Y0-BSD-663

TRANSITION METALS IN TOTAL SYNTHESIS

TRANSITION METALS IN TOTAL SYNTHESIS

Peter J. Harrington
Syntex Group Technology Center
Boulder, Colorado

A WILEY-INTERSCIENCE PUBLICATION

JOHN WILEY & SONS

New York ■ Chichester ■ Brisbane ■ Toronto ■ Singapore

Library of Congress Cataloging-in-Publication Data:

Harrington, Peter J.
Transition metals in total synthesis/Peter J. Harrington.
p. cm.
"A Wiley-Interscience publication."
Includes bibliographical references.
ISBN 0-471-61300-2
1. Organic compounds—Synthesis. 2. Transition metal compounds.
I. Title.
QD262.H27 1990
547.2—dc20

89-38139
CIP

Printed in the United States of America

10 9 8 7 6 5 4 3 2 1

To Rosi in the Rockies

FOREWORD

The use of transition metal complexes as reagents for the synthesis of complex organic molecules has been under development for at least the past several decades, and many extraordinary organic transformations of profound synthetic potential have been realized. However, incorporation of this powerful methodology into the working vocabulary of the practicing synthetic organic chemist has been exceptionally slow, and only now are transition metal reagents beginning to be used routinely in the quest for efficient synthetic approaches to structurally complex organic molecules.

This book should do much to further this endeavor. In it, Dr. Harrington presents an array of total syntheses of complex, biologically active organic molecules that rely extensively on the unique reactivity patterns of organometallic complexes for their efficiency. For each total synthesis, the biological activity of the molecule is briefly presented, followed by a discussion of the *general* principles of the organometallic processes involved in the synthesis. The total synthesis is then presented in detail, and both the power and the limitations of the organometallic processes involved are brought into sharp focus. This book will make interesting and enlightening reading for student and practicing chemist alike, and will generally give new insight into the problems associated with the total synthesis of organic compounds.

LOUIS S. HEGEDUS

Professor, Department of Chemistry
Colorado State University
Fort Collins, Colorado
1989

PREFACE

This text developed from a lecture series in Advanced Organic Synthetic Methods presented at the State University of New York at Binghamton. The original draft included nearly 40 target compounds. Of these, 20 were chosen based on the following criteria: (1) biological/pharmacological significance, (2) a novel disconnection in the retrosynthetic analysis, due to (3) application of organotransition metal chemistry in what might be considered the key step in the synthetic sequence, (4) a sufficiently broad scope in the organotransition metal chemistry to make the material both instructive and thought provoking, and (5) a recently completed total synthesis.

The presentation of the material is suitable for graduate students having some background in classical synthetic organic methods. I believe that organic chemists interested in expanding their synthetic repertoire to include new, *proven* methodology based on organotransition metal chemistry will also find it enjoyable and informative.

It is my pleasure to acknowledge the help of a great many people. Thanks to Professor Edward C. Taylor (Princeton University) and Professor Louis S. Hegedus (Colorado State University) for their enthusiastic support of this project. A special thanks to Professor Hegedus for taking the time to write the foreword and to read and critique the entire manuscript. Thanks also to Professor Susan B. Hastie (State University of New York at Binghamton), Professor Huw M. L. Davies (Wake Forest University), and Drs. Sam L. Nguyen, Jonathan C. Walker, and Robert J. Topping (Organometallic Chemistry Group at the Syntex Technology Center in Boulder) for their valuable comments and suggestions. Thanks also to Mr. Richard Hong of Hawk Scientific Systems, Inc., for providing a copy of the Molecular Presentation Graphics (MPG) software used to prepare all the figures for the text in camera-ready form. Finally, a very special thanks to my wife, Rosemarie Halchuk, who devoted a full year to converting the rough, handwritten manuscript into this finished text.

PETER J. HARRINGTON

Boulder, Colorado
December 1989

CONTENTS

TRANSITION METALS IN TOTAL SYNTHESIS

CHAPTER 1

Introduction

Organic synthetic methodology continues to develop at an exponential rate in the 1980s. The result is an ever-increasing demand on the synthetic chemist to keep abreast with new developments. Although we make a valiant effort to "keep up," we often do not have time to review previous work and get the "big picture."

We will focus on one important subsection of these new developments: the application of organotransition metal chemistry to organic synthesis. The chronology of development of most organotransition metal methods is (1) initial discovery, (2) demonstration as a reproducible and predictable method, and (3) application to a challenging synthetic problem. Thus, a good barometer of synthetic utility is *application to total synthesis*. This text provides that big picture for a number of reproducible and predictable metal-based methods in describing their application to natural product synthesis.

The text is loosely organized by metal. Chapters 2 (σ-arylpalladium complexes) and 3 (π-allylpalladium complexes) describe relatively simple organopalladium chemistry. Chapters 4 (η^4-diene iron tricarbonyl complexes); 5 (π-allyliron tricarbonyl complexes), and 6 (iron-stabilized oxallyl cation complexes) cover three important areas of organoiron chemistry. Chapters 7 (alkyne cyclotrimerization), 8 (dicobalt octacarbonyl alkyne complexes), 9 (the Khand–Pauson cyclopentenone synthesis), and 10 (phthaloyl- and maleoylcobalt complexes) provide a nearly complete picture of cobalt–alkyne chemistry. Chapters 11 (η^6-arenechromium tricarbonyl complexes) and 12 (pentacarbonylchromium carbene complexes) describe two areas of organochromium chemistry that have received considerable attention. Chapters 12 and 13 (titanium carbene complexes) provide a comparison of the synthetic application of Fischer-type and Schrock-type carbene complexes. Finally, Chapters 14 and 15 describe some advances in the field of transmetallation and return to many of the relatively simple mechanistic concepts introduced in the organopalladium chemistry chapters. Transmetallation is discussed in the last two chapters since it will have an important role in the future development of organotransition metal chemistry. Organopalladium and organonickel chemistry are intentionally deemphasized to focus on some of the less utilized metals.

Chapters 3 [($\pm$)-Ibogamine and ($\pm$)-Catharanthine] and 4 [($\pm$)-Limaspermine] provide a strategy for synthesis of the clinically important chemotherapeutic agents, vinblastine and vincristine. Chapters 2, 6, 9, and 12 have multiple targets: all targets in each chapter are accessible via the same metal-based methodology. Much of the metal chemistry of Chapters 5, 6, 9, 10, and 12 has not been thoroughly reviewed in the recent literature. The material in each of these chapters is presented more comprehensively using numerous tabulated examples. Chapters 13 and 15 illustrate the potential diversity of synthetic approaches to a single target [(±) – $\Delta^{9,12}$-Capnellane]. The metal-based methods of these two chapters and those of Chapters 6, 8, and 9 are all new methods for construction of highly functionalized five-membered rings, an important synthetic objective of the 1980s.

Each chapter consists of three sections: (1) a brief introduction to the biological activity and previous syntheses of the target compound(s), (2) the organotransition metal chemistry background necessary to understand the total synthesis, and (3) the complete total synthesis from commercially available materials to finished target. Raw material costs are given; prices quoted are obtained from current catalogs of most commonly used vendors (Aldrich Chemical Company, Fluka Chemical Corporation, Lancaster Synthesis, Ltd., and Eastman Kodak Company).

All too often the "total synthesis" is a formal total synthesis, that is, the synthesis of a key intermediate. Conversion of the key intermediate to the target by classical methods is described. The number of synthetic operations and overall yields are summarized. Particularly poor steps are noted, and, in some cases, suggestions for improvement are given. Synthetic operations for which no yield is available are noted; the yield is assumed quantitative for the overall yield calculation.

I chose to limit the presentation of the organotransition metal chemistry in each chapter; sufficient information is provided to fully understand the metal-mediated transformation(s) used in the total synthesis. The many leading references provide access to a more in-depth coverage.

As mentioned, a number of related topics were omitted. In several chapters total syntheses of alternative targets can be used to illustrate much of the same organotransition metal chemistry:

Chapter 3: (±)-Brefeldin A[1]
Chapter 4: Trichothecene analogs[2]
Chapter 5: Nocardicin derivatives[3]
Chapter 6: C Nucleosides, especially showdomycin[4]
Chapter 15: Nucleoside analogs[5]

Abbreviations Used in the Figures

AiBN	azobis(isobutyronitrile)
BQ	benzoquinone
BSA	bis-(trimethylsilyl)acetamide
DBU	1,8-diazabicyclo[5.4.0]undec-7-ene
DDQ	2,3-dichloro-5,6-dicyano-1,4-benzoquinone
DEAD	diethylazodicarboxylate
DMAP	4-dimethylaminopyridine
DME	1,2-dimethoxyethane
DMF	*N*,*N*-dimethylformamide
DMSO	dimethylsulfoxide
HMPA	hexamethylphosphoric triamide
MCPBA	*m*-chloroperoxybenzoic acid
NBS	*N*-bromosuccinimide
PCC	pyridinium chlorochromate
PDC	pyridinium dichromate

TFA	trifluoroacetic acid
TFAA	trifluoroacetic anhydride
THF	tetrahydrofuran
TMEDA	*N*,*N*,*N'*,*N'*-tetramethylethylenediamine

REFERENCES

1a. Trost, B. M. *Angew. Chem. Int. Ed. Engl.* **1986**, *25*, 1.

1b. Trost, B. M.; Lynch, J.; Renaut, P.; Steinman, D. H. *J. Am. Chem. Soc.* **1986**, *108*, 284.

1c. Trost, B. M. Pure Appl. Chem. **1981**, *53*, 2357.

2a. Pearson, A. J.; Chen, Y-S. *J. Org. Chem.* **1986**, *51*, 1939.

2b. Pearson, A. J.; Ong, C. W. *J. Am. Chem. Soc.* **1981**, *103*, 6686.

3. Hodgson, S. T.; Hollinshead, D. M.; Ley, S. V.; Low, C. M. R.; Williams, D. J. *J. Chem. Soc., Perkin Trans. 1* **1985**, 2375.

4a. Sato, T.; Hayakawa, Y.; Noyori, R. *Bull. Chem. Soc. Jpn.* **1984**, *57*, 2515.

4b. Noyori, R.; Sato, T.; Hayakawa, Y. *J. Am. Chem. Soc.* **1978**, *100*, 2561.

4c. Sato, T.; Ito, R.; Hayakawa, Y.; Noyori, R. *Tetrahedron Lett.* **1978**, 1829.

5. Bergstrom, D. E. *Nucleosides Nucleotides* **1982**, *1*, 1.

CHAPTER 2

(±)-*N*-Acetyl Clavicipitic Acid Methyl Ester and (±)-Aurantioclavine

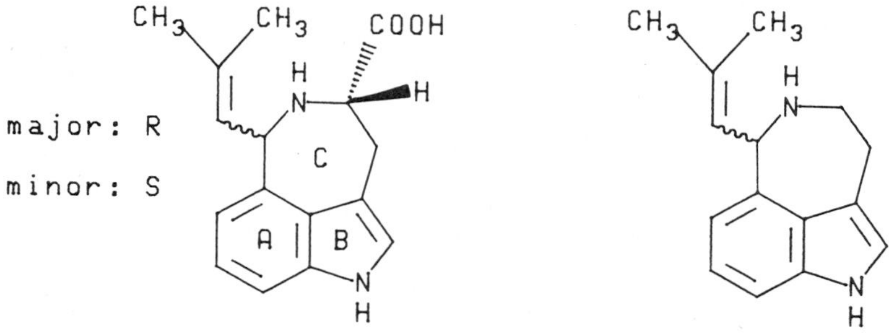

major: R

minor: S

(±)-Clavicipitic Acid (±)-Aurantioclavine

1H-Azepino[5,4,3-cd]indole-4-carboxylic acid, 3,4,5,6-tetrahydro-6-(2-methyl-1-propenyl) trans (±)[84986-03-8], cis (±)[84986-04-9] 4S-trans[33062-26-9], 4S-cis[72690-85-8]

1H-Azepino[5,4,3-cd]indole, 3,4,5,6-tetrahydro-6-(2-methyl-1-propenyl) (±)[99211-67-3], (−)[80152-02-9]

Figure 2.1

(±)-Clavicipitic acid was first isolated as a mixture of diastereoisomers from submerged cultures of *Claviceps* species, strain SD-58[1] and later from *Claviceps fusiformis* strain 139/2/1G.[2] (These two strains are probably identical.[3]) (±)-Clavicipitic acid is a derailment product in the *d*-lysergic acid biosynthetic pathway at the 4-dimethallyltryptophan stage.[4] After an initial incorrect structural assignment, mass spectral and ^{1}H nuclear magnetic resonance (NMR) studies on a derivative[2,5] and, several years later, an x-ray crystal structure of the major diastereoisomer[6] confirmed the structure as shown in Figure 2.1. (±)-Aurantioclavine was first isolated from *Penicillium aurantio-virens.*[7]

While the pharmacological activity of these ergot alkaloids has yet to be investigated, the potent psychotropic activity of another ergot, lysergic acid as its diethylamide (LSD), is common knowledge. Less well known are other closely related ergots that are effective for the treatment of prolactin-dependent disorders, postpartum hemorrhage, hypertension, poor peripheral and cerebral blood circulation, and migraine attacks. With the extremely broad spectrum of activity exhibited by the ergots, clavicipitic acid and aurantioclavine are prime synthetic targets. Although many ingenious syntheses of the ergoline framework of the more common ergots have been reported,[8] there are few reports of total synthesis of (±)-clavicipitic acid[9] and only one previous synthesis of (±)-aurantioclavine.[10]

2.1 σ-ARYLPALLADIUM COMPLEXES: SYNTHETIC APPLICATIONS[11]

σ-Arylpalladium complexes can be generated by a variety of different methods. These palladium(II) complexes can rapidly and reversibly coordinate an alkene. The π-alkene complex undergoes *migratory insertion* (i.e., migration of aryl from palladium to carbon) to produce carbon–carbon and palladium–carbon σ-bonds. The aryl group usually *migrates regioselectively* to the less substituted alkene carbon. With σ-alkylpalladium complexes possessing a hydrogen β to the metal, β-hydride elimination is rapid. Decomplexation affords the arylated alkene, palladium(0), and acid. Usually only the *E*-alkene isomer is obtained. The overall process results in generation of a new aryl–carbon bond. The entire process occurs at room temperature and tolerates a wide range of functionality (see Figure 2.2).

Arylpalladium halides have been generated by five different methods:

1. By oxidative addition of an aryl halide:

$$\mathrm{ArX + Pd(0) \longrightarrow ArPdX} \qquad (2.1)$$

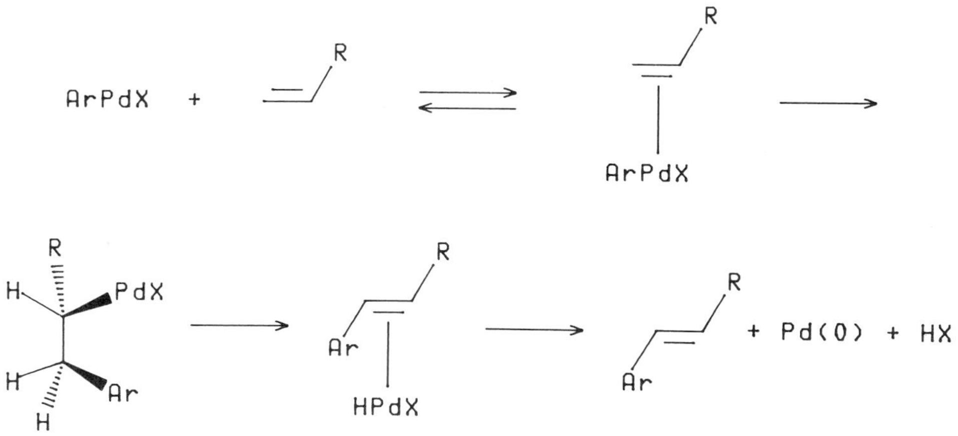

Figure 2.2

The more difficult oxidative addition with X = Br generally requires the presence of phosphines. These coordinate to the metal, increasing its electron density and, thus, lower the energy for the *oxidative* addition.

2. By transmetallation from another aryl metal species:

$$ArMX + PdX_2 \longrightarrow ArPdX + MX_2 \tag{2.2}$$

Metals that are most commonly used for transmetallation to palladium are Hg, B, and Zn. Synthetic applications of transmetallation will be discussed further in Chapters 14 and 15.

3. By direct ring metallation:

$$ArH + PdX_2 \longrightarrow ArPdX + HX \tag{2.3}$$

This is most efficient when a chelated arylpalladium complex is produced[12] (see Figure 2.3).

4. By oxidative addition of aroyl chlorides[13]:

$$ArC(O)Cl + Pd(0) \longrightarrow ArC(O)PdCl \longrightarrow ArPdCl + CO \tag{2.4}$$

5. By attack of arylsulfinates on palladium(II)[14]:

$$ArSO_2Na + PdX_2 \longrightarrow ArSO_2PdX + NaX \longrightarrow ArPdX + SO_2 \tag{2.5}$$

The first method is by far the most useful for synthetic purposes for two primary reasons: first, aryl bromides and iodides are readily available, and second, the arylpalladium halide is produced using palladium(0). This method for complex generation together with the arylation sequence constitutes a novel method for arylation that is *catalytic* in palladium(0). This method is commonly known as the "Heck arylation of alkenes" or the "Heck olefination" or simply as the "Heck reaction" (see Figure 2.4).

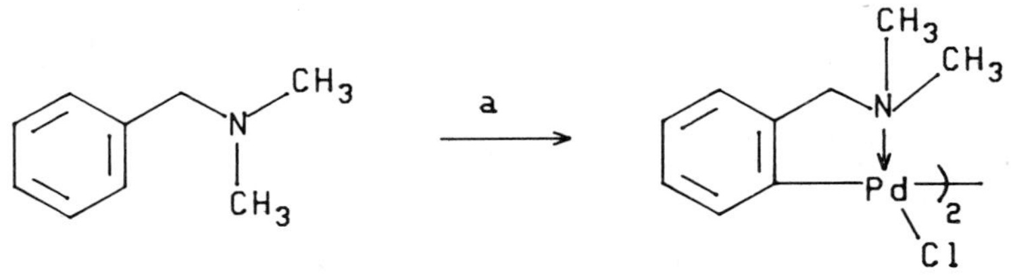

a. rt, Li_2PdCl_4, CH_3OH; 96%

Figure 2.3

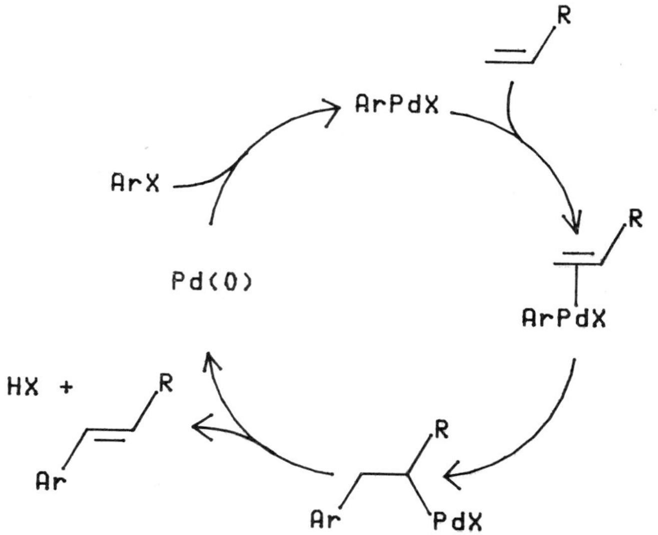

Figure 2.4

There are three significant limitations of the method:

1. Highly electron-rich aromatics undergo competitive dehalogenation[15] (see Figure 2.5).
2. Halides on π-deficient heterocycles have not been replaced with any reproducible efficiency[16] (see Figure 2.6)
3. As the number of substituents or the steric bulk of substituents on the alkene increases, coordination of the alkene to palladium becomes more difficult and the overall process becomes less efficient. The method is essentially limited to monosubstituted and some disubstituted alkenes.

Despite these limitations, the method has proven to be of tremendous synthetic value. In addition, variations on the basic theme have resulted in extensions of the

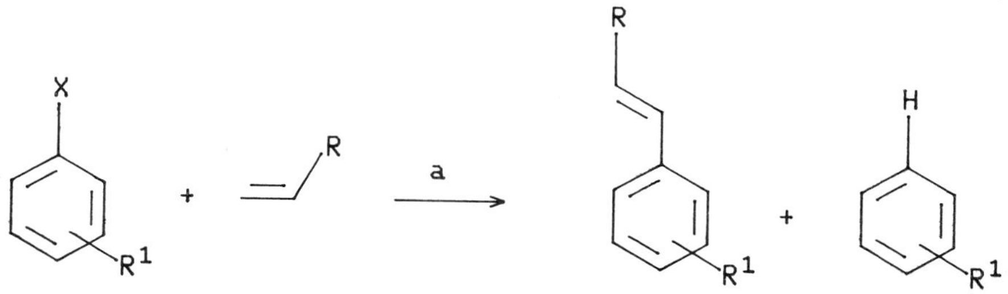

a. 100°C, Et_3N, 1 mol% $Pd(OAc)_2$, ligand (L)

Figure 2.5

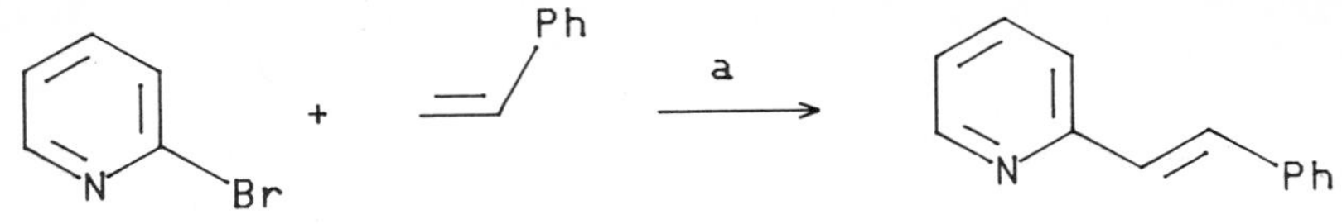

a. 100°C, Et_3N, 1 mol% $Pd(OAc)_2$,

2 mol% $(o\text{-}tol)_3P$; 19%

Figure 2.6

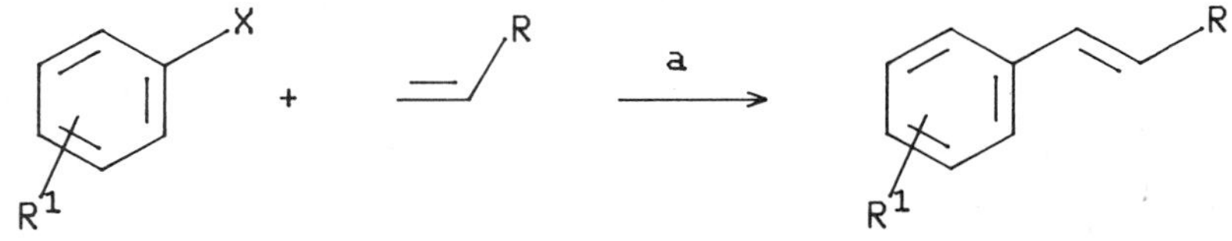

R = COOH, $COOCH_3$, COOEt, $CONH_2$, CN

a. 100°C, Et_3N, 1 mol% $Pd(OAc)_2$, ligand (L),

solvent (S)

Figure 2.7

method to produce compounds other than simple arylated alkenes. The synthetic utility of the method will be discussed in the following section.

1. The Arylation of Electron-Poor Alkenes

With R = CHO or $C(O)R^1$, polymerization of the alkene is competitive under the typical vigorous reaction conditions (see Figure 2.7).

2. Arylation of Electron-Rich Alkenes

Synthetic utility with electron-rich alkenes is limited by poor regioselectivity in migratory insertion[17–19] (see Figure 2.8).

Arylation and subsequent hydrolysis of the enol ether or enamide constitutes a novel method for synthesis of arylacetaldehydes and acetophenones.

3. Arylation of Cyclic Alkenes[20]

Stereospecific *syn*-addition of aryl and PdX to an alkene is followed by stereospecific *syn*-β-hydride elimination. The result is elimination to produce alkene away from the aromatic (see Figure 2.9).

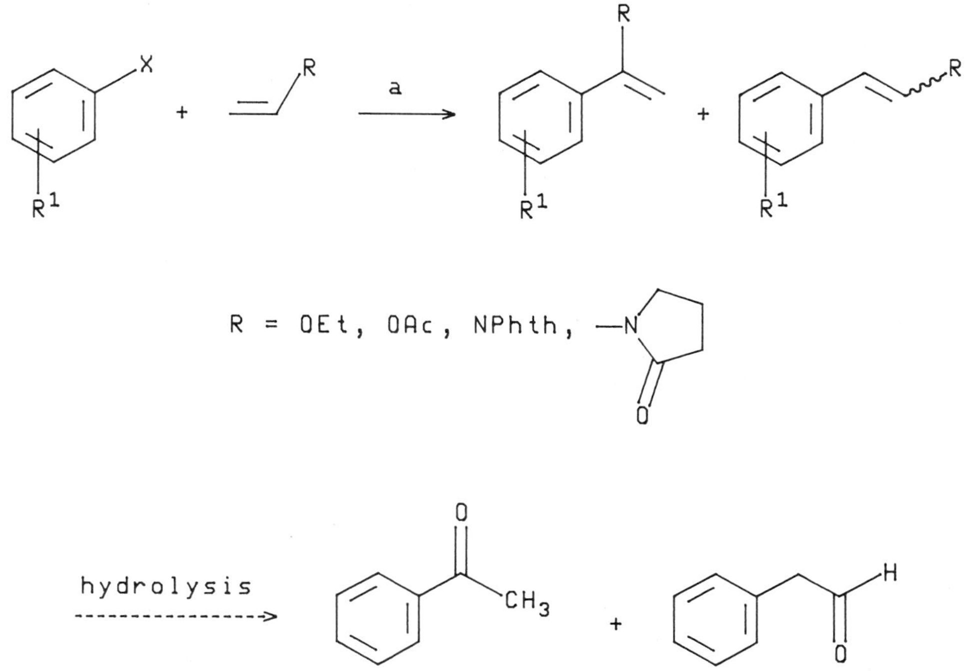

a. 100°C, Et_3N, 1 mol% $Pd(OAc)_2$, ligand (L), solvent (S)

R = OEt, OAc, NPhth, —N(2-oxopyrrolidinyl)

hydrolysis

Figure 2.8

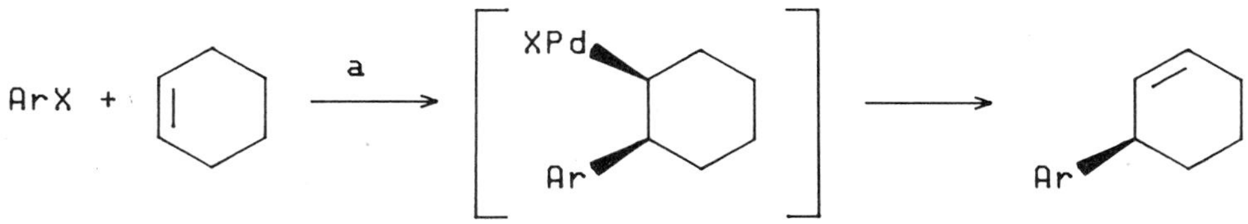

ArX + cyclohexene —a→ [XPd, Ar cyclohexane intermediate] → Ar-cyclohexene

a. 100°C, Et_3N, 1 mol% $Pd(OAc)_2$, ligand (L), solvent (S)

Figure 2.9

4. Arylation of Allylic and Homoallylic Alcohols[21]

Addition of ArPdX to allyl alcohol produces a σ-alkylpalladium complex that can undergo β-hydride elimination in two directions. Elimination is regioselective for production of an enol that, after decomplexation, rapidly tautomerizes to the keto form. Addition of ArPdX to a homoallylic alcohol suffers from poor regioselectivity. Both possible σ-alkyl complexes can β-hydride eliminate to palladium–alkene

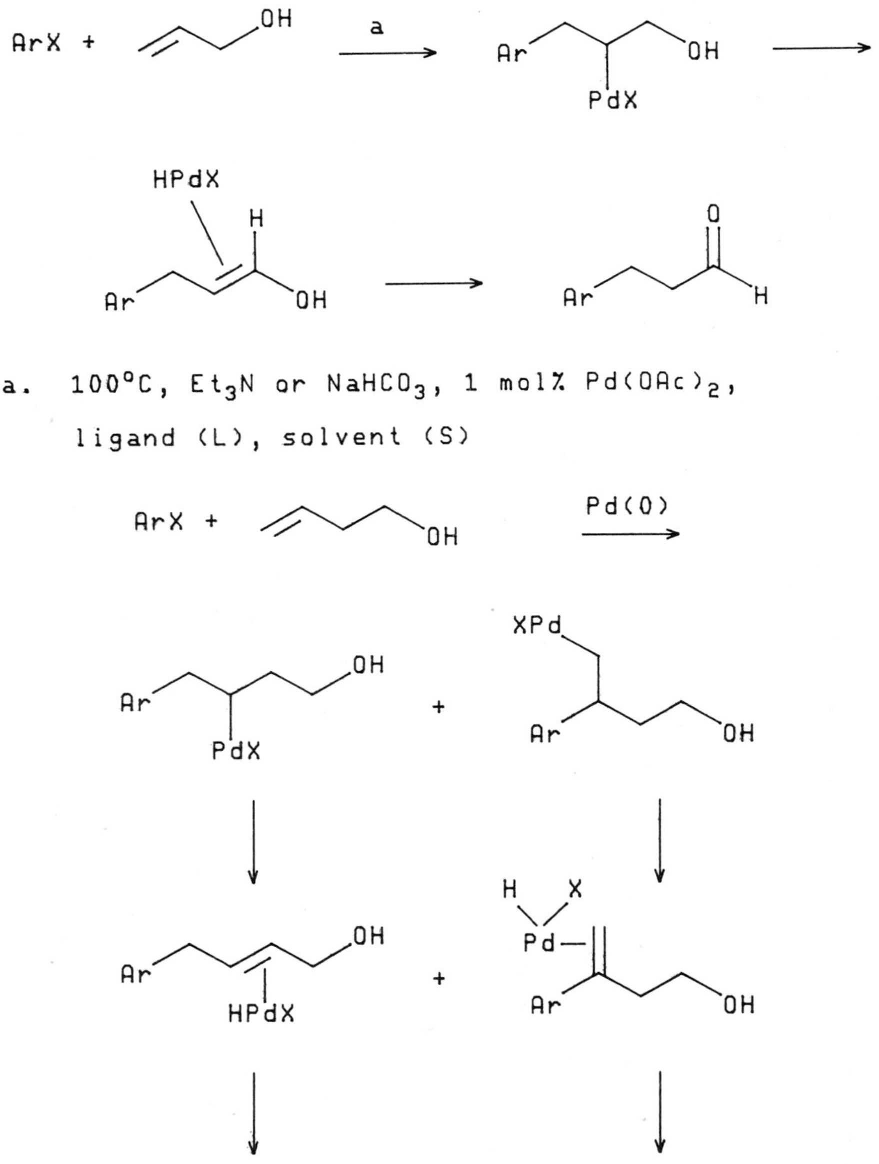

Figure 2.10

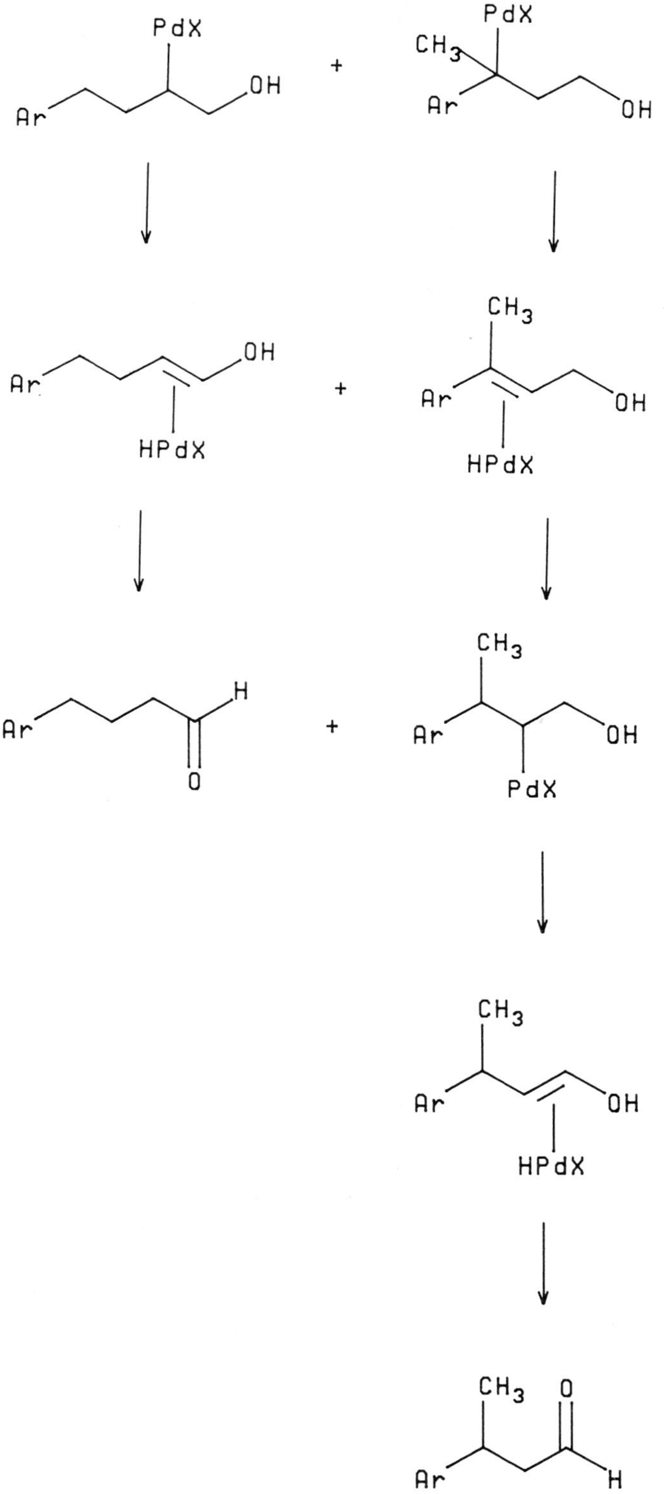

Figure 2.10 (*Continued*)

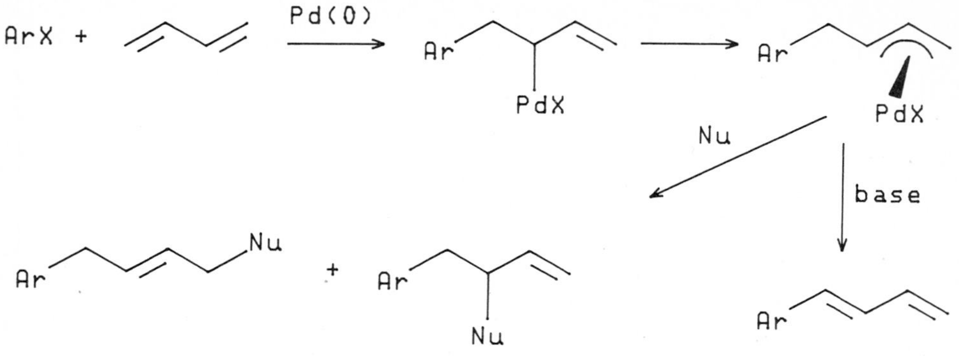

ArX + Pd(0) Ar PdX Ar PdX Nu base Ar Nu + Ar Nu Ar

Figure 2.11

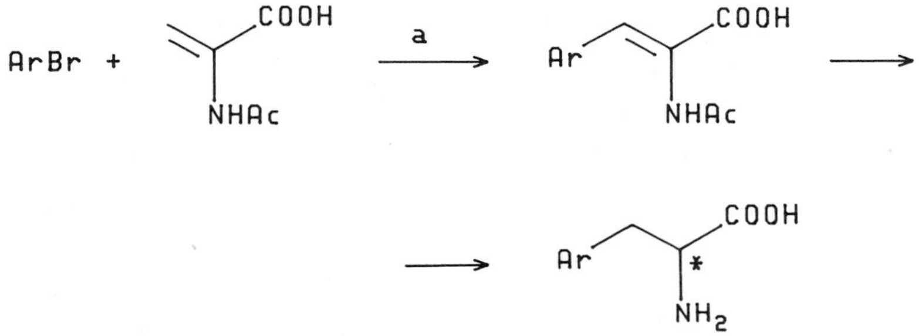

ArBr + COOH NHAc a Ar COOH NHAc Ar COOH * NH_2

a. 110°C, Et_3N or $(n\text{-}Bu)_3N$ or TMEDA, 1 mol% $Pd(OAc)_2$, 2-4 mol% PPh_3; 25-76%

Figure 2.12

complexes. Decomplexation is somewhat slower than readdition to produce new σ-alkyl complexes that β-hydride eliminate to complexed enols. Decomplexation is followed by rapid tautomerization to the keto form (see Figure 2.10).

5. Arylation of 1,3-Dienes[22]

Arylation of butadiene gives a σ-alkylpalladium complex with an alkene adjacent to the metal. This complex rapidly collapses to a π-allyl complex (see Figure 2.11). The utilization of π-allylpalladium complexes in synthesis is discussed in Chapter 3.

6. Arylation of α-Acetamidoacrylates[23,24]

α-Acetamidocinnamates can be reduced enantiospecifically and deprotected without racemization to afford phenylalanines in very high enantiomeric excess (see Figure 2.12).

7. Arylation to Produce Rings[25]

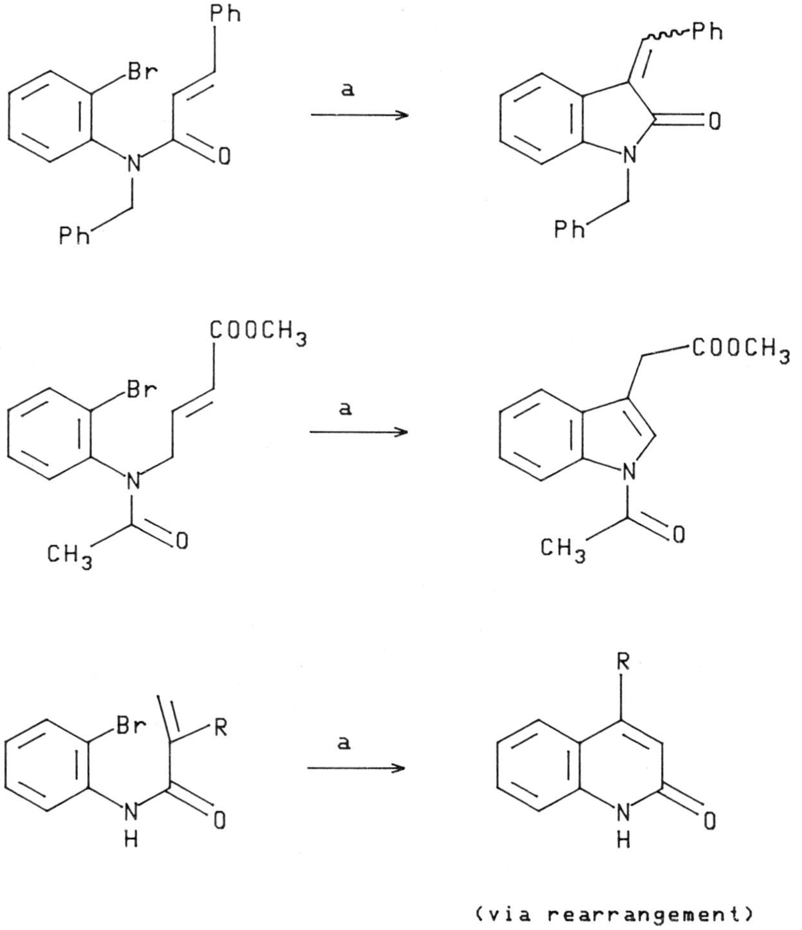

(via rearrangement)

a. 100°C, Et_3N or $(n\text{-}Bu)_3N$, 1 mol% $Pd(OAc)_2$, 1-4 mol% ligand (L), solvent (S)

Figure 2.13

8. Arylation at Room Temperature Using Phase Transfer Catalysis[26]

The polymerization of some alkenes under the relatively vigorous standard reaction conditions has already been discussed. Oxidative addition of aryl iodides proceeds *at room temperature* in the presence of tetrabutylammonium salts. Using aryl iodides and a phase transfer catalyst, arylation of thermally labile sulfolene is moderately

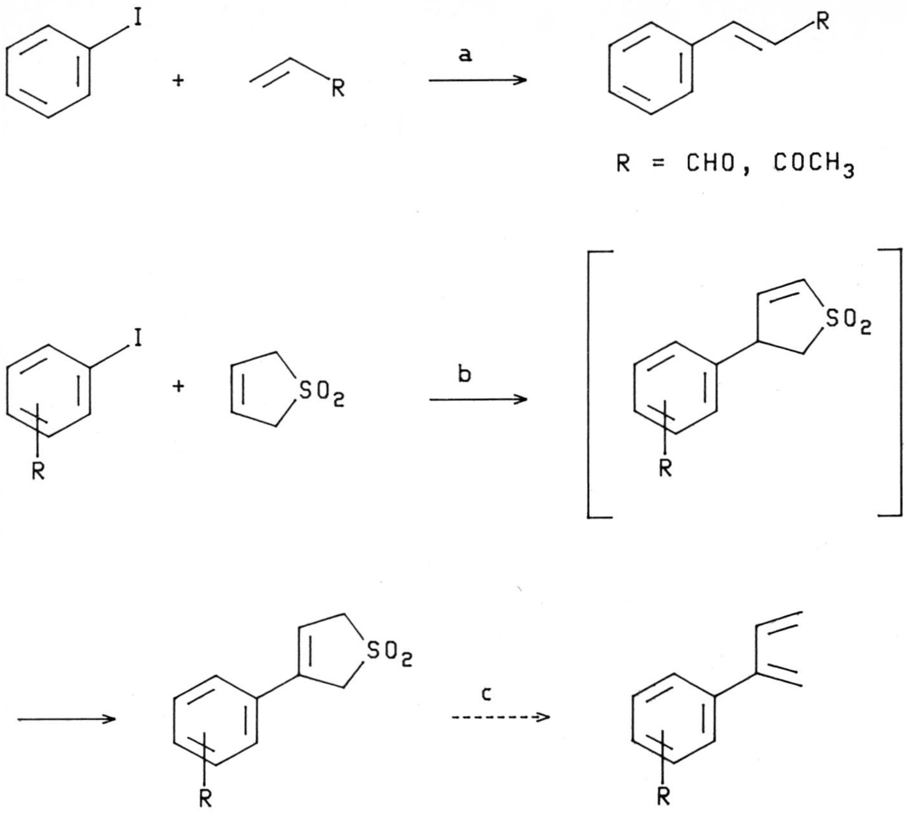

R = CHO, COCH$_3$

a. rt, $NaHCO_3$, 1 or 2 mol% $Pd(OAc)_2$, $(n\text{-}Bu)_4NCl$, DMF; >90%

b. rt, Et_3N, 5 mol% $Pd(OAc)_2$, $(n\text{-}Bu)_4NBr$, benzene; 44–70%

c. heat

Figure 2.14

efficient. The arylated sulfolenes can be converted to 2-aryl-1,3-butadienes[27] (see Figure 2.14).

2.2 TOTAL SYNTHESIS OF (±)-*N*-ACETYL CLAVICIPITIC ACID METHYL ESTER[24,28]

A retrosynthetic analysis for clavicipitic acid reveals a problem common to the synthesis of all ergots. Indole is a π-electron-rich heterocycle. As such, it undergoes facile electrophilic aromatic substitution. A map of the π-electron densities of the indole

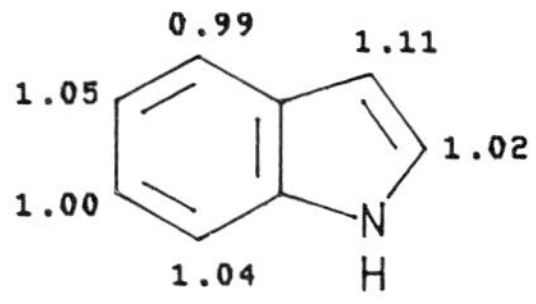

Figure 2.15 π-Electron densities of the indole ring carbons.

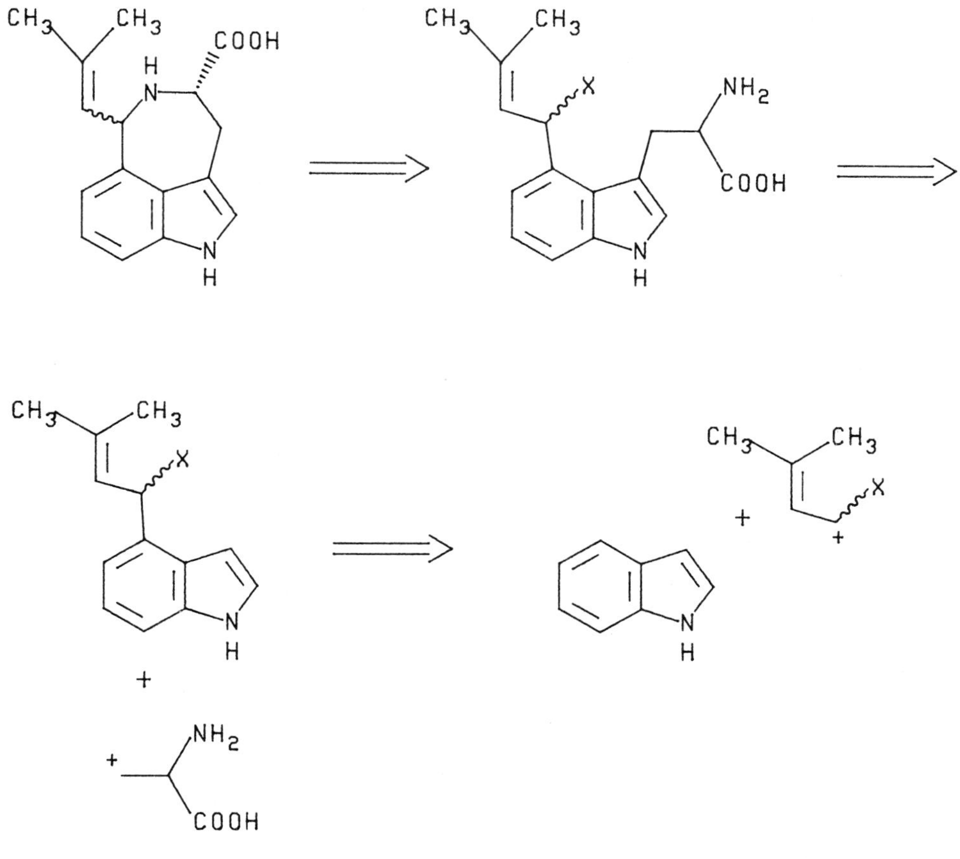

Figure 2.16

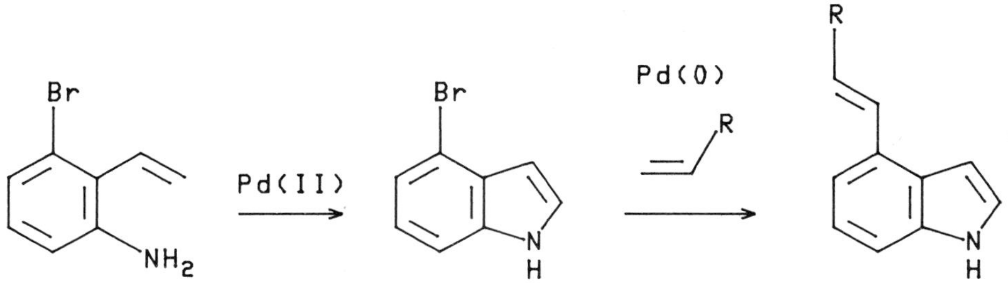

Figure 2.17

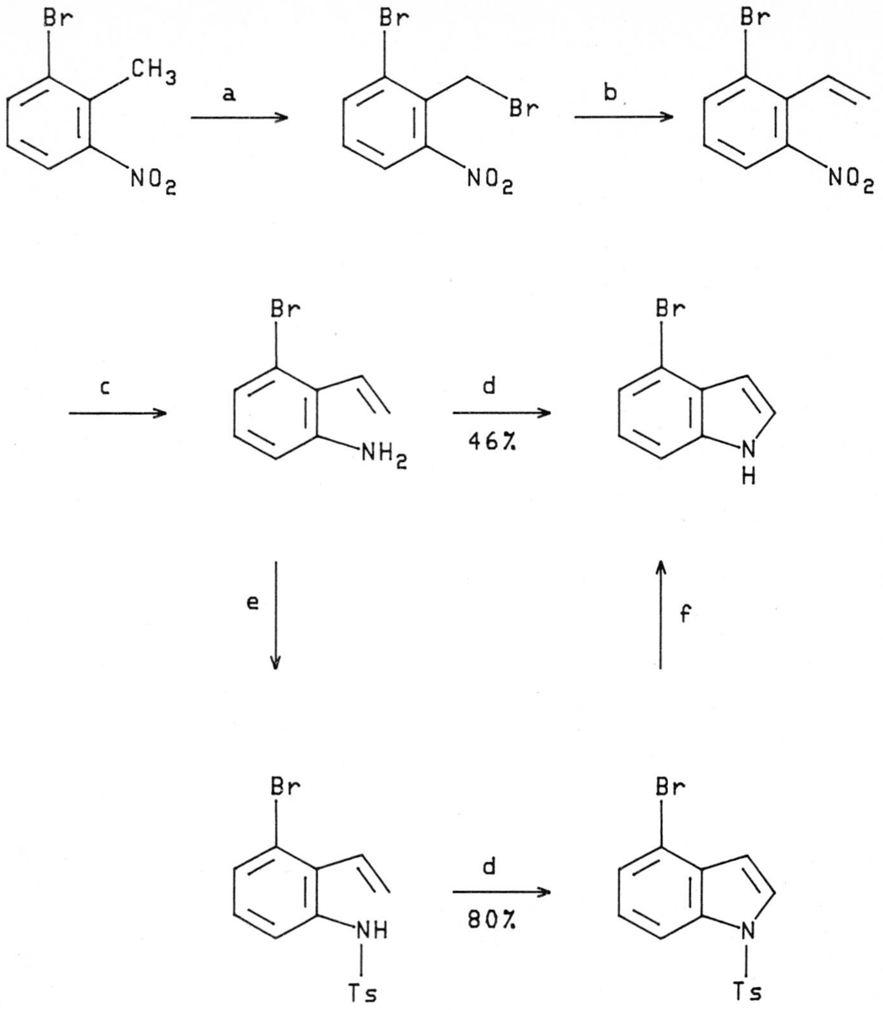

Br | CH3 | NO2 —a→ Br | Br | NO2 —b→ Br | NO2

—c→ Br | NH2 —d (46%)→ Br | N H

e ↓ ↑ f

Br | NH | Ts —d (80%)→ Br | N | Ts

a. rfx, hν, $(PhCOO)_2$, Br_2, CCl_4; 98%

b. 1) heat, PPh_3, $CHCl_3$, 2) rt, CH_2O, Et_3N, CH_2Cl_2; 85%

c. rfx, Fe powder, AcOH, EtOH; 97%

d. rfx, $PdCl_2(CH_3CN)_2$, LiCl, BQ, THF

e. rfx, p-TsCl, pyridine; 80%

f. rfx, NaOH, CH_3OH, H_2O; 98%

Figure 2.18

ring carbons reveals that the carbon at position 3 is the most electron-rich, is the most reactive, while the carbon at position 4 is the least reactive. Thus, 4-substituted indoles cannot be made by electrophilic aromatic substitution (see Figures 2.15 and 2.16).

There are two basic approaches to 4-substituted indoles[8]: (1) incorporate a 4-substituent prior to formation of the indole ring or (2) introduce a 4-substituent directly by a mechanism other than electrophilic aromatic substitution. Introduction of the substituent by *nucleophilic attack on indole* coordinated to chromium tricarbonyl has been investigated by Kozikowski and later by Semmelhack[29,30] (see Chapter 11). The approach presented in this chapter falls into the more common first category but is considerably more versatile than most previous syntheses. The substituent at the 4-position is bromine. In the synthetic strategy, oxidative addition, alkene coordination, and migratory insertion, reductive elimination (i.e., the Heck reaction) provides a mechanism for replacement of the initial 4-substituent by a wide range of mono-substituted alkenes (see Figure 2.17).

2-Bromo-6-nitrotoluene can be converted to 2-bromo-6-ethenylaniline by a series of classical steps in excellent overall yield.[31,32] Although 2-ethenylaniline could be converted to indole in good yield using a catalytic amount of palladium(II) and benzoquinone as reoxidant,[33] the cyclization of 3-bromo-2-ethenylaniline is considerably slower. Alkene polymerization and/or indole decomposition are apparently

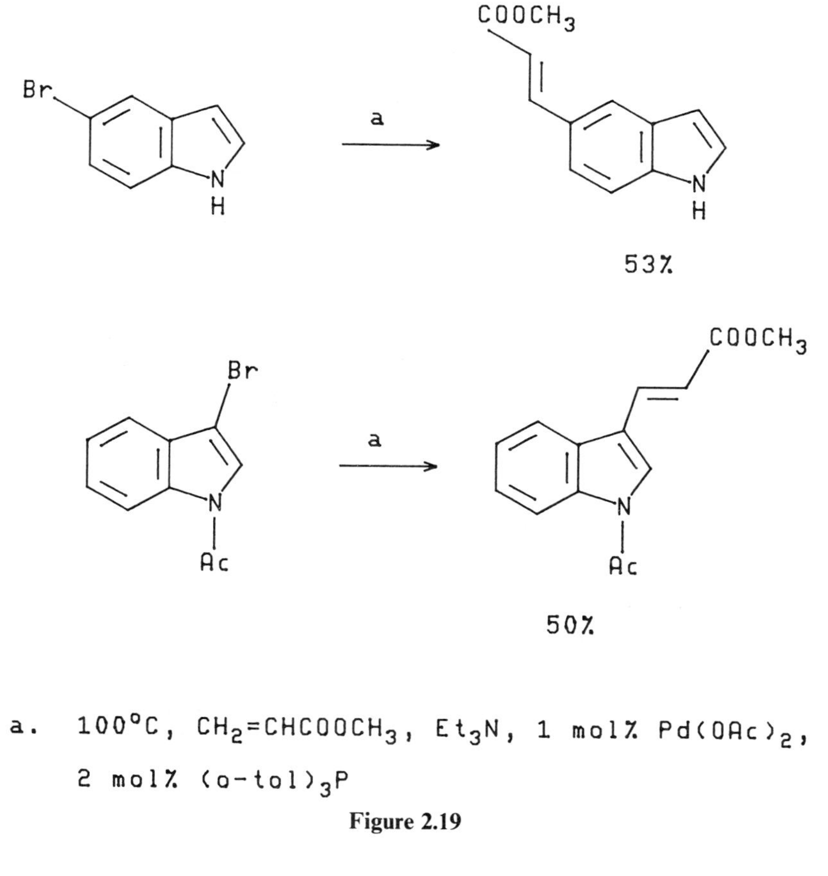

a. 100°C, CH_2=$CHCOOCH_3$, Et_3N, 1 mol% $Pd(OAc)_2$, 2 mol% $(o\text{-tol})_3P$

Figure 2.19

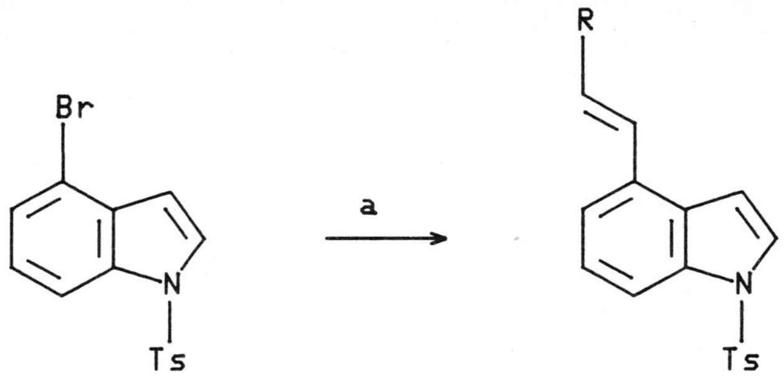

Br

a

R

N

Ts

N

Ts

R =	Yield
$COOCH_3$	86%
Ph	74%
NPhth	74%
$C(CH_3)_2OH$	97%

a. 100°C, CH_2=CHR, Et_3N, 5 mol% $Pd(OAc)_2$, 20 mol% $(o\text{-tol})_3P$

Figure 2.20

competitive. Cyclization of the acetanilide is similarly slow while cyclization of the *p*-toluenesulfonamide (tosamide) is rapid and, accordingly, efficient. Since the indoletosamide can be readily converted to the free indole,[34] the three-step route (tosylation, cyclization, detosylation) constitutes an improved route to 4-bromoindole (see Figure 2.18).

Arylation using very electron-rich aromatics is inefficient due to competitive protodehalogenation. For example, Heck has reported moderately effective arylation using 5-bromoindole and 3-bromo-1-acetylindole[16] (see Figure 2.19).

This information as well as the relative ease of manipulation and purification of the indoletosamide prompted initial investigation of the arylation sequence using the tosamide rather than the free indole (see Figure 2.20).

The product of arylation of 2-methyl-3-buten-2-ol, prepared by an alternative method, has been converted to clavicipitic acid by Natsume.[9c] The alternative functionalization of the 3-position used in this synthesis and, ultimately, an alternative synthesis of clavicipitic acid, is based on a selective arylation at the indoletosamide 3-position in the presence of the 4-bromo substituent. This is most efficiently achieved via 4-bromo-3-iodo-1-tosylindole by oxidative addition to palladium(0) in the absence of phosphine ligands. Arylation of methyl α-acetamidoacrylate produced only the *Z*-alkene isomer in 60% yield (see Figure 2.21).

Arylation of 2-methyl-3-buten-2-ol with *Z*-3-(2-acetamido-2-methoxycarbonylethenyl)-4-bromo-1-tosylindole provides a substrate suitable for investigation

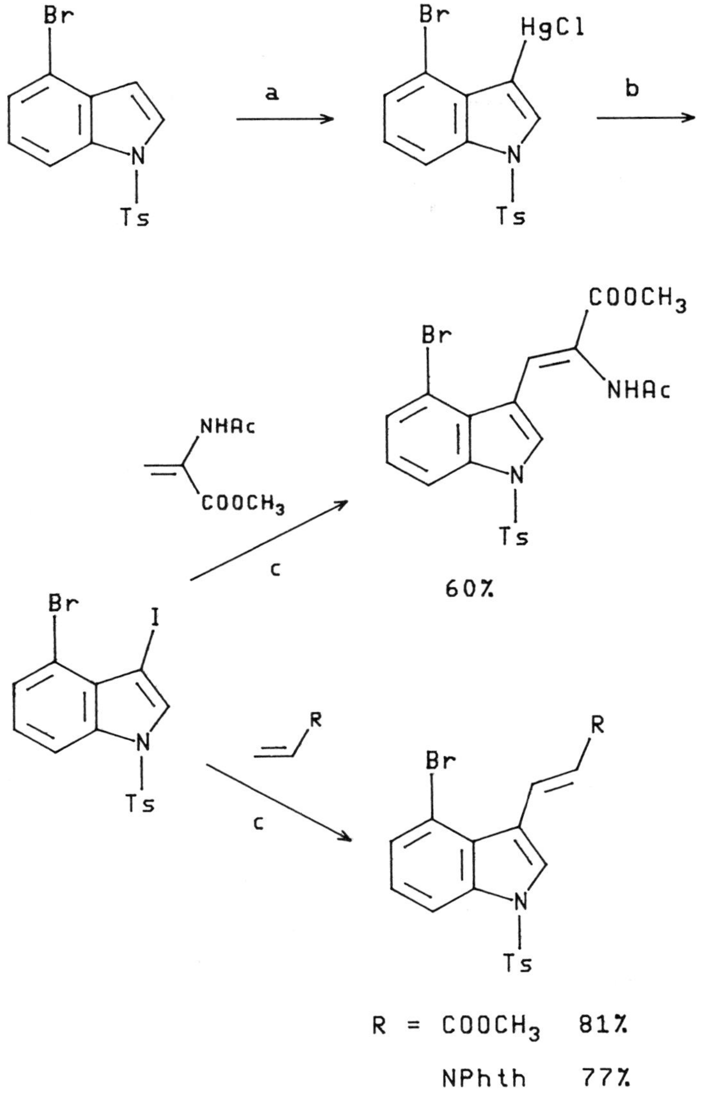

Br HgCl

a b

Ts Ts

COOCH$_3$

Br

NHAc

NHAc

COOCH$_3$

c

Ts

60%

Br I

R R

Br

Ts

c

Ts

R = COOCH$_3$ 81%

NPhth 77%

a. 1) rt, cat. $HClO_4$, $Hg(OAc)_2$, AcOH,

2) NaCl, H_2O; 100%

b. rt, I_2, $CHCl_3$; 97%

c. 100°C, Et_3N, 5 mol% $Pd(OAc)_2$

Figure 2.21

of C-ring formation. While cyclization was originally envisioned to proceed via nucleophilic attack on a π-allylpalladium complex at the less sterically encumbered terminus (see Chapter 3), the cyclization was achieved in remarkably good yields by heating the mixture in the presence of a catalytic amount of *p*-toluenesulfonic acid or, better, a catalytic amount of palladium(II). The mechanism of this interesting palladium-mediated cyclization may involve nucleophilic attack on a palladium(II)–alkene π-complex followed by preferential elimination of palladium(II) oxide. Selective

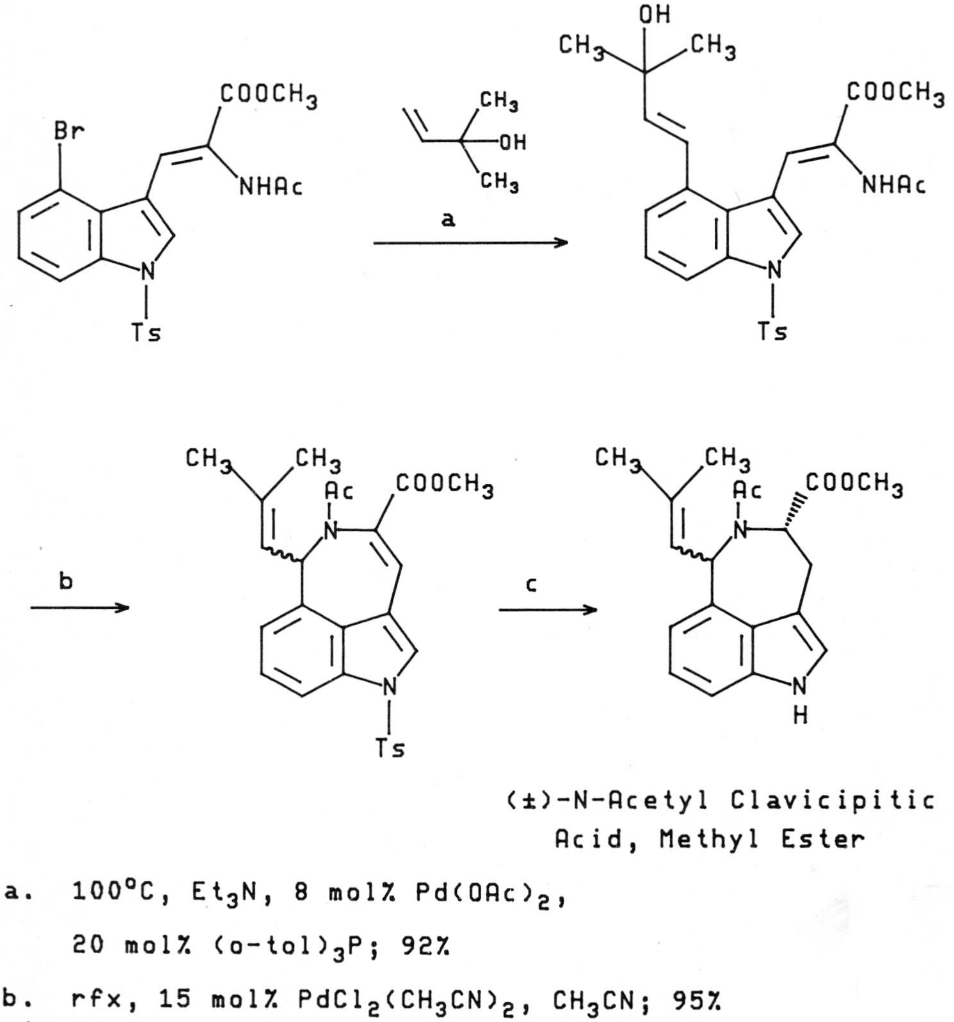

(±)-N-Acetyl Clavicipitic Acid, Methyl Ester

a. 100°C, Et_3N, 8 mol% $Pd(OAc)_2$, 20 mol% $(o\text{-tol})_3P$; 92%

b. rfx, 15 mol% $PdCl_2(CH_3CN)_2$, CH_3CN; 95%

c. −20°C, hν, $NaBH_4$, Na_2CO_3, CH_3OH, DME, H_2O;
after 10 h, 61% cis, 0% trans
20.5 h, 37% cis, 37% trans

Figure 2.22

reduction of the electron-deficient double bond *and* tosamide deprotection are simultaneously accomplished using sodium borohydride under photolytic conditions developed by Natsume.[9c] Short reaction times (10 h) result in almost exclusive formation of the *cis*-derivative (61%) corresponding to the major diastereoisomer from natural clavicipitic acid. Longer reaction times result in slightly higher yields (74%) of an approximately 1:1 mixture of *cis*- and *trans*-derivatives.

The total synthesis of the *N*-acetyl esters of (±)-clavicipitic acid was completed in 12 steps from 2-bromo-6-nitrotoluene in an overall 18% yield. "This represents an approximately forty-fold increase in yield over existing methods and provides an illustration of the efficacy of transition metals in organic synthesis"[28] (see Figure 2.22).

2.3 TOTAL SYNTHESIS OF (±)-AURANTIOCLAVINE[35]

The total synthesis of (±)-aurantioclavine requires a method for regioselective introduction of a vinyl ether at the 3-position of 4-bromo-3-iodo-1-tosylindole. Arylation of ethyl vinyl ether using a catalytic amount of palladium(0) (no phosphine ligands) would be expected to give a regioisomeric mixture of the 3-(2-ethoxyethenyl) and 3-(1-ethoxyethenyl) products. The most efficient approach is based on two different methods using organotransition metals[36]:

1. Stereo- and regioselective hydrozirconation of a triple bond (see Chapter 14): Zirconium and hydride add *cis* to a triple bond with the metal and its sterically demanding cylcopentadiencyl ligands going to the most sterically accessible carbon.

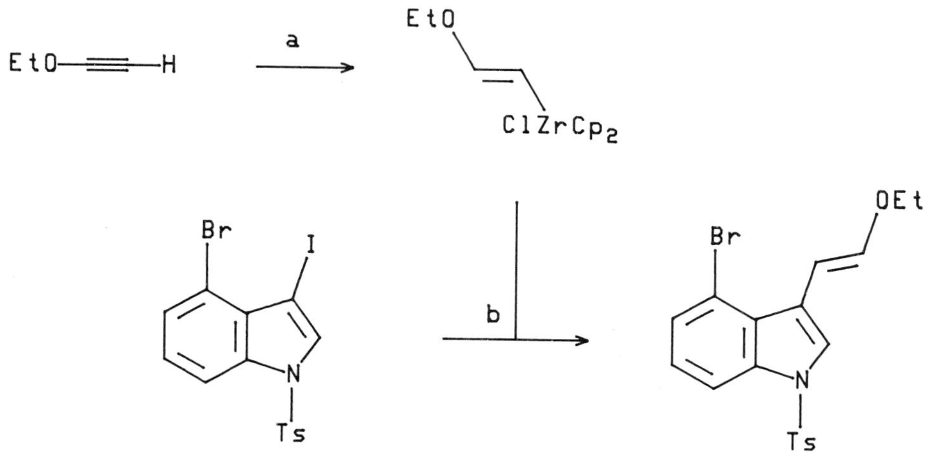

EtO—≡—H
a
EtO
ClZrCp$_2$
Br
I
N
Ts
b
OEt
Br
N
Ts

a. rt, $Cp_2Zr(H)Cl$, benzene

b. rt, $Ni(PPh_3)_4$, benzene, THF; 80%

Figure 2.23

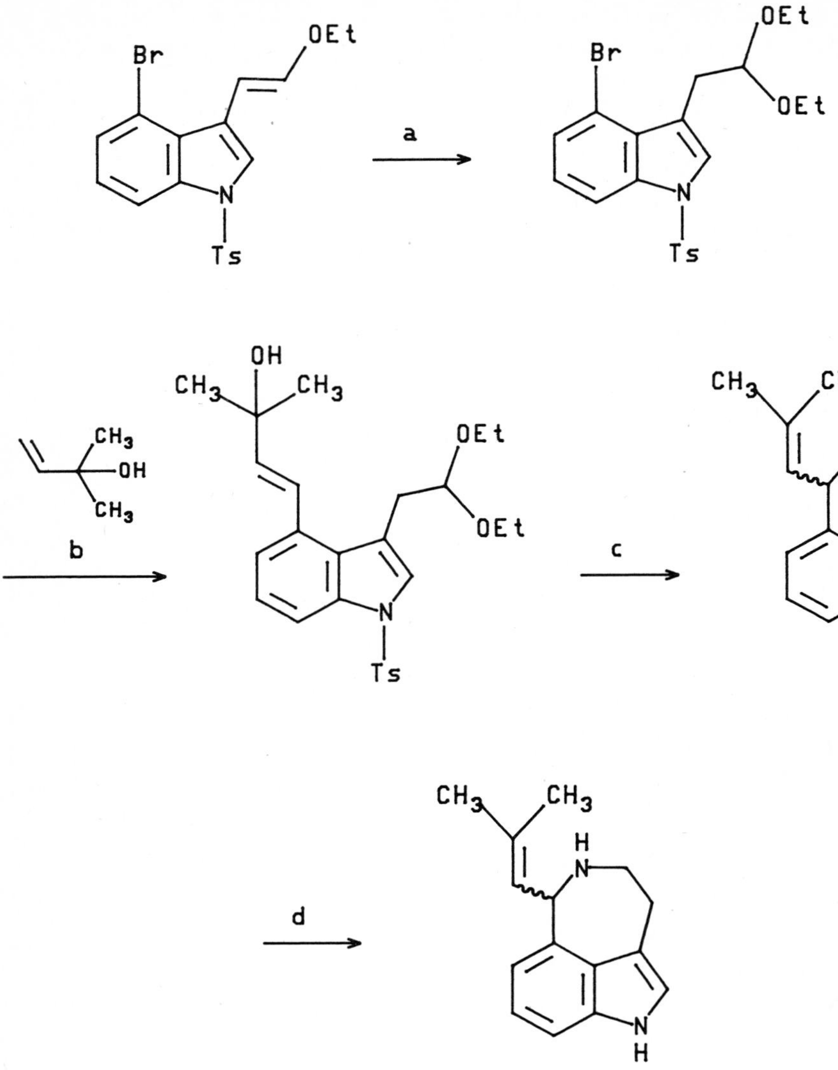

(±)-Aurantioclavine

a. 1) rt, $BBr_3 \cdot S(CH_3)_2$, CH_2Cl_2, 2) $NaHCO_3$, H_2O, EtOH; 96%

b. rfx, Et_3N, $Pd[(o\text{-}tol)_3P]_4$, CH_3CN; 92%

c. rfx, p-TsOH·H_2O, p-$TsNH_2$, CH_3CN; 56%

d. rt, hν, $NaBH_4$, CH_3OH, DME, H_2O; 85%

Figure 2.24

2. Transmetallation (see Chapters 14 and 15): Nickel(0) will oxidatively add the aryl iodide to generate an arylnickel(II) halide. The vinylzirconium complex will transmetallate to the nickel(II) complex. The resulting nickel(II) species will reductively eliminate to generate a new carbon–carbon bond (see Figure 2.23).

Prior to functionalization of the A ring, the vinyl ether is protected as an acetal. 2-Methyl-3-buten-2-ol is introduced at the 4-position, as previously described. Reaction with *p*-toluenesulfonamide and a catalytic amount of *p*-toluenesulfonic acid produces not the expected tosylimine-diene but, rather, the desired tricyclic material. Selective reduction of the enetosamide and simultaneous deprotection of both tosamides are accomplished using sodium borohydride under photolytic conditions.

The total synthesis of (±)-aurantioclavine was completed in 13 steps from 2-bromo-6-nitrotoluene in overall 23% yield (see Figure 2.24).

REFERENCES

1. Robbers, J. E.; Floss, H. G. *Tetrahedron Lett.* **1969**, 1857.
2. King, G. S.; Mantle, P. G.; Szczyrbak, C. A.; Waight, E. S. *Tetrahedron Lett.* **1973**, 215.
3. Despite a statement to the contrary: Stadler, P. A.; Stütz, P. *Alkaloids* **1975**, *15*, 5.

4a. Saini, M. S.; Cheng, M.; Anderson, J. A. *Phytochemistry* **1976**, *15*, 1497.

4b. Bajwa, R. S.; Kohler, R. D.; Saini, M. S.; Cheng, M.; Anderson, J. A. *Phytochemistry* **1975** *14*, 735.

5. King, G. S.; Waight, E. S.; Mantle, P. G.; Szczyrbak, C. A. *J. Chem. Soc., Perkin Trans. 1* **1977**, 2099.
6. Robbers, J. E.; Otsuka, H.; Floss, H. G.; Arnold, E. V.; Clardy, J. *J. Org. Chem.* **1980**, *45*, 1117.

7a. Kozlovskii, A. G.; Solov'eva. T. F.; Sakharovskii, V. G.; Adanin, V. M. *Dokl. Akad. Nauk. SSSR* **1981**, *260*, 230; *Chem. Abstr.* **1982**, *96*, 3403b.

7b. Sakharovskii, V. G.; Aripovskii, A. V.; Baru, M. B.; Kozlovskii, A. G. *Khim, Prir. Soedin* **1983**, 656.

8a. Horwell, D. C. *Tetrahedron* **1980**, *36*, 3123.

8b. Kozikowski, A. P. *Heterocycles* **1981**, *16*, 267.

9a. Kozikowski, A. P.; Okita, M. *Tetrahedron Lett.* **1985**, *26*, 4043.

9b. Kozikowski, A. P.; Greco, M. N. *J. Org. Chem.* **1984**, *49*, 2310.

9c. Muratake, H.; Takahashi, T.; Natsume, M. *Heterocycles* **1983**, *20*, 1963.

9d. Kozikowski, A. P.; Greco, M. N. *Heterocycles* **1982**, *19*, 2269.

10. Yamada, F.; Makita, Y.; Suzuki, T.; Somei, M. *Chem. Pharm. Bull.* **1985**, *33*, 2162.
11. Heck, R. F. *Organic React.* **1982**, *27*, 345.
12. Cope, A. C.; Friedrich, E. C. *J. Am. Chem. Soc.* **1968**, *90*, 909.
13. Blaser, H-U.; Spencer, A. *J. Organometal. Chem.* **1982**, *233*, 267.

14a. Garves, K. *J. Org. Chem.* **1970**, *35*, 3273.

14b. Selke, V. R.; Thiele, W. *J. Prak. Chem.* **1971**, *313*, 875.

14c. Chiswell, B.; Venanzi, L. M. *J. Chem. Soc. A* **1966**, 1246.

14d. Tamaru, Y.; Yoshida, Z-I. *Tetrahedron Lett.* **1978**, 4527.

15. Ziegler, C. B., Jr.; Heck, R. F. *J. Org. Chem.* **1978**, *43*, 2941.
16. Frank, W. C.; Kim, Y. C.; Heck, R. F. *J. Org. Chem.* **1978**, *43*, 2947.
17. Arai, I.; Daves, G. D., Jr. *J. Org. Chem.* **1979**, *44*, 21.
18. Kasahara, A.; Izumi, T.; Fukuda, N. *Bull. Chem. Soc. Jpn.* **1977**, *50*, 551.
19. Ziegler, C. B., Jr.; Heck, R. F. *J. Org. Chem.* **1978**, *43*, 2949.
20. Cortese, N. A.; Ziegler, C. B., Jr.; Hrnjez, B. J.; Heck, R. F. *J. Org. Chem.* **1978**, *43*, 2952.
21a. Chalk, A. J.; Magennis, S. A. *J. Org. Chem.* **1976**, *41*, 273.
21b. Chalk, A. J.; Magennis, S. A. *J. Org. Chem.* **1976**, *41*, 1206.
21c. Melpolder, J. B.; Heck, R. F. *J. Org. Chem.* **1976**, *41*, 265.
22a. Patel, B. A.; Dickerson, J. E.; Heck, R. F. *J. Org. Chem.* **1978**, *43*, 5018.
22b. Stakem, F. G.; Heck, R. F. *J. Org. Chem.* **1980**, *45*, 3584.
23. Cutolo, M.; Fiandanese, V.; Naso, F.; Sciacovelli, O. *Tetrahedron Lett.* **1983**, *24*, 4603.
24. Harrington, P. J.; Hegedus, L. S. *J. Org. Chem.* **1984**, *49*, 2657.
25a. Mori, M.; Ban, Y. *Tetrahedron Lett.* **1979**, 1113.
25b. Mori, M.; Chiba, K.; Ban, Y. *Tetrahedron Lett.* **1977**, 1037.
25c. Terpko, M. O.; Heck, R. F. *J. Am. Chem. Soc.* **1979**, *101*, 5281.
26. Jeffery, T. *J. Chem. Soc., Chem. Commun.* **1984**, 1287.
27. Harrington, P. J.; DiFiore, K. A. *Tetrahedron Lett.* **1987**, *28*, 495.
28. Harrington, P. J.; Hegedus, L. S.; McDaniel, K. F. *J. Am Chem. Soc.* **1987**, *109*, 4335.
29. Kozikowski, A. P.; Isobe, K. *J. Chem. Soc., Chem. Commun.* **1978**, 1076.
30. Semmelhack, M. F. *Pure Appl. Chem.* **1981**, *53*, 2379.
31. 2-Bromo-6-nitrotoluene is commercially available (Aldrich 88-89 catalog, 25 g for $50.55).
32. Alternative starting materials are shown in Figure 2.25. 2,6-Dinitrotoluene is commercially available and inexpensive (Aldrich 88–89 catalog, 100 g for $27.30). 2-Methyl-3-nitroaniline is commercially available (Aldrich 88–89 catalog, 25 g for $14.10).
32a. Lounasmaa, M. *Acta Chem. Scand.* **1968**, *22*, 2388.
32b. Gibson, C. S.; Johnson, J. D. A. *J. Chem. Soc.* **1929**, 1229.
33. Hegedus, L. S.; Allen, G. F.; Bozell, J. J.; Waterman, E. L. *J. Am. Chem. Soc.* **1978**, *100*, 5800.
34a. Sundberg, R. J.; Russell, H. F. *J. Org. Chem.* **1973**, **38**, 3324.
34b. Jones, C. D. *J. Org. Chem.* **1972**, *37*, 3624.
35. Hegedus, L. S.; Toro, J. L.; Miles, W. H.; Harrington, P. J. *J. Org. Chem.* **1987**, *52*, 3319.
36. Negishi, E-I.; Takahashi, T. *Aldrichimica Acta* **1985**, *18*, 31.

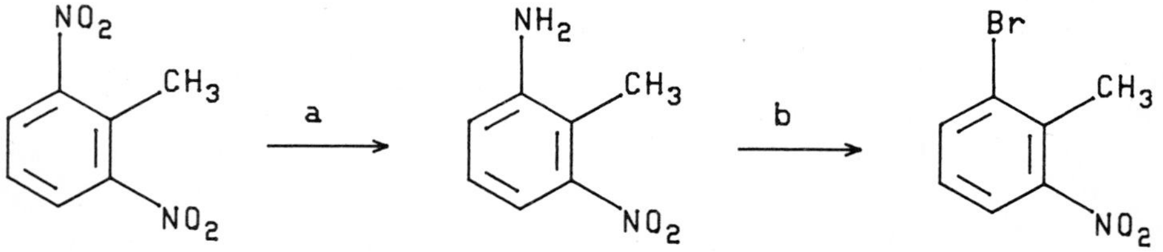

a. rt, H_2, Pd-C, EtOH; 80%

b. 1) 0°C, HONO, H_2O, 2) rt, CuBr, HBr, H_2O; 94%

Figure 2.25

CHAPTER 3

(±)-Ibogamine and (±)-Catharanthine

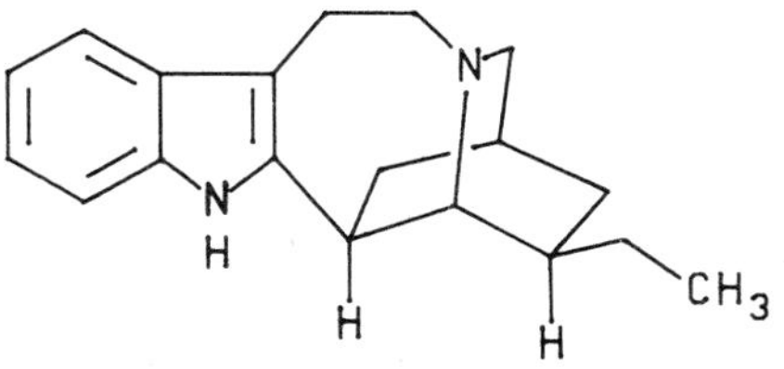

(−)-Ibogamine [481-87-8]

(±)-Ibogamine [2288-55-3]

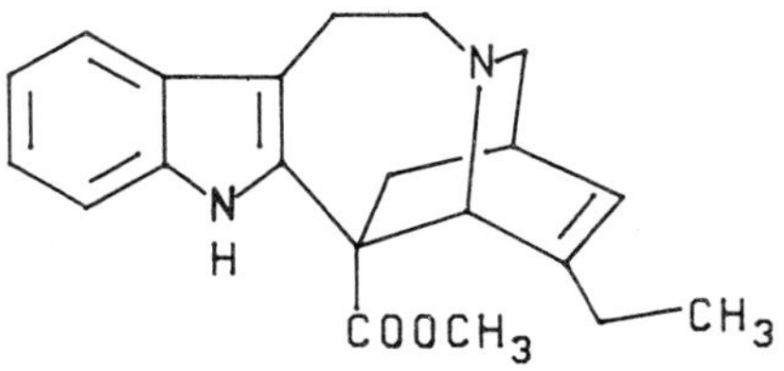

Catharanthine

Ibogamine-18-carboxylic acid,

3,4-didehydro methyl ester

(2α, 5β, 6α, 18β) [2468-21-5]

(±) [20395-98-6]

Figure 3.1

Ibogamine is a member of the *iboga* alkaloid group, common to the genera *Vocanga, Tabernaemontana, Ervatamia, Pandaca,* and *Conopharyngia* (see Figure 3.1). Other related *iboga* alkaloids are ibogaine, ibogaline, coronaridine, voacangine, and catharanthine (see Figure 3.2).[1]

Ibogamine has significant central nervous system (CNS)-stimulating and hallucinogenic activity. It is the active principle of the roots of the West African plant *Tabernathe iboga* used by hunters to combat hunger and sleep and relieve fatigue under stress, allowing them to remain motionless for long periods without loss of concentration, and by natives to induce a wild euphoric condition during ceremonial rituals. A nonamphetamine CNS-stimulating activity, as evidenced by reserpine catalepsy, is common with *iboga* alkaloids. Several *iboga* alkaloids, particularly ibogaline, cause hypotension and bradycardia in anesthetized cats. Coronaridine has analgesic, diuretic, cytotoxic,[2] and estrogenic[3] activity. Catharanthine displays hypoglycemic activity; voacangine has cytotoxic activity.[4]

The bis-indole *Catharanthus* alkaloids vinblastine (vincaleukoblastine) and vincristine (leurocristine) are two of the most effective chemotherapeutic agents for treatment of Hodgkin's disease and acute childhood leukemia. Both consist of an *Aspidosperma*

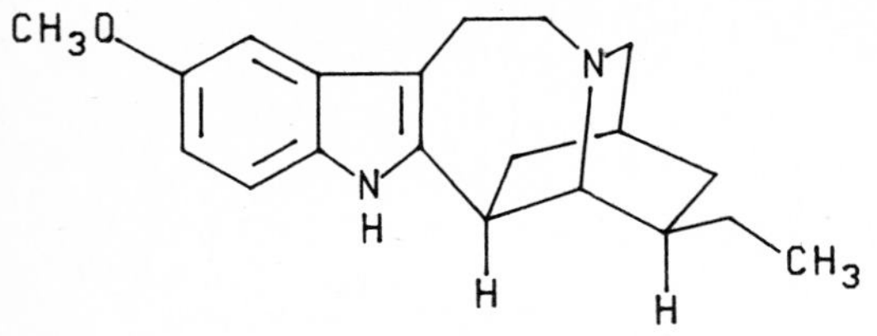

Ibogaine

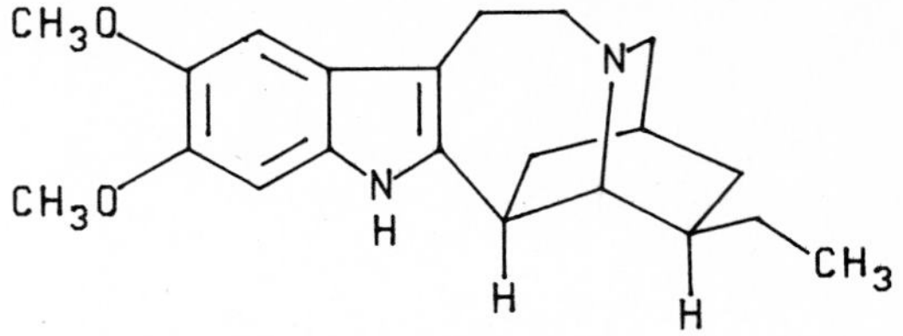

Ibogaline

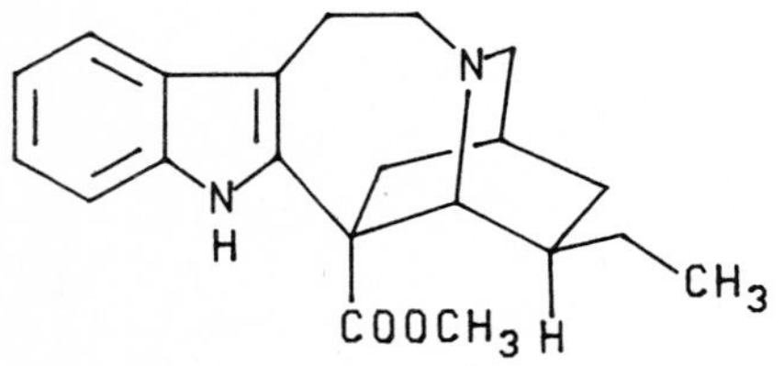

Coronaridine

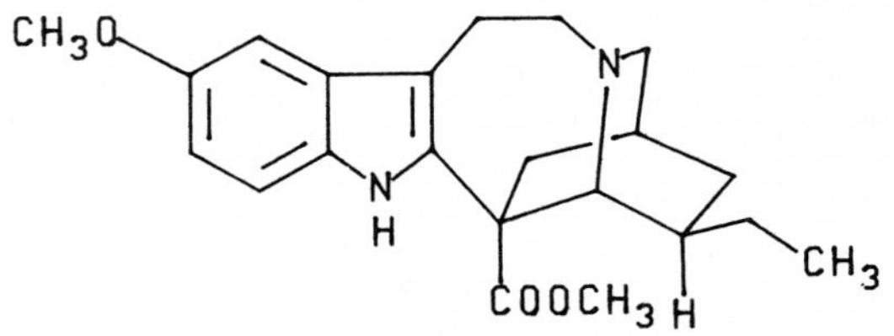

Voacangine

Figure 3.2

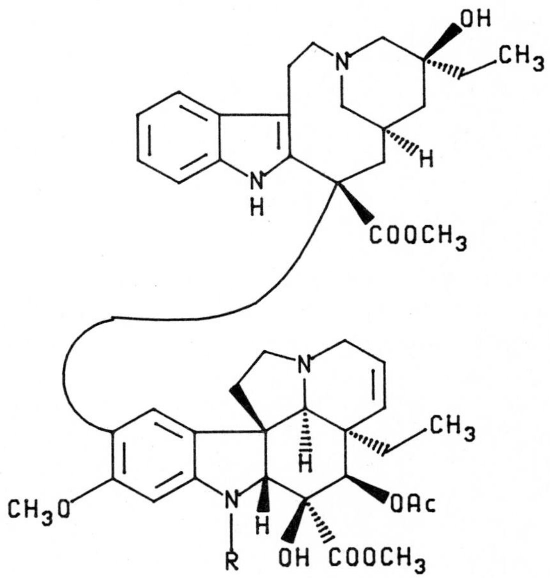

Vinblastine R = CH_3

Vincristine R = CHO

Figure 3.3

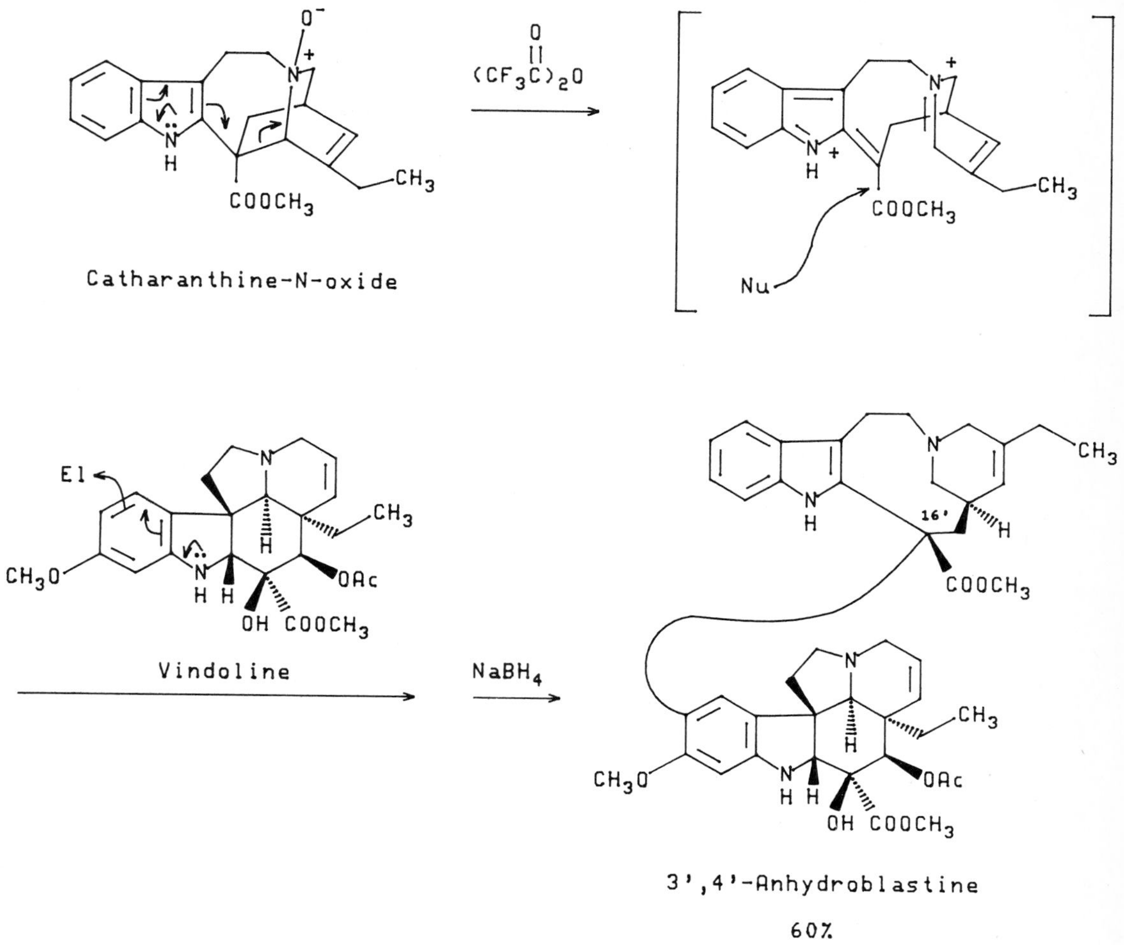

Figure 3.4

alkaloid (vindoline) and an *Iboga* alkaloid (16-β-carbomethoxyvelbanamine). Vinblastine and vincristine were first isolated from *Catharanthus roseus* by Svoboda, and today they are still produced by isolation despite typical yields of 1 g/1000 kg and 20 mg/1000 kg, respectively.[5] A nicely convergent synthetic approach might have as a final step a "biomimetic" carbon–carbon bond formation between vindoline and a nitrogen oxide of the *iboga* alkaloid catharanthine. Cell-free extracts of *Catharanthus roseus* are able to generate 3′,4′-anhydrovinblastine from catharanthine and vindoline and transform it into vinblastine.[6]

Despite the varied activity of the *iboga* family itself and the demonstrated chemotherapeutic value of the *Catharanthus* alkaloids, relatively little progress has been made toward a stereospecific synthesis of the *iboga* skeleton of ibogamine or catharanthine.[7] This chapter focuses on Trost's palladium-based approach to both racemic and optically active ibogamine; Chapter 4 will focus on Pearson's iron-carbonyl-based approach to the *Aspidosperma* alkaloids (see Figures 3.3 and 3.4).

3.1 π-ALLYLPALLADIUM COMPLEXES: INTRODUCTION

Dimeric π-allylpalladium chloride complexes are yellow-to-orange solids that are easily prepared, handled, and stored in the presence of air. They can be isolated by silica gel chromatography and purified by recrystallization from halogenated hydrocarbon–hydrocarbon mixtures. While generally insoluble as the dimer, they readily dissolve in organic solvents when the dimer is split by addition of a ligand (e.g. trialkyl- or triarylphosphines, pyridine).

π-Allylpalladium complexes have the three carbons and their substituent atoms in one plane, with the metal located above or below the plane. The terminal carbon substituents are termed *syn* or *anti*, referring to their geometric relationship with the single substituent on the central allyl carbon. In the presence of excess ligand, the π-allyl forms may equilibrate with the σ-allyl forms, permitting interconversion of *syn* and *anti* terminal substituents. Terminal substituents will prefer *syn* positions to avoid *anti–anti* steric interaction (see Figure 3.5).

Two important synthetic processes involving allylpalladium complexes are nucleophilic attack on π-allylpalladium complexes with displacement of the metal and reductive elimination resulting in carbon–carbon bond formation betwegn π-allyl and σ-allyl ligands.[8]

3.2 PREPARATION OF π-ALLYLPALLADIUM COMPLEXES

3.2.1 Method A. Alkenes and Palladium(II)

π-Allylpalladium chloride dimers can be prepared directly from alkenes using palladium(II) as the dichloride or bis-trifluoroacetate.[9] The mechanism involves rapid and reversible formation of a palladium(II)–alkene π-complex. Three mechanisms for conversion of the π-alkene complex to the π-allyl have been suggested: (1) insertion of palladium into the allylic carbon–hydrogen bond, (2) allylic proton abstraction by a chloride ligand, or (3) allylic proton abstraction by an external base. Allylic proton abstraction by a chloride ligand best explains the results of kinetic studies and deuterium isotope effects.[10] The process can be run in a variety of solvents, such as acetic acid, chloroform, dichloromethane, *N,N*-dimethylformamide, or methanol, and is facilitated by base (see Figure 3.6).

A review of the available literature reveals several trends in chemo-, regio-, and stereoselectivity:

1. More highly substituted alkenes are more reactive.
2. With two similarly substituted alkenes, the more electron rich is more reactive.
3. With unsymmetrical alkenes, mixtures are formed; the ratio of complexes produced is strongly dependent upon reaction conditions. Formation of the major product(s) can often be rationalized in terms of a Markovnikov-like addition of palladium(II) to the carbon–carbon double bond (see Figure 3.7).
4. Formation of π-allyl complexes with terminal *anti* substituents is unfavorable. These less stable complexes will form only when no other option is available.
5. Electron-withdrawing groups, which facilitate the allylic proton abstraction by stabilizing the resulting anion, facilitate π-allyl complex formation.

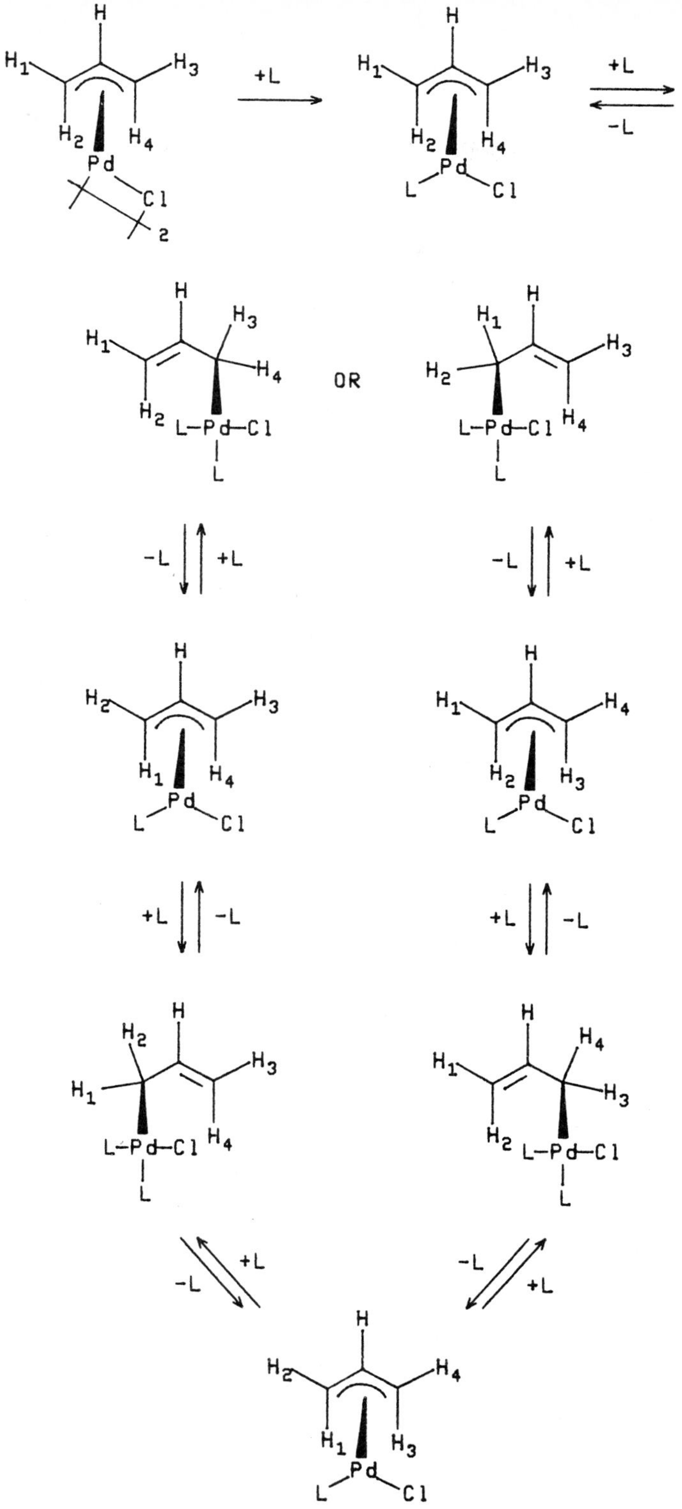

Figure 3.5

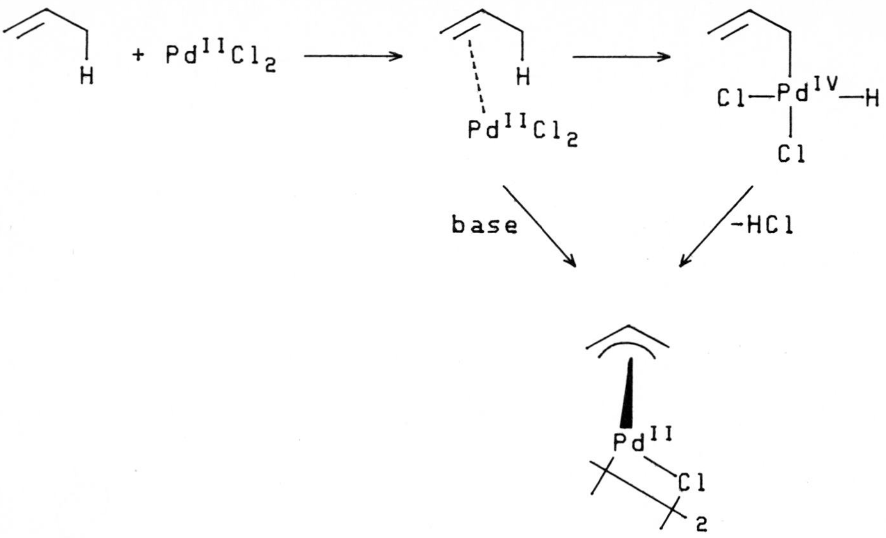

Figure 3.6

TABLE 3.1 Regioselective π-Allyl Formation Using Palladium Chloride in 50% Acetic Acid[13]

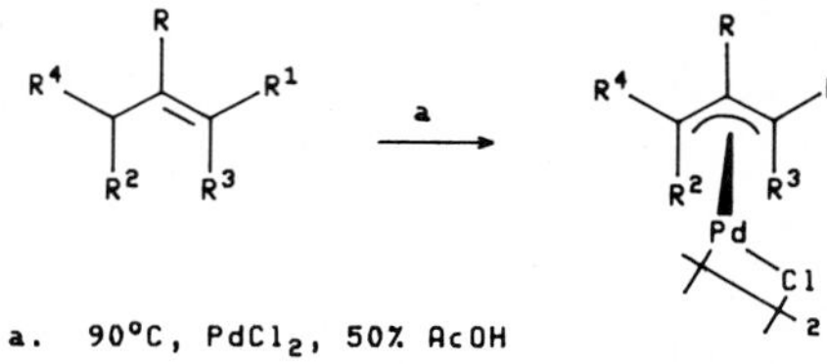

a. 90°C, $PdCl_2$, 50% AcOH

R	R^1	R^2	R^3	R^4	Yield (%)
CH_3	H	H	H	H	27
CH_3	H	H	H	CH_3	32
CH_3	CH_3	H	H	H	43
CH_3	H	H	H	Et	27
CH_3	Et	H	H	H	24
CH_3	CH_3	H	H	CH_3	68
CH_3	H	CH_3	H	CH_3	90
CH_3	CH_3	H	CH_3	H	91
CH_3	*n*-Pr	H	H	H	34
Et	CH_3	H	H	CH_3	98
CH_3	*i*-Pr	H	H	H	72
CH_3	*i*-Bu	H	H	H	30
neo-C_5H_{11}	H	H	H	H	89
CH_3	*t*-Bu	H	H	H	90
—$(CH_2)_{10}$—		H	H	H	21
—$(CH_2)_5$—		H	H	H	23
—$(CH_2)_6$—		H	H	H	91
H	H	—$(CH_2)_9$—		H	40
H	H	—$(CH_2)_5$—		H	59

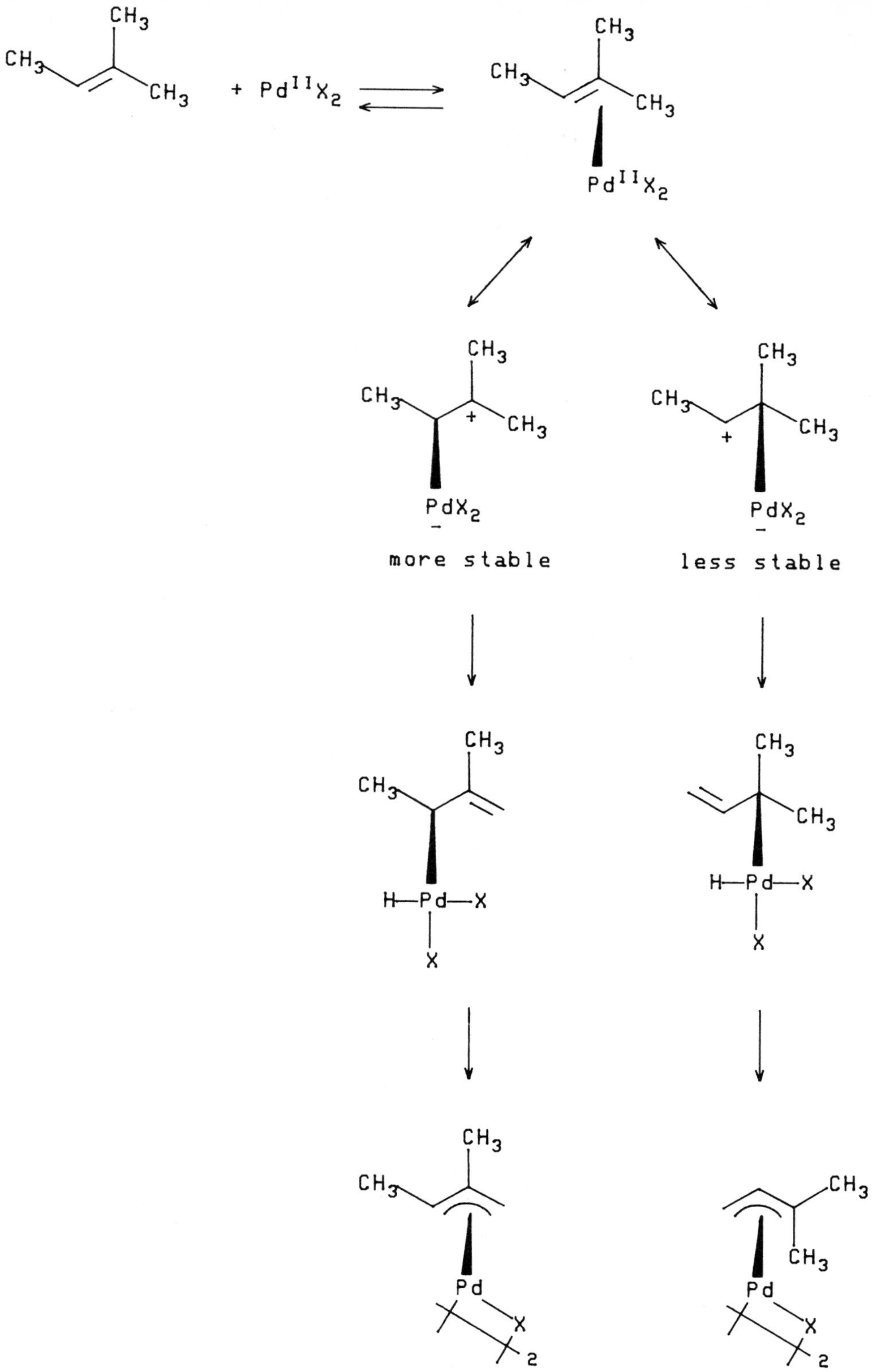

CH_3 CH_3 CH_3 + $Pd^{II}X_2$ ⇌ $Pd^{II}X_2$

PdX_2^- more stable

PdX_2^- less stable

H—Pd—X

Figure 3.7

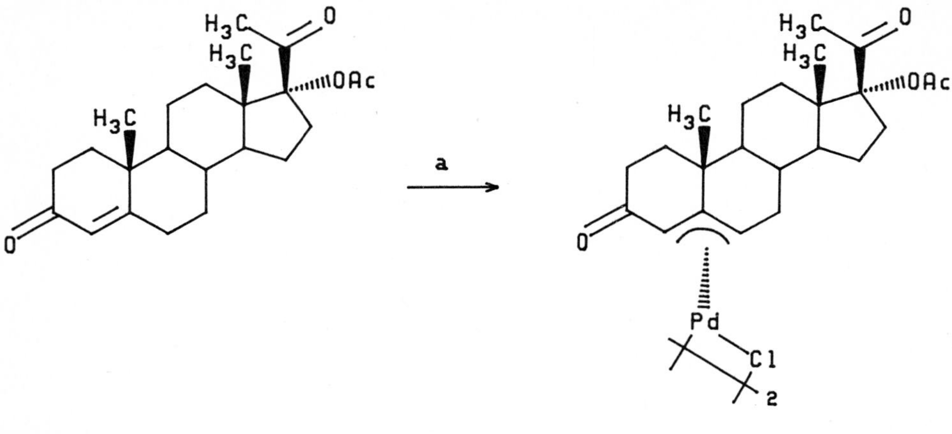

Substrates:

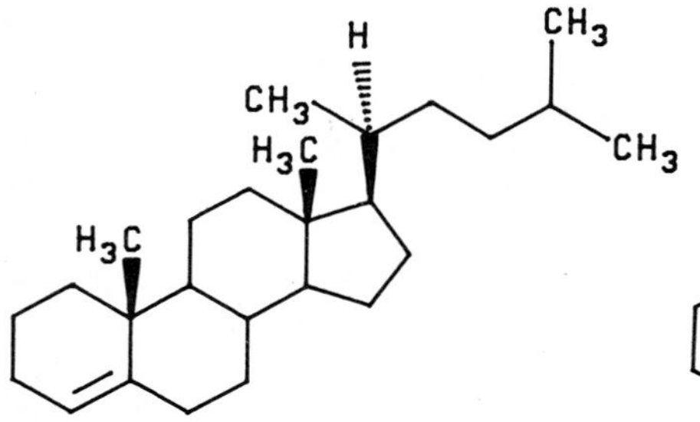

Cholest-4-ene

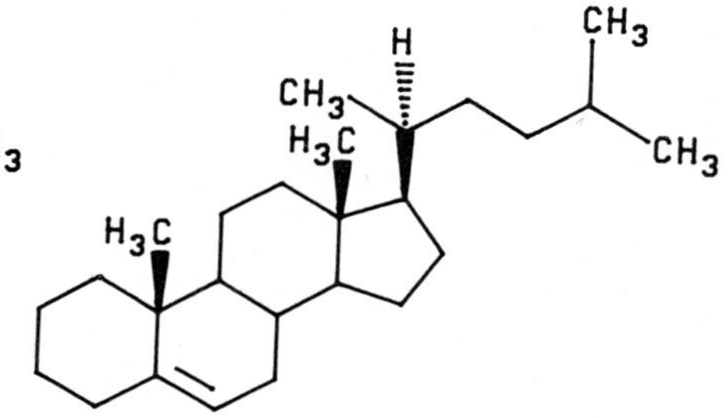

Cholest-5-ene

Products:

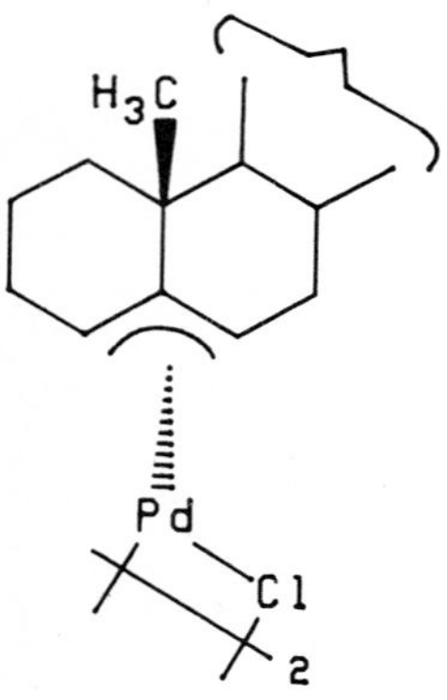

A

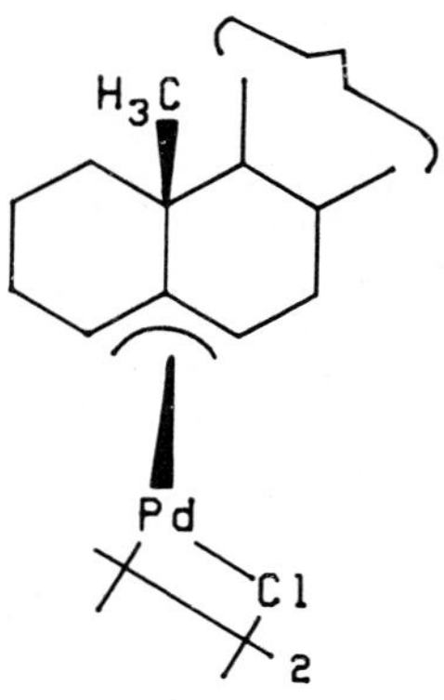

B

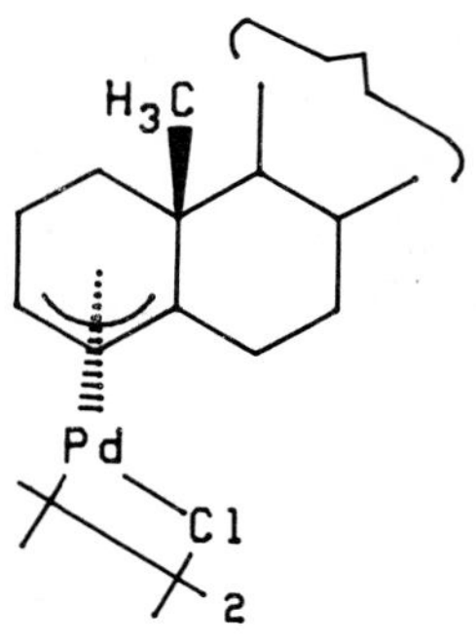

C

Figure 3.8

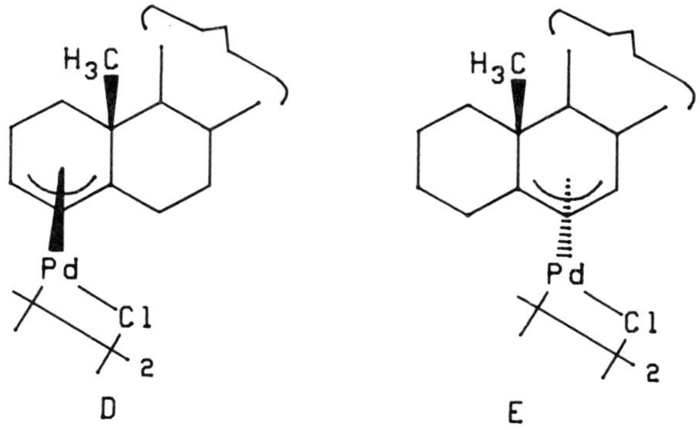

D

E

Cholest-4-ene $\xrightarrow{b}$ A + B

46% 14%

Cholest-5-ene $\xrightarrow{b}$ A

71%

Cholest-4-ene $\xrightarrow{c}$ A + B + C + D

26% 8% 5% 18%

Cholest-5-ene $\xrightarrow{c}$ A + B + E

4% 4% 12%

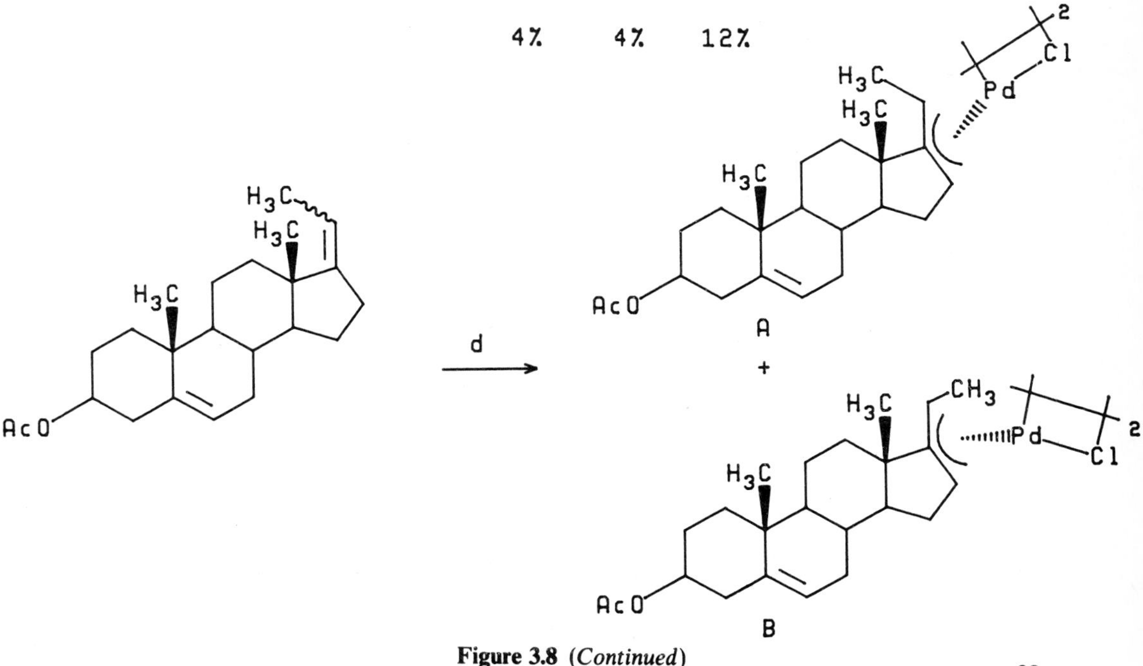

A

+

B

Figure 3.8 (*Continued*)

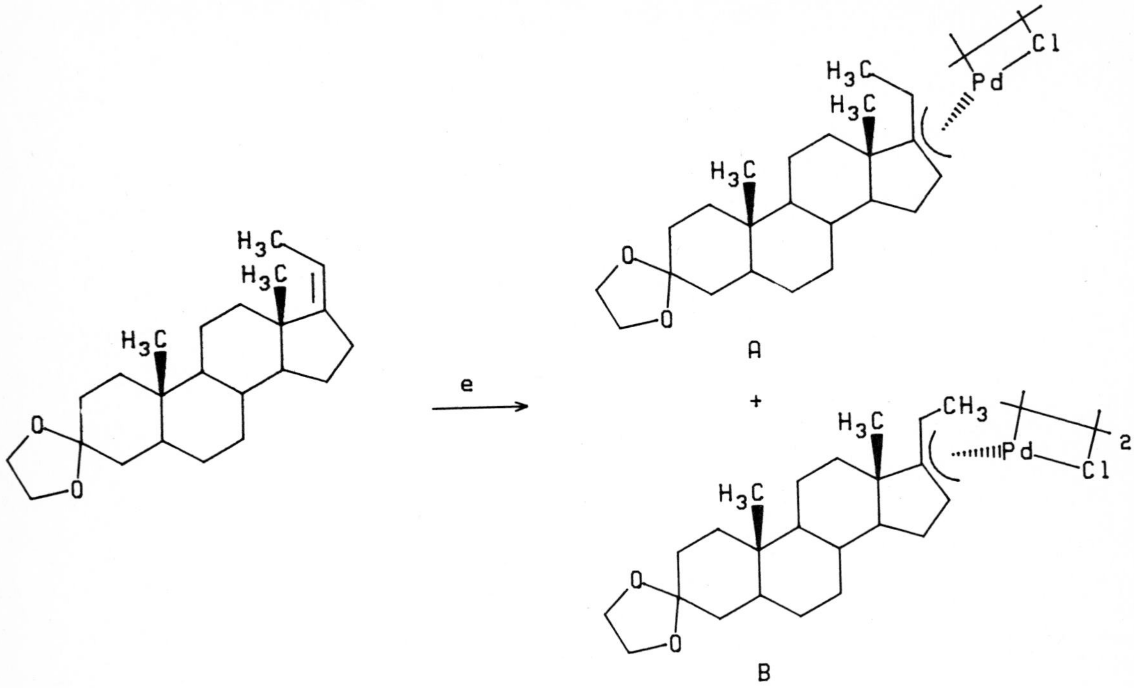

e

A

+

B

Minor product via the β-face π-allyl:

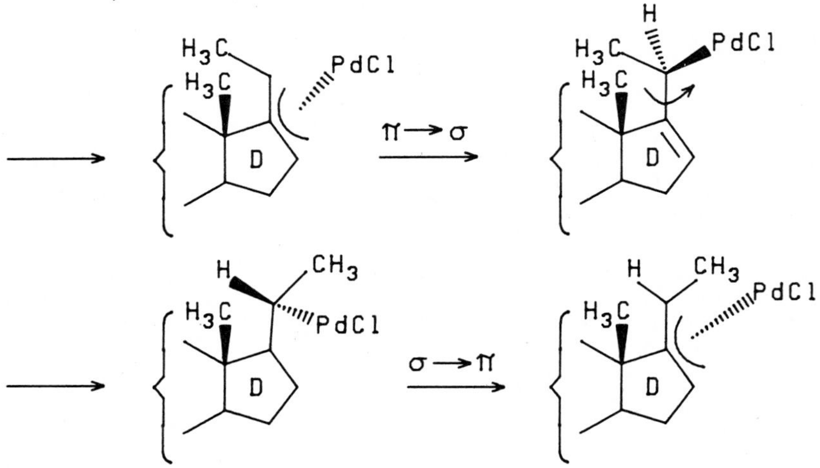

π → σ

σ → π

a. 60°C, $PdCl_2$, DMF

b. rfx, $(PhCN)_2PdCl_2$, $CHCl_3$

c. 50°C, $PdCl_2$, NaCl, NaOAc, Ac_2O, AcOH

Figure 3.8 *(Continued)*

d. $PdCl_2$, NaCl, NaOAc, $CuCl_2$, AcOH;

A:B, 93:7, 92% yield

e. rt, $(CH_3CN)_2PdCl_2$, Na_2CO_3, NaCl, CH_2Cl_2;

A:B, 95:5, 96% yield

Figure 3.8 (*Continued*)

TABLE 3.2 Regioselective π-Allyl Formation Using Palladium Chloride–Copper(II) Chloride and Sodium Acetate in Acetic Acid[11]

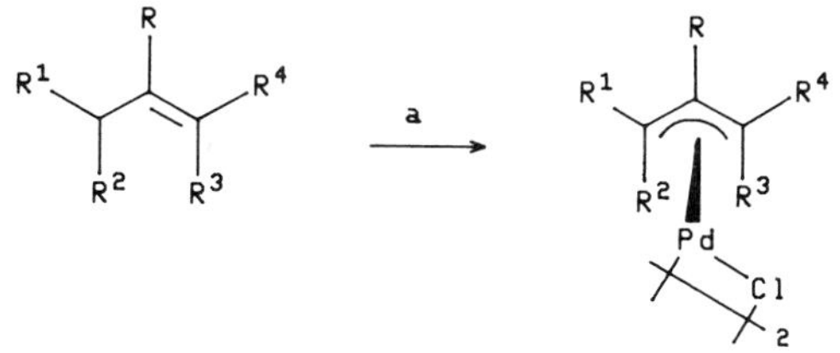

a. rt, $PdCl_2$, NaCl, NaOAc, $CuCl_2$, AcOH

R	R^1	R^2	R^3	R^4	Yield (%)
$-(CH_2)_3-$		H	H	H	66
H	H	$-(CH_2)_3-$		H	100
$-(CH_2)_4-$		H	H	H	86
$-(CH_2)_4-$		H	H	H	92
$-(CH_2)_4-$		H	H	CH_3	68
$-CH_2CH_2N(Ac)CH_2-$		H	H	CH_3	71
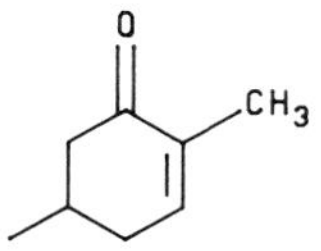 A	H	H	H	H	52
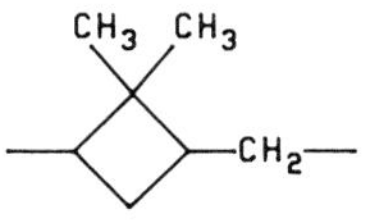 B		H	H	H	60
$-(CH_2CH_2CH(t\text{-}Bu)CH_2-$		H	H	H	90
CH_3	H	H	H	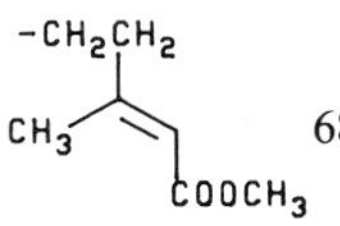 C	68

6. When diastereoisomeric π-allyl complexes can be formed, the major product will have palladium on the less sterically encumbered π-allyl face (e.g., the α face of a steroid)[11,12] (see Figure 3.8).

π-Allylpalladium chloride dimers can be formed by simple heating at 90°C of the alkene with palladium chloride in 50% acetic acid (Table 3.1).[13] Base facilitates the conversion of π-alkene complex to π-allyl complex. Near optimal conditions are achieved with base on addition of an oxidant, copper(II) chloride (Table 3.2).[11] Finally, palladium bis-trifluoroacetate is so electrophilic that π-allyl formation occurs rapidly at room temperature without base. Ligand replacement with tetrabutylammonium chloride affords π-allylpalladium chloride dimers in high yield (Table 3.3).[14]

The utility of this method is limited by (1) the use of a stoichiometric amount of

TABLE 3.3 Regioselective π-Allyl Formation Using Palladium bis-Trifluoroacetate in Various Solvents[14]

a. rt, $Pd[OC(O)CF_3]_2$, solvent

b. rt, $(n\text{-}Bu)_4NCl$

R	R^1	R^2	Yield (%) Acetone	Ethyl Acetate	THF
H	n-Pr	n-Bu	70	—	—
A	H	H	66	—	—
H	$n\text{-}C_7H_{15}$	H	68	64	50
H	$(CH_2)_7COOCH_3$	H	68	78	82
CH_3	$CH_2CH_2C(O)CH_3$	H	70	—	—
B		H	83	—	—
CH_3	H	C	83	—	—

TABLE 3.4 Effect of Reaction Conditions on Regioselectivity of π-Allyl Complex Formation[15]

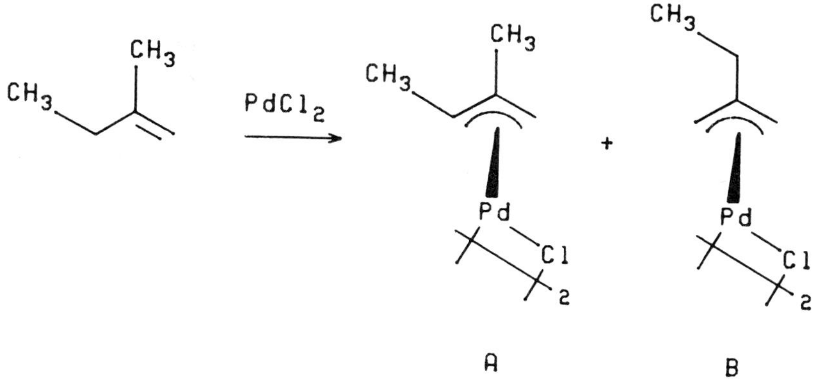

Conditions	Yield (%)	A:B Ratio
Heat, 85°C; 50% AcOH	40	90:10
Heat, 85°C; 50% AcOH + NaOH (pH 3)	69	88:12
20°C; AcOH + NaOAc + NaCl	49	71:29
Heat, 95°C; AcOH + NaOAc + CuCl	89	26:74
20°C; $CHCl_3$ + Na_2CO_3	36	70:30

palladium(II) since subsequent utilization of the π-allyl invariably releases palladium(0) and (2) the often poor regioselectivity of proton abstraction, leading to mixtures of π-allyl complexes (see Table 3.4 and Figure 3.9).[15] Regiospecific π-allyl formation is observed under mild conditions in the special case of substrates with electron-withdrawing groups capable of stabilizing an anion formed by allylic proton abstraction[16].

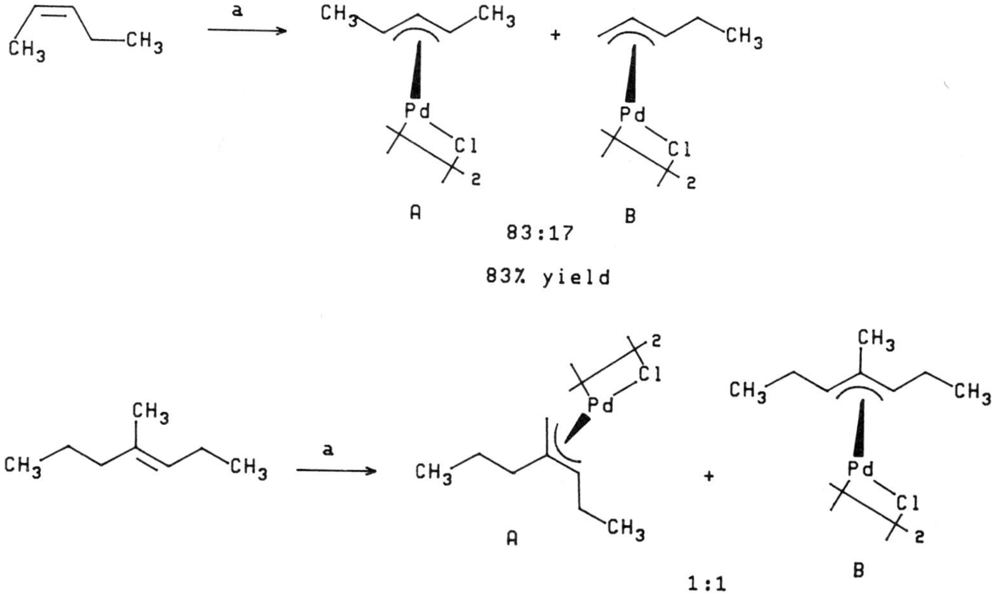

Figure 3.9

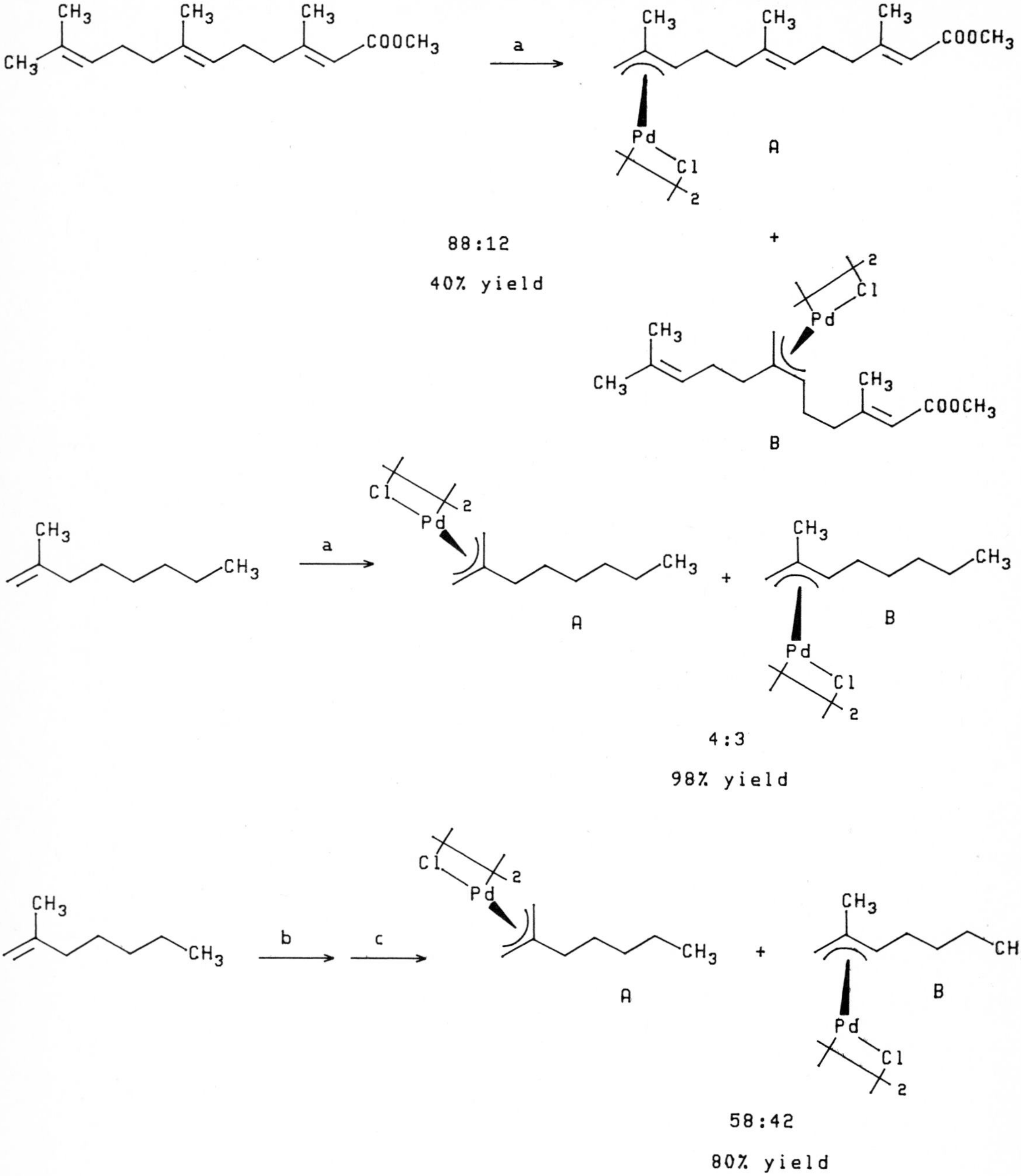

a. $PdCl_2$, $CuCl_2$, NaCl, NaOAc, AcOH; Reference 11

b. rt, $Pd[OC(O)CF_3]_2$, acetone

c. $(n\text{-}Bu)_4NCl$; Reference 14

Figure 3.9 (*Continued*)

3.2.2 Method B. Allylic Functionalized Alkenes and Palladium(0)

The most common method for preparation of π-allylpalladium complexes—specifically for use in situ—is by oxidative addition of allylic functionalized alkenes to palladium(0). There is clear evidence that the oxidative addition is an S_N2 displacement of a heteroatom substituent, typically N, O, S, or halogen, from the allylic carbon by the nucleophilic metal.[17,18] With good leaving groups, for example, Cl or Br, the palladium(0) generated in situ by palladium(II) reduction is sufficiently reactive[19,20] (see Table 3.5).

With other leaving groups, a phosphine complex of palladium(0) is generally required. The phosphine ligands serve a dual purpose: the phosphine facilitates oxidative addition by increasing the electron density on the metal and maintains the metal in a reactive *soluble* complex form. Since phosphine complexes of palladium(0) are somewhat air sensitive, π-allyl formation using a phosphine complex is carried out

TABLE 3.5 π-Allyl Formation from Allylic Chlorides and Palladium(0)[19]

Mechanism with H_2O added:

$$HO{-}Pd(Cl)(CO)(CH_2{=}CHCH_2Cl) \longrightarrow Pd(Cl)(C(O)OH)(CH_2{=}CHCH_2Cl) \longrightarrow ClPd(\pi\text{-allyl}) + CO_2 + HCl$$

a. CO as reducing agent, Na_2PdCl_4 or NaCl, $PdCl_2$, CH_3OH

A = Na_2PdCl_4, B = NaCl, $PdCl_2$

R^1	R^2	R^3	A or B	Yield (%)
H	H	H	A	84
H	H	H	B	68, $CaCl_2$ added
H	CH_3	H	A	93
H	CH_3	H	B	76, $CaCl_2$ added
H	H	CH_3	A	98
CH_3	H	H	A	84
CH_3	H	H	B	21
CH_3	H	H	B	100, H_2O aded
CH_3	H	H	B	98, $CaCl_2$ added

under an inert atmosphere. The process can be run in aromatic hydrocarbons, ethers, or highly polar solvents such as acetonitrile or dimethylsulfoxide. To minimize metal precipitation prior to high conversion to the π-allyl complex, the phosphine-to-palladium ratio is generally equal to or greater than 4:1.

The oxidative addition is more facile with complexes having chelating diphosphine ligands. A five- or six-membered ring chelate maintains two phosphines at adjacent sites in the square planar metal coordination sphere, leaving two adjacent sites vacant, and thus provides an optimal arrangement for oxidative addition[21] (see Figure 3.10).

Both E and Z alkene isomers will give the π-allyl with a terminal substituent in the more sterically favorable *syn* position. Thus, a π-allyl complex may have as many as four suitable precursors (see Figure 3.11).

There appear to be few limitations on the nature of the π-allyl substituents. Generally, at least one terminal substituent is hydrogen. (That is, at least one terminal

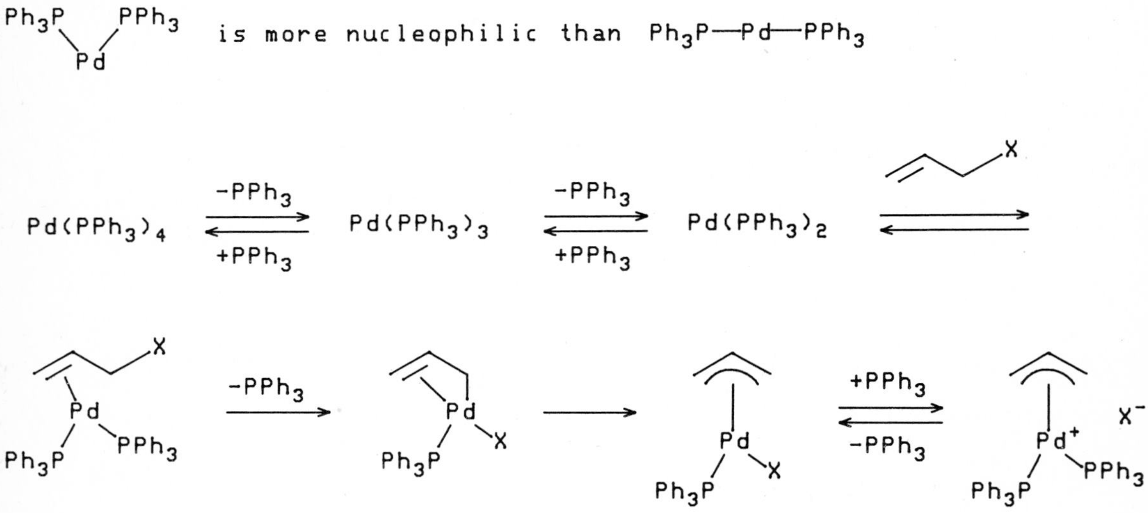

Ph3P PPh3 Pd is more nucleophilic than Ph3P—Pd—PPh3

Pd(PPh3)4 ⇌ (−PPh3 / +PPh3) Pd(PPh3)3 ⇌ (−PPh3 / +PPh3) Pd(PPh3)2 ⇌

−PPh3 → Pd X Ph3P → Pd X Ph3P ⇌ (+PPh3 / −PPh3) Pd+ PPh3 Ph3P X−

Figure 3.10

X R R1 → R R1 Pd X ← X R1 R

X R R1 → ← R X R1

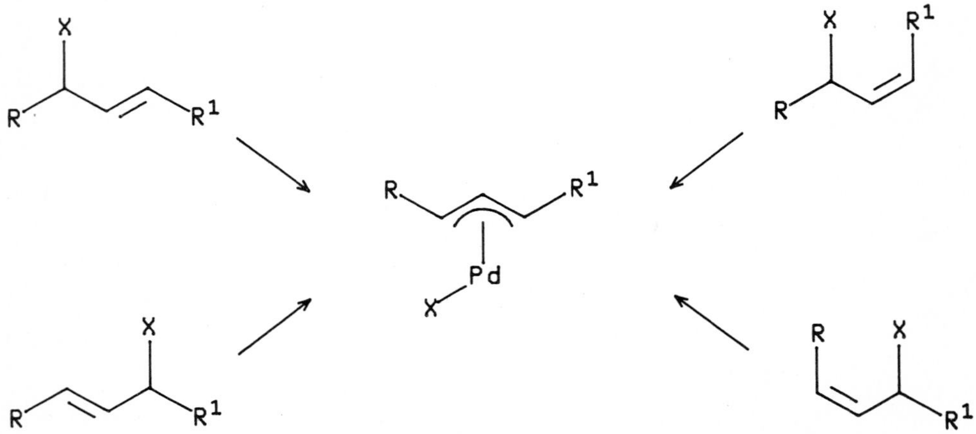

Figure 3.11

carbon of the allylic substrate is sterically accessible.) Displacement with inversion has been observed when the carbon bearing the leaving group is secondary. The stereochemistry of displacement of a leaving group from a tertiary carbon has not been determined. An $S_N'2$ mechanism would give π-allyl with retention; an S_N1 mechanism would give a mixture of π-allyls.

While the initial displacement by an S_N2 mechanism is stereospecific, the π-allyl formed may not be configurationally stable under conditions required for its generation. For example, Trost suggested that this could occur by conversion of a π-allylpalladium acetate to an inverted allyl acetate by reductive elimination followed by an S_N2 displacement of this inverted acetate.[18a] Stille suggested an alternative mechanism involving backside displacement of the metal by another metal.[17] In either case, subsequent reaction of the in-situ-generated π-allyl should be rapid to capitalize on the stereospecific inversion in initial formation of the π-allyl (see Figure 3.12).

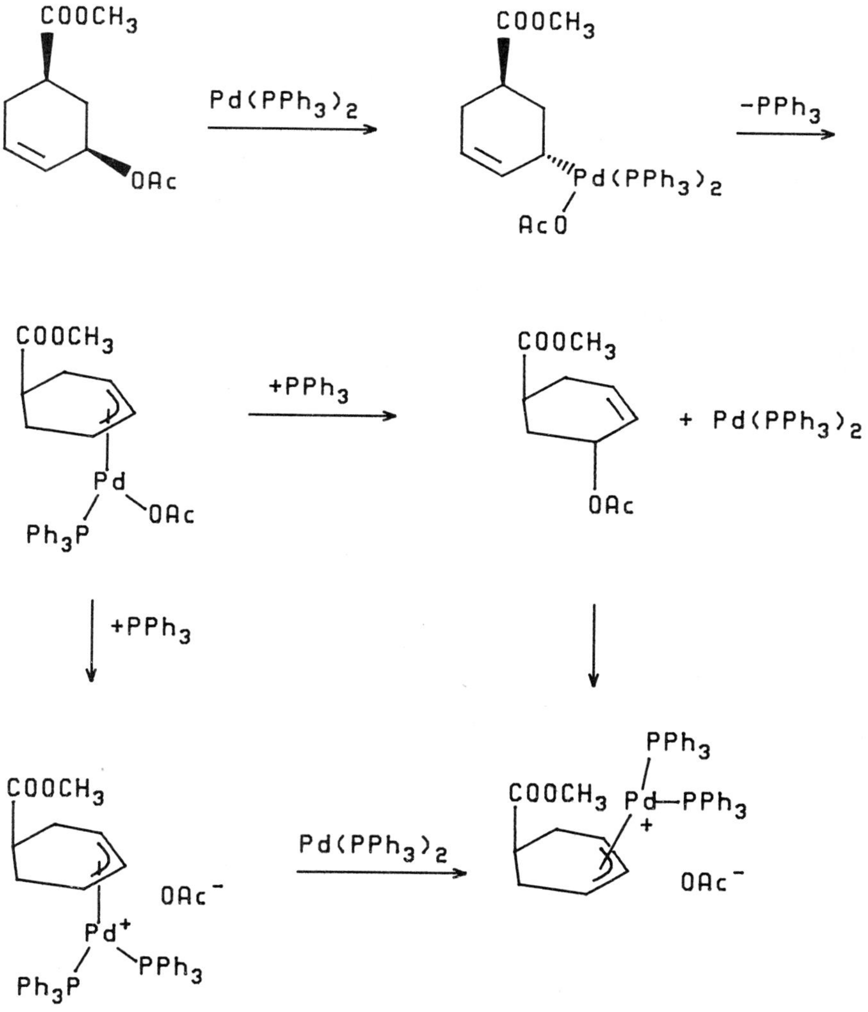

COOCH3 — OAc —[Pd(PPh3)2]→ COOCH3 — Pd(PPh3)2 / AcO —[−PPh3]→

COOCH3 — Pd / Ph3P / OAc —[+PPh3]→ COOCH3 — OAc + Pd(PPh3)2

↓ +PPh3 ↓

COOCH3 — Pd+ / Ph3P / PPh3 OAc⁻ —[Pd(PPh3)2]→ COOCH3 Pd(PPh3)3 + OAc⁻

Figure 3.12

π-Allyl formation by method B and reaction of the π-allyl—releasing palladium(0)—can be carried out in one pot using a *catalytic amount of metal*. This catalytic cycle together with regiospecificity of preparation and, if conditions are carefully chosen, stereospecificity of preparation, make method B the method of choice for π-allyl synthesis in most instances.

3.2.3 Method C. 1,3-Dienes and Palladium(II)

π-Allylpalladium complexes can be formed from 1,3-dienes and palladium(II) by two mechanisms with different stereochemical consequences. With palladium(II) halide complexes, selective π-alkene complexation with the more sterically accessible π-bond is followed by *anti* attack by a weak nucleophile (halide ion or, more typically, the solvent) at a terminal carbon to give a σ-alkyl complex capable of rapid isomerization to a π-allyl. Donati and Conti reported the preparation of the butadiene–$PdCl_2$ π-complex at low temperature and the nearly quantitative conversion to the π-allyl complex when warmed to room temperature[22] (see Figure 3.13).

Water, alcohols,[23] carboxylic acids,[24] amines,[25] and arylsulfinate salts[26] are also suitable nucleophiles (Table 3.6). While yields are not often high, the starting materials are readily available and workup is trivial. A copper(II) salt can be added to prevent palladium(0) precipitation.

With organopalladium halide complexes, selective π-alkene complexation with the more sterically accessible π-bond is followed by *syn*-migration of the σ-bonded organic ligand from the metal to the terminal diene carbon (so-called migratory insertion) to give a σ-complex capable of isomerization to a π-allyl. The organopalladium halide can be generated in two ways. Organopalladium chlorides prepared by transmetallation of organomercuric chlorides using palladium(II) at low temperature are used to prepare π-allylpalladium chloride dimers (Table 3.7).[27] Again, yields are not often high, but

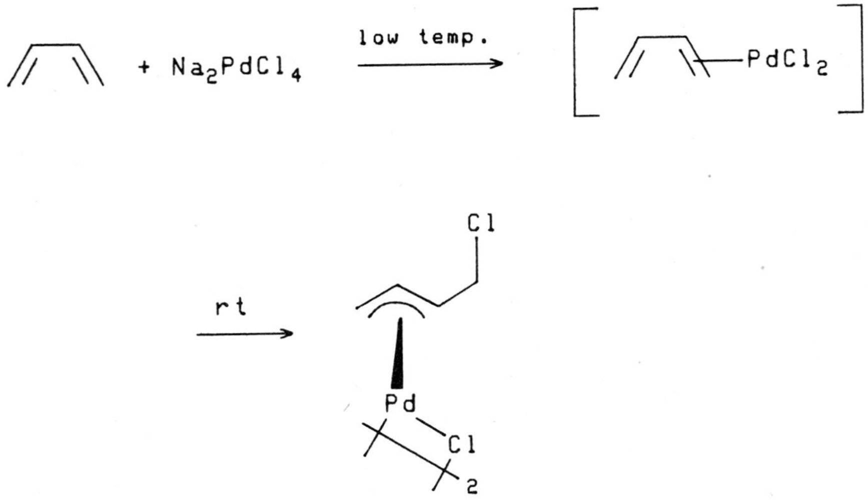

+ Na_2PdCl_4 —low temp.→ [—$PdCl_2$]

—rt→ Cl, Pd, Cl, 2

Figure 3.13

TABLE 3.6 π-Allyl Complex Formation by Nucleophilic Attack on Palladium(II)-Diene π-Complexes[23,24d]

a. rt, Cu(II), NuH

R^1	R^2	R^3	R^4	R^5	R^6	PdX_2	NuH	Cu(II)	Yield (%)
H	H	H	H	H	H	$PdCl_2$	AcOH	—	50
H	H	H	H	H	H	$PdCl_2$	AcOH	$Cu(OAc)_2$	89
H	H	CH_3	H	H	H	$PdCl_2$	AcOH	—	43
H	H	CH_3	H	H	H	$PdCl_2$	AcOH	$Cu(OAc)_2$	60
H	H	CH_3	H	H	H	PdI_2	AcOH	$Cu(OAc)_2$	52
H	H	CH_3	H	H	H	$PdBr_2$	AcOH	$Cu(OAc)_2$	56
H	H	CH_3	H	H	H	$PdCl_2$	EtCOOH	$Cu(OAc)_2$	40
H	CH_3	H	H	H	H	$PdCl_2$	AcOH	$Cu(OAc)_2$	46
H	H	H	H	H	H	Na_2PdCl_4	MeOH	—	75
H	H	CH_3	CH_3	H	H	Na_2PdCl_4	MeOH	—	76
H	H	CH_3	H	H	H	Na_2PdCl_4	MeOH	—	81
H	H	H	H	CH_3	CH_3	Na_2PdCl_4	MeOH	—	80
H	H	CH_3	H	CH_3	CH_3	Na_2PdCl_4	MeOH	—	68
CH_3	CH_3	H	H	CH_3	CH_3	Na_2PdCl_4	MeOH	—	74
H	H	CH_3	H	H	H	Na_2PdCl_4	EtOH	—	73
H	H	H	H	CH_3	CH_3	Na_2PdCl_4	EtOH	—	37
CH_3	CH_3	H	H	CH_3	CH_3	Na_2PdCl_4	EtOH	—	57

the starting materials are readily available and workup is trivial. Organopalladium bromides and iodides prepared by oxidative addition of aryl and alkenyl bromides and iodides to palladium(0) at relatively high temperatures, typically 100°C, are used in preparation of π-allylpalladium complexes for subsequent reaction in situ.[28]

The two mechanisms for π-allyl formation are illustrated in Figure 3.14 using 1,3-cyclohexadiene.

While a variety of functional groups may be compatible under mild conditions, it is clear that alternative sites of unsaturation may lead to unexpected cyclization by carbon–carbon bond formation. Ocimine gave the expected π-allyl complex with sodium tetrachloropalladate in methanol or acetic acid; isomeric myrcine afforded a π-allyl complex in methanol as only a minor product.[29] The unexpected cyclization could be inhibited by addition of hexamethylphosphoric triamide[30] (see Table 3.8 and Figure 3.15).

TABLE 3.7 π-Allyl Complex Formation from Organopalladium Halides and 1,3-Dienes[27]

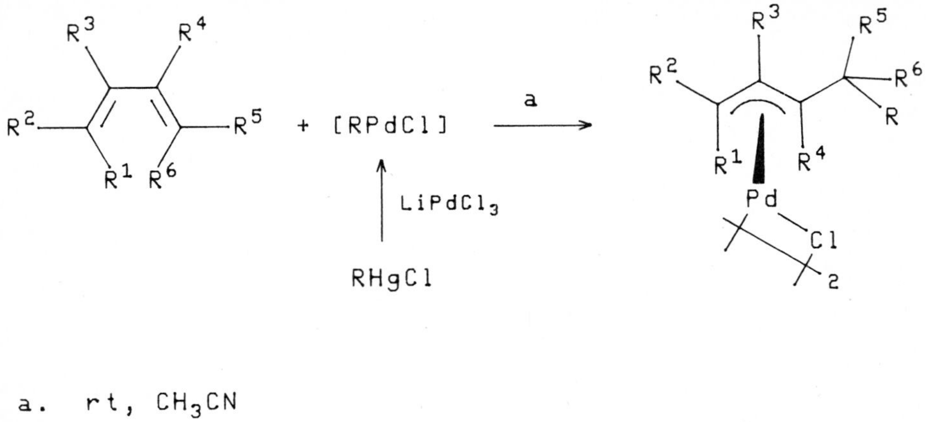

a. rt, CH_3CN

R^1	R^2	R^3	R^4	R^5	R^6	R	Yield (%)
H	H	H	H	H	H	Ph	48
H	H	CH_3	H	H	H	Ph	58
H	H	CH_3	CH_3	H	H	Ph	63
H	CH_3	H	CH_3	H	H	Ph	44
H	CH_3	CH_3	H	H	H	Ph	33
CH_3	CH_3	H	H	H	H	Ph	96
H	CH_3	H	H	CH_3	H	Ph	33
H	$(CH_2)_2CH_3$	H	H	CH_3	H	Ph	51
H	$COOCH_3$	H	H	CH_3	H	Ph	74
CH_3	CH_3	H	CH_3	H	H	Ph	49
H	H	CH_3	H	H	H	*m*-CHO-Ph	20
H	H	CH_3	H	H	H	*o,p*-$(OCH_3)_2$-Ph	22
H	H	H	H	H	H	1-Naphthyl	10
H	H	H	H	H	H	Ferrocenyl	35
H	H	CH_3	H	H	H	Ferrocenyl	24
H	H	CH_3	CH_3	H	H	Ferrocenyl	29
H	H	H	H	H	H	CH_3	30
H	H	CH_3	H	H	H	$COOCH_3$	21
H	H	CH_3	H	H	H	H[a]	39
H	CH_3	H	H	H	H	H[a]	75

[a]From β-elimination of RPdCl, R = *n*-Bu.

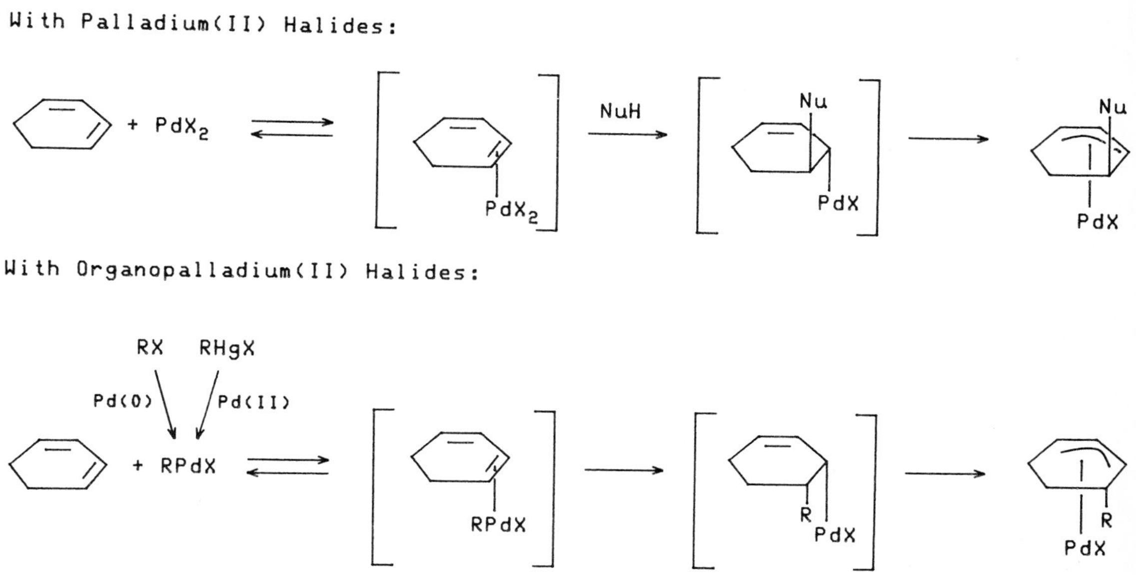

With Palladium(II) Halides:

$$\text{cyclohexene} + PdX_2 \rightleftharpoons [\pi\text{-complex}, PdX_2] \xrightarrow{NuH} [\text{Nu}, PdX] \longrightarrow \pi\text{-allyl}(Nu)PdX$$

With Organopalladium(II) Halides:

$$RX \xrightarrow{Pd(0)} ;\quad RHgX \xrightarrow{Pd(II)}$$

$$\text{cyclohexene} + RPdX \rightleftharpoons [\pi\text{-complex}, RPdX] \longrightarrow [R, PdX] \longrightarrow \pi\text{-allyl}(R)PdX$$

Figure 3.14

TABLE 3.8 Cyclizations Inhibited by Addition of Hexamethylphosphoric Triamide[30]

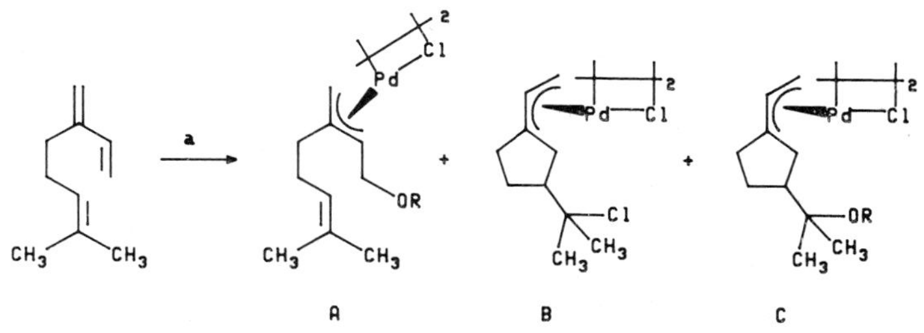

a. rt, $PdCl_2(CH_3CN)_2$, solvent

Solvent	Yield of A(%)	Yield of B(%)	Yield of C(%)
H_2O-HMPA	75	4	4
H_2O-Acetone	—	60	30
H_2O-Acetone + Li_2CO_3	25	6	30
H_2O-DMF + Li_2CO_3	33	7	6
EtOH	—	26	35
EtOH-HMPA	57	5	—
i-PrOH	—	48	43
i-PrOH-HMPA	44	15	—
AcOH	—	51	30
AcOH-HMPA + LiOAc	64	—	—

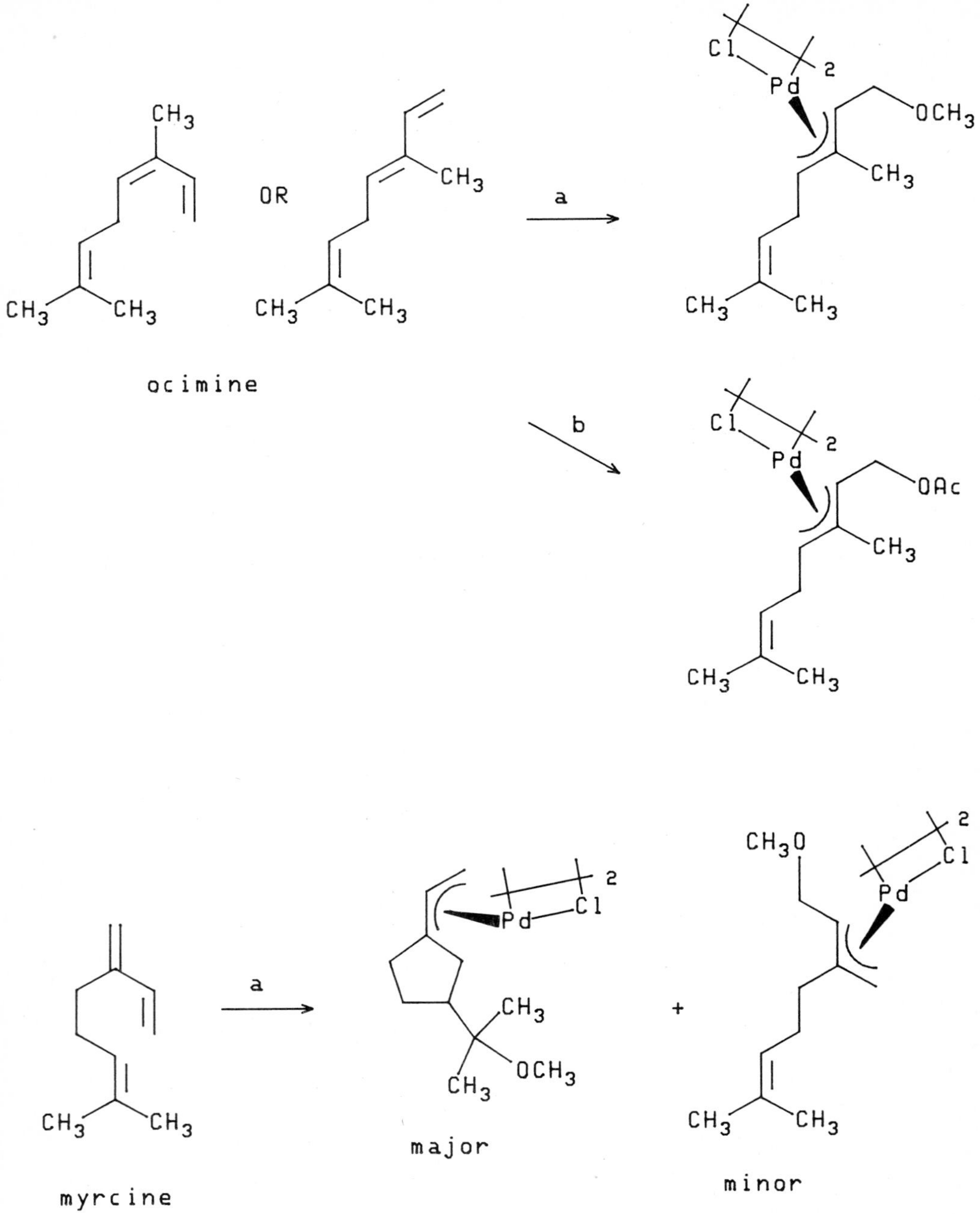

ocimine

a

b

myrcine

a

major

+

minor

Figure 3.15

Major product via:

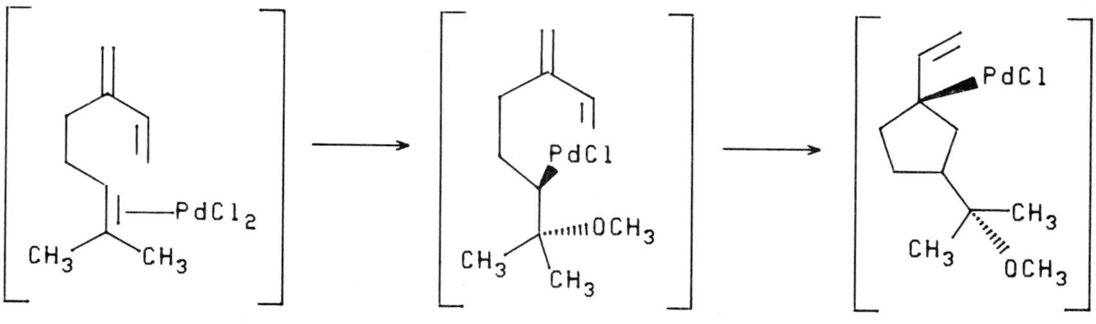

a. rt, Na_2PdCl_4, CH_3OH

b. rt, Na_2PdCl_4, AcOH, NaOAc; or $PdCl_2(PhCN)_2$, AcOH

Figure 3.15 (*Continued*)

3.2.4 Method D. Palladium Migrations: Alkenylpalladium Halides and Alkenes, Organopalladium Halides and Nonconjugated Dienes

Like arylpalladium halides (Chapter 2), alkenylpalladium halides add to alkenes, forming σ-complexes ("alkene insertion"). β-Hydride elimination affords a diene-HPdX complex. Readdition of HPdX in the opposite direction affords a σ-allyl complex that collapses to the more stable π-allyl form. Since addition of the alkenylpalladium halide, β-elimination of HPdX, and readdition of HPdX are all *syn* processes, only two diastereoisomeric complexes are possible (see Figure 3.16).

That the diene-HPdX π-complex may cleave and reform prior to HPdX decomposition is clearly indicated in several examples of π-allyl formation using cyclic alkenes.[31] Base facilitates HPdX decomposition, leading to formation of 1,4-dienes (see Figure 3.17).

Two routes to the vinylpalladium halide complexes have been used: (1) transmetallation of vinylmercuric halide[32] to palladium(II) is the initial step in an efficient synthesis of π-allylpalladium chloride dimers[33] (Table 3.9) and (2) the oxidative addition of vinyl halides to palladium(0) is used to prepare π-allyl complexes for subsequent reaction in situ.[34] Transmetallation is accomplished under quite mild conditions, 0°C to room temperature, while oxidative addition of an alkenyl iodide or bromide often requires temperatures in excess of 100°C. In both cases the alkenylpalladium halide is formed with retention of precursor *E*, *Z* stereochemistry (see Figure 3.18).

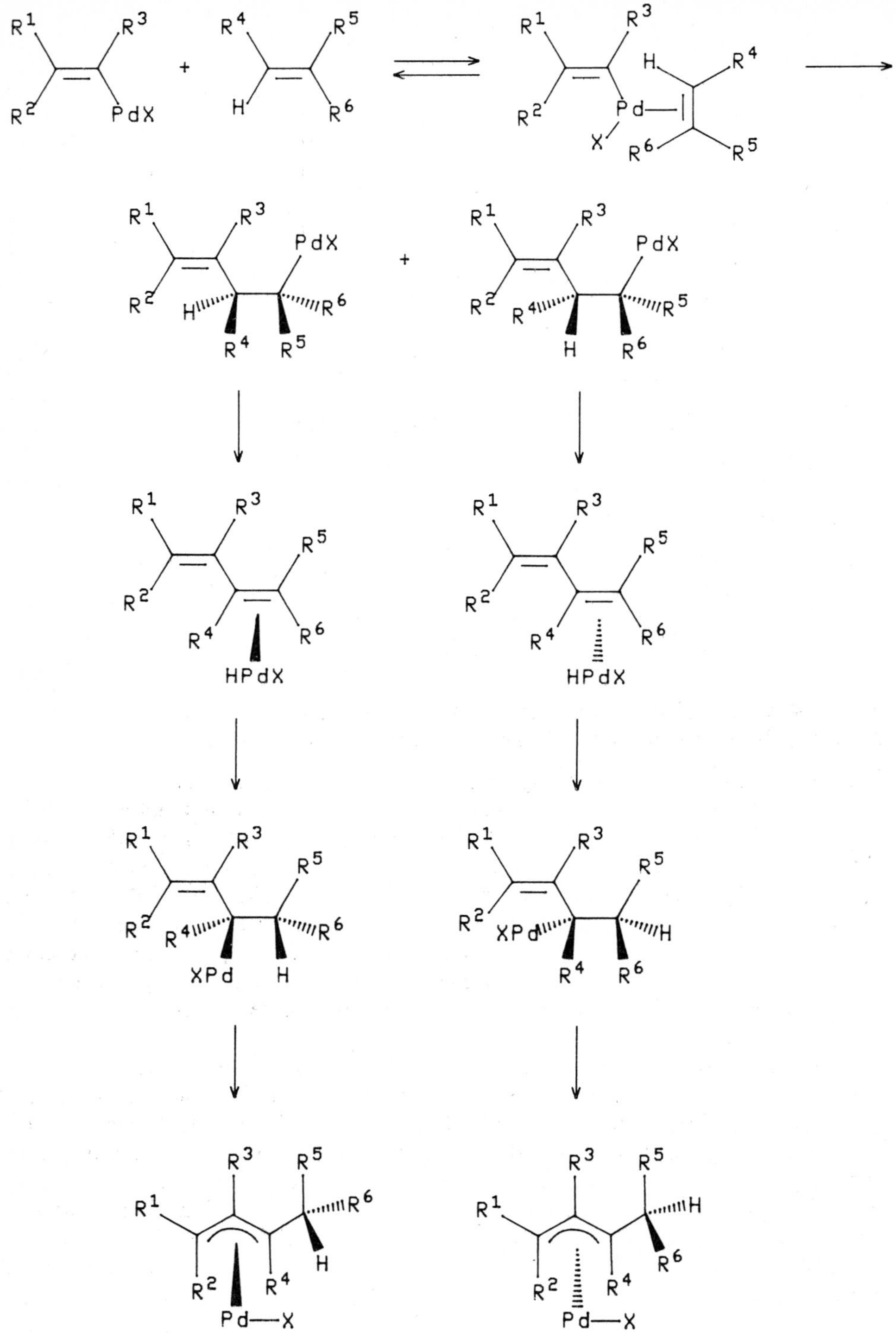

Figure 3.16

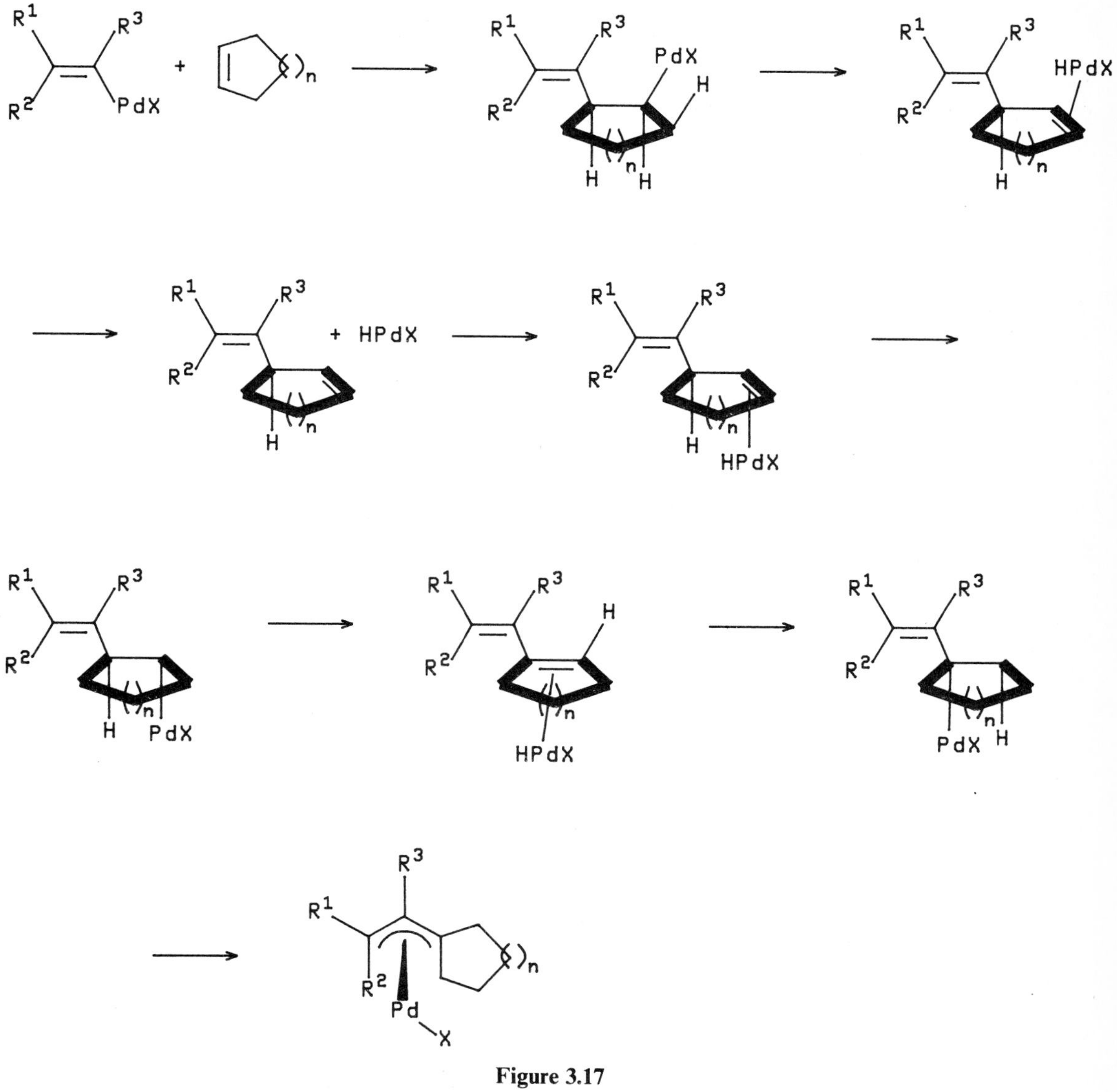

Figure 3.17

The method is limited by (1) reversibility of coordination of the alkene to the vinylpalladium halides (as the number or steric bulk of alkene substituents increases, the equilibrium favors free alkene and the π-allyl formation becomes less efficient); (2) poor regioselectivity of alkene insertion in cases where R^4, R^5, and R^6 do not provide a pronounced steric or electronic differentiation of the alkene carbons; and (3) formation of *syn-anti* mixtures when terminal π-allyl substituents are similar in size (see Figures 3.19 and 3.20).

Larock has extended the potential utility of this method considerably by demonstrating efficient HPdX migration several carbons along a chain. Thus, organopalladium chlorides (from organomercuric chlorides by transmetallation) add to

TABLE 3.9 π-Allylpalladium Chloride Dimer Formation by Method D[33]

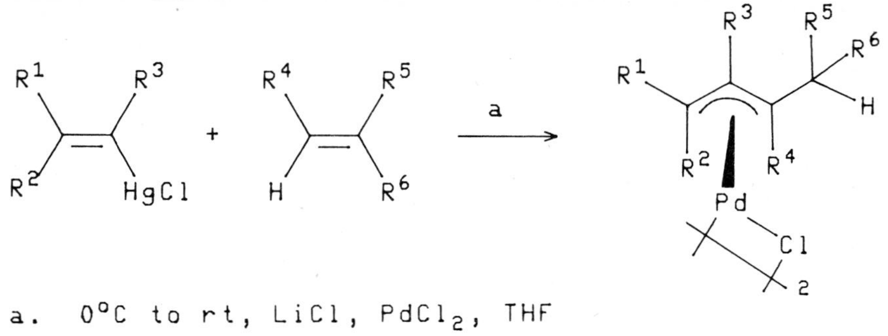

a. 0°C to rt, LiCl, $PdCl_2$, THF

R^1	R^2	R^3	R^4	R^5	R^6	Yield (%) Crude	Recrystallized
$CH_3(CH_2)_3$	H	H	H	COOEt	H	66	—
Ph	H	H	H	COOEt	H	100	58
t-Bu	H	CH_3	H	COOEt	H	97	83
t-Bu	H	H	H	COOEt	H	90	82
t-Bu	H	H	H	CN	H	87	71
t-Bu	H	H	H	$COCH_3$	H	100	67
t-Bu	H	H	H	$COOCH_3$	CH_3	29	22
t-Bu	H	H	H	H	H	92	—
t-Bu	H	H	H	CH_3	CH_3	41	21

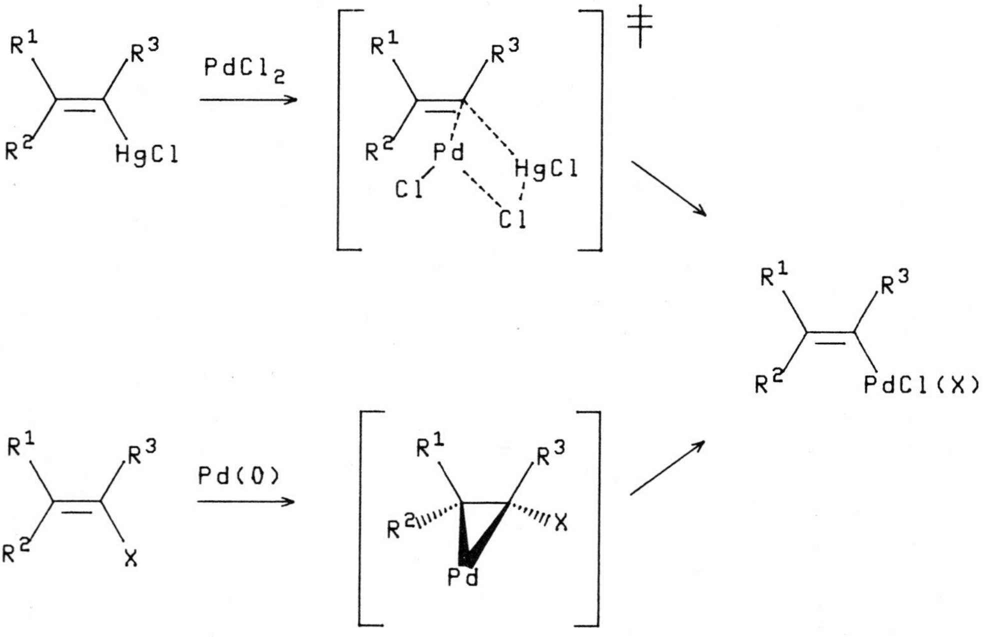

Figure 3.18

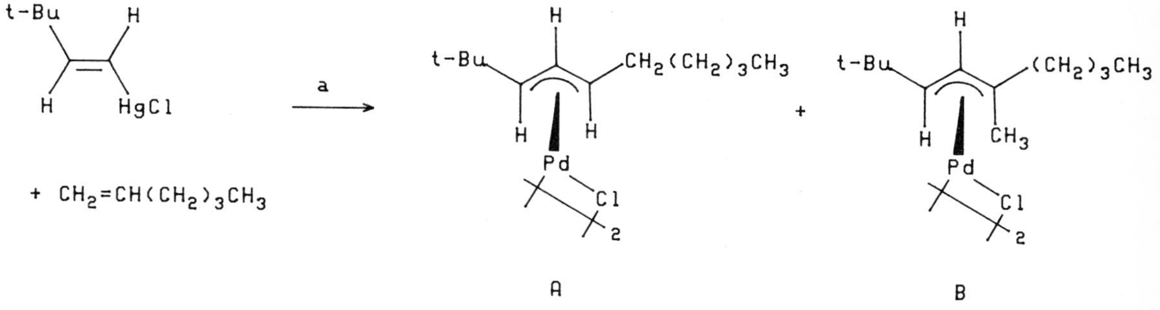

a. 0°C to rt, LiCl, $PdCl_2$, THF; A:B, 5:1, 67% crude yield

Figure 3.19

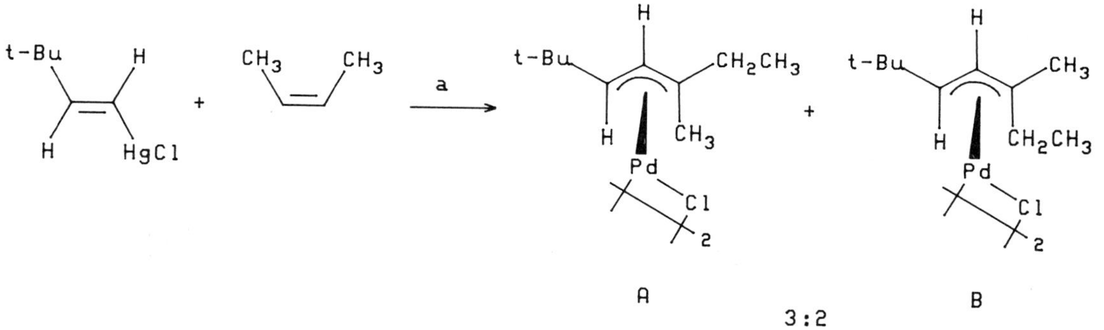

3:2

59% crude yield

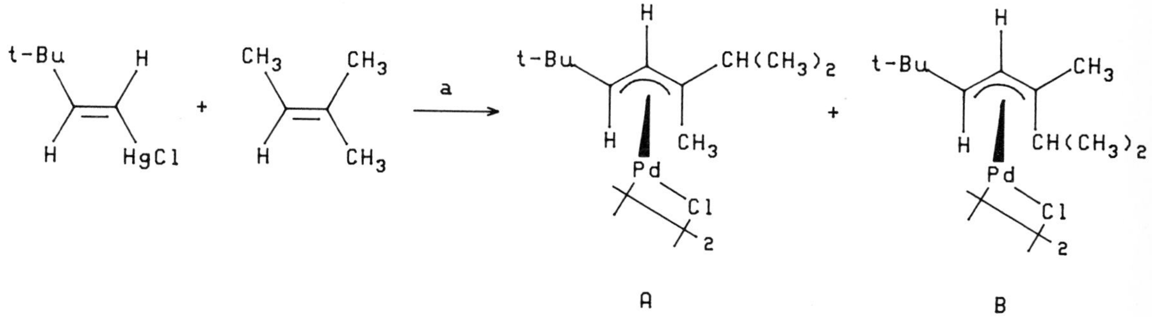

7:3

19% crude yield

a. 0°C to rt, LiCl, $PdCl_2$, THF

Figure 3.20

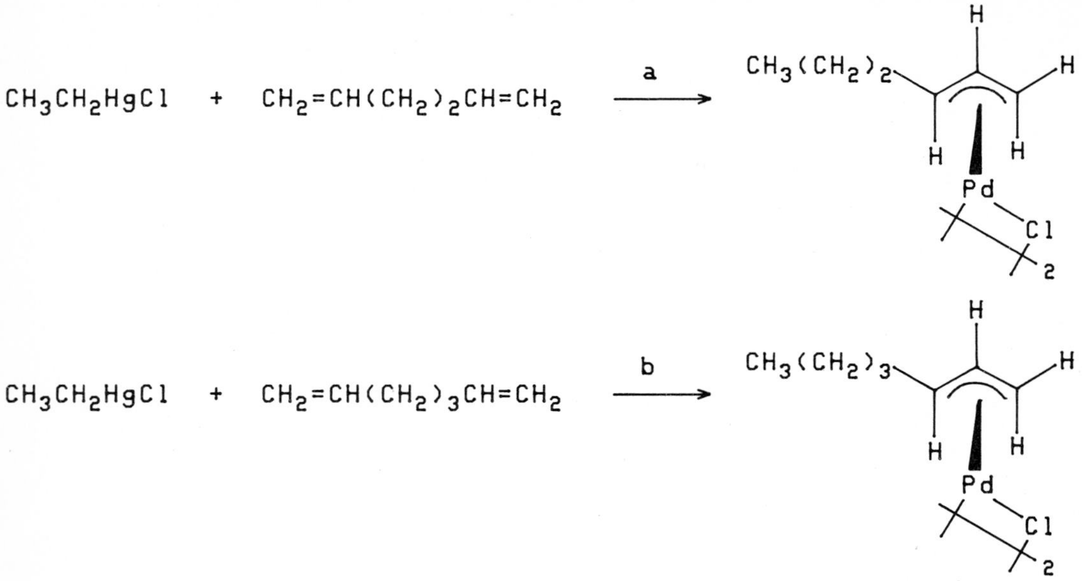

CH_3CH_2HgCl + $CH_2{=}CH(CH_2)_2CH{=}CH_2$ $\xrightarrow{a}$ [π-allylpalladium chloride dimer: $CH_3(CH_2)_2$–C(H)–C(H)–C(H)–H, PdCl, 2]

CH_3CH_2HgCl + $CH_2{=}CH(CH_2)_3CH{=}CH_2$ $\xrightarrow{b}$ [π-allylpalladium chloride dimer: $CH_3(CH_2)_3$–C(H)–C(H)–C(H)–H, PdCl, 2]

a. −78°C to rt, LiCl, $PdCl_2$, THF; 55% crude yield

b. 0°C to rt, LiCl, $PdCl_2$, THF; 61% crude yield

Figure 3.21

nonconjugated dienes to give moderate to excellent yields of π-allylpalladium chloride dimers. The HPdX must remain strongly coordinated until the migration is complete since a single π-allyl complex is isolated in many cases[35] (see Figure 3.21).

Other general methods for synthesis of π-allylpalladium complexes of less utility primarily due to the relative inaccessibility of starting materials are:

Method E. From Main-Group Metal–Allyl Complexes and Palladium(II)[36,37]
Method F. From Cyclopropanes or Cyclopropenes and Palladium(II)[38]

3.3 π-ALLYLPALLADIUM COMPLEXES: REACTIONS

3.3.1 Palladium Displacement by Nucleophiles

The most synthetically significant reaction of π-allyl complexes is metal displacement by nucleophiles. The π-allyl complex must first be activated by conversion to a cationic complex. Activation has been accomplished by halide precipitation with silver tetrafluoroborate[24a,25,39] or by addition of good ligands for palladium, at least two equivalents per metal atom (see Figure 3.22).

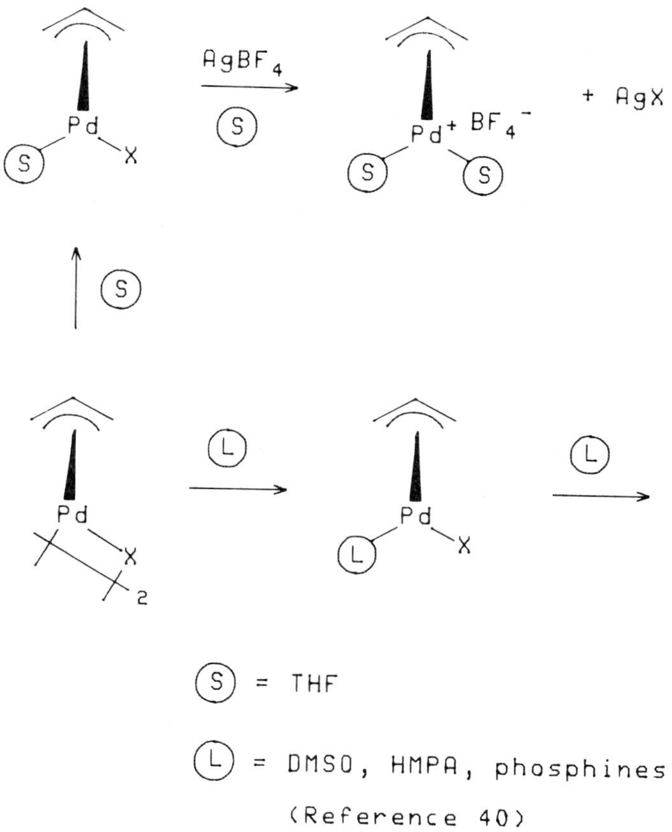

(Reference 40)

Figure 3.22

Since π-allyl preparation by method B is carried out using a phosphine-activated and soluble form of palladium(0), preparation by method B in the presence of a nucleophile results in π-allyl formation, activation, and metal displacement in one pot usng a catalytic amount of metal (see Figure 3.23).

The allylic leaving group can be halogen or some oxygen, nitrogen, or sulfur functionality (typically, OR, OAr, OC(O)R, OC(O)OR, NR_2, SO_2R, SO_2Ar, NO_2). While allylic halogen can be displaced by a classical nucleophilic substitution mechanism, the oxygen, nitrogen and sulfur functionalities cannot. In addition, classical substitution and substitution via π-allylpalladium chemistry may have opposite stereochemical consequences. Stereochemistry of substitution via π-allylpalladium complexes is primarily controlled by the "hard' or "soft" character of the nucleophile. Finally, the regioselectivity of nucleophilic substitution using unsymmetrical allylic reagents is quite different for classical substitution and substitution via π-allylpalladium complexes. The regioselectivity of substitution via π-allylpalladium complexes is affected by (1) the structure of the substrate, (2) the structure of the nucleophile, and (3) the choice of π-allyl-activating ligands.

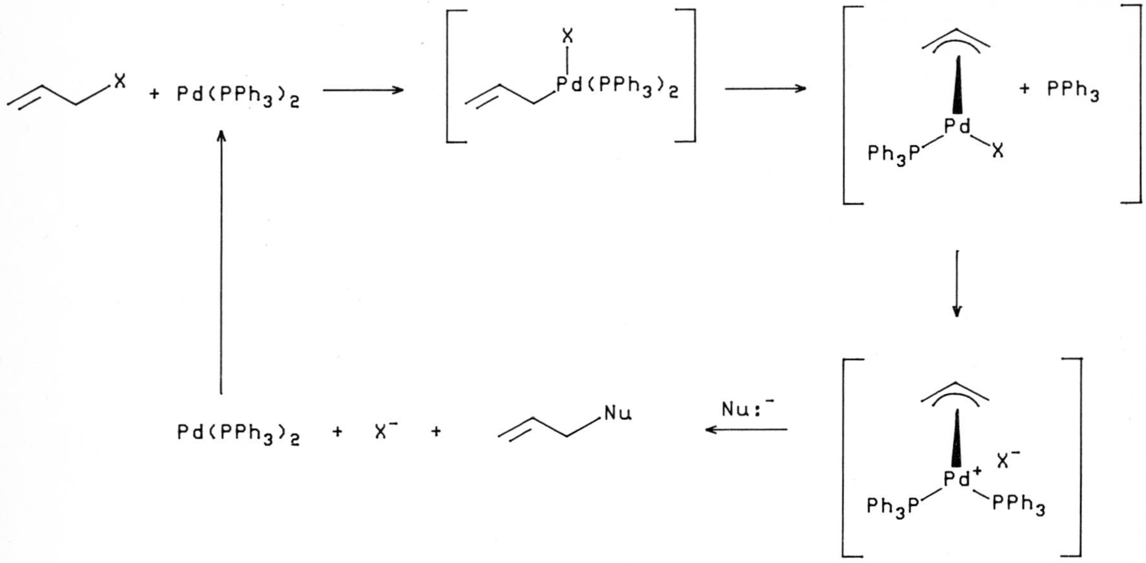

Figure 3.23

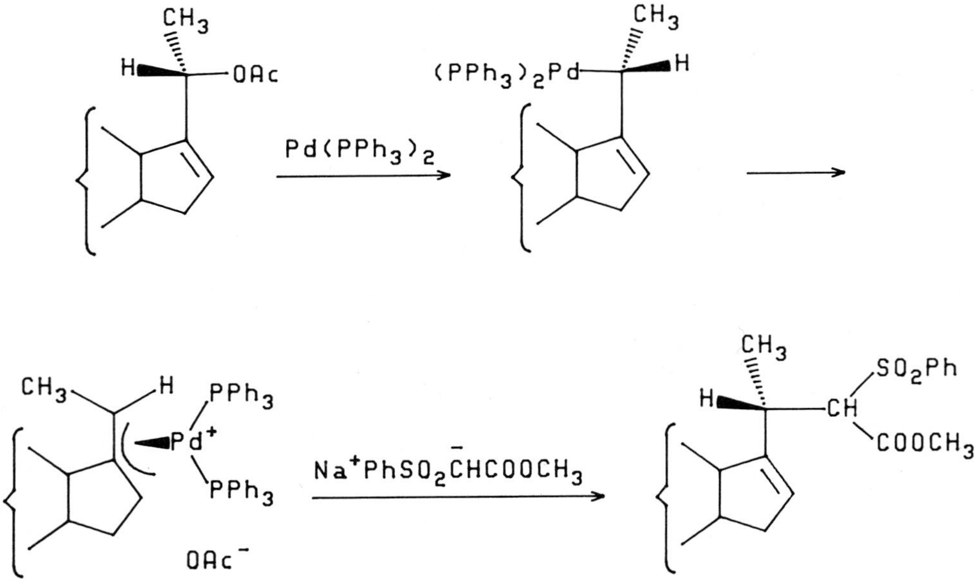

Figure 3.24

3.3.2 Stereochemistry and Mechanism

Displacement by "soft" carbon nucleophiles occurs by attack on the π-allyl at one of the terminal carbons from the side opposite the metal (backside displacement with inversion). Combining backside displacement (inversion) of the allylic functionality in

π-allyl preparation method B with backside displacement of the metal by the nucleophile, the allylic functionality is replaced by the nucleophile with overall retention. From the earlier discussion of preparation method B, nucleophilic displacement of the metal should be fast to capitalize on the short-lived stereochemical integrity of the π-allyl complex[18c,d] (see Figure 3.24).

With "hard," more strongly basic carbon nucleophiles and π-allyl complexes having adjacent hydrogens, proton abstraction leads to protodemetallation, affording a diene and palladium(0). When the π-allyl cannot eliminate to a diene, attack apparently occurs at the metal center. Reductive elimination of the σ-π-complex results in replacement of the metal with retention of configuration[41] (see Figure 3.25).

With nucleophiles capable of β-elimination, the relative rates of β-elimination and reductive elimination will determine which product is formed. If β-elimination is faster, the metal will be replaced by hydrogen; if reductive elimination is faster, the metal will be replaced by the nucleophile.

Hegedus has reported a most unusual nucleophilic attack at the central carbon of a π-allylpalladium complex by a branched ester enolate.[42] Nucleophilic attack occurs at a terminal carbon with phosphines as activating ligands. With hexamethylphosphoric triamide and triethylamine as potential activating ligands, attack occurs at the central carbon to give a metallocyclobutane that reductively eliminates. It is not possible to decide between mechanisms involving initial attack on carbon or on the metal without additional information. The intermediacy of an electron-rich 18-electron cationic complex, $[\pi\text{-allyl PdL}_3]^+$ (L = hexamethylphosphoric triamide, triethylamine) has been suggested[43] (see Figure 3.26).

While the mechanism and thus stereochemical outcome is fairly well established using soft and hard carbon nucleophiles, nitrogen and oxygen nucleophiles can apparently displace metal by either mechanism with comparable facility.[24b,c,27a,44] An increase in the ligand–metal ratio or effective steric bulk of metal ligands can result in a decrease in availability of a vacant site in the metal coordination sphere; attack on carbon will predominate. To illustrate, a pool of chloride ions effectively blocks attack

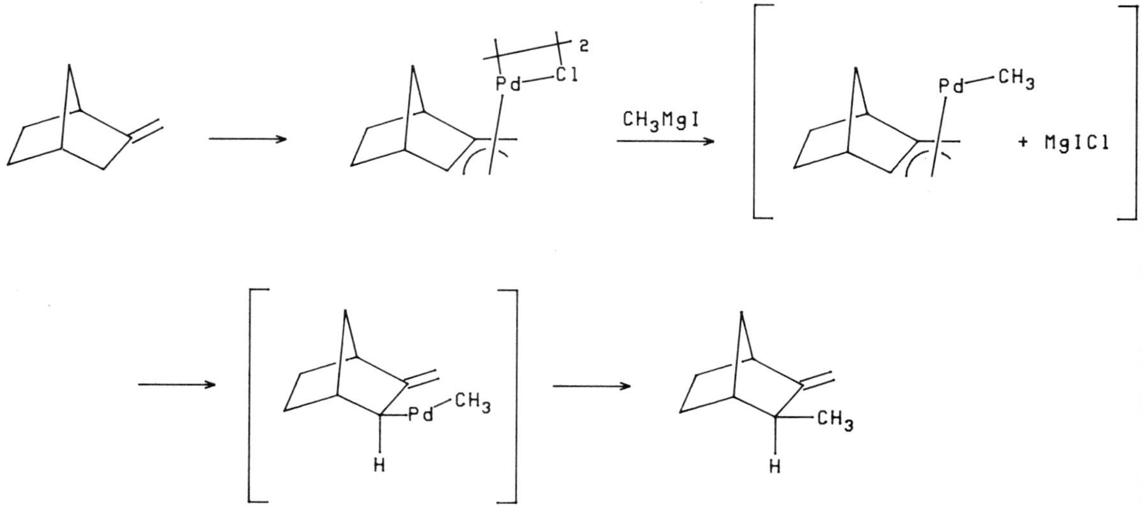

Figure 3.25

Li^+ ^-O OCH_3 a + Pd b

a inversion

$COOCH_3$ Pd (L) (L)

$-Pd(L)_2$

$COOCH_3$

$-Pd(L)_2$

retention b

CH_3O O Pd (L)

CH_3OOC Pd (L) (L)

(L) = HMPA, Et_3N

Figure 3.26

OCH_3 a OAc OCH_3

Pd Cl 2

b

OCH_3 c OCH_3

Pd OAc 2 AcO

a. BQ, LiOAc, LiCl, AcOH; >95% cis

b. AgOAc, AcOH

c. BQ, AcOH; >98% trans

Figure 3.27

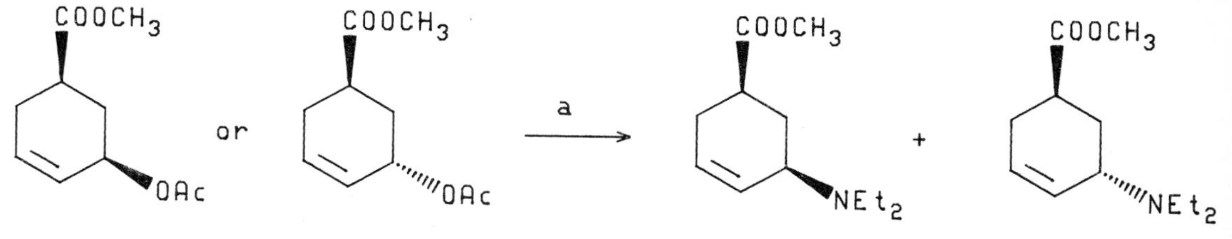

		C:D Ratio	Yield (%)
A	$Pd(PPh_3)_4$	67:33	85
	(P)-Pd	100:0	83
B	$Pd(PPh_3)_4$	35:65	60
	(P)-Pd	0:100	86

a. Pd(0), Et_2NH, THF

Figure 3.28

by acetate on the metal[24b,c], and metal coordination to phosphine bound to a polymer backbone effectively blocks attack by an amine on the metal[44e] (see Figures 3.27 and 3.28).

On the other hand, Larock reported several efficient intramolecular nucleophilic displacements apparently involving exclusive nucleophilic attack on the metal[27a] (see Figure 3.29).

3.3.3 Regioselectivity

Allylic alkylation is essentially irreversible with soft carbon nucleophiles. Thus, product ratios reflect relative rates of product formation. In general, the major product results from nucleophilic attack on the less substituted end of the π-allyl. However, several factors play a role in determining the relative rates: (1) choice of activating ligands, (2) nature of the nucleophile, (3) electronic effects of π-allyl substituents, (4) choice of π-allyl precursor using preparation method B, and (5) conformational rigidity (when nucleophilic displacement is intramolecular). The strong *syn* preference for terminal allyl substituents is translated into a preference for *trans* (*E*) disposition of alkene substituents arising from the carbon attacked by the nucleophile and the other terminal carbon substituent. Note, however, that this selectivity is nucleophile dependent[40a] (see Figure 3.30).

1. Choice of Activating Ligands

The effect of activating ligands is most clearly illustrated in alkylation using di-μ-chloro-bis-1,2-tetramethylene-π-allyl)dipalladium[40a] (see Table 3.10). With the not-n=1 51%

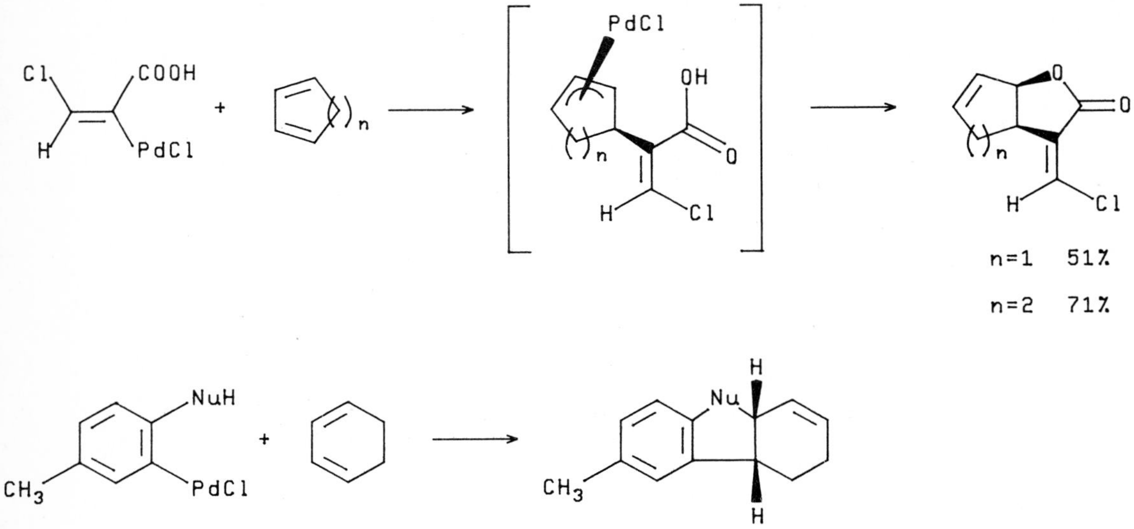

n=2 71%

Nu=O 86%

Nu=NAc 74%

Figure 3.29

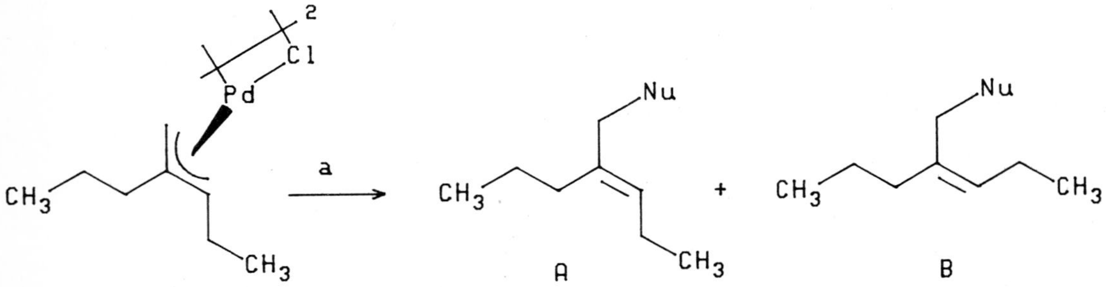

Nu = $CHCOOCH_3$ (SO_2CH_3): A:80%, B not observed

Nu = $CH(COOCH_3)_2$: A:B, 1.1:6, 68% yield

a. rt, PPh_3, NaNu, THF

Figure 3.30

TABLE 3.10 Effect of Ligand on Regioselectivity of Nucleophilic Attack[40a]

2
Cl
Pd
SO_2CH_3
$COOCH_3$
a
A
+
SO_2CH_3
$COOCH_3$
B

a. rt, L, $Na^+CH_3SO_2\bar{C}HCOOCH_3$, DMSO

Ligand L	Yield (%)	A:B Ratio
PPh_3	75	62:38
$(n\text{-Bu})_3P$	100	100:0
dppe	64	76:24
dppp	100	75:25
$(o\text{-tol})PPh_2$	66	54:46
$(o\text{-tol})_3P$	90	18:82
$(o\text{-}CH_3OC_6H_4)_3P$	80	70:30
$(o\text{-}CF_3C_6H_4)_3P$	79	49:51
HMPA	58	100:0
$(CH_3O)_3P$	91	87:13
$(PhO)_3P$	86	93:7

able exception of tri-*o*-tolylphosphine as ligand, the major product results from attack at the less substituted terminal carbon atom. One explanation for the regioselectivity recognizes that asymmetry in the π-allyl ligand will result in asymmetry in bonding of the allyl to the metal. The more electron-rich (more highly substituted) terminal carbon should bond more tightly to the metal. Attack by the nucleophile occurs at the less substituted carbon, breaking the weaker carbon–metal bond. When tri-*o*-tolylphosphine is used, the metal with two such ligands will have a much larger effective size. Steric repulsion will weaken the bond to the more highly substituted carbon (see Figure 3.31).

Another explanation for the ligand effect considers the initial products of alkylation: the palladium–alkene π-complexes. π-Complexes become less stable as the number of alkyl substituents on the alkene increases due both to unfavorable steric interactions and decreased metal–alkene backbonding. As the transition state becomes more productlike, the relative stabilities of the π-alkene complexes become more important, and attack on the more substituted carbon will be favored (see Figure 3.32).

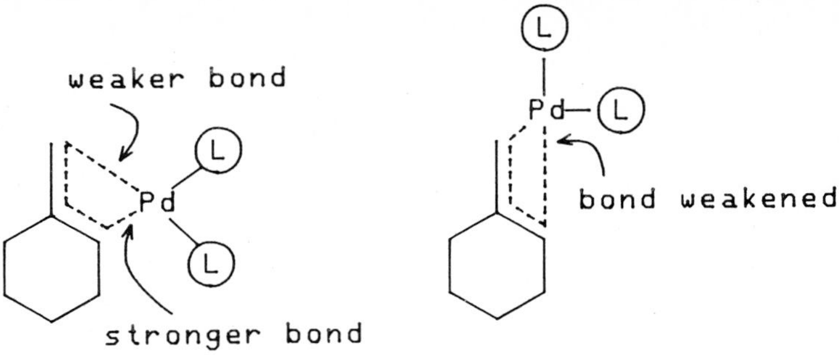

weaker bond

Pd

L

L

stronger bond

L

Pd—L

bond weakened

Figure 3.31

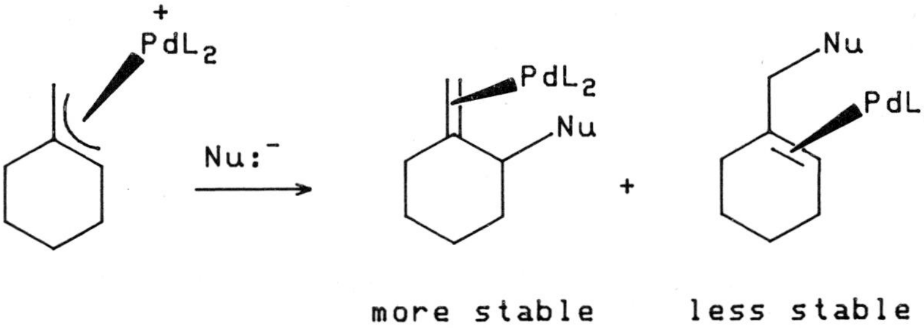

$\overset{+}{PdL_2}$

$Nu:^-$

PdL_2

Nu

Nu

PdL_2

+

more stable

less stable

Figure 3.32

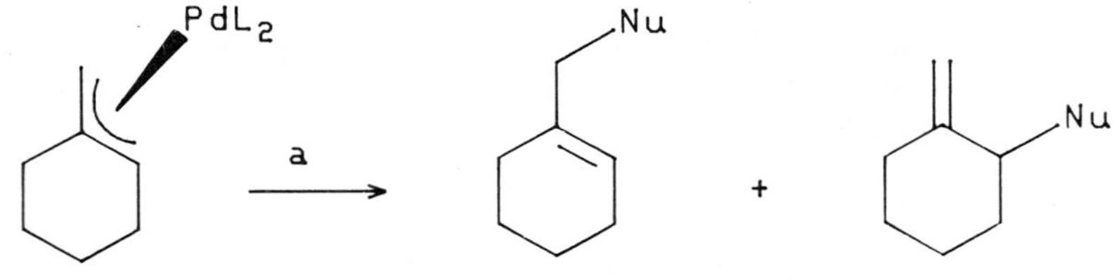

PdL_2

a

Nu

Nu

+

L = HMPA

A

B

Nu = $CH(COOCH_3)_2$: A:B, 79:21, 70% yield

Nu = $CHCOOCH_3$: A:B, 100:0, 58-90% yield
$|$
SO_2CH_3

a. NaNu:, THF or DMSO

Figure 3.33

2. Nature of the Nucleophile

The regioselectivity of alkylation is dependent on the nature of the nucleophile. Since at least one of the electron-withdrawing groups on a soft carbon nucleophile is often removed after carbon–carbon bond formation, the choice of nucleophile can be dictated by regioselectivity of attack (see Figure 3.33).

3. Electronic Character of π-Allyl Substituents

Strongly electron-withdrawing or electron-releasing substituents capable of conjugation to the π-allyl system exert a controlling influence on regiochemistry of nucleophilic attack. Electron-withdrawing groups (COOR, COR, or CN) direct to the distant terminal carbon[45]; electron-releasing groups (OR) direct to the adjacent terminal carbon.[46] A weaker donor such as acetate apparently does not have a pronounced directing effect[47] (see Figure 3.34).

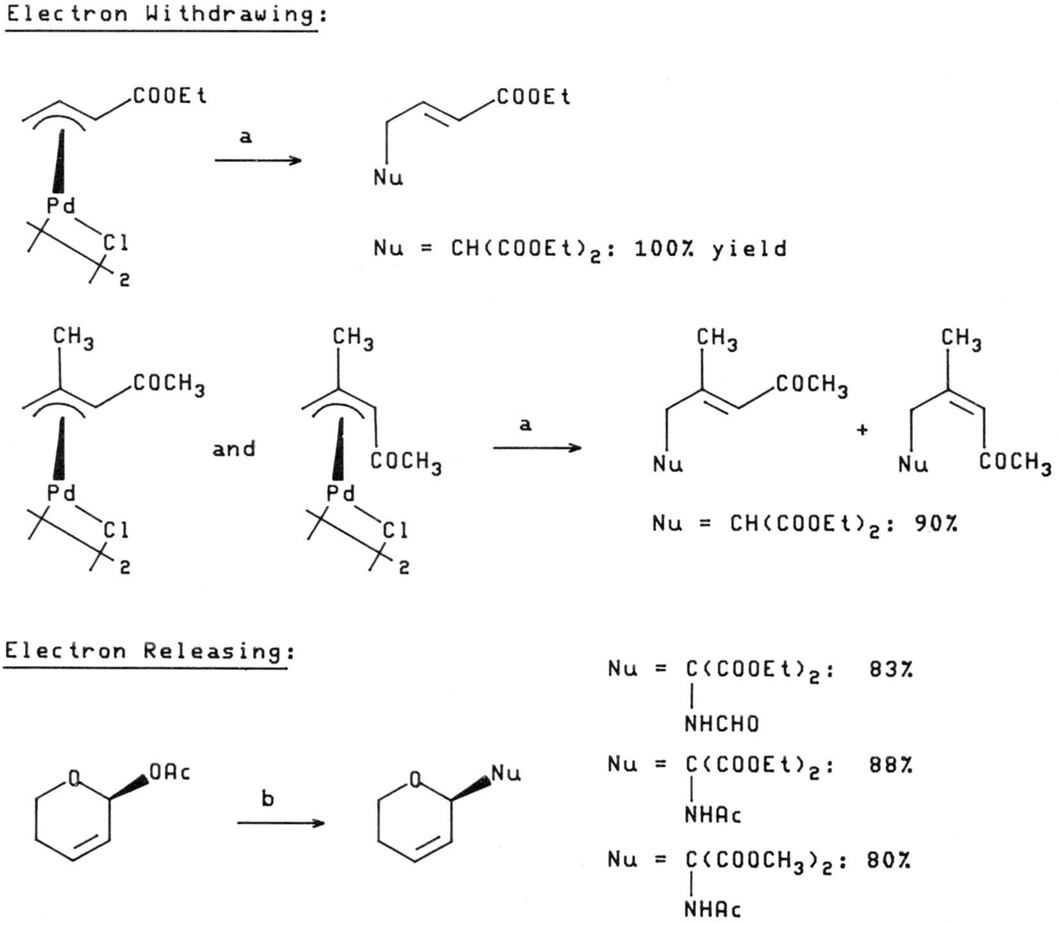

Electron Withdrawing:

COOEt, Pd, Cl, 2 —a→ COOEt, Nu

Nu = $CH(COOEt)_2$: 100% yield

CH_3, $COCH_3$, Pd, Cl, 2 and CH_3, $COCH_3$, Pd, Cl, 2 —a→ CH_3, $COCH_3$, Nu + CH_3, Nu, $COCH_3$

Nu = $CH(COOEt)_2$: 90%

Electron Releasing:

O, OAc —b→ O, Nu

Nu = $C(COOEt)_2$(NHCHO): 83%

Nu = $C(COOEt)_2$(NHAc): 88%

Nu = $C(COOCH_3)_2$(NHAc): 80%

Figure 3.34

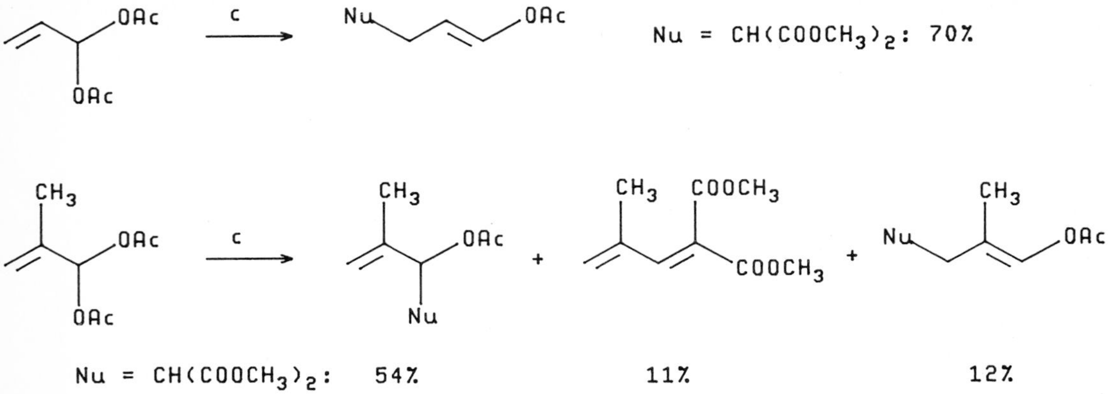

Nu = $CH(COOCH_3)_2$: 70%

Nu = $CH(COOCH_3)_2$: 54% 11% 12%

a. rt, NaNu, DMSO with THF or DMF

b. rt, NaNu, $Pd(PPh_3)_4$, PPh_3, DMF

c. rfx, BSA, NaNu, $Pd(PPh_3)_4$, PPh_3, THF

Figure 3.34 (*Continued*)

4. Choice of π-Allyl Precursor Using Preparation Method B

When the π-allyl precursor using preparation method B is a vinyl epoxide, nucleophilic attack is regiospecific for the terminal carbon distant from the hydroxyl group formed. This is true even if there is a steric or electronic bias in favor of the alternative regioisomer.[48] Note that the alkoxide formed on epoxide ring opening serves as the base to generate the nucleophilic anion (see Table 3.11 and Figure 3.35).

The situation may be complicated by reversibility of alkylation with oxygen or nitrogen nucleophiles. A kinetic mixture of regioisomers would be expected under conditions minimizing reversibility (a large difference in leaving-group ability of the original leaving group and the oxygen or nitrogen nucleophile, low temperature, short reaction time, poor activating ligands). A thermodynamic mixture of regioisomers would be expected under conditions facilitating conversion of the alkylation products back to the π-allylpalladium complex. Again, in general, this methodology should be useful for preparation of the regioisomer resulting from nucleophilic attack at the less substituted terminal carbon.

3.3.4 Regioselectivity in Intramolecular Alkylations

5. Conformational Rigidity for Intramolecular Nucleophilic Displacement

A more rigid arrangement obtained by connecting the π-allyl and nucleophile using a short chain may provide some additional control over the regioselectivity of attack. Three ring products are possible, depending on the *syn–anti* location of the nucleophile and the length of the connecting chain (see Figure 3.36).

TABLE 3.11 Regioselectivity of Nucleophilic Attack on π-Allyl Complexes Generated from Vinyl Epoxides[48]

Vinyl Epoxide	Product	Nucleophile (NuH)	Yield (%)
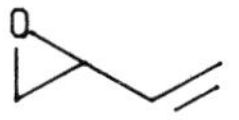	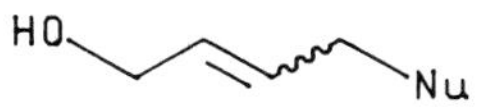	$CH_2(COOCH_3)_2$	75
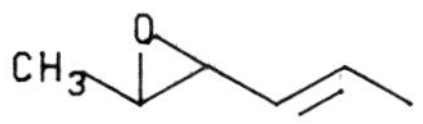	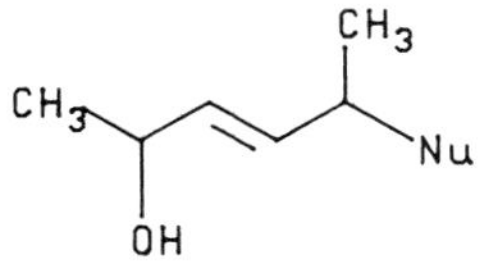	$CH_2(COOCH_3)_2$	76
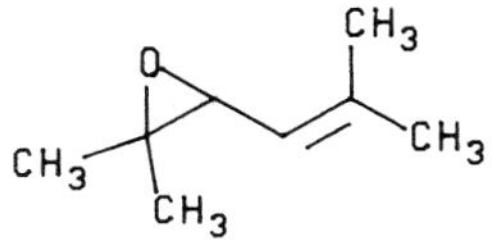	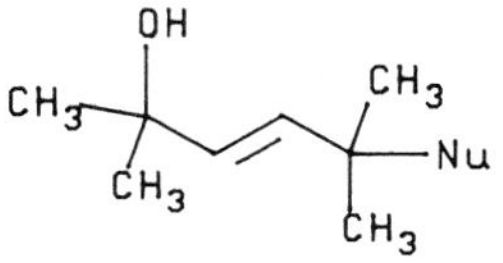	$CH_2(COOCH_3)_2$	85
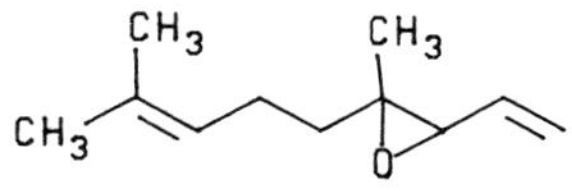	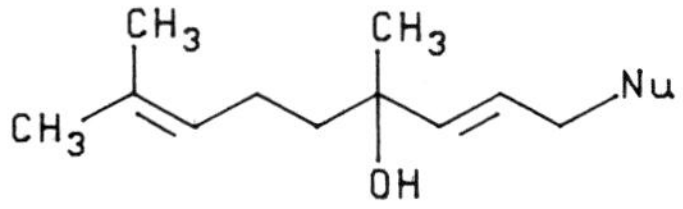	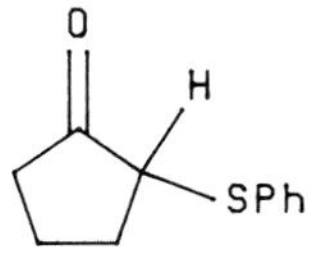	72
	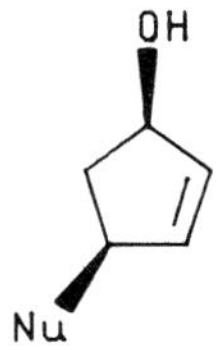	$CH_2(COOCH_3)_2$	57
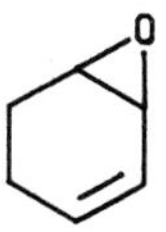		$CH_2(COOCH_3)_2$	74
		$PhOC(O)CH_2COOCH_3$	61
		$CH_2(SO_2Ph)_2$	85
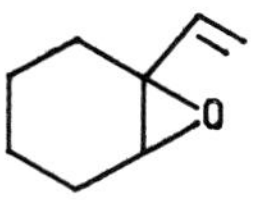	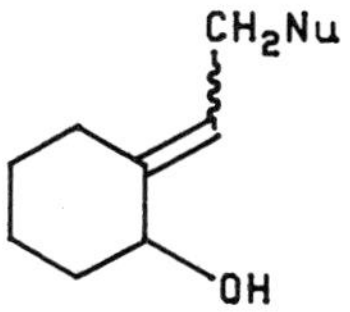	$CH_2(COOCH_3)_2$	76

(continued)

TABLE 3.11 (*Continued*)

Vinyl Epoxide	Product	Nucleophile (NuH)	Yield (%)
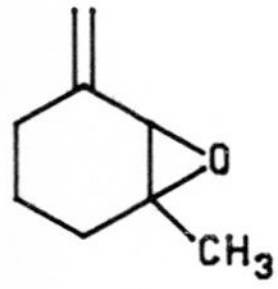 O, CH_3	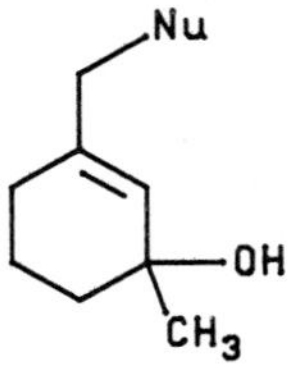 Nu, OH, CH_3	$CH_2(COOCH_3)_2$	71
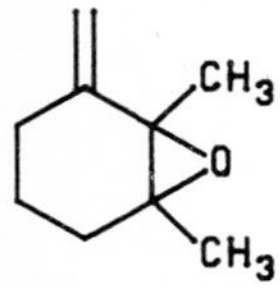 CH_3, O, CH_3	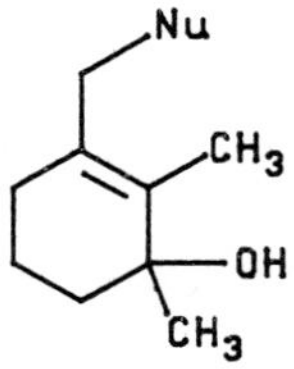 Nu, CH_3, OH, CH_3	$CH_2(COOCH_3)_2$	75
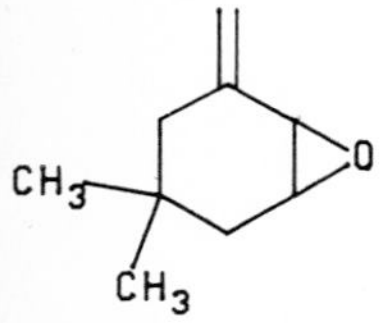 O, CH_3, CH_3	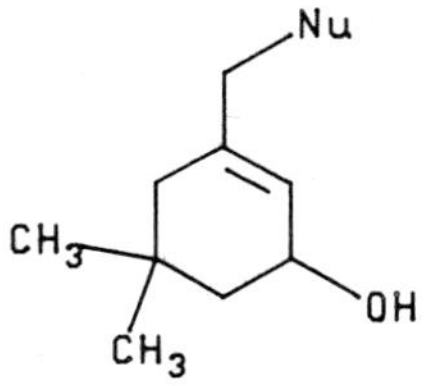 Nu, OH, CH_3, CH_3	$CH_2(COOCH_3)_2$	85
		$PhSO_2CH_2COOCH_3$	81
		$C_2H_5CH(COOCH_3)_2$	73
		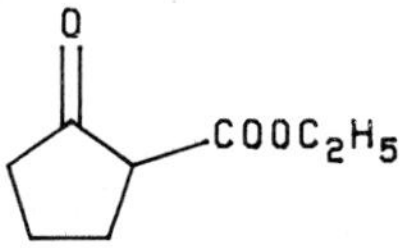 O, $COOC_2H_5$	73
		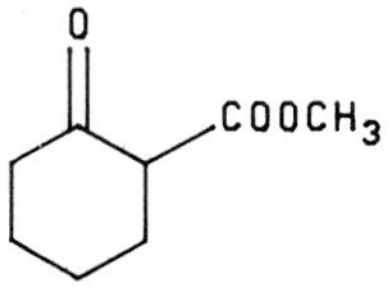 O, $COOCH_3$	79
		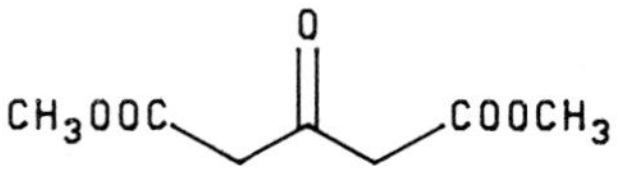 CH_3OOC, O, $COOCH_3$	74
		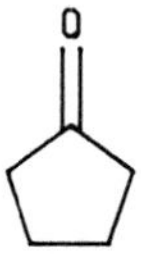 O	29

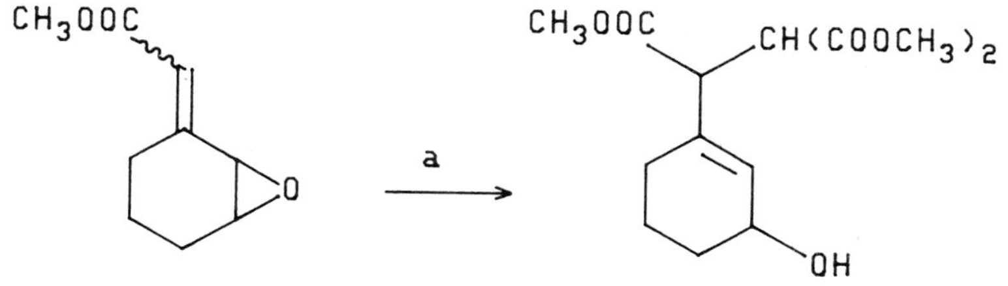

a. 40°C, $Pd(PPh_3)_4$, $CH_2(COOCH_3)_2$, THF

Figure 3.35

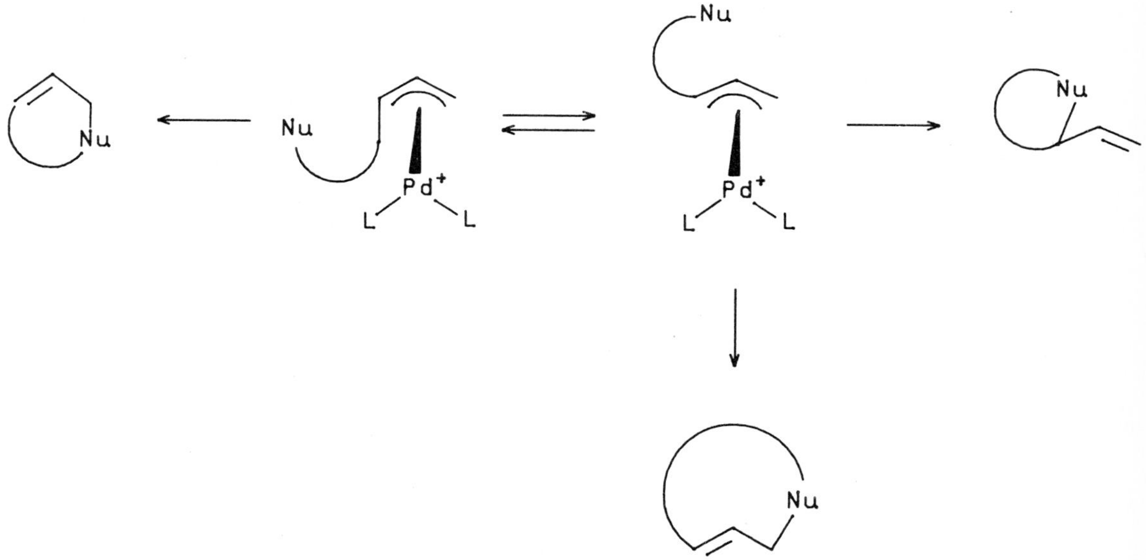

With connecting chains of 5 or more atoms

Figure 3.36

A bicyclo[4.2.0] system can be formed by attack of the nucleophile attached to a pseudoequatorial substituent in a half-chair conformation; the alternative bicyclo-[2.2.2] system can be formed by attack of the nucleophile attached to a pseudoaxial substituent in a twist-boat conformation. The kinetic product, the bicyclo[4.2.0] system, predominates with a soft carbon nucleophile.[49] With a nitrogen nucleophile,

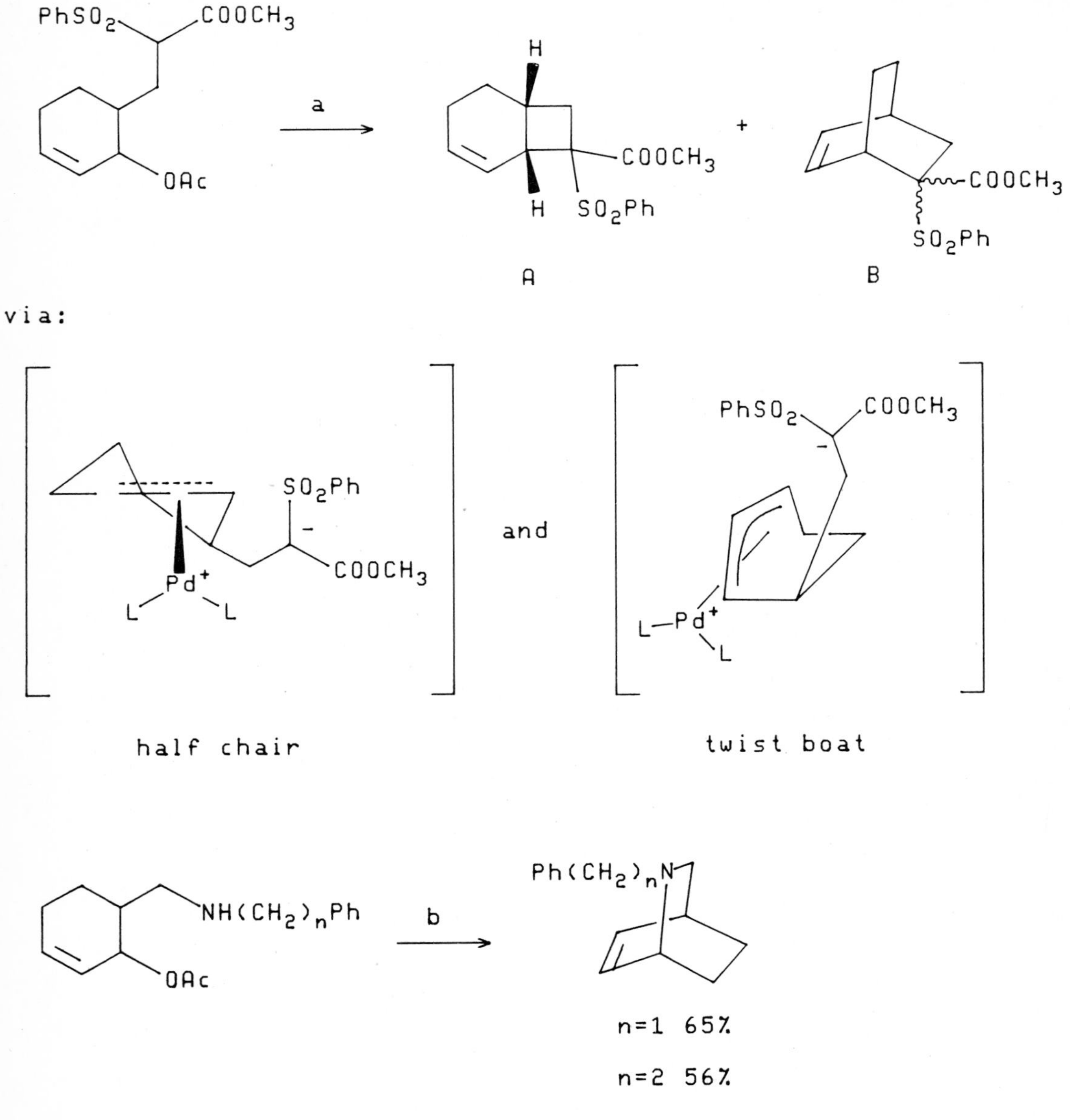

$PhSO_2$ $COOCH_3$

OAc

a

H

$COOCH_3$

H SO_2Ph

A

$COOCH_3$

SO_2Ph

B

via:

SO_2Ph

$COOCH_3$

Pd^+

L L

and

$PhSO_2$ $COOCH_3$

$L—Pd^+$

L

half chair

twist boat

$NH(CH_2)_nPh$

OAc

b

$Ph(CH_2)_nN$

n=1 65%

n=2 56%

a. 1) NaH, 2) rfx, $Pd(PPh_3)_4$, THF; A:80%, B:20%

b. 70°C, Et_3N, $Pd(PPh_3)_4$, PPh_3, CH_3CN

Figure 3.37

nucleophilic attack on the π-allyl is reversible, and the thermodynamic product, the bicyclo[2.2.2] system, is the only one isolated[50] (see Figure 3.37).

In this particular case the nucleophile is "locked" in an *anti* position. In cases where the substituent with nucleophile attached can be *syn* or *anti*, the substituent, as always, prefers to be *syn*. This preference might be translatable into a kinetic preference for formation of three- over five-membered rings, four- over six-membered rings, and so forth. With oxygen or nitrogen nucleophiles the smaller ring kinetic products may revert back to π-allyl, driven by relief of ring strain. *Syn–anti* isomerization and nucleophilic displacement would then afford the thermodynamic larger ring products. Unfortunately, vigorous conditions are required for *syn–anti* isomerization. Under these vigorous conditions, elimination to give diene can be competitive. To illustrate, Bäckvall and Nystrom have reported condensation of 1,3-dienes with a primary amine to give pyrroles[51] (see Figure 3.38).

Condensation of an arylpalladium halide with a 3,5-hexadienylamine at low temperature affords an azetidine. At higher temperatures a unique migration of the π-allyl one carbon along the chain followed by nucleophilic displacement affords a pyrrolidine[52] (see Figures 3.39 and 3.40).

In a competition between five- and seven-membered ring formation using a soft carbon nucleophile, diphos as activating ligand and relatively vigorous conditions promote rapid *syn–anti* isomerization; the seven-membered ring is formed exclusively. A polymer-supported palladium catalyst affords a five-membered ring exclusively[53] (see Figure 3.41). No *syn–anti* isomerization is observed under the exceedingly mild conditions suitable for π-allyl formation from a vinyl epoxide; a five-membered ring product is formed exclusively[54] (see Figure 3.42).

In a six–eight competition using a soft carbon nucleophile and diphos as activating ligand, relatively vigorous conditions promote *syn–anti* isomerization; the major product is the eight-membered ring.[55] In contrast, Tsuji set up an analogous competition, generating the π-allyl from a vinyl epoxide. The mild conditions together with the directing effect of epoxide opening on regioselectivity of nucleophilic attack combine to afford only the six-membered ring[56] (see Figures 3.43 and 3.44).

A seven–nine competition is complicated by the possibility of larger ring formation via the *syn*-π-allyl. Here again, disphos as activating ligand promotes formation of the larger ring. Other regioselectivity factors, such as the nature of the nucleophile, can also be used to advantage[55] (see Figure 3.45).

Other factors influencing regioselectivity of attack (activating ligands, nature of the nucleophile, electronic effects of π-allyl substituents, and π-allyl precursor) can be incorporated in the design of efficient intramolecular nucleophilic displacements to produce still larger rings.[57]

3.3.5 Enantioselectivity

Enantioselectivity of nucleophilic displacement of palladium based on the use of chiral ancillary ligands has recently received attention. The applicability of enantioselectivity depends on the substitution pattern of the π-allyl. In the absence of strong substituent electronic effects, the regioselectivity for attack on a less substituted terminal carbon makes a π-allyl with a single terminal substituent or two terminal substituents on the same carbon a poor substrate. Again, in the absence of strong substituent electronic effects, poor regioselectivity is expected using a π-allyl having one terminal

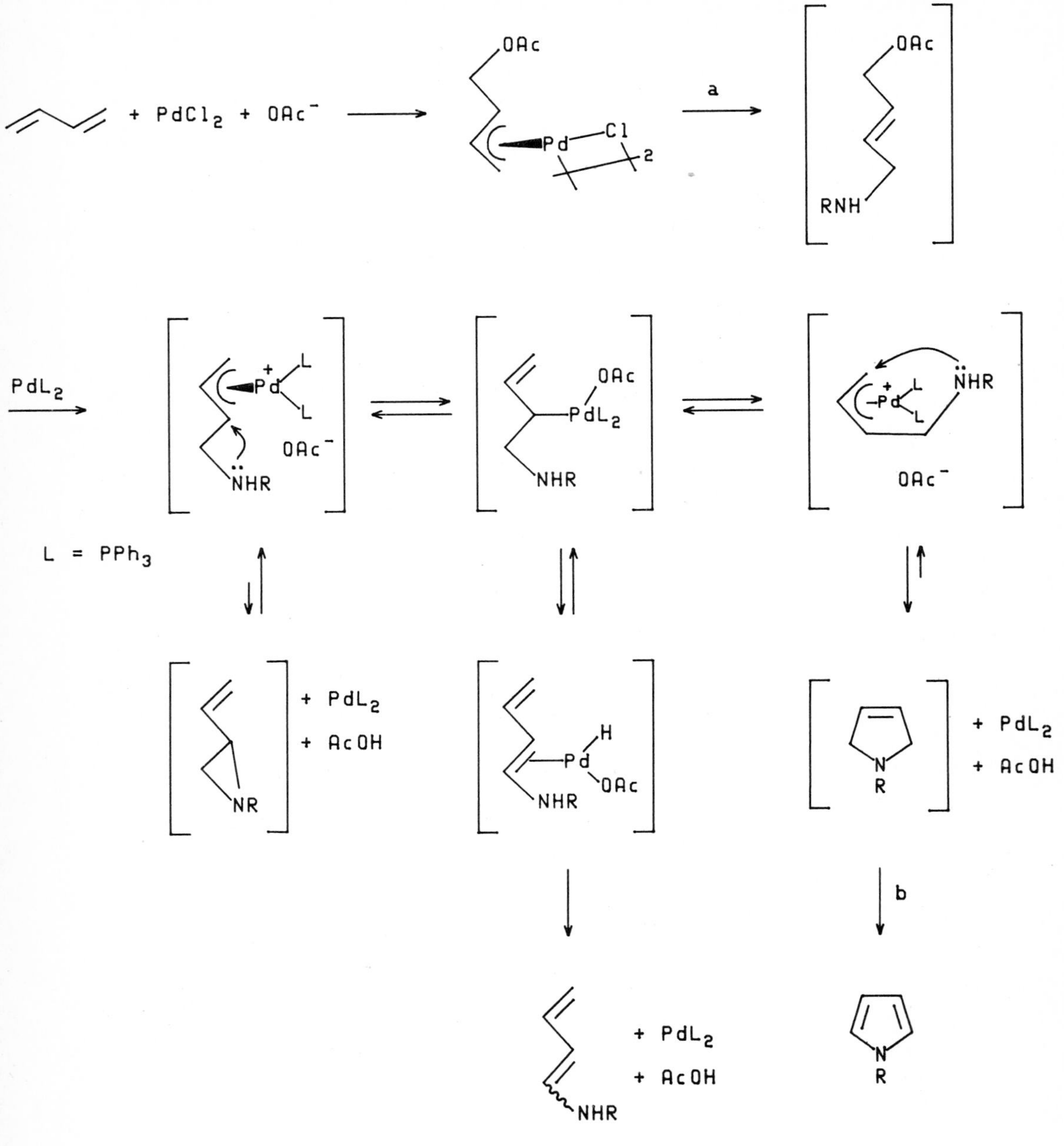

a. PPh_3, Cu(II), RNH_2

b. [O]; 14-40%

Figure 3.38

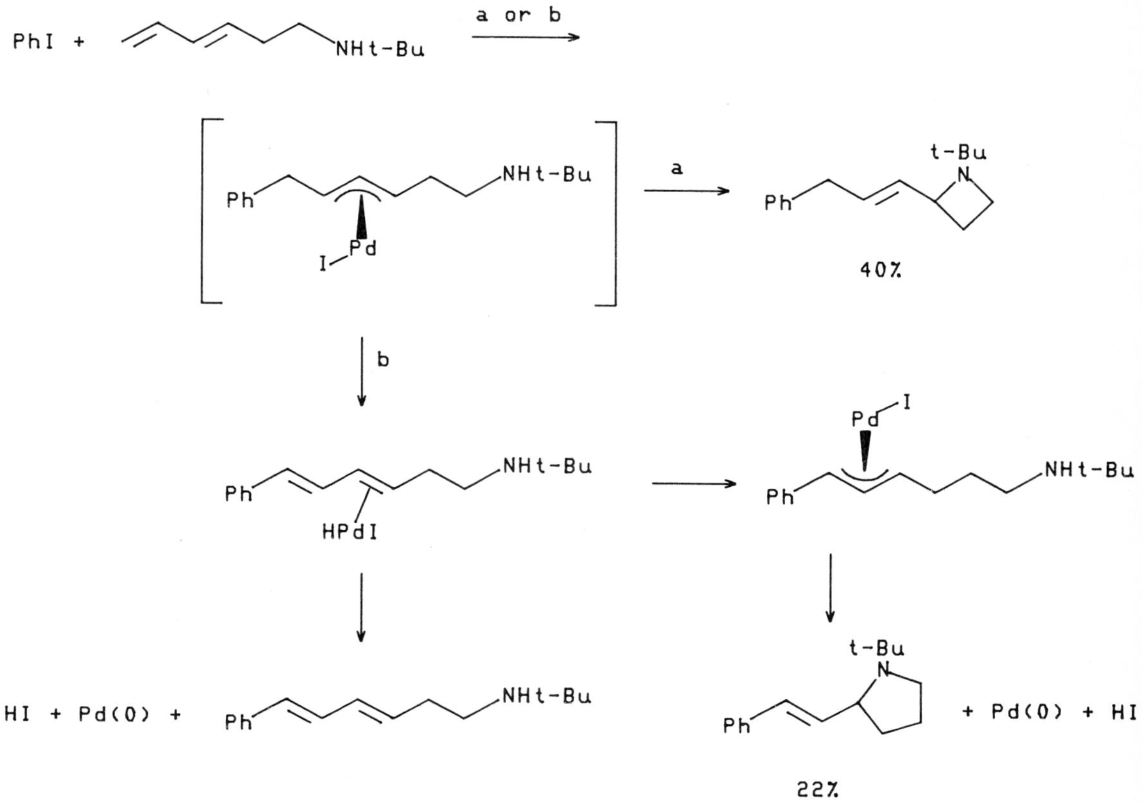

PhI + NHt-Bu $\xrightarrow{\text{a or b}}$

[Ph ... NHt-Bu, I–Pd π-allyl complex] $\xrightarrow{\text{a}}$ Ph ... (N-t-Bu azetidine) 40%

↓ b

Ph ... NHt-Bu (HPdI) ⟶ Ph ... NHt-Bu (Pd–I π-allyl)

HI + Pd(0) + Ph ... NHt-Bu

Ph ... (N-t-Bu pyrrolidine) + Pd(0) + HI

22%

a. 1 mmol PhI, 1.25 mmol diene, 1.25 mmol Et_3N, 5 mol% $Pd(OAc)_2$, 10 mol% PPh_3, 1 mL DMF, 60°C for 66 h

b. 1 mmol PhI, 1 mmol diene, 5 mol% $Pd(OAc)_2$, 20 mol% PPh_3, 100°C for 19 h

Figure 3.39

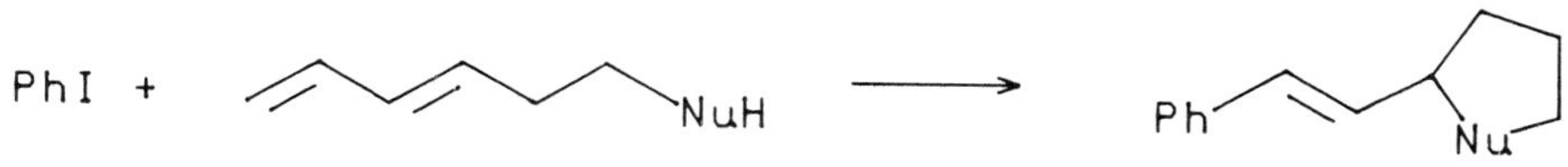

PhI + NuH ⟶ Ph ... Nu

<u>Nu = O</u>

Run 1: 1 mmol PhI, 1.25 mmol diene, 1.25 mmol Et_3N, 5 mol% $Pd(OAc)_2$, 10 mol% PPh_3, 1 mL CH_3CN, 100°C for 19 h; 23%

Figure 3.40

Nu = NTs

Run 1: 1 mmol PhI, 1.25 mmol diene, 1.25 mmol Et_3N, 5 mol% $Pd(OAc)_2$, 20 mol% PPh_3, 1 mL DMF, 100°C for 24 h; 25%

Run 2: 1 mmol PhI, 1.25 mmol diene, 1.25 mmol $NaHCO_3$, 5 mol% $Pd(OAc)_2$, 20 mol% PPh_3, 1 mL DMF, 100°C for 21 h; 52%

Nu = $C(COOEt)_2$

Run 1: 1 mmol PhI, 1.10 mmol diene, 1.25 mmol $NaHCO_3$, 5 mol% $Pd(OAc)_2$, 20 mol% PPh_3, 1 mL DMF, 100°C for 24 h; 40%

Run 2: 1 mmol PhI, 1.10 mmol diene, 1.25 mmol MgO, 5 mol% $Pd(OAc)_2$, 20 mol% PPh_3, 1 mL DMF, 100°C for 60 h; 60%

Figure 3.40 (*Continued*)

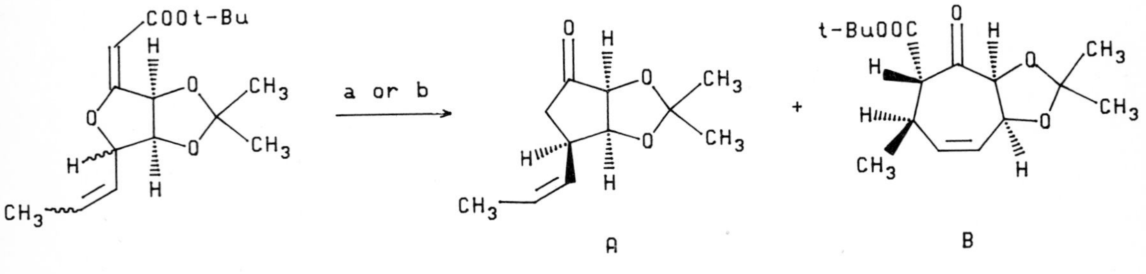

a. 100°C, 6 mol% $Pd(dppe)_2$, DMSO; B:64%

b. 70°C, 3 mol% Pd-Ⓟ, DMSO; A:82%

Figure 3.41

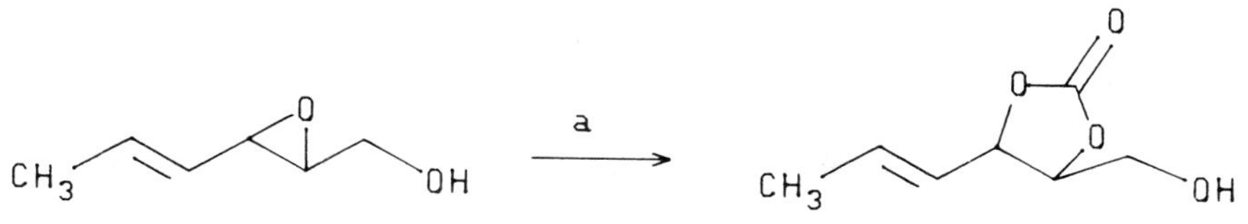

a. rt, Pd(0) catalyst, CO_2, THF; 71%

Pd(0) catalyst is formed from $Pd(OAc)_2$ and excess $P(i\text{-}PrO)_3$ at rt on addition of n-butyllithium.

Figure 3.42

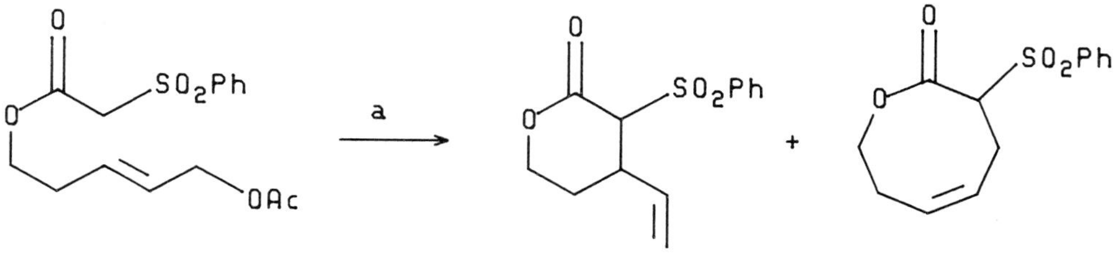

a. 1) NaH, 2) rfx, 5 mol% $Pd(PPh_3)_4$, 10 mol% dppe, THF; A:B, 6:94, 73% yield

Figure 3.43

substituent on each end. When the two substituents are different, facile *syn–anti* isomerism makes several products accessible. When the two substituents are identical, the analysis is greatly simplified (see Figures 3.46 and 3.47).

Nucleophilic displacement of palladium from a *syn,syn*-1,3-dimethyl-π-allylpalladium complex having a chiral ancillary ligand with diethylsodiomalonate at low temperatures gives optical yields as high as 25%[64] (see Table 3.12).

Oxidative addition of racemic 1-(1-acetoxyethyl)cyclopentene to palladium(0) having chiral phosphine ligands affords two diastereoisomeric π-allyl complexes (see Table 3.13). Under the reaction conditions (high phosphine–palladium ratio) nucleophilic displacement is apparently slow relative to interconversion of the diastereoisomeric π-allyls by either a Trost or Stille mechanism. Thus, the high optical yields observed, even at the reflux temperatures of tetrahydrofuran or 1,2-dimethoxyethane, are due to a difference in the thermodynamic stabilities of the diastereoisomeric π-allyls.[18a]

Bosnich has studied a very similar system having one distinct advantage: the π-allyl has 1,1-diphenyl substitution and undergoes epimerization of the diastereoisomeric

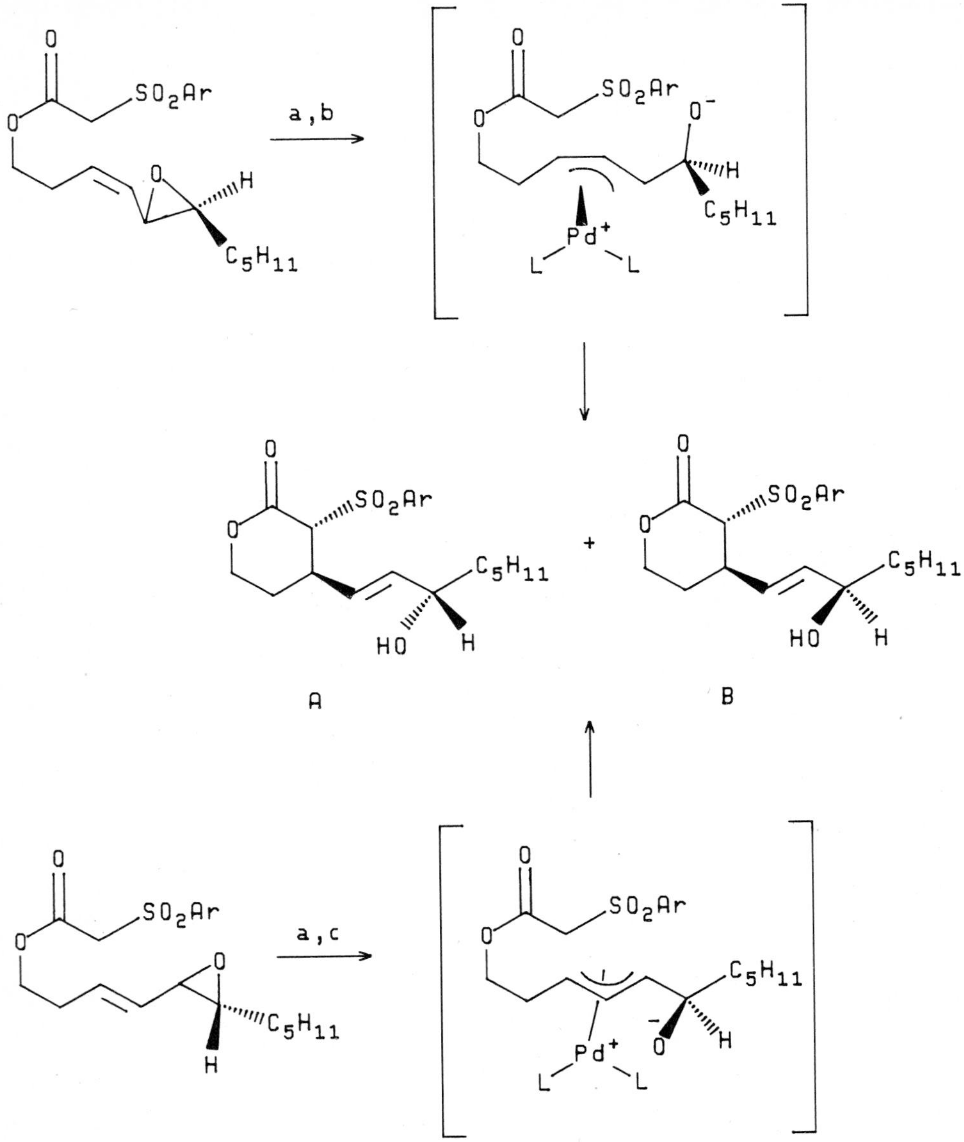

a,b

A + B

a,c

a. rt, $Pd_3(TBAA)_3 \cdot CHCl_3$, excess [bicyclic phosphite P(OCH₂)₃C–CH₃], THF

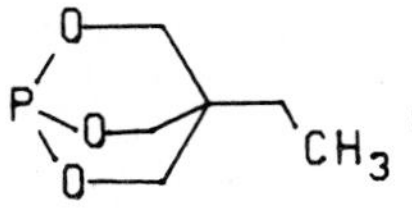

b. A:B, 92:8, 60-70% yield

c. A:B, 5:95, 60-70% yield

Figure 3.44

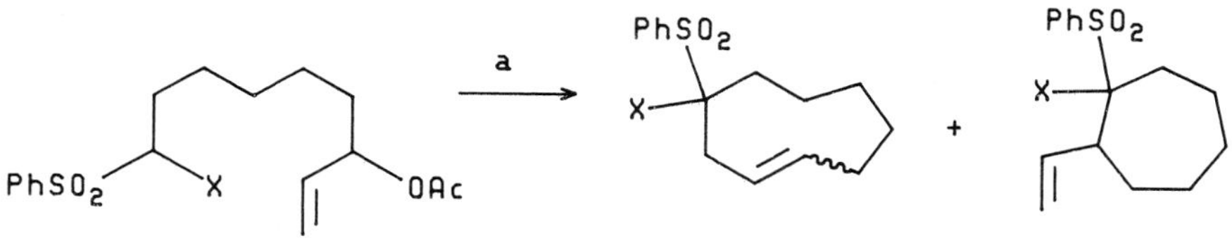

$X = COOCH_3$ with dppe A:B = 37:63

without dppe 7:93

$X = PhSO_2$ with dppe 73:27

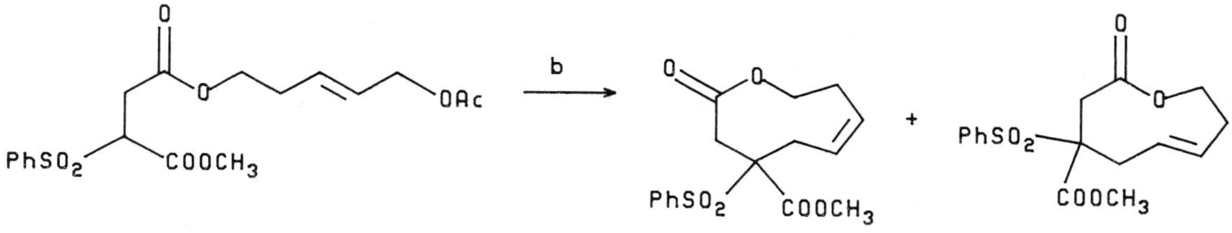

a. 1) NaH, 2) rfx, $Pd(PPh_3)_4$, THF

b. 1) NaH, 2) rfx, 10 mol% $Pd(PPh_3)_4$, 10 mol% dppe, THF; A:B, 92:8, 60% yield

Figure 3.45

(+)-DiOP = (+)-2,3-O-isopropylidene-2,3-dihydroxy-1,4-bis-(diphenylphoshino)butane[58]

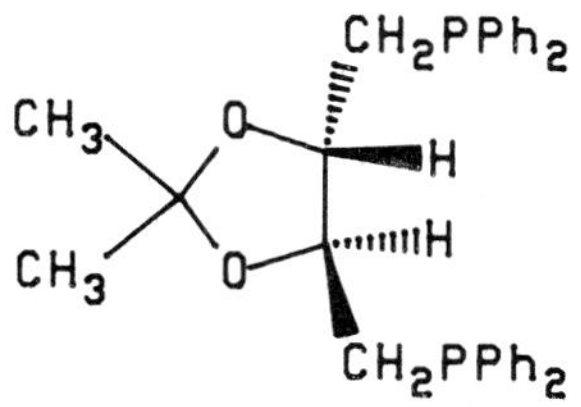

Figure 3.46 Ligands used in asymmetric allylic alkylation.

(+)-ACMP = (+)-o-anisylcyclohexylmethylphosphine[59]

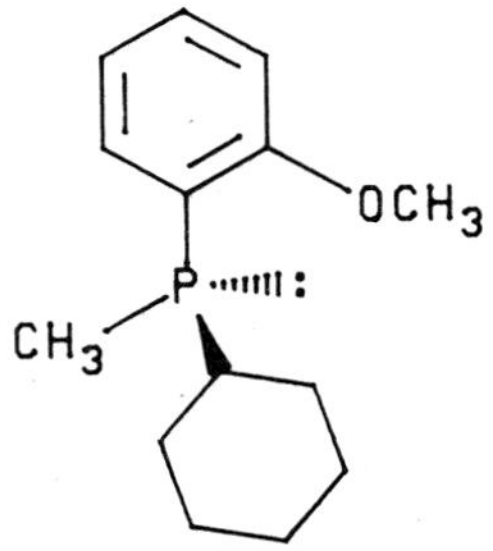

(−)-DMiP = (−)-dimenthylisopropylphosphine[60]

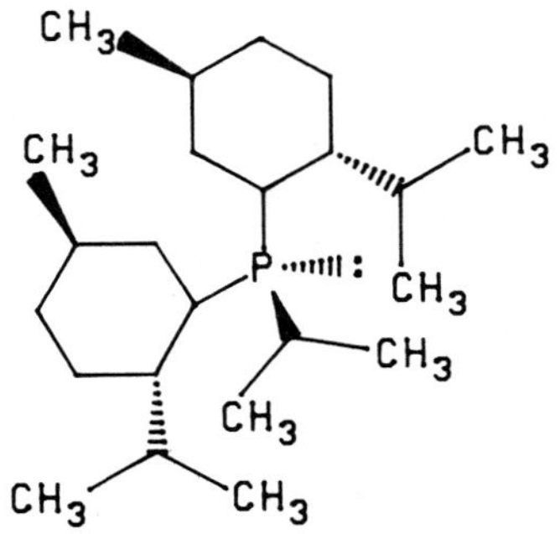

(−)-Sparteine

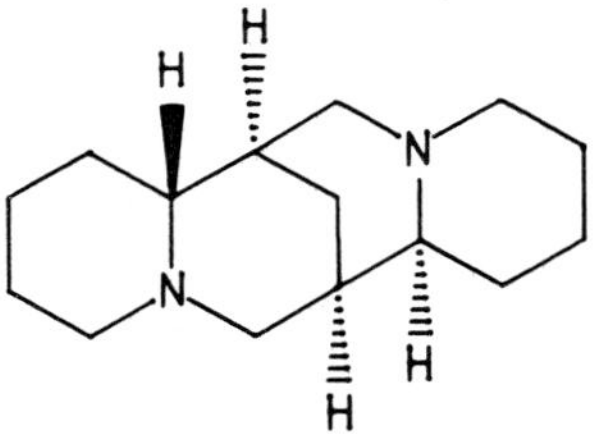

Figure 3.46 (*Continued*)

(−)-DiPAMP = (−)-1,2-bis-(o-anisylphenyl-phosphinoethane)[61]

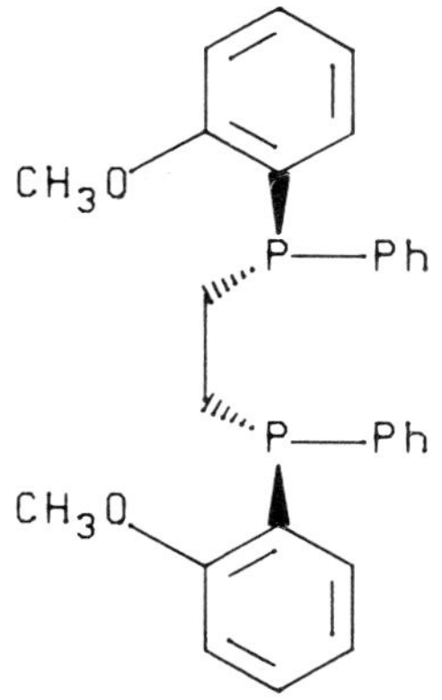

CH_3O P—Ph P—Ph CH_3O

(+)-CAMPHOS = (+)-1,2,2-trimethyl-cis-1,3-bis-(diphenylphosphinomethyl)cyclopentane[62]

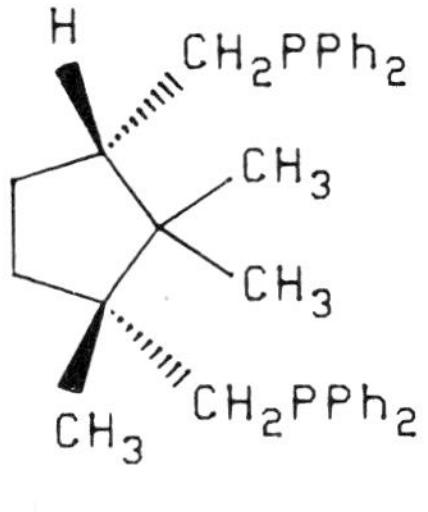

H CH_2PPh_2 CH_3 CH_3 CH_2PPh_2 CH_3

(−)-Chiraphos = (−)-(2S,3S)-2,3-bis-(diphenylphosphino)butane

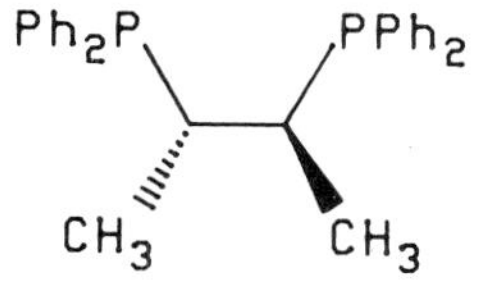

Ph_2P PPh_2 CH_3 CH_3

Figure 3.46 (*Continued*)

(−)-BiNAP = (−)-2,2'-bis-(diphenylphosphino)-1,1'-binaphthyl

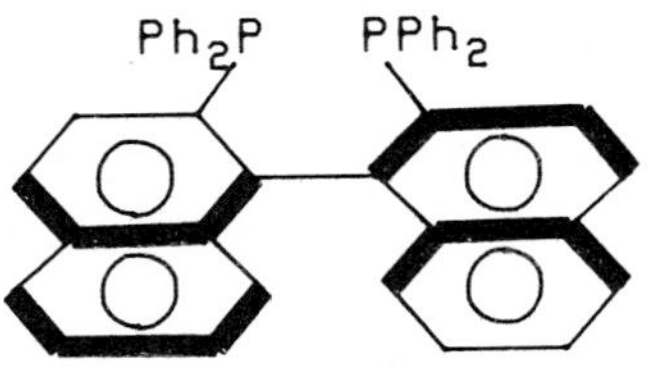

(−)-BiNAPO = (−)-1,1'-bi-2-naphthylbis-(diphenylphosphinite)[63]

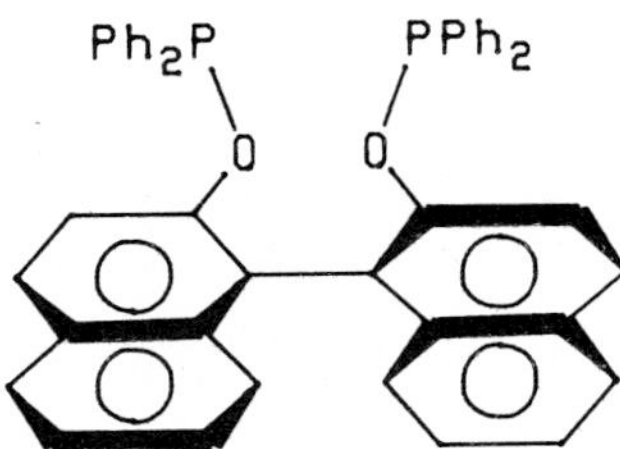

Figure 3.46 *(Continued)*

π-allyl complexes by a much more rapid π–σ–π sequence. Thus, a single enantiomer or a racemic mixture of the π-allyl precursor will give the same intermediate mixture and optical yield. The optical yields accurately reflect a difference in thermodynamic stabilities of the diastereoisomeric π-allyls[65] (see Figure 3.48).

The distance separating nucleophile trajectory (backside on terminal carbon) and the chirality associated with an ancillary ligand places a severe limitation on optical efficiency (see Table 3.14). Trost has recently developed a working model based on the concept of "chiral pockets" that may provide a solution. Enlarging the "bite" of a bidentate ligand should lead to a decrease in the distance separating the phosphine aryl substituents and π-allyl terminal carbons. Since the aryl rings are fixed in a conformation minimizing aryl–aryl steric interactions, the π-allyl complex has "chiral pockets" enveloping the terminal carbons.

Returning to an achiral π-allyl having two identical substituents, one on each terminal carbon, nucleophilic displacement of palladium by bis-(benzenesulfonyl)-methane in the presence of a chiral ancillary ligand affords product with enantiomeric excesses as high as 38% at the reflux temperature of tetrahydrofuran. *Ortho* or *meta* substituents on the aryl rings should enhance the difference in accessibility of the two terminal carbons and improve optical efficiency. In fact, 69% *enantiomeric excess* is

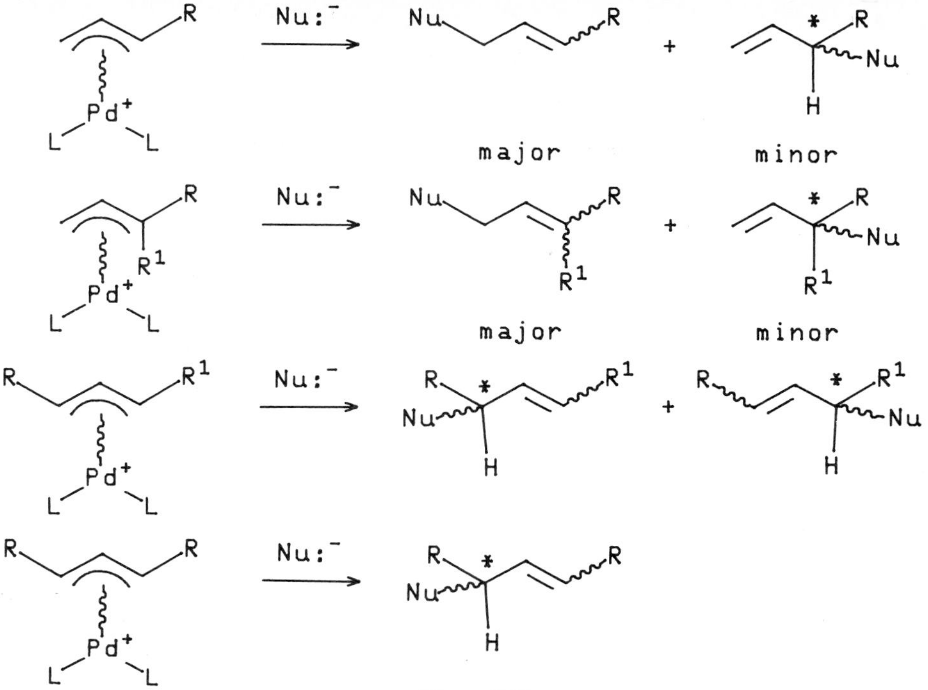

Nu:⁻
major
minor
Nu:⁻
major
minor
Nu:⁻
Nu:⁻

Figure 3.47

TABLE 3.12 Asymmetric Alkylation of π-Allyl Complexes Having Chiral Monodentate Ancillary Ligands[64]

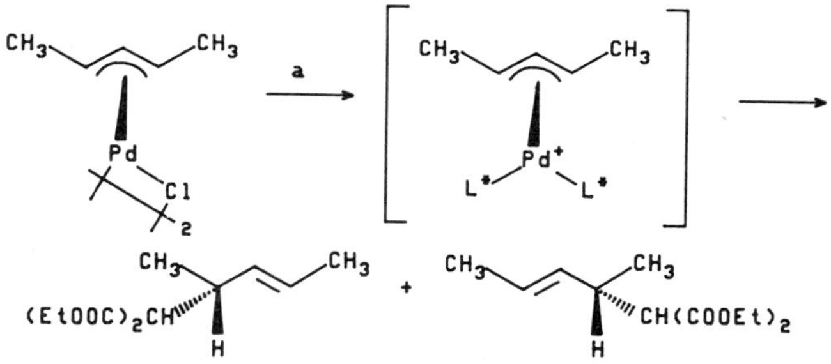

a. $NaCH(COOEt)_2$, L*, THF; 66-88% yield

L*	Reaction Temperature (°C)	Enantiomeric Excess (%)
(+)–Diop	0	12.2 ± 0.8
	–40	22.4 ± 2.2
	–78	17.9 ± 1.8
(+)-ACMP	25	17.9 ± 1.8
	–40	24.4 ± 1.6
	–78	22.4 ± 2.8
(–)-Sparteine	25	20.2 ± 2.1
(–)-DMIP	0	2.0 ± 0.3

TABLE 3.13 Asymmetric Alkylation of π-Allyl Complexes Having Chiral Bidentate Ancillary Ligands[18a]

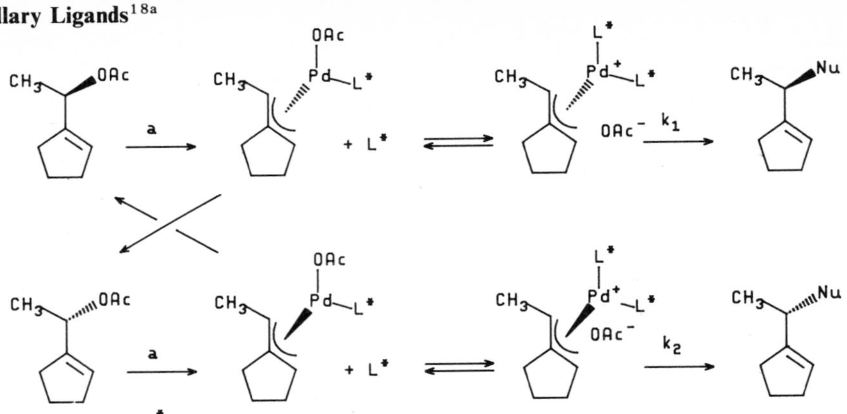

CH3
OAc
a
Pd
L*
+ L*
Pd+
OAc−
k1
k2
Nu
a. rfx, PdL*2, NaNu, THF or DME

L* (bidentate)	NaNu	Yield (%)	Enantiomeric Excess (%)
(+)-Diop	$NaCH(COOCH_3)_2$	57	21
(+)-Diop	$NaCH(COOCH_3)_2$	82	38 (without PPh_3 present)
(−)-DiPamP	$NaCH(COOCH_3)_2$	62	16
(+)-CAMPHOS	$NaCH(COOCH_3)_2$	99	37
(+)-CAMPHOS	$NaCHSO_2Ph$ | $COOCH_3$	76	39
(+)-Diop	$NaCHSO_2Ph$ | $COOCH_3$	84	46

TABLE 3.14 Asymmetric Alkylations of π-Allyl Complexes Having Chiral Bidentate Ancillary Ligands[66]

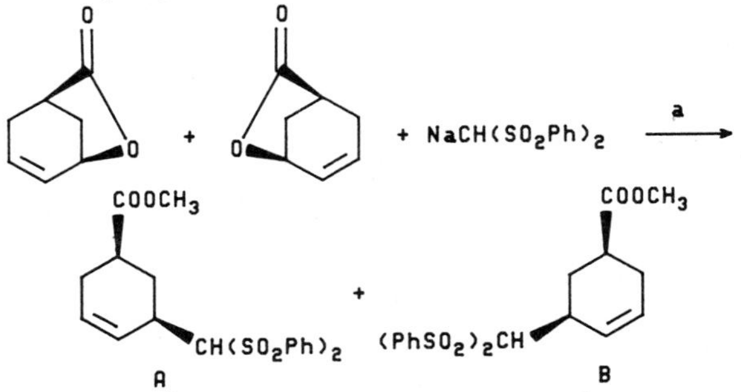

O
+ NaCH(SO2Ph)2
a
COOCH3
CH(SO2Ph)2
(PhSO2)2CH
A
B
a. rfx, PdL*n, THF, then CH2N2

L*	Yield (%)	A:B Ratio	Enantiomeric Excess (%)
(−)-Chiraphos	73	59:41	18
(+)-Diop	82	58:42	16
(−)-BiNAP	92	65:35	31
(−)-BiNAPO	66	69:31	38
BiNAPO (*m*-$Si(CH_3)_3)_8$	82	82:15	69

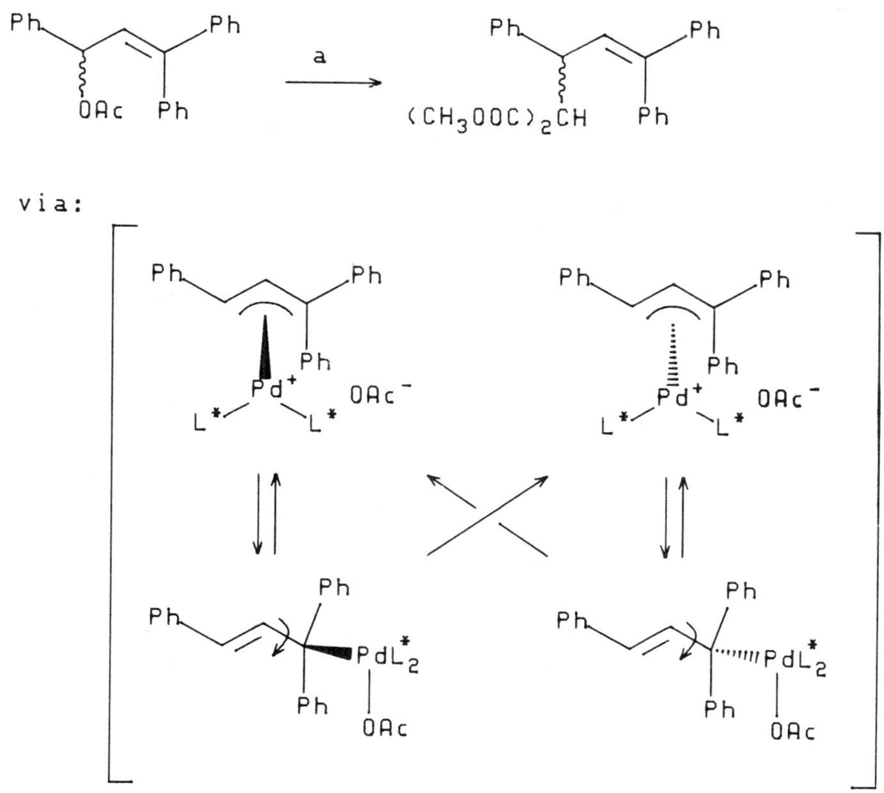

a. rt, $NaCH(COOCH_3)_2$, 5 mol% ClO_4^-, THF; Pd^+(S,S-chiraphos)

100% (84% optical yield)

Figure 3.48

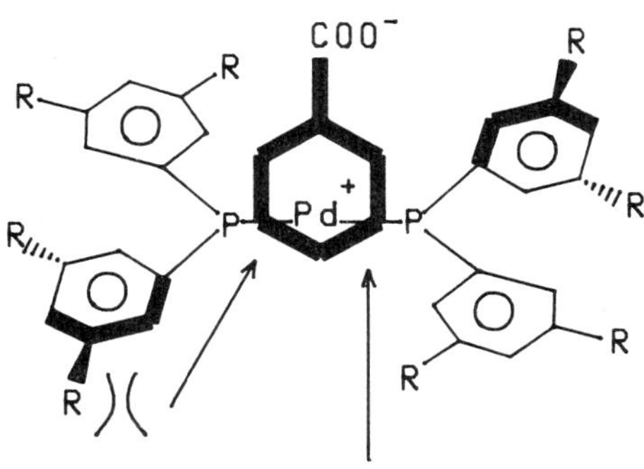

Figure 3.49

observed using a catalyst with *meta*-trimethylsilyl groups on a BiNAPO-type ligand[66] (see Figure 3.49).

These results indicate that high asymmetric induction (90 to 100% enantiomeric excess) in carbon–carbon bond formation by catalytic allylic alkylation is just a matter of time.

3.4 TOTAL SYNTHESIS OF (±)-IBOGAMINE[67]

A mixture of geometric isomers of a dienol acetate is prepared from *trans*-2-hexenal.[68–70] The *E*,*E*, -isomer is best able to adopt the *S*-*cis* conformation necessary

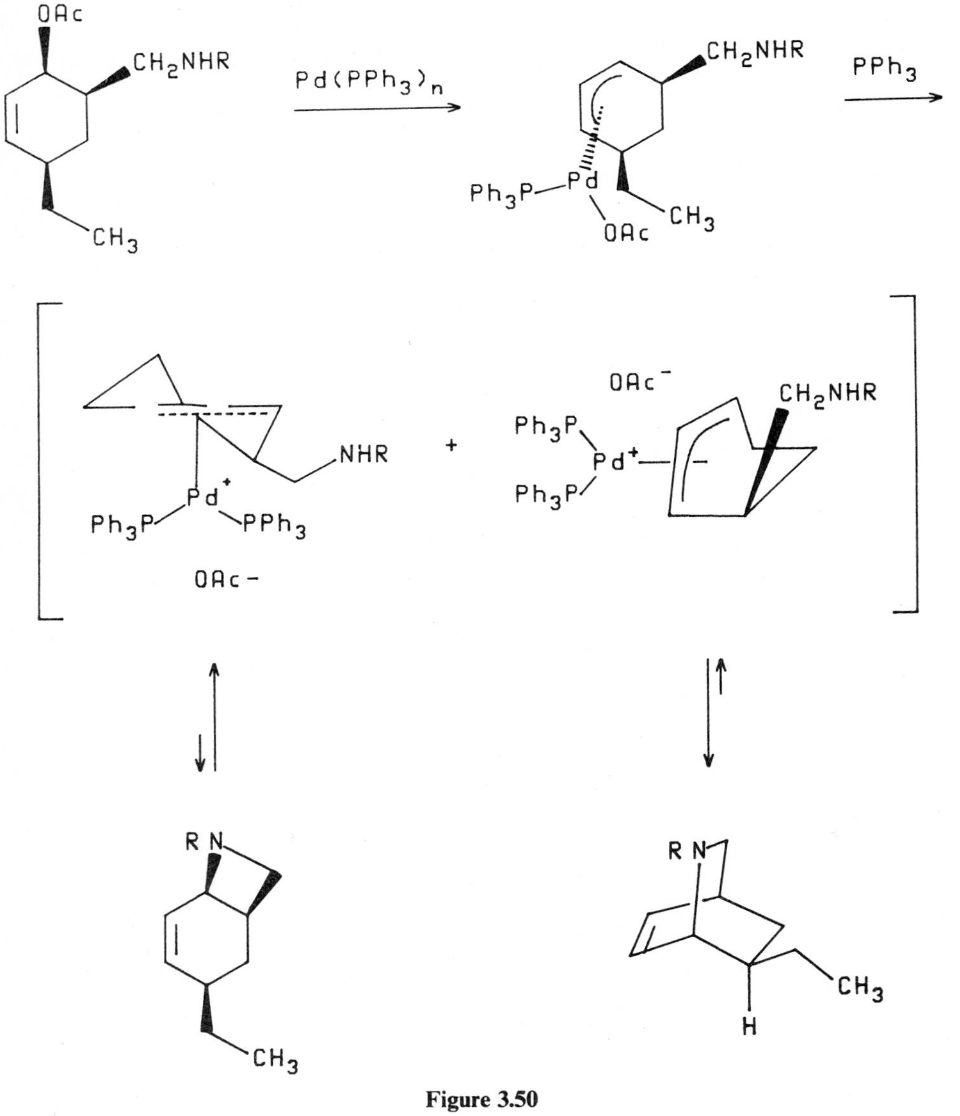

OAc
CH_2NHR
CH_3
$Pd(PPh_3)_n$
CH_2NHR
Ph_3P—Pd
OAc
CH_3
PPh_3
NHR
Ph_3P
Pd^+
PPh_3
OAc^-
+
OAc^-
CH_2NHR
Ph_3P
Pd^+
Ph_3P
R N
CH_3
R N
H
CH_3

Figure 3.50

for Diels–Alder cycloaddition. In fact, cycloaddition with acrolein[71] under mild conditions affords exclusively the adduct from the *E,E*-isomer in excellent yield. Aldehyde condensation with tryptamine[72] followed by reduction of the imine introduces a nucleophilic secondary amine. The allylic acetate is converted to a π-allyl *with inversion* using a palladium(0)–phosphine complex. Subsequent rapid intramolecular palladium displacement by the secondary amine also occurs with inversion; the result is replacement of acetate by secondary amine with retention of configuration. Thus, the Diels–Alder cycloaddition establishes the relationship of acetate, ethyl, and nucleophile substituents. This relationship is maintained in the palladium(0)-catalyzed cyclization. Note that no bicyclo[4.2.0] product is isolated. Attack by an amine nucleophile on a π-allylpalladium complex is reversible. Thus, any [4.2.0] product formed can be converted back to π-allyl (driven by relief of ring strain) and then onto the more stable bicyclo[2.2.2] product (see Figure 3.50).

The final operation is carbon–carbon bond formation by coupling the 2-position of the indole ring with the alkene. A strong palladium(II) electrophile is prepared by chloride abstraction with silver tetrafluoroborate. A heteroarylpalladium chloride produced by electrophilic substitution at the indole 2-position coordinates and inserts

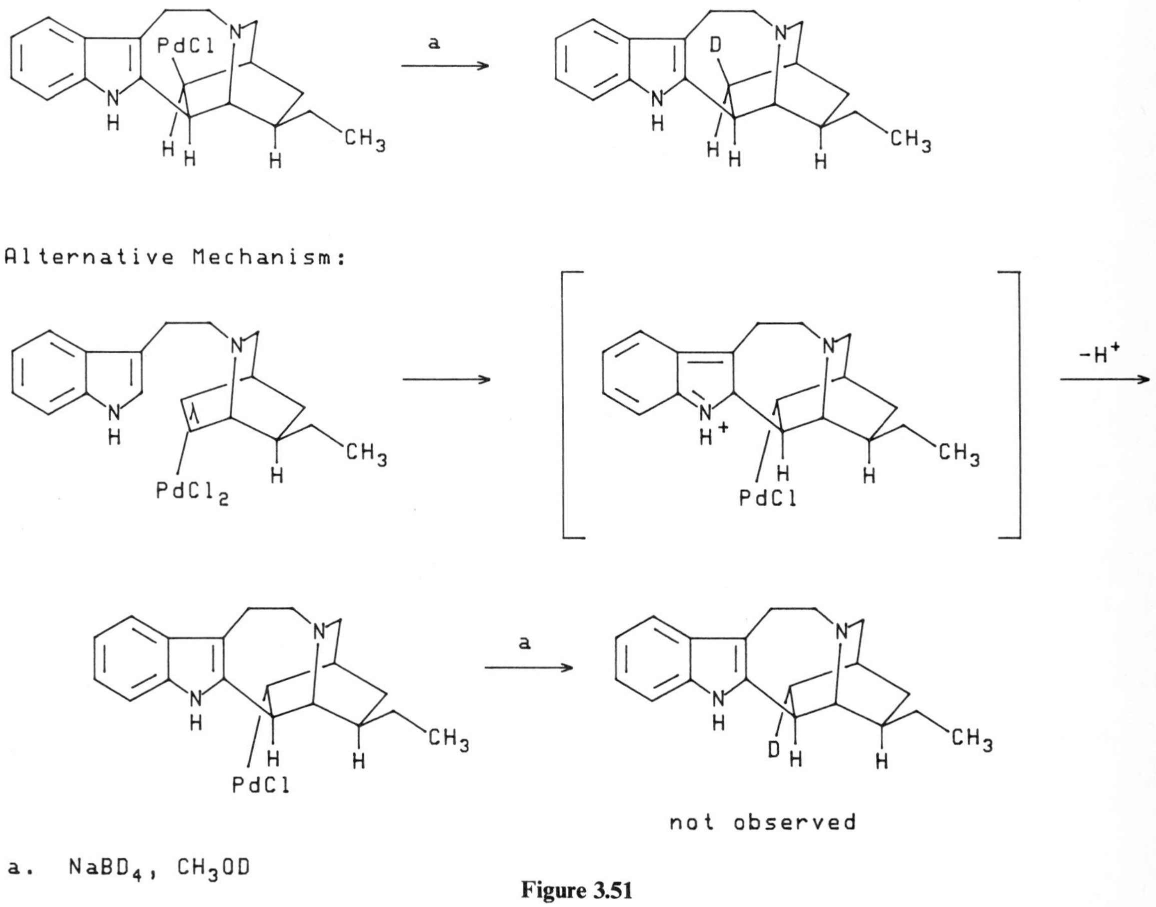

Alternative Mechanism:

not observed

a. $NaBD_4$, CH_3OD

Figure 3.51

the alkene (note the *syn* addition of the heterocycle and metal to the alkene; see Chapter 2), affording a σ-alkylpalladium complex that cannot β-hydride eliminate. Reductive demetallation completes the synthesis.

An alternative mechanism for this carbon–carbon bond formation via nucleophilic attack on a palladium(II)–alkene π-complex would result in *anti* addition of the heterocycle and the metal to the alkene. This mechanism can be disproven by reductive demetallation with sodium borodeuteride (see Figure 3.51).

The synthesis of (±)-ibogamine from 1-acetoxy-1,3-hexadiene and acrolein was completed in just six steps in overall 17% yield (see Figure 3.52).

3.5 TOTAL SYNTHESIS OF OPTICALLY ACTIVE IBOGAMINE[67]

A total synthesis of optically active ibogamine can be achieved by replacing the acetoxy group on the diene by a chiral acyloxy group. The adduct from Diels–Alder cycloaddition of cyclopentadiene[73] and maleic anhydride[74,75] is converted to a diester. While the standard acyloin condensation is generally not suitable for synthesis of four-membered rings, a modification using chlorotrimethylsilane affords a bis-(trimethylsiloxy)cyclobutene in excellent yield.[76] Hydrolysis produces an α-hydroxycyclobutanone.[77] The hydroxyl group is removed by tosylation and reductive cleavage.[78] The enolate from deprotonation with lithium diisopropylamide alkylates exclusively on the *exo* face. Carbonyl reduction with borohydride delivers hydride to the more accessible *exo* face. Acylation with (*S*)-(+)-α-methoxyphenylacetyl chloride[79,80] produces two diastereoisomers. A retro Diels–Alder at 460°C is followed by conrotatory ring opening of the *trans*-disubstituted cyclobutene. Diels–Alder cycloaddition of the chiral diene with acrolein at low temperature affords the (3*R*, 4*S*, 6*R*) and (3*S*, 4*R*, 6*S*) diastereoisomers in a 4:1 ratio. This mixture is then converted to a 4:1 mixture of (+)- and (−)-ibogamine by the same sequence as used in the racemic systhesis.

The chiral diene was prepared from cyclopentadiene and maleic anhydride in ten steps in 18% overall yield (Figure 3.53).

3.6 TOTAL SYNTHESIS OF (±)-CATHARANTHINE[81]

The total synthesis of (±)-catharanthine utilizes the same approach for construction of the requisite diene as the synthesis of optically active ibogamine. Pivaloylation of the *endo*-α-hydroxy ketone is followed by a base-induced epimerization. Ketone reduction with borohydride again delivers hydride on the *exo* face. The new (*endo*) hydroxy is also pivaloylated. A retro Diels–Alder occurs on heating from 25 to 240 °C in a quartz tube; the *trans*-disubstituted cyclobutene produced undergoes conrotatory ring opening.

Diels–Alder cycloaddition with acrolein at low temperature affords the *endo* adduct in excellent yield. Reductive amination is accomplished in two steps: imine formation with tryptamine[72] and then reduction with borohydride. Since both pivaloyloxy groups are allylic, both can oxidatively add to palladium(0) to produce a π-allyl complex. In fact, reaction with *tetrakis*-(triphenylphosphine)palladium(0) proceeds via the more stable transition state having the nucleophile-containing substituent pseudoequatorial. Intramolecular nucleophilic displacement of palladium then occurs

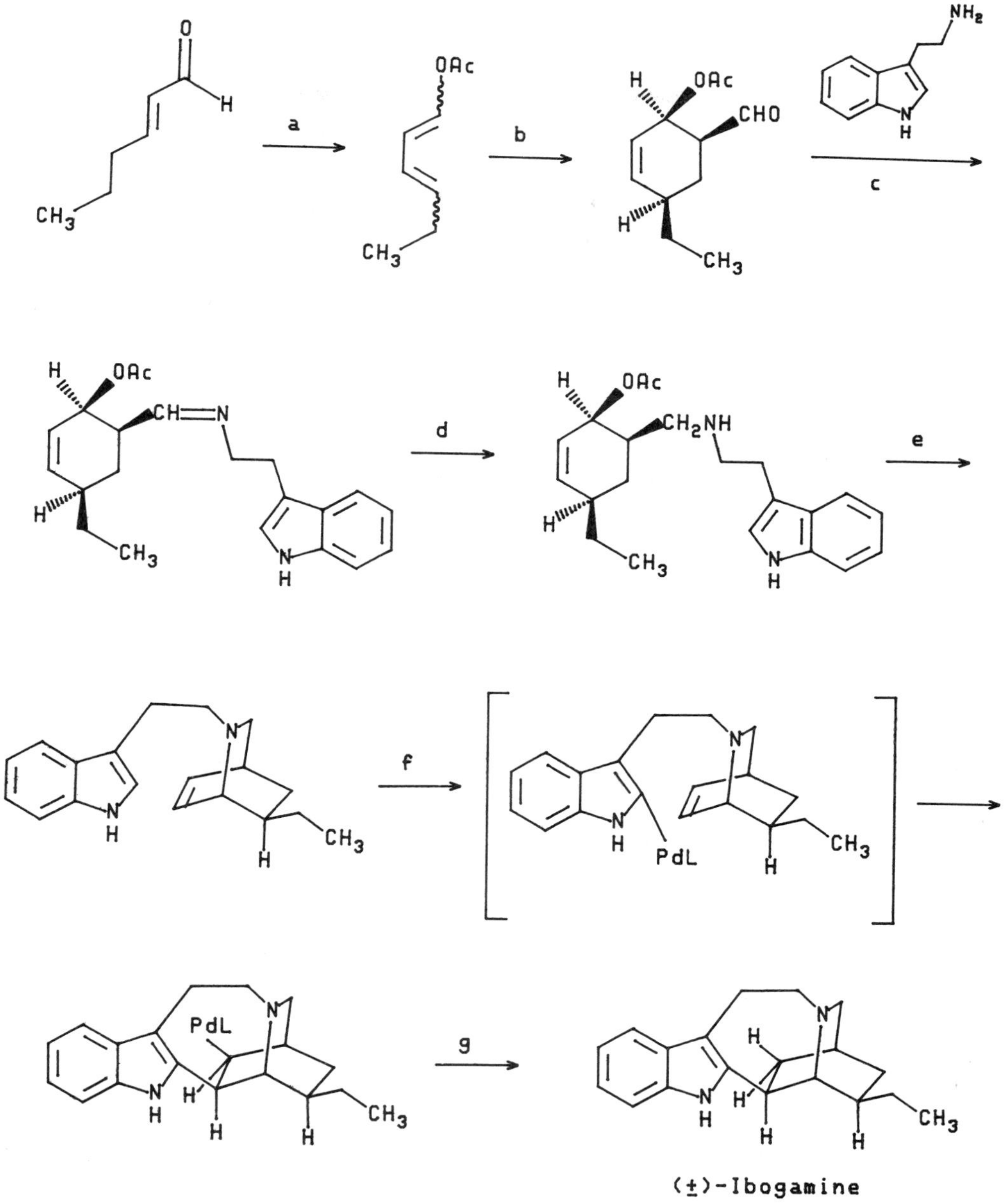

(±)-Ibogamine

a. p-TsOH, $CH_2=C(OAc)CH_3$; or 50°C, Ac_2O, Et_3N, DMAP (85%)

b. -10°C, $CH_2=CHCHO$, toluene; 90%

c. -10 to -5°C, $MgSO_4$, toluene

d. 0°C, $NaBH_4$, CH_3OH; 93% for two steps

Figure 3.52

e. 70°C, 3-6 mol% $Pd(PPh_3)_4$, CH_3CN; 45%

f. rt to 75°C, $AgBF_4$, 1 eq $PdCl_2(CH_3CN)_2$, Et_3N

g. 0°C, $NaBH_4$, CH_3OH; 45% for two steps

Figure 3.52 (*Continued*)

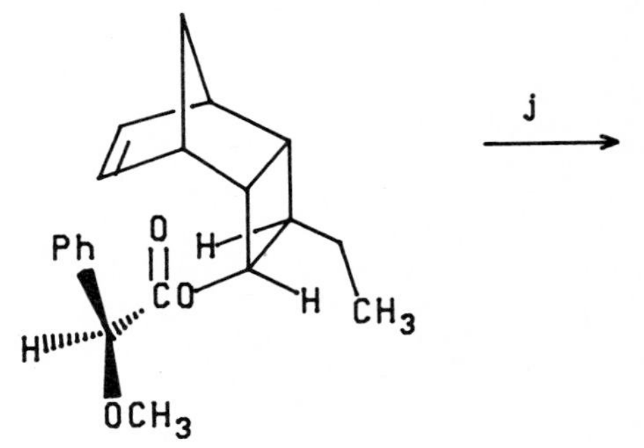

Figure 3.53

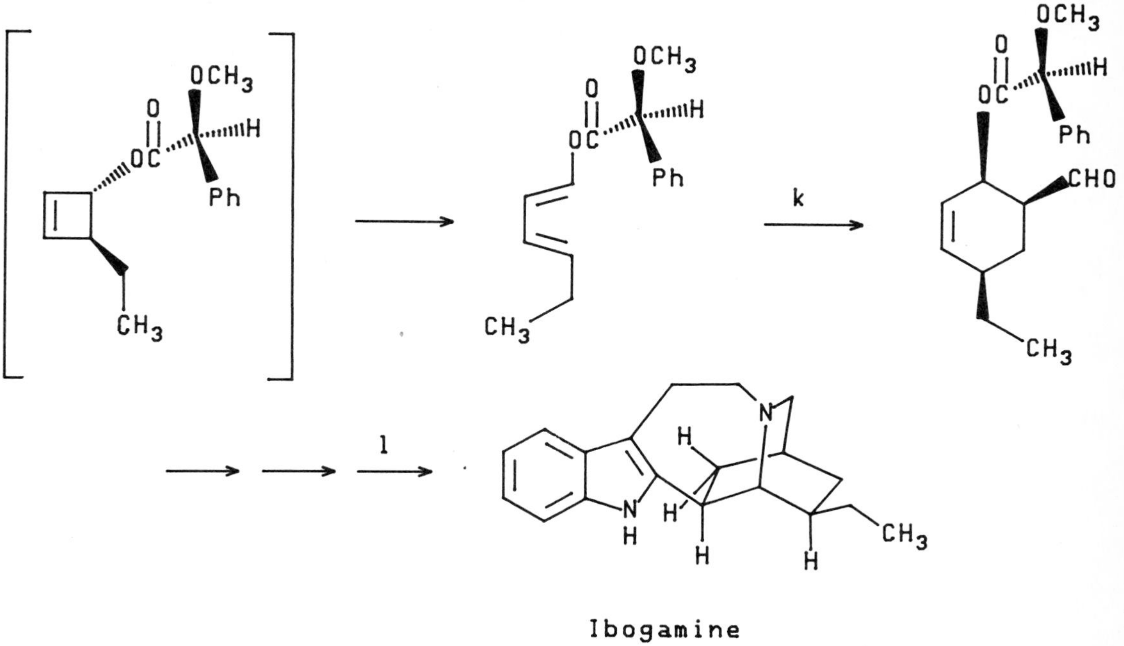

Ibogamine

a. $NaOCH_3$, CH_3OH

b. $CH_3OSO_2OCH_3$; 87%

c. rfx, Na dispersion, $(CH_3)_3SiCl$, toluene

d. rfx, 1 M HCl, THF, H_2O; 90% for two steps

e. rt, p-TsCl, pyridine; 84%

f. 60°C, under N_2, $CrCl_2$, acetone; 100%

g. 1) -78°C, $LiN(i\text{-}Pr)_2$, THF, 2) -30°C to rt, EtI; 52%

h. 0°C, $NaBH_4$, EtOH; 98%

i. rt, pyridine, benzene; 55%

j. 460°C, quartz tube; 100%

k. -10°C, CH_2=CHCHO, $BF_3 \cdot Et_2O$, toluene; 93%, 80:20 mixture of (3R,4S,6R) and (3S,4R,6S) diastereoisomers

l. Same sequence as for racemic synthesis; 80:20 mixture of (+) and (-) diastereoisomers

Figure 3.53 *(Continued)*

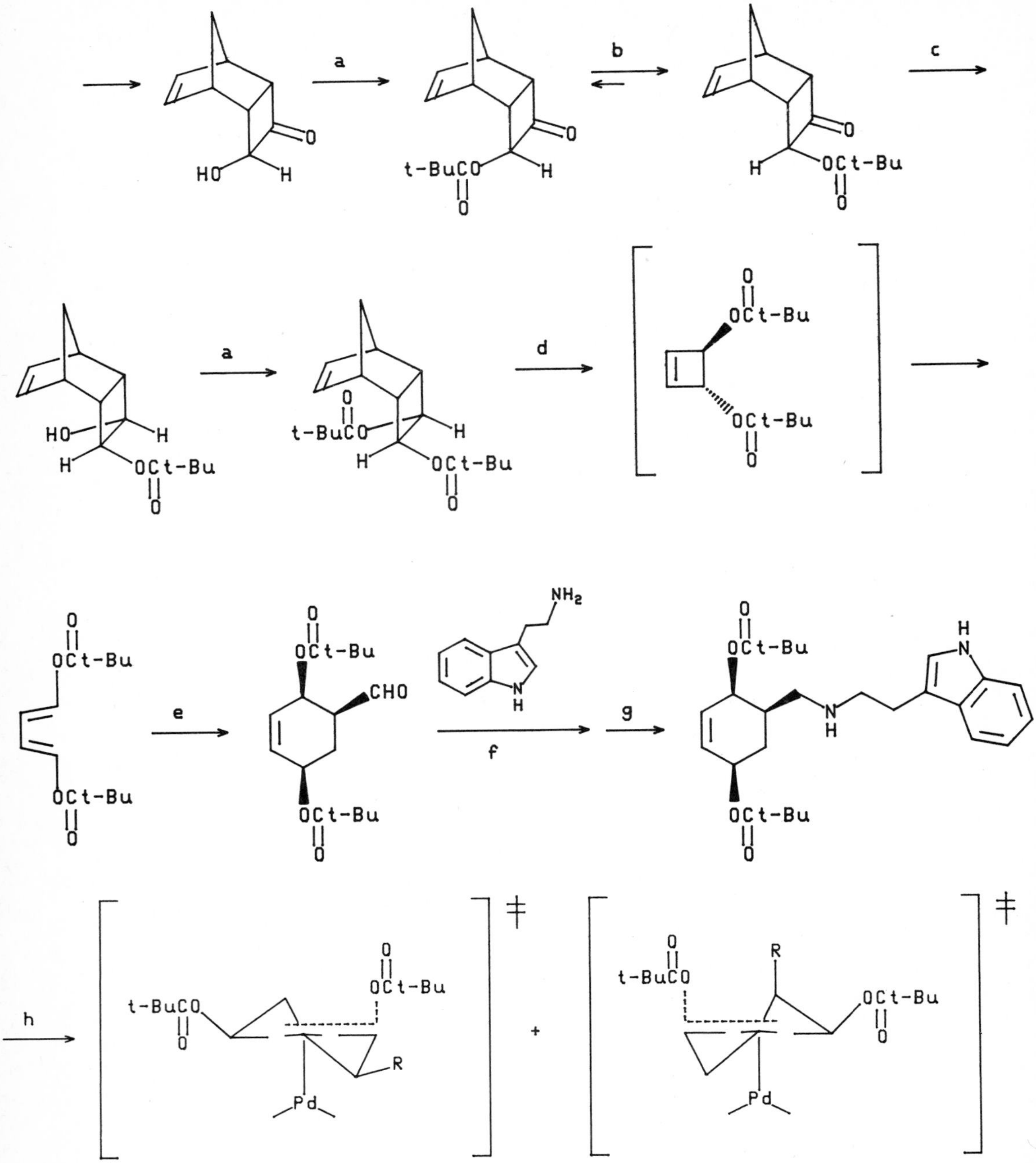

Lower energy transition state: Less steric interaction when R is pseudoequatorial

Higher energy transition state: more steric interaction when R is pseudoaxial

Figure 3.54

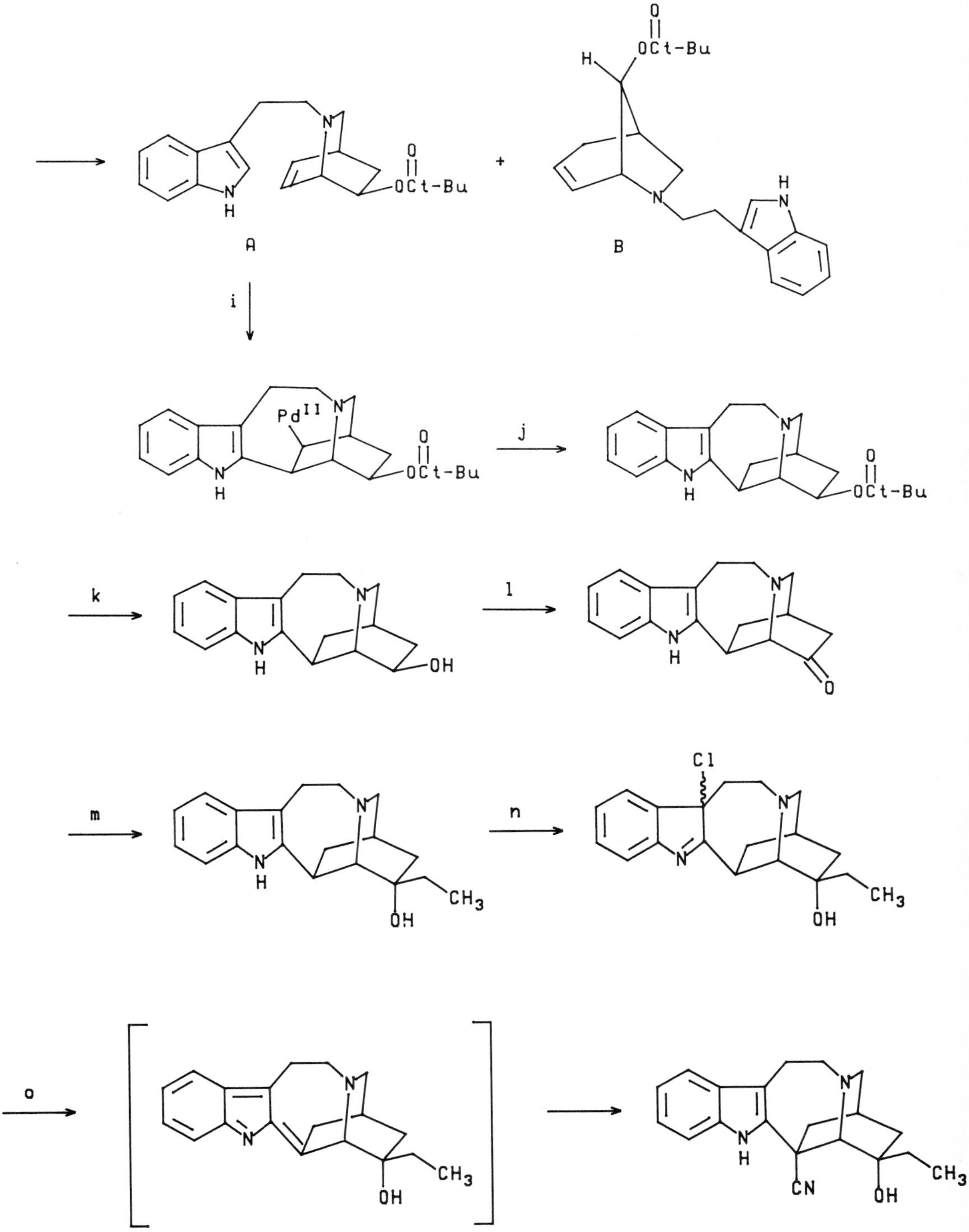

Figure 3.54 (*Continued*)

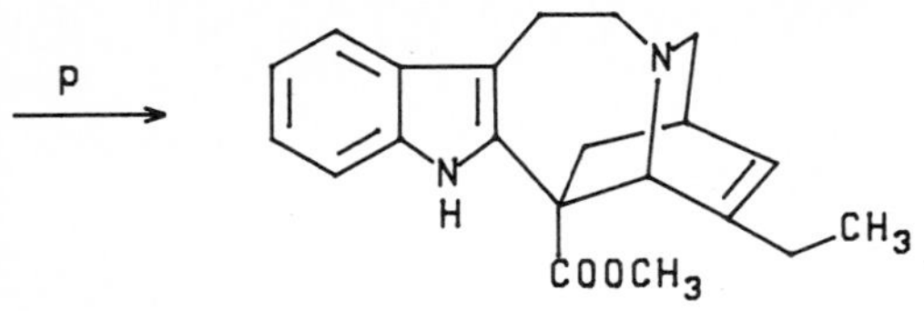

(±)-Catharanthine

a. t-BuC(O)Cl, pyridine

b. DBU, THF

c. 0°C, $NaBH_4$, EtOH

d. 25 to 240°C, quartz tube; for the five step preparation of (E,E)-1,4-diacetoxy-1,3-butadiene, the yields are 84, 83, 95, 90 and 89%, or 53% overall

e. -30 to -10°C, CH_2=CHCHO, $BF_3 \cdot Et_2O$, toluene; 90%

f. -5°C, $MgSO_4$, toluene

g. -5°C, $NaBH_4$, CH_3OH; 64% for two steps

h. 75°C, 5 mol% $Pd(PPh_3)_4$, CH_3CN

i. 25 to 67°C, Et_3N, $AgBF_4$, $PdCl_2(CH_3CN)_2$, CH_3CN; A:32%, B:"variable yield"

j. 0°C, $NaBH_4$, CH_3OH; 22%

k. rt, CH_3Li, Et_2O

l. rt, Et_3N, pyridine$\cdot SO_3$, DMSO; 88% for two steps

m. -78 to -10°C, EtMgBr, THF, Et_2O; 85%

n. -20°C, Et_3N, t-BuOCl, CH_2Cl_2, CCl_4

o. 70°C, under N_2, KCN, $CH_3C(O)N(CH_3)_2$; 22% for two steps

p. 1) 0°C to rt, 100% H_2SO_4, 2) 150°C, 20% KOH, diethylene glycol, 3) 0°C, HCl, CH_3OH, 4) rt, CH_2N_2, Et_2O, 5) repeat 3) and 4); 29% for five steps

Figure 3.54 *(Continued)*

at a terminal carbon to give a modest yield of the bicyclo[2.2.2] product and poorly reproducible yields of the alternative bicyclo[3.2.1] product. Intramolecular coupling of the indole ring with the alkene is accomplished using the palladium(II) methodology developed for ibogamine synthesis.

Depivaloylation with methyllithium followed by Moffatt oxidation affords a ketone. After introduction of an ethyl group using ethylmagnesium bromide, the remaining steps result in introduction of an ester group and dehydration.[82] Chlorination of the indole ring with *t*-butylhypochlorite occurs at the 3-position. Heating liberates HCl, producing an electron-deficient imine that can add cyanide. Dehydration occurs under relatively mild acidic conditions while nitrile hydrolysis is accomplished using strong base at high temperatures. Esterification using diazomethane completes the synthesis.

The total synthesis of (±)-catharanthine from cyclopentadiene and maleic anhydride was completed in 25 steps in overall 0.076% yield (assuming yields in the synthesis of 1,4-diacetoxy-1,3-butadiene and 1,4-dipivaloyloxy-1,3-butadiene were comparable). A related approach to optically active ibogamines has the potential to be both more versatile and significantly more efficient[83] (see Figure 3.54).

REFERENCES

1a. Danieli, B.; Palmisano, G. *Alkaloids* **1986**, *27*, 1.

1b. Taylor, W. I. *Alkaloids* **1968**, *11*, 79.

1c. Taylor, W. I. *Alkaloids* **1965**, *8*, 203.

1d. Cordell, G. A. *Introduction to Alkaloids: A Biogenetic Approach*; Wiley: New York, 1981, pp 761–771.

2. ED_{50} (μg/mL against P_{388} cells) = 0.43: Gunasekera, S. P.; Cordell, G. A.; Farnsworth, N. R. *Phytochemistry* **1980**, *19*, 1213.

3. Coronaridine partially prevents pregnancy in adult female rats: Meyer, W. E.; Coppola, J. A.; Goldman, L. *J. Pharm. Sci.* **1973**, *62*, 1199.

4. ED_{50} (μg/mL against P_{388} cells) = 6.8, (μg/mL against KB cells) = 5100: Kingston, D. G. I. *J. Pharm. Sci.* **1978**, *67*, 272.

5. Cordell, G. A.; Saxton, J. E. *Alkaloids* **1981**, *20*, 1.

6a. Potier, P. *Pure Appl. Chem.* **1986**, *58*, 737.

6b. Kutney, J. P. *Pure Appl. Chem.* **1984**, *56*, 1011.

7. See the following references for the most recent synthetic efforts.

7a. Huffman, J. W.; Shanmugasundaram, G.; Sawdaye, R.; Raveendranath, P. C.; Desai, R. C. *J. Org. Chem.* **1985**, *50*, 1460.

7b. Kuehne, M. E.; Reider, P. J. *J. Org. Chem.* **1985**, *50*, 1464.

7c. Imanishi, T.; Yagi, N.; Shin, H.; Hanaoka, M. *Tetrahedron Lett.* **1981**, *22*, 4001.

7d. Atta-ur-Rahman; Beisler, J. A.; Harley-Mason, J. *Tetrahedron* **1980**, *36*, 1063.

7e. Nagata, W.; Hirai, S. U. S. Patent 3 716 528, 13 Feb 1973; *Chem. Abstr.* **1973**, *78*, P136493f.

7f. Rosenmund, P.; Haase, W. H.; Bauer, J.; Frische, R. *Chem. Ber.* **1975**, *108*, 1871.

7g. Sallay, S. I. *J. Am. Chem. Soc.* **1967**, *89*, 6762.

7h. Hirai, S.; Kawata, K.; Nagata, W. *Chem. Commun.* **1968**, 1016.

7i. Ikezaki, M.; Wakamatsu, T.; Ban, Y. *Chem. Commun.* **1969**, 88.

7j. Kutney, J. P., Cretney, W. J.; Le Quesne, P.; McKague, B.; Piers, E. *J. Am. Chem. Soc.* **1970**, *92*, 1712.

8. See the following references for two excellent reviews on palladium reagents in organic synthesis:
8a. Tsuji, J. *Organic Synthesis with Palladium Compounds*; Springer-Verlag: New York, 1980.
8b. Heck, R. F. *Palladium Reagents in Organic Syntheses*; Academic Press: Orlando, FL, 1985.
9. Palladium bis-trifluoroacetate is prepared conveniently from $Pd(OAc)_2$ and trifluoroacetic acid: Stephenson, T. A.; Morehouse, S. M.; Powell, A. R.; Heffer, J. P.; Wilkinson, G. *J. Chem. Soc.* **1965**, 3632.
10. Chrisope, D. R.; Beak, P.; Saunders, W. H., Jr. *J. Am. Chem. Soc.* **1988**, *110*, 230.
11. Trost, B. M.; Strege, P. E.; Weber, L.; Fullerton, T. J.; Dietsche, T. J. *J. Am. Chem. Soc.* **1978**, *100*, 3407.
12a. Harrison, I. T.; Kimura, E.; Bohme, E.; Fried, J. H. *Tetrahedron Lett.* **1969**, 1589.
12b. Jones, D. N.; Knox, S. D. *J. Chem. Soc., Chem. Commun.* **1975**, 165.
12c. Riediker, M.; Schwartz, J. *Tetrahedron Lett.* **1981,** *22*, 4655.
12d. Temple, J. S.; Riediker, M.; Schwartz, J. *J. Am. Chem. Soc.* **1982**, *104*, 1310.
13. Hüttel, R.; Christ, H. *Chem. Ber.* **1964**, *97*, 1439.
14. Trost, B. M.; Metzner, P. J. *J. Am. Chem. Soc.* **1980**, *102*, 3572.
15. Hüttel, R.; McNiff, M. *Chem. Ber.* **1973**, *106*, 1789.
16a. Henderson, K.; McQuillin, F. J. *J. Chem. Soc., Chem. Commun.* **1978**, 15.
16b. Ukhim, L. Yu.; Il'in, V. I.; Orlova, Zh. I.; Bokii, N. G.; Struchkov, Yu. T. *J. Organometal. Chem.* **1976**, *113*, 167.
16c. Itoh, K.; Nishiyama, H.; Ohnishi, T.; Ishii, Y. *J. Organometal. Chem.* **1974**, *76*, 401.
16d. Howsam, R. W.; McQuillin, F. J. *Tetrahedron Lett.* **1968**, 3667.
16e. Kasahara, A.; Tanaka, K.; Asamiya, K. *Bull. Chem. Soc. Jpn.* **1967**, *40*, 351.
16f. Tsuji, J.; Imamura, S. *Bull. Chem. Soc. Jpn.* **1967**, *40*, 197.
16g. Tsuji, J.; Imamura, S.; Kiji, J. *J. Am. Chem. Soc.* **1964**, *86*, 4491.
16h. Parshall, G. W.; Wilkinson, G. *Inorg. Chem.* **1962**, *1*, 896.
16i. Parshall, G. W.; Wilkinson, G. *Chem. Ind.* **1962**, 261.
17. For a discussion of mechanisms for oxidative addition, see Stille, J. K.; Lau, K. S. Y. *Acc. Chem. Res.* **1977**, *10*, 434.
18. See the following references for examples of S_N2 displacement of allylic acetates.
18a. Trost, B. M.; Strege, P. E. *J. Am. Chem. Soc.* **1977**, *99*, 1649.
18b. Trost, B. M.; Matsumura, Y. *J. Org. Chem.* **1977**, *42*, 2036.
18c. Trost, B. M.; Verhoeven, T. R. *J. Am. Chem. Soc.* **1976**, *98*, 630.
18d. Trost, B. M.; Verhoeven, T. R. *J. Am. Chem. Soc.* **1978**, *100*, 3435.
19. Dent, W. T.; Long, R.; Wilkinson, A. J. *J. Chem. Soc.* **1964**, 1585.
20. Nicholson, J. K.; Powell, J.; Shaw, B. L. *Chem. Commun.* **1966**, 174.
21. Fiaud, J. C.; Malleron, J. L. *Tetrahedron Lett.* **1980**, *21*, 4437.
22a. Donati, M.; Conti, F. *Tetrahedron Lett.* **1966**, 1219.
22b. Shaw, B. L. *Chem. Ind. (London)* **1962**, 1190.
23. Robinson, S. D.; Shaw; B. L. *J. Chem. Soc.* **1963**, 4806.
24a. Bäckvall, J-E.; Nyström, J-E. *J. Chem. Soc., Chem. Commun.* **1981**, 59.
24b. Bäckvall, J-E.; Nordberg, R. E. *J. Am. Chem. Soc.* **1981**, *103*, 4959.
24c. Bäckvall, J-E.; Nordberg, R. E.; Björkman, E. E.; Moberg, C. *J. Chem. Soc., Chem. Commun.* **1980**, 943.
24d. Rowe, J. M.; White, D. A. *J. Chem. Soc. (A)* **1967**, 1451.

25. Åkermark. B.; Bäckvall, J-E; Löwenborg, A.; Zetterberg, C. *J. Organometal. Chem.* **1979**, *166*, C33.
26. Julia, M.; Nel, M.; Saussine, L. *J. Organometal. Chem.* **1979**, *181*, C17.
27a. Larock, R. C.; Harrison, L. W.; Hsu, M. H. *J. Org. Chem.* **1984**, *49*, 3662.
27b. Stakem, F. G.; Heck, R. F. *J. Org. Chem.* **1980**, *45*, 3584.
27c. Kasahara, A.; Izumi, T. *Bull. Chem. Soc. Jpn.* **1972**, *45*, 1256.
27d. Heck, R. F. *J. Am. Chem. Soc.* **1968**, *90*, 5542.
28a. O'Connor, J. M.; Stallman, B. J.; Clark, W. G.; Shu, A. Y. L.; Spada, R. E.; Stevenson, T. M.; Dieck, H. A. *J. Org. Chem.* **1983**, *48*, 807.
28b. Patel, B. A.; Kao, L-C.; Cortese, N. A.; Minkiewicz, J. V.; Heck, R. F. *J. Org. Chem.* **1979**, *44*, 918.
28c. Patel, B. A.; Dickerson, J. E.; Heck, R. F. *J. Org. Chem.* **1978**, *43*, 5018.
29. Dunne, K.; McQuillin, F. J. *J. Chem. Soc. (C)* **1970**, 2196.
30. Takahashi, M.; Suzuki, H.; Moro-Oka, Y.; Ikawa, T. *Chem. Lett.* **1979**, 53.
31. Larock, R. C.; Tagaki, K.; Hershberger, S. S.; Mitchell, M. A. *Tetrahedron Lett.* **1981**, *22*, 5231.
32. Larock, R. C.; Narayanan, K. *J. Org. Chem.* **1984**, *49*, 3411.
33. Larock, R. C.; Mitchell, M. A. *J. Am. Chem. Soc.* **1978**, *100*, 180.
34a. Dieck, H. A.; Heck, R. F. *J. Org. Chem.* **1975**, *40*, 1083.
34b. Patel, B. A.; Heck, R. F. *J. Org. Chem.* **1978**, *43*, 3898.
34c. Patel, B. A.; Kim, J-I. I.; Bender, D. D.; Kao, L-C.; Heck, R. F. *J. Org. Chem.* **1981**, **46**, 1061.
34d. Kim, J-I. I.; Patel, B. A.; Heck, R. F. *J. Org. Chem.* **1981**, *46*, 1067.
35a. Larock, R. C.; Takagi, K. *Tetrahedron Lett.* **1983**, *24*, 3457.
35b. Larock, R. C.; Takagi, K. *J. Org. Chem.* **1984**, *49*, 2701.
36. Powell, J.; Shaw, B. L. *J. Chem. Soc (A)* **1968**, 774.
37a. Itoh, K.; Fukui, M.; Kurachi, Y. *J. Chem. Soc., Chem. Commun.* **1977**, 500.
37b. Kliegman, J. M. *J. Organomètal. Chem.* **1971**, *29*, 73.
38a. Larock, R. C.; Varaprath, S. *J. Org. Chem.* **1984**, *49*, 3432.
38b. Battiste, M. A.; Friedrich, L. E.; Fiato, R. A. *Tetrahedron Lett.* **1975**, 45.
38c. Fiato, R. A.; Mushak, P.; Battiste, M. A. *J. Chem. Soc., Chem. Commun.* **1975**, 869.
38d. Ketley, A. D.; Braatz, J. A.; Craig, J. *Chem. Commun.* **1970**, 1117.
38e. Mushak, P.; Battiste, M. A. *J. Organometal. Chem.* **1969**, *17*, P46.
38f. Noyori, R.; Takaya, H. *Chem. Commun.* **1969**, 525.
38g. Shono, T.; Yoshimura, T.; Matsumura, Y.; Oda, R. *J. Org. Chem.* **1968**, *33*, 876.
38h. Ketley, A. D.; Braatz, J. A. *Chem. Commun.* **1968**, 959.
38i. Ketley, A. D.; Braatz, J. A. *J. Organometal. Chem.* **1967**, *9*, P5.
39. Åkermark, B.; Zetterberg, K. *Tetrahedron Lett.* **1975**, 3733.
40a. Trost, B. M.; Weber, L.; Strege, P. E.; Fullerton, T. J.; Dietsche, T. J. *J. Am. Chem. Soc.* **1978**, *100*, 3416.
40b. Trost, B. M..; Weber, L.; Strege, P.; Fullerton, T. J.; Dietsche, T. J. *J. Am. Chem. Soc.* **1978**, *100*, 3426.
40c. Jackson, W. R.; Strauss, J. U. *Aust. J. Chem.* **1978**, *31*, 1073.
40d. Collins, D. J.; Jackson, W. R.; Timms, R. N. *Tetrahedron Lett.* **1976**, 495.
40e. Jackson, W. R.; Strauss, J. U. G. *Tetrahedron Lett.* **1975**, 2591.

40f. Tsuji, J. *Bull. Chem. Soc. Jpn.* **1973**, *46*, 1896.
40g. Tsuji, J.; Takahashi, H.; Morikawa, M. *Tetrahedron Lett.* **1965**, 4387.
41a. Castanat, Y.; Petit, F. *Tetrahedron Lett.* **1979**, 3221.
41b. Numata, S.; Kurosawa, H.; Okawara, R. *J. Organometal. Chem.* **1975**, *102*, 259.
42. Hegedus, L. S.; Darlington, W. H.; Russell, C. E. *J. Org. Chem.* **1980**, *45*, 5193.
43. Green, M. L. H.; Mitchard, L. C.; Silverthorn, W. E. *J. Chem. Soc., Dalton Trans.* **1973**, 2177.
44a. Stanton, S. A.; Felman, S. W.; Parkhurst, C. S.; Godleski, S. A. *J. Am. Chem. Soc.* **1983**, *105*, 1964.
44b. Trost, B. M.; Verhoeven, T. R. *J. Am. Chem. Soc.* **1980**, *102*, 4730.
44c. Trost, B. M.; Verhoeven, T. R.; Fortunak, J. M. *Tetrahedron Lett.* **1979**, 2301.
44d. Trost, B. M.; Keinan, E. *J. Org. Chem,* **1979**, *44*, 3451.
44e. Trost, B. M.; Keinan, E. *J. Am. Chem. Soc.* **1978**, *100*, 7779.
45a. Tsuji, J.; Ueno, H.; Kobayashi, Y.; Okumoto, H. *Tetrahedron Lett.* **1981**, 22, 2573.
45b. Collins, D. J.; Jackson, W. R.; Timms, R. N. *Aust. J. Chem.* **1977**, *30*, 2167.
45c. Jackson, W. R.; Strauss, J. U. *Aust. J. Chem.* **1977**, *30*, 553.
45d. Collins, D. J.; Jackson, W. R.; Timms, R. N. *Tetrahedron Lett.* **1976**, 495.
45e. Jackson, W. R.; Strauss, J. U. G. *Tetrahedron Lett.* **1975**, 2591.
46a. Rajanbabu, T. V. *J. Org. Chem.* **1985**, *50*, 3642.
46b. Dunkerton, L. V.; Serino, A. J. *J. Org. Chem.* **1982**, *47*, 2812.
47. Trost, B. M.; Vercauteren, J. *Tetrahedron Lett.* **1985**, *26*, 131.
48a. Trost, B. M.; Molander, G. A. *J. Am. Chem. Soc.* **1981**, *103*, 5969.
48b. Tsuji, J.; Kataoka, H.; Kobayashi, Y. *Tetrahedron Lett.* **1981**, *22*, 2575.
49. Trost, B. M.; Verhoeven, T. R *J. Am. Chem. Soc.* **1977**, *99*, 3867.
50. Trost, B. M.; Genet, J. P. *J. Am. Chem. Soc.* **1976**, *98*, 8516.
51. Bäckvall, J-E.; Nystrom, J-E. *J. Chem. Soc., Chem. Commun.* **1981**, 59.
52. Harrington, P. J.; DiFiore, K. A., unpublished results.
53a. Trost, B. M.; Runge, T. A. *J. Am. Chem. Soc.* **1981**, *103*, 7559.
53b. Trost, B. M.; Runge, T. A. *J. Am. Chem. Soc.* **1981**, *103*, 7550.
54. Trost, B. M.; Angle, S. R. *J. Am. Chem. Soc.* **1985**, *107*, 6123.
55. Trost, B. M.; Verhoeven, T. R. *J. Amer. Chem. Soc.* **1979**, *101*, 1595.
56. Takahashi, T.; Kataoka, H.; Tsuji, J. *J. Am. Chem. Soc.* **1983**, *105*, 147.
57a. Trost , B. M.; Warner, R. W. *J. Am. Chem. Soc.* **1983**, *105*, 5940.
57b. Trost, B. M.; Brickner, S. J. *J. Am. Chem. Soc.* **1983**, *105*, 568.
57c. Trost, B. M.; Warner, R. W. *J. Am. Chem. Soc.* **1982**, *104*, 6112.
57d. Kitagawa, Y.; Itoh, A.; Hashimoto, S.; Yamamoto, H.; Nozaki, H. *J. Am. Chem. Soc.* **1977**, *99*, 3864.
58. Kagan, H. B.; Dang, T-P. *J. Am. Chem. Soc.* **1972, 94**, 6429.
59. Knowles, W. S.; Sabacky, M. J.; Vineyard, B. D. *J. Chem. Soc., Chem. Commun.* **1972**, 10.
60a. Morrison, J. D.; Burnett, R. E.; Aguiar, A. M.; Morrow, C. J.; Phillips C. *J. Am. Chem. Soc.* **1971**, *93*, 1301.
60b. Bogdanvoić, B.; Henc, B.; Meister, B.; Pauling, H.; Wilke, G. *Angew. Chem. Int. Ed. Engl.* **1972**, *11*, 1023.
61. Knowles, W. S.; Sabacky, M. J.; Vineyard, B. D.; Weinkauff, D. J. *J. Am. Chem. Soc.* **1975**, *97*, 2567.

62. Morrison, J. D.; Masler, W. F.; Neuberg, M. K. *Adv. Catal.* **1976**, *25*, 81.

63a. Brussee, J.; Jansen, A. C. A. *Tetrahedron Lett.* **1983**, *24*, 3261.

63b. Grubbs, R. H.; DeVries, R. A. *Tetrahedron Lett.* **1977**, 1879.

64. Trost, B. M.; Dietsche, T. J. *J. Am. Chem. Soc.* **1973**, *95*, 8200.

65. Bosnich, B.; Mackenzie, P. B. *Pure Appl. Chem.* **1982**, *54*, 189.

66. Trost, B. M.; Murphy, D. J. *Organometallics* **1985**, *4*, 1143.

67. Trost, B. M.; Godleski, G. A.; Genet, J. P. *J. Am. Chem. Soc.* **1978**, *100*, 3930.

68. *Trans*-2-hexenal is commercially available (Fluka 88–89 catalog, 100 mL, $d = 0.853$ g/mL, for $40.50).

69. Deghenghi, R.; Engel, C. R. *J. Am. Chem. Soc.* **1960**, *82*, 3201.

70. Cecchi, R.; Favara, D.; Omodei-Salé, A.; Depaoli, A.; Consonni, P. *Gazz. Chim. Ital.* **1984**, *114*, 225.

71. Acrolein is commercially available and inexpensive (Fluka 88–89 catalog, 1 L, $d = 0.845$ g/mL, for $27.00).

72. Tryptamine is commercially available (Lancaster Synthesis 89–90 catalog, 50 g for $53.70).

73a. Dicyclopentadiene is commercially available and inexpensive (Lancaster Synthesis 89–90 catalog, 1 kg for $11.00).

73b. Cyclopentadiene is prepared by thermal cracking of dicyclopentadiene: Korach, M.; Nielsen, D. R.; Rideout, W. H. *Org. Syn. Coll.* **1973**, *V*, 414.

74. Maleic anhydride is commercially available and inexpensive (Aldrich 88–89 catalog, 5 kg for $22.90).

75. Jung, M. E. *J. Chem. Soc., Chem. Commun.* **1974**, 956.

76a. Bloomfield, J. J. *Tetrahedron Lett.* **1968**, 587.

76b. For a review, see Bloomfield, J. J.; Owsley, D. C.; Nelke, J. M. *Org. React.* **1976**, *23*, 259.

77. Schräpler, U.; Rühlmann, K. *Chem. Ber.* **1964**, *97*, 1383.

78. Trost, B. M.; Godleski, S. A.; Ippen, J. *J. Org. Chem.* **1978**, *43*, 4559.

79. (*S*)-(+)-α-Methoxyphenylacetic acid is commercially available (Aldrich 88–89 catalog, 1 g for $23.70). The acid chloride was prepared using thionyl chloride.

80. Alternatively, (*S*)-(+)-α-methoxyphenylacetic acid can be prepared from (*S*)-(+)-mandelic acid by methylation [diazomethane, $(t\text{-BuO})_3Al$ catalyst], hydrolysis of the methyl ester with dilute sodium hydroxide, precipitation of the sodium salt with sodium chloride solution, and then recrystallization of the acid from petroleum ether-benzene: Raznikiewicz, T. *Acta Chem. Scand.* **1962**, *16*, 1097. (*S*)-(+)-Mandelic acid is commercially available (Lancaster Synthesis 88–89 catalog, 25 g for $32.50).

81. Trost, B. M.; Godleski, S. A.; Belletire, J. L. *J. Org. Chem.* **1979**, *44*, 2052.

82. Büchi, G.; Kulsa, P.; Ogasawara, K.; Rosati, R. L. *J. Am. Chem. Soc.* **1970**, *92*, 999

83. Trost, B. M.; Romero, A. G. *J. Org. Chem.* **1986**, *51*, 2332.

CHAPTER 4

(±)-Limaspermine

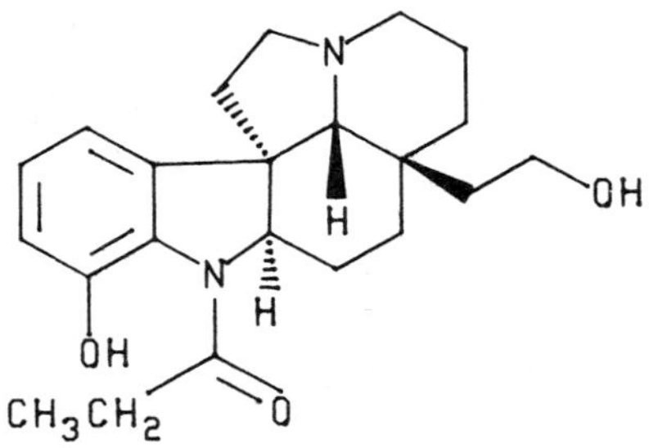

(±)-Limaspermine

Aspidospermidine-17,21-diol,

1-(1-oxopropyl)

(±)[80656-03-7]

(+)[5516-64-3]

Figure 4.1

The *Aspidosperma* alkaloid (+)-limaspermine was first isolated from *Aspidosperma limae* and assigned a structure in 1962.[1] The absolute stereochemistry was later determined by correlations with (−)-aspidospermine, an alkaloid of known configuration[2] (see Figure 4.2). (−)-Vindoline has a more highly functionalized skeleton (see Figure 4.3).

While these three related alkaloids are generally considered to be devoid of pharmacological activity,[3] the bis-indole *Catharanthus* alkaloids vinblastine (vincaleukoblastine) and vincristine (leurocristine) are two of the most effective chemotherapeutic agents for the treatment of Hodgkin's disease and acute childhood leukemia. Both contain the *Aspidosperma* alkaloid vindoline and an *iboga* alkaloid (16-β-carbomethoxyvelbanamine). Vinblastine and vincristine were first isolated from *Catharanthus roseus* by Svoboda and are still produced by isolation today despite typical yields of 1 g and 20 mg per 1000 kg, respectively.[4] A nicely convergent synthetic approach might have as a final step a biomimetic carbon–carbon bond formation between vindoline and an nitrogen oxide of the *iboga* alkaloid catharanthine. (Cell-free

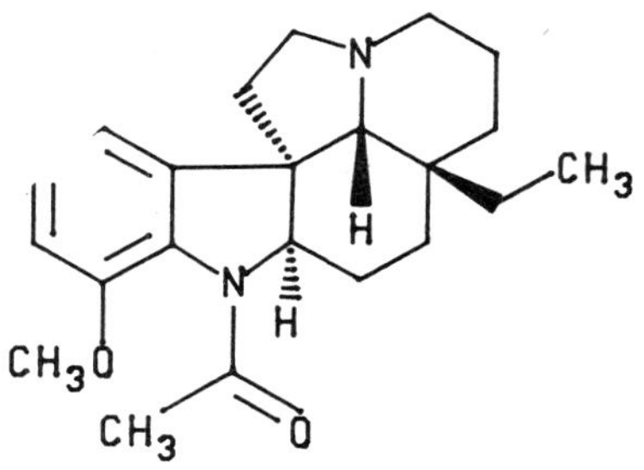

(-)-Aspidospermine

Figure 4.2

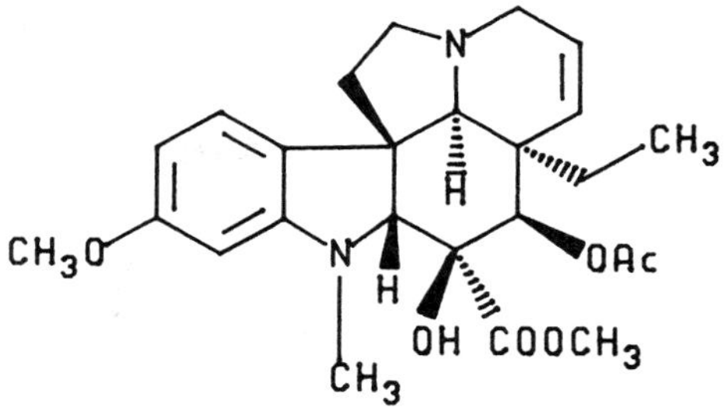

(-)-Vindoline

Figure 4.3

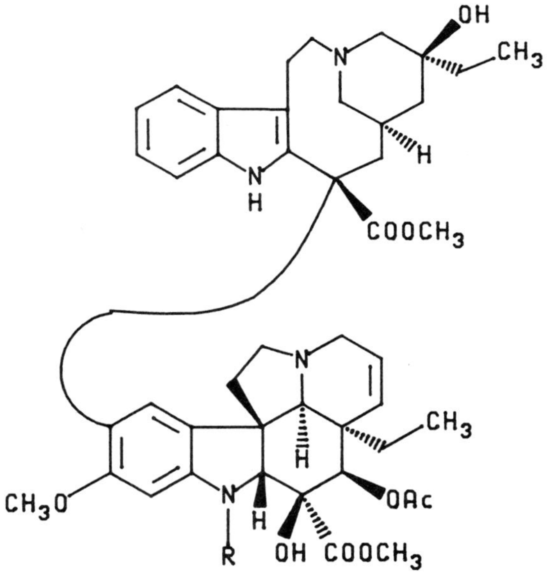

Vinblastine R = CH_3

Vincristine R = CHO

Figure 4.4

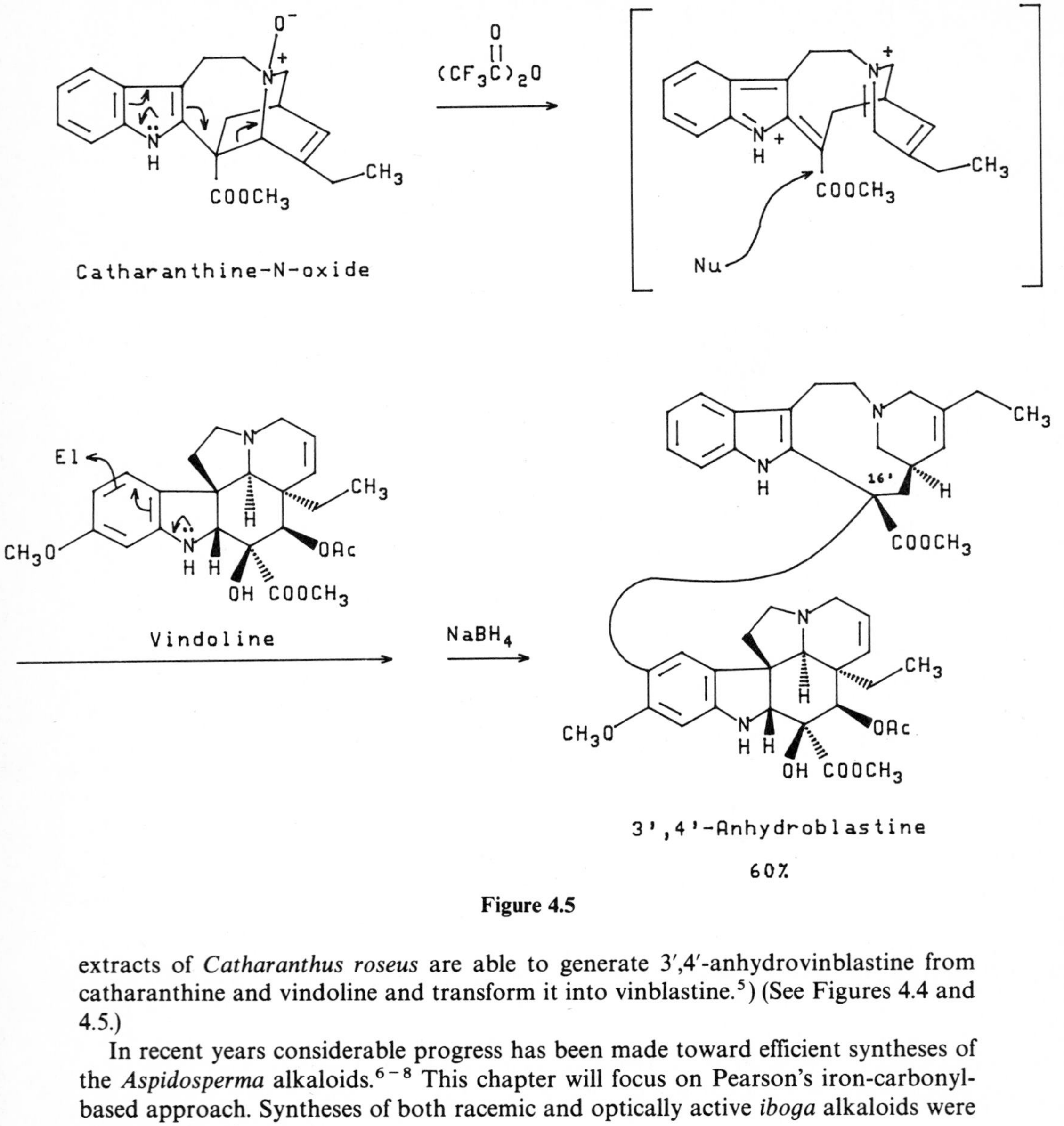

Figure 4.5

extracts of *Catharanthus roseus* are able to generate 3′,4′-anhydrovinblastine from catharanthine and vindoline and transform it into vinblastine.[5]) (See Figures 4.4 and 4.5.)

In recent years considerable progress has been made toward efficient syntheses of the *Aspidosperma* alkaloids.[6–8] This chapter will focus on Pearson's iron-carbonyl-based approach. Syntheses of both racemic and optically active *iboga* alkaloids were discussed in Chapter 3.

4.1 PREPARATION OF η^4-DIENE IRON TRICARBONYL COMPLEXES

4.1.1 Methods A and B. Conjugated Dienes and Iron Pentacarbonyl: Thermal or Photochemical Initiation

Diene iron tricarbonyl complexes are most often prepared by reaction of a conjugated diene in the *S-cis* conformation with commercially available, inexpensive iron

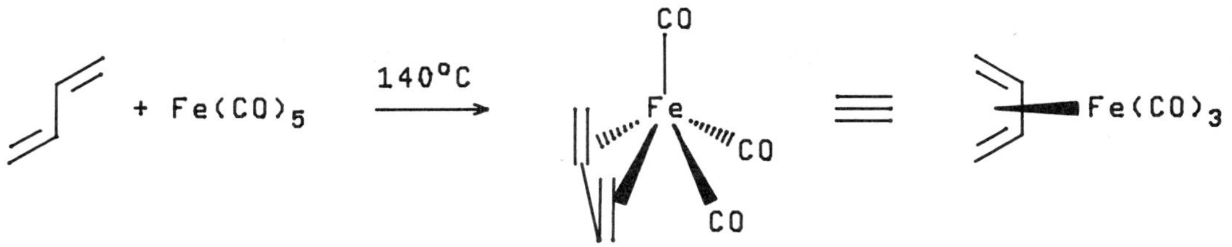

Figure 4.6

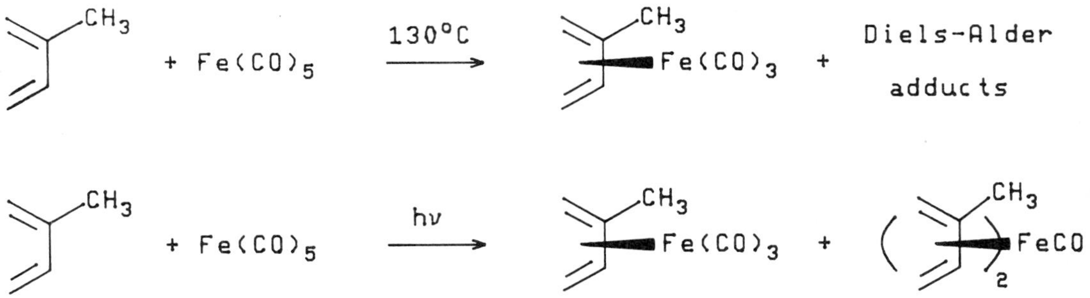

Figure 4.7

pentacarbonyl (Caution: volatile and toxic). Since iron pentacarbonyl is a coordinatively saturated, 18-electron complex, ejection of a carbon monoxide must precede formation of the initial π-alkene complex. Ejection can be accomplished by either heating at high temperature, typically 130 to 140 °C in di-*n*-butyl ether (method A) or by photolysis (method B).

The first complex of this type was prepared from butadiene by method A.[9,10] The tetrahapto structure was suggested 30 years later. The complex has a square pyramidal geometry with a carbonyl in the apical position[11,12] (see Figure 4.6).

Heating isoprene and iron pentacarbonyl at high temperature results in inefficient complex formation due to competitive Diels–Alder dimerization. Despite some bis-diene iron carbonyl complex formation on prolonged irradiation, the photochemical method (method B) proves to be superior (see Figure 4.7). Other methods of complex formation often have advantages.

4.1.2 Method C. Nonconjugated Dienes and Iron Pentacarbonyl

A nonconjugated diene can be converted to the complex of an isomeric conjugated diene under the vigorous conditions of method A. The nonconjugated diene may be more synthetically accessible (e.g., 1,4-cyclohexadienes are readily prepared by Birch reduction of aromatics).

4.1.3 Methods D and E. Conjugated Dienes and Triiron Dodecacarbonyl; Conjugated Dienes and Diiron Nonacarbonyl

A more reactive iron carbonyl will complex a conjugated diene in the *S-cis* conformation at significantly lower temperatures. For example, triiron dodecacarbonyl forms complexes in refluxing benzene[13] (method D), and diiron nonacarbonyl forms complexes at temperatures as low as 40°C (method E).[14]

4.1.4 Method F. Conjugated Dienes and Iron Pentacarbonyl: Oxidative Initiation

A mild oxidant such as trimethylamine-*N*-oxide will convert iron pentacarbonyl to the reactive tetracarbonyl. Complexation of a conjugated diene in the *S-cis* conformation will occur at or below room temperature.[15]

4.1.5 Methods G and H. α, β- and β, γ-Unsaturated Alcohols and Diiron Nonacarbonyl

Dehydration of α,β- (method G) or β,γ-unsaturated alcohols (method H) with cupric sulfate in the presence of diiron nonacarbonyl produces diene iron tricarbonyl complexes directly. The alcohols may be more accessible and, perhaps, more stable on storage than the corresponding diene.[16]

4.1.6 Methods I and J. 1,4-Dihaloalkenes and Diiron Nonacarbonyl

Reduction of 1,2-bis-bromomethyl aromatics (method I) and 3,4-dichlorocyclobutene (method J) with diiron nonacarbonyl results in complexes of what would be very unstable dienes[17,18] (see Figure 4.8).

4.1.7 Method K and Discussion. Methylenecyclopropanes and Diiron Nonacarbonyl

Ring opening of methylenecyclopropanes with diiron nonacarbonyl produces diene iron tricarbonyl complexes stereospecifically[18] (see Figure 4.9).

Four different routes to *trans*-1,3-pentadiene iron tricarbonyl and one route to the *cis*-isomer illustrate the propensity for rearrangements in complex formation under the vigorous conditions of method A as well as the advantageous use of diiron nonacarbonyl in method E (see Figure 4.10).

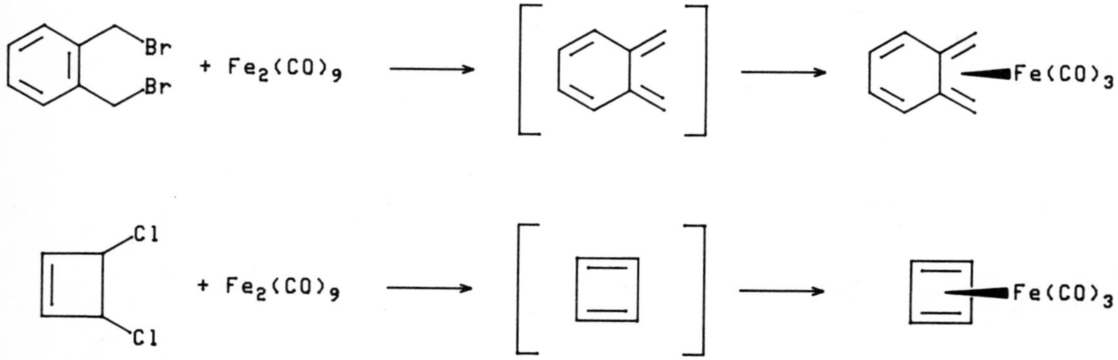

Br
Br
+ $Fe_2(CO)_9$ ⟶ [ ] ⟶ $Fe(CO)_3$

Cl
Cl
+ $Fe_2(CO)_9$ ⟶ [ ] ⟶ $Fe(CO)_3$

Figure 4.8

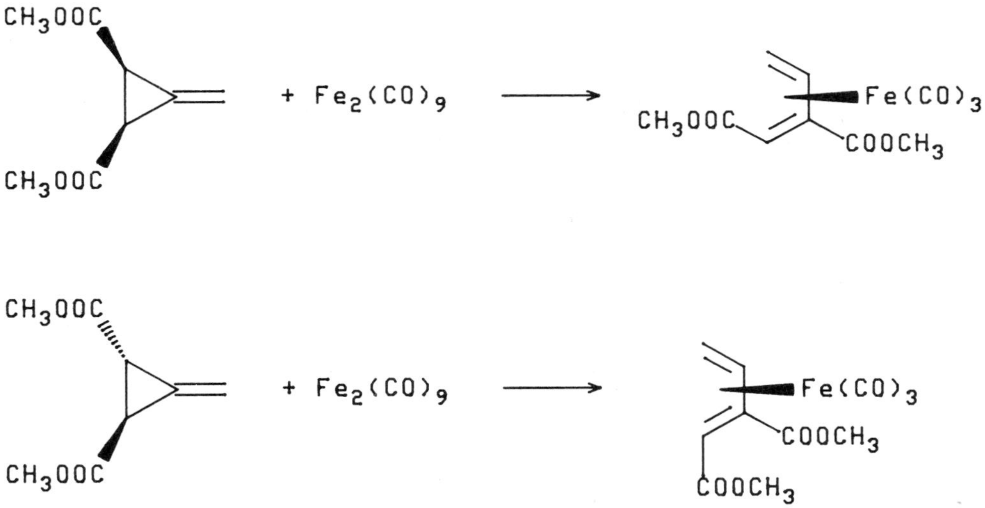

CH3OOC + Fe2(CO)9 → CH3OOC Fe(CO)3 COOCH3
CH3OOC

CH3OOC + Fe2(CO)9 → Fe(CO)3 COOCH3 COOCH3
CH3OOC

Figure 4.9

Complexation of acyclic dienes is limited to those that can adopt a *cisoid* conformation. In cyclopenta-, cyclohepta-, and cyclohexadienes, the ring size constrains the diene to be *cisoid*. Cyclopentadiene reacts with iron pentacarbonyl under the vigorous conditions of method A to produce an η^5-cyclopentadienyl complex by loss of hydrogen. The diene iron tricarbonyl complex can be prepared by reaction with diiron nonacarbonyl in refluxing diethyl ether.[19,20] Cycloheptatriene reacts with iron pentacarbonyl to give a mixture of iron tricarbonyl complexes of cycloheptatriene and 1,3-cycloheptadiene[21] (see Figure 4.11).

Complexes of cyclohexadienes can be prepared from 1,3-dienes or the more

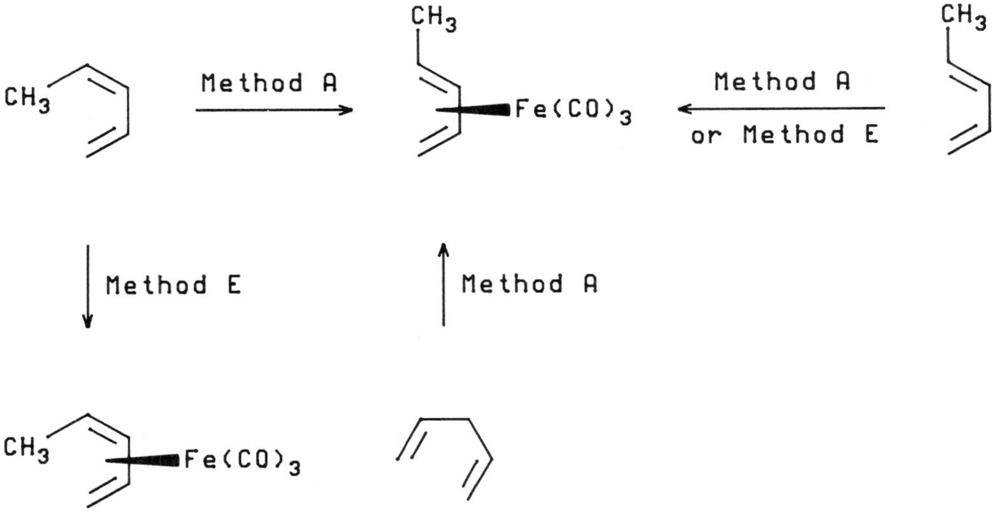

CH3 Method A → CH3 Fe(CO)3 ← Method A or Method E CH3

Method E ↓ ↑ Method A

CH3 Fe(CO)3

Figure 4.10

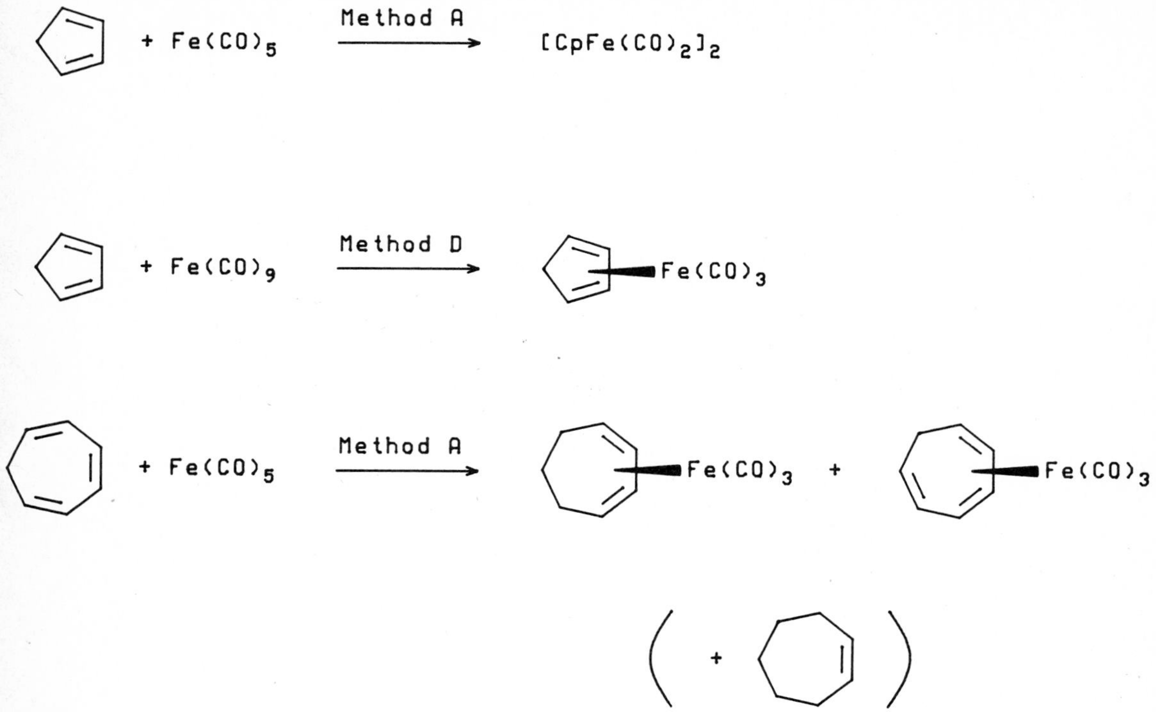

$$\text{C}_5\text{H}_6 + Fe(CO)_5 \xrightarrow{\text{Method A}} [CpFe(CO)_2]_2$$

$$\text{C}_5\text{H}_6 + Fe_2(CO)_9 \xrightarrow{\text{Method D}} (\eta^4\text{-}C_5H_6)Fe(CO)_3$$

$$\text{C}_7\text{H}_8 + Fe(CO)_5 \xrightarrow{\text{Method A}} (C_7H_{10})Fe(CO)_3 + (C_7H_8)Fe(CO)_3$$

$$(+ \ C_7H_{12})$$

Figure 4.11

$$\text{anisole} \longrightarrow \text{1-methoxy-1,4-cyclohexadiene} \xrightarrow{\text{Method C}} (\text{OCH}_3\text{-}C_6H_7)Fe(CO)_3$$

$$\text{1-methoxy-1,4-cyclohexadiene} \longrightarrow \text{1-methoxy-1,3-cyclohexadiene} \xrightarrow{\text{Method E}} (\text{OCH}_3\text{-}C_6H_7)Fe(CO)_3$$

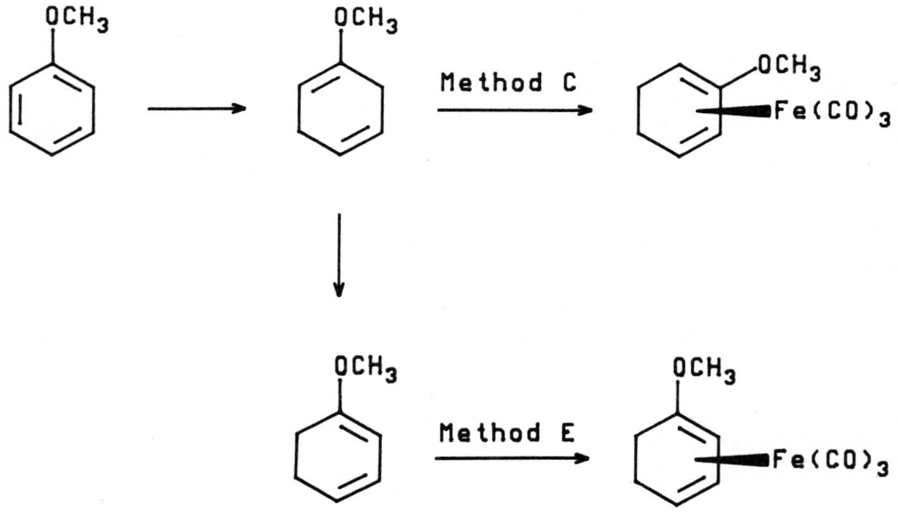

Figure 4.12

accessible 1,4-dienes. For example, Birch reduction of anisole affords 1-methoxy-1,4-cyclohexadiene. This can be isomerized to a mixture containing mostly the 1-methoxy-1,3-diene by base, charge transfer complexes, or rhodium or chromium complexes. Complexation of the 1,4-diene using iron pentacarbonyl (method C) is a convenient route to the 2-methoxy-1,3-diene complex.[22] Complexation of the 1-methoxy-1,3-diene under the mild conditions of method E affords the isomeric complex (see Figure 4.12).

4.2 GENERATION OF η^5-CATIONIC COMPLEXES

4.2.1 Cation Method A. From Diene Complexes via Hydride Abstraction

When a diene iron tricarbonyl complex has a terminal CH-substituent *anti* to the adjacent substituent, hydride abstraction by triphenylmethyl cation can produce an η^5-cationic complex. Few cationic complexes have been prepared from the relatively

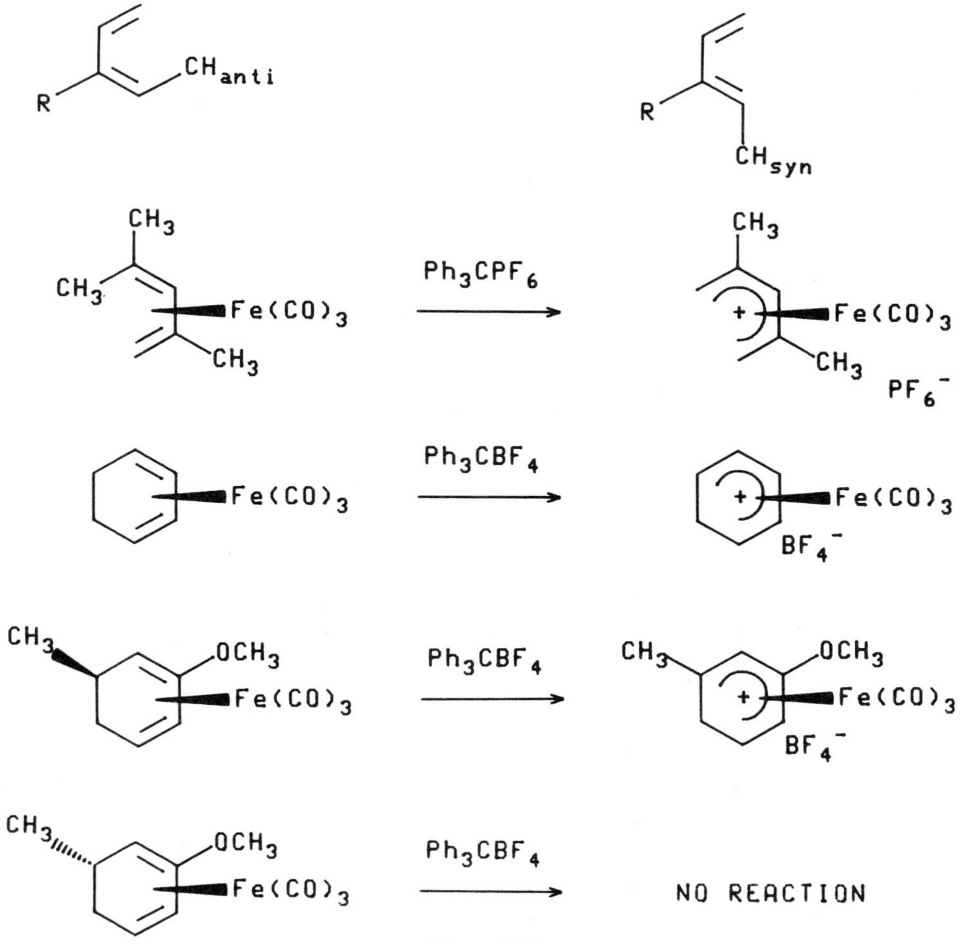

Figure 4.13

TABLE 4.1 Regioselectivity of Hydride Abstraction from Cyclohexadienyl Iron Tricarbonyl Complexes: (1-Methoxy)[9b]

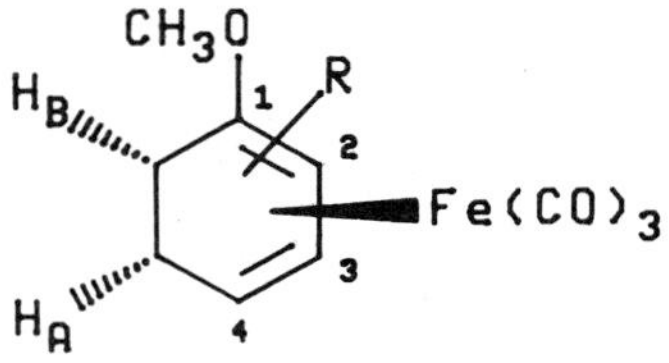

R	Percentage of Abstraction of H_A	Percentage of Abstraction of H_B	Yield of η^5-Cationic Complexes (%)
H	80	20	98
2-$COOCH_3$	0	0	0
3-$COOCH_3$	100	0	39(40% recovered starting material)
4-$COOCH_3$	0	0	0
3-CH_3	100	0	100
4-CH_3	10	90	90
3-OCH_3	44	56	84
4-*i*-PrO	50	50	95
4-Morpholino	0	100	90

TABLE 4.2 Regioselectivity of Hydride Abstraction from Cyclohexadienyl Iron Tricarbonyl Complexes: (2-Methoxy)[9b]

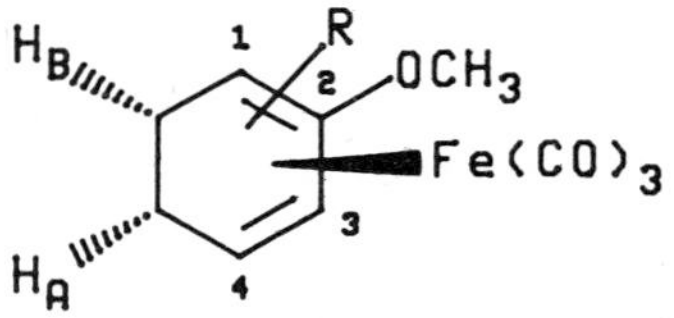

R	Percentage of Abstraction of H_A	Percentage of Abstraction of H_B	Yield of η^5-Cationic Complexes (%)
H	90	10	95
1-$COOCH_3$	100	0	73
4-$COOCH_3$	0	100	32(50% recovered starting material)
4-CH_3	0	100	98
4-OCH_3	56	44	84

inaccessible acyclic diene iron tricarbonyls while a great many cationic complexes have been prepared from cyclohexadiene iron tricarbonyls. Abstraction is stereospecific; only a hydride on the *exo* face, the face opposite the metal, is removed (see Figure 4.13).

Abstraction is also regioselective, as indicated by the results presented in Tables 4.1 and 4.2. The preferred cation does not correspond to the more stable uncomplexed cation. Pearson has suggested that the preferred cation, with a high highest occupied molecular orbital (HOMO) energy level and a low lowest unoccupied molecular orbital (LUMO) energy level, has a stronger synergic interaction with the metal.

A few factors affecting reactivity of a diene iron tricarbonyl complex toward hydride abstraction are apparent from the data presented in Tables 4.1 and 4.2. First, electron-withdrawing substituents inhibit η^5-cation formation. Second, sterically demanding substituents adjacent to the *exo* hydrogen block access by the triphenylmethyl cation. The use of less sterically demanding hydride abstracting reagents circumvents this difficulty.

4.2.2 Cation Method B. From Alkoxy-Substituted Diene Complexes Using Strong Acid

While hydride abstraction provides the most reliable method for η^5-cation complex formation, the reactions of 1- and 2-methoxy-substituted diene complexes with strong acids does provide ready access to some η^5-complexes difficult to prepare by hydride abstraction. The conversion involves stereospecific *endo* protonation[23] of a diene terminal position (perhaps via initial protonation of the metal) to give a cationic π-allyl complex followed by either *anti* 1,2-elimination (route A) or *anti* 1,5-elimination (route B) of methanol. The same sequence carried out with D_2SO_4 provides evidence for a 1,4-migration of an *endo* deuterium (hydrogen) from one terminal carbon of the diene to the

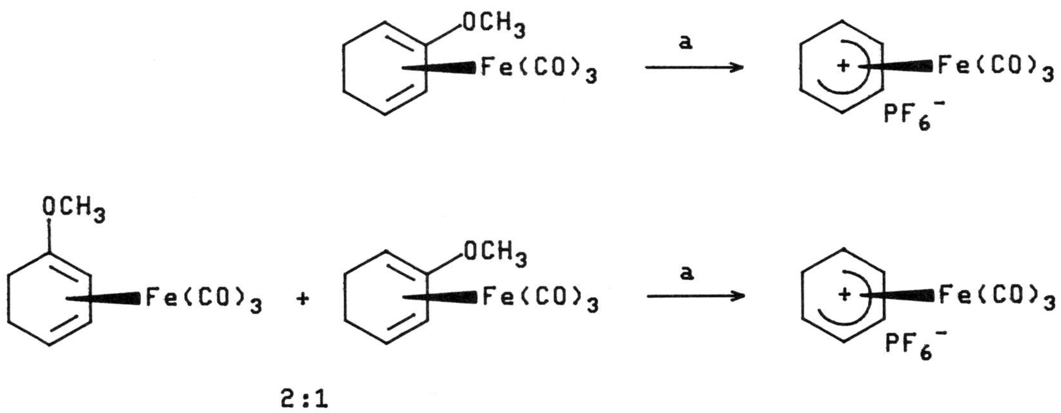

OCH3 Fe(CO)3 —a→ + Fe(CO)3 PF6⁻

OCH3 Fe(CO)3 + OCH3 Fe(CO)3 —a→ + Fe(CO)3 PF6⁻

2:1

a. 1) rt, H_2SO_4, 2) Et_2O, 3) 0°C, NH_4PF_6, H_2O; 70%

Figure 4.14

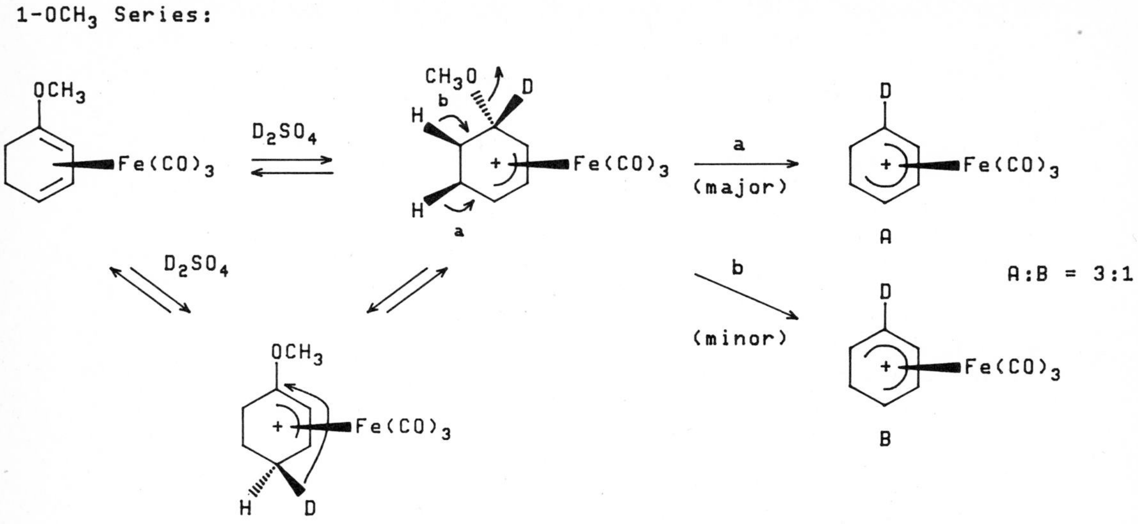

1-OCH_3 Series:

A:B = 3:1

Figure 4.15

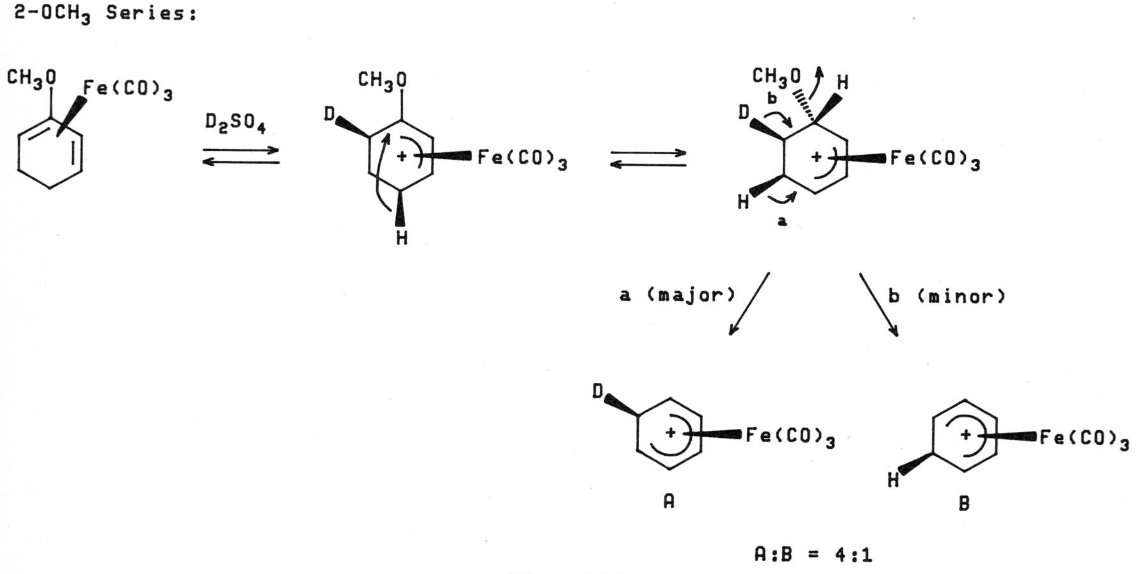

2-OCH_3 Series:

A:B = 4:1

Figure 4.16

other. With either substrate the major product has the η^5-system terminating at a carbon originally having a methoxyl group[24] (see Figures 4.14–4.16).

The 3-methoxy cation can be conveniently prepared by the method[25] shown in Figure 4.17. Two major limitations of this preparative method are (1) the propensity for acid-catalyzed isomerization of the 1,3-diene iron tricarbonyl complex *prior* to η^5-complex formation and (2) the unpredictable competition between 1,2-*anti* (route A)

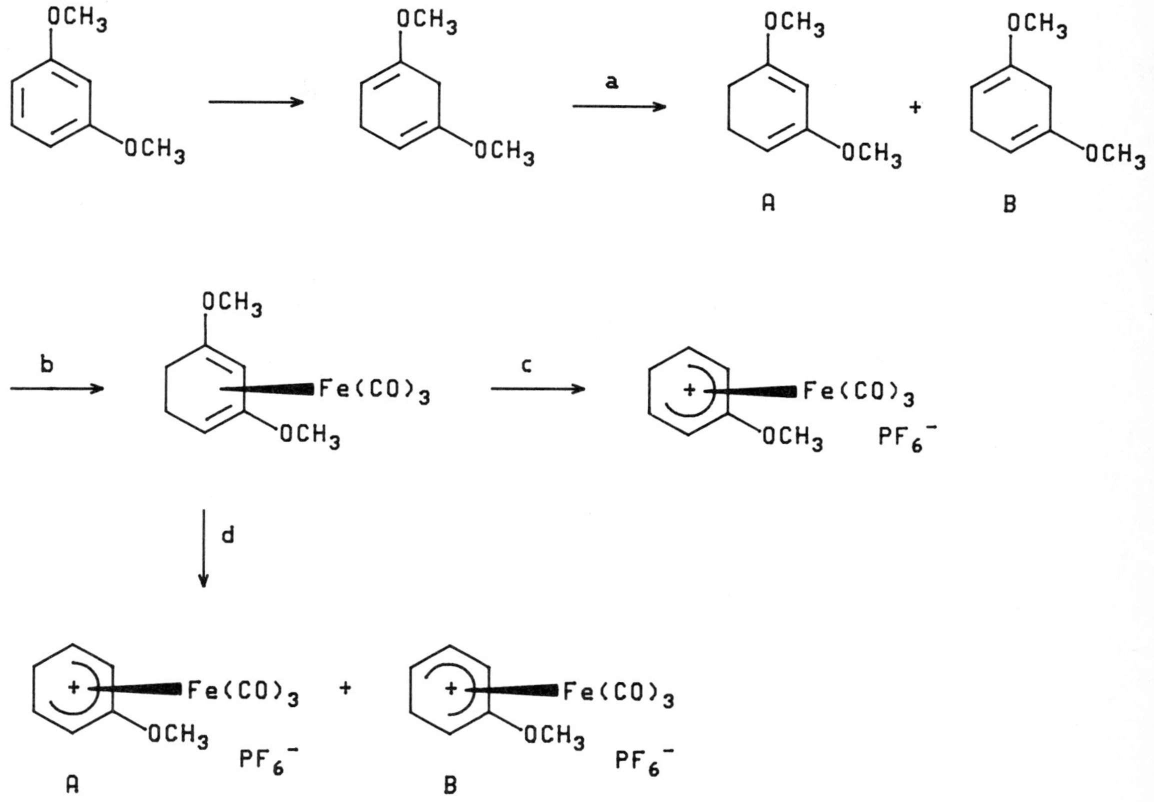

a. 65–75°C, t-BuOK, DMSO; A:B, 3:1

b. 135–145°C, $Fe(CO)_5$, $(n\text{-}Bu)_2O$; 44%

c. 0 to 10°C, 1) TFA, 2) NH_4PF_6, H_2O; 70%

d. 1) rt, H_2SO_4, 2) Et_2O, 3) 0°C, NH_4PF_6, H_2O;
A:80%, B:5–10%

Figure 4.17

and 1,5-*anti* (route B) elimination of methanol, resulting in mixtures of isomeric complexes[24b] (see Figure 4.18).

4.2.3 Cation Method C. From 1,3,5-Triene Complexes via Protonation

Protonation of the terminal position of an uncomplexed alkene in a 1,3,5-triene iron tricarbonyl complex provides an alternative route to some cyclic and acyclic cation complexes.[26,27] Protonation is apparently stereospecific, occurring on the *exo* face

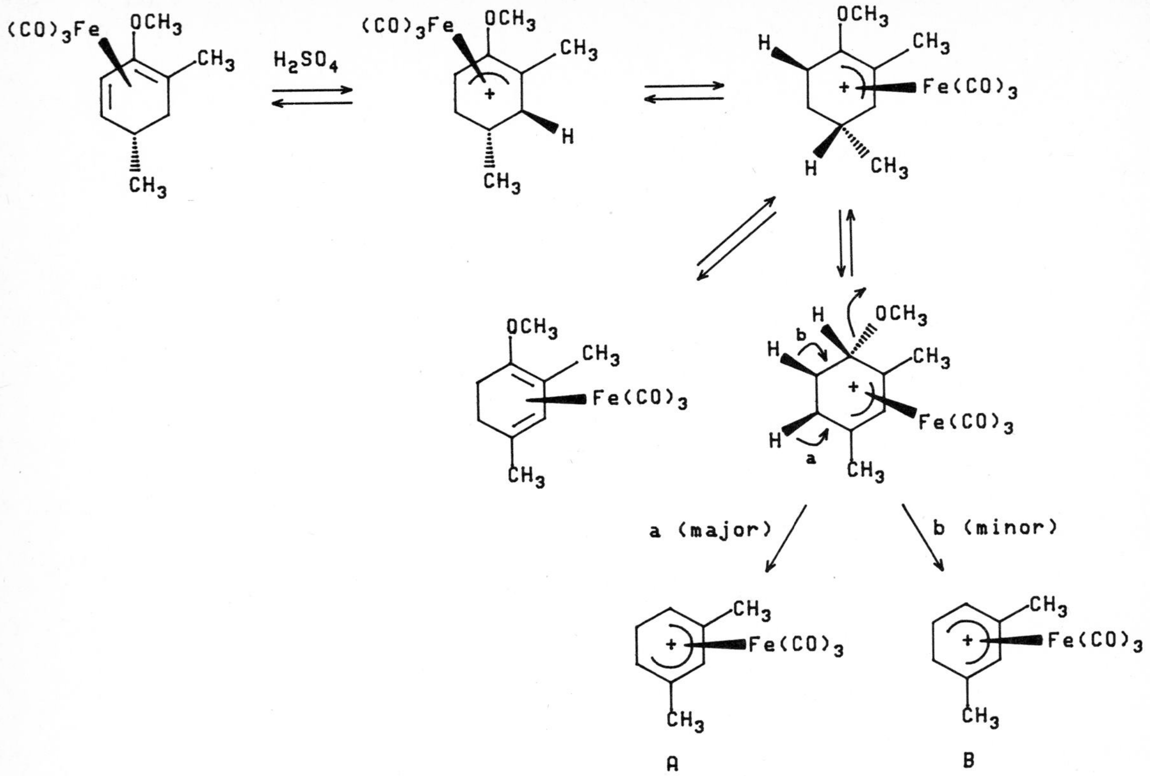

$(CO)_3Fe$ OCH_3 CH_3 H_2SO_4 CH_3 ...

A:B = 3:1

Figure 4.18

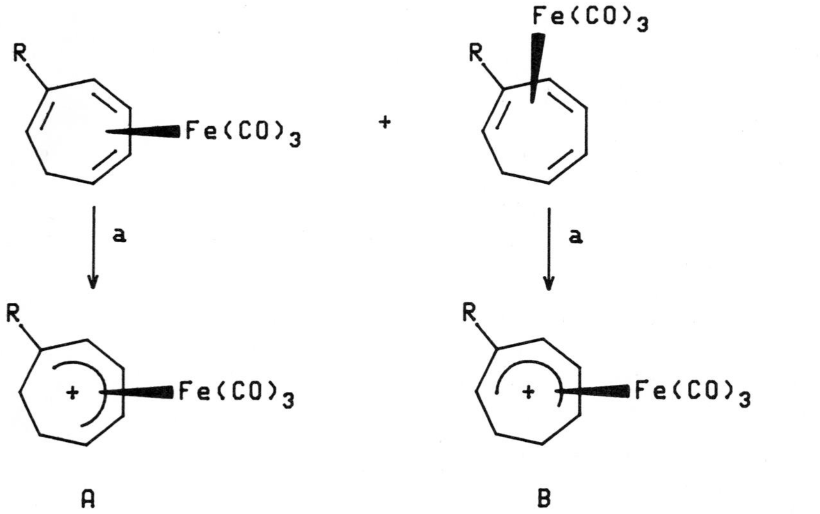

a. rt, $CDCl_3$, CF_3COOH; A:B, 1:1 for R = Ph

1.6:1 for R = p-$CH_3C_6H_4$

Figure 4.19

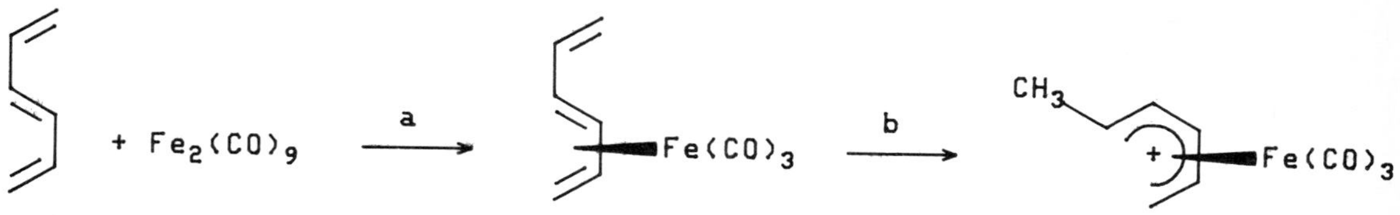

+ $Fe_2(CO)_9$ —a→ $Fe(CO)_3$ —b→ CH_3 $Fe(CO)_3$

a. rfx, pentane then benzene; 60%

b. 5 to 10°C, HBF_4, $[CH_3CH_2C(O)]_2O$; 74%

Figure 4.20

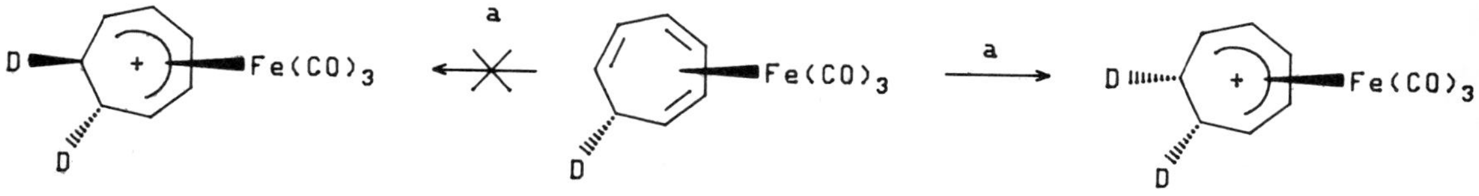

D D $Fe(CO)_3$ ←a✕— $Fe(CO)_3$ D —a→ D D $Fe(CO)_3$

a. −78°C, CF_3COOD, CD_2Cl_2

Figure 4.21

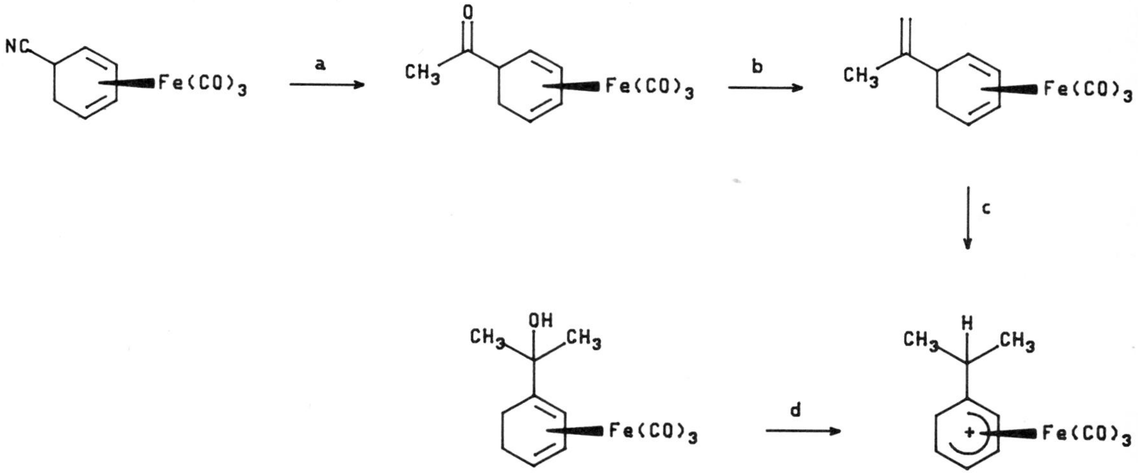

NC $Fe(CO)_3$ —a→ O CH_3 $Fe(CO)_3$ —b→ CH_3 $Fe(CO)_3$ —c→ OH CH_3 CH_3 $Fe(CO)_3$ —d→ H CH_3 CH_3 $Fe(CO)_3$

a. CH_3MgBr

b. rt, n-BuLi, Ph_3PCH_3I, hexane

c. rt, HBF_4, CH_3OH, H_2O

d. −120 to 0°C, FSO_3H, SO_2ClF

Figure 4.22

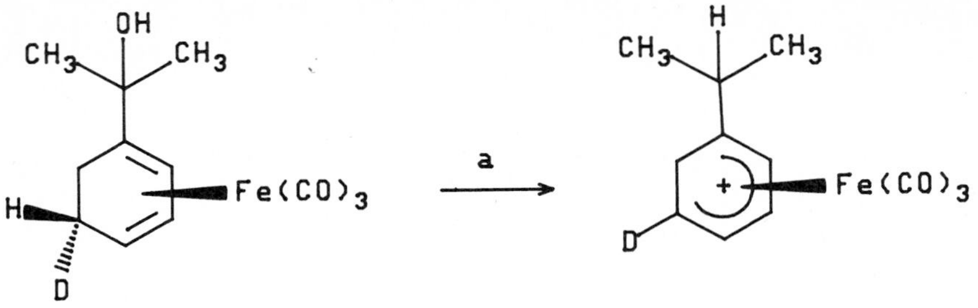

a. −120°C, FSO_3H, SO_2ClF, $CDCl_3$

Figure 4.23

exclusively[28] (see Figures 4.19–4.21). The difficulty in preparing regioisomerically pure precursors limits the utility of cation method C.

4.2.4 Cation Method D. From Diene Complexes via Remote Cation Migration

Protonation of a remote double bond followed by a hydride shift produces a cation complex.[29] A similar hydride shift is observed with a cation generated from an alcohol. The hydride shift occurs stereospecifically on the *endo* face[30] (see Figures 4.22 and 4.23).

Other examples of pentadienyl cation formation include (1) protonation of an acyclic 2,4-pentadien-1-ol complex[31] and (2) geometric isomerization of a 2,4-hexadien-1-one complex in the presence of strong acid.[32]

4.3 NUCLEOPHILIC ATTACK ON η^5-CATIONIC COMPLEXES

η^5-Cationic complexes are attacked by a range of nucleophiles. Carbon–carbon bond formation is accomplished using cyanide,[25] ketones,[33] enamines,[33] enol silyl ethers,[34a] silyl ketene acetals,[34b] allyl trimethylsilanes and -stannanes,[25,34a] lithium alkyls,[35a] organozinc and organocadmium reagents,[35b] stabilized enolates,[25,36] and activated aromatics.[37] Other effective nucleophiles include borohydride,[33,38] amines,[33,37,39] water,[33] methoxide,[38] and tertiary phosphines, tertiary phosphites, and arsines.[40] Two major concerns that must be addressed in utilization of this method of diene functionalization are (1) regioselectivity of attack and (2) reversibility of attack (even, in some cases, where a carbon–carbon bond is generated).

There are four potential sites for nucleophilic attack: (1) attack on the metal with loss of carbon monoxide, (2) attack on coordinated carbon monoxide to give an acyl complex, and (3) and (4) attack on either of the terminal carbons of the η^5-cationic system to give a new η^4-diene iron complex. We will focus on the synthetic utilization of attack on a terminal carbon of the η^5-cationic system.

4.3.1 Regioselectivity

The nucleophile–cation coupling is stereospecific; the nucleophile enters on the more accessible *exo* face. The nucleophile–cation coupling is also regioselective. Three of the factors determining regioselectivity of attack on a terminal carbon in an unsymmetrical η^5-cationic complex are (1) steric bulk of the terminal carbon substituents, (2) the resonance effect (electron-withdrawing or electron-releasing) of the η^5-cation substituents, and (3) the nature of the nucleophile.

1. Steric Bulk of Terminal Carbon Substituents

As expected, as a terminal carbon becomes more sterically accessible, attack on that carbon becomes more favorable[41] (see Figure 4.24).

2. Resonance Effect of η^5-Cation Substituents

While no clear relationship between substituent resonance effects and regioselectivity has been established, a few cases are illustrative (see Figure 4.25).

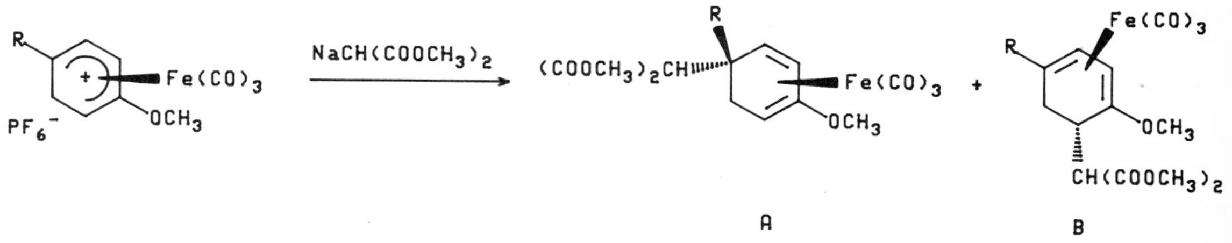

R, +, $Fe(CO)_3$, OCH_3, PF_6^- —[$NaCH(COOCH_3)_2$]→ $(COOCH_3)_2CH$, R, $Fe(CO)_3$, OCH_3 (A) + R, $Fe(CO)_3$, OCH_3, $CH(COOCH_3)_2$ (B)

R =		A:B =
	CH_2CH_2OAc	50:50
	$CH_2CH_2OCH_3$	72:28
	CH_2CH_3	82:18
	CH_3	95:5

Figure 4.24

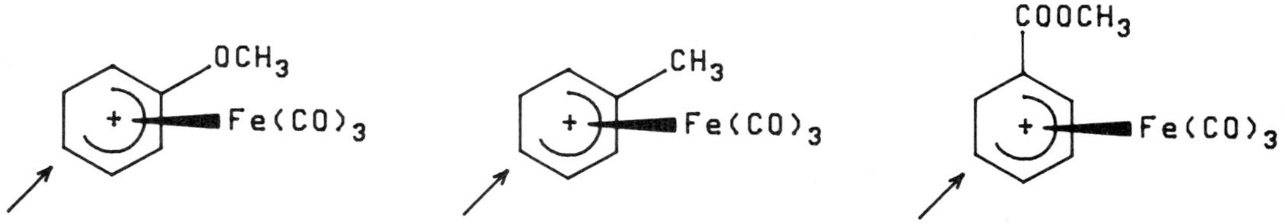

OCH_3, +, $Fe(CO)_3$ CH_3, +, $Fe(CO)_3$ $COOCH_3$, +, $Fe(CO)_3$

The 2-OCH_3, 2-CH_3 and 1-$COOCH_3$ cations are all attacked at position 5.

Figure 4.25

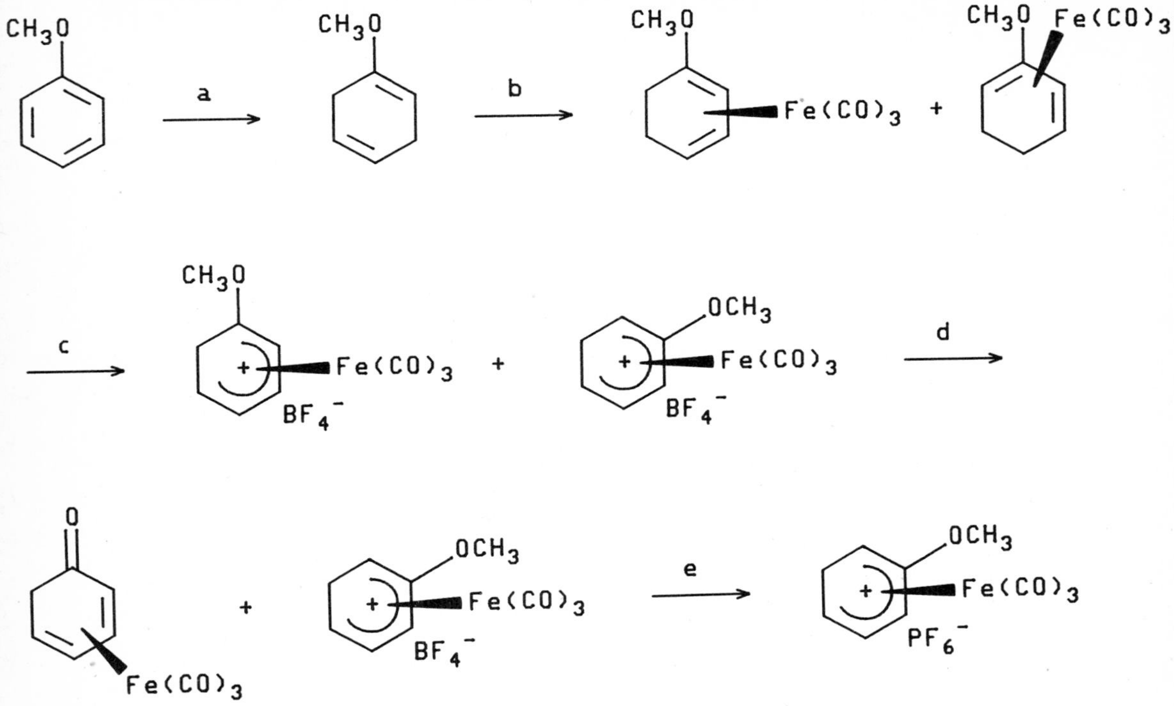

A

B

a. rfx, Li, NH_3, t-BuOH, THF; 75%

b. rfx, $Fe(CO)_5$, $(n\text{-}Bu)_2O$; 78%

c. rt, Ph_3CBF_4, CH_2Cl_2; 91%

d. rfx, H_2O; A:51%

e. rt, NH_4PF_6, H_2O; 44%

Figure 4.26

In a mixture of 1- and 2-methoxy cations, the 1-methoxy cation is attacked by water at position 1 to give a cyclohexadienone complex. This difference in reactivity is the basis of a separation of the cation complex mixture and, thus, a preparation of pure 2-methoxy cation[22] (see Figure 4.26).

3. The Nature of the Nucleophile

The nucleophile effect is not primarily steric since, for example, attack on the 2-methoxy cation is exclusively at position 5 using methyllithium while attack occurs 40% at position 1 using *t*-butyllithium.[35a]

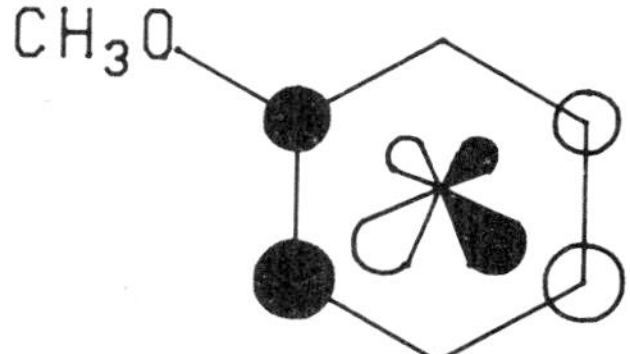

Figure 4.27

With enolates derived from dimethylmalonate and methyl acetoacetate, an increase in the proportion of product resulting from attack at C-5 is observed with increasing nucleophile–counterion association.[36] The complex LUMO is an antibonding combination of the dienyl ψ_2 and iron hybrid orbitals (see Figure 4.27).

A frontier molecular orbital interaction will favor attack at position 1. The tighter the nucleophile–counterion association, the lower the energy of the nucleophile HOMO and the stronger the HOMO–LUMO interaction. While the coulombic effects

TABLE 4.3 Effect of Enolate Counterion on Regioselectivity of Nucleophilic Addition to an η^5-Cationic Complex[36]

OR^1, R^2, $Fe(CO)_3$ + M^+Nu^- ⟶ A (OR^1, $Fe(CO)_3$, R^2, Nu) + B (OR^1, Nu, $Fe(CO)_3$, R^2)

R^1	R^2	M^+	Nu^-	A:B Ratio	Yield (%)
CH_3	CH_2CH_3	Li	$CH(COOCH_3)_2$	75:25	78
CH_3	CH_2CH_3	Na	$CH(COOCH_3)_2$	82:18	77
CH_3	CH_2CH_3	K	$CH(COOCH_3)_2$	85:15	96
i-Pr	CH_2CH_3	Li	$CH(COOCH_3)_2$	89:11	81
i-Pr	CH_2CH_3	Na	$CH(COOCH_3)_2$	94:6	77
i-Pr	CH_2CH_3	K	$CH(COOCH_3)_2$	100:0	97
CH_3	$CH_2CH_2COOCH_3$	Li	$CH(COOCH_3)_2$	68:32	86
CH_3	$CH_2CH_2COOCH_3$	Na	$CH(COOCH_3)_2$	79:21	75
CH_3	$CH_2CH_2COOCH_3$	Na[a]	$CH(COOCH_3)_2$	85:15	85
CH_3	$CH_2CH_2COOCH_3$	K	$CH(COOCH_3)_2$	85:15	81
CH_3	$CH_2CH_2COOCH_3$	Li	$CH(COOCH_3)COCH_3$	—	—[b]
CH_3	$CH_2CH_2COOCH_3$	Na	$CH(COOCH_3)COCH_3$	33:67	73
CH_3	$CH_2CH_2COOCH_3$	K	$CH(COOCH_3)COCH_3$	55:45	65
CH_3	$(CH_2)_3NPhth$	Li	$CH(COOCH_3)_2$	60:40	92
CH_3	$(CH_2)_3NPhth$	Na	$CH(COOCH_3)_2$	74:26	85
CH_3	$(CH_2)_3NPhth$	Na[a]	$CH(COOCH_3)_2$	81:19	82
CH_3	$(CH_2)_3NPhth$	K	$CH(COOCH_3)_2$	82:18	81

[a]18-crown-6 added.
[b]No addition products isolated.

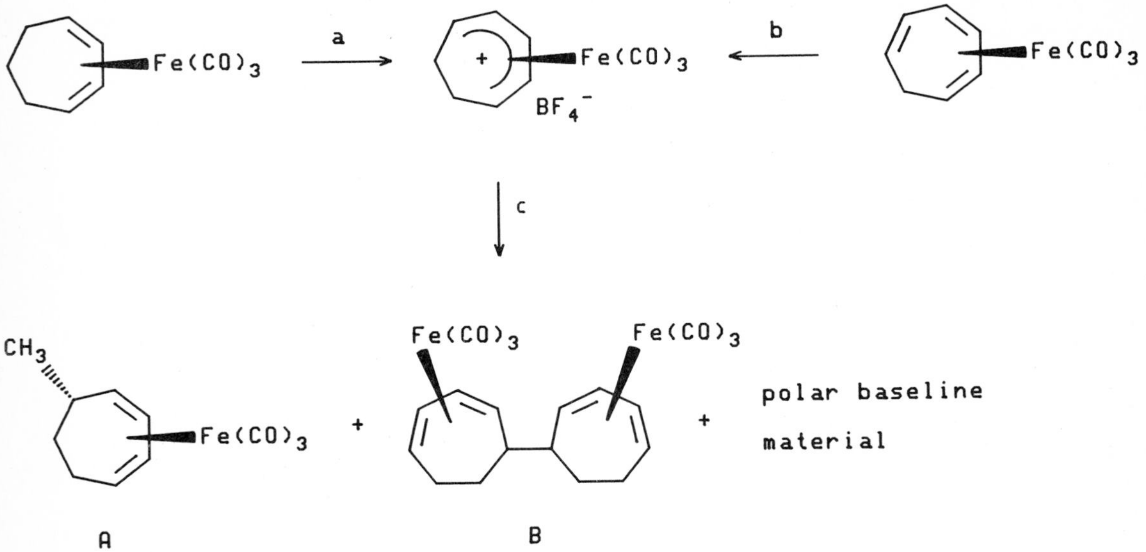

a. rt, Ph_3CBF_4, CH_2Cl_2

b. HBF_4, Ac_2O

c. 0°C, $(CH_3)_2CuLi$, Et_2O; A:15-20%, B:20%

Figure 4.28

dominate in almost all cases, the frontier orbital effects become significant with the more strongly associated enolates (see Table 4.3).

Counterion chelation is apparently required since little or no change in the A:B ratio is observed with anions derived from malononitrile and methyl cyanoacetate.

4. Ancillary Metal Ligands

Finally, while most work has been done using cationic iron tricarbonyl complexes, modification of the ancillary metal ligands can significantly alter regioselectivity of nucleophilic attack. In developing a strategy for functionalization of cycloheptadienyl cation complexes, Pearson found that changing one ancillary carbon monoxide ligand to a phosphine reduced both competitive attack on coordinated carbon monoxide and competitive reductive dimerization[42] (see Figure 4.28).

4.4 REMOVAL OF THE METAL

After introduction of the nucleophile, oxidation can provide a metal-free functionalized organic fragment suitable for further elaboration. Coupling oxidative demetallation with subsequent oxidation of the organic fragment can increase synthetic efficiency by

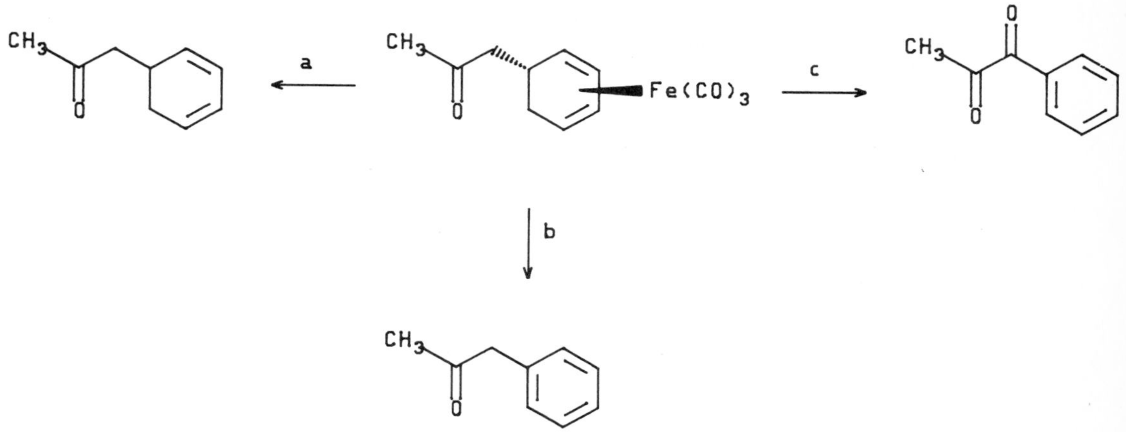

a. rt, $FeCl_3$, conc. HCl, EtOH; 70%

b. rt, $Pb(OAc)_4$, AcOH; 55%

c. rfx, $Tl[OC(O)CF_3]_3$, CCl_4; 50%

Figure 4.29

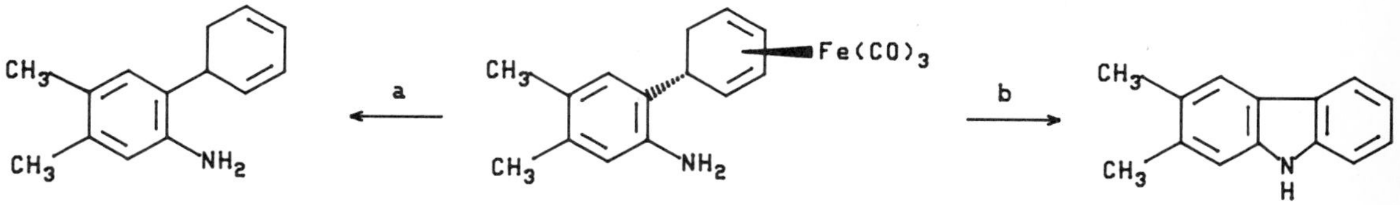

a. warm, $FeCl_3$, AcOH, H_2O

b. rt, I_2, pyridine; 32%

Figure 4.30

streamlining a synthetic sequence. To illustrate, 5-acetonylcyclohexa-1,3-diene iron tricarbonyl was demetallated with iron(III) chloride, lead(IV) acetate, and thallium(III) trifluoroacetate to produce 5-acetonylcyclohexa-1,3-diene, 1-phenylpropan-2-one, and 1-phenylpropan-1,2-dione, respectively[33] (see Figure 4.29). 5-(2-Anilino)-cyclohexa-1,3-diene iron tricarbonyl complexes can be demetallated with iron(III) chloride to afford the free dienes or demetallated and cyclized with iodine in pyridine to afford carbazoles (see Figure 4.30).

The reagent of choice for oxidative demetallation is trimethylamine-*N*-oxide. Diene is produced in high yields using either trimethylamine-*N*-oxide hydrate in warm *N*,*N*-

dimethylacetamide or dry trimethylamine-*N*-oxide in refluxing toluene or benzene; the anhydrous procedure is preferred for methoxy-substituted complexes where enol ether hydrolysis is a problem.

The unpredictable regioselectivity of nucleophilic attack on highly functionalized η^5-cationic complexes suggests that this method of functionalization should be used in the early stages of a total synthesis. This is elegantly illustrated in Pearson's synthesis of (±)-limaspermine.[43,44]

4.5 TOTAL SYNTHESIS OF (±)-LIMASPERMINE

One iron-mediated approach to (±)-limaspermine begins with the condensation of malonic acid and *p*-anisaldehyde.[45] Birch reduction of *p*-methoxycinnamic acid[46] affords a 1,4-diene with a saturated side chain in excellent yield. Esterification followed by acid-catalyzed rearrangement produces a 1,3-diene suitable for complex formation with iron pentacarbonyl by method A. The side chain ester is reduced to a primary alcohol that is then converted to a tosylate. Nucleophilic displacement by potassium phthalimide introduces a protected primary amine.

Hydride abstraction (cation method A) affords a single 2-methoxy-substituted η^5-cation complex. Nucleophilic attack by malonate occurs stereospecifically on the *exo* face but with poor (4.6:1) regioselectivity for the 5-position. The desired regioisomeric adduct can be obtained in pure form by recrystallization (see Figure 4.31).

Oxidative demetallation using trimethylamine-*N*-oxide affords the free diene. The primary amino group is unmasked with hydrazine hydrate. Enol ether hydrolysis is followed by conjugate addition to close the second ring. Both the amine and ketone are then protected to allow synthetic manipulation of the malonate substituent.

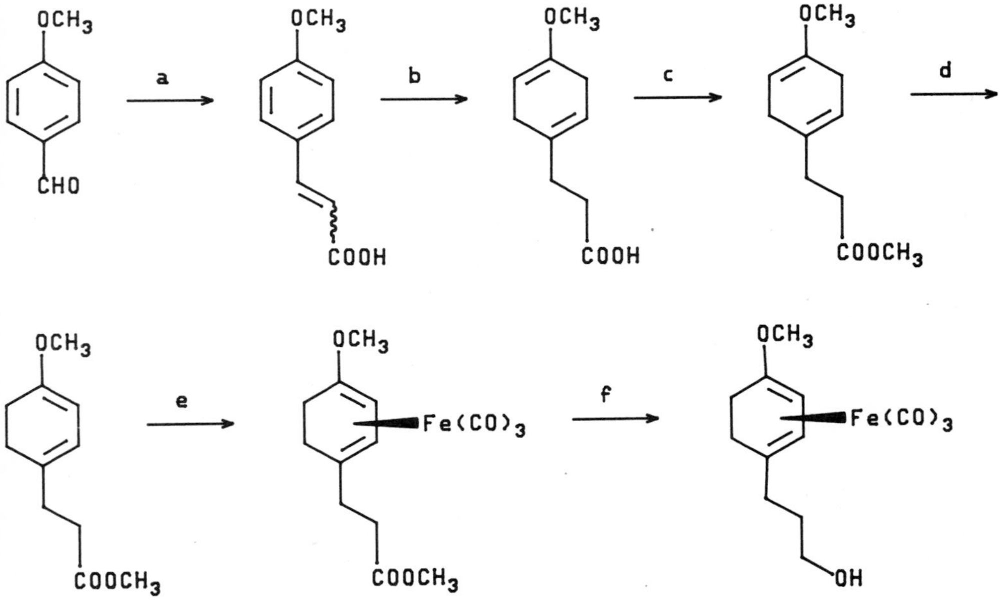

OCH_3 / CHO —a→ OCH_3 / $COOH$ —b→ OCH_3 / $COOH$ —c→ OCH_3 / $COOCH_3$ —d→

OCH_3 / $COOCH_3$ —e→ OCH_3, $Fe(CO)_3$ / $COOCH_3$ —f→ OCH_3, $Fe(CO)_3$ / OH

Figure 4.31

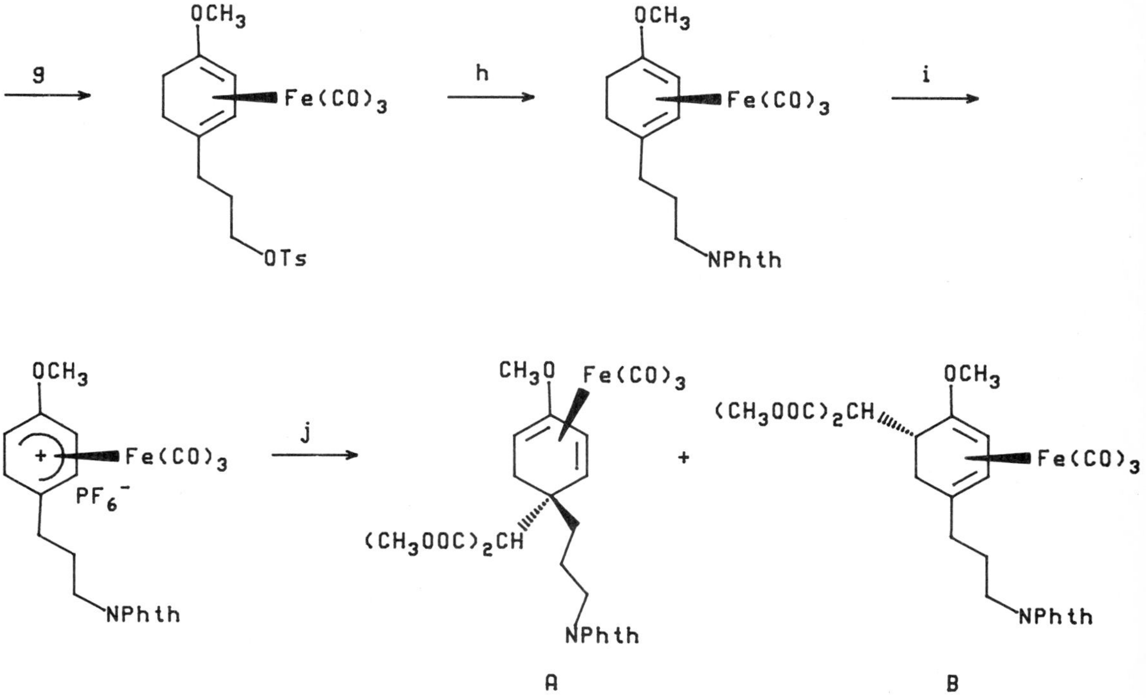

a. 100°C, $CH_2(COOH)_2$, piperidine, pyridine; 99%

b. 1) rfx (-33°C), Li, NH_3, t-BuOH, 2) NH_4Cl; 98%

c. rfx, $CH_3OSO_2OCH_3$, acetone; 93%

d. 80°C, cat. p-TsOH; 99%

e. rfx, $Fe(CO)_5$, $(n\text{-}Bu)_2O$; 55% (Reference 54)

f. -78°C to rt, $(i\text{-}Bu)_2AlH$, THF, hexane; 97%

g. -10 to 0°C, p-TsCl, pyridine; 100%

h. rt to 40°C, KNPhth, DMF; 90%

i. rfx, Ph_3CPF_6, CH_2Cl_2; 88%

j. rt, $KCH(COOCH_3)_2$, THF; A:B, 4.6:1, 68% yield of the desired isomer after recrystallization

Figure 4.31 *(Continued)*

Heating with cyanide in dimethylsulfoxide affects decarbomethoxylation. The monoester is reduced and the alcohol protected as a methyl ether. The acetamide is cleaved and a new amide generated using chloroacetyl chloride. The ketal is cleaved, providing an enolic position for condensation to close the third ring. The ketone is then reprotected as a ketal. Hydride reduces the amide to an amine. The ketone is again deprotected, then employed in a classical Fisher indole synthesis with *o*-methoxyphenylhydrazine hydrochloride.[47] Imine reduction, acylation, and methyl

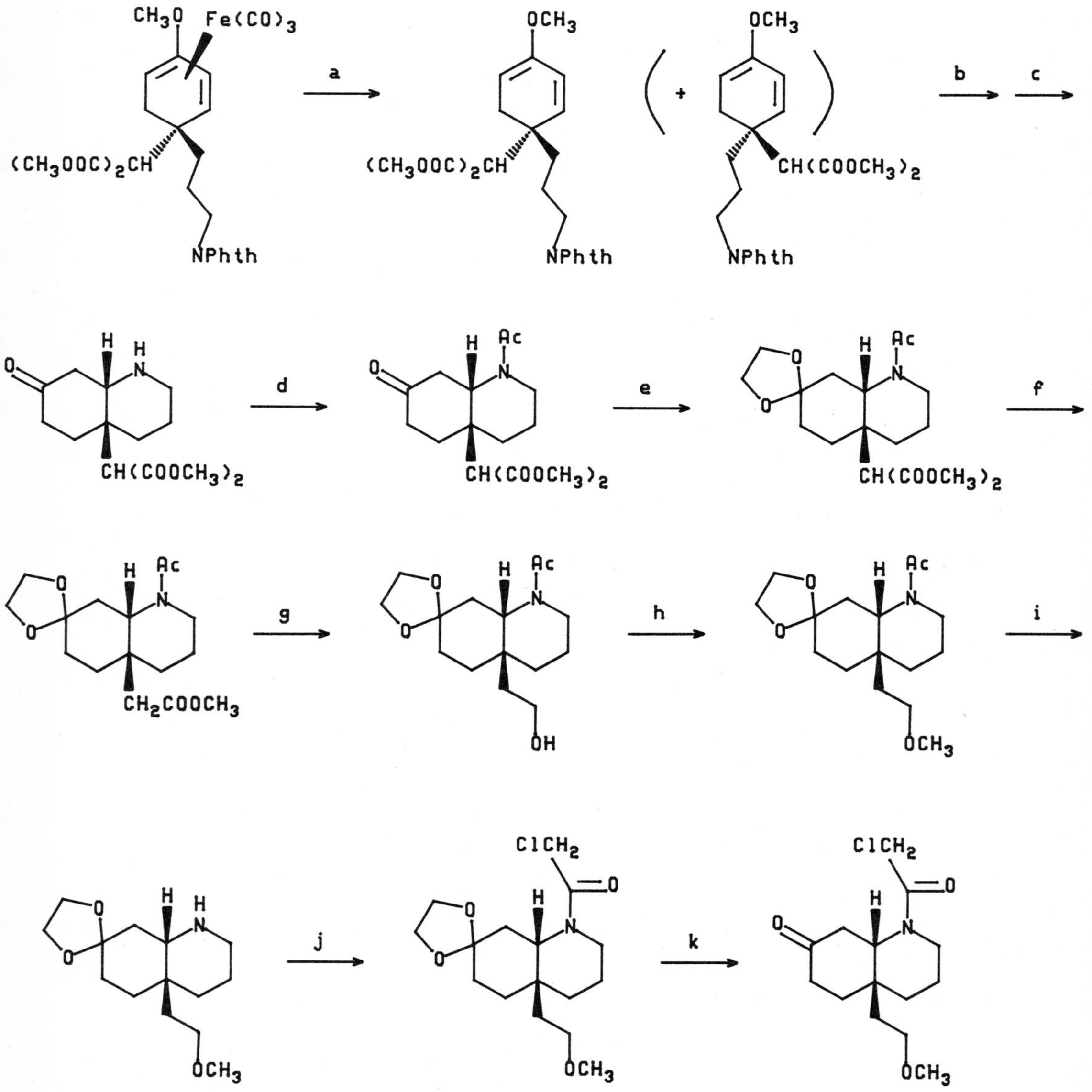

Figure 4.32

a. 50°C, $(CH_3)_3NO$, benzene; 85%

b. 40°C, $NH_2NH_2 \cdot H_2O$, CH_3OH

c. 1) rt, cat. HOOCCOOH, 2) $NaHCO_3$, H_2O, CH_3OH;
99% for two steps

d. 0°C, Ac_2O, CH_2Cl_2, pyridine; 67%

e. rfx, cat. p-TsOH, $HOCH_2CH_2OH$, benzene; 100%

f. 118°C, NaCN, DMSO; 87%

g. rt, $LiBH_4$, THF; 77%

h. 1) rt, NaH, 2) CH_3I, THF; 100%

i. 1) rfx, NH_3, 2) Ca, DME, EtOH; 96%

j. 10°C, $ClCH_2C(O)Cl$, pyridine, benzene; 84%

k. 76°C, cat. HCl, H_2O, EtOH; 95%

Figure 4.32 (*Continued*)

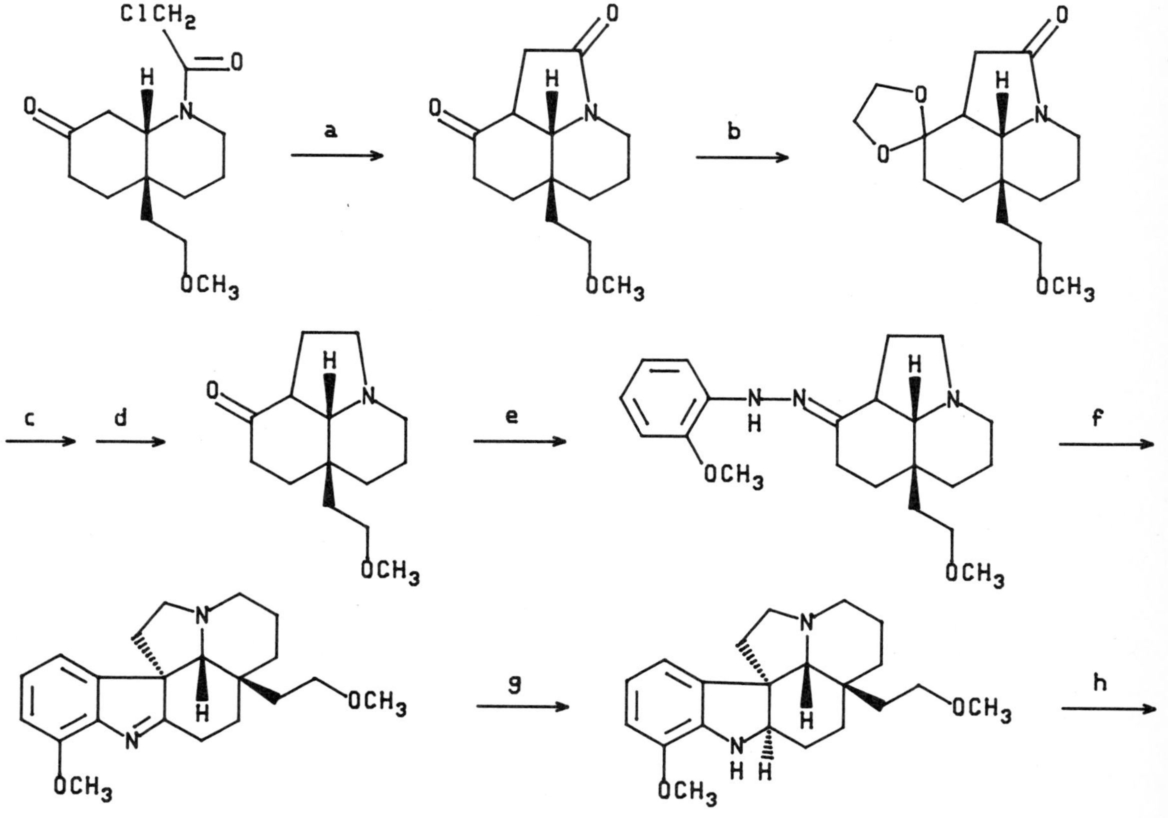

ClCH2 H O N O OCH3 a O H N O OCH3 b O O H N O OCH3

c d O H N OCH3 e N H N H N OCH3 OCH3 f

N H N OCH3 OCH3 g N H N H H OCH3 OCH3 h

Figure 4.33

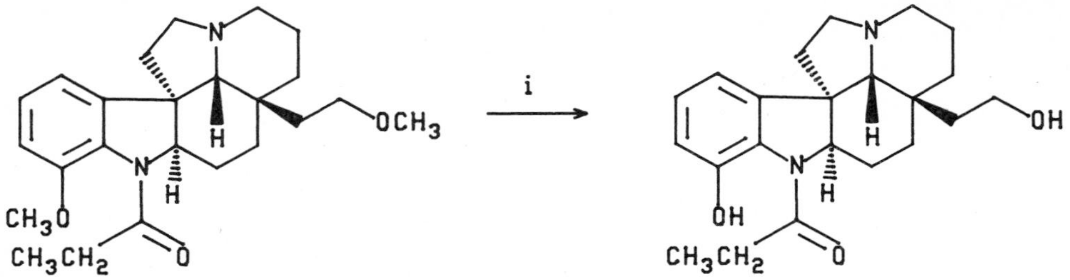

i

(±)-Limaspermine

a. rt, t-BuOK, benzene, t-BuOH; 100%

b. rfx, cat. p-TsOH, $HOCH_2CH_2OH$, benzene; 100%

c. rt, $LiAlH_4$, THF

d. 90°C, HCl, H_2O; 33% for two steps

e. rfx, (2-methoxyphenyl)$-NHNH_2 \cdot HCl$, EtOH

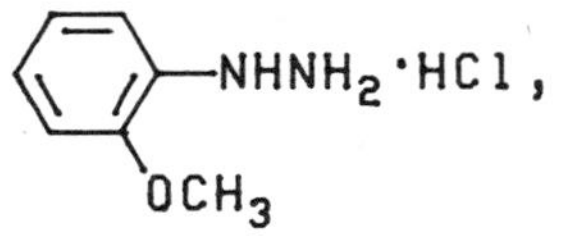

f. 95°C, AcOH

g. rt, $LiAlH_4$, THF; 39% for three steps

h. 10°C, $CH_3CH_2C(O)Cl$, pyridine, benzene; 95%

i. 60°C, $(CH_3)_3SiI$, pyridine, $CHCl_3$; 50% (unoptimized)

Figure 4.33 (*Continued*)

ether cleavage with iodotrimethylsilane completes the synthesis (see Figures 4.32 and 4.33).

A shorter route to a key precursor begins with *p*-hydroxyphenylacetic acid.[48,49] Phenoxide alkylation is followed by hydride reduction of the side chain acid. The alcohol is protected as a methyl ether. Birch reduction affords a 1,4-diene in excellent yield. Acid-catalyzed isomerization produces a mixture of 1,4- and 1,3-dienes suitable for η^4-diene iron tricarbonyl complex formation using method E. Hydride abstraction (cation method A) affords a single 2-methoxy-substituted η^5-cation complex. Nucleophilic attack by malonate ion is stereospecifically *exo*. The larger isopropyl ether substituent at position 2 provides some steric encumbrance to attack at position 1 (8.8:1 ratio of regioisomers). The major regioisomer is obtained in pure form by flash chromatography.

Decarbomethoxylation by cyanide is followed by hydride reduction of the monoester. The alcohol is converted to a tosylate, then the tosylate is displaced by

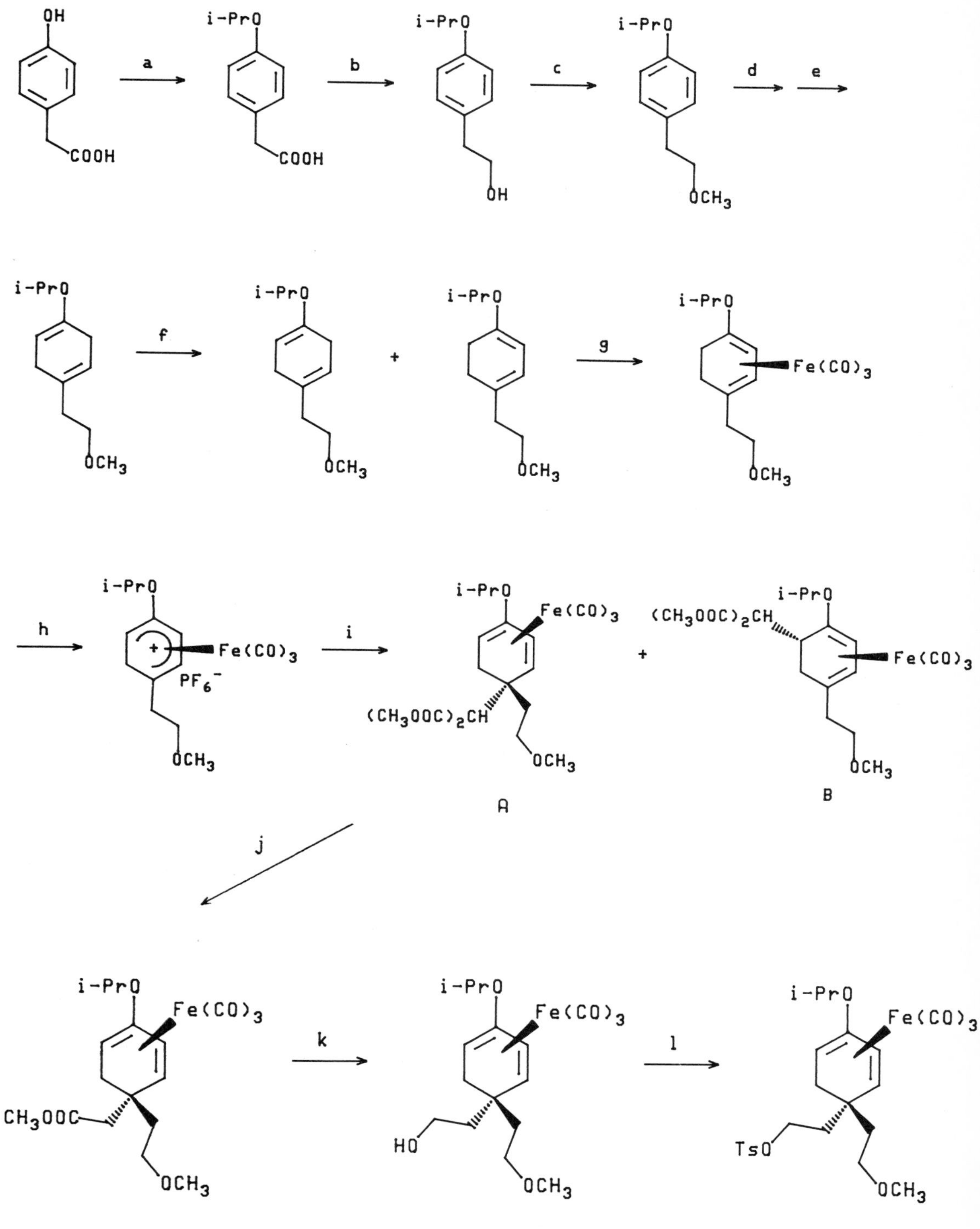

a, b, c, d, e, f, g, h, i, j, k, l

i-PrO, OH, COOH, OCH$_3$, Fe(CO)$_3$, PF$_6^-$, (CH$_3$OOC)$_2$CH, CH$_3$OOC, HO, TsO

A

B

Figure 4.34

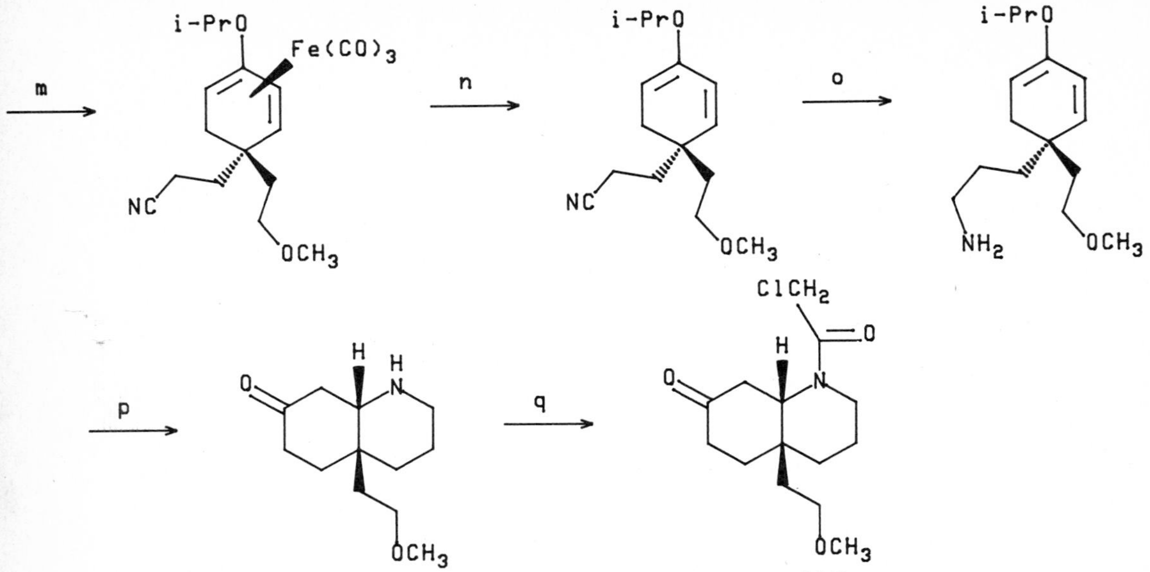

a. rfx, NaOH, $(CH_3)_2CHBr$, i-PrOH, H_2O; 62%

b. 0°C to rt, $BH_3 \cdot THF$, THF; 100%

c. 1) NaH, 2) CH_3I, THF; 92%

d. Li, NH_3, THF, EtOH

e. CH_3OH, H_2O; 93% for two steps

f. p-TsOH

g. 30°C, $Fe_2(CO)_9$, acetone; 50% for two steps

h. rfx, Ph_3CPF_6, CH_2Cl_2; 85%

i. rt, $KCH(COOCH_3)_2$, THF; A:70%, B:8%

j. 100 to 120°C, KCN, DMSO, H_2O; 62%

k. -78°C to rt, $(i\text{-}Bu)_2AlH$, THF; 80%

l. 0°C, p-TsCl, pyridine; 96%

m. 55°C, NaCN, HMPA; 94%

n. 46°C, $(CH_3)_3NO$, benzene; 84%

o. 0°C, $LiAlH_4$, Et_2O; 79%

p. 1) rt, cat. HOOCCOOH, 2) $NaHCO_3$, CH_3OH, H_2O; 100%

q. 5 to 10°C, $ClCH_2C(O)Cl$, pyridine; 70%

Figure 4.34 (*Continued*)

cyanide. Oxidative demetallation using trimethylamine-*N*-oxide affords the free diene. Lithium aluminum hydride converts the nitrile to a primary amino group. Enol ether hydrolysis results in conjugate addition to close the second ring. Finally, reaction with chloroacetyl chloride produces the amide (see Figure 4.34).

(±)-Limaspermine was prepared from *p*-anisaldehyde in 30 steps in overall 0.45% yield and from *p*-hydroxyphenylacetic acid in 25 steps in overall 0.20% yield. Particularly inefficient operations common to both routes are the amide reduction and Fisher indole synthesis.

REFERENCES

1. Pinar, M.; von Phillipsborn, W.; Vetter, W.; Schmid, H. *Helv. Chim. Acta* **1962**, *45*, 2260.
2. Cava, M. P.; Talapatra, S. K.; Nomura, K.; Weisbach, J. A.; Douglas, B.; Shoop, E. C. *Chem. Ind. (London)* **1963**, 1242.
3. (−)-Aspidospermine has shown some promise as an antiarrhythmic or bronchodilator: Rojas-Martinez, R. L.; Perez-Souto, N.; Cantero-Martinez, A.; Diaz-Rondon, B.; Arruzazabala-Valmana, M. L. *Rev. Cubana Farm.* **1980**, *14*, 289.
4. Cordell, G. A.; Saxton, J. E. *Alkaloids*, **1981**, *20*, 1.

5a. Potier, P. *Pure Appl. Chem.* **1986**, *58*, 737.

5b. Kutney, J. P. *Pure Appl. Chem.* **1984**, *56*, 1011.

6. For an attempted synthesis of (±)-limaspermine, see Ban, Y.; Iijima, I.; Inoue, I.; Akagi, M.; Oishi, T. *Tetrahedron Lett.* **1969**, 2067.
7. See the following references for synthesis of (±)-aspidospermine:

7a. Martin, S. F.; Desai, S. R.; Phillips, G. W.; Miller, A. C. *J. Am. Chem. Soc.* **1980**, *102*, 3294.

7b. Stevens, R. V.; Fitzpatrick, J. M.; Kaplan, M.; Zimmerman, R. L. *J. Chem. Soc., Chem. Commun.* **1971**, 857.

7c. Kuehne, M. E.; Bayha, C. *Tetrahedron Lett.* **1966**, 1311.

7d. Ban, Y.; Sato, Y.; Inoue, I.; Nagai, M.; Oishi, T.; Terashima, M.; Yonemitsu, O.; Kanaoka, Y. *Tetrahedron Lett.* **1965**, 2261.

7e. Stork, G.; Dolfini, J. E. *J. Am. Chem. Soc.* **1963**, *85*, 2872.

8. See the following references for syntheses of (−)- and (±)-vindoline:

8a. Feldman, P. L.; Rapoport, H. *J. Am. Chem. Soc.* **1987**, *109*, 1603.

8b. Andriamialisoa, R. Z.; Langlois, N.; Langlois, Y. *J. Org. Chem.* **1985**, *50*, 961.

8c. Danieli, B.; Lesma, G.; Palmisano, G.; Riva, R. *J. Chem. Soc., Chem. Commun.* **1984**, 909.

8d. Kutney, J. P.; Bunzli-Trepp, U.; Chan, K. K.; de Souza, J. P.; Fujise, Y.; Honda, T.; Katsube, J.; Klein, F. K.; Leutwiler, A.; Morehead, S.; Rohr, M.; Worth, B. R. *J. Am. Chem. Soc.* **1978**, *100*, 4220.

8e. Takano, S.; Shishido, K.; Sato, M.; Yuta, K.; Ogasawara, K. *J. Chem. Soc., Chem. Commun.* **1978**, 943.

8f. Ban, Y.; Sekine, Y.; Oishi. T. *Tetrahedron Lett.* **1978**, 151.

8g. Takano, S.; Shishido, K.; Sato, M.; Ogasawara, K. *Heterocycles* **1977**, *6*, 1699.

8h. Ando, M.; Büchi, G.; Ohnuma, T. *J. Am. Chem. Soc.* **1975**, *97*, 6880.

9. See the following references for recent reviews of the chemistry of diene iron tricarbonyl complexes:

9a. King, R. B. *The Organic Chemistry of Iron*, Koerner, E. A.; von Gustorf, F. W.; Grevels, F. W.; Fischler, I., Eds.; Academic Press: New York, 1978; Vol. 1.
9b. Birch, A. J.; Kelly, L. F. *J. Organometal. Chem.* **1985**, *285*, 267.
9c. Pearson, A. J. *Science* **1984**, *223*, 895.
9d. Pearson, A. J. *Pure Appl. Chem.* **1983**, *55*, 1767.
10. Reihlen, H.; Gruhl, A.; v. Hessling, G.; Pfrengle, O. *Justus Liebigs Ann. Chem.* **1930**, *482*, 161.
11. Hallam, B. F.; Pauson, P. L. *J. Chem. Soc.* **1958**, 642.
12a. Mills, O. S.; Robinson, G. *Proc. Chem. Soc.* **1960**, 421.
12b. Mills, O. S.; Robinson, G. *Acta Cryst.* **1963**, *16*, 758.
13. King, R. B.; Manuel, T. A.; Stone, F. G. A. *J. Inorg. Nucl. Chem.* **1961**, *16*, 233.
14. Murdoch, H. D.; Weiss, E. *Helv. Chim. Acta* **1962**, *45*, 1156.
15. Shvo, Y.; Hazum, E. *J. Chem. Soc., Chem. Commun.* **1975**, 829.
16. Nametkine, N. S.; Tyurine, V. D.; Nekhaev, A. I.; Ivanov, V. I.; Bayaouova, F. S. *J. Organometal Chem.* **1976**, *107*, 377.
17. Emerson, G. F.; Watts, L.; Pettit, R. *J. Am. Chem. Soc.* **1965**, *87*, 131.
18. Whitesides, T. H.; Slaven, R. W. *J. Organometal. Chem.* **1974**, *67*, 99.
19. Piper, T. S.; Cotton, F. A.; Wilkinson, G. *J. Inorg. Nucl. Chem.* **1955**, *1*, 165.
20. Kochhar, R. K.; Pettit, R. *J. Organometal. Chem.* **1966**, *6*, 272.
21a. Burton, R.; Pratt, L.; Wilkinson, G. *J. Chem. Soc.* **1961**, 594.
21b. Dauben, H. J., Jr.; Bertelli, D. J. *J. Am. Chem. Soc.* **1961**, *83*, 497.
22. Birch, A. J.; Chamberlain, K. B. *Org. Syn.* **1977**, *57*, 107.
23. Whitesides, T. H.; Arhart, R. W. *J. Am. Chem. Soc.* **1971**, *93*, 5296.
24a. Birch, A. J.; Haas, M. A. *J. Chem. Soc. (C)* **1971**, 2465.
24b. Birch, A. J.; Chauncy, B.; Kelly, L. F.; Thompson, D. J. *J. Organometal. Chem.* **1985**, *286*, 37.
25. Birch, A. J.; Kelly, L. F.; Thompson, D. J. *J. Chem. Soc., Perkin Trans. I* **1981**, 1006.
26. Cunningham, D.; McArdle, P.; Sherlock, H.; Johnson, B. F. G.; Lewis, J. *J. Chem. Soc., Dalton Trans.* **1977**, 2340.
27. Johnson, B. F. G.; Lewis, J.; Parker, D. G.; Postle, S. R. *J. Chem. Soc., Dalton Trans.* **1977**, 794.
28. Brookhart, M.; Karel, K. J.; Nance, L. E. *J. Organometal. Chem.* **1977**, *140*, 203.
29. Johnson, B. F. G.; Lewis, J.; Parker, D. G.; Stephenson, G. R. *J. Organometal. Chem.* **1980**, *194*, C14.
30. Jablonski, C. R.; Sorensen, T. S. *Can. J. Chem.* **1974**, *52*, 2085.
31. Mahler, J. E.; Pettit, R. *J. Am. Chem. Soc.* **1963**, *85*, 3955.
32. Brookhart, M.; Harris, D. L. *J. Organometal. Chem.* **1972**, *42*, 441.
33. Birch, A. J.; Chamberlain, K. B.; Haas, M. A.; Thompson, D. J. *J. Chem. Soc., Perkin Trans. I* **1973**, 1882.
34a. Birch, A. J.; Kelly, L. F.; Narula, A. S. *Tetrahedron* **1982**, *38*, 1813.
34b. Pearson, A. J.; O'Brien, M. K. *Tetrahedron Lett.* **1988**, *29*, 869.
35a. Bandara, B. M. R.; Birch, A. J.; Khor, T-C. *Tetrahedron Lett.* **1980**, *21*, 3625.
35b. Birch, A. J.; Pearson, A. J. *Tetrahedron Lett.* **1975**, 2379.
36. Pearson, A. J.; Ham, P.; Ong, C. W.; Perrior, T. R.; Rees, D. C. *J. Chem. Soc., Perkin Trans. I* **1982**, 1527.
37a. Kane-Maguire, L. A. P.; Odiaka, T. I.; Turgoose, S.; Williams. P. A. *J. Chem. Soc., Dalton Trans.* **1981**, 2489.
37b. Birch, A. J.; Liepa, A. J.; Stephenson. G. R. *Tetrahedron Lett.* **1979**, 3565.

38. Birch, A. J. ; Westerman, P. W.; Pearson, A. J. *Aust. J. Chem.* **1976**, *29*, 1671.
39. Odiaka, T. I.; Kane-Maguire, L. A. P. *J. Chem. Soc., Dalton Trans.* **1981**, 1162.
40a. Birch, A. J.; Jenkins, I. D.; Liepa, A. J. *Tetrahedron Lett.* **1975**, 1723.
40b. John, G. R.; Kane-Maguire, L. A. P. *J. Chem. Soc., Dalton Trans.* **1979**, 873.
41. Pearson, A. J.; Chandler, M. *J. Chem. Soc., Perkin Trans. I* **1980,** 2238.
42. Pearson, A. J.; Kole, S. L.; Chen, B. *J. Am. Chem. Soc.* **1983**, *105*, 4483.
43. Pearson, A. J.; Rees, D. C. *J. Chem. Soc., Perkin Trans. I* **1982**, 2467.
44. Pearson, A. J.; Rees, D. C. *J. Chem. Soc., Perkin Trans. I* **1983**, 619.
45. *p*-Anisaldehyde is commercially available and inexpensive (Lancaster Synthesis 89–90 catalog, 250 g for $13.80).
46. Johnson, W. S.; Shelberg, W. E. *J. Am. Chem. Soc.* **1945**, *67*, 1853.
47a. *o*-Methoxyphenylhydrazine hydrochloride is prepared by diazotization and reduction of *o*-anisidine (34% yield): Bergmann, E. D.; Hoffmann, E. *J. Chem. Soc.* **1962**, 2827.
47b. *o*-Anisidine is commercially available and inexpensive (Fluka 88–89 catalog, 250 mL, $d = 1.09$ g/mL, for $10.80).
48. *p*-Hydroxyphenylacetic acid is commercially available (Lancaster Synthesis 89–90 catalog, 100 g for $59.20).
49a. Pearson, A. J.; Ham, P.; Rees, D. C. *J. Chem. Soc., Perkin Trans. I* **1982**, 489.
49b. Pearson, A. J. *J. Chem. Soc., Perkin Trans. I* **1979**, 1255.

CHAPTER 5

(+)-Thienamycin

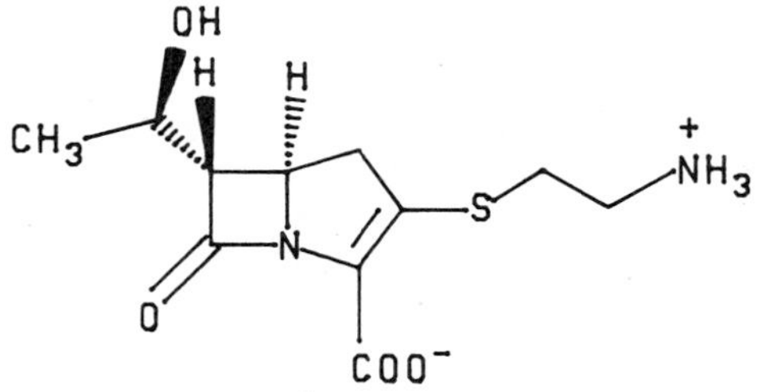

(+)-Thienamycin

1-Azabicyclo[3.2.0]hept-2-ene-2-carboxylic acid,

2-[(2-aminoethyl)thio]-6-(1-hydroxyethyl)-7-oxo

[5R-[5α,6α(R*)]] [59995-64-1]

Figure 5.1

(+)-Thienamycin, obtained from cultures of *Streptomyces cattleya*,[1] is an extremely potent broad-spectrum antibiotic.[2] Several groups have been working toward the development of an asymmetric synthesis of (+)-thienamycin since (1) production by *Streptomyces* is inefficient and (2) only the diastereoisomer with the natural configuration has significant bactericidal activity.[3]

5.1 π-ALLYLIRON TRICARBONYL LACTONE COMPLEXES

5.1.1 Preparation

In Chapter 3 π-allylmetal complexes were prepared by oxidative addition of an allylic-functionalized alkene to palladium(0) (see Figure 5.2). The allylic leaving groups most commonly involved are chloride, carboxylate, and alkoxide. A more difficult yet analogous oxidative addition to iron(0) is possible using 1,3-diene monoepoxides[4,5] (see Figure 5.3).

With iron pentacarbonyl (method A), a coordination site is accessed by photolytic ejection of a carbon monoxide ligand. The volatility and high toxicity of the

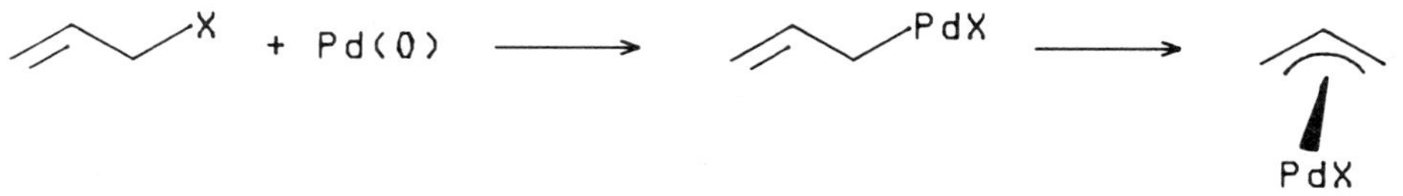

X + Pd(0) ⟶ PdX ⟶ PdX

Figure 5.2

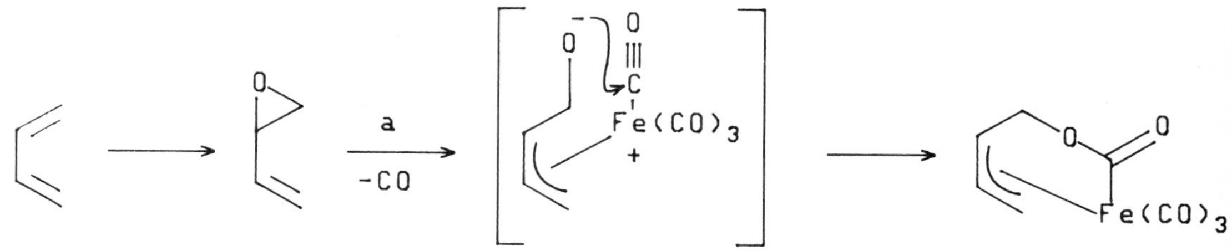

a
-CO

O O C Fe(CO)$_3$ +

O O Fe(CO)$_3$

a. hν, Fe(CO)$_5$, benzene; 90%

Figure 5.3

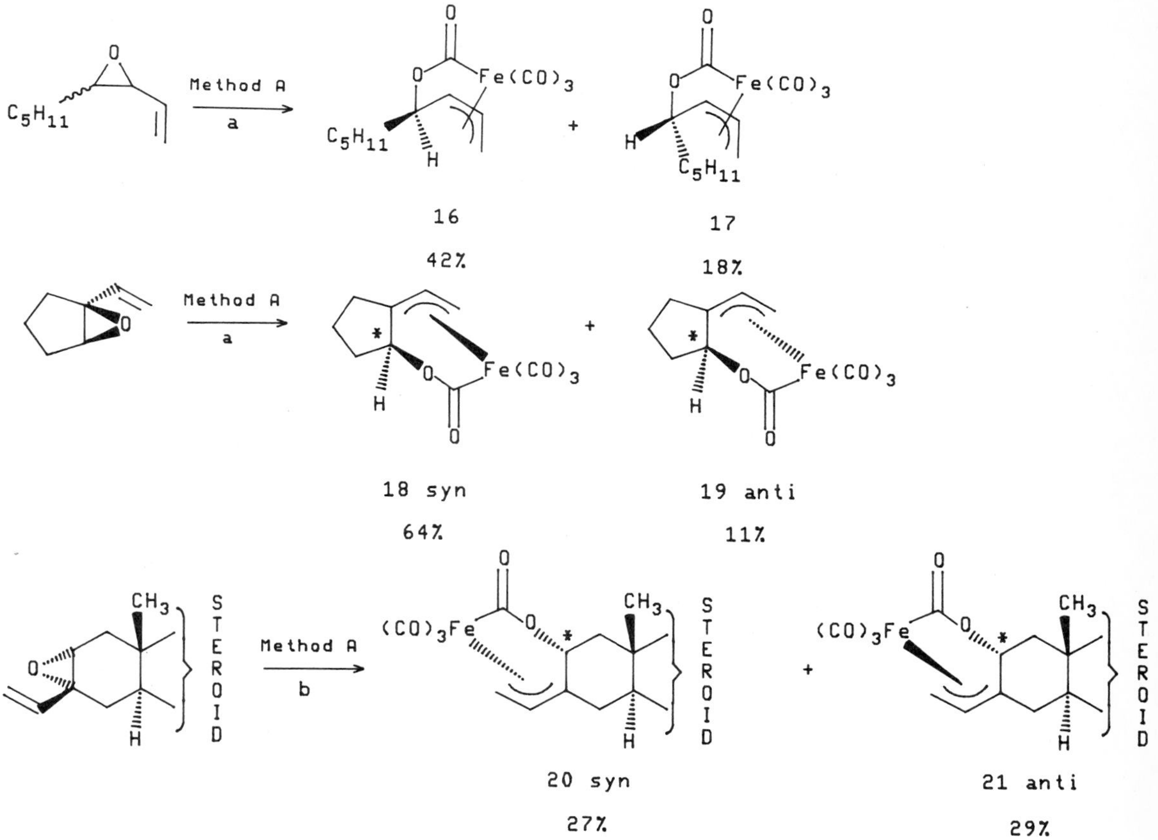

C_5H_{11} O Method A a

O O Fe(CO)$_3$ C_5H_{11} H + O O Fe(CO)$_3$ H C_5H_{11}

16 42%   17 18%

O Method A a

Fe(CO)$_3$ O O H + Fe(CO)$_3$ O O H

18 syn 64%   19 anti 11%

CH_3 O H STEROID Method A b

O (CO)$_3$Fe O CH_3 H STEROID + O (CO)$_3$Fe O CH_3 H STEROID

20 syn 27%   21 anti 29%

a. Products separated on florisil

b. Products separated on silica gel

Figure 5.4

TABLE 5.1 Preparation of π-Allyliron Tricarbonyl Lactone Complexes[4,5b,6]

		Yields (%)		
	Methods:	A	B	C
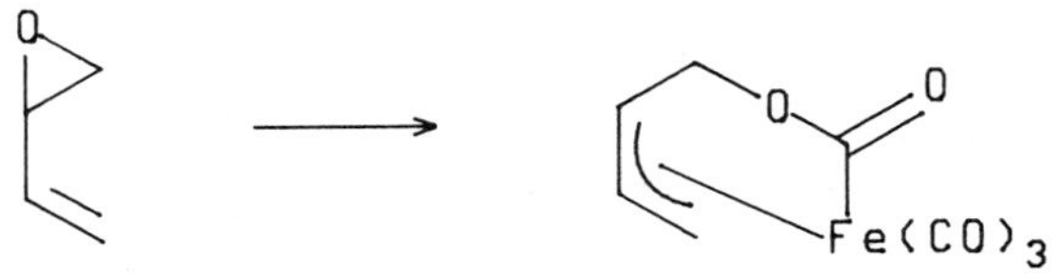 1		90	–	–
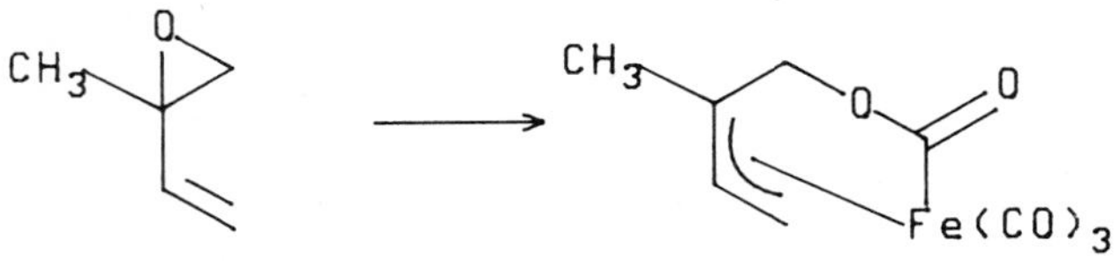 2		89	94	76
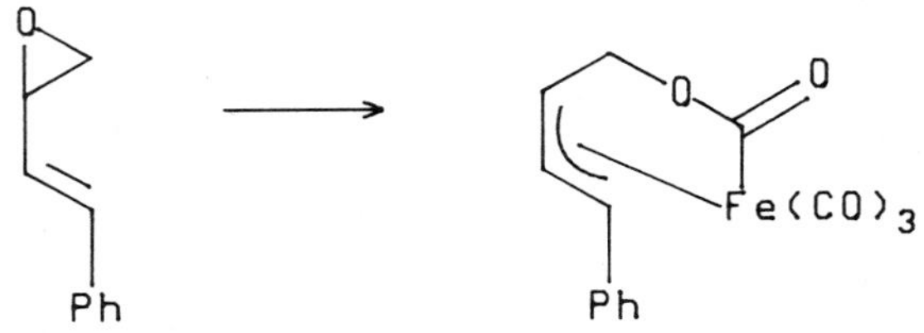 3		–	100	68
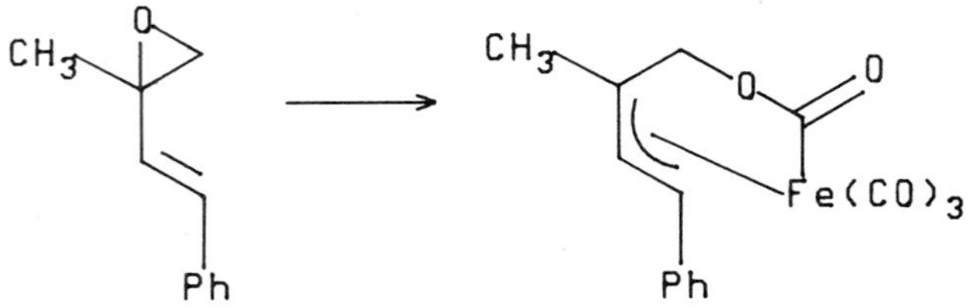 4		–	64	63

	Yields (%)		
Methods:	A	B	C
5	–	59	65
6	–	56	66
7	65	42	44
8	72	–	–

(continued)

	Yields (%)		
Methods:	A	B	C
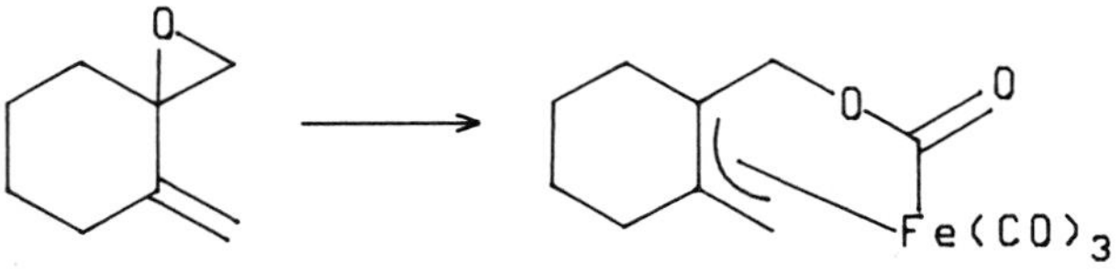 9	29	–	–
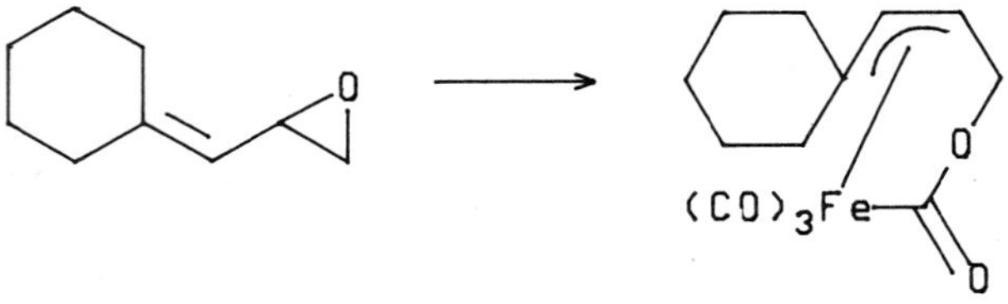 10	46	–	–
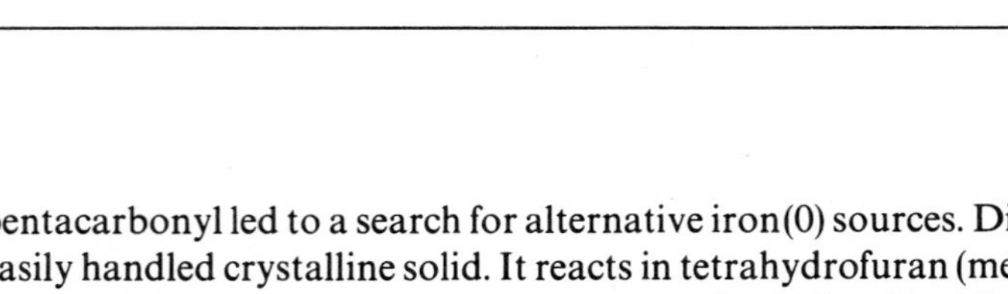 11	49	–	50

pentacarbonyl led to a search for alternative iron(0) sources. Diiron nonacarbonyl is an easily handled crystalline solid. It reacts in tetrahydrofuran (method B) or even in lower polarity solvents when ultrasonicated (method C) with 1,3-diene monoepoxides to produce the same π-allyliron tricarbonyl complexes in moderate to excellent yields[6] (see Table 5.1).

Diastereoisomers are produced when the organic portion of the lactone product contains a chiral center. These are often separated by column chromatography on silica gel or florisil (see Table 5.2 and Figure 5.4).

TABLE 5.2 Preparation of Diastereoisomeric π-Allyliron Tricarbonyl Lactone Complexes[4,5b,6]

		Yields (%)		
	Methods:	A	B	C
(epoxycyclopentene) → $(CO)_3Fe$ lactone complex	12	48	–	–
(epoxycyclohexene) → $(CO)_3Fe$ lactone complex	13	60	–	34
(epoxide, CH_2OBz, CH_3) → lactone complex (CH_2OBz, $Fe(CO)_3$, CH_3)	14	–	34	79
(epoxide, C_9H_{19}, CH_2OH, CH_3) → $(CO)_3Fe$ lactone complex (C_9H_{19}, CH_2OH, CH_3)	15	–	71	–

5.1.2 Demetallation

Demetallation of a π-allyliron tricarbonyl lactone complex by oxidation with ceric ammonium nitrate at low temperature produces a β-lactone (see Table 5.3). Demetallation of the lactone complex at high temperature under a high pressure of carbon monoxide produces a δ-lactone.[7] Oxidation of the *syn* lactone complex (oxygen and the metal are *syn*) **18** affords a separable mixture of *cis*-fused β-lactone and δ-lactone. Oxidation of the *anti* complex **19** affords only the δ-lactone since a *trans*-fused β-lactone would be impossibly strained[8] (see Figure 5.5).

The mechanism for demetallation may involve a tautomeric equilibrium between two σ-alkyliron intermediates via a π-allyl complex. Metal oxidation produces radical cations. Metal extrusion (a reductive elimination) from the σ-alkyliron radical cation at low temperature or from the σ-alkyliron complex at high temperature results in

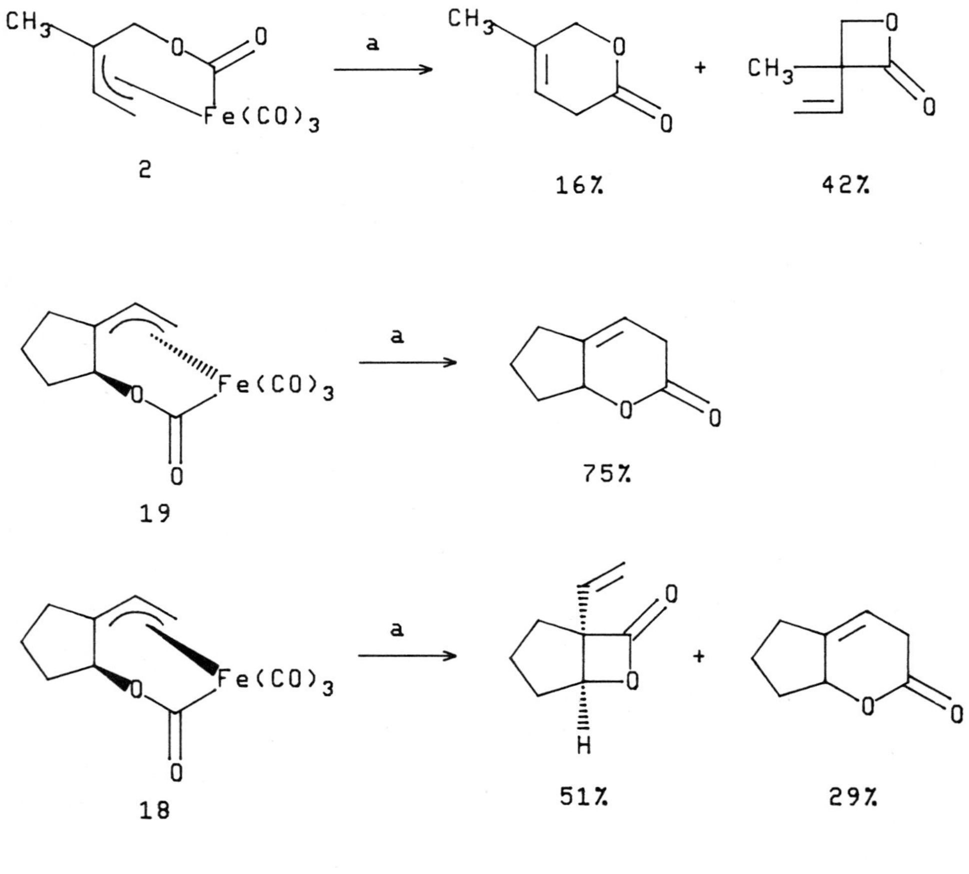

CH3 O O Fe(CO)3 2 —a→ CH3 O O 16% + CH3 O O 42%

Fe(CO)3 O O 19 —a→ O O 75%

Fe(CO)3 O O 18 —a→ O O H 51% + O O 29%

a. rt, $(NH_4)_2Ce(NO_3)_6$, CH_3CN

Figure 5.5

TABLE 5.3 Demetallation by Oxidation[5b]

Iron Complex	β-Lactone Yield (%)	δ-Lactone Yield (%)	Solvent
7	—	38	EtOH-H_2O
7	42	16	CH_3CN
8	38	—	Anhydrous EtOH
9	15	55	CH_3CN
10	52	—	CH_3CN
11	52	—	CH_3CN
13	86	—	Anhydrous EtOH
16	68	—	CH_3CN (*trans*-4-pentyl-3-vinyl)
17	64	—	CH_3CN (*cis*-4-pentyl-3-vinyl)
18	51	29	CH_3CN
19	—	75	CH_3CN
20	12	27	CH_3CN-benzene
21	—	25	CH_3CN-benzene

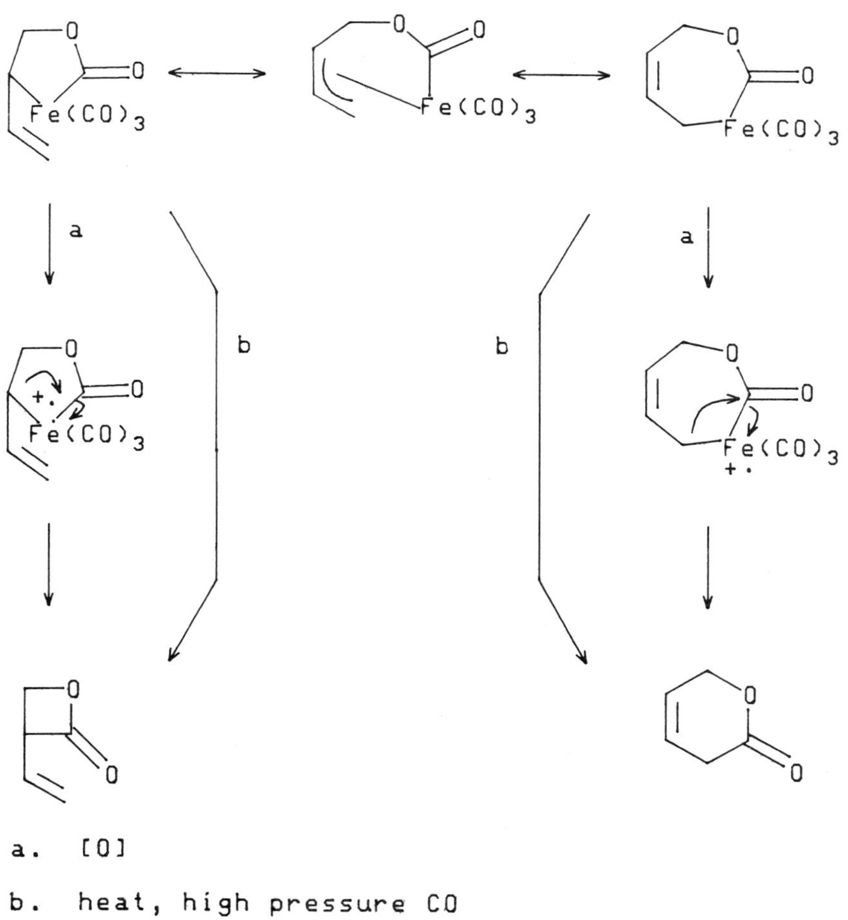

Figure 5.6

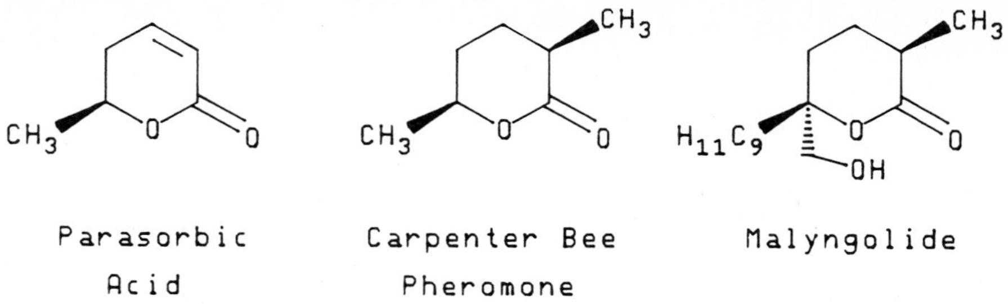

Parasorbic Acid | Carpenter Bee Pheromone | Malyngolide

Figure 5.7

β-lactone from the five-membered ring tautomer and δ-lactone from the seven-membered ring tautomer (see Figure 5.6).

The new methodology for δ-lactone synthesis has been applied to total syntheses of (1) parasorbic acid, a natural δ-lactone from the tree *Sorbus acucuparia*; (2) the carpenter bee pheromone; and (3) malyngolide, an unusual δ-lactone antibacterial agent[7] (see Figure 5.7).

Monoepoxidation of 1,3-pentadiene affords an inseparable 3 : 1 mixture of regioisomers. The mixture is converted to a separable mixture of π-allyliron tricarbonyl lactone complexes with diiron nonacarbonyl in tetrahydrofuran (method B). The major complex, corresponding to epoxidation at the more substituted alkene bond, reductively eliminates at high temperature and under a high pressure of carbon monoxide (see Figure 5.8).

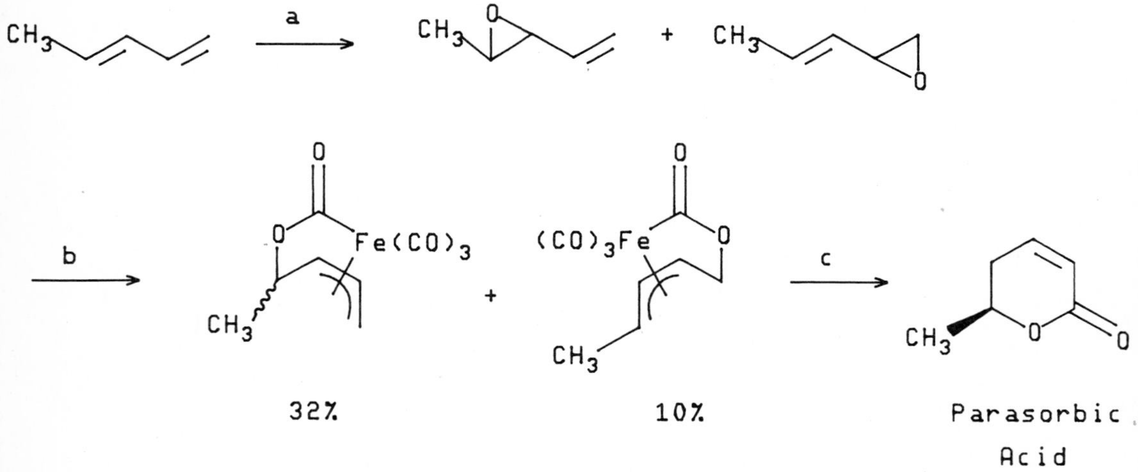

32% | 10% | Parasorbic Acid

a. $-5^{\circ}C$, Na_2CO_3, CH_3COOH, CH_2Cl_2

b. $Fe_2(CO)_9$, THF

c. $195^{\circ}C$, 60 atm CO, benzene; 73% (racemic)

Figure 5.8

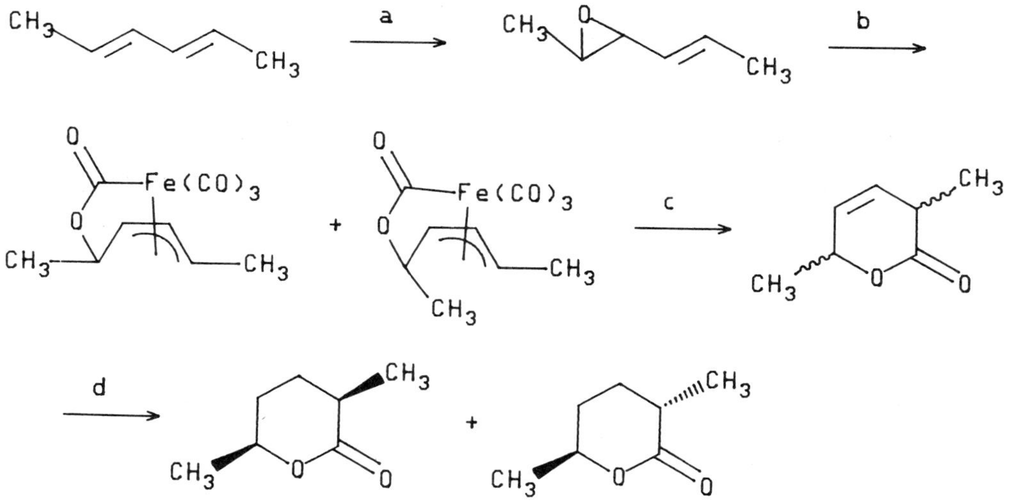

Carpenter Bee Pheromone

a. 0°C, Na_2CO_3, CH_3COOH, CH_2Cl_2

b.))), $Fe_2(CO)_9$, benzene

c. 90°C, 300 atm CO, benzene; 89% (1:1 mixture)

d. H_2, Pd-C, EtOAc; 92% (1:1 mixture, separable by GLC)

Figure 5.9

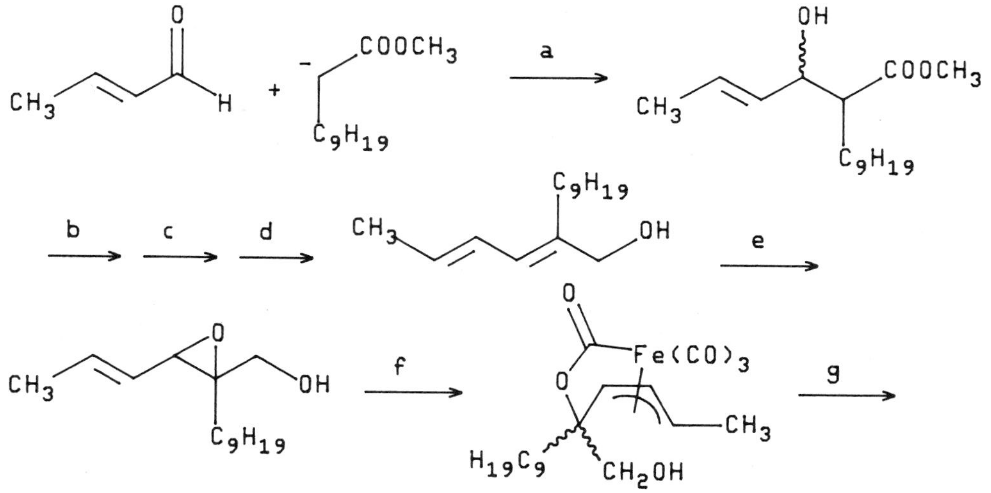

Figure 5.10

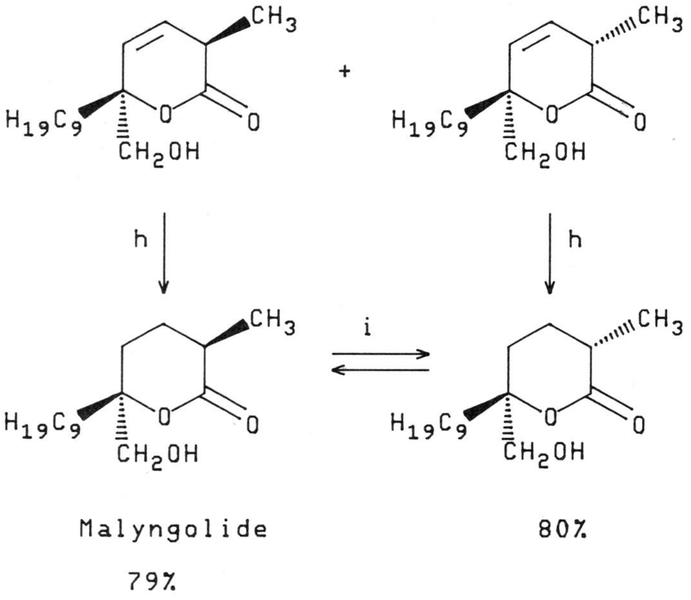

Malyngolide

79%

80%

a. 88%

b. heat, $KHSO_4$, toluene

c. rt, DBU

d. 0°C, $LiAlH_4$, Et_2O; 71% for three steps

e. heat, t-BuOOH, $VO(acac)_2$, benzene; 88%

f. rt, $Fe_2(CO)_9$, THF; 71%

g. 90°C, 300 atm CO, benzene; 74% (separable)

h. H_2, PtO_2, CH_3OH

i. -78°C, $LiN(i\text{-}Pr)_2$

Figure 5.10 (*Continued*)

The monoepoxide from 1,3-pentadiene affords two diastereoisomeric lactone complexes with diiron nonacarbonyl in benzene when ultrasonicated (method C). Demetallation occurs at high temperature under a high pressure of carbon monoxide. Carpenter bee pheromone is prepared from 2,4-hexadiene by a similar sequence (see Figure 5.9).

The synthesis of malyngolide begins with the condensation of methyl undecanoate with crotonaldehyde. Alcohol dehydration and ester reduction affords a dienyl alcohol. Epoxidation under Sharpless conditions produces a monoepoxide that is converted to a mixture of diastereoisomeric lactone complexes with diiron nonacarbonyl in tetrahydrofuran (method B). Demetallation at high temperature and under a high pressure of carbon monoxide produces separable diastereoisomeric lactones in 74% yield. Catalytic reduction completes the synthesis (see Figure 5.10).

5.2 π-ALLYLIRON TRICARBONYL LACTAM COMPLEXES

The unexpected isolation of significant amounts of β-lactones from oxidative demetallation of π-allyliron tricarbonyl lactone complexes led to an investigation of a potential route to β-lactams by synthesis and oxidative demetallation of π-allyliron

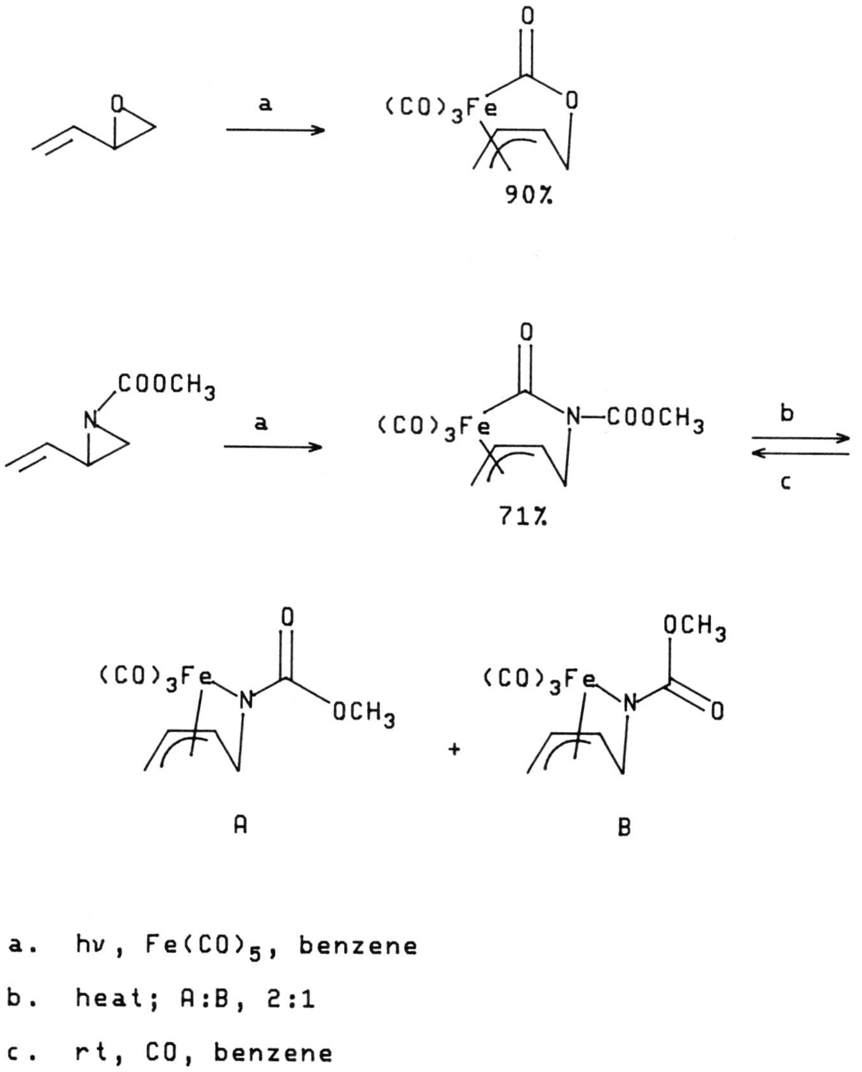
a
(CO)3Fe
90%
COOCH3
a
(CO)3Fe
N—COOCH3
b
c
71%
(CO)3Fe
OCH3
(CO)3Fe
OCH3
+
A
B
a. hν, Fe(CO)5, benzene
b. heat; A:B, 2:1
c. rt, CO, benzene

Figure 5.11

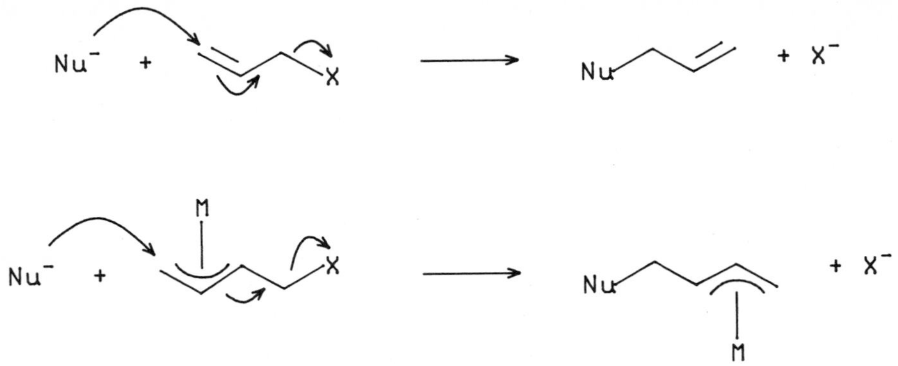

Figure 5.12

tricarbonyl lactam complexes. The oxidative addition of monoaziridines of 1,3-dienes to iron(0) does produce π-allyliron tricarbonyl lactams.[4] However, the method is limited by the relative inaccessibility of the monoaziridines (see Figure 5.11). Alternatively, Shvo reported the $S_N'2$ attack by nucleophiles on π-allylmetal complexes possessing α-leaving groups[9,10] (see Figure 5.12).

π-Allyliron tricarbonyl lactone complexes are converted to the *transposed* lactams in reasonably good yields by reaction with an amine and a mild Lewis acid catalyst. A reasonable mechanism for the lactone-to-lactam conversion involves (see Figure 5.13):

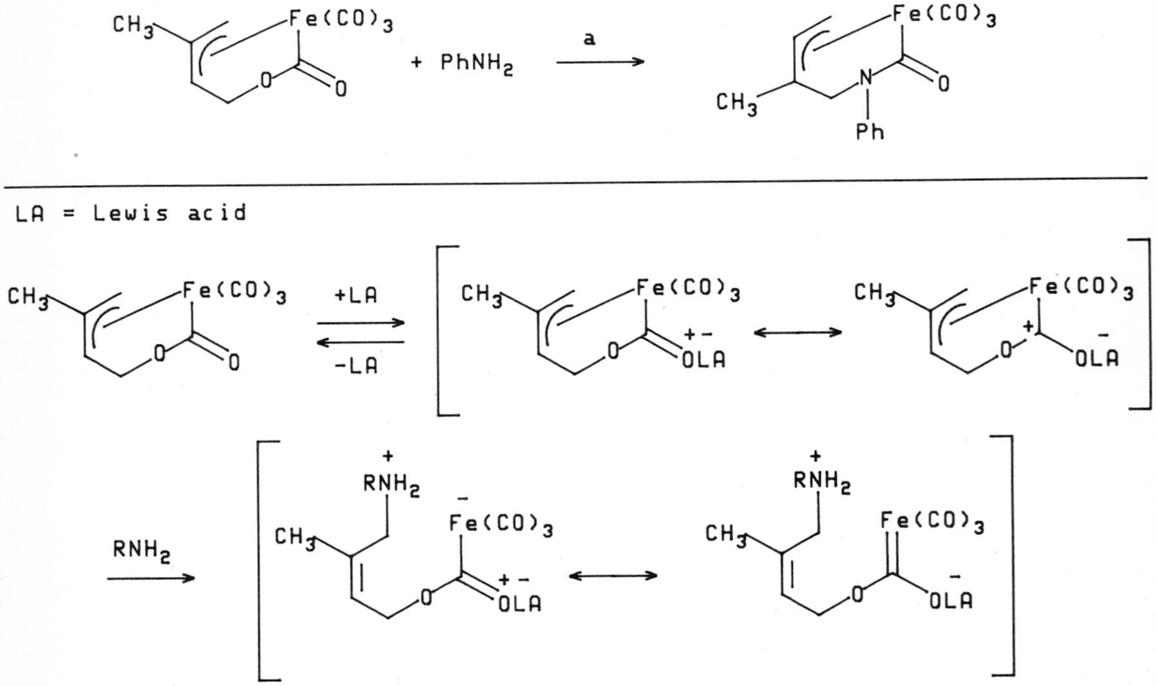

LA = Lewis acid

Figure 5.13

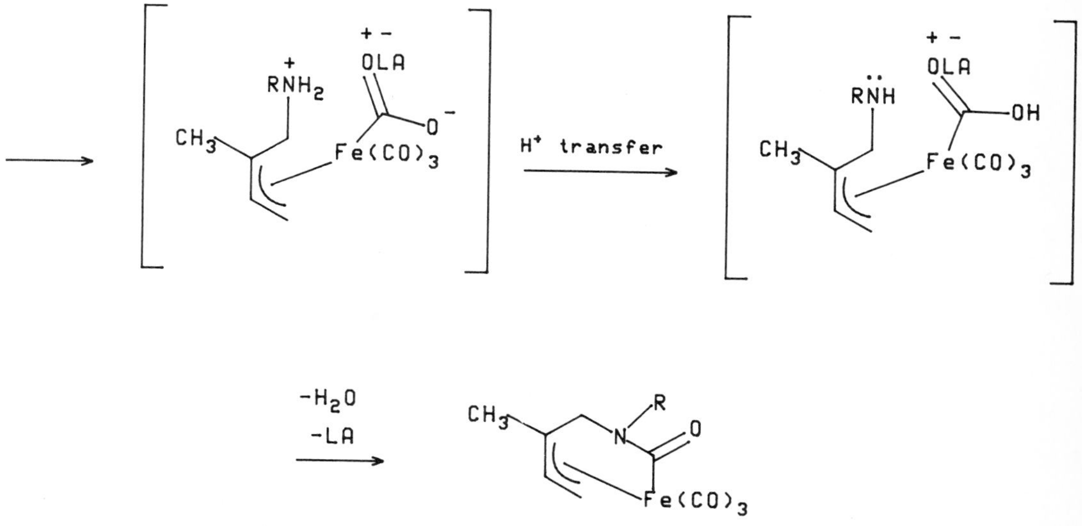

$\xrightarrow{H^+ \text{ transfer}}$

$\xrightarrow[-LA]{-H_2O}$

a. rt, Al_2O_3, benzene; 50%

Figure 5.13 (*Continued*)

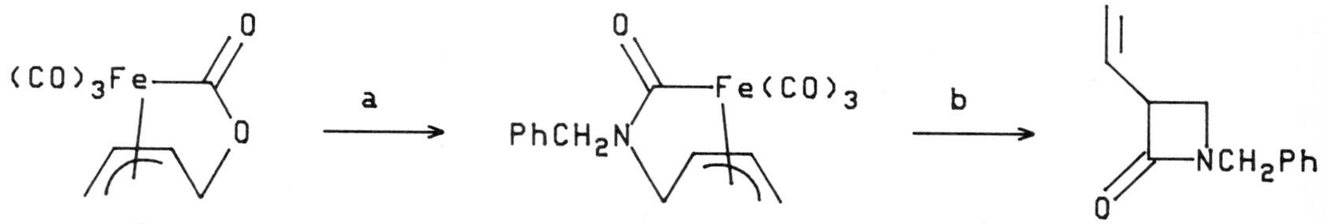

$(CO)_3Fe$ 1 $\xrightarrow{a}$ $PhCH_2N$ $Fe(CO)_3$ 1a $\xrightarrow{b}$ NCH_2Ph 1b

a. rt, $ZnCl_2$, $PhCH_2NH_2$, Et_2O; 82%

b. −30°C to rt, $(NH_4)_2Ce(NO_3)_6$, anh. EtOH; 72%

Figure 5.14

1. Lewis acid complexation with the lactone carbonyl oxygen, producing a more electron-deficient metal and, thus, a π-allylmetal complex more prone to nucleophilic attack
2. Nucleophilic attack by amine on the terminal π-allyl carbon distant from oxygen
3. Intramolecular oxidative addition of the allylic ester to the iron carbonyl anion
4. A proton transfer followed by Lewis-acid-promoted ring closure with loss of H_2O

TABLE 5.4 **Preparation of β-Lactams by Oxidative Demetallation of π-Allyliron Tricarbonyl Lactam Complexes**[11,12]

Lactone Complex	Lactam Complex	Yield (%)	Catalyst	β-Lactam	Yield (%)
1	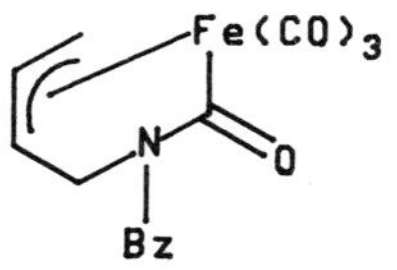 1a	82	$ZnCl_2$	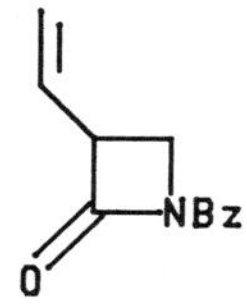 1b	75
2	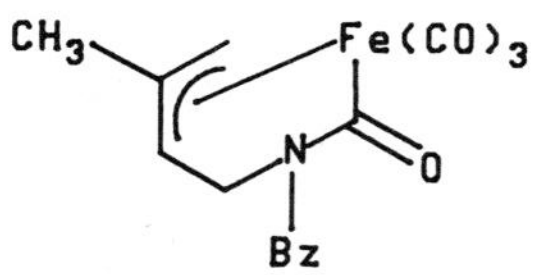 2a	95	$ZnCl_2$	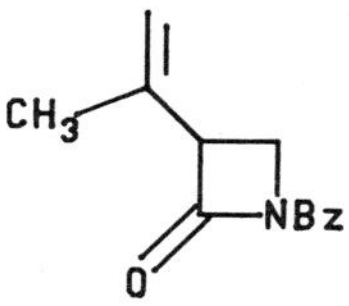 2b	88
2	2a	100	$ZnCl_2 \cdot TMEDA$		
2	2a	98	Et_2AlCl		
3	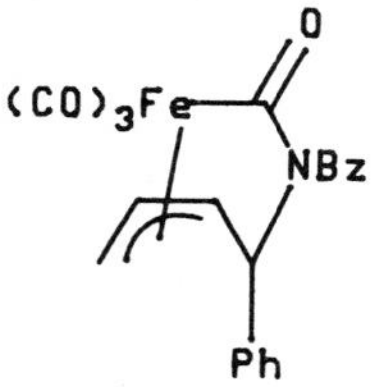 3a	80	Et_2AlCl	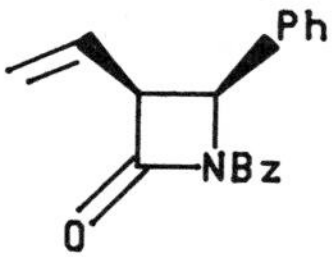 3b	78
7	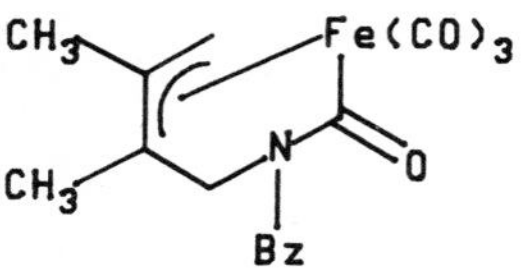 7a	75	$ZnCl_2$	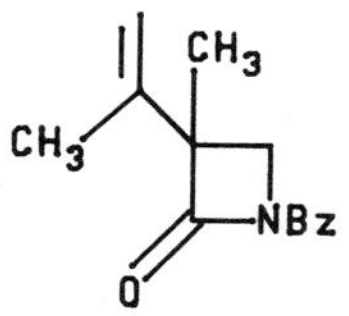 7b	34 (56% lactone)

Lactone Complex	Lactam Complex	Yield (%)	Catalyst	β-Lactam	Yield (%)
8	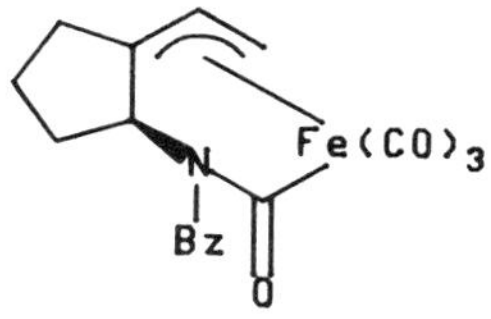 Fe(CO)3, N, Bz, O 8a	23	$ZnCl_2$	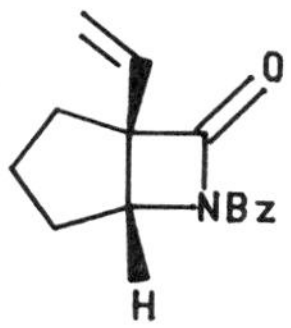 O, NBz, H 8b	75
	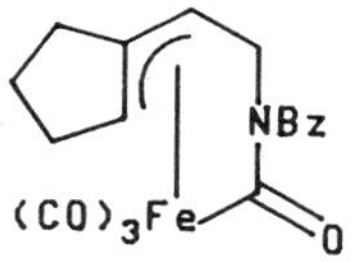 NBz, (CO)3Fe, O 18a	12			
18,19	18a	48	$ZnCl_2$	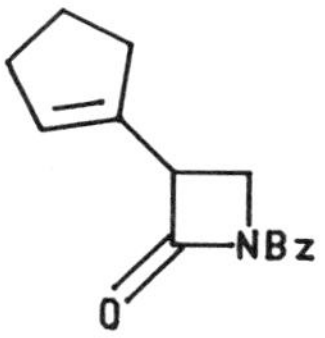 NBz, O 18b	84
18,19	18a	68	Et_2AlCl		
16,17	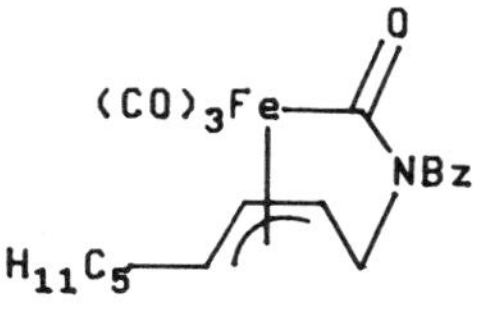 O, (CO)3Fe, NBz, $H_{11}C_5$ 16a	82	$ZnCl_2$	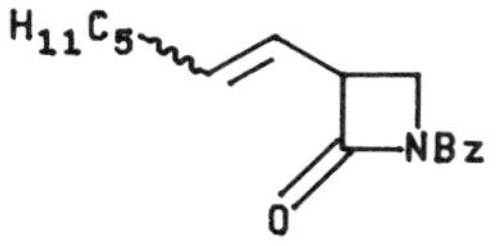 $H_{11}C_5$, NBz, O 16b	64

Ley demonstrated that (1) π-allyliron tricarbonyl lactam synthesis using benzylamine and a Lewis acid catalyst is both efficient and general and (2) oxidative demetallation provides the desired β-lactams in excellent yield[11,12] (see Figure 5.14 and Table 5.4). This methodology for the synthesis of 3-alkenyl-2-azetidinones, which are otherwise difficult to access, can be applied to the synthesis of (+)-thienamycin.[13,14]

5.3 TOTAL SYNTHESIS OF (+)-THIENAMYCIN

A key intermediate, 3-(1-hydroxyethyl)-4-(2,2-dimethoxyethyl)-2-azetidinone, is first synthesized in racemic form. The starting diene monoepoxide is prepared in two steps: (1) a modified Wittig condensation of 3,3-dimethoxypropanal[15] with a phosphonate ester to afford an α,β-unsaturated ketone and (2) reaction with dimsyl anion to convert the ketone to an epoxide. The π-allyliron tricarbonyl lactone complex is efficiently prepared using either iron pentacarbonyl under photolytic conditions (method A) or using diiron nonacarbonyl in tetrahydrofuran (method B). Nucleophilic attack by benzylamine produces a mixture of diastereoisomeric lactam complexes. Oxidation of the mixture with ceric ammonium nitrate gives both β-lactam and δ-lactam. When the β-lactam alkene is ozonolyzed, the resulting ketone epimerizes to the *trans*-substituted isomer. The desired 3-(1-hydroxyethyl) substituent is available by ketone reduction using K-Selectride. Benzylamine hydrogenolysis is best done with sodium at low temperature in an ammonia–ethanol mixture (see Figure 5.15).

The corresponding chiral material is prepared by use of a chiral auxiliary and separation of diastereoisomeric lactam complexes. Nucleophilic attack by (*S*)-(−)-α-methylbenzylamine[16] on the π-allyliron tricarbonyl lactone complex affords diastereoisomeric lactam complexes in an approximately 1:1 ratio and a small amount of an iron diene complex. The mixture is separable by careful column chromatography on silica gel. Oxidation of the π-allyliron tricarbonyl lactams produces the β-lactams in excellent yields. When the diastereoisomer with the 4-*R* configuration is ozonolyzed, the ketone produced epimerizes. K-Selectride reduction affords an 11:2 mixture of diastereoisomeric alcohols in 85% yield. Benzylamine hydrogenolysis is again efficient using sodium at low temperature in an ammonia–ethanol mixture (see Figure 5.16).

The remaining transformations are done on racemic material to produce (±)-thienamycin by Kametani[17,18] and the Merck group.[19] The optically active material can be converted in an analogous fashion. The alcohol is protected as a carbonate and the acetal is converted to a dithioacetal and then to a thioenol ether. Condensation of

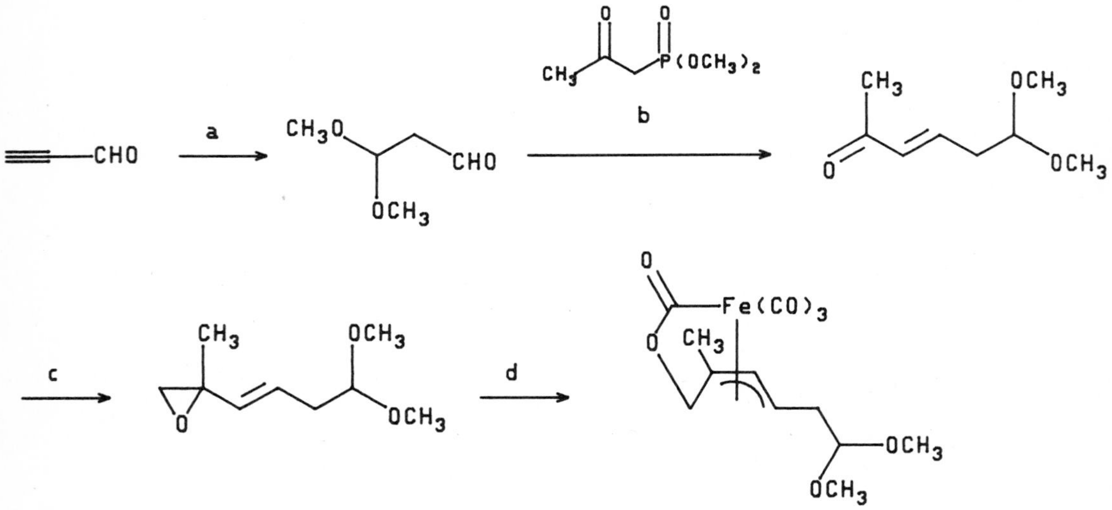

CHO a CH3O CHO OCH3 b P(OCH3)2 CH3 OCH3 O OCH3 c CH3 OCH3 O OCH3 d Fe(CO)3 CH3 OCH3 OCH3

Figure 5.15

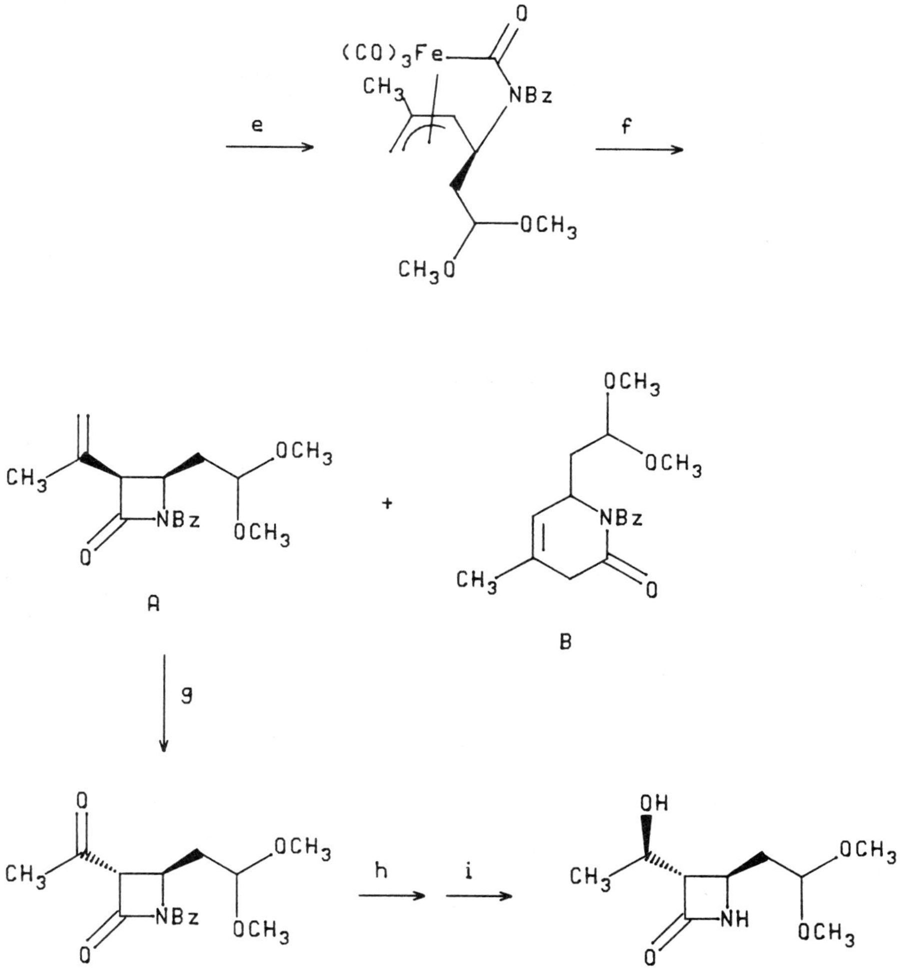

a. $0^{\circ}C$, K_2CO_3, CH_3OH; 55%

b. K_2CO_3, benzene; 84%

c. $-17^{\circ}C$ to rt, NaH, DMSO, THF; 66%

d. rt, hν, $Fe(CO)_5$, benzene; 91% (Method B: 84%)

e. rt, $ZnCl_2$, $BzNH_2$, Et_2O, THF; 57%

f. $-30^{\circ}C$ to rt, $(NH_4)_2Ce(NO_3)_6$, CH_3OH; A:64%, B:24%

g. 1) $-78^{\circ}C$, O_3, 2) CH_3SCH_3; 84%

h. rt, K-Selectride®, THF, Et_2O

i. $-78^{\circ}C$, Na, EtOH, NH_3; 74% for two steps

Figure 5.15 *(Continued)*

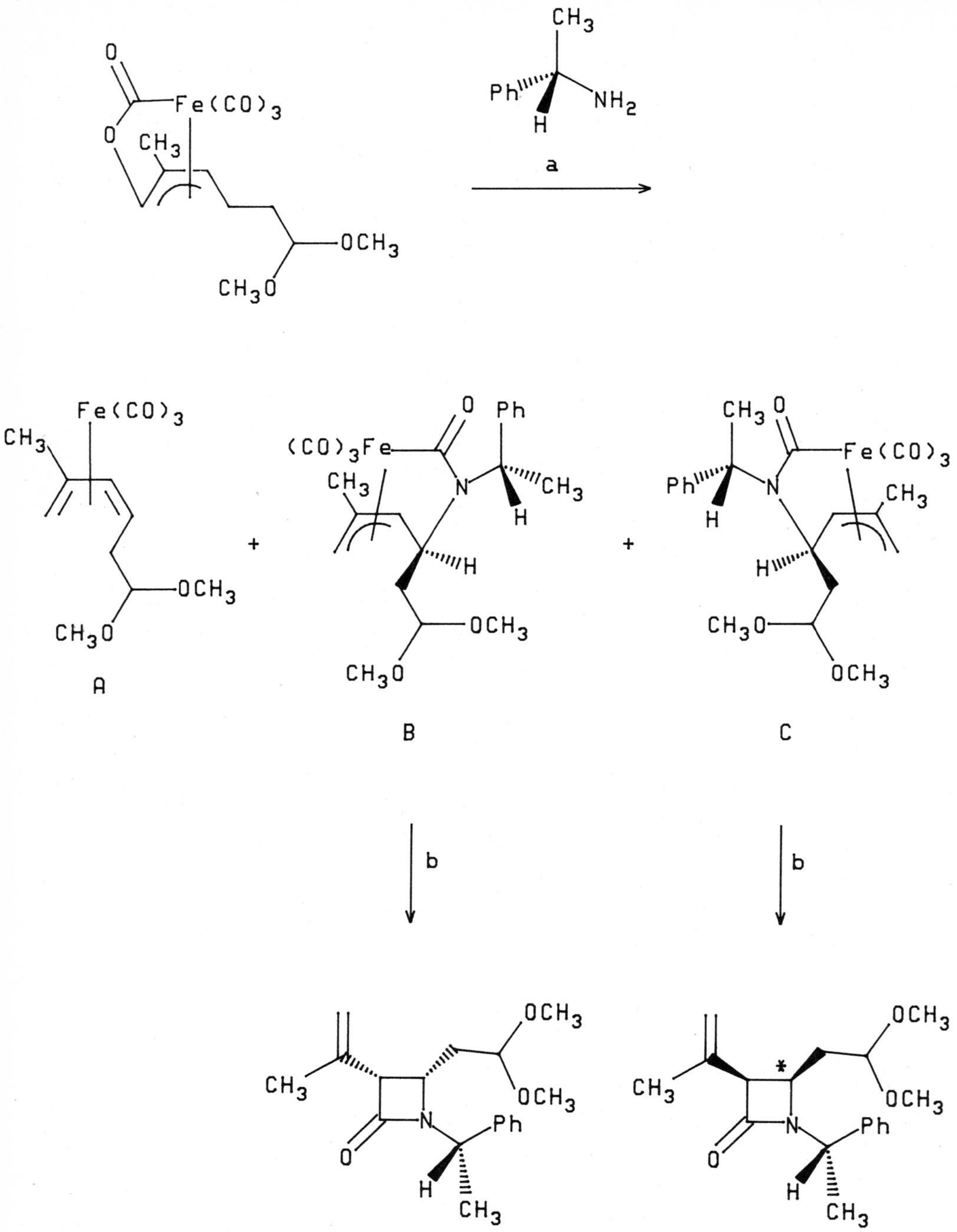

Figure 5.16

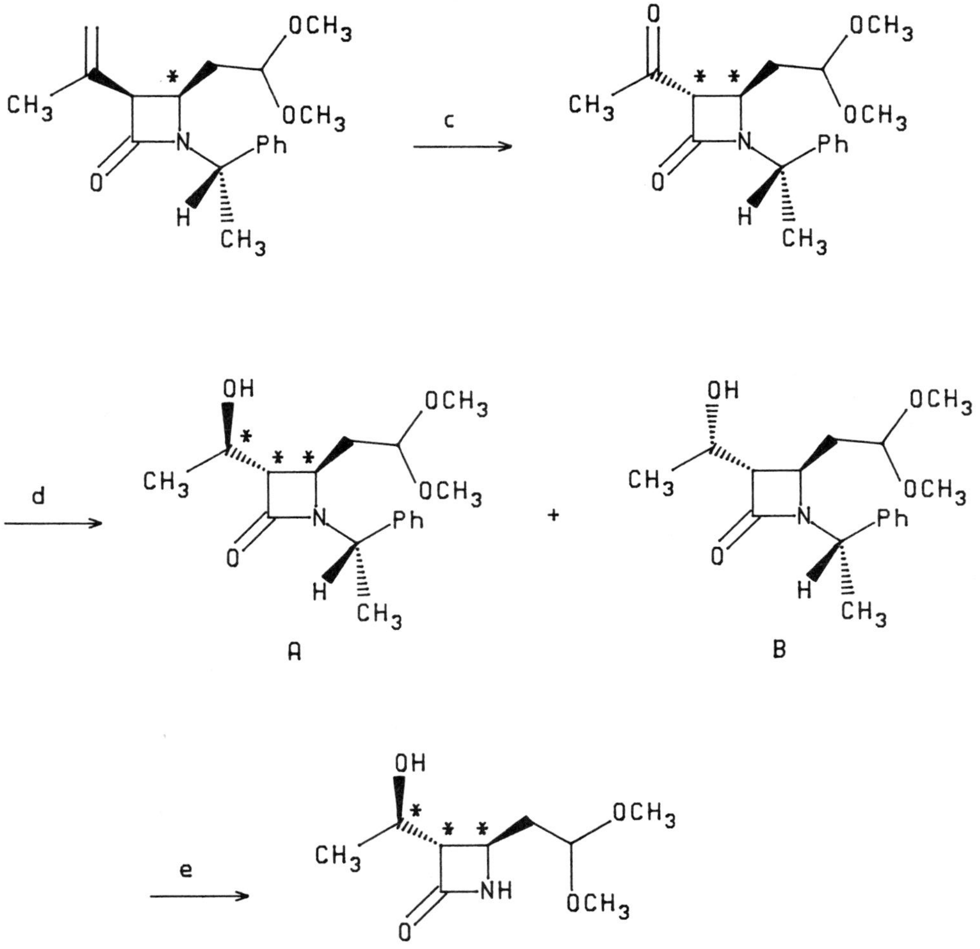

C

d

A

B

e

a. rt, $ZnCl_2 \cdot TMEDA$, Et_2O, THF; A:16%, B:30%, C:29%, mixture separable by chromatography

b. -30°C to rt, $(NH_4)_2Ce(NO_3)_6$, CH_3OH; 88%

c. 1) -78°C, O_3, 2) CH_3SCH_3; 81%

d. 0°C, KI, K-Selectride®, Et_2O; A:B, 11:2, 85% yield

e. -78°C, Na, EtOH, NH_3; 83%

Figure 5.16 *(Continued)*

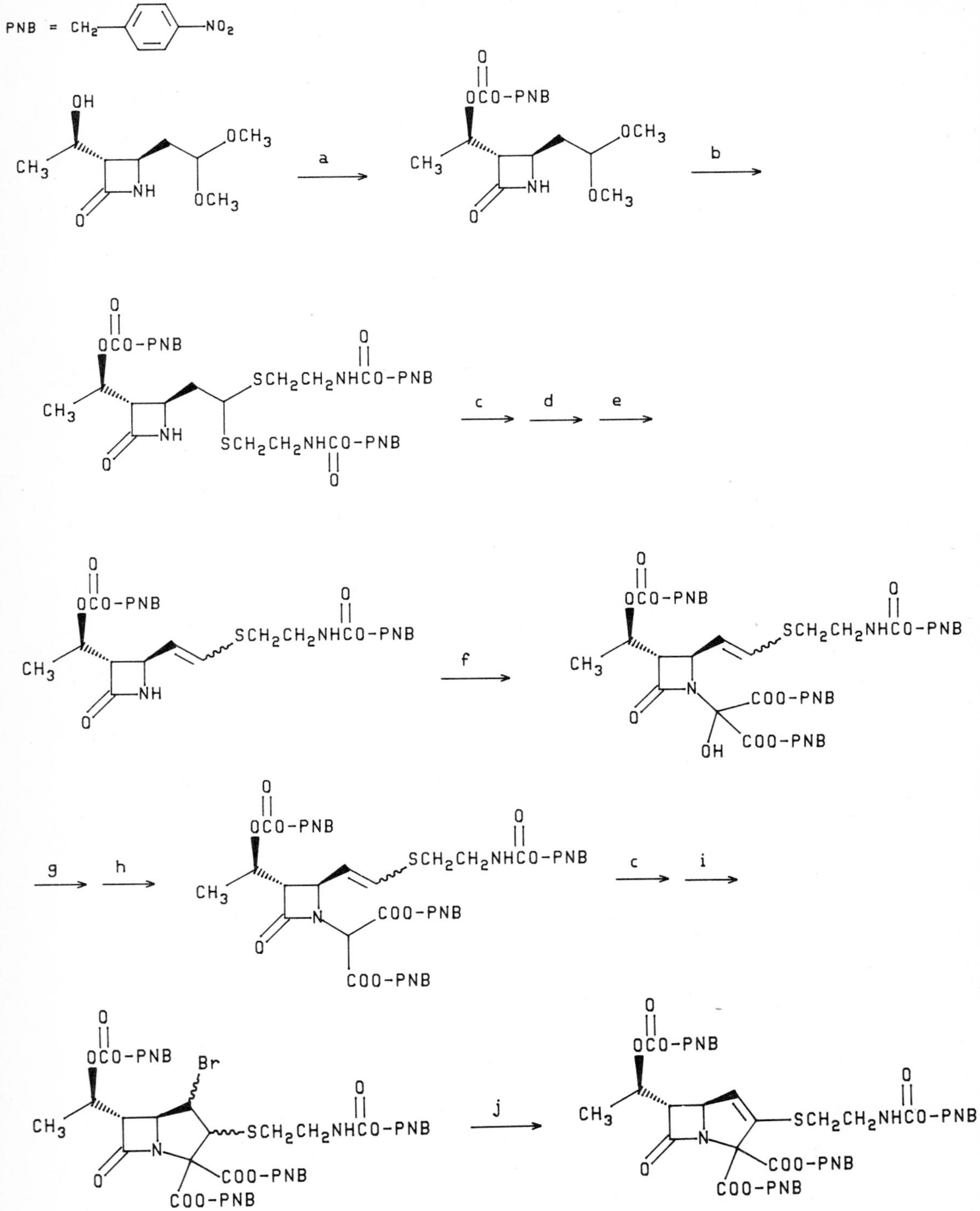

Figure 5.17

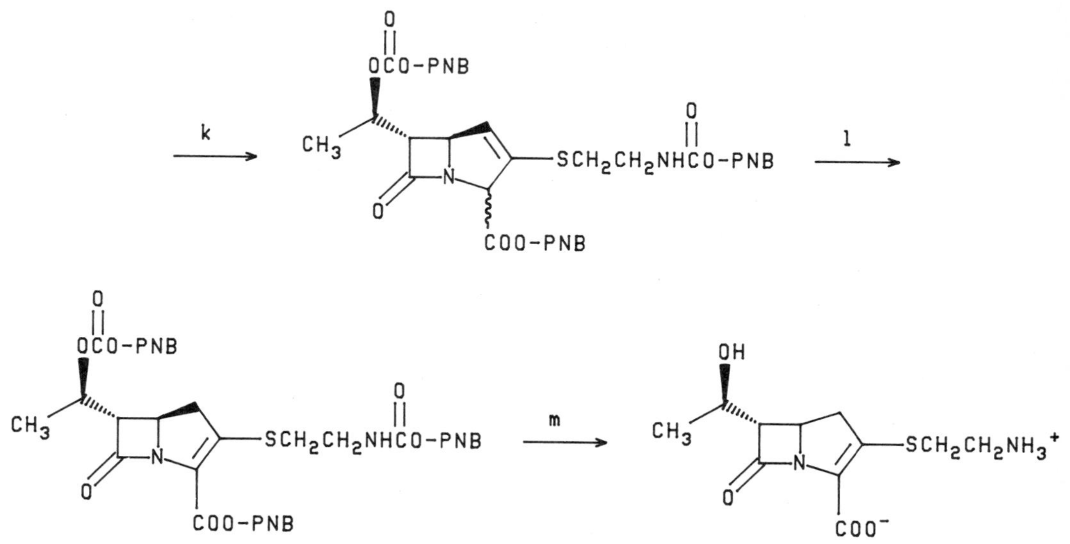

a. $0^{\circ}C$, DMAP, ClC(O)O-PNB, CH_2Cl_2; 85%

b. rt, $HSCH_2CH_2NHC(O)O$-PNB, CF_3COOH; 90%

c. $0^{\circ}C$, Br_2, Et_2O, THF

d. $0^{\circ}C$, cyclohexene

e. rt, Et_3N, DMF; 87% for three steps

f. rfx, PNB-OC(O)C(O)C(O)O-PNB, toluene; 44%

g. -20 to $20^{\circ}C$, pyridine, $SOCl_2$, THF

h. rt, K_2HPO_4, P(n-Bu)$_3$, DMF, H_2O; 60% for two steps

i. 1) $0^{\circ}C$, Br_2, Et_2O, THF, 2) rt, Et_3N, DMF; 60% for two steps

j. rt, AgF, pyridine; 68%

k. $120^{\circ}C$, LiI, collidine; 47%

l. rt, $(i\text{-}Pr)_2NH$, DMSO

m. K_2HPO_4, H_2, 10% Pd-C, H_2O, dioxane, EtOH; 23% for two steps

Figure 5.17 (*Continued*)

the amide nitrogen with ketomalonate affords an alcohol. The hydroxyl is removed by conversion to a chloride, then chloride reduction. Bromination of the thioenol ether at low temperature is followed by cyclization via a facile nucleophilic displacement of bromide by malonate. Dehydrobromination to a thioenol ether, decarboalkoxylation and base-mediated isomerization affords the properly functionalized bicyclic system. Hydrogenolysis of the *p*-nitrobenzyl protecting groups completes the synthesis.

The synthesis of (+)-thienamycin was completed in 23 steps from propiolaldehyde in overall 0.025% yield. The synthesis of the chiral intermediate, 3-(1-hydroxyethyl)-4-(2,2-dimethoxyethyl)-2-azetidinone, from propiolaldehyde was completed in nine steps in overall 3.4% yield (see Figure 5.17).

REFERENCES

1a. Kahan, J. S.; Kahan, F. M.; Goegelman, R.; Currie, S. A.; Jackson, M.; Stapley, E. O.; Miller, T. W.; Miller, A. K.; Hendlin, D.; Mochales, S.; Hernandez, S.; Woodruff, H. B.; Birnbaum, J. *J. Antibiot.* **1979**, *32*, 1.

1b. Albers-Schönberg, G.; Arison, B. H.; Hensens, O. D.; Hirshfield, J.; Hoogsteen, K.; Kaczka, E. A.; Rhodes, R. E.; Kahan, J. S.; Kahan, F. M.; Ratcliffe, R. W.; Walton, E.; Ruswinkle, L. J.; Morin, R. B.; Christensen, B. G. *J. Am. Chem. Soc.* **1978**, *100*, 6491.

1c. For a recent review on the biosynthetic pathway for (+)-thienamycin, see Williamson, J. M. *CRC Crit. Rev. Biotechnol.* **1986**, *4*, 111.

2a. Neu, H. C. *Symp. Front. Pharmacol.* **1982**, *1* (New β-Lactam Antibiot.), 85; *Chem. Abstr.* **1982**, *97*, 141359p.

2b. Ratcliffe, R. W.; Albers-Schönberg, G. *Chem. Biol. β-Lactam Antibiot.* **1982**, *2*, 227; *Chem. Abstr.* **1983**, *98*, 16441d.

2c. Leanza, W. J.; Wildonger, K. J.; Hannah, J.; Shih, D. H.; Ratcliffe, R. W.; Barash, L.; Walton, E.; Firestone, R. A.; Patel, G. F.; et al. *Spec. Publ. R. Soc. Chem.* **1981**, (Publ. *1980*) *38* (*Recent* Adv. Chem. β-Lactam Antibiot.) 240; *Chem. Abstr.* **1981**, *95*, 34942k.

2d. Mingoia, Q. *Cron. Farm.* **1980**, *23*, 20; *Chem. Abstr.* **1980**, *93*, 89273g.

2e. Weaver, S. S.; Bodey, G. P.; LeBlanc, B. M. *Antimicrob. Agents Chemother.* **1979**, *15*, 518.

2f. Tally, F. P.; Jacobus, N. V.; Gorbach, S. L. *Antimicrob. Agents Chemother.* **1978**, *14*, 436.

3. See the following references for recent syntheses of (+)-thienamycin:

3a. Buynak, J. D.; Mathew, J.; Rao, M. N. *J. Chem. Soc., Chem. Commun.* **1986**, 941.

3b. Maruyama, H.; Shiozaki, M.; Hiraoka, T. *Bull. Chem. Soc. Jpn.* **1985**, *58*, 3264.

3c. Hart, D. J.; Ha, D-C. *Tetrahedron Lett.* **1985**, *26*, 5493.

3d. Chiba, T.; Nakai, T. *Tetrahedron Lett.* **1985**, *26*, 4647.

3e. Melillo, D. G.; Cvetovich, R. J.; Ryan, K. M.; Sletzinger, M. *J. Org. Chem.* **1986**, *51*, 1498.

3f. Georg, G. I.; Gill, H. S. *J. Chem. Soc., Chem. Commun.* **1985**, 1433.

3g. Maruyama, H.; Hiraoka, T. *J. Org. Chem.* **1986**, *51*, 399.

3h. Chiba, T.; Nagatsuma, M.; Nakai, T. *Chem. Lett.* **1985**, 1343.

3i. Iimori, T.; Shibasaki, M. *Tetrahedron Lett.* **1985**, *26*, 1523.

3j. Kametani, T.; Nagahara, T.; Honda, T. *J. Org. Chem.* **1985**, *50*, 2327.

4. Aumann, R.; Fröhlich, K.; Ring, H. *Angew. Chem. Int. Ed. Engl.* **1974**, *13*, 275.

5a. Annis, G. D.; Ley, S. V. *J. Chem. Soc., Chem. Commun.* **1977**, 581.

5b. Annis, G. D.; Ley, S. V.; Self, C. R.; Sivaramakrishnan, R. *J. Chem. Soc., Perkin Trans. 1* **1981**, 270,

6. Horton, A. M.; Hollinshead, D. M.; Ley, S. V. *Tetrahedron* **1984**, *40*, 1737.
7. Horton, A. M.; Ley, S. V. *J. Organometal. Chem.* **1985**, *285*, C17.
8. Annis, G. D.; Ley, S. V.; Sivaramakrishnan, R.; Atkinson, A. M.; Rogers, D.; Williams, D. J. *J. Organometal. Chem.* **1979**, *182*, C11.
9. Becker, Y.; Eisenstadt, A.; Shvo, Y. *Tetrahedron* **1976**, *32*, 2123.
10. Becker, Y.; Eisenstadt, A.; Shvo, Y. *Tetrahedron* **1974**, *30*, 839.
11. Annis, G. D.; Hebblethwaite, E. M.; Ley, S. V. *J. Chem. Soc., Chem. Commun.* **1980**, 297.
12. Annis, G. D.; Hebblethwaite, E. M.; Hodgson, S. T.; Hollinshead, D. M.; Ley, S. V. *J. Chem. Soc., Perkin Trans. 1* **1983**, 2851.
13. Hodgson, S. T.; Hollinshead, D. M.; Ley, S. V. *J. Chem. Soc., Chem. Commun.* **1984**, 494.
14. Hodgson, S. T.; Hollinshead, D. M.; Ley, S. V. *Tetrahedron* **1985**, *41*, 5871.
15. Propiolaldehyde can be prepared by oxidation of the alcohol with chromium trioxide: Sauer, J. C. *Org. Syn., Coll.* **1963**, *IV*, 813. The alcohol is commercially available and inexpensive (Aldrich 88–89 catalog, 1 kg for $17.30).
16. (*S*)-(−)-α-Methylbenzylamine is commercially available (Fluka 88–89 catalog, 50 mL, $d = 0.953$ g/mL, for $30.60).
17. Kametani, T.; Nagahara, T.; Ihara, M. *J. Chem. Soc., Perkin Trans. 1* **1981**, 3048.
18. Kametani, T.; Huang, S-P.; Yokohama, S.; Suzuki, Y.; Ihara, M. *J. Am. Chem. Soc.* **1980**, *102*, 2060.
19. Johnston, D. B. R.; Schmitt, S. M.; Bouffard, F. A.; Christensen, B. G. *J. Am. Chem. Soc.* **1978**, *100*, 313.

CHAPTER 6

The Tropane Alkaloids: *l*-Hyoscyamine and Atropine

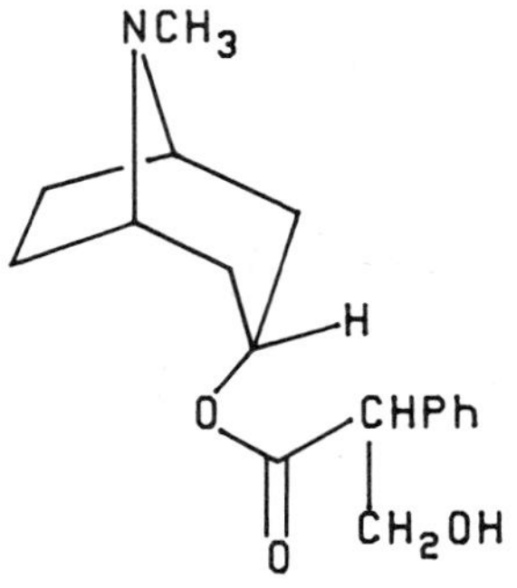

Atropine (dl), Hyoscyamine (l)

Benzeneacetic acid, α-(hydroxymethyl)-8-methyl-

8-azabicyclo[3.2.1]octan-3-yl ester, endo (±)

[51-55-8], [3(S) endo] [101-31-5]

Figure 6.1

l-Hyoscyamine, the most common of the naturally occurring tropane alkaloids, is often converted to the racemic form, atropine, in isolation procedures. Atropine was first isolated from the leaves of *Atropabelladonna* in 1833. The tropane alkaloids and atropine, in particular, have a broad spectrum of pharmacological activity. The most significant activity of atropine is parasympathetic inhibition, acting as a muscarinic receptor antagonist. These muscarinic receptors are involved in constriction of the pupil of the eye, vasodilation, slowing of the heart, and stimulation of secretion. Thus, clinical uses based on these effects are:

1. To induce mydriasis in eye examinations
2. To reduce salivary and bronchial secretions by smooth muscle relaxation of bronchi
3. To alter heart rate in the initial treatment of a myocardial infarction or high-grade atioventricular block

4. To reduce gastric secretion in treatment of peptic ulcers
5. To increase motility in the gastrointestinal tract as part of the treatment of severe infectious diarrhea in infants

Other tropane alkaloids include scopine, valerine or tropanediol, teloidine, and the infamous cocaine. Such broad and potent activity prompted the design and development of many versatile syntheses of the 8-azabicyclo[3.2.1]octan-3-ol skeleton[1] (see Figure 6.2).

6.1 IRON-STABILIZED OXALLYL CATIONS: PREPARATION

Noyori's synthesis of atropine is based on the formation of an iron-stabilized oxallyl cation by reaction of an α,α'-dibromoketone[2] with diiron nonacarbonyl.[3–6] The oxidative addition of an organic halide to a low-valent metal carbonyl results in an unstable organometallic intermediate. Simple aliphatic halides and aliphatic acid chlorides require highly reactive anionic metal complexes such as sodium cobalt tetracarbonyl or lithium acylmetal carbonylates. α,β-Unsaturated halides are activated; oxidative addition to neutral nickel carbonyl and iron pentacarbonyl is facile. The successful preparation of an iron carbonyl complex of trimethylenemethane[7] and

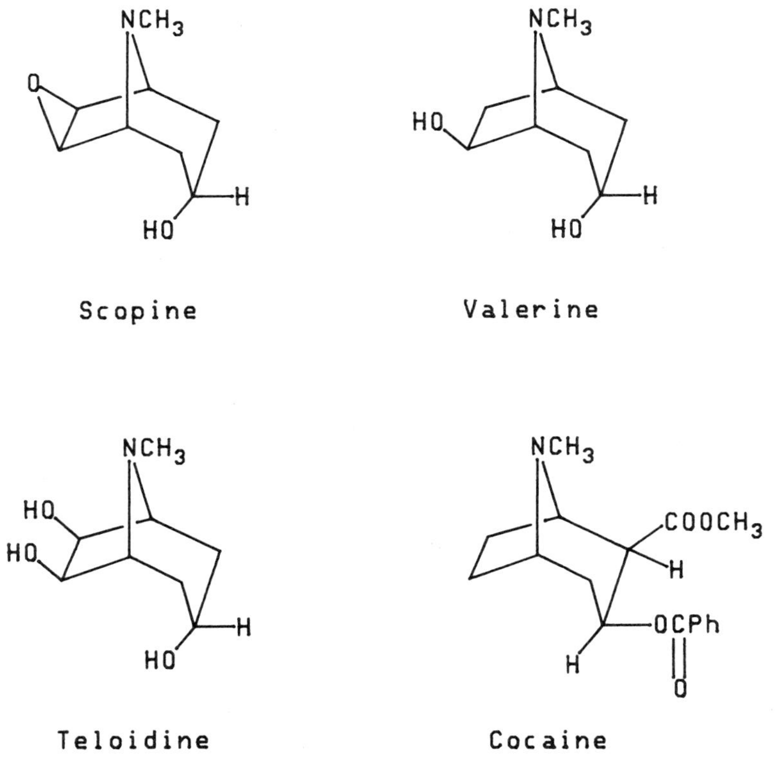

NCH_3 HO H — Scopine

NCH_3 HO HO H — Valerine

NCH_3 HO HO HO H — Teloidine

NCH_3 $COOCH_3$ H OCPh O H — Cocaine

Figure 6.2

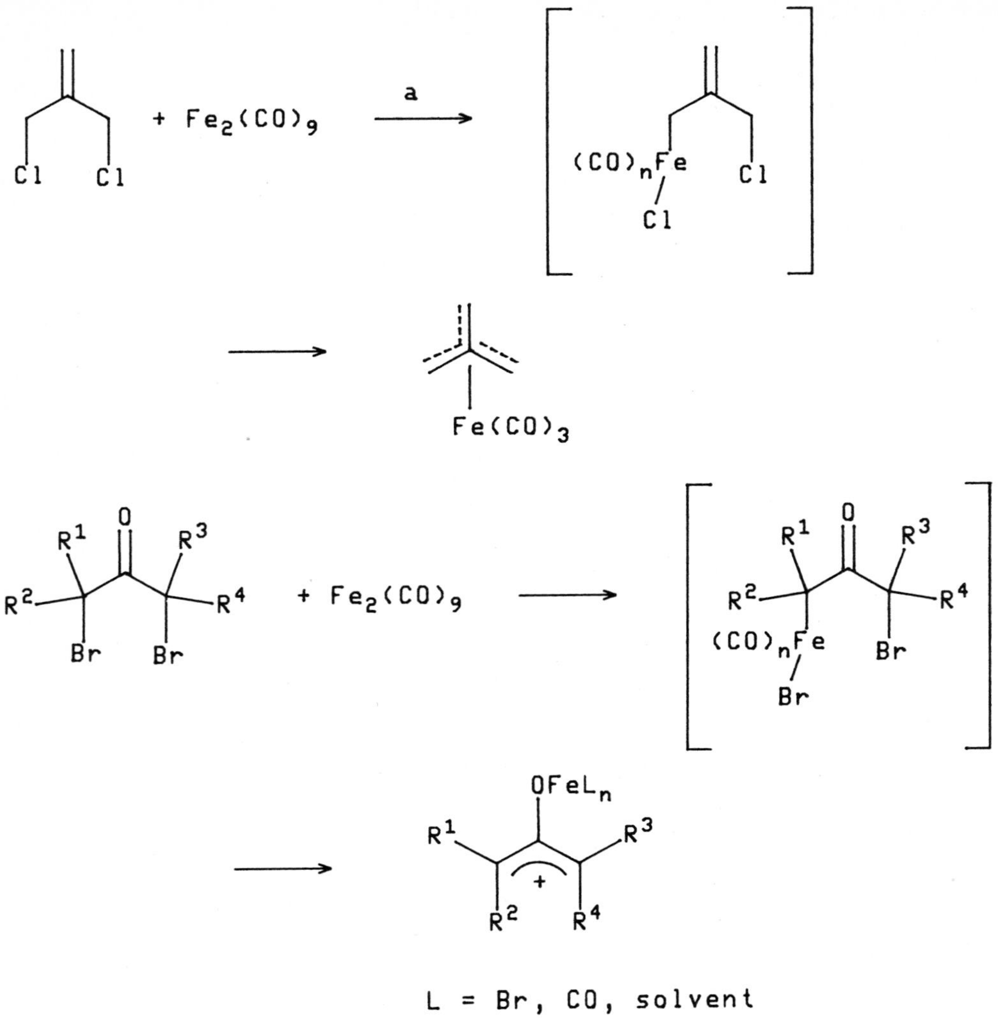

+ $Fe_2(CO)_9$ →(a) [$(CO)_nFe$, Cl] → $Fe(CO)_3$

+ $Fe_2(CO)_9$ → [$(CO)_nFe$, Br] → $OFeL_n$

L = Br, CO, solvent

a. rt, Et_2O; 30%

Figure 6.3

several diiron nonacarbonyl-mediated rearrangements of dibromoketones[8,9] led to investigation of methods for trapping iron-stabilized oxallyl cations with nuleophiles, alkenes, and dienes (see Figures 6.3 and 6.4).

6.2 TRAPPING BY NUCLEOPHILES

While reaction of 2,6-dibromo-2,6-diisopropylcyclohexanone with diiron nonacarbonyl in benzene, tetrahydrofuran, or *N*,*N*-dimethylformamide results in the enone as the

$+ Fe_2(CO)_9 \xrightarrow{a}$

70%

$+ Fe_2(CO)_9 \xrightarrow{a}$

95%

$+ Fe_2(CO)_9 \xrightarrow{a}$

78%

$+ Fe_2(CO)_9 \xrightarrow{a}$ +

80% 3%

a. rt, benzene

Figure 6.4

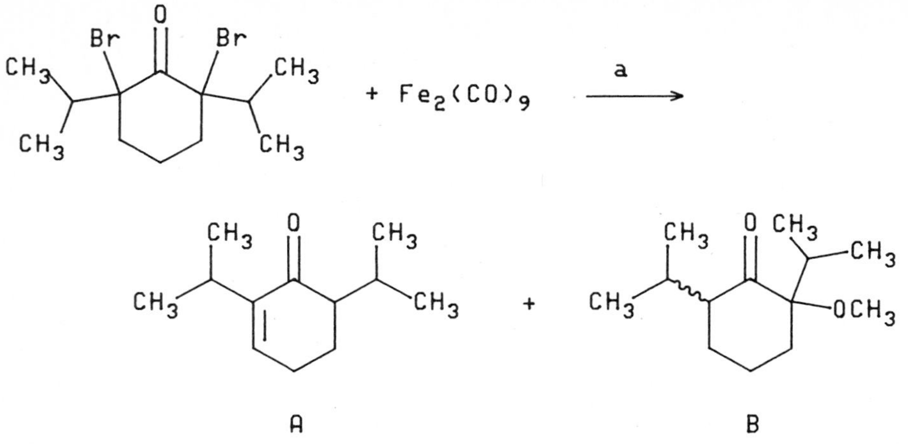

a. rt, CH_3OH; A:21%, B:61% (cis:trans, 69:31)

Figure 6.5

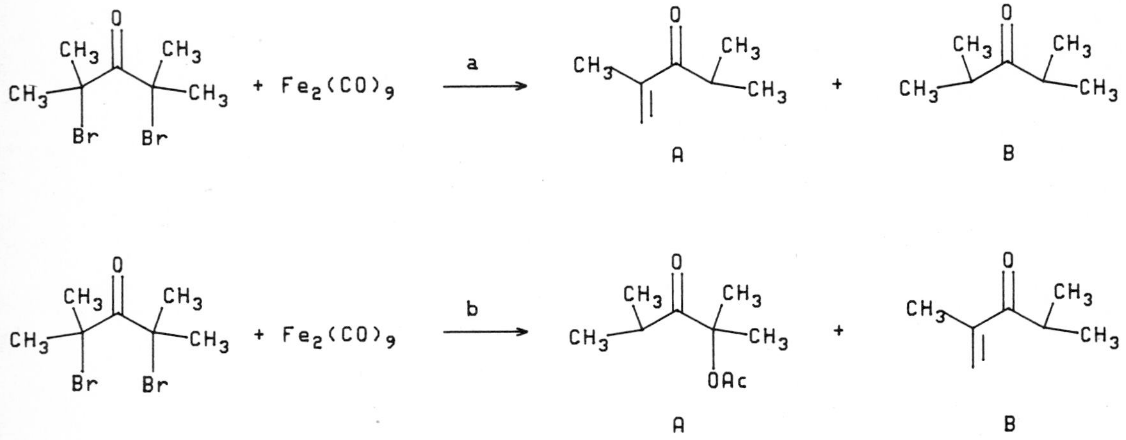

a. 30°C, DMF; A:80%, B:9%

b. 30°C, NaOAc, DMF; A:60%, B:20%

Figure 6.6

primary product, reaction in methanol affords the methoxy cyclohexanone (see Figure 6.5).

2,4-Dibromo-2,4-dimethyl-3-pentanone is converted to enone with diiron nonacarbonyl in *N*,*N*-dimethylformamide. When the reaction is carried out in the presence of a large excess of sodium acetate, the acetoxy-substituted ketone is the major product (see Figure 6.6).

a. CH_3OH; A:B, 90:10, 45% yield

b. NaOAc, DMF; A:B, 69:31, 46% yield

Figure 6.7

With unsymmetrical dibromoketones some measure of regioselectivity of attack by nucleophiles is possible. Attack predominates at the more substituted carbon, the carbon best able to accommodate a positive charge (or the carbon with the largest coefficient in the lowest unoccupied molecular orbital (see Figure 6.7).

6.3 TRAPPING BY ALKENES AND AMIDES: [3 + 2]-CYCLOADDITIONS

Iron-stabilized oxallyl cations generated in situ from α,α'-dibromoketones and diiron nonacarbonyl react with simple alkenes such as isobutylene to afford adducts typical of an ene reaction as well as [3 + 2]-adducts[10] (see Figure 6.8).

Cyclopentanones are efficiently produced using aromatic-substituted alkenes. The conversion is generally run in benzene at 50 to 60°C with the dibromide as limiting reagent (dibromide–alkene–diiron nonacarbonyl, 1 : 4 : 1.2).[11–13] General utility of the method is limited by the formation of open-chain 1 : 1 adducts and 2-alkylidenetetrahydrofurans in certain cases (see Figure 6.9).

Cyclopentanones are also produced en route to cyclopentenones using enamines. The conversion is generally run in benzene at room temperature with the dibromide as limiting reagent (dibromide–enamine–diiron nonacarbonyl, 1 : 2.5–3.0 : 1.2). The same process can also be carried out, although considerably less efficiently, using enol ethers[14–16] (see Figure 6.10).

3-Furanones are similarly produced by [3 + 2]-cycloaddition with an amide carbonyl followed by loss of amine. The reaction is run in amide as solvent at room

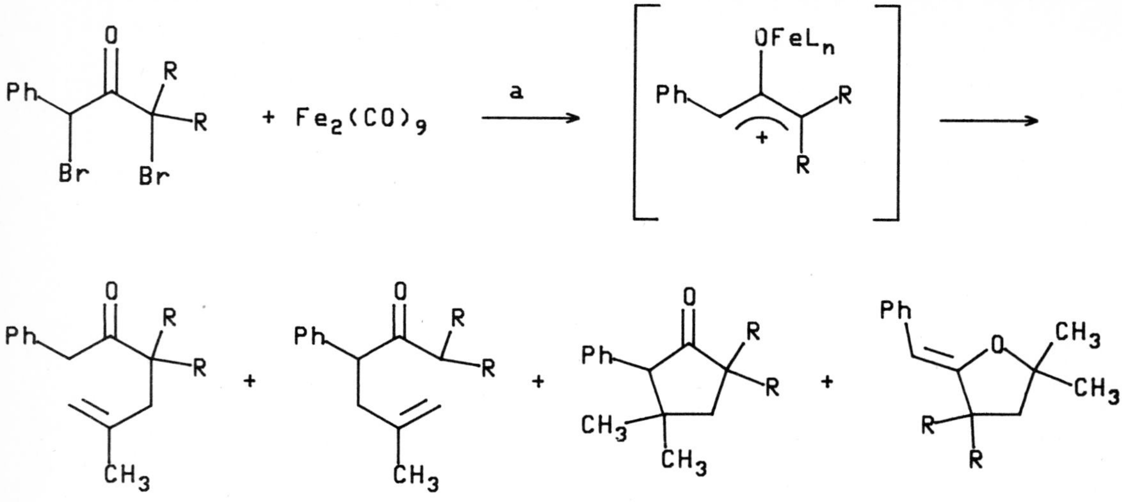

$R = CH_3$: A:B:C:D = 52:5:14:29, 35% yield

R = H: A:C = 9:91, 6% yield

a. 55°C, benzene, $CH_2{=}C(CH_3)_2$

Figure 6.8

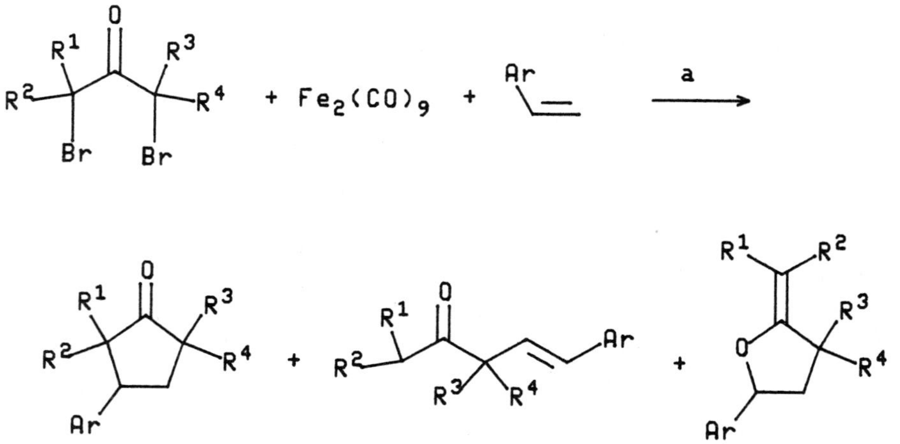

a. 50–60°C, benzene

Figure 6.9

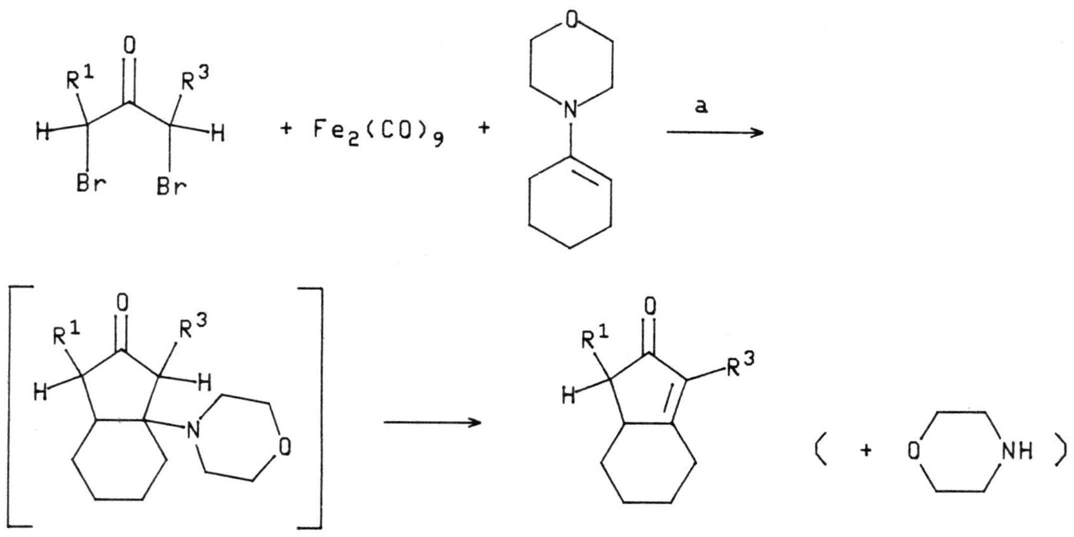

a. rt, benzene

Figure 6.10

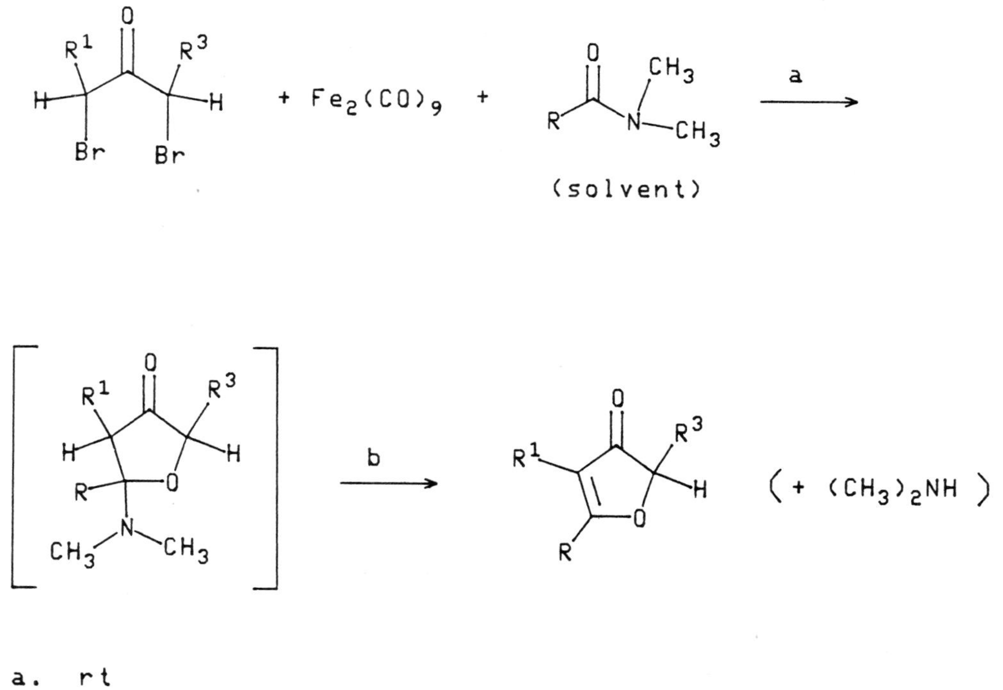

a. rt

b. rt or heat

Figure 6.11

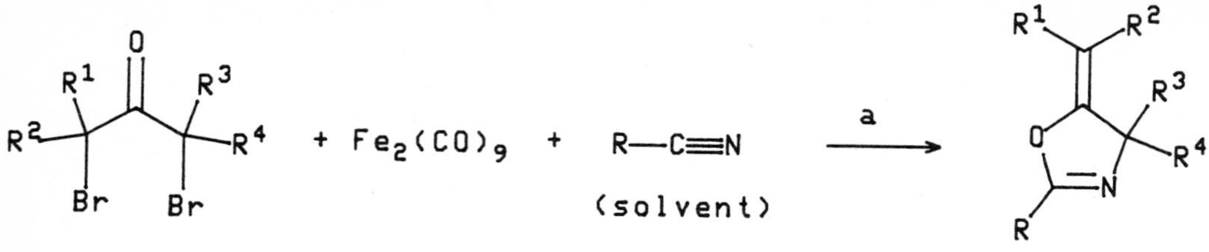

R1, R2, R3, R4-substituted dibromo ketone (Br, Br) + $Fe_2(CO)_9$ + R—C≡N (solvent) —a→ oxazoline product (R1, R2, R3, R4, O, N, R)

a. rt to 40°C

Figure 6.12

temperature with the dibromide as limiting reagent (dibromide–amide–diiron nonacarbonyl, 1–large excess–1.2). Amine elimination is considerably slower when the dibromide substituents are isopropyl or *t*-butyl; the isopropyl cycloadduct affords 3-furanone by brief heating at 110°C[17,18] (see Figure 6.11).

Finally, a few unpublished examples of [3 + 2]-cycloaddition using nitriles have appeared in a recent review.[5] The oxallyl cation provides a C—C—O, not C—C—C, fragment for the new ring. All [3 + 2]-cycloaddition data are collected in Table 6.1 (see also Figure 6.12).

These [3 + 2]-cycloadditions, formally $\pi^2 + \pi^2$ processes, are symmetry forbidden. Conversion is *assumed* stepwise via zwitterionic intermediates. Noyori has uncovered some evidence for the involvement of Z- and U-shaped zwitterionic intermediates.[19]

TABLE 6.1 [3 + 2]-Cycloadditions Using Iron-Stabilized Oxallyl Cations[3,5,12–18]

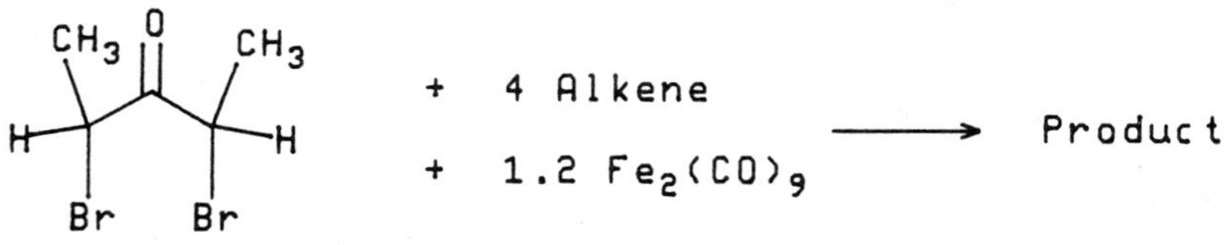

CH_3, H, Br / O / CH_3, H, Br dibromo ketone + 4 Alkene + 1.2 $Fe_2(CO)_9$ ⟶ Product

Solvent = Benzene

Temperature = 50-60°C

Alkene	Product	Yield (%)
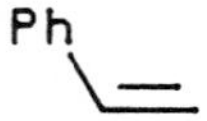 Ph (styrene)	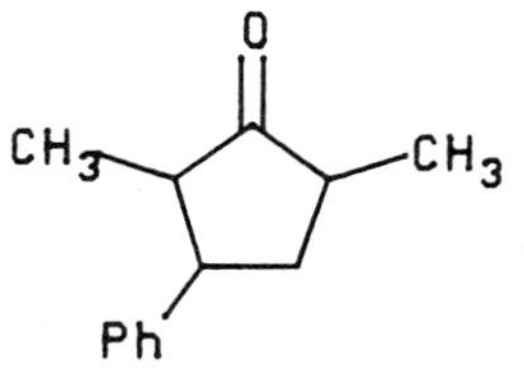 2,5-dimethyl-3-phenylcyclopentanone (O, CH_3, CH_3, Ph)	65

Alkene	Product	Yield (%)
Ph, Ph	O, CH_3, CH_3, Ph, Ph	70
Ph, CH_3	O, CH_3, CH_3, CH_3, Ph	70
Ph	O, CH_3, CH_3, Ph	95
Ph, Ph	O, CH_3, CH_3, Ph, Ph	30
	O, CH_3, CH_3	45

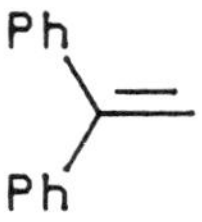

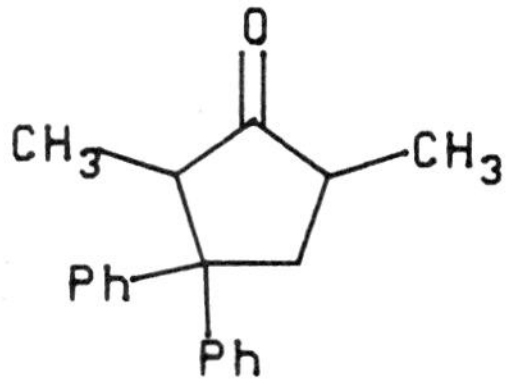

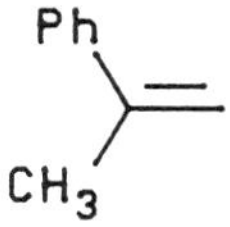

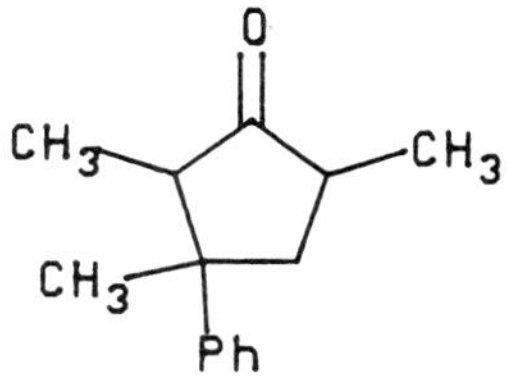

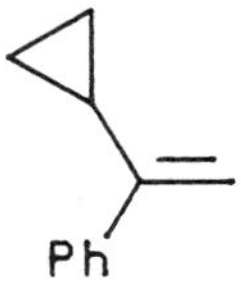

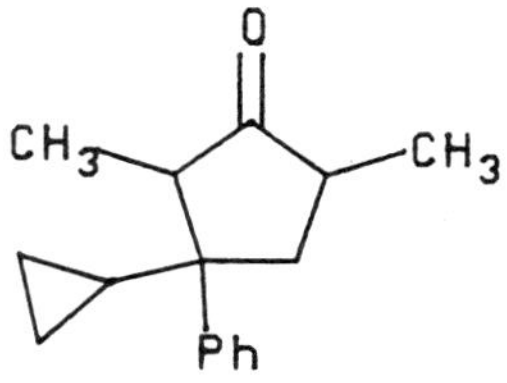

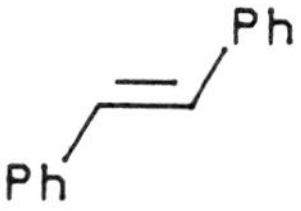

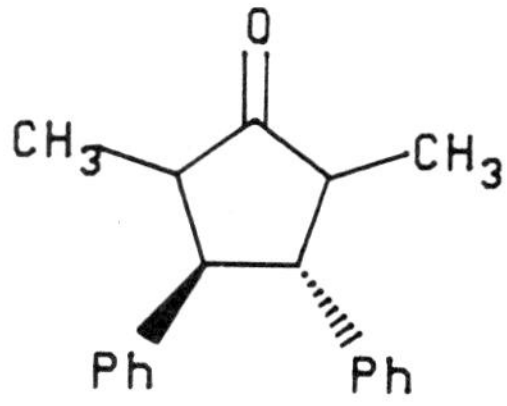

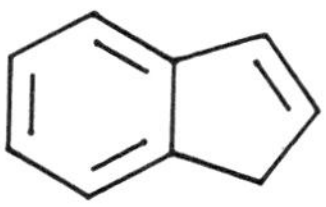

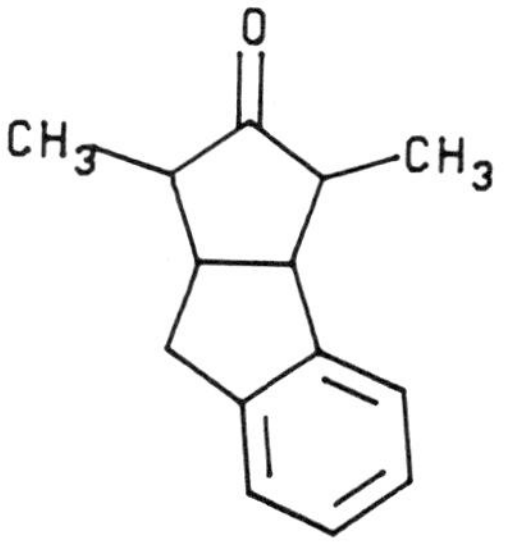

(*continued*)

TABLE 6.1 *(Continued)*

Alkene	Product	Yield (%)
Vinylferrocene	2,5-Dimethyl-3-ferrocenylcyclopentanone	30
Ph–CH=CH–CH$_3$	2,5-Dimethyl-3-phenyl-4-methylcyclopentanone (trans Ph/CH$_3$)	20

CH$_3$CH(CH$_3$)CHBrC(O)CHBrCH(CH$_3$)CH$_3$ + 4 Alkene + 1.2 $Fe_2(CO)_9$ ⟶ Product

Solvent = Benzene

Temperature = 50–60°C

Alkene	Product	Yield (%)
Ph$_2$C=CH$_2$	2,5-Diisopropyl-3,3-diphenylcyclopentanone	44

(CH$_3$)$_2$CBrC(O)CH$_2$Br + 4 Alkene + 1.4 $Fe_2(CO)_9$ ⟶ Product

Solvent = Benzene

Temperature = 50–60°C

Alkene	Product	Yield (%)
[structure]	[structure] +	18
	[structure]	1

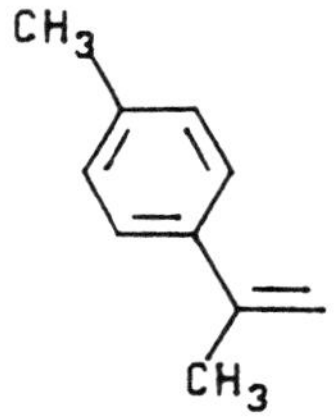

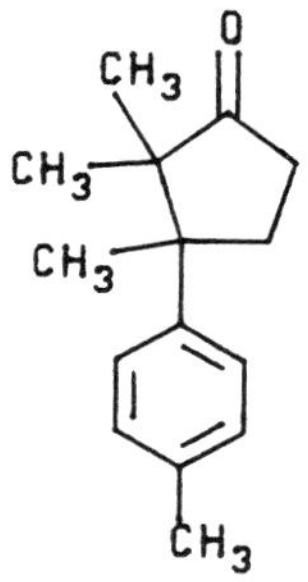

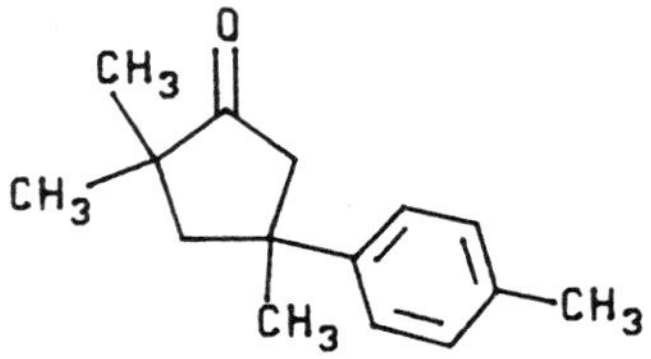

[Reaction scheme: 2,4-dibromo-2,4-dimethyl-3-pentanone] + 4 Alkene + 1.2 $Fe_2(CO)_9$ ⟶ Product

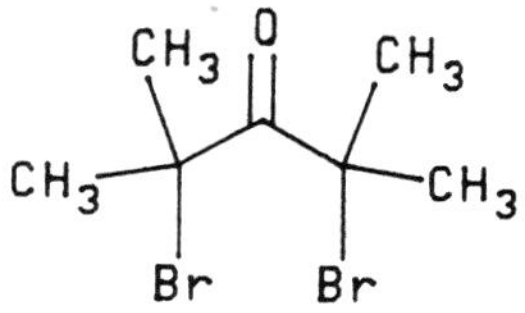

Solvent = Benzene

Temperature = 50–60°C

Alkene	Product	Yield (%)
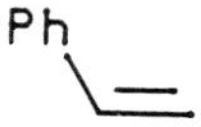 Ph (styrene)	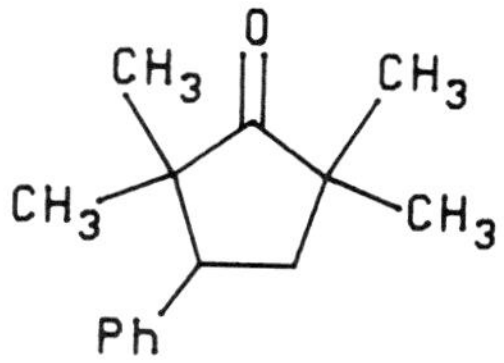 [structure]	23
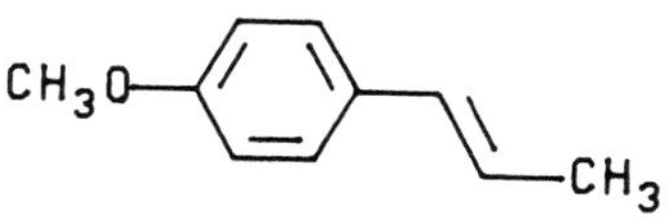 [structure]	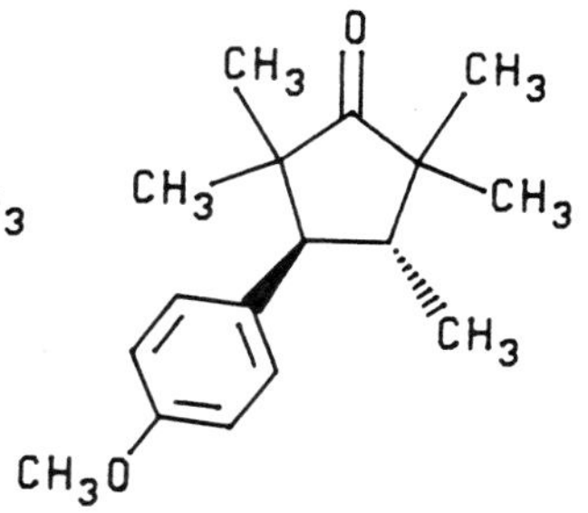 [structure]	55

(continued)

TABLE 6.1 (*Continued*)

Alkene	Product	Yield (%)
Ph(CH₃)C=CH₂	2,2,4,5,5-pentamethyl-4-phenylcyclopentanone	5

$CH_3CH(Br)C(O)CH(Br)CH_3$ + 2.5-3 Enamine + 1.2 $Fe_2(CO)_9$ ⟶ Product

Solvent = Benzene

NR_2 = morpholine

Temperature = 25-30°C

Enamine	Product	Yield (%)
$R_2N(H)C{=}C(CH_3)_2$	2,4,4,5-tetramethylcyclopent-2-enone	83
$R_2N(Ph)C{=}CH_2$	2,5-dimethyl-3-phenylcyclopent-2-enone	91
$R_2N(CH_3CH_2)C{=}CHCH_3$	2,5-dimethyl-3-ethyl-4-methylcyclopent-2-enone	73
1-(NR_2)cyclopentene	1,3-dimethyl bicyclic (cyclopenta-fused) cyclopentenone	75

Enamine	Product	Yield (%)
NR_2	O, CH_3, CH_3	100
NR_2	O, CH_3, CH_3	100
NR_2	O, CH_3, CH_3	90
NR_2	CH_3, O, CH_3	71
NR_2	CH_3, O, CH_3	65

(continued)

TABLE 6.1 (*Continued*)

3,5-Dibromo-4-heptanone ($CH_3CH_2CHBr-CO-CHBrCH_2CH_3$) + 2.5-3 Enamine + 1.2 $Fe_2(CO)_9$ ⟶ Product

NR_2 = morpholine

Solvent = Benzene

Temperature = 32°C

Enamine	Product	Yield (%)
R_2N, Ph	Ph	64
R_2N, CH_3, CH_3	CH_3, CH_3, CH_3, CH_3, O	70
NR_2	CH_3, CH_3, O	89

2,6-Dimethyl-3,5-dibromo-4-heptanone + 2.5-3 Enamine + 1.2 $Fe_2(CO)_9$ ⟶ Product

NR_2 = morpholine

Solvent = Benzene

Temperature = 32°C

Enamine	Product	Yield (%)
R_2N–C(Ph)=CH_2	2,5-diisopropyl-3-phenylcyclopent-2-enone	72
1-(NR_2)cyclohexene	2,5-diisopropyl bicyclic enone (fused cyclohexane)	77

$$PhCH(Br)C(=O)CH(Br)Ph \; + \; 2.5\text{-}3 \; \text{Enamine} \; + \; 1.2 \; Fe_2(CO)_9 \longrightarrow \text{Product}$$

Solvent = Benzene

NR_2 = morpholine

Temperature = 32°C

Enamine	Product	Yield (%)
1-(NR_2)cyclopentene	2,5-diphenyl bicyclic enone (fused cyclopentane)	66

$$CH_3CH(Br)C(=O)CH(Br)CH_3 \; + \; \text{Amide (solvent)} \; + \; 1.2 \; Fe_2(CO)_9 \longrightarrow \text{Product}$$

Temperature = 20-25°C

(continued)

TABLE 6.1 *(Continued)*

Amide	Product	Yield (%)
$HC(O)N(CH_3)_2$	2,4-dimethyl-3(2H)-furanone (CH_3, CH_3, H)	53
$CH_3C(O)N(CH_3)_2$	2,4,5-trimethyl-3(2H)-furanone (CH_3, CH_3, CH_3)	21*
$PhC(O)N(CH_3)_2$	2,4-dimethyl-5-phenyl-3(2H)-furanone (CH_3, CH_3, Ph)	25*

* NaH_2 edta added

$CH_3CH_2CH(Br)C(O)CH(Br)CH_2CH_3$ + Amide (solvent) + 1.2 $Fe_2(CO)_9$ ⟶ Product

Temperature = 20–25°C

Amide	Product	Yield (%)
$HC(O)N(CH_3)_2$	2,4-diethyl-3(2H)-furanone (CH_2CH_3, CH_2CH_3, H)	64

Amide	Product	Yield (%)
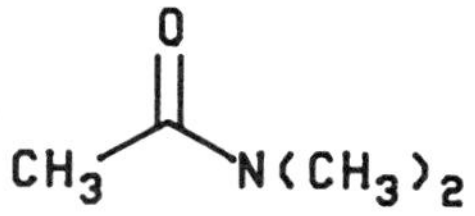 $CH_3C(O)N(CH_3)_2$	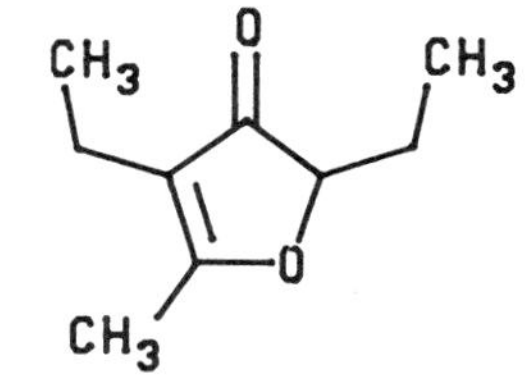 CH_3, O, CH_3, O, CH_3	51*
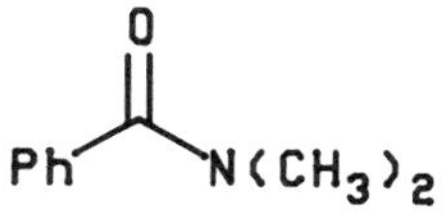 $PhC(O)N(CH_3)_2$	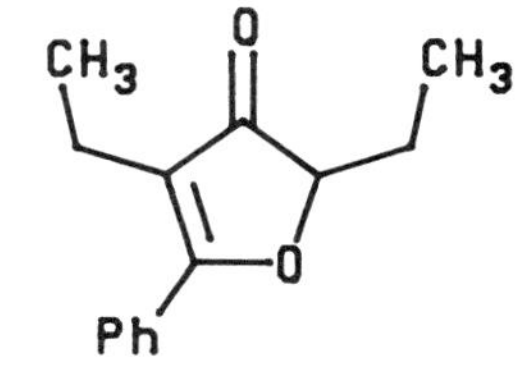 CH_3, O, CH_3, O, Ph	42*

* NaH_2 edta added

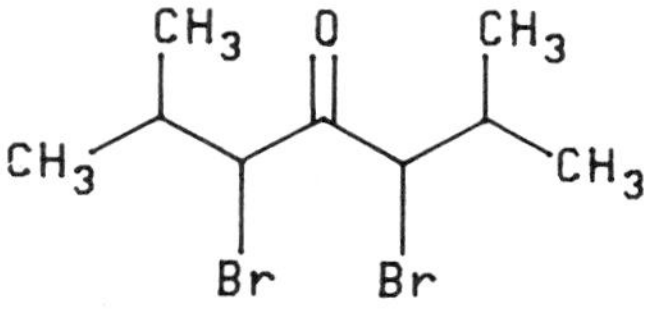

$CH_3CH(CH_3)CH(Br)C(O)CH(Br)CH(CH_3)CH_3$ + Amide (solvent) + 1.2 $Fe_2(CO)_9$ $\longrightarrow$ Product

Temperature = 20-25°C

Amide	Product	Yield (%)
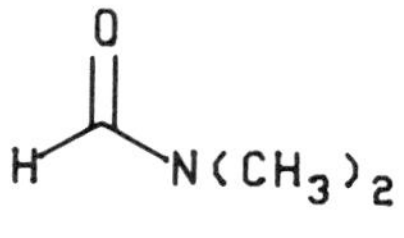 $HC(O)N(CH_3)_2$	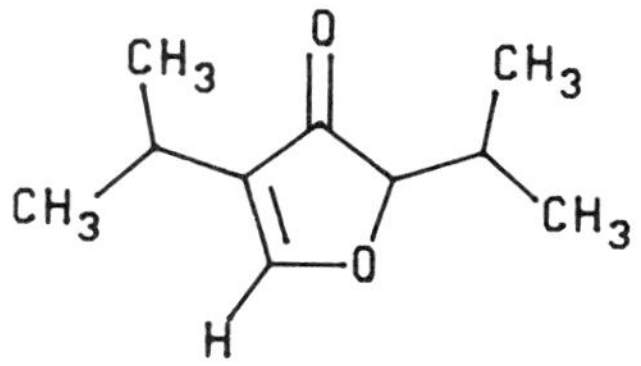 CH_3, CH_3, O, CH_3, CH_3, O, H	90*
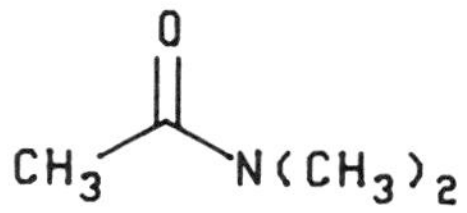 $CH_3C(O)N(CH_3)_2$	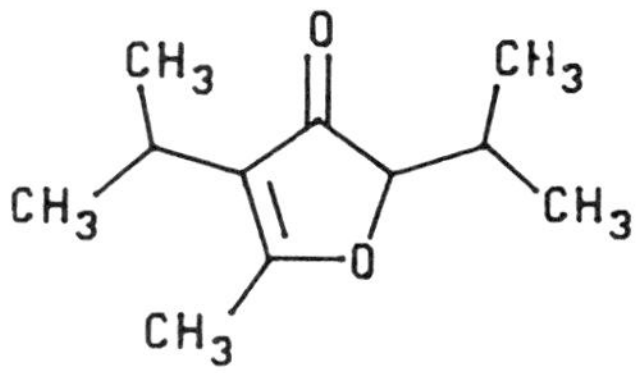 CH_3, CH_3, O, CH_3, CH_3, O, CH_3	87*

*Requires heating to 110°C after ring formation to complete amine elimination

(continued)

TABLE 6.1 (*Continued*)

Amide	Product	Yield (%)
O, Ph, $N(CH_3)_2$	CH_3, O, CH_3, CH_3, CH_3, O, Ph	83
O, N, CH_3	CH_3, O, CH_3, CH_3, CH_3, O, CH_3N, H	26

CH_3 O CH_3 / CH_3 CH_3 / CH_3 CH_3 / Br Br + Amide (solvent) + 1.2 $Fe_2(CO)_9$ ⟶ Product

Temperature = 20–25°C

Amide	Product	Yield (%)
O, H, $N(CH_3)_2$	O, CH_3, CH_3, CH_3, CH_3, CH_3, CH_3, H, O, $N(CH_3)_2$	98

CH_3 O CH_3 / CH_3 CH_3 / Br Br + Amide (solvent) + 1.2 $Fe_2(CO)_9$ ⟶ Product

Temperature = 20–25°C

Amide	Product	Yield (%)
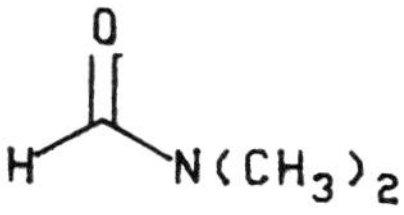 $HC(O)N(CH_3)_2$	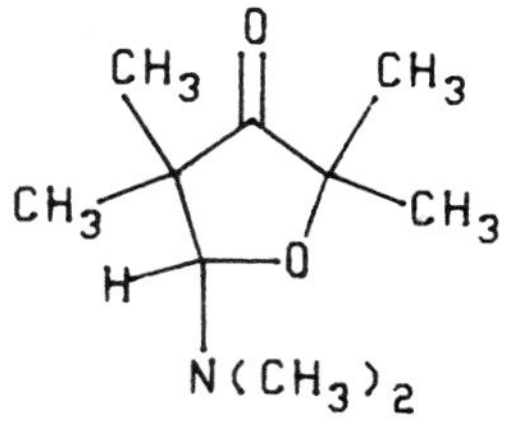 [structure]	3

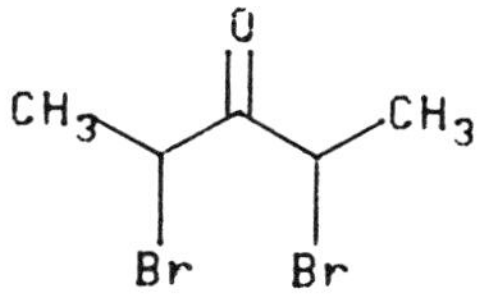

$CH_3CH(Br)C(O)CH(Br)CH_3$ + Enol Ether or Nitrile + 1.2 $Fe_2(CO)_9$ ⟶ Product

Solvent = benzene

Temperature = rt

Enol Ether or Nitrile	Product	Yield (%)
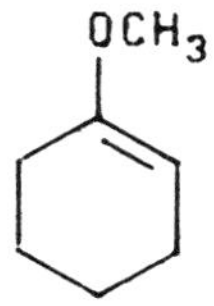 [structure]	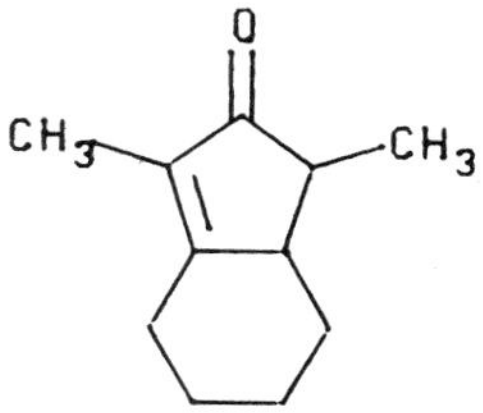 [structure]	12
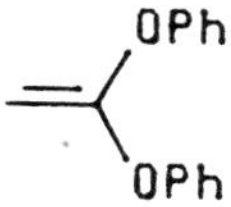 [structure]	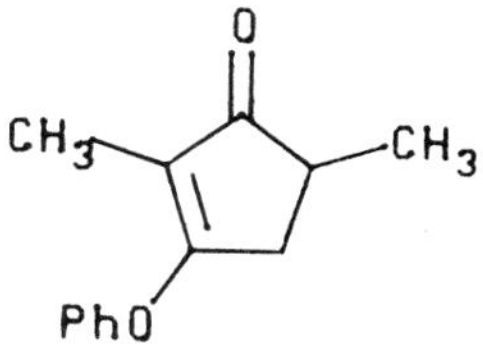 [structure]	8

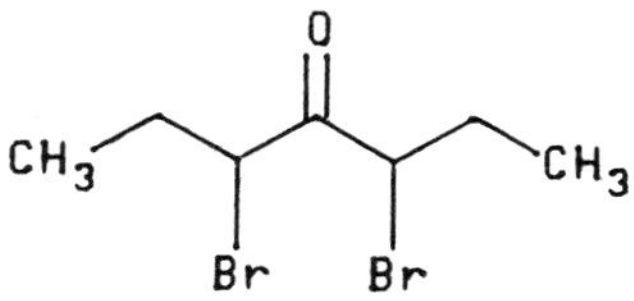

$CH_3CH_2CH(Br)C(O)CH(Br)CH_2CH_3$ + Enol Ether or Nitrile + 1.2 $Fe_2(CO)_9$ ⟶ Product

Solvent = benzene

Temperature = 40°C

(continued)

TABLE 6.1 (*Continued*)

Enol Ether or Nitrile	Product	Yield (%)
CH_3CN (excess)	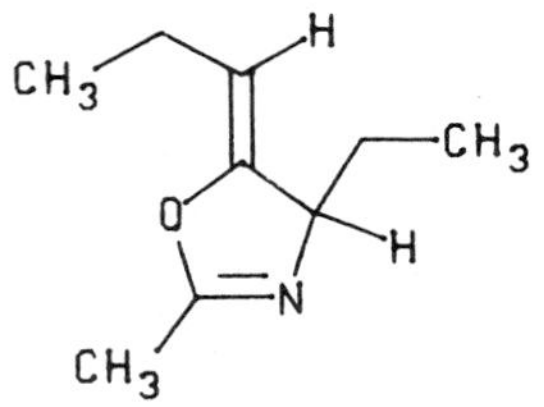 [structure: 2-methyl-4-ethyl-5-ethylidene-oxazoline; labels CH3, H, CH3, H, O, N, CH3]	44

$CH_3C(Br)(CH_3)C(=O)C(Br)(CH_3)CH_3$ + Enol Ether or Nitrile + 1.2 $Fe_2(CO)_9$ ⟶ Product

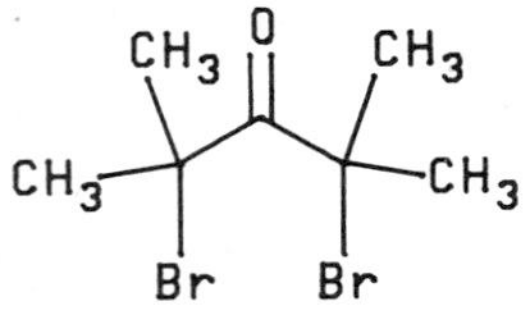

Solvent = benzene

Temperature = rt

Enol Ether or Nitrile	Product	Yield (%)
CH_3CN (excess)	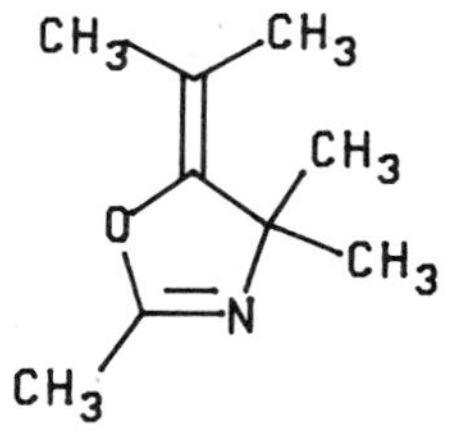 [structure: 2,4,4-trimethyl-5-isopropylidene-oxazoline; labels CH3, CH3, CH3, CH3, O, N, CH3]	50

The oxallyl cation from 2,4-dibromo-2,4-dimethyl-3-pentanone and diiron nonacarbonyl cycloadds *Z*-α-deuteriostyrene to afford a cyclopentanone retaining the relationship between phenyl and deuterium. Also produced is a 1 : 1 open-chain adduct of E-alkene configuration (see Figure 6.13).

Stereospecific formation of the cyclopentanone involves the U-shaped zwitterion. A strong attractive interaction between the enolate and carbocation prevents rotation about the Ph—C—C bond prior to closing the ring. Formation of the 1 : 1 open-chain adduct involves the Z-shaped zwitterion, which is able to rotate freely about the Ph—C—C bond. A 60° rotation results in retention of deuterium adjacent to phenyl. A 120° rotation results in relocation of deuterium. Other rotamers, less stable because phenyl and R are in close proximity, are not observed (see Figures 6.14 and 6.15).

Regioselectivity in the [3 + 2]-cycloaddition of an unsymmetrical oxallyl cation and unsymmetrical alkene is the basis of a synthesis of cuparane-type sesquiterpenes.[12]

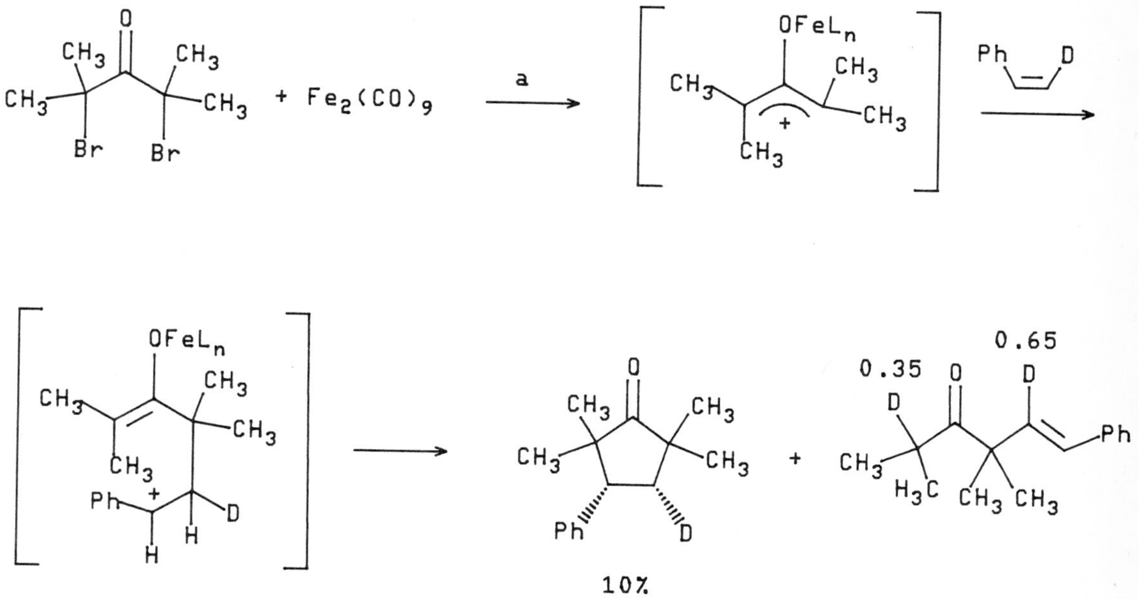

a. 55-60°C, benzene

Figure 6.13

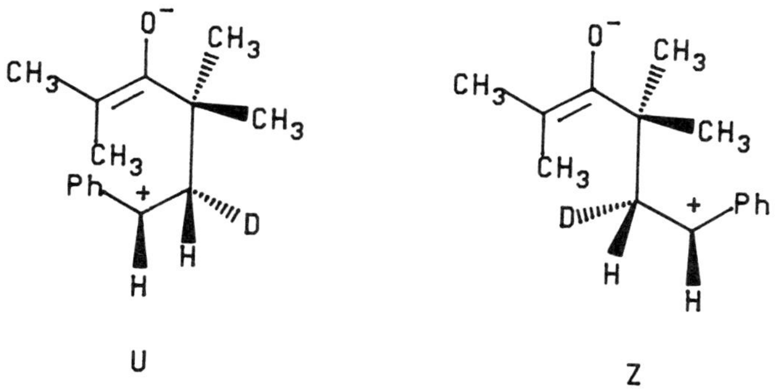

Figure 6.14

The major product is formed from the most stable zwitterionic intermediate[20] (see Figure 6.16 and Table 6.2).

The [3 + 2]-cycloaddition with amides was applied to muscarine alkaloid synthesis.[21] The adduct from 2,4-dibromo-3-pentanone and *N*,*N*-dimethylformamide is acetylated using acetyl chloride in 1,2-dimethoxyethane. A dimethylaminomethyl group is introduced at the 5-position via the Mannich reaction. Hydrolysis of the

Figure 6.15

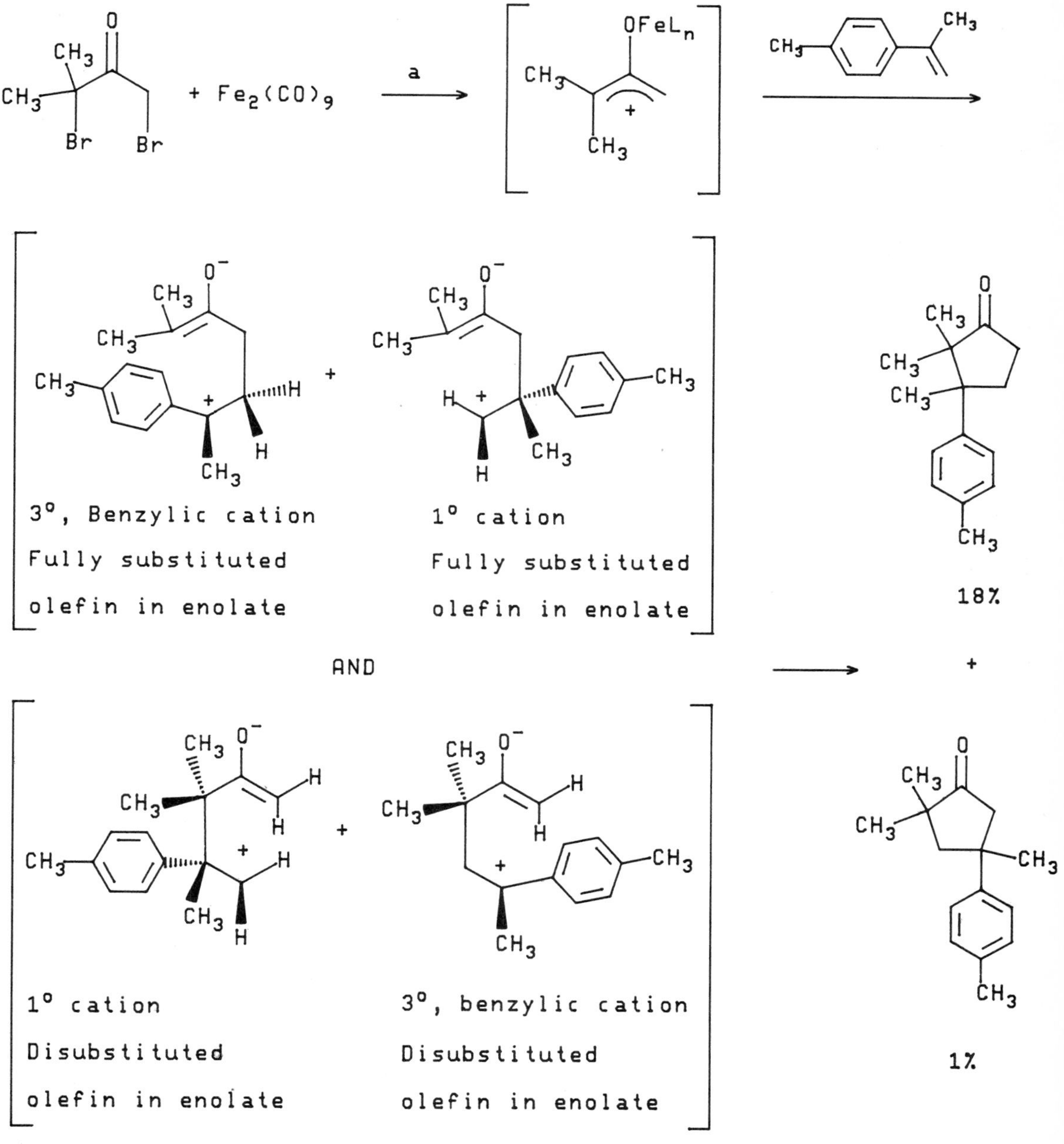

CH_3 CH_3 Br Br O + $Fe_2(CO)_9$ —a→ [CH_3 $OFeL_n$ + CH_3] CH_3 CH_3 →

[CH_3 CH_3 O^- CH_3 + H H CH_3

3°, Benzylic cation

Fully substituted

olefin in enolate

+ CH_3 CH_3 O^- H + CH_3 CH_3 H

1° cation

Fully substituted

olefin in enolate]

CH_3 CH_3 CH_3 O CH_3

18%

AND →

+

[CH_3 CH_3 O^- H H CH_3 + H CH_3 H

1° cation

Disubstituted

olefin in enolate

+ CH_3 CH_3 O^- H H + CH_3 CH_3

3°, benzylic cation

Disubstituted

olefin in enolate]

CH_3 CH_3 O CH_3 CH_3

1%

a. 55-60°C, benzene

Figure 6.16

TABLE 6.2 **Regioselectivity in [3 + 2]-Cycloadditions of Unsymmetrical Iron-Stabilized Oxallyl Cations and Unsymmetrical Alkenes**[20]

Dibromide + Alkene ⟶ Product A + Product B

Dibromide	Alkene	Product A	Product B	A:B	Yield (%)
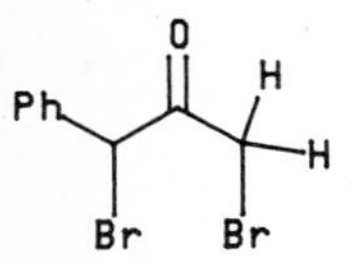 PhCHBr–CO–CH₂Br	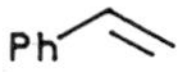 PhCH=CH₂	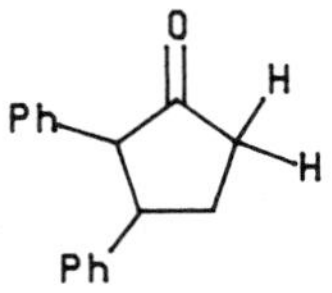	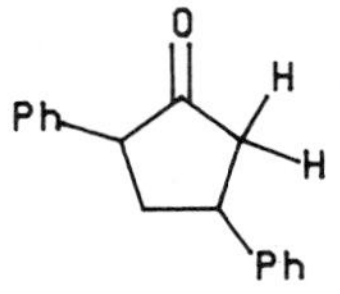	100:0	44
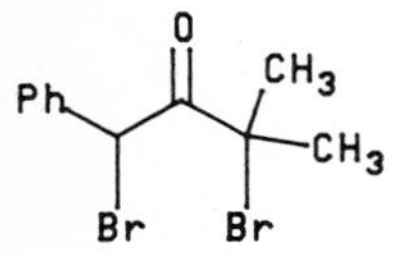 PhCHBr–CO–CBr(CH₃)₂	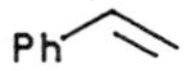 PhCH=CH₂	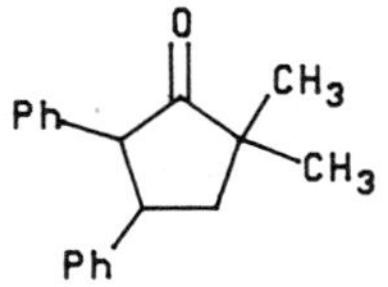	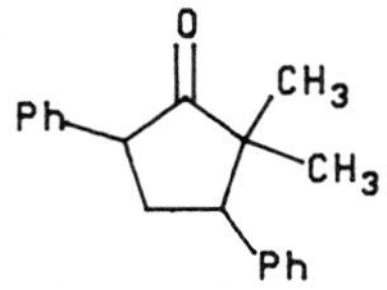	100:0	49
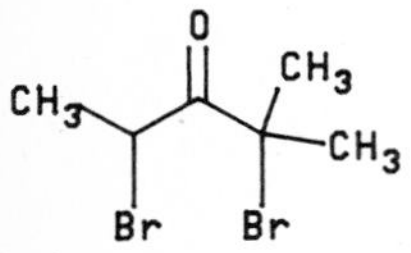 CH₃CHBr–CO–CBr(CH₃)₂	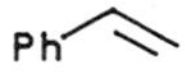 PhCH=CH₂	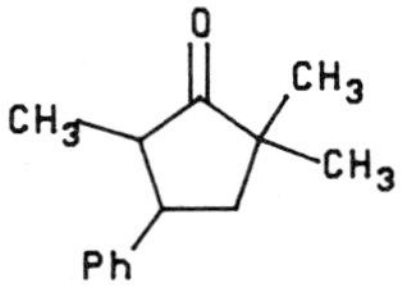	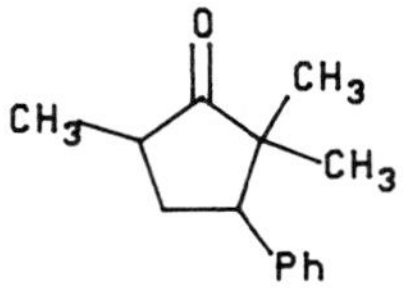	6:94	31

Dibromide + Enamine ⟶ Product A + Product B

Dibromide	Enamine	Product A	Product B	A:B	Yield (%)
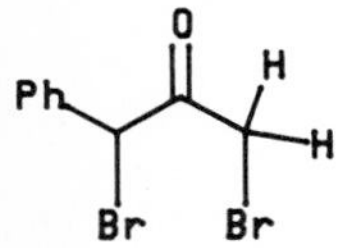	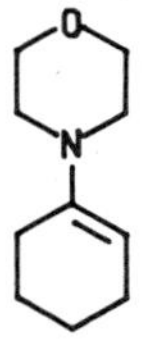	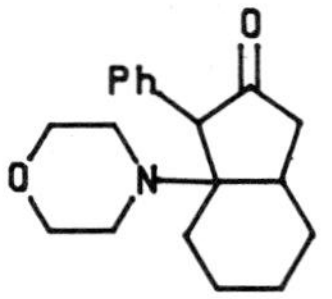	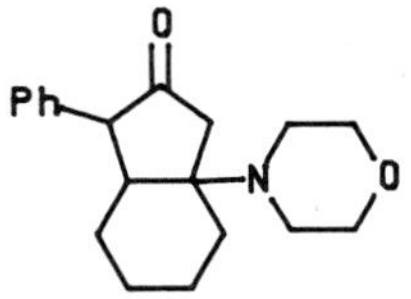	76:24	22
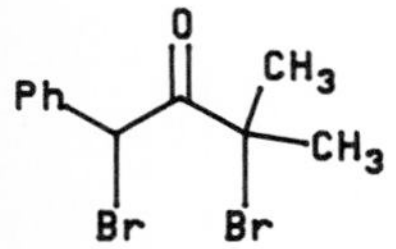	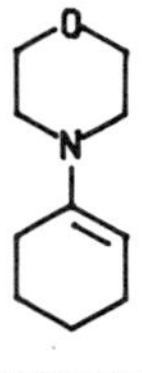	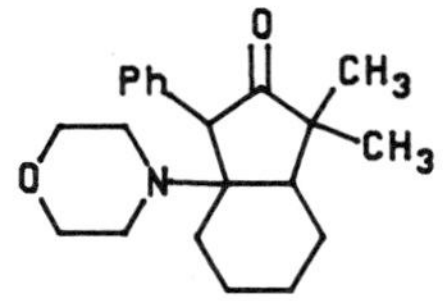	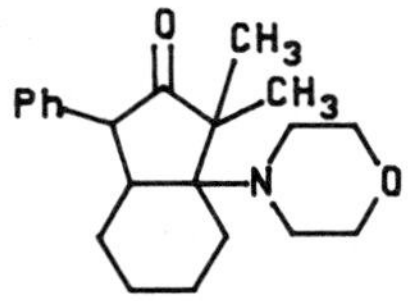	75:25	61
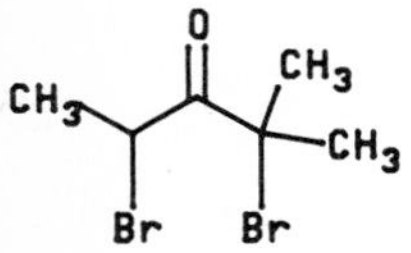	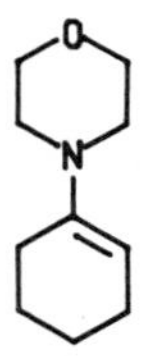	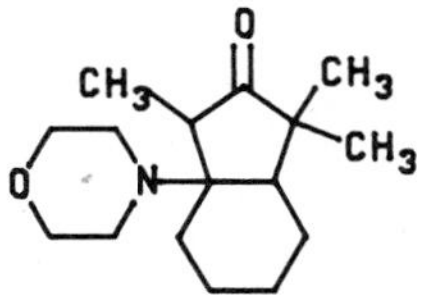	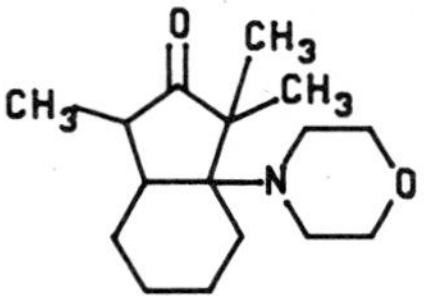	41:59	86

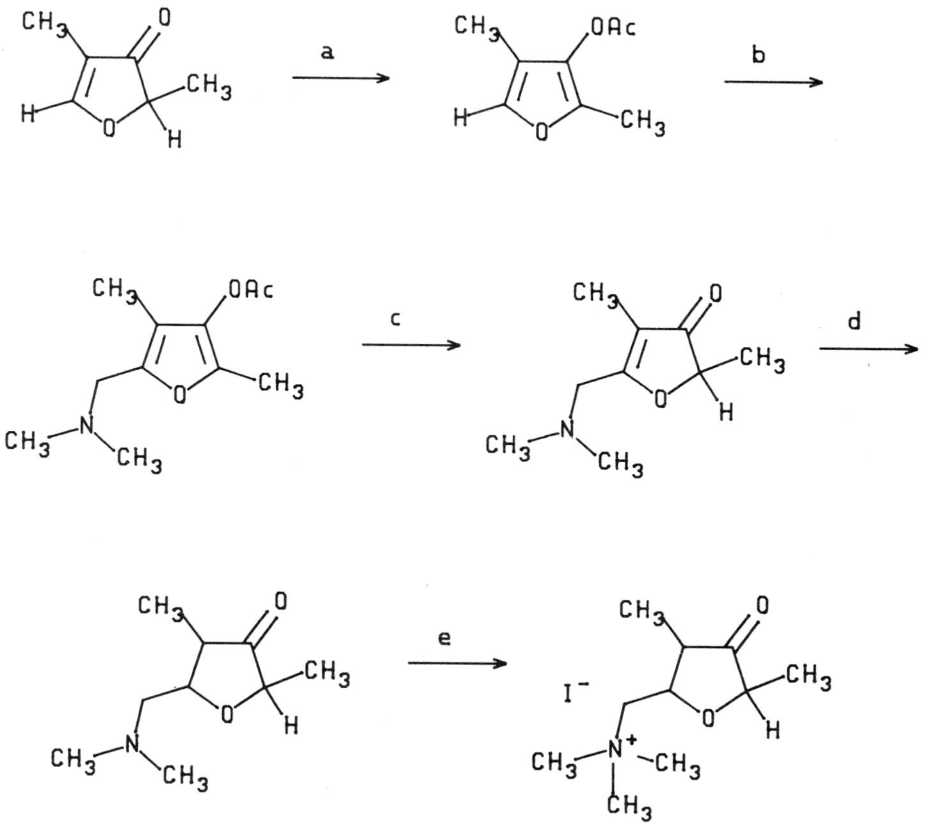

a. 55°C, $CH_3C(O)Cl$, DME; 70%

b. 70°C, CH_2O, $(CH_3)_2NH$, CH_3COOH; 80%

c. 50°C, 70% $HClO_4$, DME; 80%

d. −35°C, Li, CH_3OH, NH_3; 60%

e. CH_3I

Figure 6.17

labile acetate followed by Birch reduction and quaternization with methyl iodide affords two isomeric 4-methylmuscarine iodides[17] (see Figure 6.17).

Biosynthesis of camphor from geraniol may proceed via a carbocation-alkene [3 + 2]-cyclization. A laboratory synthesis of (±)-camphor involves conversion of geraniol to an α,α′-dibromoketone, then a similar intramolecular [3 + 2]-cycloaddition[22] (see Figures 6.18 and 6.19). The intermediate zwitterion is far less stable

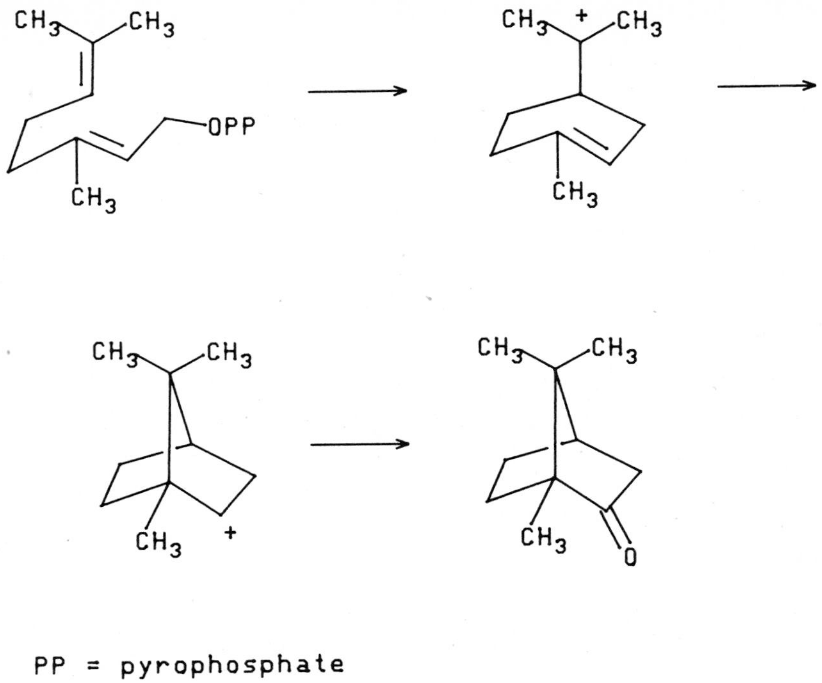

PP = pyrophosphate

Figure 6.18

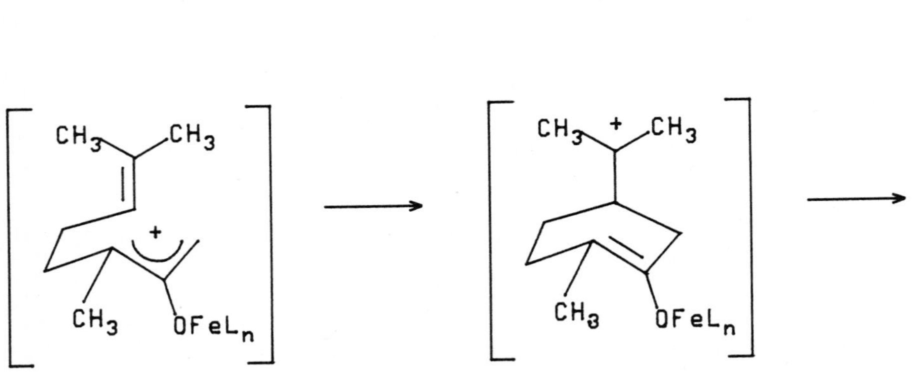

a

Figure 6.19

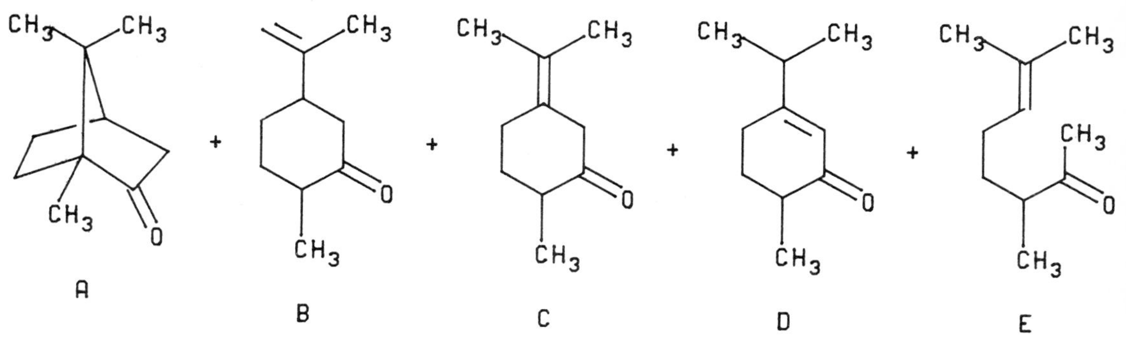

a. 100–110°C, $Fe_2(CO)_9$, benzene;

A:B:C:D:E = 54:20:4:10:7, 70% yield

Figure 6.19 (*Continued*)

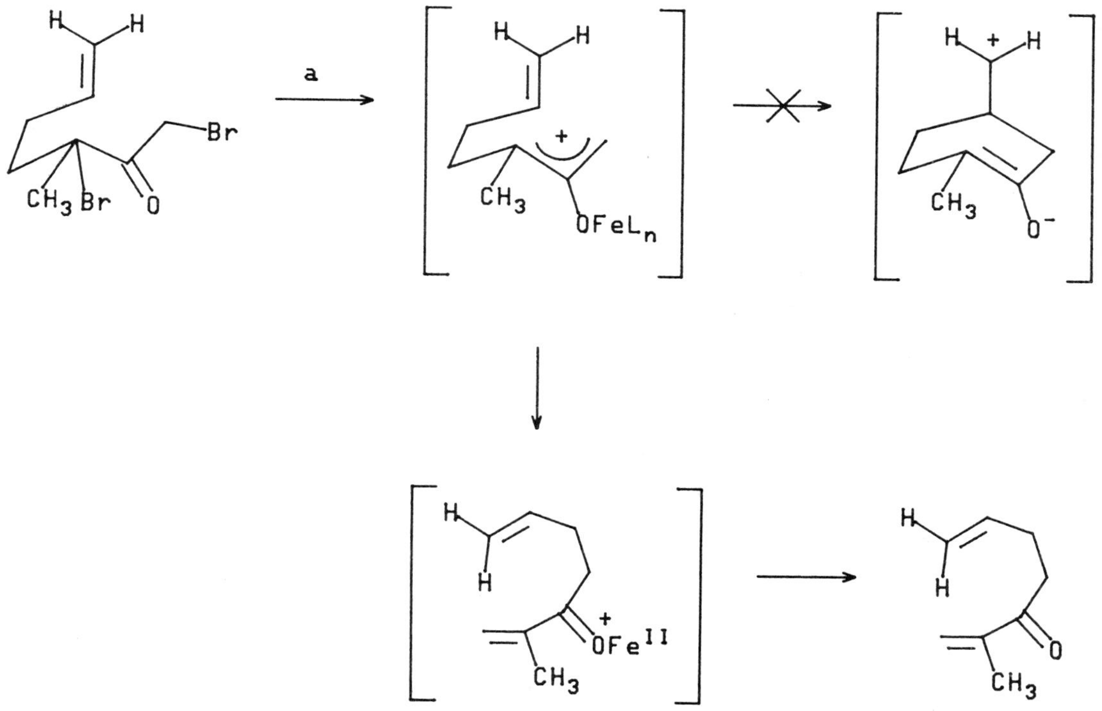

a. 80°C, $Fe_2(CO)_9$, benzene; 80% (glc)

Figure 6.20

when the alkene is monosubstituted. In this case the only isolable product is an enedione from a unique cationic [3 + 4]-sigmatropic rearrangement[23] (see Figure 6.20).

6.4 TRAPPING BY DIENES, FURANS, AND PYRROLES: [3 + 4]-CYCLOADDITIONS

Trapping of the oxallyl cation intermediate with dienes, furans, and certain pyrroles is a convenient route to several versatile seven-membered ring ketones. The cycloaddition with diene can be run (1) in benzene at 60–80°C using dibromide, diene, and diiron nonacarbonyl with dibromide as limiting reagent (1–large excess–1.2); (2) in benzene at room temperature with irradiation using dibromide, diene, and diiron nonacarbonyl with dibromide as limiting reagent (1–large excess–1.2); and (3) in benzene at 80 to 120°C using dibromide and η^4-diene iron tricarbonyl complex (1 : 1.3). The third method is apparently the most efficient. There are two limitations: (1) with acyclic dienes, the diene must exist to a considerable extent in the *S-cis* configuration and (2) the α,α'-dibromoketone must have at least one alkyl or aryl or a second halogen substituent on both the α and α' carbons[24–26] (see Figure 6.21).

Unsuccessful [3 + 4]-cycloaddition using α,α'-dibromoacetone is attributed to the high reactivity of the intermediate unsubstituted iron enolate and the instability of the unsubstituted oxallyl cation. Increasing steric requirements by substitution with alkyl groups decreases the enolate reactivity. Substitution with alkyl groups also increases stability of the oxallyl cation by increasing steric requirements and by inductive electron release. An unsubstituted cycloadduct can be prepared using $\alpha,\alpha,\alpha',\alpha'$-tetrabromoacetone since (1) a bromine substituent has the same stabilizing effect and (2) after cycloaddition, the bromine can be efficiently removed using Zn–Cu couple.

Trapping of the oxallyl cation with furan is run in refluxing furan (32°C) using dibromide and diiron nonacarbonyl (1 : 1.2). With relatively expensive substituted furans, the experiment is run in benzene at 60 to 80°C using dibromide and diiron

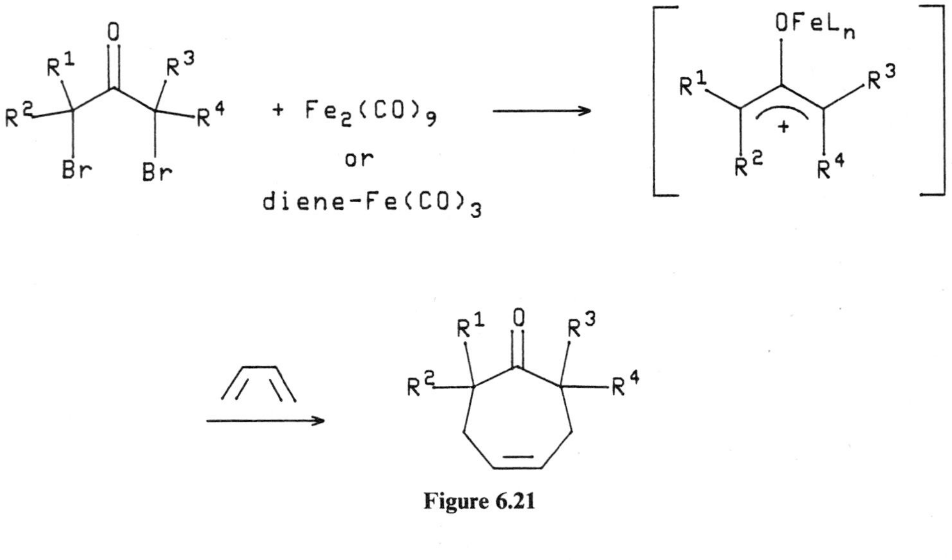

Figure 6.21

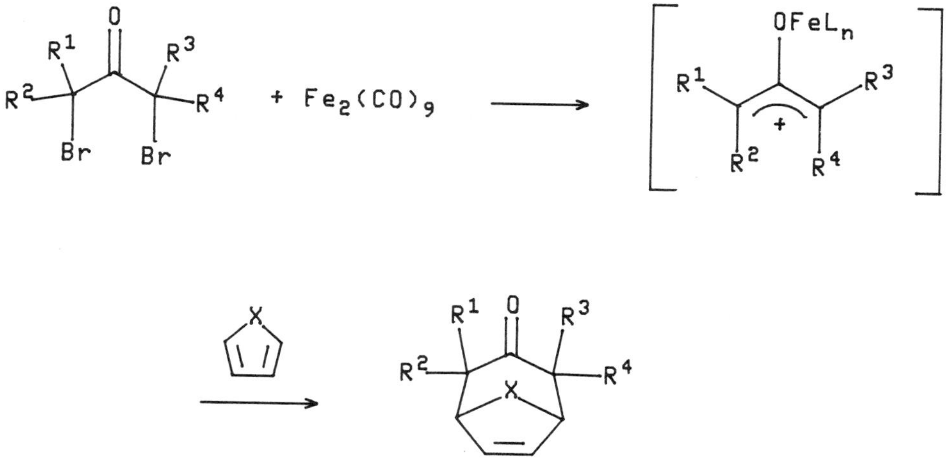

X = O, $NCOOCH_3$, $NCOCH_3$

Figure 6.22

nonacarbonyl with the furan as limiting reagent (dibromide–substituted furan–diiron nonacarbonyl, 2.5 : 1 : 1.2).[27–29] While the oxallyl cation undergoes nucleophilic addition with *N*-methylpyrrole, *N*-methoxycarbonylpyrroles and *N*-acetylpyrroles are trapped by [3 + 4]-cycloaddition. These reactions are run in benzene at 40 to 50°C using dibromide, substituted pyrrole, and diiron nonacarbonyl (3 : 1 : 1.5)[30,31] (see Figure 6.22). Data on [3 + 4]-cycloadditions are collected in Table 6.3.

The symmetry-forbidden [3 + 2]-cycloadditions ($\pi^2 + \pi^2$) proceed stepwise with regioselectivity best interpreted in terms of stability of zwitterionic intermediates.[32]

TABLE 6.3 [3 + 4]-Cycloadditions Using Iron-Stabilized Oxallyl Cations[24–30,37]

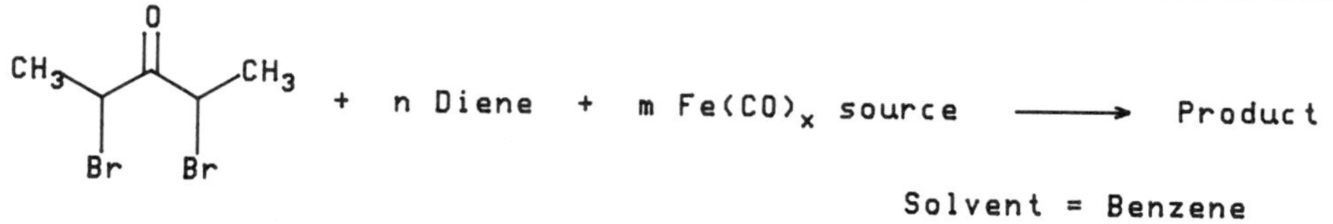

$CH_3CH(Br)C(O)CH(Br)CH_3$ + n Diene + m $Fe(CO)_x$ source ⟶ Product

Solvent = Benzene

Diene (n)	$Fe(CO)_x$ Source (m)	Temperature (°C)	Product	Yield (%)
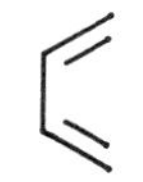 butadiene (excess)	$Fe_2(CO)_9$ (1.2)	60	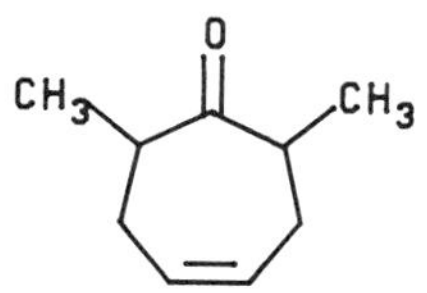 2,7-dimethylcyclohept-4-enone	44

(continued)

TABLE 6.3 (*Continued*)

Diene (n)	$Fe(CO)_x$ Source (m)	Temperature (°C)	Product	Yield (%)
Butadiene–$Fe(CO)_3$ (1.3)		60-80	2,7-dimethylcyclohept-4-enone (CH_3, CH_3, O)	55
2-methylbutadiene (CH_3) (excess)	$Fe_2(CO)_9$ (1.3)	25 (hν)	2,5,7-trimethylcyclohept-4-enone (CH_3, CH_3, CH_3, O)	36
2-methylbutadiene (CH_3)–$Fe(CO)_3$ (1.3)		120	2,5,7-trimethylcyclohept-4-enone (CH_3, CH_3, CH_3, O)	43
2,3-dimethylbutadiene (CH_3, CH_3) (excess)	$Fe_2(CO)_9$ (1.2)	60	2,4,5,7-tetramethylcyclohept-4-enone (CH_3, CH_3, CH_3, CH_3, O)	47
2,3-dimethylbutadiene (CH_3, CH_3)–$Fe(CO)_3$ (1.3)		120	2,4,5,7-tetramethylcyclohept-4-enone (CH_3, CH_3, CH_3, CH_3, O)	33
cyclopentadiene (excess)	$Fe_2(CO)_9$ (1.2)	60	2,4-dimethylbicyclo[3.2.1]oct-6-en-3-one (CH_3, CH_3, O)	81

Diene (n)	$Fe(CO)_x$ Source (m)	Temperature (°C)	Product	Yield (%)
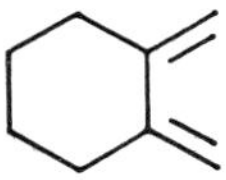 (excess)	$Fe_2(CO)_9$ (1.2)	63	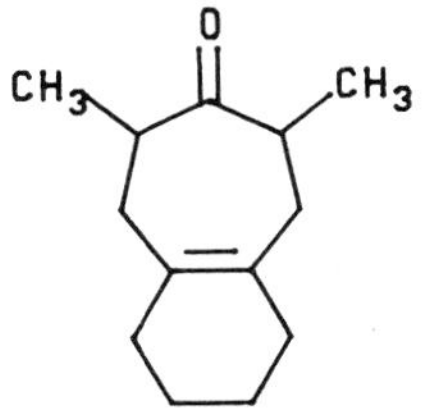 O, CH_3, CH_3	80

$$CH_3C(CH_3)(Br)C(=O)C(CH_3)(Br)CH_3 + n \text{ Diene} + m\ Fe(CO)_x \text{ source} \longrightarrow \text{Product}$$

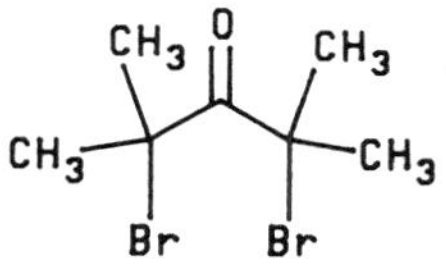

Solvent = Benzene

Diene (n)	$Fe(CO)_x$ Source (m)	Temperature (°C)	Product	Yield (%)
 (excess)	$Fe_2(CO)_9$ (1.2)	60	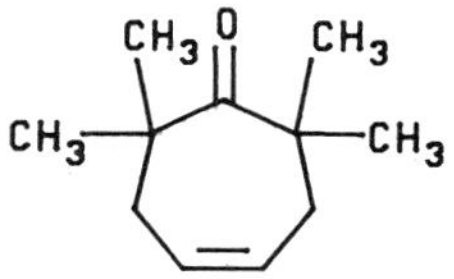 CH_3, O, CH_3, CH_3, CH_3	33
 $Fe(CO)_3$ (1.3)		80	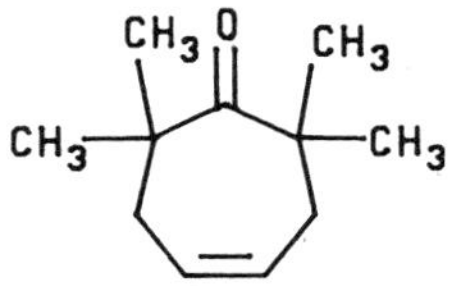 CH_3, O, CH_3, CH_3, CH_3	90
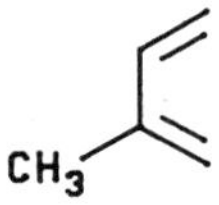 CH_3 (excess)	$Fe_2(CO)_9$ (1.2)	65	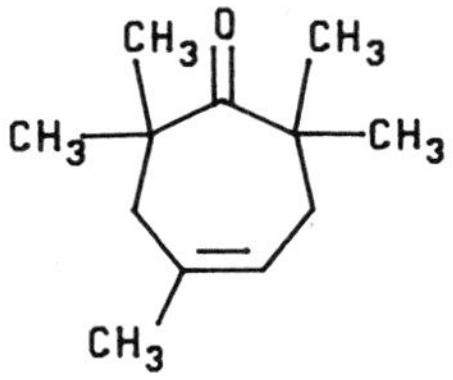 CH_3, O, CH_3, CH_3, CH_3, CH_3	47
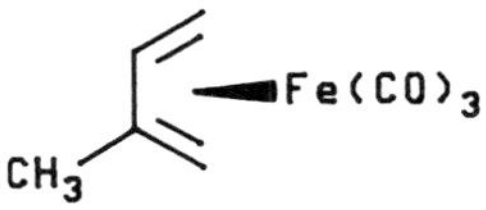 CH_3 $Fe(CO)_3$ (1.3)		87	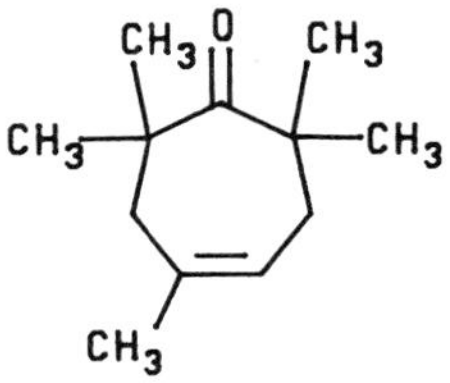 CH_3, O, CH_3, CH_3, CH_3, CH_3	70

(continued)

TABLE 6.3 *(Continued)*

Diene (n)	$Fe(CO)_x$ Source (m)	Temperature (°C)	Product	Yield (%)
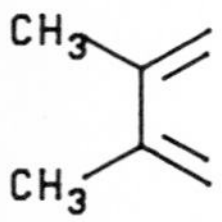 CH_3 CH_3 (excess)	$Fe_2(CO)_9$ (1.2)	60	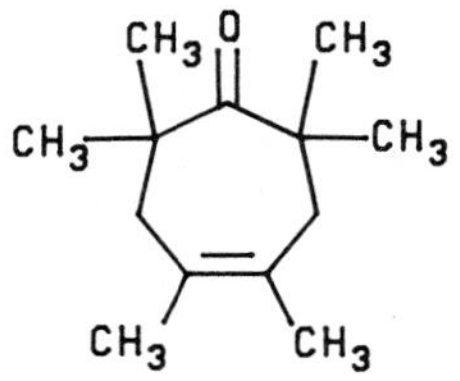 CH_3 O CH_3 CH_3 CH_3 CH_3 CH_3	71
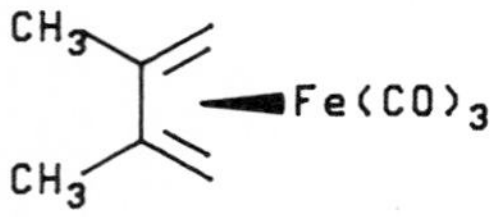 CH_3 CH_3 $Fe(CO)_3$ (1.3)		80	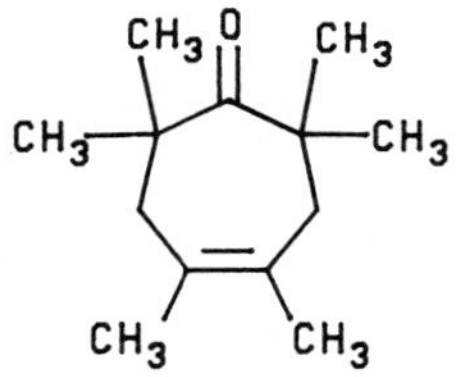 CH_3 O CH_3 CH_3 CH_3 CH_3 CH_3	100
 (excess)	$Fe_2(CO)_9$ (1.2)	60	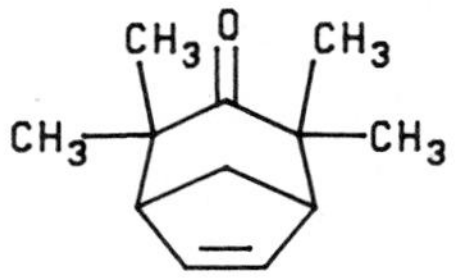 CH_3 O CH_3 CH_3 CH_3	82

CH_3 O CH_3 CH_3 CH_3 Br Br + n Diene + m $Fe(CO)_x$ source ⟶ Product

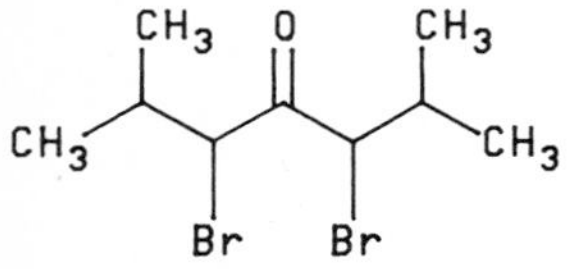

Solvent = Benzene

Diene (n)	$Fe(CO)_x$ Source (m)	Temperature (°C)	Product	Yield (%)
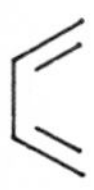 (excess)	$Fe_2(CO)_9$ (1.2)	90	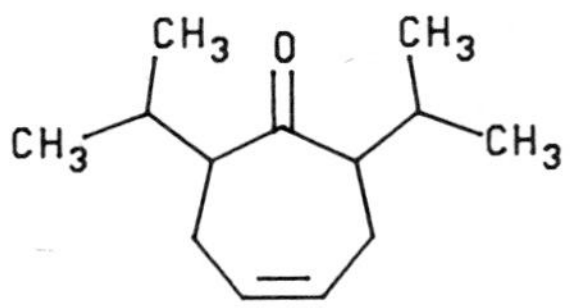 CH_3 O CH_3 CH_3 CH_3	44
 $Fe(CO)_3$ (1.3)		70	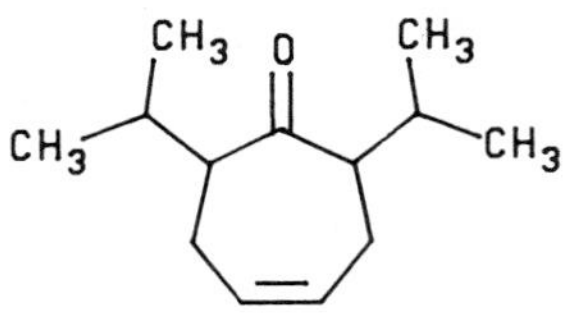 CH_3 O CH_3 CH_3 CH_3	77

Diene (n)	$Fe(CO)_x$ Source (m)	Temperature (°C)	Product	Yield (%)
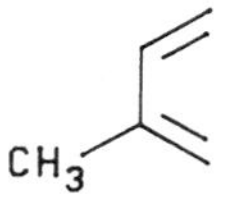 CH_3 (excess)	$Fe_2(CO)_9$ (1.2)	25 (hν)	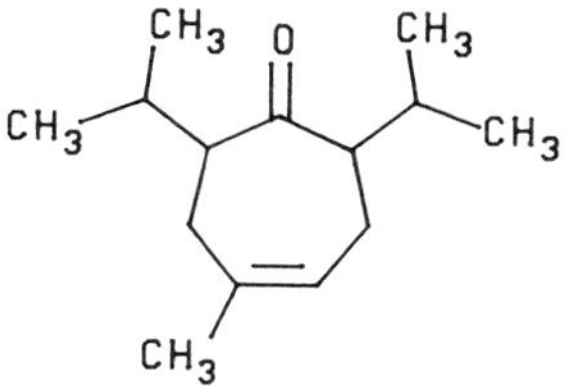 CH_3 O CH_3 CH_3 CH_3 CH_3	31
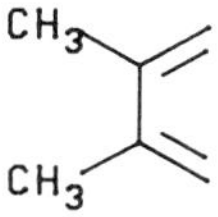 CH_3 CH_3 (excess)	$Fe_2(CO)_9$ (1.2)	60	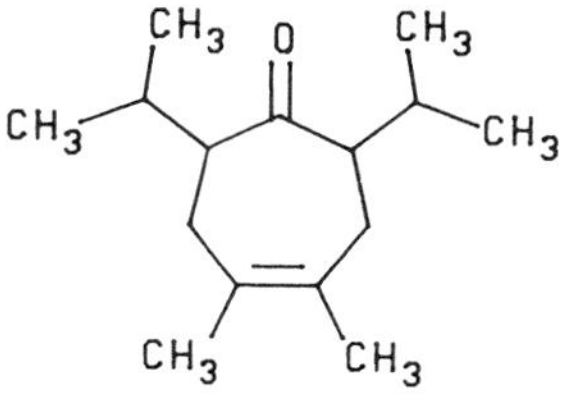 CH_3 O CH_3 CH_3 CH_3 CH_3 CH_3	50
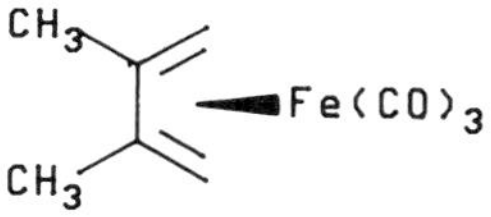 CH_3 CH_3 —$Fe(CO)_3$ (1.3)		120	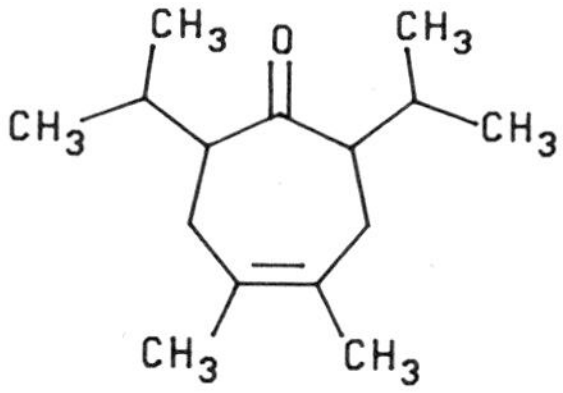 CH_3 O CH_3 CH_3 CH_3 CH_3 CH_3	55
 (excess)	$Fe_2(CO)_9$ (1.2)	90	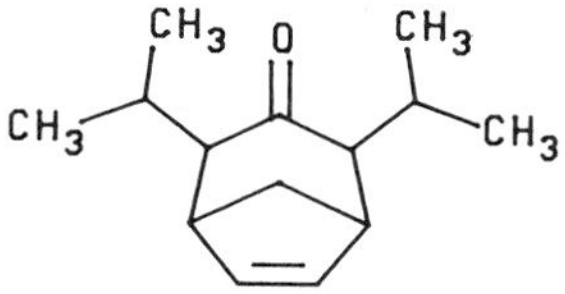 CH_3 O CH_3 CH_3 CH_3	93

CH_3 O Br CH_3 H Br Br + n Diene + m $Fe(CO)_x$ source ⟶ Product

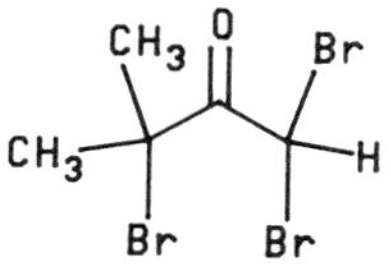

Solvent = Benzene

Diene (n)	$Fe(CO)_x$ Source (m)	Temperature (°C)	Product	Yield (%)
 (excess)	$Fe_2(CO)_9$ (1.2)	80	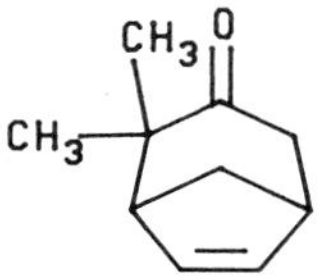 CH_3 O CH_3	66 after Zn-Cu reduction

(continued)

TABLE 6.3 (*Continued*)

y $Br_2CH-CO-CHBr_2$ + n Heterocycle + m $Fe(CO)_x$ source ⟶ Product

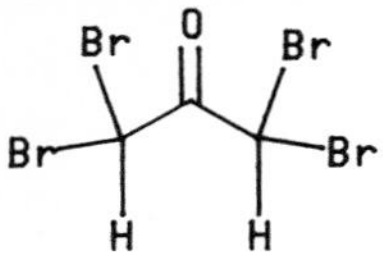

Solvent = Benzene

Heterocycle (n)	$Fe(CO)_x$ Source (m)	Temperature (°C)	Product	Yield (%)
y = 1  furan (excess)	$Fe_2(CO)_9$ (1.2)	32 (no solvent)	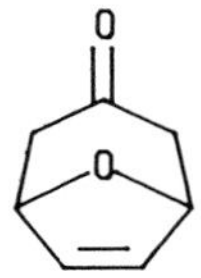	60 after Zn-Cu reduction
y = 2.5 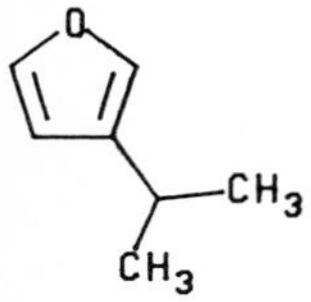 3-isopropylfuran (CH_3, CH_3) (1)	$Fe_2(CO)_9$ (2)	60	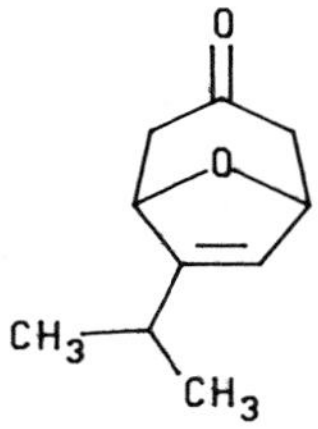 (CH_3, CH_3)	71 after Zn-Cu reduction
y = 2 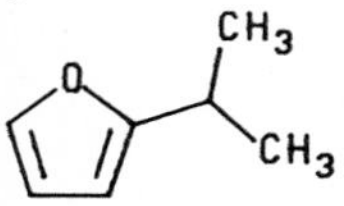 2-isopropylfuran (CH_3, CH_3) (1)	$Fe_2(CO)_9$ (1.5)	60	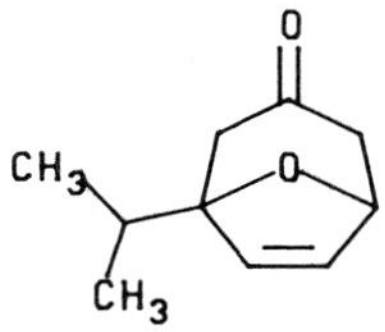 (CH_3, CH_3)	47 after Zn-Cu reduction
y = 3  N-R-pyrrole (1) R = $COOCH_3$	$Fe_2(CO)_9$ (1.5)	50	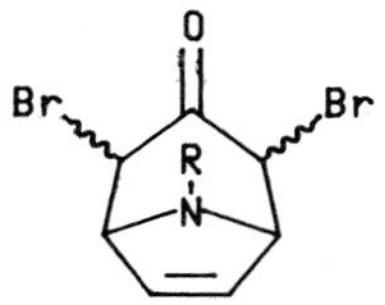 (Br, Br, R–N)	52

$BrCH(Br)-CO-C(CH_3)Br_2$ + n Heterocycle + m $Fe(CO)_x$ source ⟶ Product

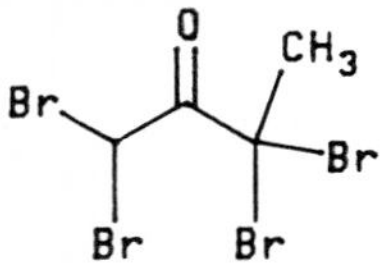

Solvent = None

Heterocycle (n)	$Fe(CO)_x$ Source (m)	Temperature (°C)	Product	Yield (%)
 (excess)	$Fe_2(CO)_9$ (1.2)	32	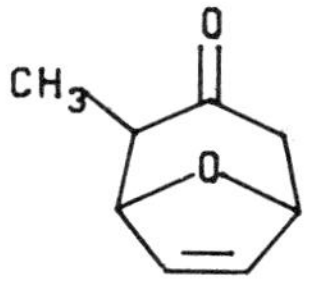 CH_3 O O	63 after Zn-Cu reduction

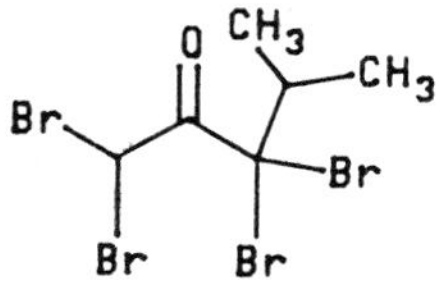

Br, Br, Br, Br, O, CH_3, CH_3 + n Heterocycle + m $Fe(CO)_x$ source ⟶ Product

Solvent = None

Heterocycle (n)	$Fe(CO)_x$ Source (m)	Temperature (°C)	Product	Yield (%)
 (excess)	$Fe_2(CO)_9$ (1.2)	32	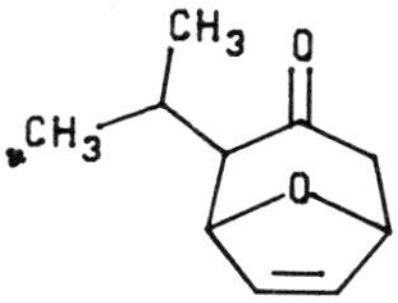 CH_3 O CH_3 O	35 after Zn-Cu reduction

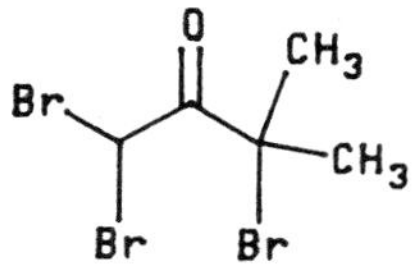

Br, Br, Br, O, CH_3, CH_3 + n Heterocycle + m $Fe(CO)_x$ source ⟶ Product

Solvent = None

Heterocycle (n)	$Fe(CO)_x$ Source (m)	Temperature (°C)	Product	Yield (%)
 (excess)	$Fe_2(CO)_9$ (1.2)	32	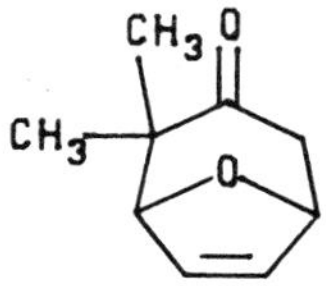 CH_3 O CH_3 O	87 after Zn-Cu reduction

(continued)

TABLE 6.3 (*Continued*)

$CH_3CH(Br)C(=O)CH(Br)CH_3$ + n Heterocycle + m $Fe(CO)_x$ source ⟶ Product

Solvent = Benzene

Heterocycle (n)	$Fe(CO)_x$ Source (m)	Temperature (°C)	Product	Yield (%)
Furan (excess)	$Fe_2(CO)_9$ (1.2)	32 (no solvent)	2,4-dimethyl-8-oxabicyclo[3.2.1]oct-6-en-3-one	90 after Zn-Cu reduction
2,5-Dimethylfuran (3)	$Fe_2(CO)_9$ (1.2)	80	1,2,4,5-tetramethyl-8-oxabicyclo[3.2.1]oct-6-en-3-one	78
N-R-pyrrole (1) R = $COOCH_3$	$Fe_2(CO)_9$ (1.2)	25 (hν)	2,4-dimethyl-8-R-8-azabicyclo[3.2.1]oct-6-en-3-one	60

2,6-Dibromocyclohexanone + n Heterocycle + m $Fe(CO)_x$ source ⟶ Product

Solvent = None

Heterocycle (n)	$Fe(CO)_x$ Source (m)	Temperature (°C)	Product	Yield (%)
Furan (excess)	$Fe_2(CO)_9$ (1.2)	32	tricyclic oxa-bridged ketone	35

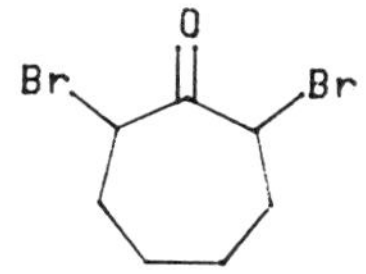

Br-(2,7-dibromocycloheptanone) + n Heterocycle + m $Fe(CO)_x$ source ⟶ Product

Solvent = None

Heterocycle (n)	$Fe(CO)_x$ Source (m)	Temperature (°C)	Product	Yield (%)
 furan (excess)	$Fe_2(CO)_9$ (1.2)	32	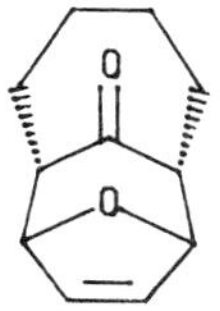	61

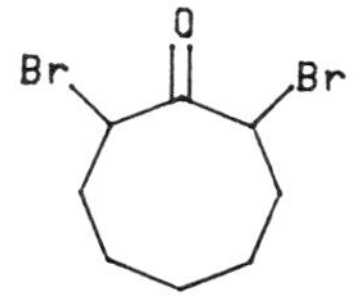

Br-(2,8-dibromocyclooctanone) + n Heterocycle + m $Fe(CO)_x$ source ⟶ Product

Solvent = None

Heterocycle (n)	$Fe(CO)_x$ Source (m)	Temperature (°C)	Product	Yield (%)
 furan (excess)	$Fe_2(CO)_9$ (1.2)	32	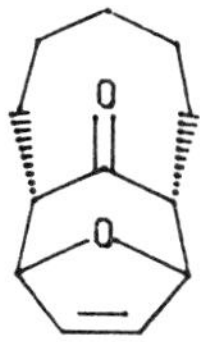	60

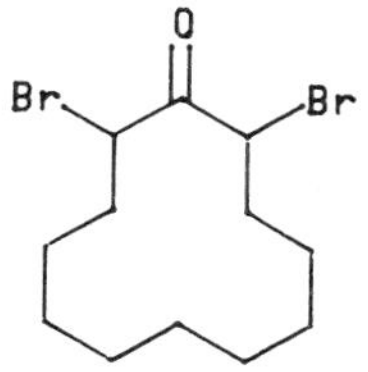

Br-(2,12-dibromocyclododecanone) + n Heterocycle + m $Fe(CO)_x$ source ⟶ Product

Solvent = None

Heterocycle (n)	$Fe(CO)_x$ Source (m)	Temperature (°C)	Products	Yield (%)
 furan (excess)	$Fe_2(CO)_9$ (1.2)	32	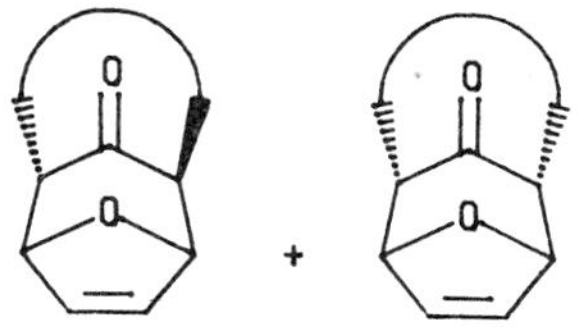 +	52

(continued)

TABLE 6.3 (*Continued*)

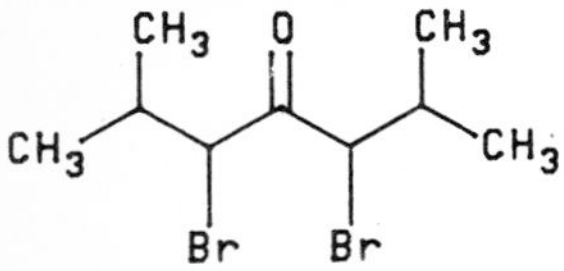

Reactant: 2,4-dibromo-1,5-dimethyl... ketone (CH_3, CH_3 / C=O / Br, Br) + n Heterocycle + m $Fe(CO)_x$ source ⟶ Product

Solvent = None

Heterocycle (n)	$Fe(CO)_x$ Source (m)	Temperature (°C)	Product	Yield (%)
 furan (excess)	$Fe_2(CO)_9$ (1.2)	32	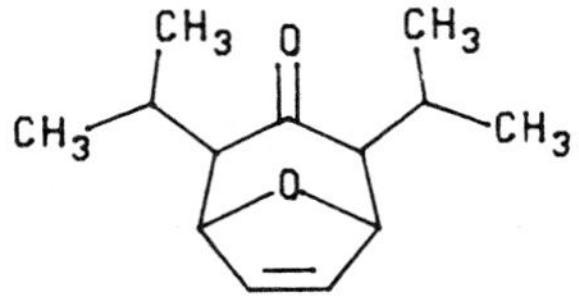 2,4-diisopropyl-8-oxabicyclo[3.2.1]oct-6-en-3-one	97

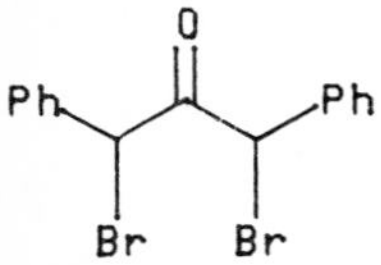

Reactant: Ph–CH(Br)–C(=O)–CH(Br)–Ph + n Heterocycle + m $Fe(CO)_x$ source ⟶ Product

Solvent = None

Heterocycle (n)	$Fe(CO)_x$ Source (m)	Temperature (°C)	Product	Yield (%)
 furan (excess)	$Fe_2(CO)_9$ (1.2)	32	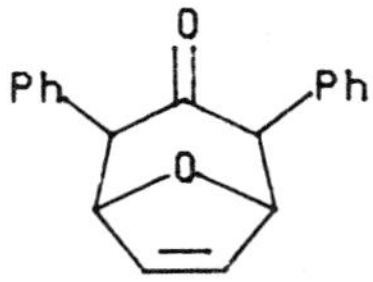 2,4-diphenyl-8-oxabicyclo[3.2.1]oct-6-en-3-one	90

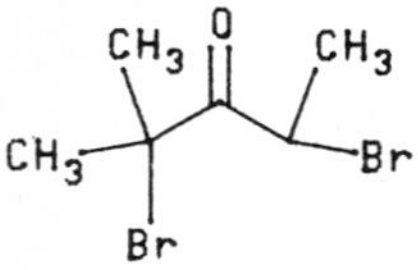

Reactant: $(CH_3)_2C(Br)$–C(=O)–CH(Br)CH_3 + n Heterocycle + m $Fe(CO)_x$ source ⟶ Product

Solvent = None

Heterocycle (n)	$Fe(CO)_x$ Source (m)	Temperature (°C)	Product	Yield (%)
 furan (excess)	$Fe_2(CO)_9$ (1.2)	32	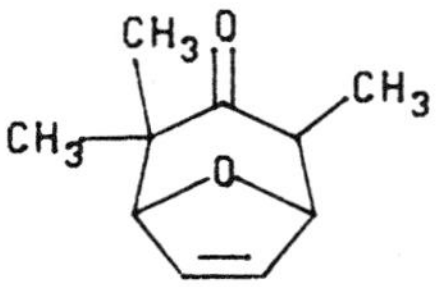 2,2,4-trimethyl-8-oxabicyclo[3.2.1]oct-6-en-3-one	84

$$y\ CH_3C(Br)(CH_3)-CO-C(Br)(CH_3)CH_3 + n\ \text{Heterocycle} + m\ Fe(CO)_x\ \text{source} \longrightarrow \text{Product}$$

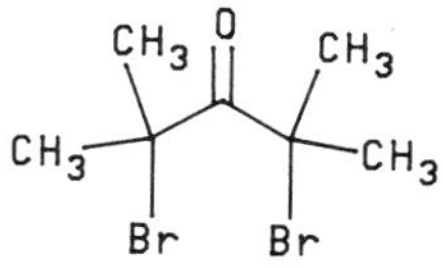

Solvent = Benzene

Heterocycle (n)	$Fe(CO)_x$ Source (m)	Temperature (°C)	Product	Yield (%)
y = 1  furan (excess)	$Fe_2(CO)_9$ (1.2)	32 (no solvent)	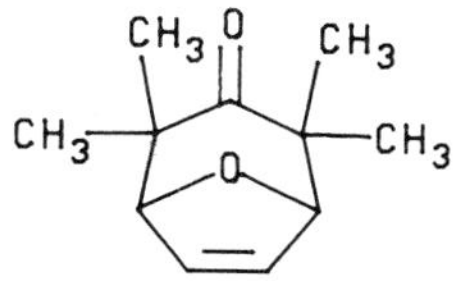 2,2,4,4-tetramethyl-8-oxabicyclo[3.2.1]oct-6-en-3-one	96
y = 3 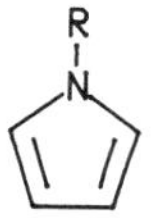 N-R pyrrole (1) R = $COCH_3$	$Fe_2(CO)_9$ (1.5)	40-50	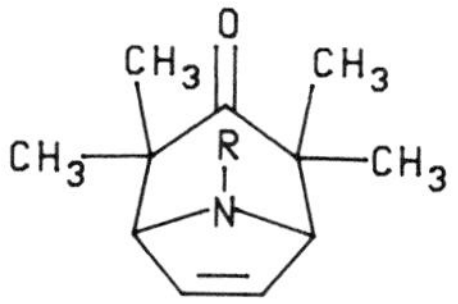 2,2,4,4-tetramethyl-8-R-8-azabicyclo[3.2.1]oct-6-en-3-one	70

Primary and Secondary MO Interactions in a [3+4] Cycloaddition

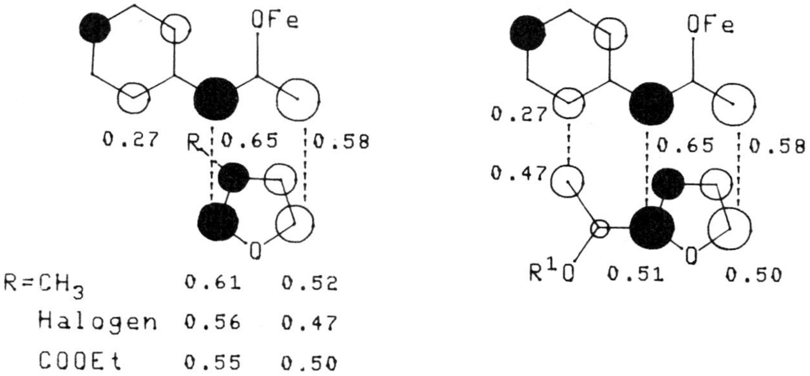

R		
R=CH_3	0.61	0.52
Halogen	0.56	0.47
COOEt	0.55	0.50

Figure 6.23

The symmetry-allowed [3 + 4]-cycloadditions ($\pi^2 + \pi^4$) are concerted with regioselectivity controlled by both primary and secondary frontier molecular orbital interactions.[20] The primary orbital interactions for combination of oxallyl cation LUMO with furan HOMO are shown in Figure 6.23. The similarity of furan HOMOs for 3-methyl, 3-halogeno-, 3-ethoxycarbonyl, and 2-ethoxycarbonylfurans results in similar regioselectivity of cycloaddition with an unsymmetrical oxallyl cation (see Table 6.4).

TABLE 6.4 Regioselectivity in [3 + 4]-Cycloadditions of Unsymmetrical Iron-Stabilized Oxallyl Cations and Unsymmetrical Dienes[20]

+ Heterocycle $\xrightarrow{Fe_2(CO)_9}$ [OFeL$_n$ cation]

⟶ Product A + Product B
(Major) (Minor)

Temperature = 30°C

Heterocycle	Product A	Product B	A:B	Yield (%)
3-methylfuran (CH_3)	CH_3	CH_3	56:44	48
3-bromofuran (Br)	Br	Br	65:35	73
3-ethoxycarbonylfuran (EtOOC)	EtOOC	COOEt	55:45	52

Heterocycle	Product A	Product B	A:B	Yield (%)
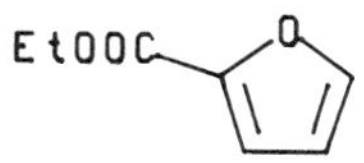 EtOOC-furan	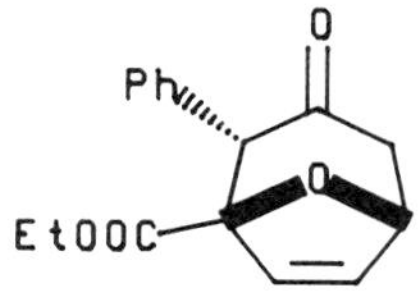 Ph, O, O, EtOOC	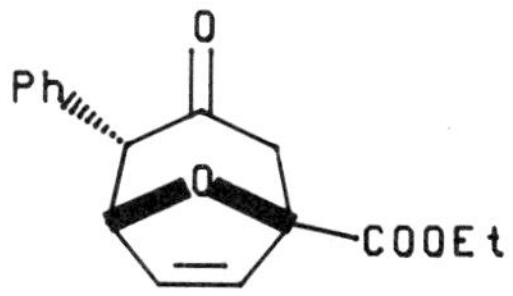 Ph, O, O, COOEt	90:10	56
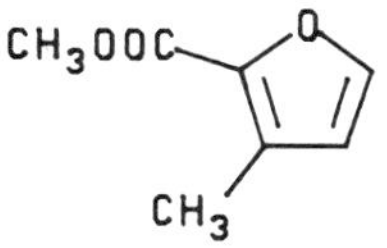 CH_3OOC-furan, CH_3	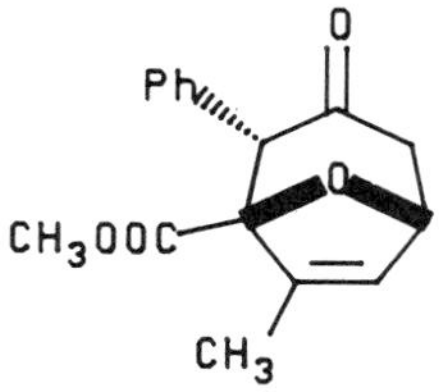 Ph, O, O, CH_3OOC, CH_3	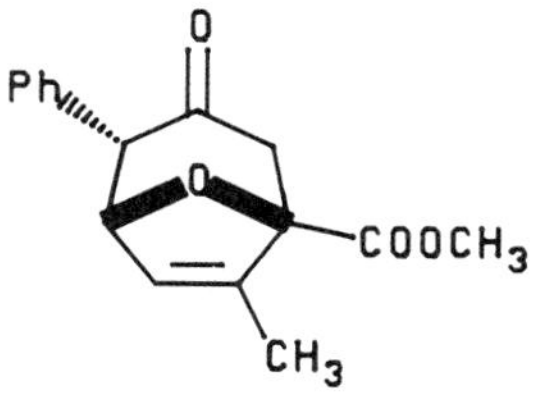 Ph, O, O, $COOCH_3$, CH_3	92:8	63
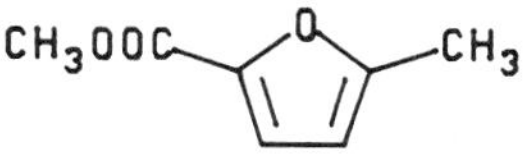 CH_3OOC-furan-CH_3	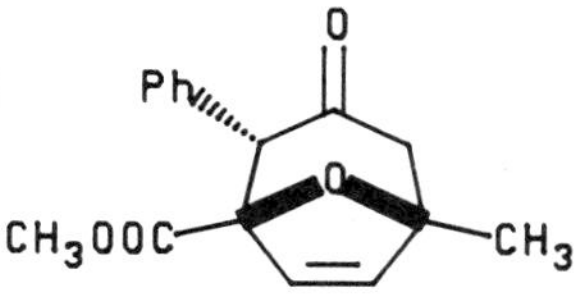 Ph, O, O, CH_3OOC, CH_3	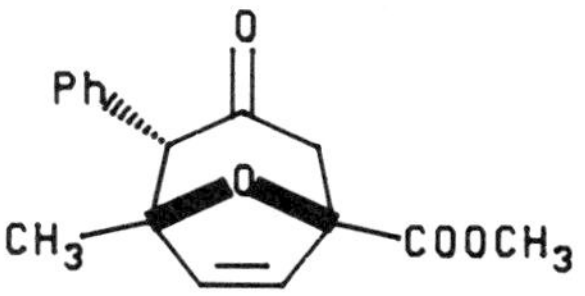 Ph, O, O, CH_3, $COOCH_3$	65:35	65
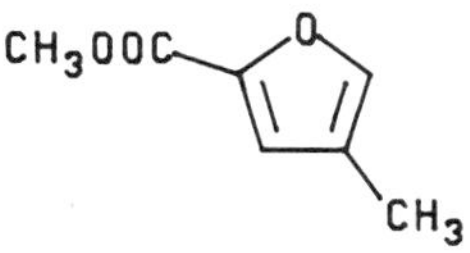 CH_3OOC-furan, CH_3	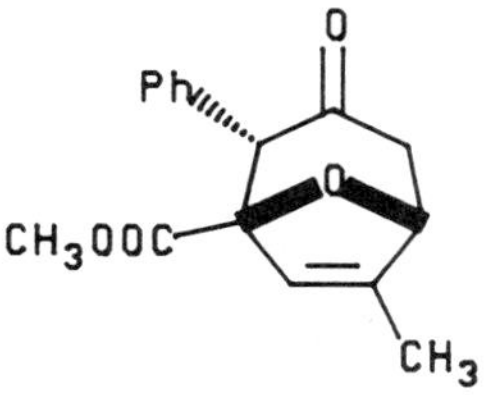 Ph, O, O, CH_3OOC, CH_3	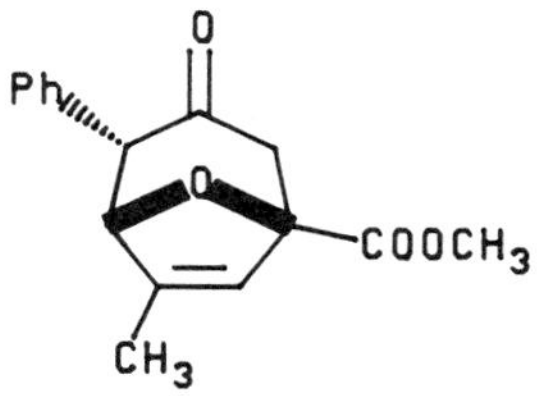 Ph, O, O, $COOCH_3$, CH_3	76:24	68

The greatly enhanced regioselectivity with 2-ethoxycarbonylfuran is a result of secondary molecular orbital interaction between the *ortho* carbon of the aromatic ring and the ester carbonyl.

This versatile and often efficient [3 + 4]-cycloaddition is the key step in several syntheses, including syntheses of the tropane alkaloids. In these, in the less involved preliminary syntheses of carbocamphelinone,[33] thujaplicins,[28] nezukone,[29,34] and the bridge ring system of the lolium alkaloids[35], a seven-membered ring ketone is

Carbocamphenilone

a. 80°C, benzene, THF

b. Zn-Cu, NH_4Cl, CH_3OH; 66%

c. H_2, Pd-C, EtOAc; 80%

d. 140°C, SeO_2, xylene; 100%

Figure 6.24

constructed at an early stage. Subsequent elaboration to target involves just a few well-precedented steps. Cycloaddition of the oxallyl cation from 1,1,3-tribromo-3-methylbutan-2-one and cyclopentadiene affords the seven-membered ring ketone for carbocamphelinone synthesis. Catalytic hydrogenation and selenium dioxide oxidation to the α-diketone completes the synthesis (see Figure 6.24).

Cycloaddition of the oxallyl cation from α,α,α′,α′-tetrabromoacetone with 3-isopropylfuran affords the seven-membered ring ketone for nezukone synthesis. Hydrogenation followed by acid-induced dehydration gives the cross-conjugated dienone. Dehydrogenation with 2,3-dichloro-5,6-dicyano-*p*-benzoquinone (DDQ) completes the synthesis of nezukone. Cycloaddition of the same oxallyl cation with 2-isopropylfuran affords the seven-membered ring ketone for β-thujaplicin synthesis. Hydrogenation followed by treatment first with boron trifluoride etherate and acetic anhydride and then with basic alumina affords two compounds, both readily converted to the tropone. Treatment with hydrazine hydrate and subsequent hydrolysis with alcoholic potassium hydroxide completes the synthesis of β-thujaplicin. Cycloaddition of the oxallyl cation from 1,1,3-tribromo-4-methylpenten-2-one and furan provides the seven-membered ring for α-thujaplicin synthesis. The remaining steps are similar to those for β-thujaplicin (see Figures 6.25 to 6.27).

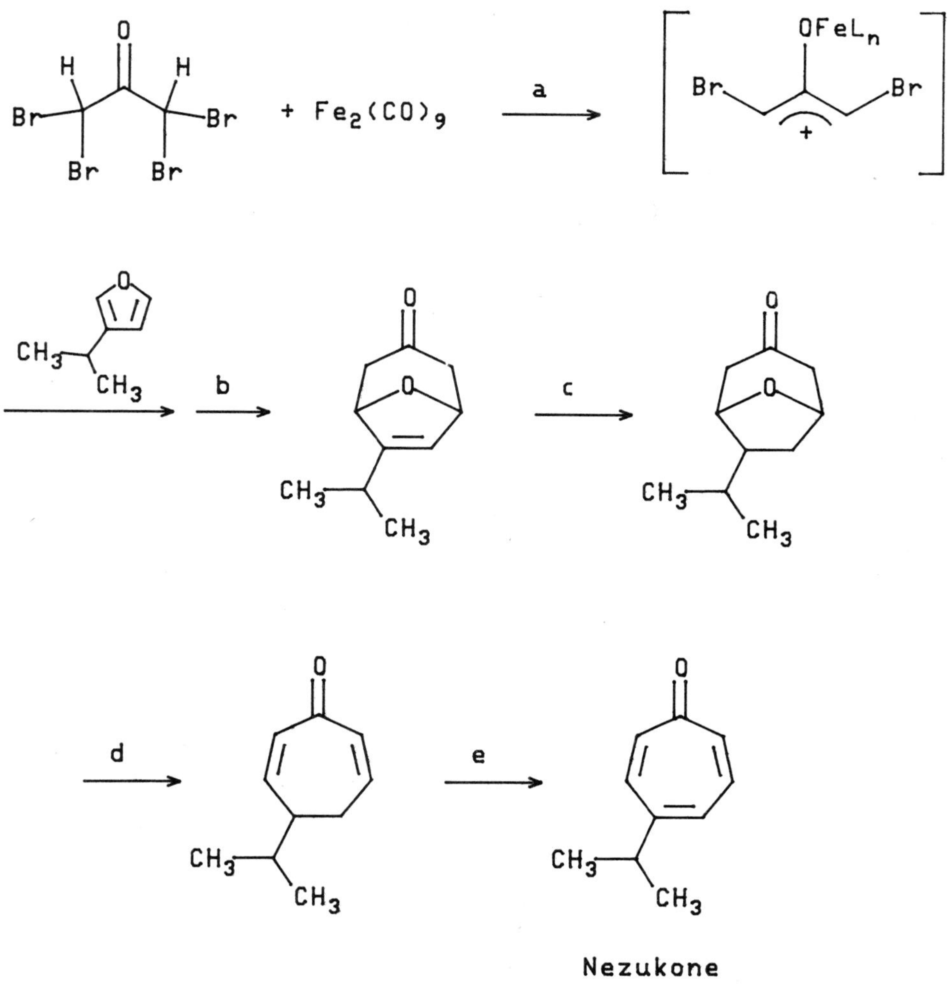

a. 60°C, benzene

b. rt, Zn-Cu, NH_4Cl, benzene, CH_3OH; 57% for two steps

c. rt, H_2, Pd-C, $NaHCO_3$, CH_3OH; 77%

d. rt, FSO_3H, CH_2Cl_2; 59%

e. 100°C, DDQ, p-TsOH, benzene; 54%

Nezukone

Figure 6.25

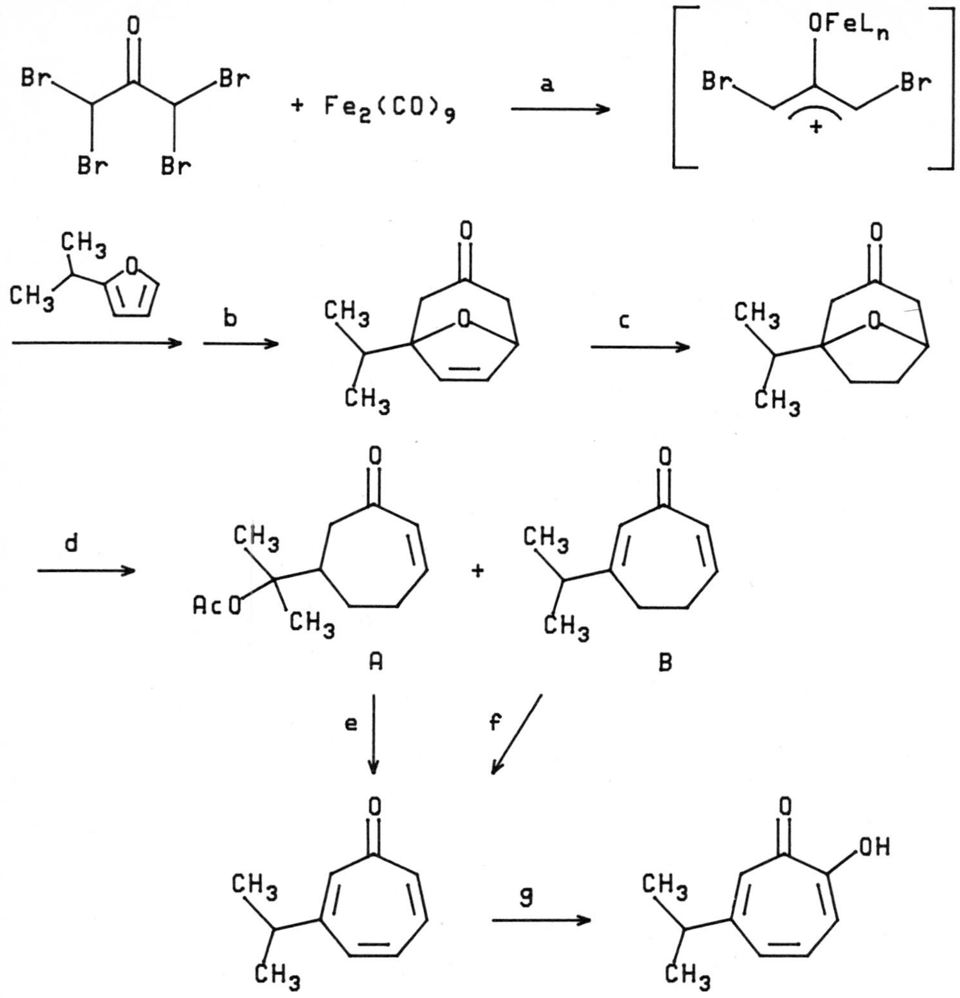

$$\text{Br}_2\text{CH–CO–CHBr}_2 + Fe_2(CO)_9 \xrightarrow{a} [\text{OFeL}_n \text{ oxyallyl cation}] \xrightarrow{\text{2-isopropylfuran}} \xrightarrow{b} \xrightarrow{c} \xrightarrow{d} A + B; \quad A \xrightarrow{e},\ B \xrightarrow{f} \xrightarrow{g} \beta\text{-Thujaplicin}$$

A | B

β-Thujaplicin

a. 60°C, benzene

b. rt, Zn-Cu, CH_3OH; 47% for two steps

c. rt, H_2, Pd-C, $NaHCO_3$, EtOH; 96%

d. 1) -20 to -10°C, $BF_3 \cdot Et_2O$, Ac_2O, 2) rt, Al_2O_3, EtOAc; A:33%, B:25%

e. 1) 80°C, NBS, AiBN, CCl_4, 2) 130-140°C, LiCl, Li_2CO_3, DMF; 77%

f. 100°C, DDQ, p-TsOH, benzene; 80%

g. 1) rt, $NH_2NH_2 \cdot H_2O$, EtOH, 2) 100°C, KOH, H_2O, EtOH; 100%

Figure 6.26

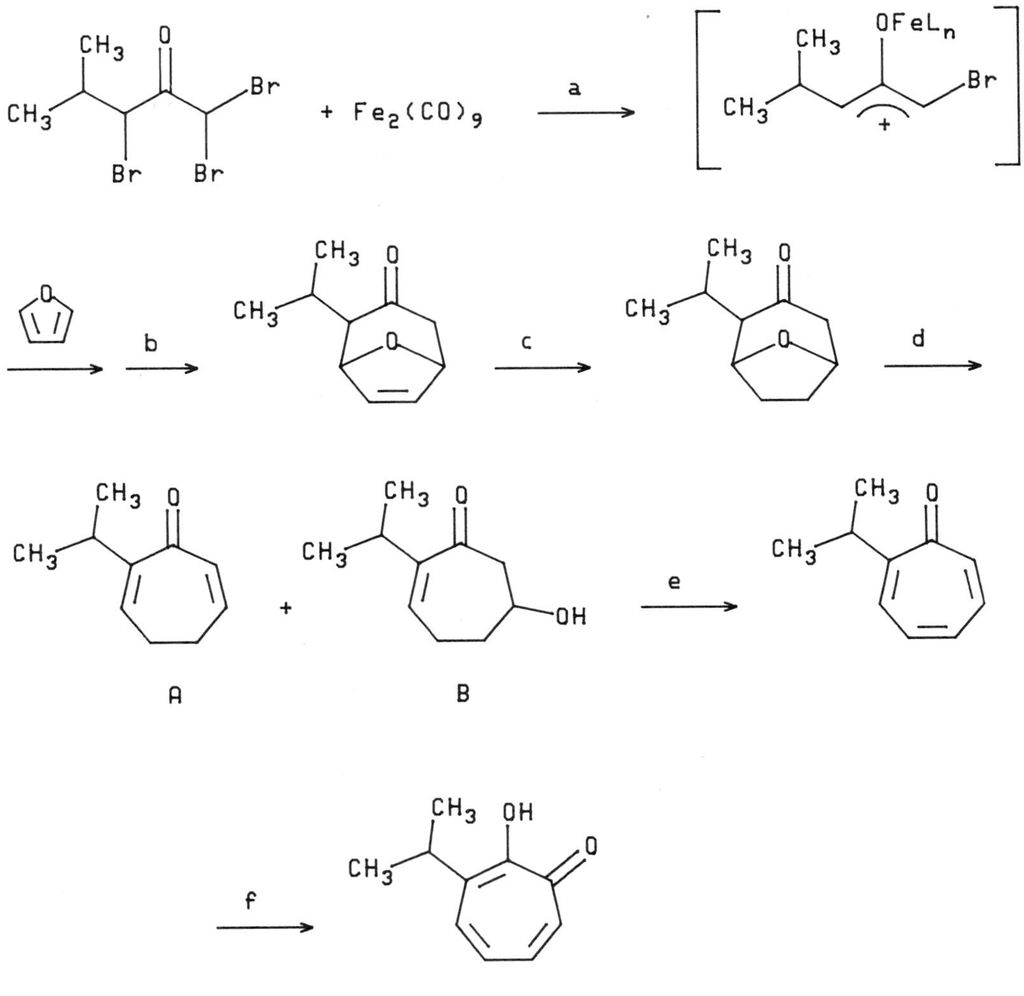

α-Thujaplicin

a. rfx, neat furan

b. rt, Zn-Cu, NH_4Cl, CH_3OH; 35% for two steps

c. rt, H_2, Pd-C, EtOH; 96%

d. 0°C to rt, FSO_3H, CH_2Cl_2; A:11%, B:57%

e. 160°C, DDQ, p-TsOH, benzene; 55% for two steps

f. 1) rt, $NH_2NH_2 \cdot H_2O$, EtOH, 2) 100°C, KOH, H_2O, EtOH

Figure 6.27

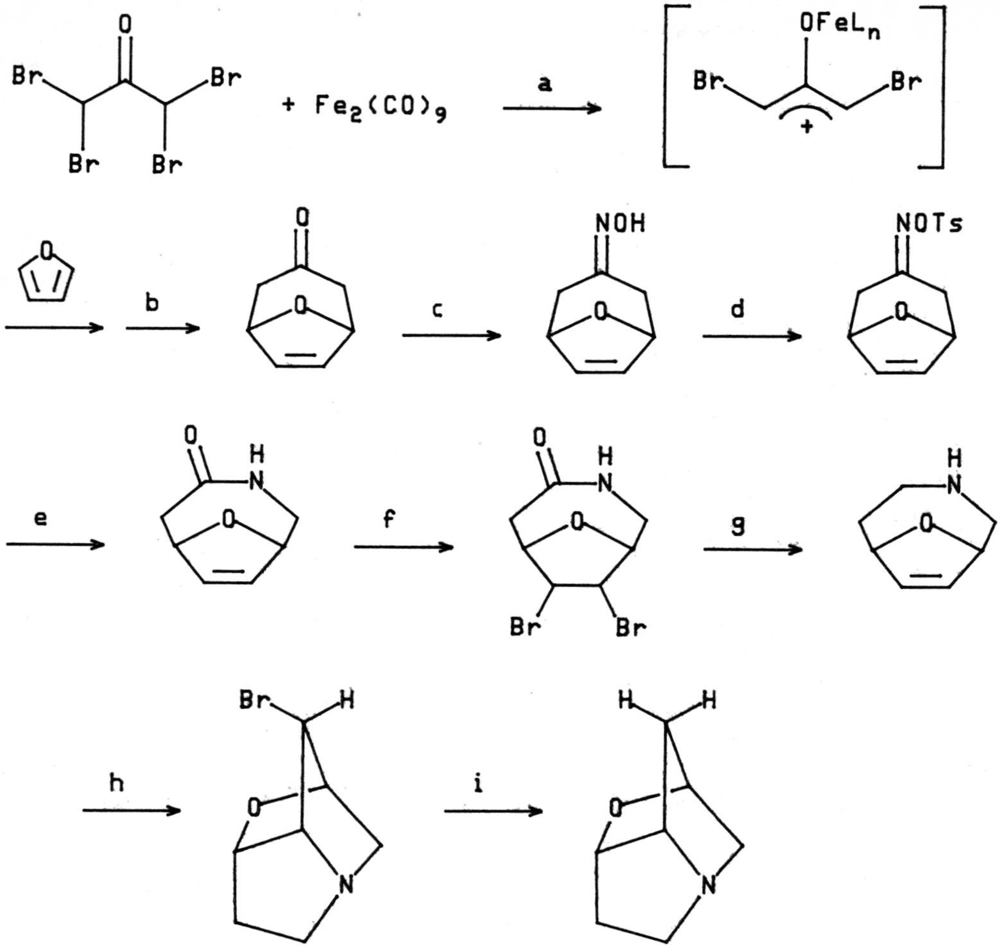

a. rfx, neat furan

b. rt, Zn-Cu, NH_4Cl, CH_3OH; 57% for two steps

c. rfx, $NH_2OH{\cdot}HCl$, $NaHCO_3$, CH_3OH; 81%

d. -10°C to rt, p-TsCl, pyridine, CH_2Cl_2; 91%

e. rt, K_2CO_3, H_2O, THF; 85%

f. rt, Br_2, CH_2Cl_2; 58%

g. rt to rfx, $LiAlH_4$, THF, Et_2O; 65%

h. rt, Br_2, Et_2O, CH_2Cl_2; 99%

i. rfx, $LiAlH_4$, Et_2O; 65%

Figure 6.28

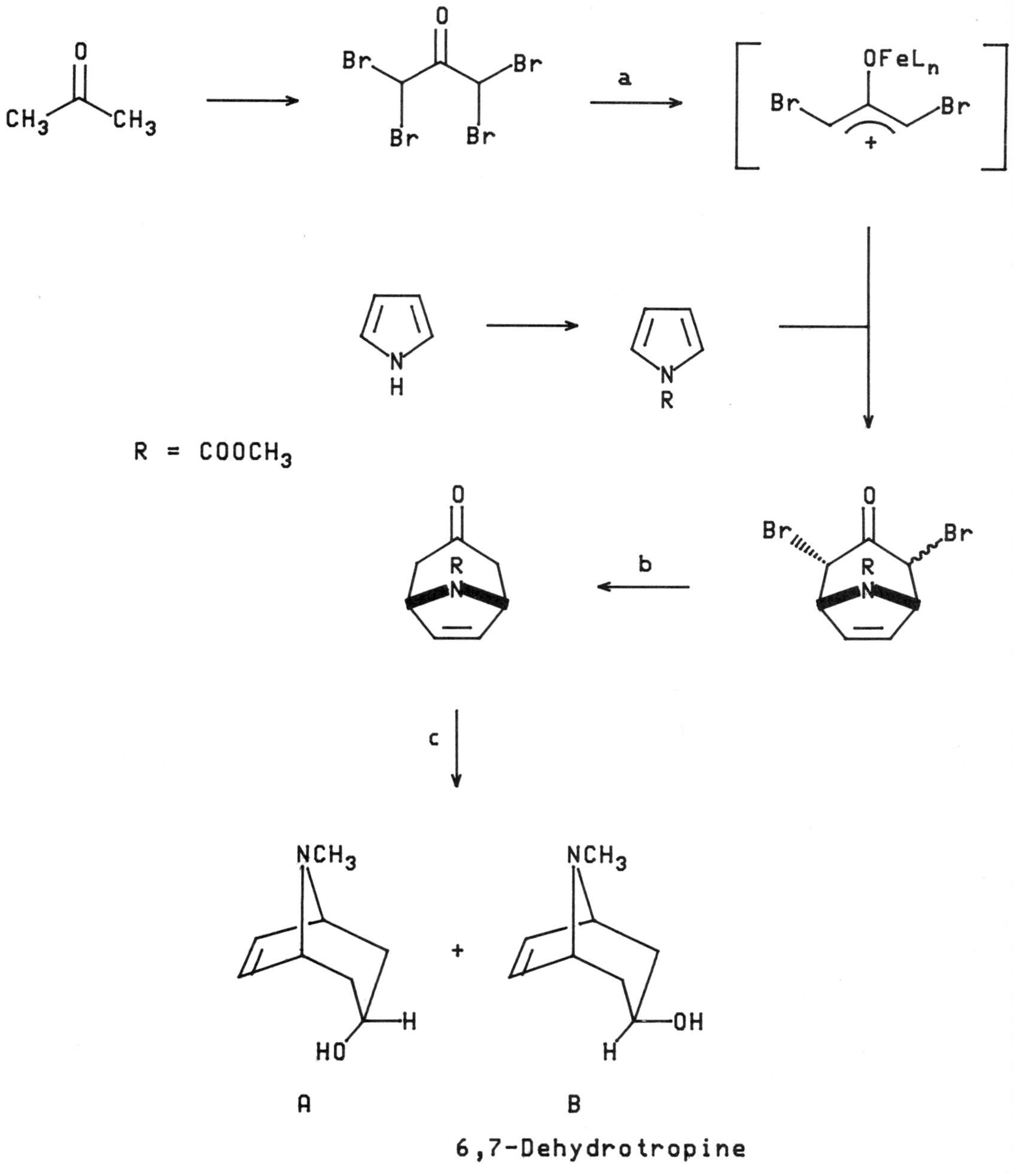

R = COOCH$_3$

A B

6,7-Dehydrotropine

a. 50°C, $Fe_2(CO)_9$, benzene; 70%

b. rt, Zn-Cu, NH_4Cl, CH_3OH; 100%

c. -78°C, $(i\text{-}Bu)_2AlH$, THF; A:87%, B:5%

Figure 6.29

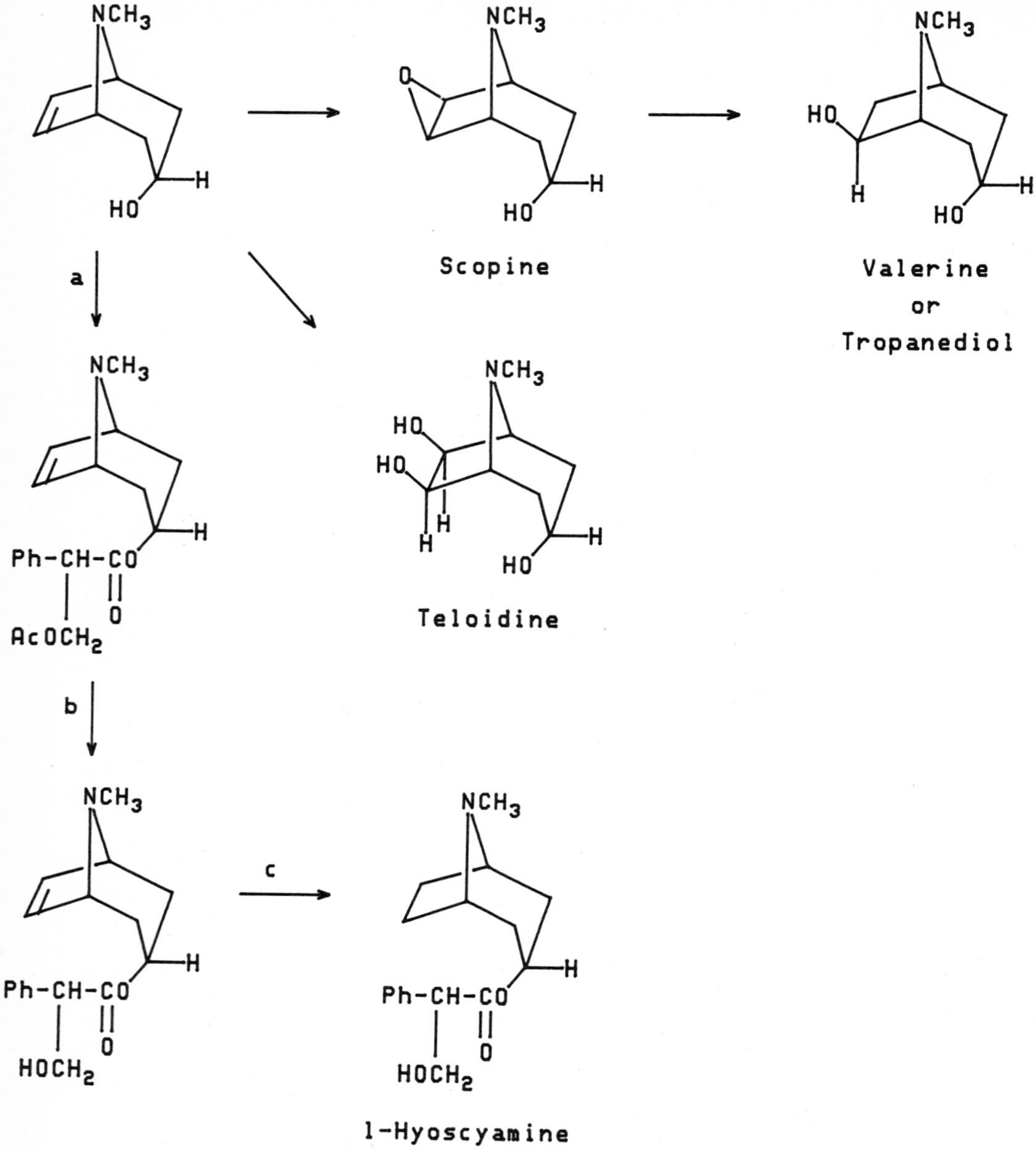

Scopine

Valerine
or
Tropanediol

Teloidine

1-Hyoscyamine

a. p-TsOH, ClC(O)CH(Ph)CH_2OAc

b. HCl; 88% for two steps

c. rt, H_2, Pd-C, EtOH; 99%

Figure 6.30

Synthesis of the lolium alkaloids begins with a [3 + 4]-cycloaddition of the oxallyl cation from $\alpha,\alpha,\alpha',\alpha'$-tetrabromoacetone and furan. The ketone is converted to an oxime. Tosylation and Beckman rearrangement afford the amide. Bromination and reduction produces the amine, which cyclizes to the bromoalkaloid on treatment with bromine. Dehalogenation is accomplished with lithium aluminum hydride (see Figure 6.28).

6.5 TOTAL SYNTHESIS OF *l*-HYOSCYAMINE[30,36,37]

A tropane skeleton can be constructed in just a few short steps from very inexpensive materials: acetone and pyrrole. Acetone is first converted to $\alpha,\alpha,\alpha',\alpha'$-tetrabromoacetone.[2] Pyrrole is converted to an *N*-methoxycarbonyl derivative.[38] A [3 + 4]-cycloaddition of the oxallyl cation from tetrabromoacetone produces two diastereoisomers. After reduction with Zn-Cu couple, reduction with excess di-*i*-butylaluminum hydride in tetrahydrofuran at $-78°C$ beautifully accomplishes three synthetic objectives in one operation; (1) retention of the alkene, (2) conversion of *N*-methoxycarbonyl to *N*-methyl, and (3) stereoselective reduction of the ketone to the α-alcohol (see Figure 6.29).

6,7-Dehydrotropine is a relay compound for syntheses of several more highly functionalized tropane alkaloids. Scopine is synthesized by alkene epoxidation with trifluoroperacetic acid, hydrogen peroxide–formic acid, or hydrogen peroxide–tungstic acid.[39–42] Catalytic hydrogenolysis of the epoxide in scopine affords valerine (or tropanediol).[41,43] Teloidine is synthesized by alkene dihydroxylation with potassium permanganate. *l*-Hyoscyamine is synthesized by esterification of the *p*-toluenesulfonic acid salt of 6,7-dehydrotropine with (−)-*O*-acetyltropyl chloride. The acetyl protecting group is removed with hydrochloric acid.[44] Finally, catalytic alkene hydrogenation affords the target in an excellent 88% yield for the three steps.

l-Hyoscyamine was prepared from $\alpha,\alpha,\alpha',\alpha'$-tetrabromoacetone, *N*-methoxycarbonylpyrrole and (−)-*O*-acetyltropyl chloride in six steps in overall 53% yield (see Figure 6.30).

REFERENCES

1. The chemistry, biochemistry, and pharmacology of tropane alkaloids have been extensively reviewed:

1a. Fodor, G.; Dharanipragada, R. *Nat. Prod. Rep.* **1986**, *3*,181.

1b. Swan, G. A. *An Introduction to the Alkaloids*; Wiley: New York, 1967, p. 50.

1c. Pelletier, S. W. *Chemistry of the Alkaloids*; Van Nostrand Reinhold: New York 1970, p. 431.

1d. Dalton, D. R. *The Alkaloids*; Marcel Dekker: New York, 1979, p. 65.

2. The α,α'-dibromoketones are prepared by the method of Rappe: Rappe, C. *Acta Chem. Scand.* **1962**, *16*, 2467.
3. Noyori, R. *Ann. N.Y. Acad. Sci.* **1977**, *295*, 225.
4. Noyori, R. *Acc. Chem. Res.* **1979**, *12*, 61.
5. Noyori, R.; Hayakawa, Y. *Org. React.* **1983**, *29*, 163.

6. Noyori, R.; Hayakawa, Y. *Tetrahedron* **1985**, *41*, 5879.
7. Emerson, G. F.; Ehrlich, K.; Giering, W. P.; Lauterbur, P. C. *J. Am. Chem. Soc.* **1966**, *88*, 3172
8. Noyori, R.; Hayakawa, Y.; Funakura, M.; Takaya, H.; Murai, S.; Kobayashi, R-I., Tsutsumi, S. *J. Am. Chem. Soc.* **1972**, *94*, 7202.
9. Noyori, R.; Hayakawa, Y.; Takaya, H.; Murai, S.; Kobayashi, R.; Sonoda, N. *J. Am. Chem. Soc.* **1978**, *100*, 1759.
10. Noyori, R.; Shimizu, F.; Hayakawa, Y. *Tetrahedron Lett.* **1978**, 2091
11. Noyori, R.; Yokoyama, K.; Hayakawa, Y. *J. Am. Chem. Soc.* **1973**, 95, 2722.
12. Hayakawa, Y.; Shimizu, F.; Noyori, R. *Tetrahedron Lett.* **1978**, 993.
13. Hayakawa, Y.; Yokoyama, K.; Noyori R. *J. Am. Chem. Soc.* **1978**, *100*, 1791.
14. Noyori, R.; Yokoyama, K.; Makino, S.; Hayakawa, Y. *J. Am. Chem. Soc.* **1972**, *94*, 1772.
15. Hayakawa, Y.; Yokoyama, K.; Noyori, R. *J. Am. Chem. Soc.* **1978**, *100*, 1799.
16. Noyori, R.; Yokoyama, K.; Hayakawa, Y., *Org. Syn.* **1978**, *58*, 56.
17. Noyori, R.; Hayakawa, Y.; Makino, S.; Hayakawa, N.; Takaya, H. *J. Am. Chem. Soc.* **1973**, *95*, 4103.
18. Hayakawa, Y.; Takaya, H.; Makino, S.; Hayakawa, N.; Noyori, R. *Bull. Chem. Soc. Jpn.* **1977**, *50*, 1990.
19. Hayakawa, Y.; Yokoyama, K.; Noyori, R. *Tetrahedron Lett.* **1976**, 4347.
20. Noyori, R.; Shimizu, F.; Fukuta, K.; Takaya, H.; Hayakawa, Y. *J. Am. Chem. Soc.* **1977**, *99*, 5196.
21. Wang, P. C.; Joullié, M. M. *Alkaloids* **1984**, *23*, 327.
22. Noyori, R.; Nishizawa, M.; Shimizu, F.; Hayakawa, Y.; Maruoka, K.; Hashimoto, S.; Yamamoto, H.; Nozaki, H. *J. Am. Chem. Soc.* **1979**, *101*, 220.
23. Nishizawa, M.; Noyori, R. *Bull Chem. Soc. Jpn.* **1981**, *54*, 2233.
24. Noyori, R.; Makino, S.; Takaya, H. *J. Am. Chem. Soc.* **1971**, *93*, 1272.
25. Noyori, R.; Souchi, T.; Hayakawa, Y. *J. Org. Chem.* **1975**, *40*, 2681.
26. Takaya, H.; Makino, S.; Hayakawa, Y.; Noyori, R. *J. Am. Chem. Soc.* **1978**, *100*, 1765.
27. Noyori, R.; Baba, Y.; Makino, S.; Takaya, H. *Tetrahedron Lett.* **1973**, 1741.
28. Noyori, R.; Makino, S.; Okita, T.; Hayakawa, Y. *J. Org. Chem.* **1975**, *40*, 806.
29. Takaya, H.; Hayakawa, Y.; Makino, S.; Noyori, R. *J. Am. Chem. Soc.* **1978**, *100*, 1778.
30. Noyori, R.; Makino, S.; Baba, Y.; Hayakawa, Y. *Tetrahedran Lett.* **1974**, 1049.
31. Noyori, R.; Baba, Y.; Hayakawa, Y. *J. Am. Chem. Soc.* **1974**, *96*, 336.
32a. Huisgen, R.; Steiner, G. *J. Am. Chem. Soc.* **1973**, *95*, 5055.
32b. Huisgen, R.; Schug, R. *J. Am. Chem. Soc.* **1976**, *98*, 7819.
33. Noyori, R.; Souchi, T.; Hayakawa, Y. *J. Org. Chem.* **1975**, *40*, 2681.
34. Hayakawa, Y.; Sakai, M.; Noyori, R. *Chem. Lett.* **1975**, 509.
35a. Wilson, S. R.; Sawicki, R. A. *Tetrahedron Lett.* **1978**, 2969.
35b. Wilson, S. R.; Sawicki, R. A.; Huffman, J. C. *J. Org. Chem.* **1981**, *46*, 3887.
36. Hayakawa, Y.; Noyori, R. *Bull. Chem. Soc. Jpn.* **1974**, *47*, 2617.
37. Hayakawa, Y.; Baba, Y.; Makino, S.; Noyori, R. *J. Am. Chem. Soc.* **1978**, *100*, 1786.
38. Acheson, R. M.; Vernon, J. M. *J. Chem. Soc.* **1961**, 457.
39. Dobo, P.; Fodor, G.; Janzso, G.; Koczor, I.; Toth, J.; Vincze, I. *J. Chem. Soc.* **1959**, 3461.

40. Fodor, G.; Toth, J.; Romeike, A.; Vincze, I.; Dobo, P.; Janzso, G. *Angew. Chem.* **1957**, *69*, 678.

41. Fodor, G.; Toth, J.; Koczor, I.; Dobo, P.; Vincze, I. *Chem. Ind. (London)* **1956**, 764.

42. Fodor, G.; Kiss, S.; Rákoczi, *J. Chem. Ind. (Paris)* **1963**, *90*, 225.

43. Fodor, G.; Kovács. O.; Mészáros, L. *Research (London)* **1952**, *5*, 534.

44. Fung, V. A.; DeGraw, J. I. Synthesis **1976**, 311.

CHAPTER 7

(±)-Estrone

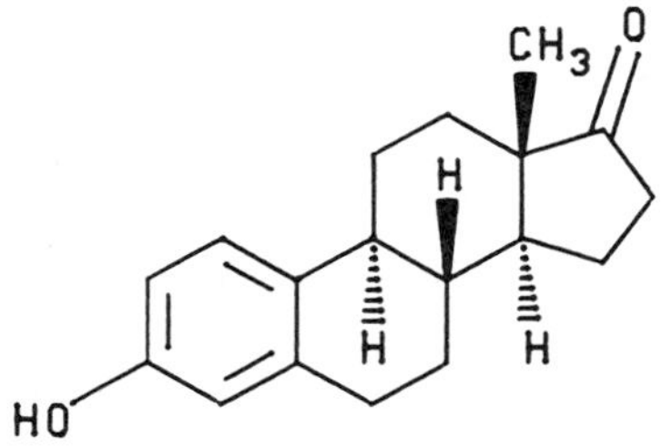

(±)-Estrone

estra-1,3,5(10)trien-17-one, 3-hydroxy [53-16-7]
(±) [19973-76-3]

Figure 7.1

A major research objective of the late 1920s was the isolation and characterization of substances associated with sexuality and reproduction. The first such substance isolated, estrone, was reported independently by Butenandt and by Doisy in 1929.[1,2] Estrone is an important member of a group of steroids classified as estrogens or estranes, which are responsible for the development of female sex characteristics. Yet another member of this group is 17-β-estradiol, which is produced in cells lining the follicles of the ovary and released under the control of pituitary glycoproteins LH (leutenizing hormone) and FSH (follicle-stimulating hormone).

A key point in the human reproductive cycle occurs about 10 days after the end of menses when a growing young follicle reaches a certain size and then secretes a surge of estradiol. A feedback mechanism causes the release of LH from the anterior portion of the pituitary gland, and the surge in LH in turn induces ovulation.[3]

With this close association of estrogens and ovulation, it is not surprising that many oral contraceptives contain an estrogen component. Two possible mechanisms for contraception are (1) blockage of ovulation by prevention of an LH surge and (2) inhibition of the release of FSH, necessary for the initial growth of the follicle. Thus, in both cases the immediate effect is on the pituitary gland, not directly on the ovary. Table 7.1 contains relative estrogenic potencies of various estrogens based on the Allen–Doisy test in rats[4] (see Figures 7.2 and 7.3).

TABLE 7.1 Biological Activity of Various Estrogens[a]

	Relative Potency	
Estrogen Type	Subcutaneous	Oral
17β-Estradiol	1	1
17α-Ethynylestradiol	1	12
Estrone	0.12	0.9
Equilin	0.08	1.2
Estriol	0.01	5
Equilenin	0.006	

[a]Used with permission from John Wiley & Sons, Inc.

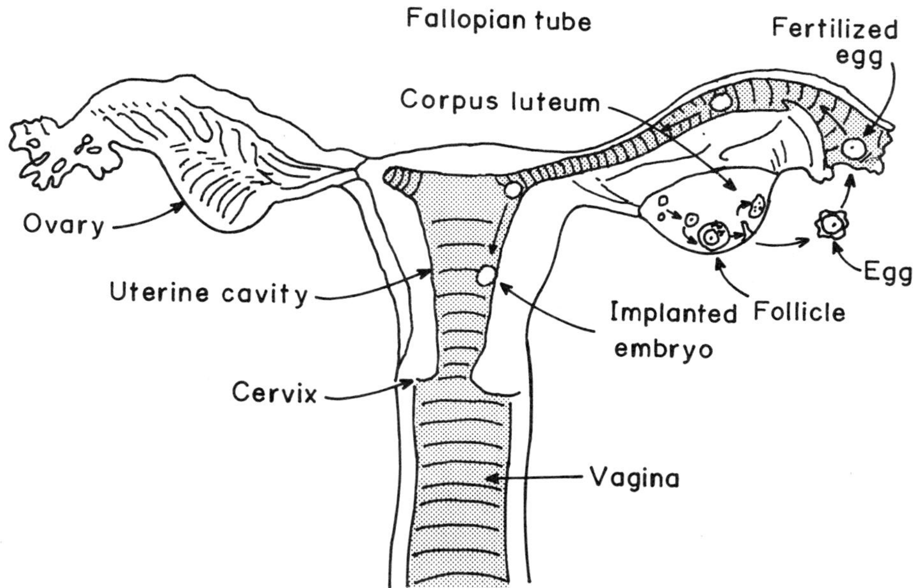

Figure 7.2 Egg maturation in the female reproductive tract. (Used with permission from John Wiley & Sons, Inc.)

While serious side effects have been associated with long-term administration of exogenous estrogens (as in oral contraception), estrogens still have important clinical applications. Estrogens are used in the treatment of menstrual irregularities, failure of ovarian development, prostatic carcinoma, and hormonal deficiencies in menopausal and postmenopausal women.[5]

The elucidation of specific biochemical roles and pharmacological applications for estrogens and steroids in general has encouraged a significant synthetic effort that continues to the present day[6,7] with the elegant work from K. P. C. Vollhardt's group.

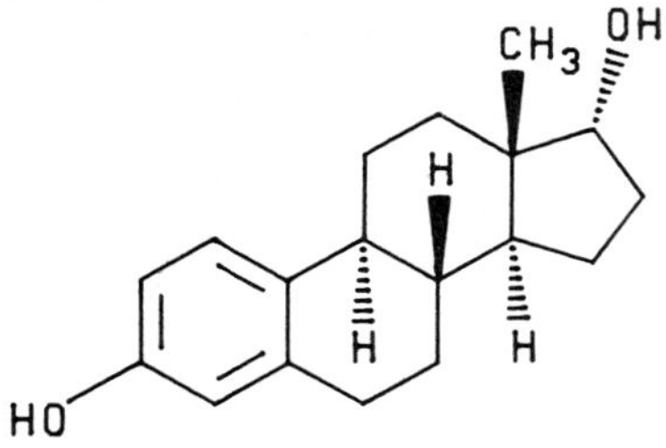

17-β-Estradiol

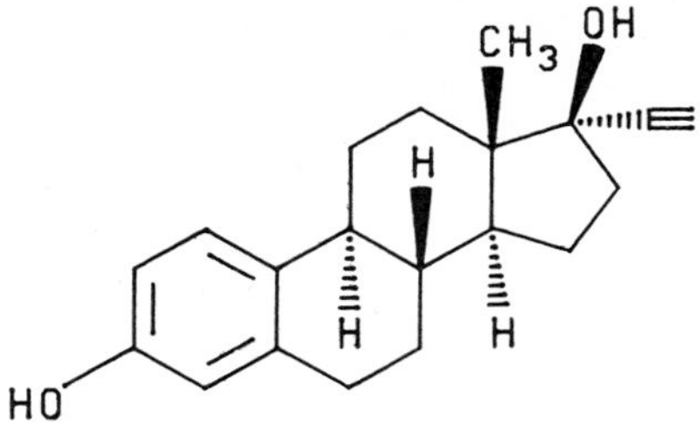

17-α-Ethynylestradiol

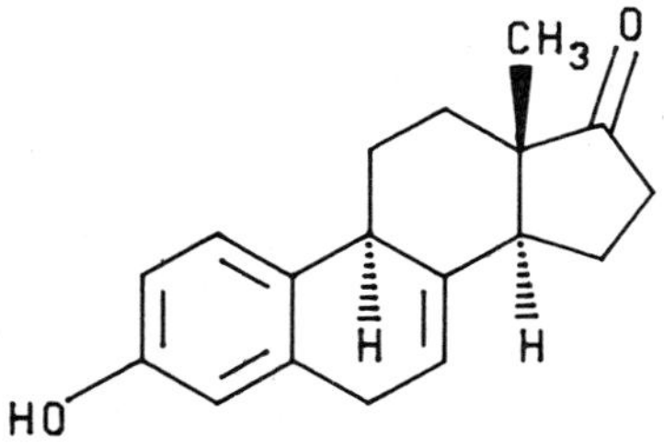

Equilin

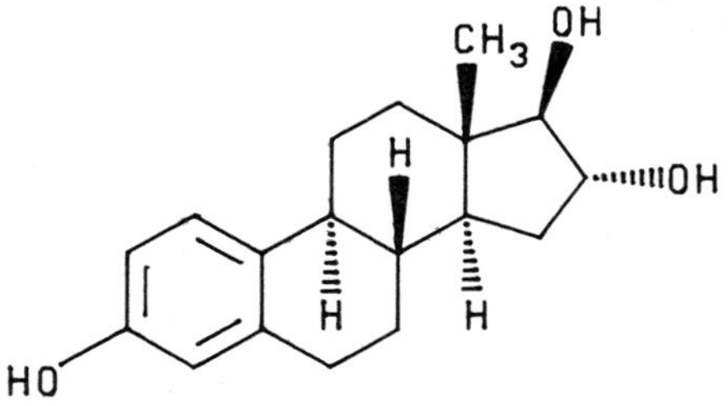

Estriol

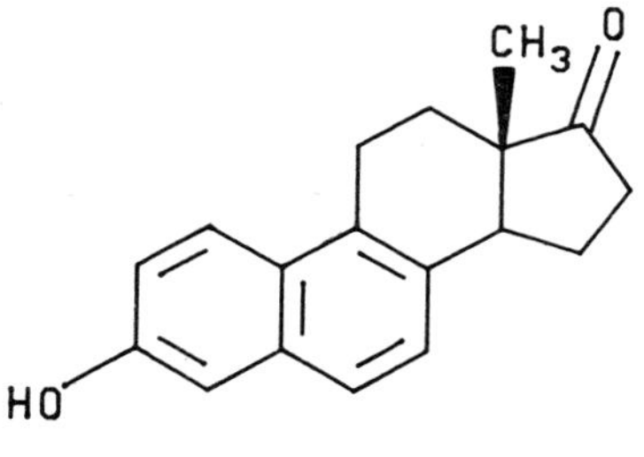

Equilenin

Figure 7.3

7.1 THE TRANSITION-METAL-CATALYZED TRIMERIZATION OF ALKYNES: INTRODUCTION

The oligomerization of acetylenes using Group VIII metals is well known to convert acetylene to cyclooctatetraene and monosubstituted acetylenes to tetrasubstituted cyclooctatetraenes. Disubstituted acetylenes do not cyclotetramerize but can be cooligermized with acetylene. Added phosphine ligands compete for coordination sites with alkyne; this competition results in conversion of a monosubstituted acetylene to

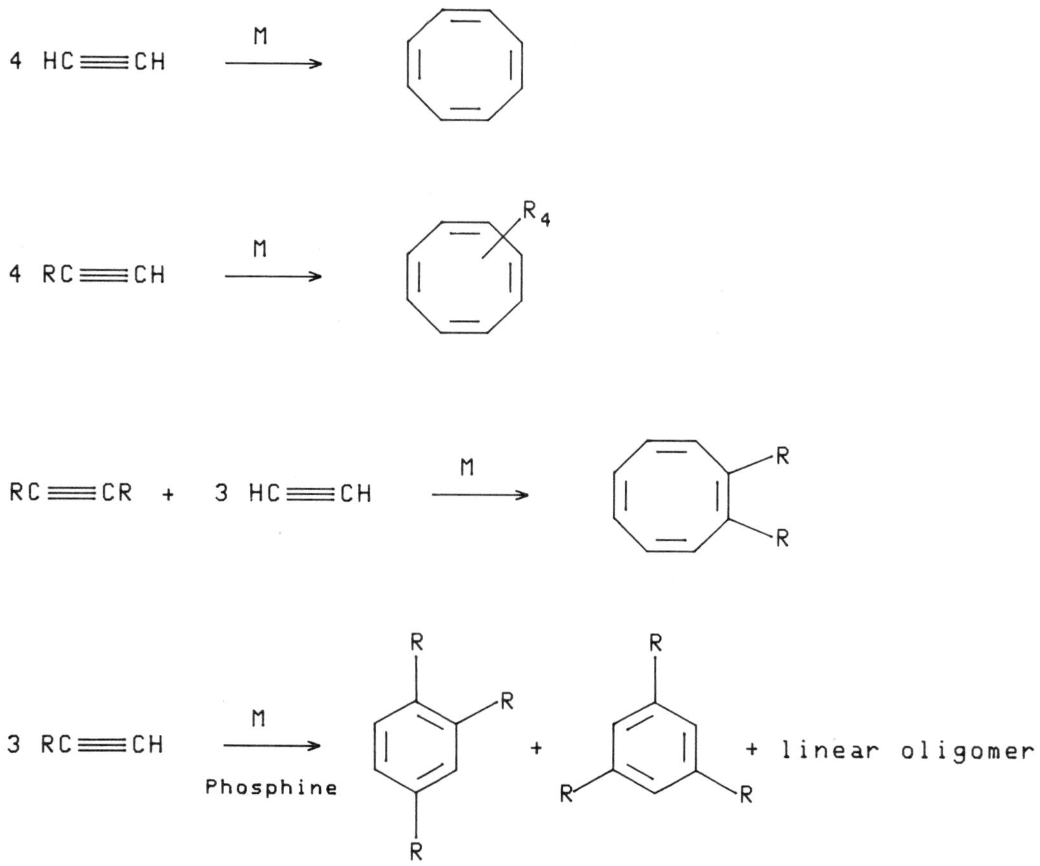

$$4\ HC{\equiv}CH \xrightarrow{M}$$

$$4\ RC{\equiv}CH \xrightarrow{M}$$

$$RC{\equiv}CR + 3\ HC{\equiv}CH \xrightarrow{M}$$

$$3\ RC{\equiv}CH \xrightarrow[\text{Phosphine}]{M}$$

+ linear oligomer

Figure 7.4

1,2,4- and 1,3,5-trisubstituted benzenes and linear oligomers (see Figure 7.4). Such oligomerizations are not restricted to the Group VIII metals but are common throughout transition metal–alkyne chemistry. Extensive studies on the cyclotrimerization of alkynes by η^5-cyclopentadienylcobalt dicarbonyl have resulted in a process suitable for synthetic application and extrapolation (see Figure 7.5).

A mechanism for cobalt-catalyzed cyclotrimerization of alkynes begins with the replacement of two carbon monoxides by alkynes. Oxidative coupling of the dialkynecobalt(I) complex produces the key cobaltacyclopentadiene (see Figure 7.6 and Table 7.2). Coordination of another alkyne can be followed by (1) insertion into a metal–carbon σ-bond to give a cobaltacycloheptatriene or (2) Diels–Alder reaction in the coordination sphere of the metal to give a cobaltabicyclo [2.2.1] heptadiene. Either can reductively eliminate cobalt(I) to produce the aromatic (see Table 7.3). In support of this mechanism, Wakatsuki and Yamazaki[8,9] converted bis-(triphenylphosphine)cyclopentadienylcobalt(I) in a stepwise fashion to isolable monoalkyne complexes, isolable cobaltacyclopentadienes, and aromatics.

Some regioselectivity in oxidative cyclization to a cobaltacyclopentadiene is clearly indicated by the results in Table 7.2. Both electronic and steric directing effects have

$CpCo(CO)_2 + RC{\equiv}CR \longrightarrow$ CpCo(RC≡CR)(CO) $\xrightarrow{RC{\equiv}CR}$ $Cp\overset{I}{Co}$(RC≡CR)$_2$

$\longrightarrow$ $Cp\overset{III}{Co}$ (metallacyclopentadiene, R$_4$) $\xrightarrow{RC{\equiv}CR}$ CpCo(metallacyclopentadiene)(RC≡CR)

$\longrightarrow$ CpCo (metallacycloheptatriene, R$_6$) or CpCo (bicyclic, R$_6$)

$\xrightarrow{-CpCo}$ C_6R_6

Figure 7.5

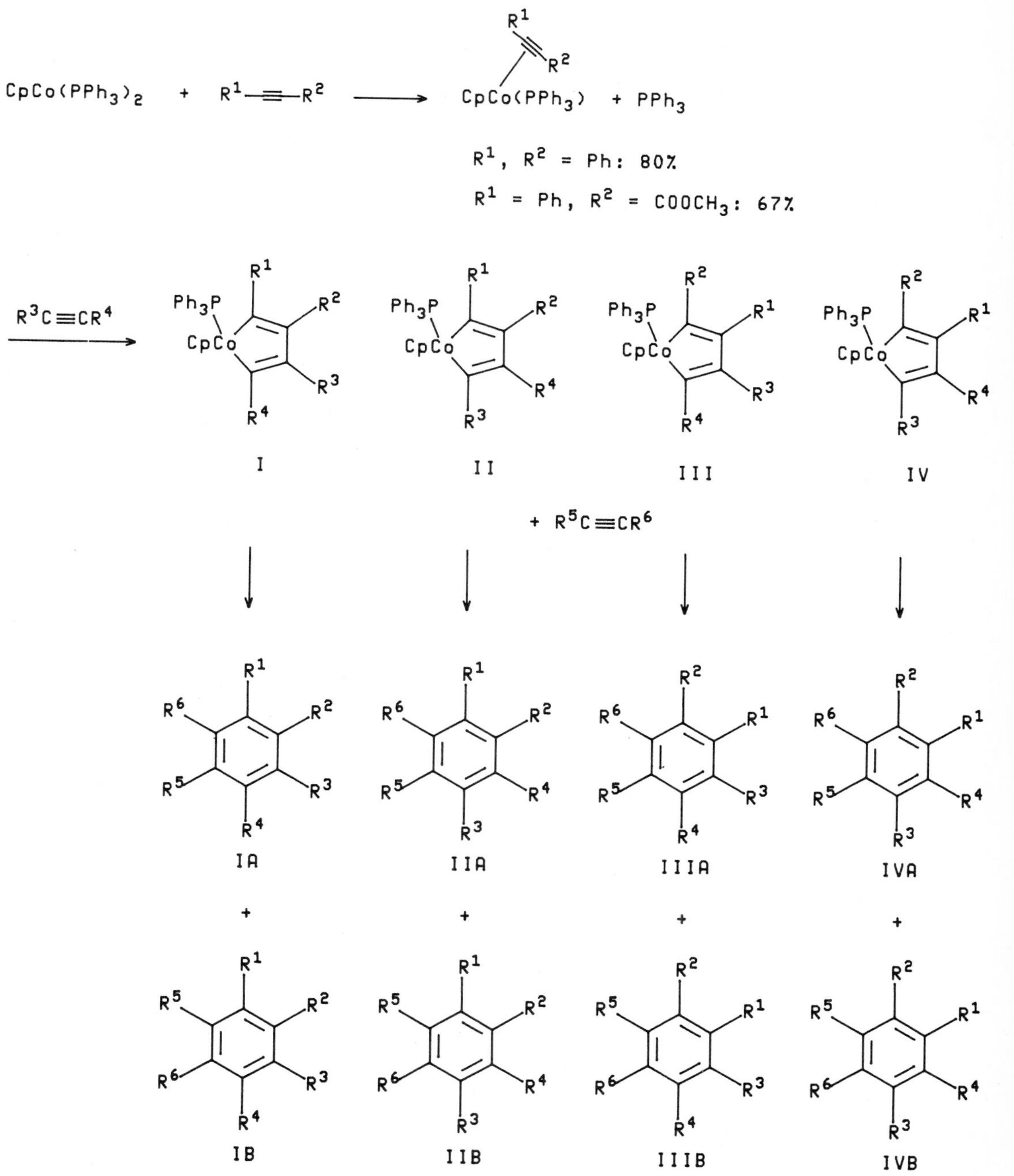

$CpCo(PPh_3)_2$ + $R^1—\equiv—R^2$ ⟶ $CpCo(PPh_3)(R^1C\equiv CR^2)$ + PPh_3

$R^1, R^2 = Ph$: 80%

$R^1 = Ph, R^2 = COOCH_3$: 67%

$\xrightarrow{R^3C\equiv CR^4}$ I + II + III + IV

+ $R^5C\equiv CR^6$

IA + IB; IIA + IIB; IIIA + IIIB; IVA + IVB

Figure 7.6

TABLE 7.2 Preparation of Cobaltacyclopentadienes[8,9]

R^1	R^2	R^3	R^4	Isolated Yields (%) I	II	III	IV
Ph	Ph	Ph	Ph	88	—	—	—
$COOCH_3$	$COOCH_3$	$COOCH_3$	$COOCH_3$	14	—	—	—
CH_2OCH_3	CH_2OCH_3	CH_2OCH_3	CH_2OCH_3	36	—	—	—
Ph	Ph	$COOCH_3$	$COOCH_3$	48	—	—	—
Ph	Ph	CH_2OCH_3	CH_2OCH_3	40	—	—	—
Ph	Ph	CH_3	$COOCH_3$	68	—	—	—
Ph	Ph	H	$COOCH_3$	48	—	—	—
Ph	Ph	H	Ph	27	—	—	—
Ph	Ph	H	p-$CH_3C_6H_4$	55	—	—	—
Ph	Ph	CH_3	Ph	67	—	—	—
Ph	CH_3	CH_3	Ph	54	—	—	—
Ph	Ph	Ph	$COOCH_3$	5	43	—	—
Ph	$COOCH_3$	Ph	$COOCH_3$	29	26	—	—
$COOCH_3$	CH_3	CH_3	$COOCH_3$	9	50	—	—
$COOCH_3$	CH_3	Ph	$COOCH_3$	9	39	—	—

TABLE 7.3 Preparation of Aromatics from Cobaltacyclopentadienes[a]

$$\mathrm{Ph_3P(CpCo)C_4R^1R^2R^3R^4} + R^5C{\equiv}CR^6 \longrightarrow C_6R^1R^2R^3R^4R^5R^6$$

R^1	R^2	R^3	R^4	R^5	R^6	Product Yield (%)
Ph	Ph	Ph	Ph	$COOCH_3$	$COOCH_3$	17
Ph	Ph	Ph	Ph	H	$COOCH_3$	22
Ph	CH_3	CH_3	Ph	$COOCH_3$	$COOCH_3$	13
Ph	CH_3	CH_3	Ph	H	$COOCH_3$	26
Ph	CH_3	CH_3	Ph	Ph	$COOCH_3$	59
$COOCH_3$	CH_3	$COOCH_3$	CH_3	Ph	Ph	16
Ph	$COOCH_3$	CH_3	$COOCH_3$	Ph	Ph	36

[a]Used with permission from Pergamon Press.

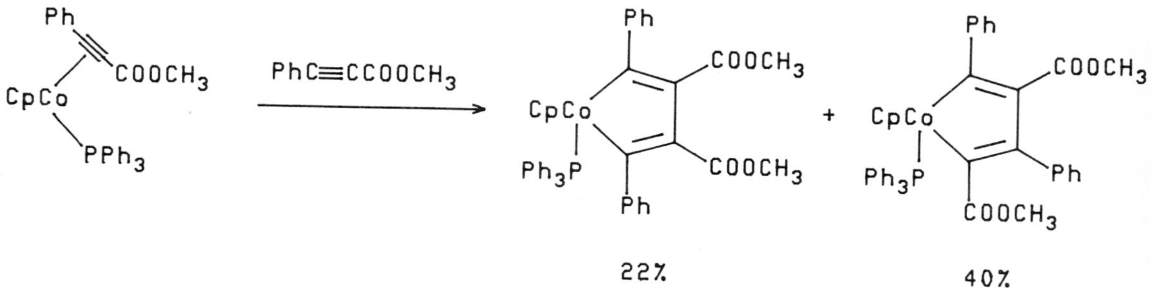

Figure 7.7

been suggested. The controlling orbital in oxidative cyclization is the HOMO of the complex, the ligand part of which is dominated by π^*. Polarity in the acetylene π^* then results in regioselectivity. The electronic effect should result in cobaltacyclopentadienes with electron-withdrawing acetylene substituents adjacent to the metal.[10] The data in Table 7.2 are more compatible with steric direction. The effect results in cobaltacyclopentadienes with bulky substituents adjacent to the metal in order to minimize interaction with other ring substituents[11] (see Figure 7.7).

7.2 USE OF α,ω-DIYNES

The stepwise approach can produce aromatics from three different alkynes, often with good regioselectivity. The catalytic cyclotrimerization produces mixtures that are often difficult to separate. The synthesis of aromatics using a catalytic amount of cobalt(I) can be made efficient by tethering two alkynes together and adding the diyne slowly via a syringe pump to a non-self-trimerizable monoalkyne such as bis-trimethylsilylacetylene as solvent (see Table 7.4). The excess bistrimethylsilylacetylene can be distilled from the reaction mixture and recycled without buildup of impurities. Benzocyclobutenes, indanes, and tetralins can be made; extension to higher ring sizes is not viable.[12–15]

7.3 MECHANISM WITH α,ω-DIYNES

Two mechanisms are proposed for cotrimerization of an α,ω-diyne with a monoacetylene. In mechanism A the α,ω-diyne coordinates to cobalt via one acetylene, then the other. The diacetylene complex then oxidatively cyclizes to a fused cobaltacyclopentadiene. This then coordinates the monoacetylene en route to a fused aromatic. In mechanism B the α,ω-diyne coordinates the cobalt via one acetylene. The monoacetylene then enters. This diacetylene complex then oxidatively cyclizes to a cobaltacyclopentadiene. Intramolecular cobalt coordination of the remaining acetylene then leads to a fused aromatic (see Figures 7.8 and 7.9).

Mechanism B provides a very clear explanation of the preferred formation of more sterically hindered benzocyclobutene products. Cotrimerization of 1-trimethylsilyl-1,5-hexadiyne and trimethylsilylacetylene can produce as many as eight cobaltacyclopentadienes. Apparently the complex with minimal steric interaction is preferred.

TABLE 7.4 [2 + 2 + 2] Alkyne + Alkyne + Alkyne Cyclotrimerizations Using α,ω-Diynes[12–15]

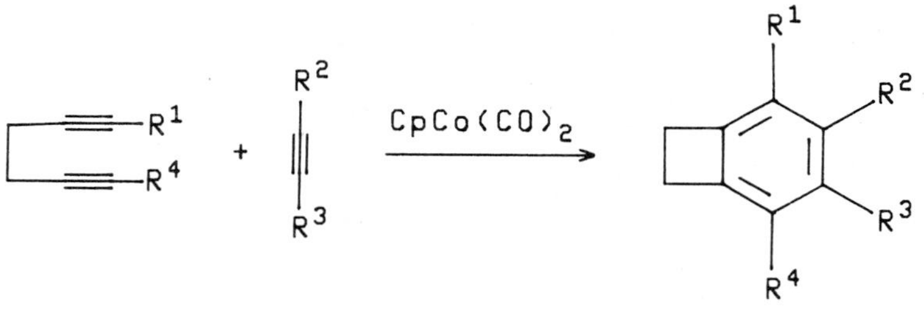

R^1	R^2	R^3	R^4	Yield (%)	Note
H	Ph	Ph	H	48	Diyne–monoaklyne, 1:1
H	$COOCH_3$	$COOCH_3$	H	14	Diyne–monoaklyne, 1:1
H	H	Ph	H	17	Diyne–monoaklyne, 1:1
H	H	CH_2OH	H	14	Excess monoalkyne
H	$Si(CH_3)_3$	$Si(CH_3)_3$	H	> 60	Excess monoalkyne
H	$CH_2Si(CH_3)_3$	$Si(CH_3)_3$	H	64	Excess monoalkyne
H	CH_2OCH_3	$Si(CH_3)_3$	H	55	Excess monoalkyne
CH_2OCH_3	$Si(CH_3)_3$	H	CH_2OCH_3	26	Excess monoalkyne
$Si(CH_3)_3$	$Si(CH_3)_3$	H	H	22	Excess monoalkyne
$Si(CH_3)_3$	$Si(CH_3)_3$	H	$Si(CH_3)_3$	2	Excess monoalkyne
$Si(CH_3)_3$	$Si(CH_3)_3$	CH_2OCH_3	H	63	Excess monoalkyne

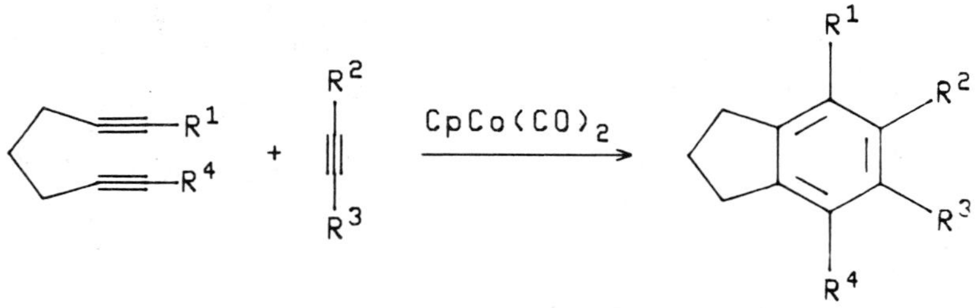

R^1	R^2	R^3	R^4	Yield (%)	Note
H	$Si(CH_3)_3$	$Si(CH_3)_3$	H	80	Excess monoalkyne
H	$COOCH_3$	$COOCH_3$	H	20	Diyne–monoaklyne, 1:1
H	Ph	H	H	26	Diyne–monoaklyne, 1:1
H	Ph	Ph	H	24	Diyne–monoaklyne, 1:2
H	n-C_6H_{13}	H	H	14	Diyne–monoaklyne, 1.2:1

TABLE 7.4 (*Continued*)

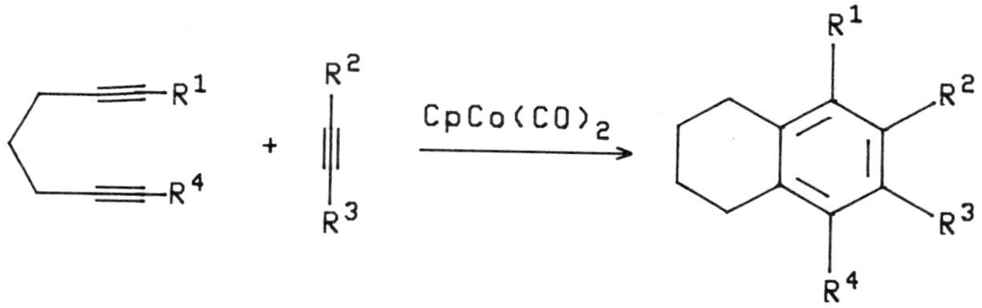

R^1	R^2	R^3	R^4	Yield (%)	Note
H	$Si(CH_3)_3$	$Si(CH_3)_3$	H	83	Excess monoalkyne
H	$COOCH_3$	$COOCH_3$	H	26	Diyne–monoaklyne, 1:1
H	Ph	H	H	18	Diyne–monoaklyne, 1:1
H	Ph	Ph	H	21	Diyne–monoaklyne, 1:1
H	n-C_6H_{13}	H	H	14	Diyne–monoaklyne, 1:1
H	$Si(CH_3)_3$	CH_3	H	34	Diyne–monoaklyne, 1:2

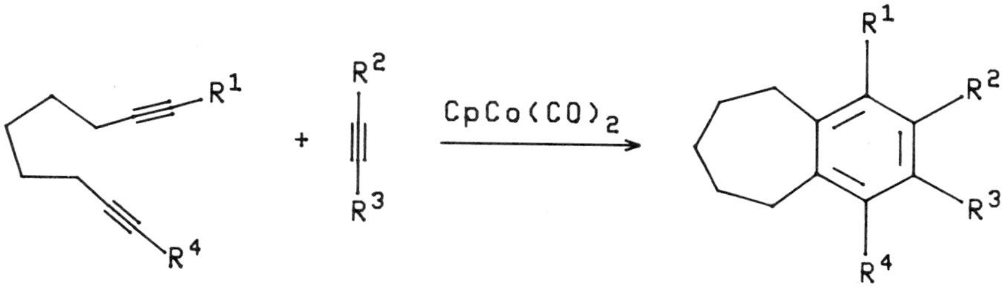

R^1	R^2	R^3	R^4	Yield (%)
H	$Si(CH_3)_3$	$Si(CH_3)_3$	H	3

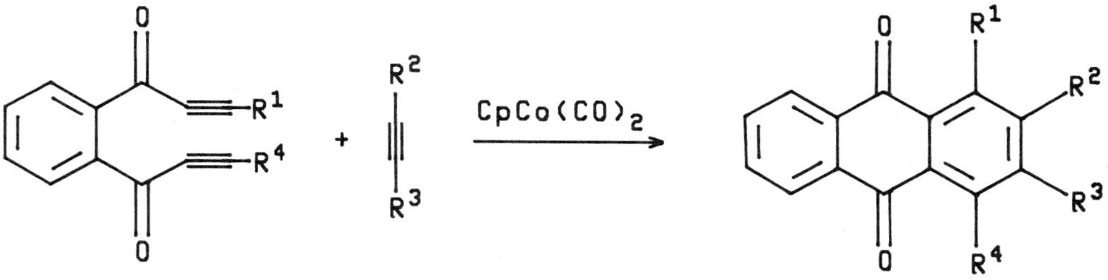

R^1	R^2	R^3	R^4	Yield (%)
H	$Si(CH_3)_3$	$Si(CH_3)_3$	H	15
$Si(CH_3)_3$	n-C_6H_{13}	H	$Si(CH_3)_3$	21

Mechanism A:

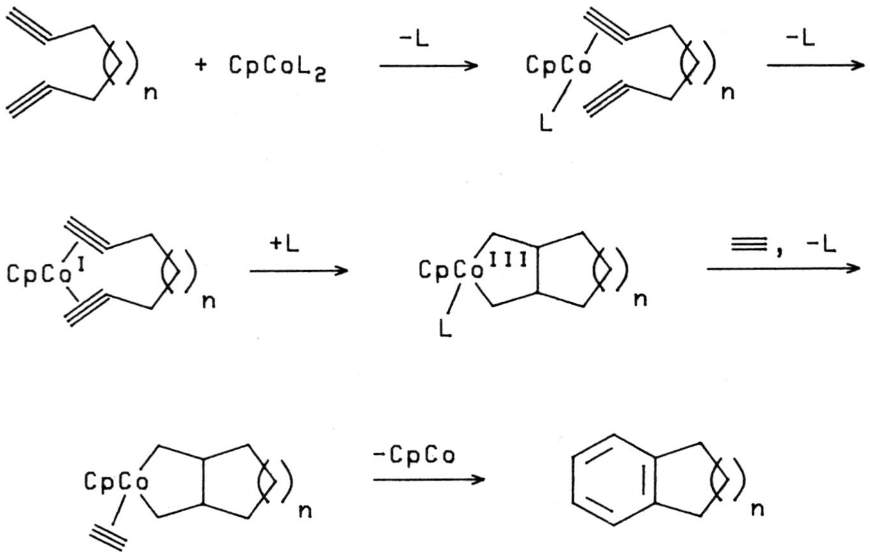

Figure 7.8

Mechanism B:

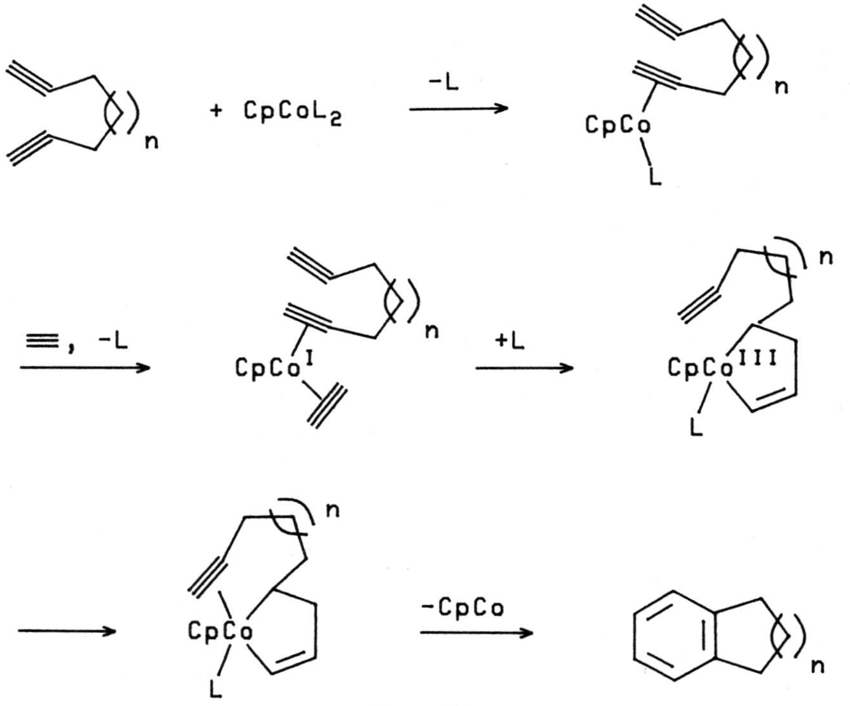

Figure 7.9

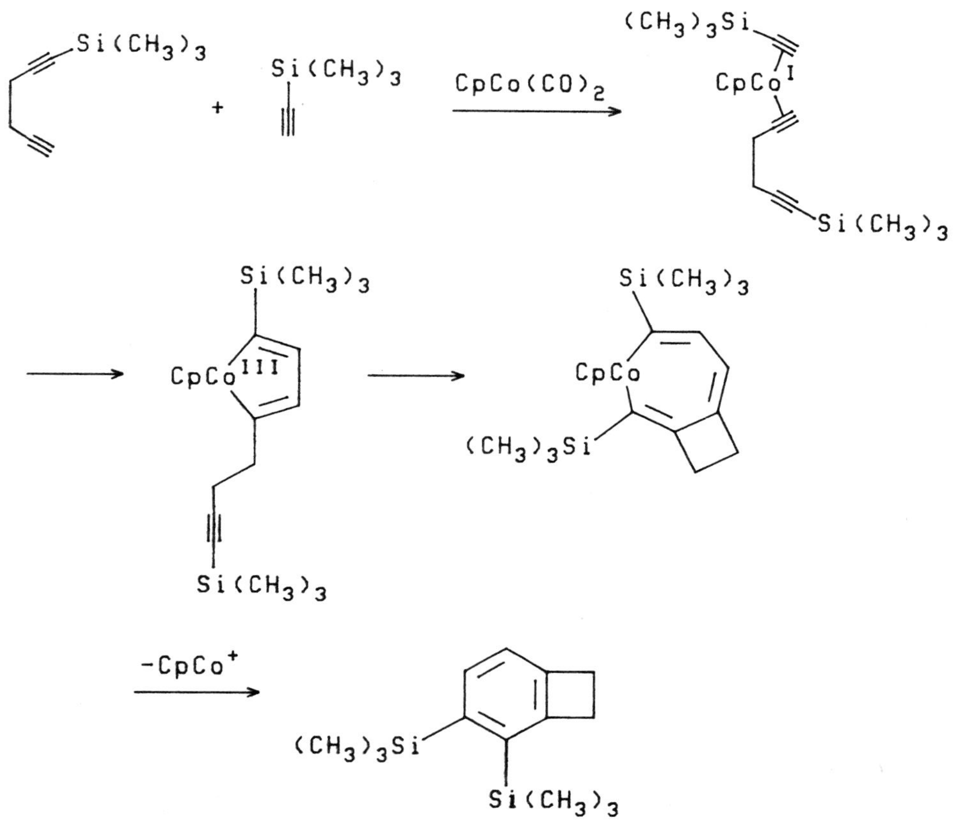

Figure 7.10

The considerable steric interaction produced by subsequent aromatic formation is more than offset by the increase in resonance stabilization (see Figure 7.10).

An efficiency-limiting side reaction in the cotrimerization of α,ω-diynes with silylated acetylenes is catalyst depletion by valence tautomerization of intermediate cobaltacyclopentadienes to *catalytically inactive* η^4-cyclobutadiene complexes.[16] Three different types of η^4-cyclobutadiene complexes are observed (see Figure 7.11).

R. G. Bergman was able to isolate fused cobaltacyclopentadienes from 1,6-heptadiynes and 1,7-octadiynes and $CpCo(PPh_3)_2$. Diynes with shorter or longer connecting chains gave more complex reactions. From this we suggest the following:

1. Mechanism B is preferred with 1,5-hexadiynes since it avoids a highly strained fused cobaltacyclopentadiene.
2. Mechanism A is preferred with 1,6-heptadiynes and 1,7-octadiynes since formation of the metallacycle is facile.
3. Both mechanisms may operate with diynes with still larger connecting chains.[17]

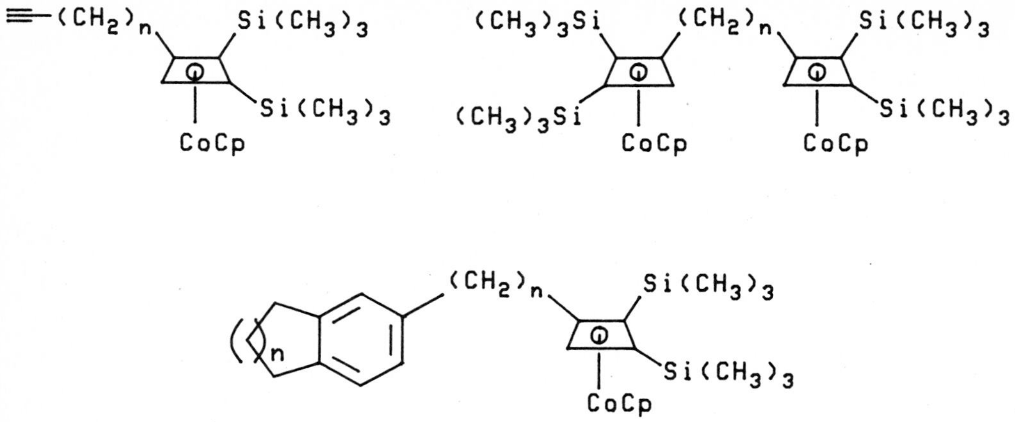

Figure 7.11

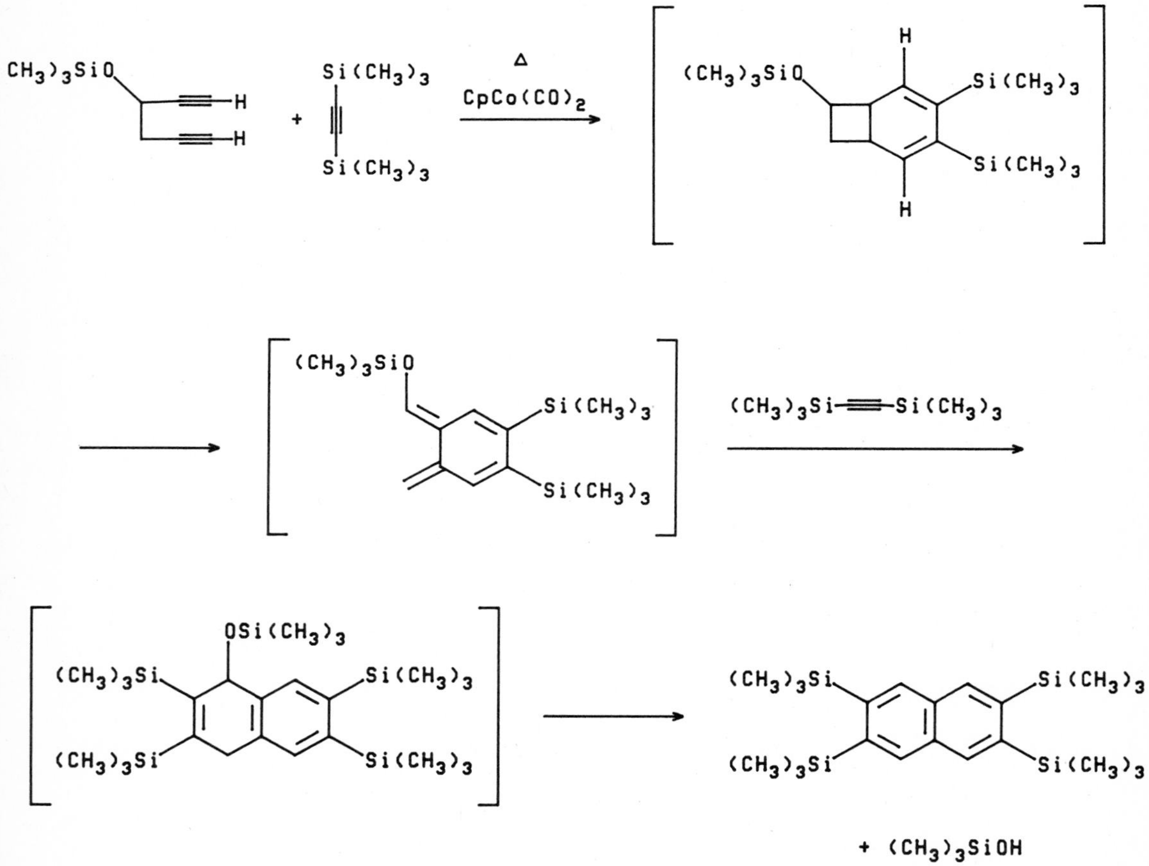

Figure 7.12

7.4 BENZOCYCLOBUTENES TO o-XYLYLENES

Thermal ring opening of a benzocyclobutene affords a highly reactive *o*-xylylene capable of Diels–Alder cycloaddition. Thus, prolonged heating of a mixture of 3-trimethylsilyloxyhexa-1,5-diyne, cyclopentadienylcobalt dicarbonyl (5 mol%), and excess bis-trimethylsilylacetylene results first in benzocyclobutene, then *o*-xylylene, and then Diels–Alder adduct. Aromatization by loss of alcohol occurs either during the reaction or on chromatographic purification (see Figure 7.12).

3-Trimethylsiloxyhexa-1,5-diyne is prepared from the corresponding alcohol.[18] Alternative Williamson ether synthesis with 5-bromo-1-pentene affords a diyne with side chain capable of intramolecular Diels–Alder capture of the *o*-xylylene[19,20] (see Figure 7.13).

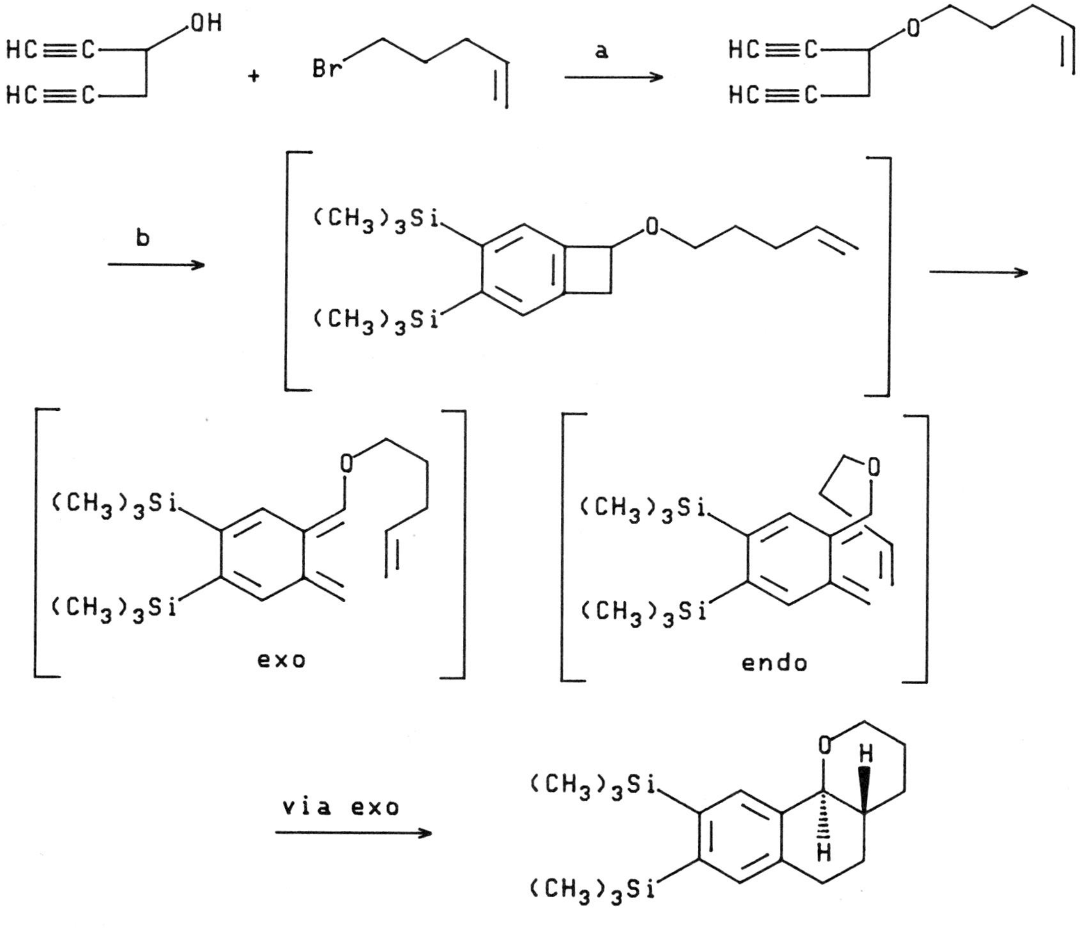

a. 1) rt, NaH, 2) 45°C, THF; 70%

b. rfx, $CpCo(CO)_2$, $(CH_3)Si$—≡—$Si(CH_3)_3$, octane; 60%

Figure 7.13

Since no cobalt-catalyzed *cis–trans* isomerization can be demonstrated in analogous systems, *trans*-naphthopyran is formed stereospecifically via an *exo* transition state. Other examples of intramolecular Diels–Alder trapping of an *o*-xylylene are in Figures 7.14 and 7.15. Note that regioisomers of the expected cycloadduct are often observed when there is sufficient flexibility (i.e., a longer connecting chain between diene and dienophile) in the transition state for *o*-xylylene cycloaddition or when electronic effects operate (dienophile C=O).

o-Xylylene formation is considerably more difficult with 2-alkyl-substituted benzocyclobutenes.[21] The benzocyclobutene can be isolated and then converted to tricyclic material in a separate step. The requisite 3-alkylated-1,5-diynes are prepared from 1,5-diynes via the trilithio derivatives (see Figure 7.16).

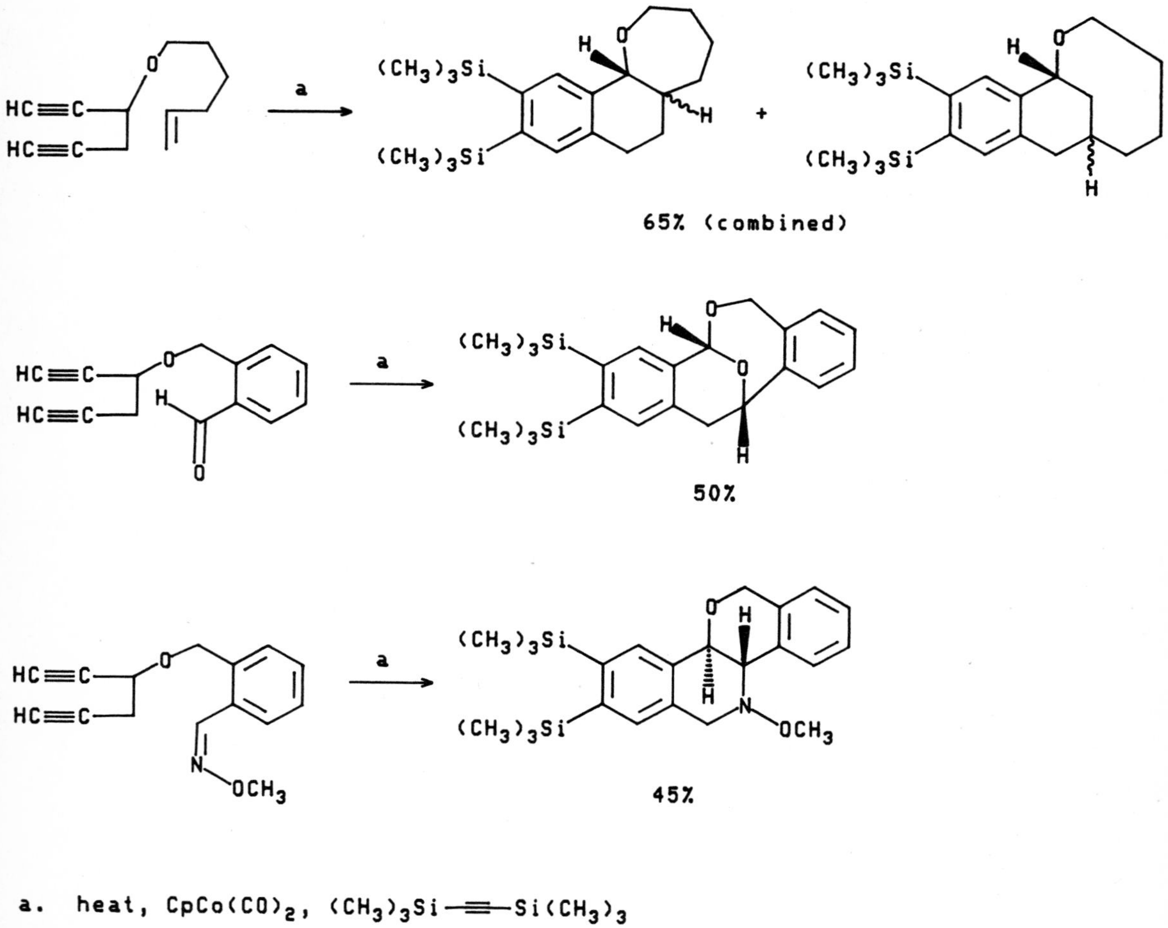

65% (combined)

50%

45%

a. heat, $CpCo(CO)_2$, $(CH_3)_3Si{-}{\equiv}{-}Si(CH_3)_3$

Figure 7.14

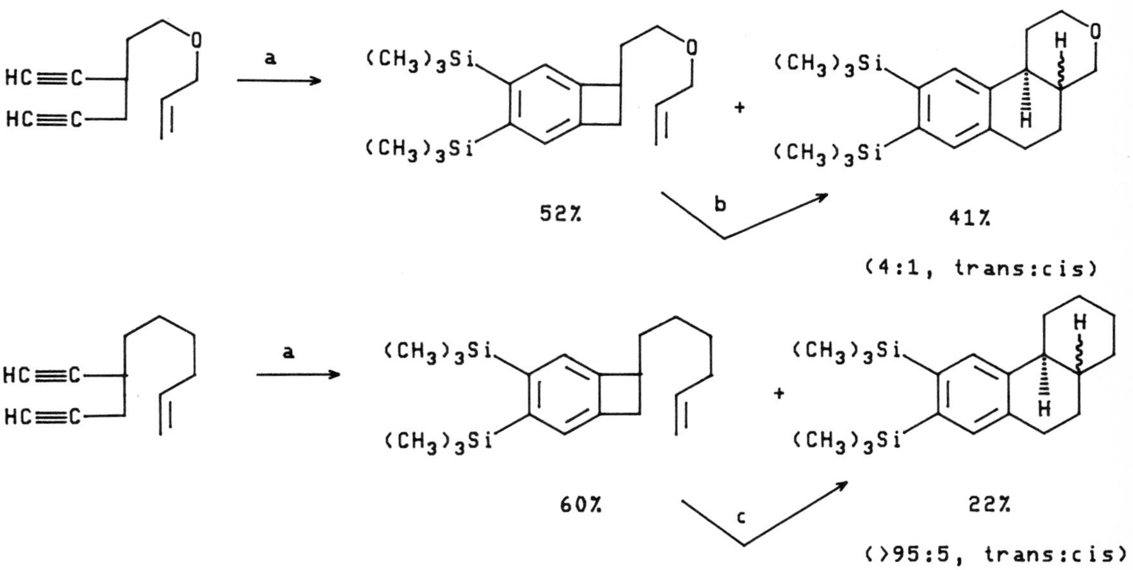

$(CH_3)_3Si$ $(CH_3)_3Si$ $HC{\equiv}C$ a b 52% 41% (4:1, trans:cis)

$(CH_3)_3Si$ $(CH_3)_3Si$ a c 60% 22% (>95:5, trans:cis)

a. heat, $CpCo(CO)_2$, $(CH_3)_3Si{\equiv}Si(CH_3)_3$

b. rfx, decane; 93%

c. rfx, decane; 97%

Figure 7.15

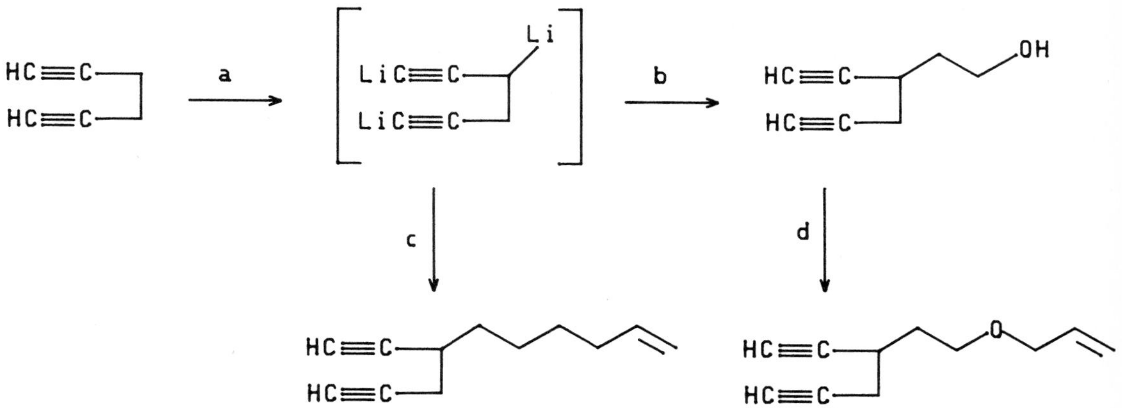

a b c d

a. −30°C, TMEDA, 3 n-BuLi, THF, hexane

b. −65 to −10°C, ethylene oxide; 65%

c. −50°C to rt, $CH_2{=}CH(CH_2)_4Cl$; 84%

d. 1) rt, NaH, THF, 2) rt to 50°C, $CH_2{=}CHCH_2Br$; 86%

Figure 7.16

7.5 SYNTHETIC MANIPULATION OF THE AROMATIC TRIMETHYLSILYL SUBSTITUENTS

Vollhardt demonstrated that the most efficient cyclotrimerizations, those using bis-trimethylsilylacetylene, are synthetically viable and even versatile by (1) replacing the trimethylsilyl ring substituents with electrophiles, (2) quantitatively replacing one of the *ortho*-situated trimethylsilyl groups while leaving the other in place, and (3) regioselectively replacing one trimethylsilyl group in an unsymmetrical cyclotrimer.

Silicon is less electronegative than carbon (electronegativities are 1.8 and 2.5, respectively). Thus silylated aromatics are inductively activated toward attack by electrophiles, and attack results in clean displacement of the relatively stable trimethylsilyl cation (see Figure 7.17).

Steric interaction between the two *ortho*-situated trimethylsilyl substituents results in *steric acceleration*[22] of the replacement of the first silicon. (Rate enhancements of 30 to 40 times are common). This provides a mechanism, then, for quantitative replacement of one silicon. Synthetic manipulation of 2,3,6,7-*tetrakis*-(trimethylsilyl)naphthalene illustrates the power of this ring functionalization methodology.[20,23] Coupling replacement by halide with palladium(0)-catalyzed alkene arylations and carbonylations provides a wide range of regioselectively substituted aromatics (see Figure 7.18).

Protodesilylation is run in trifluoroacetic acid-rich mixtures. Use of dilute acid results in *ortho* to *meta* silicon migration, driven by relief of steric interaction.[24] The *meta*-bis-trimethylsilyl aromatics produced open new possibilities for regioselective ring functionalization. Trimethylsilyl migration and replacements on 2,3-bis-trimethylsilylbenzocyclobutene are illustrated[15] in Figures 7.19 and 7.20.

Having demonstrated replacement and stepwise replacement of trimethylsilyl substituents, the remaining question is regioselectivity with unsymmetrical cyclotrimers. Regioselectivity is observed when replacement of one silicon will significantly decrease steric interactions[15] (see Figure 7.21).

Without this steric effect one must rely on some electronic preference: a greater

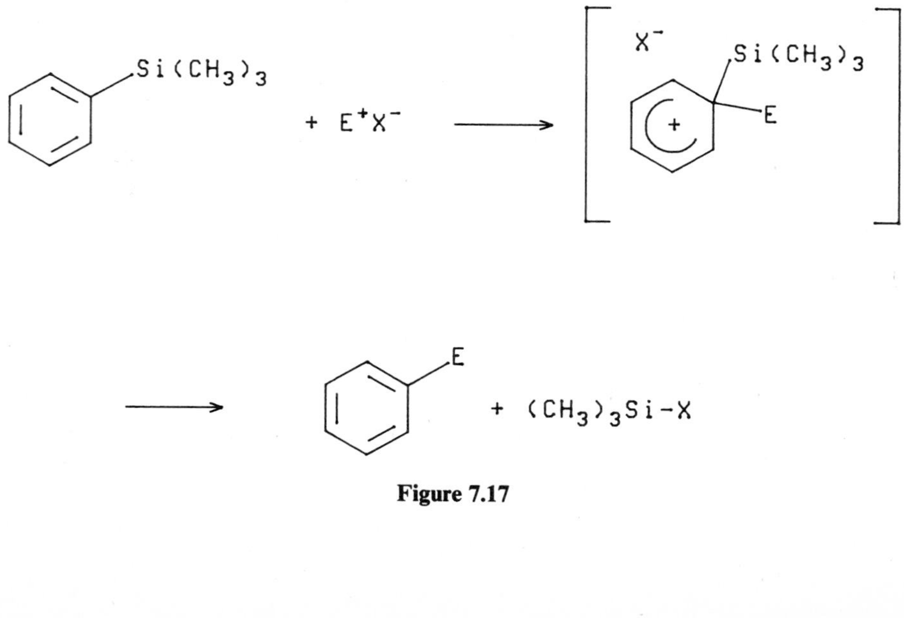

Figure 7.17

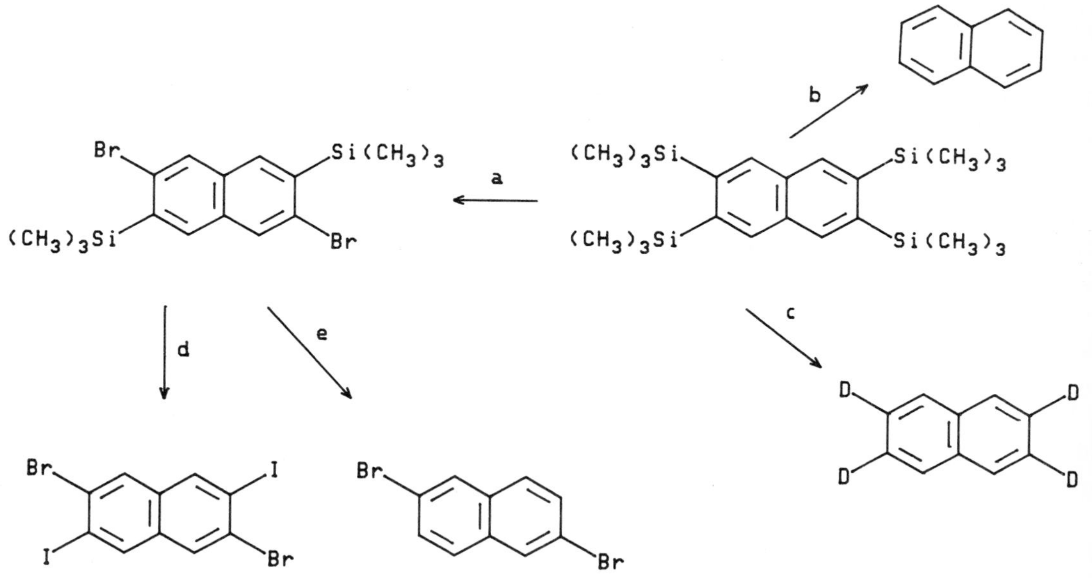

a. cold to rt, Br_2, pyridine, CCl_4; 89%

b. rt, CF_3COOH, CCl_4; 100%

c. rt, CF_3COOD, CCl_4; 95%

d. rt, ICl, $CHCl_3$; 88%

e. rt, CF_3COOH, CCl_4; 95%

Figure 7.18

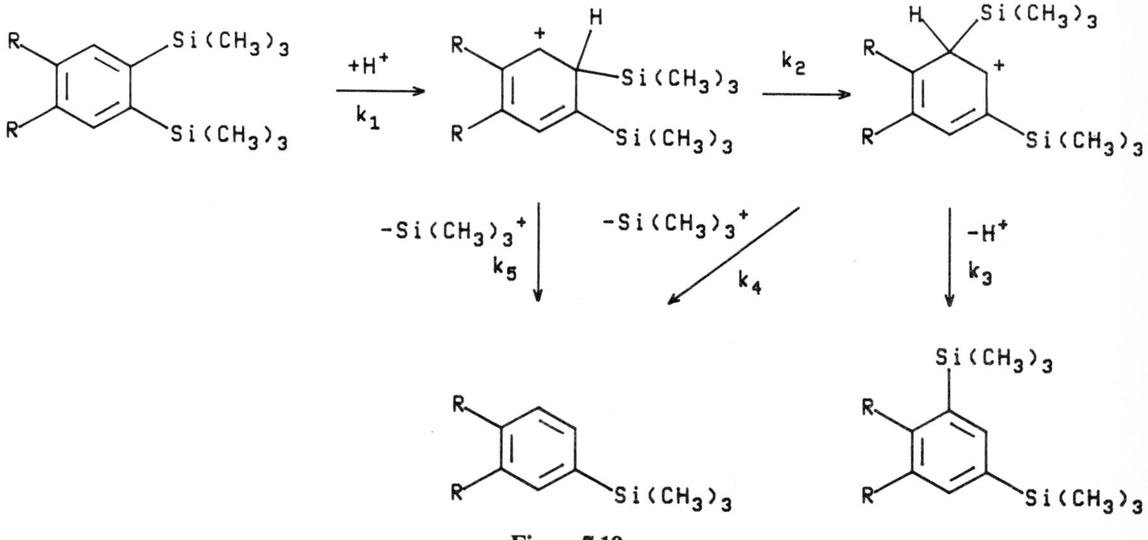

Figure 7.19

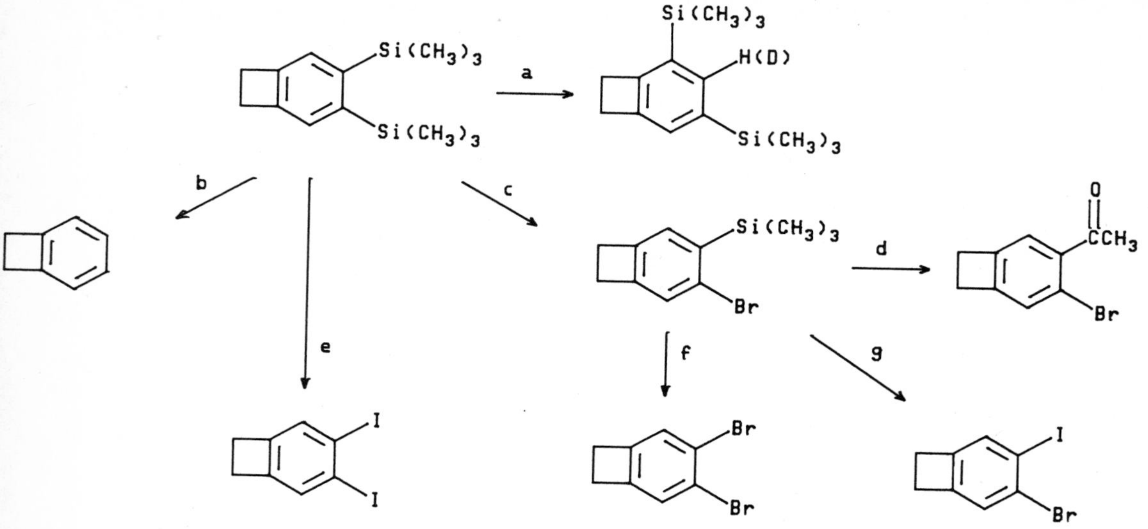

a. dilute $CF_3COOH(D)$; H, 65% by NMR

b. rt, CF_3COOH, CCl_4; >90%

c. cold, Br_2, pyridine, CCl_4; 92%

d. rt, AcCl, $AlCl_3$, CS_2; 56%

e. rt, ICl, CCl_4; 94%

f. rt, Br_2, CCl_4; 95%

g. rt, ICl, CCl_4; 95%

Figure 7.20

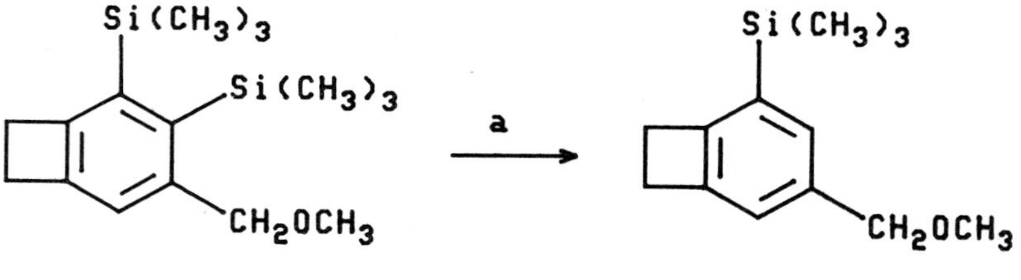

a. rt, CF_3COOH, CH_3COOH, CCl_4; 78%

Figure 7.21

reactivity toward electrophiles of one silicon-bearing ring carbon. An electronic effect is observed in bromodesilylation of a steriod model[20] and an estrone precursor. Excellent regioselectivity can be achieved at low temperature in a protodesilylation en route to estrone. Note the efficient replacement of the remaining trimethylsilyl by oxygen using lead tetratrifluoroacetate[25] (see Figure 7.22).

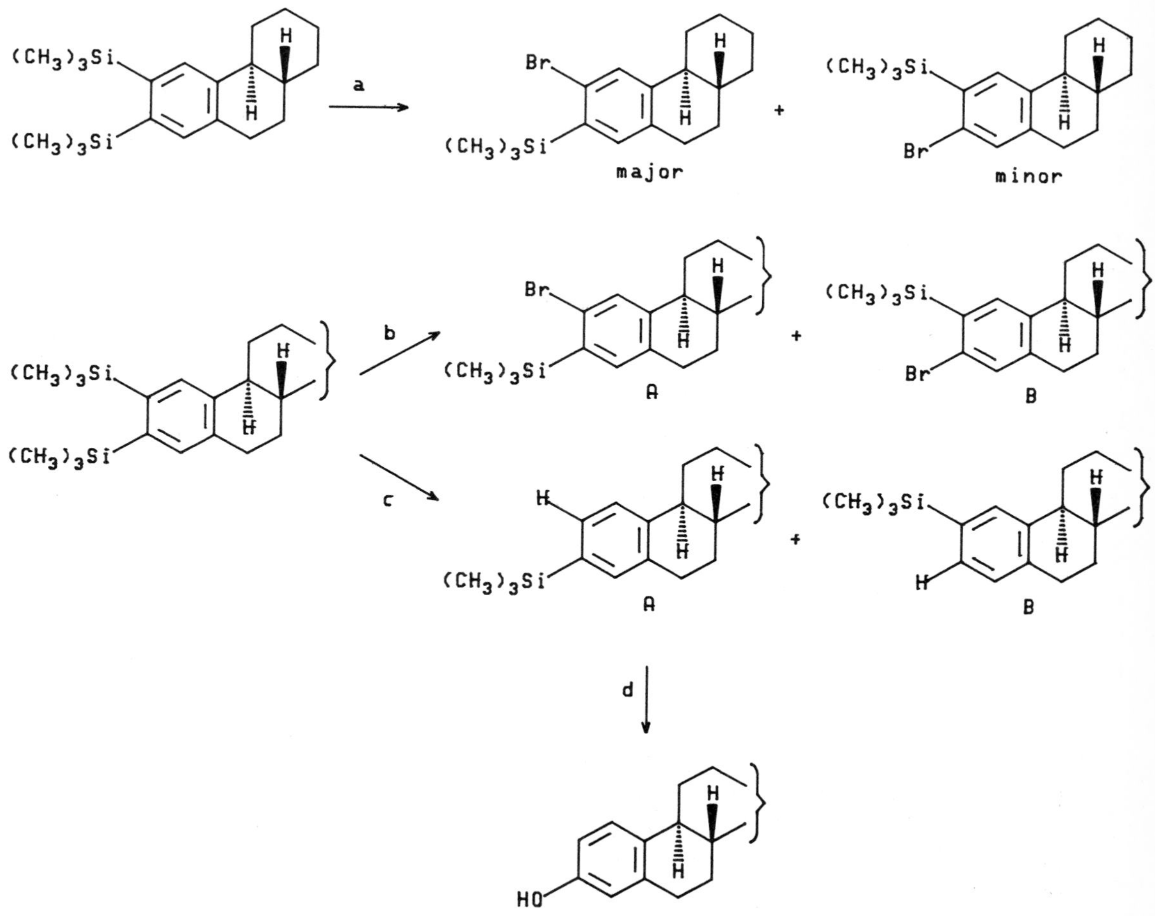

a. rt, 2 Br_2, pyridine, CCl_4; 93%

b. rt, 2 Br_2, pyridine, CCl_4; A:B, 4:1

c. −30°C, CF_3COOH; A:B, 9:1

d. rt, $Pb[OC(O)CF_3]_4$, CF_3COOH

Figure 7.22

7.6 [2 + 2 + 2] ALKYNE + ALKYNE + ALKENE

The [2 + 2 + 2]-cyclotrimerization of two alkynes and one alkene produces synthetically versatile η^4-cyclohexadienylcobalt complexes. Two approaches minimize or eliminate alternative oligomerizations of the alkynes and alkene. Method A involves stepwise construction of the η^4-cyclohexadienyl complex. Method B involves construction of a precursor containing both acetylenes and the alkene suitably positioned for cyclotrimerization. Wakatsuki's stepwise method (A) involves three steps: (1) preparation of a monoacetylene complex, (2) preparation of a diacetylene complex that oxidatively cyclizes to a cobaltacyclopentadiene, and (3) reaction with an alkene to produce the η^4-cyclohexadienyl complex.[26] The third step involves prior coordination of the alkene to the metal since the reaction is inhibited by addition of triphenylphosphine or replacement of triphenylphosphine by the stronger ligand diphenylmethylphosphine. Conversion of the π-alkene complex to product may involve either alkene insertion to give a cobaltacycloheptadiene or Diels–Alder cycloaddition in the coordination sphere of the metal. Cyclotrimerization is *exo*-stereoselective when the terminal substituents on the cobaltacyclopentadiene are both phenyl (see Table 7.5 and Figure 7.23.)

Vollhardt's method (B) begins with construction of an enediyne set for cobalt-mediated cyclotrimerization.[27] Enediynes with an internal alkene are prepared by Wittig reaction. *Cis–trans* alkene interconversion is effected by irradiation, and pure alkene isomers are obtained by preparative gas chromatography. Cyclotrimerization

TABLE 7.5 Preparation of η^4-Cyclohexadienylcobalt Complexes from Colbaltacyclopentadienes and Alkenes[26]

R¹	R²	R³	R⁴	R⁵	R⁶	*exo–endo* Ratio	Yield (%)
Ph	Ph	$COOCH_3$	$COOCH_3$	H	H	—	23
Ph	$COOCH_3$	Ph	$COOCH_3$	H	H	—	39
Ph	Ph	Ph	Ph	CH_3	H	ND	36
Ph	Ph	Ph	Ph	Ph	H	ND	9
Ph	CH_3	CH_3	Ph	$COOCH_3$	H	ND	50
Ph	Ph	H	$COOCH_3$	$COOCH_3$	$COOCH_3$	ND	75
Ph	Ph	Ph	Ph	$COOCH_3$	$COOCH_3$	10:90	40
Ph	Ph	$COOCH_3$	Ph	$COOCH_3$	$COOCH_3$	30:70	67
Ph	Ph	CH_3	$COOCH_3$	$COOCH_3$	$COOCH_3$	60:40	95
Ph	$COOCH_3$	Ph	$COOCH_3$	$COOCH_3$	$COOCH_3$	36:64	80
Ph	$COOCH_3$	CH_3	$COOCH_3$	$COOCH_3$	$COOCH_3$	60:40	95

ND = not determined.

$CpCo(PPh_3)_2 \xrightarrow[-PPh_3]{R^1C\equiv CR^2} \xrightarrow{R^3C\equiv CR^4}$

CpCo Ph3P R1 R2 R3 R4

R5 R6

CpCo R1 R2 R3 R4 R5 R6

CpCo R1 R2 R3 R4 H R6 R5 H

+

CpCo R1 R2 R3 R4 R6 H H R5

CpCo R2 R1 R6 R5 R3 R4 H H

exo

+

CpCo R2 R1 H H R3 R4 R5 R6

endo

R1 R6 R2 CoCp R5 R3 R4

R1 R6 R2 CoCp R5 R3 R4

Figure 7.23

TABLE 7.6 Preparation of η^4-Cyclohexadienylcobalt Complexes from Enediynes (Internal Alkene)[28,30]

Starting Material	$\xrightarrow{a}$ Product Complex	$\xrightarrow{b}$ Product
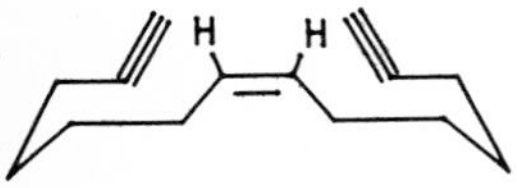 H H	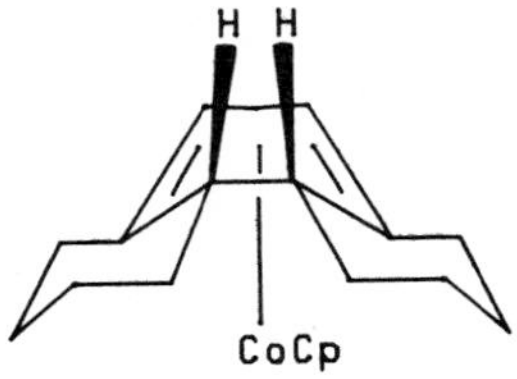 H H, CoCp 63%	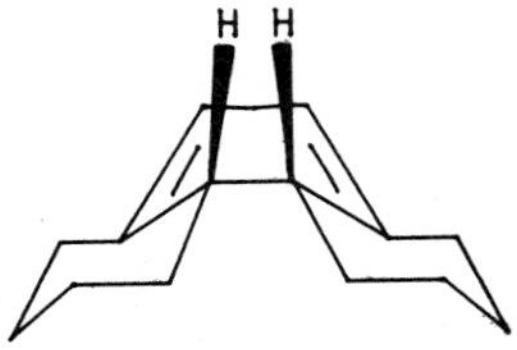 H H 93%
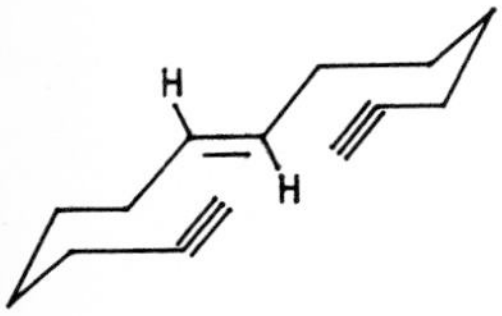 H, H	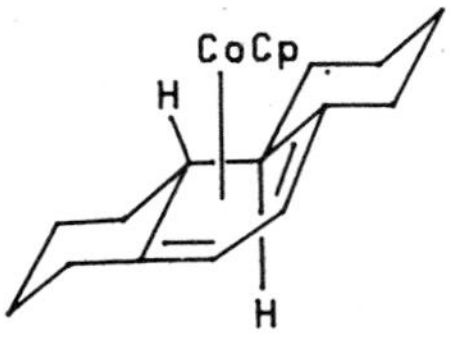 CoCp, H, H 63%	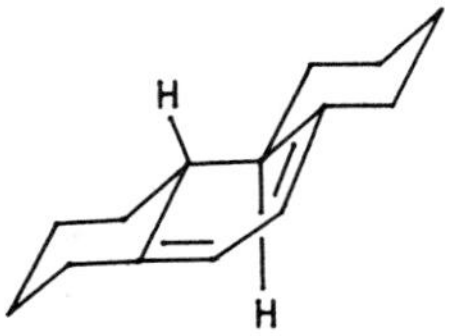 H, H 96%
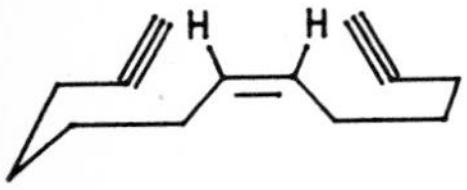 H H	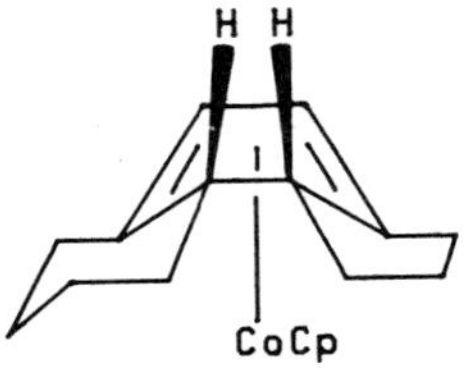 H H, CoCp 65%	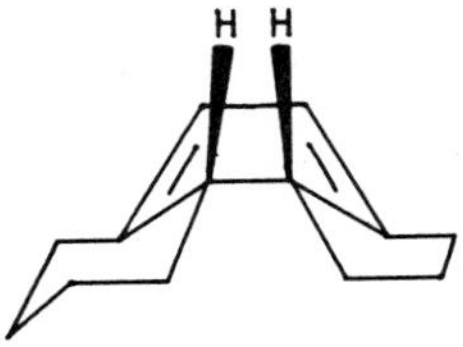 H H 86%
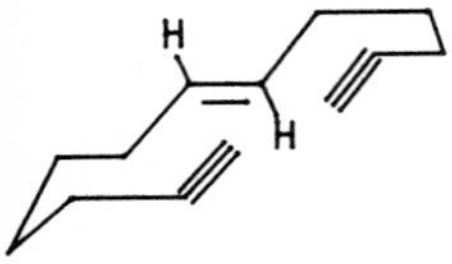 H, H	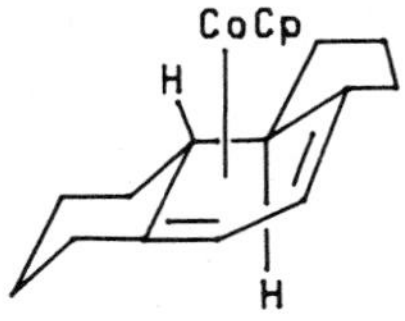 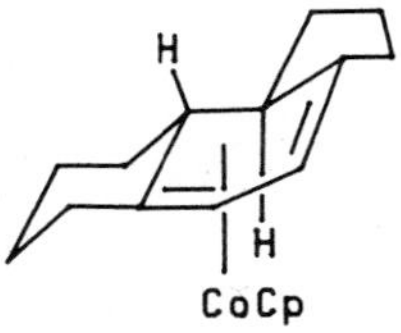 CoCp, H, H + H, H, CoCp (3:2 ratio) 63%	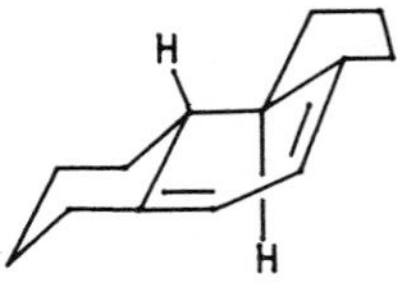 H, H 89%

TABLE 7.6 (*Continued*)

Starting Material $\xrightarrow{a}$	Product Complex $\xrightarrow{b}$	Product
CH_3 H	CoCp H_3C H 74%	H_3C H 84%
CH_3 H	CoCp H_3C H + CH_3 H CoCp (3:1 ratio) 76%	CH_3 H 92%
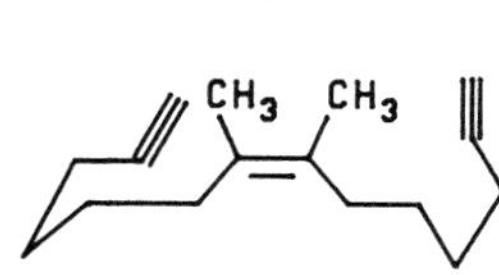	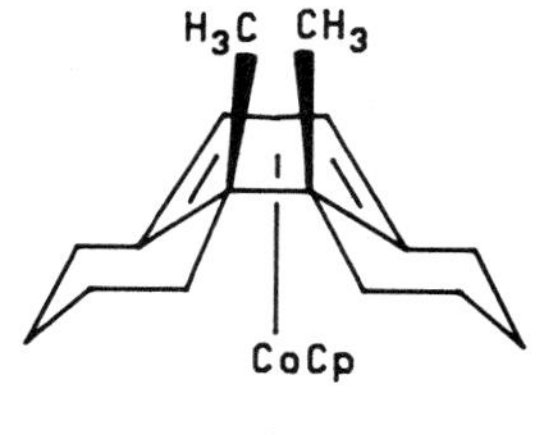 60%	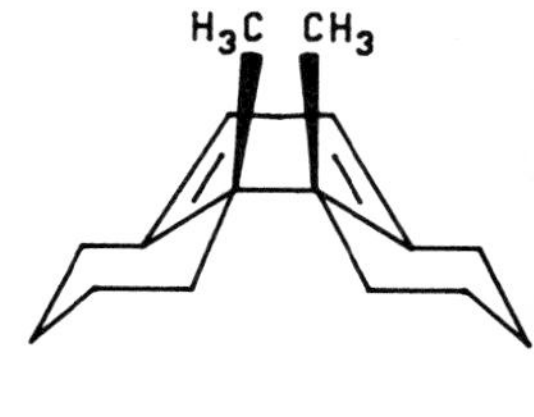 61%

Note: Demetallation on filtration through silica gel

a. heat, hν, $CpCo(CO)_2$, m-xylene

b. 0°C, $CuCl_2 \cdot 2H_2O$, Et_3N, CH_3CN

preserves the stereochemical relationship of alkene substituents, proceeds with high stereoselectivity with respect to cobalt, and is remarkably efficient even in the construction of two adjacent quaternary carbons![28] Oxidative demetallation produces the free dienes[29] (see Table 7.6).

Enediynes with a terminal alkene prepared by alkylation of diynes, also cyclotrimerize efficiently[30] (see Table 7.7).

Further manipulation of the η^4-cyclohexadienylcobalt complexes can be accomplished by (1) application of well-established chemistry of analogous η^4-cyclohexadienyliron complexes or (2) replacement of trimethylsilyl by electrophiles in the little known dienylsilanes obtained by mild oxidative demetallation. Poor *endo–exo* stereoselectivity with respect to cobalt in cyclotrimerization may complicate utilization of the first option. To illustrate, in one case where cyclotrimerization is

TABLE 7.7 Preparation of η^4-Cyclohexadienylcobalt Complexes from Enediynes (Terminal Alkene)[30]

Starting Material	$\xrightarrow{a}$ Product Complex	$\xrightarrow{b \text{ or } c}$ Product
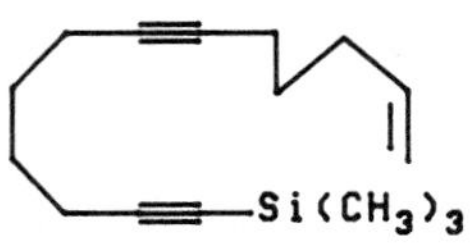 Enediyne, $Si(CH_3)_3$	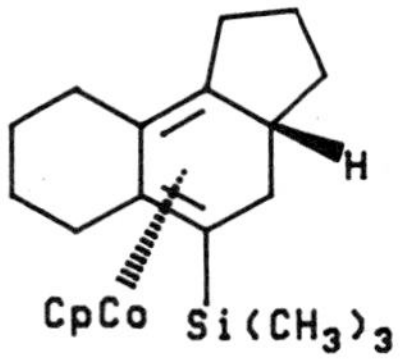 Complex (CpCo, $Si(CH_3)_3$, H) 85%	(b) Diene ($Si(CH_3)_3$, H) 85% 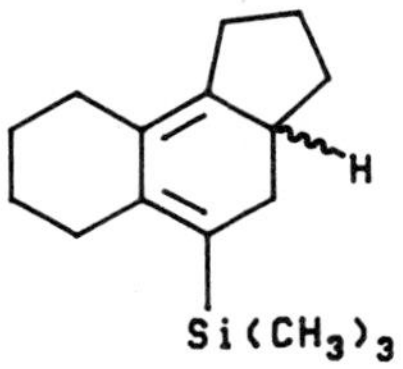
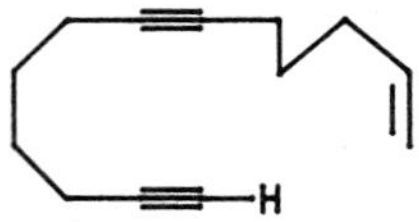 Enediyne, H	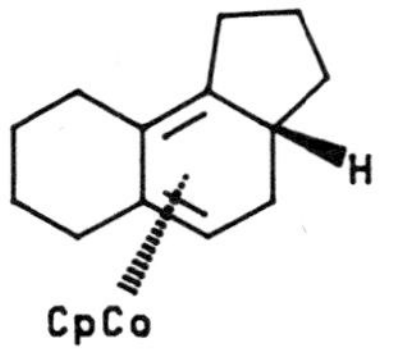 exo (CpCo, H) + endo (CpCo, H) 66%	Demetallation not reported

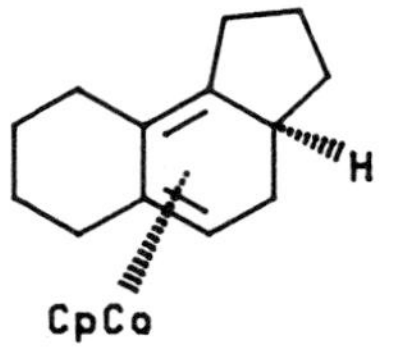

TABLE 7.7 *(Continued)*

Starting Material $\xrightarrow{a}$ Product Complex $\xrightarrow{\text{b or c}}$ Product

exo

CpCo Si(CH$_3$)$_3$

Si(CH$_3$)$_3$

+

(b)

Si(CH$_3$)$_3$

endo

CpCo Si(CH$_3$)$_3$

80%

55:45 ratio, 92%

exo

CpCo

+

(c)

H

endo

67%

CpCo

+ another isomer

2:1.5:1 ratio, 76%

O Si(CH$_3$)$_3$

O CpCo Si(CH$_3$)$_3$

34%

(b)

O Si(CH$_3$)$_3$

93%

a. rfx, CpCo(CO)$_2$, isooctane

b. 0°C, $CuCl_2 \cdot 2H_2O$, Et_3N, CH_3CN

c. rt, $(NH_4)_2Ce(NO_3)_6$, CH_3CN

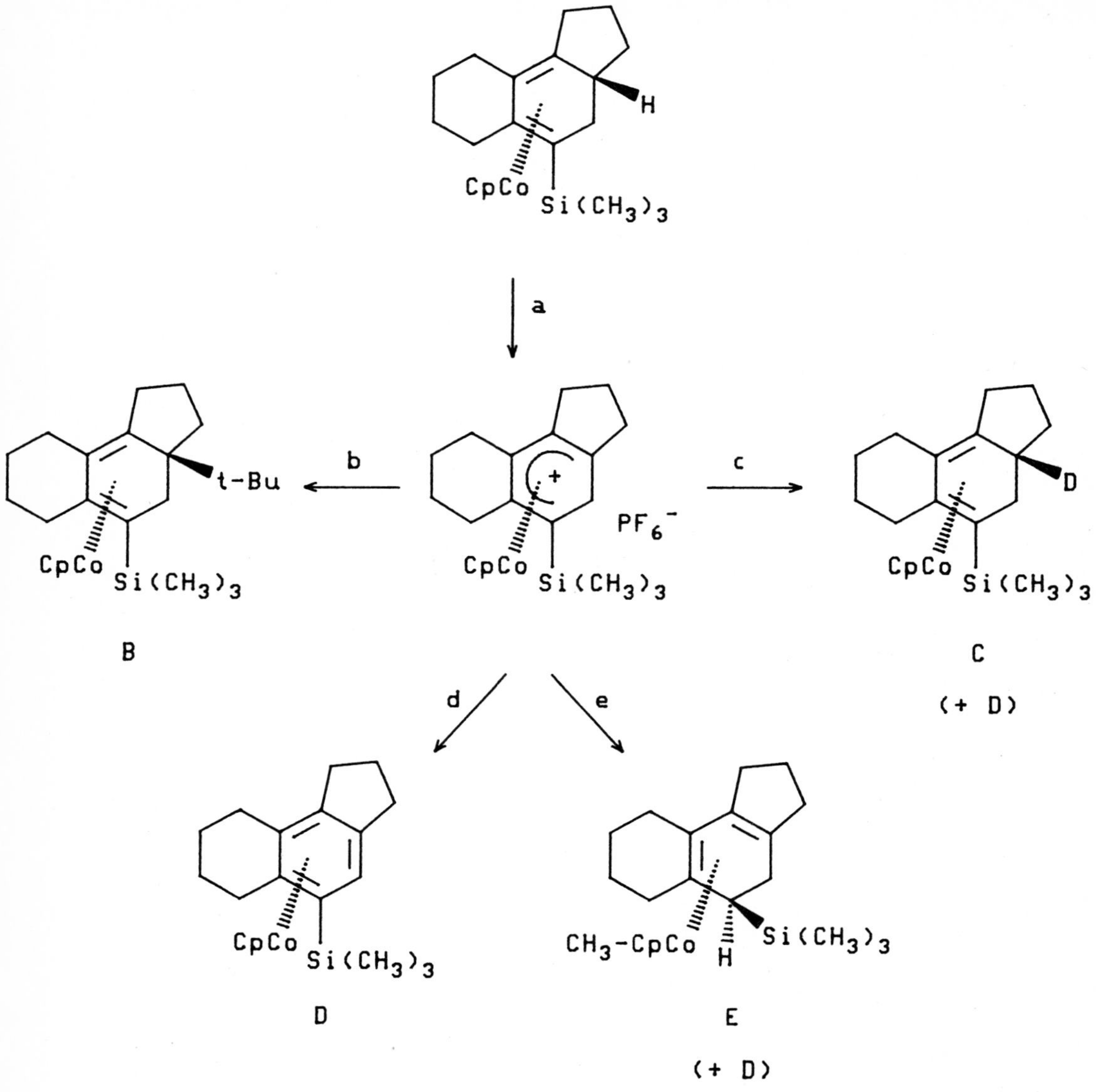

CpCo
Si(CH3)3
H

a

b

t-Bu

CpCo
Si(CH3)3

B

CpCo
Si(CH3)3
PF6−

c

D

CpCo
Si(CH3)3

C

(+ D)

d

e

CpCo
Si(CH3)3

D

CH3-CpCo
H
Si(CH3)3

E

(+ D)

a. rt, Ph_3CPF_6, CH_2Cl_2; 70%

b. −78°C, t-BuLi, THF; 77%

c. −78°C to rt, $NaBD_4$, CH_3OH, petroleum ether; C:D, 1:1

d. 0°C, K_2CO_3, CH_3OH, petroleum ether; 85%

e. −78°C, CH_3Li, THF; E:60% (+ D:15%)

Figure 7.24

efficient and *exo*-stereospecific, the resulting η^4-cyclohexadienyl complex is converted to a single η^5-cyclohexadienyl cation by regio- and stereospecific *exo*-hydride abstraction. *Exo*-stereospecificity is well known with the analogous iron complexes (see Chapter 4). Regiospecificity may be due to steric bulk of the trimethylsilyl group. Reaction with sodium borodeuteride and with *t*-butyllithium affords the *exo*-nucleophilic addition product. Deprotonation affords an unusual and rather unstable η^4-arene complex. Finally, methyllithium initially attacks the cyclopentadienyl ligand. *Endo*-hydride addition (perhaps by hydride transfer from Cp to cobalt to the η^5-cyclohexadienyl system) affords the observed product (see Figure 7.24).

The other option is oxidative demetallation, then manipulation of the dienylsilane. To illustrate, treatment of the dienylsilane with *m*-chloroperbenzoic acid affords an epoxide that rapidly ring opens to a tertiary carbocation. Silicon migration and

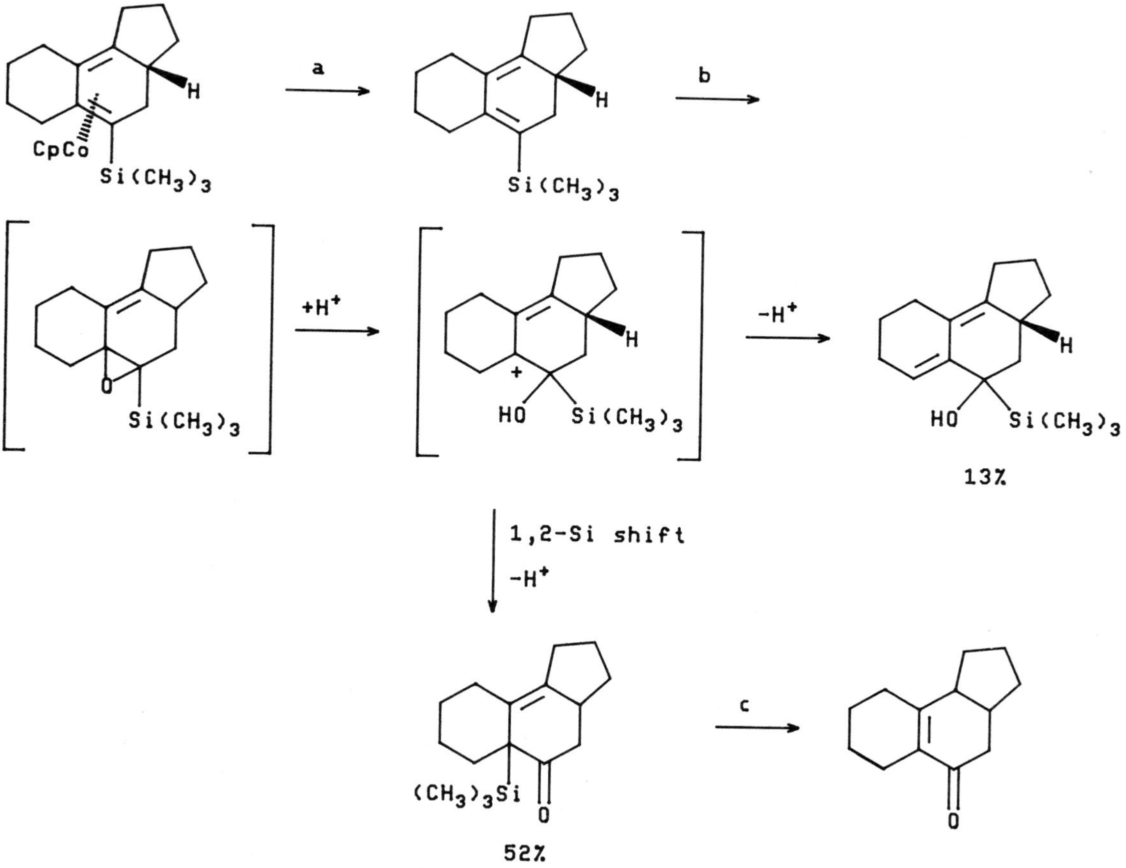

CpCo H $Si(CH_3)_3$ → a → H $Si(CH_3)_3$ → b →

[O $Si(CH_3)_3$] → $+H^+$ → [+ H HO $Si(CH_3)_3$] → $-H^+$ → H HO $Si(CH_3)_3$ 13%

↓ 1,2-Si shift $-H^+$

$(CH_3)_3Si$ O 52% → c → O

a. 0°C, $CuCl_2 \cdot 2H_2O$, Et_3N, CH_3CN; 85%

b. 0°C, MCPBA, CH_2Cl_2

c. 0°C, CF_3COOH, CH_2Cl_2; 69%

Figure 7.25

subsequent proton loss affords a ketone as major product. Simple proton loss affords the silyl alcohol as minor product. The α-trimethylsilyl-β,γ-unsaturated ketone is converted to an α,β-unsaturated ketone by treatment with trifluoroacetic acid (see Figure 7.25).

7.7 [2 + 2 + 2] ALKYNE + ALKYNE + NITRILE AND [2 + 2 + 2] ALKYNE + ALKYNE + ISOCYANATE

Two synthetically important extensions of alkyne trimerization not pertinent to steroid synthesis are the cobalt-catalyzed [2 + 2 + 2]-cyclotrimerization of two alkynes with

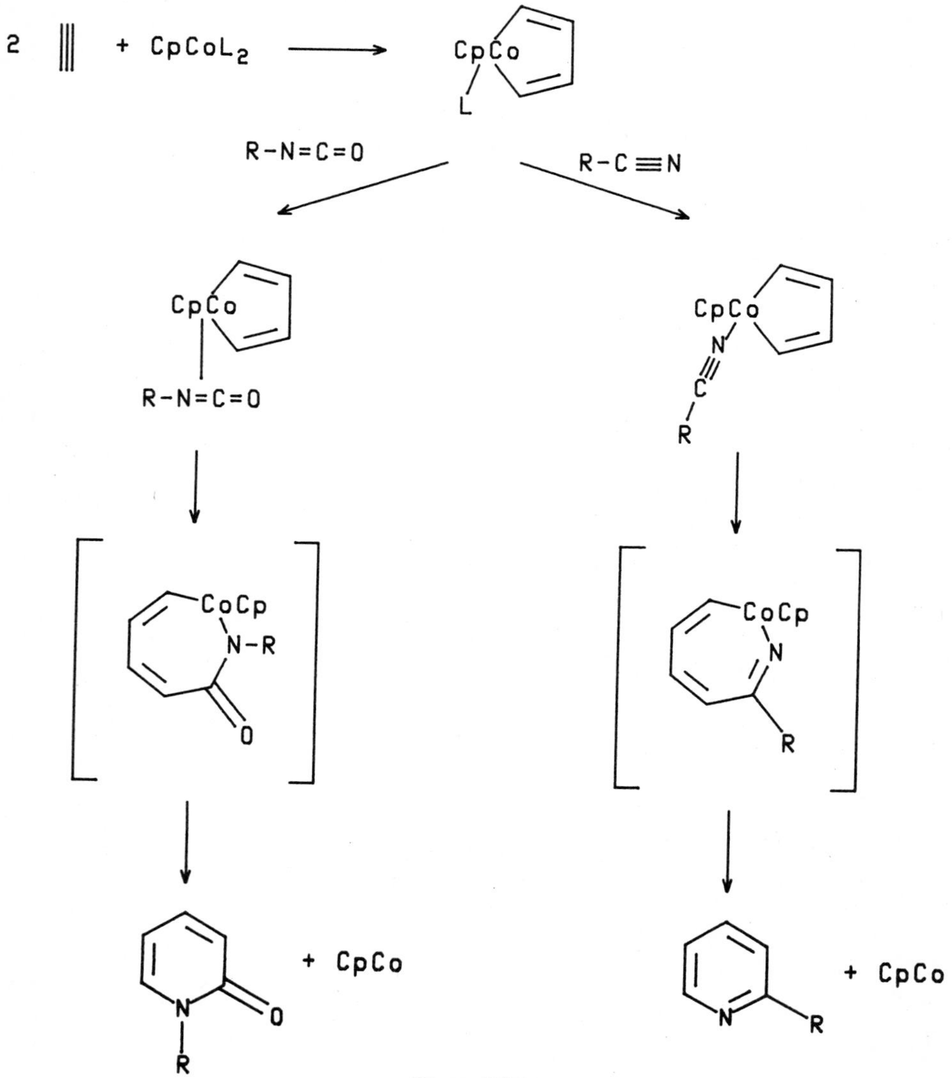

2 ||| + CpCoL2 → CpCo(L)

R−N=C=O
R−C≡N

CpCo
R−N=C=O

CpCo
N
C
R

CoCp
N−R
O

CoCp
N
R

N
R
O
+ CpCo

N
R
+ CpCo

Figure 7.26

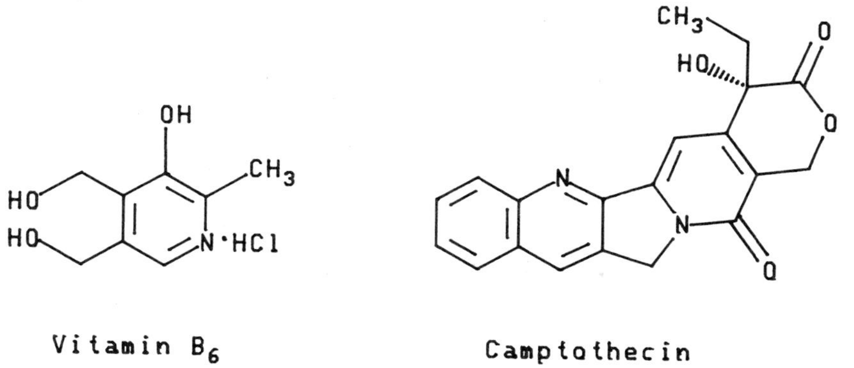

Figure 7.27

one nitrile and two alkynes with one isocyanate to give pyridines and 2-pyridones, respectively.[31–36] Both apparently involve an intermediate cobaltacyclopentadiene (see Figure 7.26).

Cocyclization of a diyne with a nitrile is the key step in a cobalt-based total synthesis of vitamin B_6.[37] Cocyclization of an yne-isocyanate with an alkyne is the key step in a cobalt-based total synthesis of the antitumor alkaloid camptothecin[38] (Figure 7.27).

7.8 TOTAL SYNTHESIS OF (±)-ESTRONE

7.8.1 Synthesis A. From an Intact D-Ring Precursor[39–41]

Synthesis A of (±)-estrone begins with an intact D ring. Synthesis of the ABC portion is based on construction of the aromatic A ring by [2 + 2 + 2]-cyclotrimerization of alkynes. Synthesis of the steroid framework can be broken down into three operations. First, a functionalized diyne is prepared. Treatment of 1,5-hexadiyne[42] with three equivalents of *n*-butyllithium affords the trilithio derivative. This is trapped with ethylene oxide. The resulting alcohol is converted to a tosylate, and the tosylate replaced by iodide. Second, the diyne is attached to an intact and functionalized D ring. Michael addition of vinyl cuprate to 2-methyl-2-cyclopentenone[43] followed by treatment with trimethylsilyl chloride affords the silyl enol ether. Clean conversion of the silyl enol ether back to enolate is followed by a stereoselective alkylation using the iodide. Finally, the diyne is efficiently cotrimerized with bis-trimethylsilylacetylene by slow addition (using a syringe pump, over 35 h) to a refluxing mixture of the cobalt catalyst in bis-trimethylsilylacetylene.[44] The intermediate benzocyclobutene slowly ring opens to an *o*-xylylene, which then undergoes an intramolecular *exo*-stereospecific Diels–Alder cycloaddition to give the steroid framework in the correct *trans–anti–trans*-configuration in a remarkable 71% yield! (See Figure 7.28.)

The *trans* C–D ring juncture is set in construction of the diyne precursor. An *exo*-transition state for the Diels–Alder cycloaddition, either in pseudochair or pseudoboat form (as observed in model studies), leads to the trans BC ring juncture. The pseudochair would afford the *trans-anti-trans* product, the pseudoboat *trans–syn–*

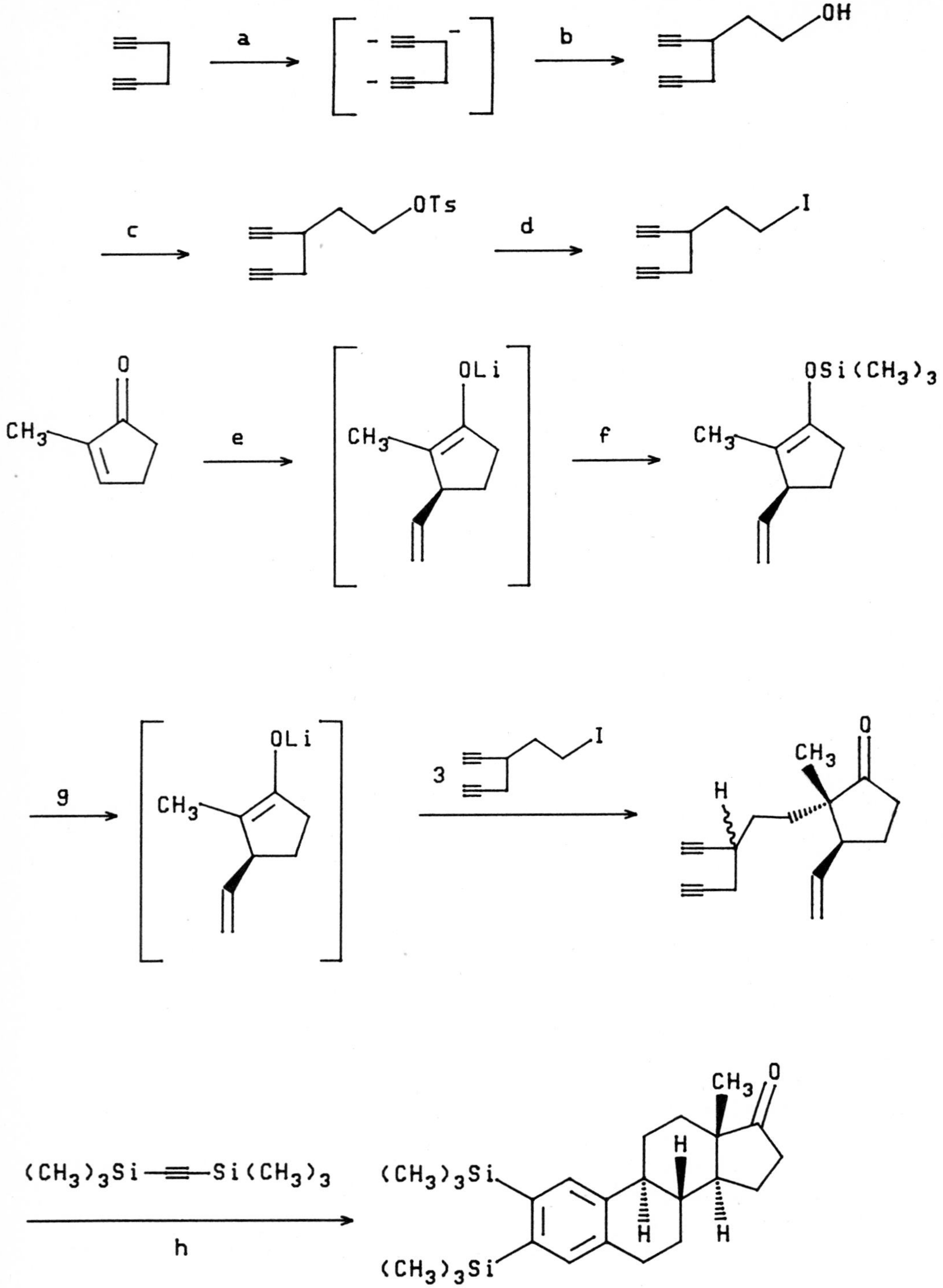

Figure 7.28

a. −30°C, 3 n-BuLi, THF

b. −65 to −10°C, ethylene oxide; 65%

c. 0°C, p-TsCl, pyridine; 100%

d. 45°C, KI, acetone; 96%

e. −60 to −40°C, CH_2=CHMgBr, CuI, THF

f. −40 to rt, Et_3N, $(CH_3)_3SiCl$, HMPA; 89%

g. $LiNH_2$, NH_3, THF; 64% (mixture of diastereoisomers)

h. 5 mol% $CpCo(CO)_2$, syringe pump addition over 35 h; 71%

Figure 7.28 (*Continued*)

trans. Exclusive formation of the *trans–anti–trans* product is a result of steric crowding in the pseudoboat (see Figure 7.29).

Alternative cotrimerization of the diyne with trimethylsilylmethoxyacetylene is inefficient due to catalyst depletion by competing formation of η^4-cyclobutadiene complexes and poor (2:1) regioselectivity (see Figure 7.30).

Conversion of the 2,3-bis-trimethylsilyl-substituted steroid to (±)-estrone requires selectivity in reaction of the trimethylsilyl substituents with electrophiles. Fortunately, the 2-position of analogous steroids is known to be more reactive toward electrophiles[45]: treatment with trifluoroacetic acid at −30°C results in highly selective monodesilylation (steric acceleration) of the 2-position. Subsequent replacement of the remaining trimethylsilyl group by lead tetratrifluoroacetate affords (±)-estrone (see Figure 7.31).

7.8.2 Synthesis B. From an Intact A-Ring Precursor[46,47]

Synthesis B begins with an intact A ring. Synthesis of the BCD portion is based on construction of the C ring (and, coincidentally the B and D rings) by [2 + 2 + 2]-cyclotrimerization of an enediyne. Again, this nicely convergent synthesis of the steroid framework can be broken down into three operations. First, an appropriately functionalized A ring is prepared. Treatment of 3-methoxybenzyl alcohol[48] with acetic anhydride in pyridine affords the acetate. Subsequent Friedel–Crafts acetylation affords a mixture of ketones. The desired 2-acetoxymethyl-4-methoxyacetophenone is converted to a vinyl chloride with phosphorus pentachloride–phosphorus oxychloride. Note that the ester is hydrolyzed also. Dehydrohalogenation of the vinyl chloride affords 2-hydroxymethyl-4-methoxyphenylacetylene. The benzylic alcohol is converted to a benzylic bromide with triphenylphosphine dibromide (see Figure 7.32).

Second, an eneyne is prepared from 4-pentyn-l-ol.[49] Treatment with pyridinium chlorochromate produces the aldehyde. Addition of allyl Grignard followed by alcohol

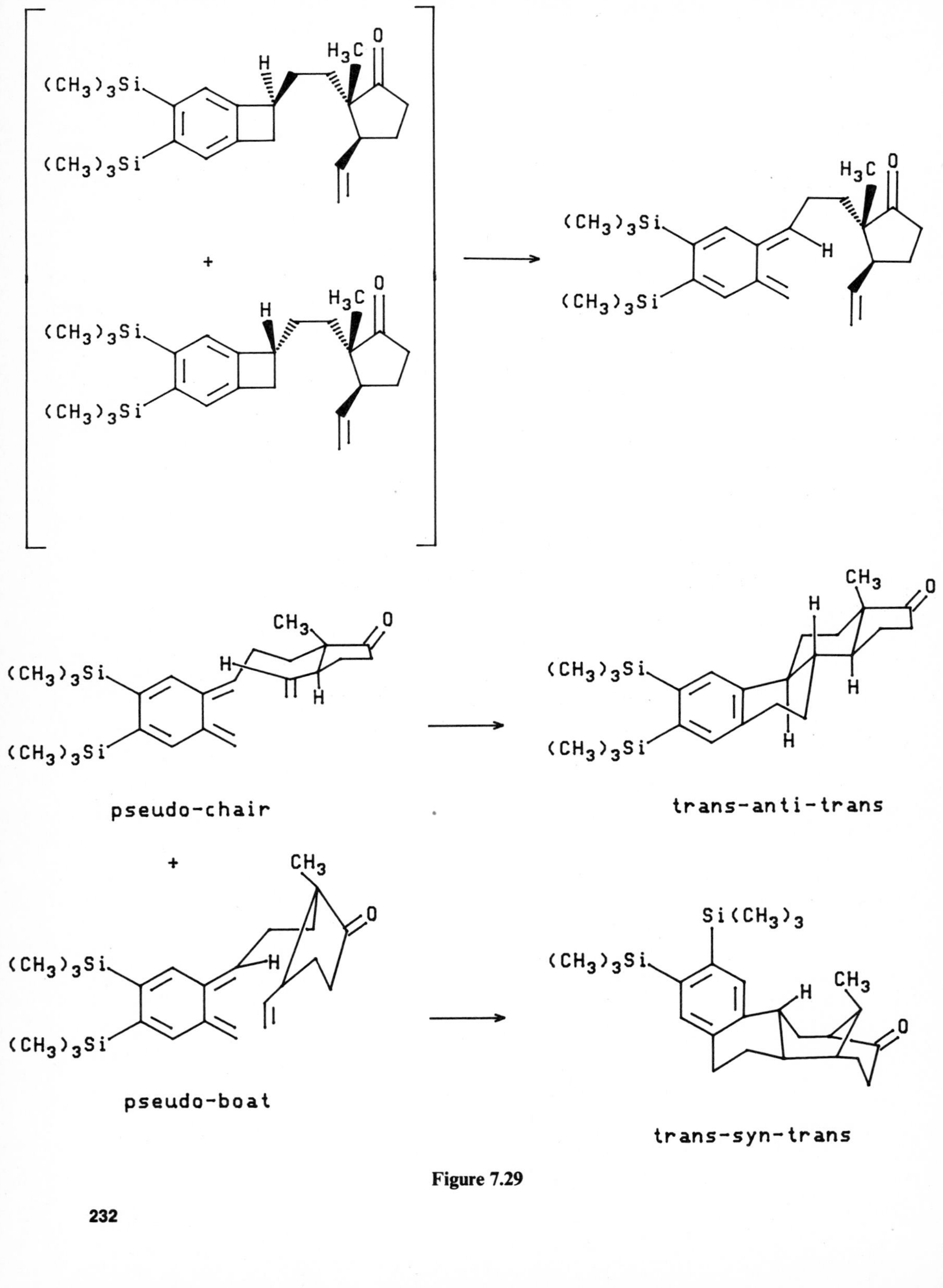

$(CH_3)_3Si$
$(CH_3)_3Si$
H
H_3C
O

+

$(CH_3)_3Si$
$(CH_3)_3Si$
H
H_3C
O

$(CH_3)_3Si$
$(CH_3)_3Si$
H_3C
O
H

$(CH_3)_3Si$
$(CH_3)_3Si$
CH_3
O
H
H

pseudo-chair

$(CH_3)_3Si$
$(CH_3)_3Si$
CH_3
O
H
H
H

trans-anti-trans

+

$(CH_3)_3Si$
$(CH_3)_3Si$
CH_3
O
H

pseudo-boat

$Si(CH_3)_3$
$(CH_3)_3Si$
H
CH_3
O

trans-syn-trans

Figure 7.29

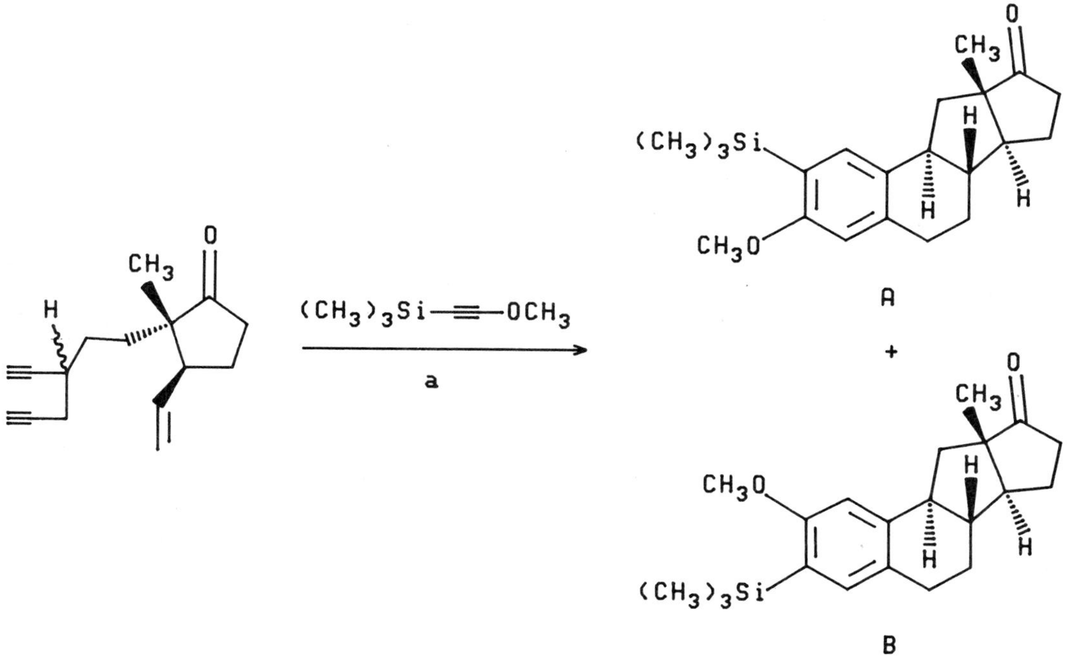

$(CH_3)_3Si$—≡—OCH_3

a

A

+

B

a. $CpCo(CO)_2$; A:B, 2:1, 31% yield

Figure 7.30

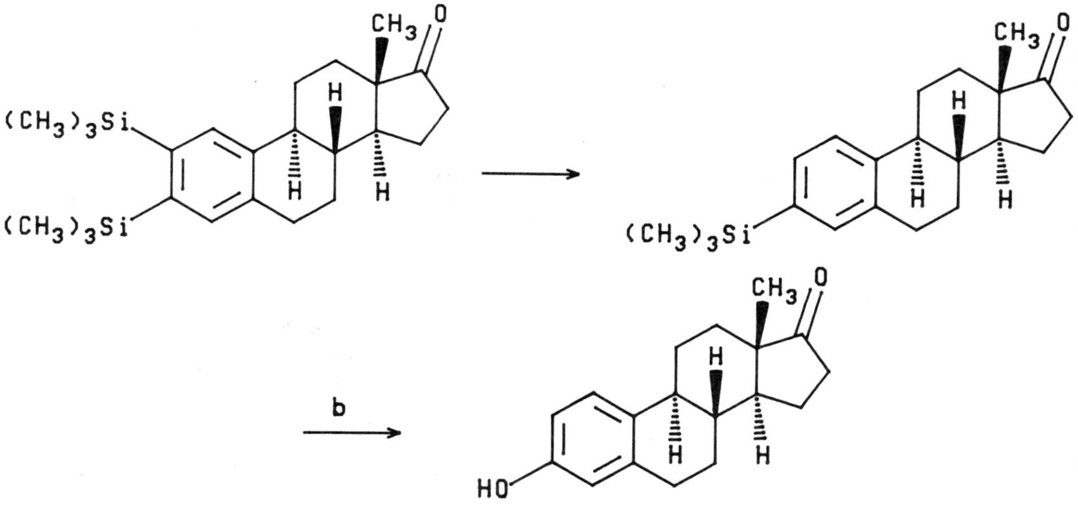

b

(±)-Estrone

a. −30°C, CF_3COOH; 90% (+10% isomer)

b. rt, $Pb[OC(O)CF_3]_4$, CF_3COOH; 100%

Figure 7.31

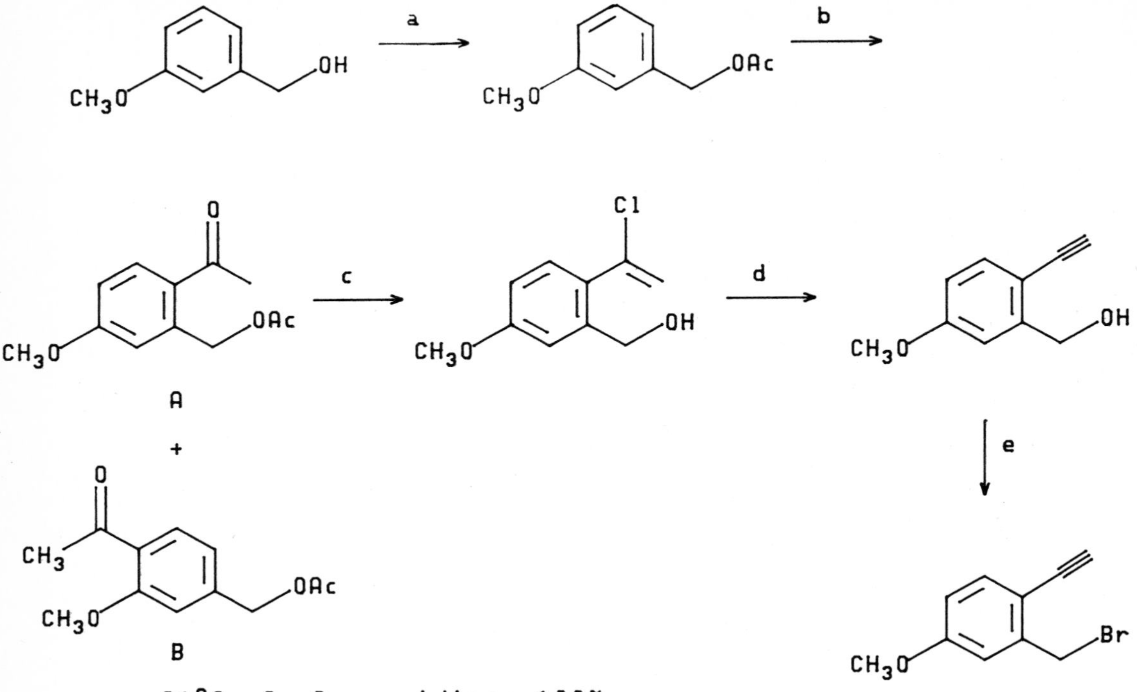

a. 21°C, Ac_2O, pyridine; 100%

b. 0°C, AcCl, $AlCl_3$, CS_2; A:B, 3:1, 68% yield of pure desired isomer

c. 40 to 50°C, PCl_5-$POCl_3$; 92%

d. 40 to 50°C, $NaNH_2$, HMPA, THF; 88%

e. 21°C, $Ph_3P{\cdot}Br_2$, collidine, CH_2Cl_2; 94%

Figure 7.32

oxidation with pyridinium chlorochromate affords the α,β-unsaturated ketone. The ketone is protected as a ketal. The acetylide from deprotonation with *n*-butyllithium is trapped with formaldehyde. The resulting propargylic alcohol is converted to a propargylic bromide with triphenylphosphine dibromide (see Figure 7.33).

Finally, the two bromides prepared are coupled using either the Wurtz procedure or a superior sulfur template approach.[50,51] Cobalt-catalyzed cyclization of the enediyne affords the η^4-cyclohexadiene cobalt complex with an *exo*-methyl group. Mild oxidative demetallation followed by ketal deprotection and double-bond migration produces the Torgov intermediate en route to (±)-estrone.[52] Hydrogenation using Raney nickel selectively reduces the D ring double bond and creates a *trans* CD ring juncture. Reduction with potassium in liquid ammonia affords a mixture of ketone and alcohol, both with the correct *trans–anti–trans* configuration. Chromic anhydride

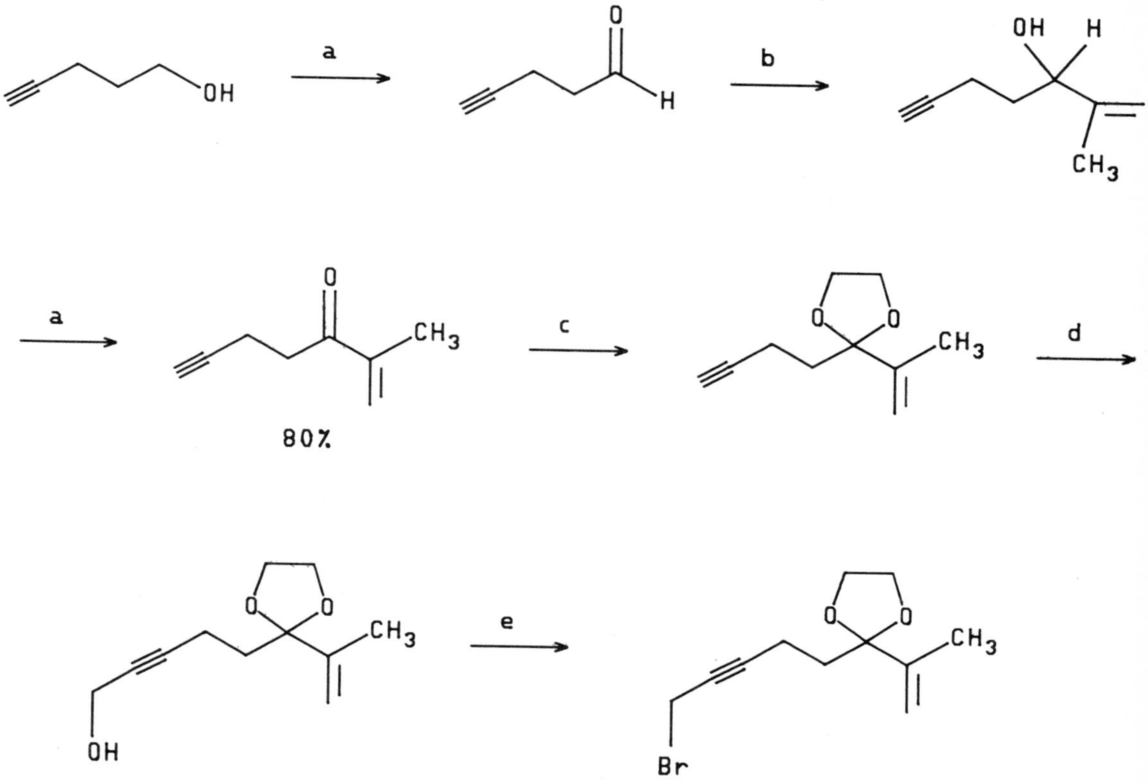

a. PCC, NaOAc, CH_2Cl_2

b. $CH_2=C(CH_3)CH_2MgBr$, THF; 85% for two steps

c. $HOCH_2CH_2OH$, p-TsOH, benzene; 52%

d. −78 to 21°C, n-BuLi, CH_2O, THF; 94%

e. 0°C, $Ph_3P \cdot Br_2$, collidine, CH_2Cl_2; 54%

Figure 7.33

oxidation converts the mixture to pure ketone. Finally, demethylation with pyridine hydrochloride at high temperature affords (±)-estrone[53] (see Figures 7.34 to 7.37).

The A synthesis for (±)-estrone, completed in just five steps in overall 24% yield from 2-methyl-2-cyclopentenone (just seven steps in overall 17% yield from 1,5-hexadiyne), is a beautiful illustration of the power of cobalt-mediated [2+2+2]-cyclotrimerization for synthesis of complex polycyclic molecules.

The B synthesis of (±)-estrone was completed in 13 steps in overall 3.4% yield from

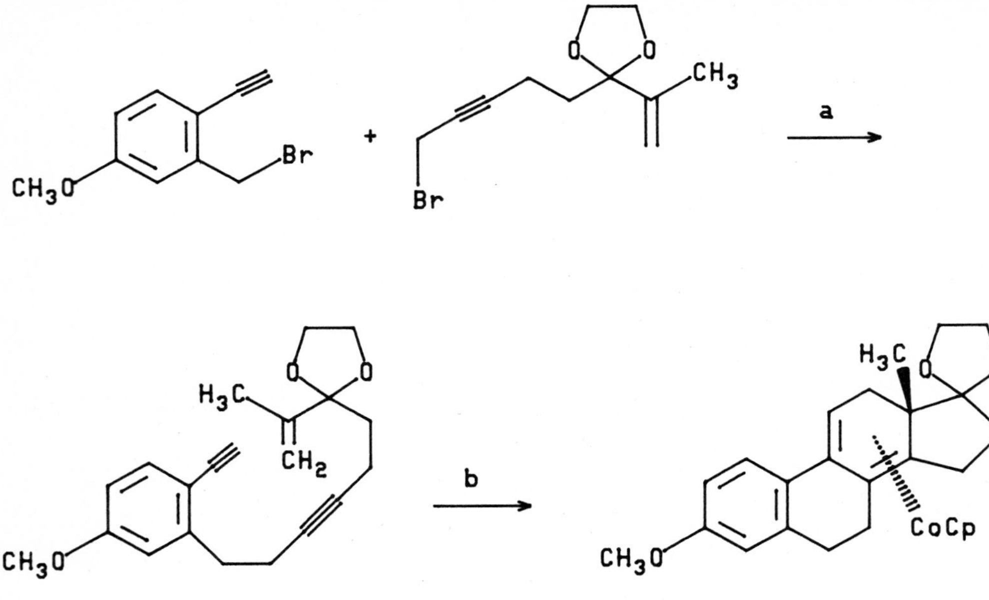

a. −78°C, t-BuLi, THF or see Figure 7.36; 25% or 65%

b. heat, $CpCo(CO)_2$, isooctane; 65%

Figure 7.34

Via:

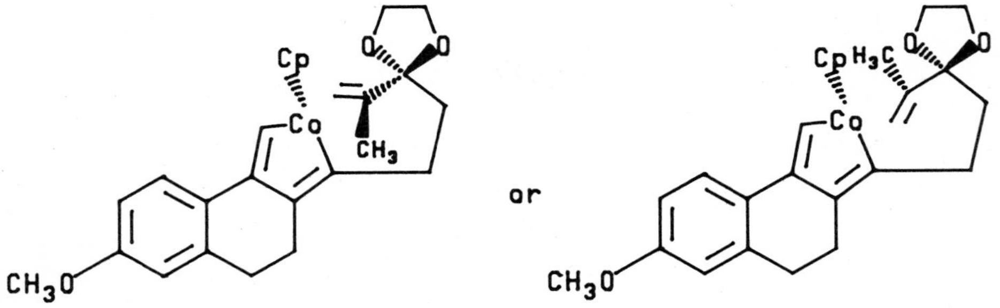

Preferred

or

Figure 7.35

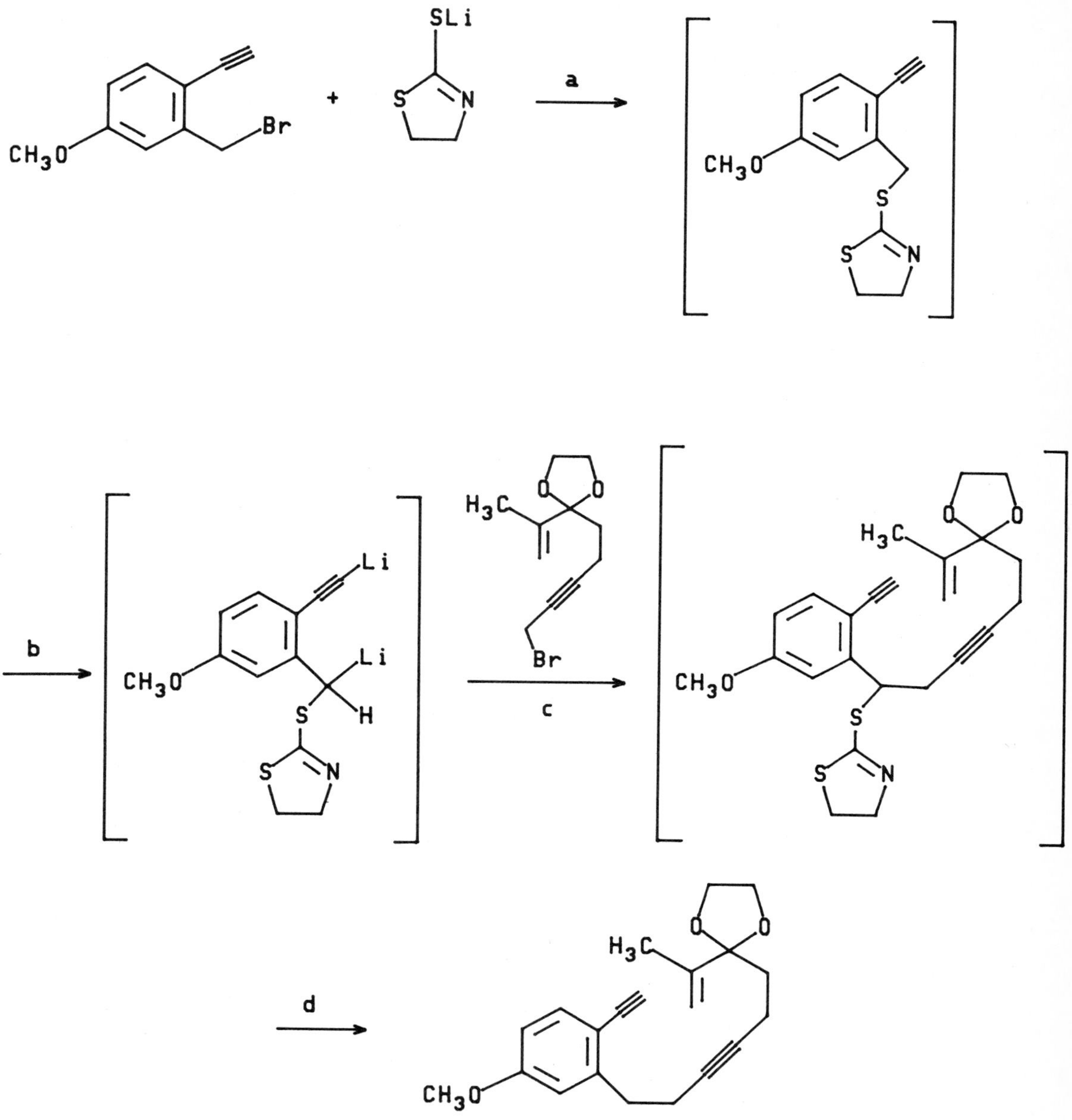

a. -78 to 0°C, THF

b. -78 to -25°C, 2 n-BuLi, THF

c. -78 to 0°C, THF

d. rt, Al(Hg), THF, H_2O; 65% for four steps

Figure 7.36

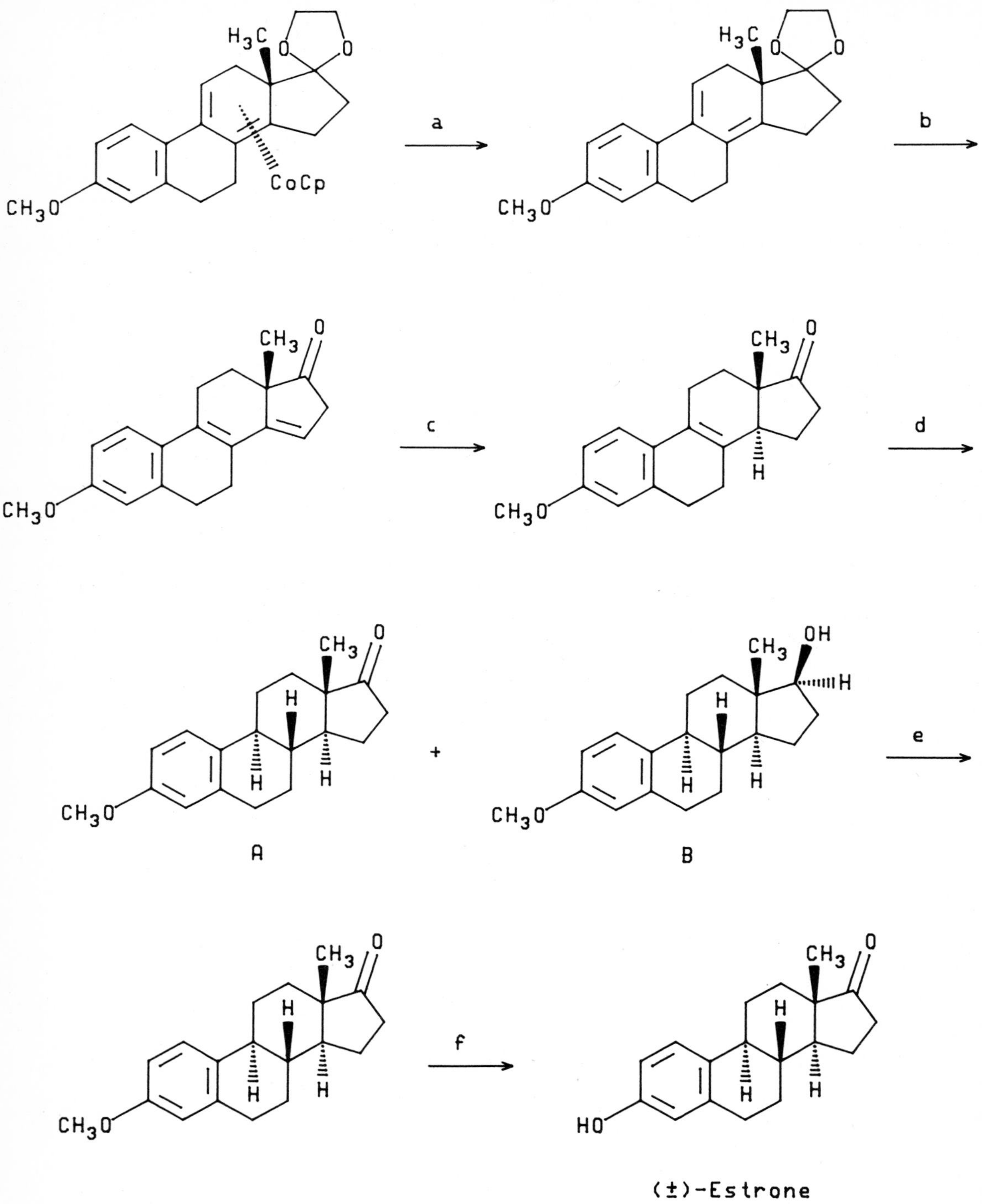

CH_3O ... H_3C ... CoCp —a→ CH_3O ... H_3C —b→

CH_3O ... CH_3 O —c→ CH_3O ... CH_3 O H —d→

CH_3O ... CH_3 O H H H + CH_3O ... CH_3 OH H H H H —e→

A B

CH_3O ... CH_3 O H H H —f→ HO ... CH_3 O H H H

(±)-Estrone

Figure 7.37

a. −35 to 0°C, $FeCl_3$, CH_3CN; 77%

b. rt, p-TsOH, THF, H_2O; 85%

c. H_2, Raney Ni, THF; 70%

d. K, NH_3; A:B, 1:1

e. CrO_3; 40% for two steps

f. heat, pyridine.HCl; 85%

Figure 7.37 (*Continued*)

3-methoxybenzyl alcohol (14 steps in overall 1.2% yield from 4-pentyn-l-ol). The weak steps are bromide–bromide coupling and stereoselective reduction of the C- and D-ring alkenes to give the *trans–anti–trans* configuration. While considerably less efficient than the A synthesis, the B synthesis does provide an intermediate with a handle on the C ring, making the synthesis of C-ring oxygenated steroids a future possibility.

REFERENCES

1. Butenandt, A. *Naturwissenschaften*, **1929**, *17*, 879.
2. Doisy, E. A.; Veler, C. D.; Thayer, S. A. *Am. J. Physiol.* **1929**, *90*, 329.
3. Dence, J. B. *Steroids and Peptides*; Wiley-Interscience: New York, 1980, pp. 49–51.

4a. Fieser, L. F.; Fieser, M. *Steroids*; Reinhold: New York, 1959, p. 477.

4b. Allen, E.; Doisy E. A. *J. Am. Med. Assoc.* **1923**, *81*, 819.

5. Lednicer, D.; Mitscher, L. A. *The Organic Chemistry of Drug Synthesis*; Wiley: New York, 1980, Vol. 2, p.136
6. Akhrem, A. A.; Titov, Yu. A. *Total Steroid Synthesis*; Plenum: New York, 1970.
7. Blickenstaff, R. T.; Ghosh, A. C.; Wolf, G. C. *Total Synthesis of Steroids*; Academic: New York, 1974.
8. Wakatsuki, Y.; Kuramitsu, T.; Yamazaki, H. *Tetrahedron Lett.* **1974**, 4549.
9. Yamazaki, H.; Wakatsuki, Y. *J. Organometal. Chem.* **1977**, *139*, 157.
10. Stockis, A.; Hoffmann, R. *J. Am. Chem. Soc.* **1980**, *102*, 2952.
11. Wakatsuki, Y.; Nomura, O.; Kitaura, K.; Morokuma, K.; Yamazaki, H. *J. Am. Chem. Soc.* **1983**, *105*, 1907.
12. Vollhardt, K. P. C.; Bergman, R. G. *J. Am. Chem. Soc.* **1974**, *96*, 4996.
13. Aalbersberg, W. G. L.; Barkovich, A. J.; Funk, R. L; Hillard, R. L., III; Vollhardt, K. P. C. *J. Am. Chem. Soc.* **1975**, *97*, 5600.
14. Gesing, E. R. F.; Sinclair, J. A.; Vollhardt, K. P. C. *J. Chem. Soc., Chem. Comm.* **1980**, 286.
15. Hillard, R. L., III; Vollhardt, K. P. C. *J. Am. Chem. Soc.* **1977**, *99*, 4058.
16. See the following references for a study of the reversibility of η^4-cyclobutadiene complex formation from cobaltacyclopentadiene:

16a. Ville, G.; Vollhardt, K. P. C.; Winter, M. J. *J. Am. Chem. Soc.* **1981**, *103*, 5267.

16b. Ville, G.; Vollhardt, K. P. C.; Winter, M. J. *Organometallics* **1984**, *3*, 1177.

17. McDonnell Bushnell, L. P.; Evitt, E. R.; Bergman, R. G. *J. Organometal. Chem.* **1978**, *157*, 445.

18. Sondheimer, F.; Amiel, Y.; Gaoni, Y. *J. Am. Chem. Soc.* **1962**, *84*, 270.
19. Funk, R. L.; Vollhardt, K. P. C. *J. Am. Chem. Soc.* **1976**, *98*, 6755.
20. Funk, R. L.; Vollhardt, K. P. C. *J. Am. Chem. Soc.* **1980**, *102*, 5245.
21. Kametani, T.; Kajiwara, M.; Takahashi, T.; Fukumoto, K. *Tetrahedron* **1975**, *31*, 949.
22. Eaborn, C.; Walton, D. R. M.; Young, D. J. *J. Chem. Soc., B* **1969**, 15.
23. Funk, R. L.; Vollhardt, K. P. C. *J. Chem. Soc., Chem. Comm.* **1976**, 833.
24. Seyferth, D.; White, D. L. *J. Am. Chem. Soc.* **1972**, *94*, 3132.
25. Kalman, J. R.; Pinhey, J. T.; Sternhell, S. *Tetrahedron Lett.* **1972**, 5369.
26. Wakatsuki, Y.; Yamazaki, H. *J. Organometal. Chem.* **1977**, *139*, 169.
27. Sternberg, E. D.; Vollhardt, K. P. C. *J. Am. Chem. Soc.* **1980**, *102*, 4839.
28. Malacria, M.; Vollhardt, K. P. C. *J. Org. Chem.* **1984**, *49*, 5010.
29. Gadek, T. R.; Vollhardt, K. P. C. *Angew. Chem. Int. Ed. Engl.* **1981**, *20*, 802.
30. Sternberg, E. D.; Vollhardt, K. P. C. *J. Org. Chem.* **1984**, *49*, 1564.
31. Wakatsuki, Y.; Yamazaki, H. *Tetrahedron Lett.* **1973**, 3383.
32. Bönnemann, H.; Brinkmann, R.; Schenkluhn, H. *Synthesis* **1974**, 575.
33. Naiman, A.; Vollhardt, K. P. C. *Angew. Chem. Int. Ed. Engl.* **1977**, *16*, 708.
34. Bönnemann, H. *Angew. Chem. Int. Ed. Engl.* **1978**, *17*, 505.
35. Hong, P.; Yamazaki, H. *Synthesis* **1977**, 50.
36. Hong, P.; Yamazaki, H. *Tetrahedron Lett.* **1977**, 1333.
37. Parnell, C. A.; Vollhardt, K. P. C. *Tetrahedron* **1985**, *41*, 5791.
38a. Earl, R. A.; Vollhardt, K. P. C. *J. Am. Chem. Soc.* **1983**, *105*, 6991.
38b. Earl, R. A.; Vollhardt, K. P. C. *J. Org. Chem.* **1984**, *49*, 4786.
39. Funk, R. L.; Vollhardt, K. P. C. *J. Am. Chem. Soc.* **1977**, *99*, 5483.
40. Funk, R. L.; Vollhardt, K. P. C. *J. Am. Chem. Soc.* **1979**, *101*, 215.
41. Vollhardt, K. P. C. *Ann. N.Y. Acad. Sci.* **1980**, *333*, 241.
42. 1,5-Hexadiyne is commercially available (Lancaster Synthesis 89–90 catalog, 10 g for $113.50).
43. 2-Methyl-2-cyclopentenone can be prepared from 2-methyl-1,3-cyclopentanedione by conversion to 3-ethoxy-2-methyl-2-cyclopentenone, di-*i*-butylaluminum hydride reduction, and acidic workup. 2-Methyl-1,3-cyclopentanedione is commercially available (Fluka 88–89 catalog, 25 g for $63.00).
44. Bis-trimethylsilylacetylene is commercially available (Lancaster Synthesis 89–90 catalog, 50 g for $86.00).
45a. Nambara, T.; Honma, S.; Akiyama, S. *Chem. Pharm. Bull.* **1970**, *18*, 474.
45b. Nambara, T.; Numazawa, M.; Akiyama, S. *Chem. Pharm. Bull.* **1971**, *19*, 153.
46. Sternberg, E. D.; Vollhardt, K. P. C. *J. Org. Chem.* **1982**, *47*, 3447.
47. Sternberg, E. D.; Vollhardt, K. P. C. *J. Org. Chem.* **1984**, 49, 1574.
48. 3-Methoxybenzyl alcohol is commercially available (Fluka 88-89 catalog, 25 mL, $d =$ 1.113 g/mL, for $14.40).
49. 4-Pentyn-1-ol is commercially available (Lancaster Synthesis 89–90 catalog, 25 g for $79.60).
50. Hirai, K.; Kishida, Y. *Tetrahedron Lett.* **1972**, 2117.
51. Hirai, K.; Iwano, Y.; Kishida, Y. *Tetrahedron Lett.* **1977**, 2677.
52. Torgov, I. V. *Pure Appl. Chem.* **1963**, *6*, 525.
53a. Cole, J. E., Jr.; Johnson, W. S.; Robins, P. A.; Walker, J. *Proc. Chem. Soc.* **1958**, 114 g.
53b. Cole, J. E., Jr.; Johnson, W. S.; Robins, P. A.; Walker, J. *J. Chem. Soc.* **1962**, 244.

CHAPTER 8

(±)-Cyclocolorenone

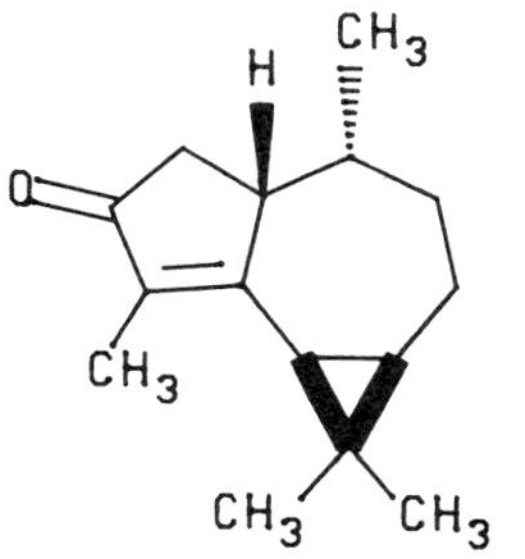

(±)-Cyclocolorenone

6H-cycloprop[e]azulen-6-one,

1,1a,2,3,4,4a,5,7b-octahydro-

1,1,4,7-tetramethyl

[1aR-(1aα, 4α, 4aβ, 7bα)] [489-45-2]

[1aS-(1aα, 4α, 4aβ, 7bα)] [53584-68-2]

Figure 8.1

(±)-Cyclocolorenone is a sesquiterpene first isolated from *Pseudowintera colorata*, a shrub endemic to New Zealand.[1] Although little is known about the pharmacological properties of cyclocolorenone, research on related sesquiterpenoids of the *guaiane* and *pseudoguainane* families reveals potential cytotoxic and antitumor activity.[2] The small number of successful syntheses of (±)-cyclocolorenone reported is indicative of the challenging functionalized perhydroazulene framework. Previous syntheses generate the skeleton by rearrangement of decalones or start with natural products already possessing the bicyclo[5.3.0] ring system.[3]

8.1 DICOBALT HEXACARBONYL ALKYNE COMPLEXES: SYNTHETIC APPLICATIONS

The cobalt alkyne complexes involved in Vollhardt's cyclotrimerizations (Chapter 7) are highly reactive. In contrast, stable cobalt alkyne complexes are prepared using dicobalt octacarbonyl. These complexes have two cobalt atoms capping the acetylene triple bond, forming a bimetallo-cluster with tetrahedral geometry. Since the alkyne

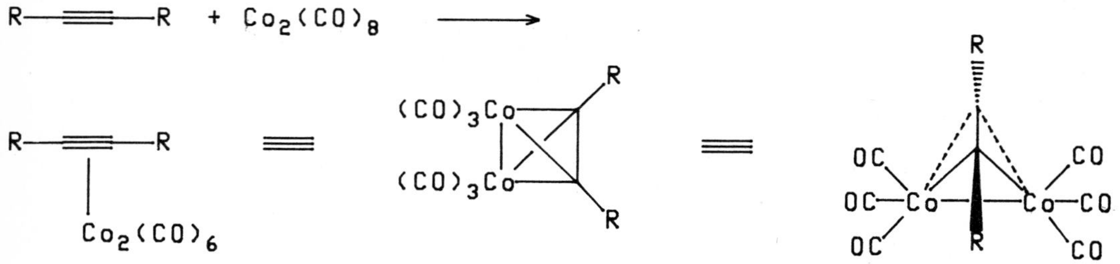

R—≡—R + $Co_2(CO)_8$ ⟶

R—≡—R ($Co_2(CO)_6$) ≡ $(CO)_3Co$ / $(CO)_3Co$ cluster ≡ OC, OC—Co—Co—CO, OC, CO, CO

Figure 8.2

R^3—≡—$C(R^1)(R^2)$—OH + $Co_2(CO)_8$ —a→ $(CO)_3Co$ / $(CO)_3Co$ complex (R^3; $C(R^1)(R^2)OH$)

—b→ —c→ $(CO)_3Co$ / $(CO)_3Co$ complex, BF_4^- (R^3; $C^+(R^1)(R^2)$) —d→ $(CO)_3Co$ / $(CO)_3Co$ complex (R^3; $C(R^1)(R^2)$Nu)

—e→ R^3—≡—$C(R^1)(R^2)$—Nu

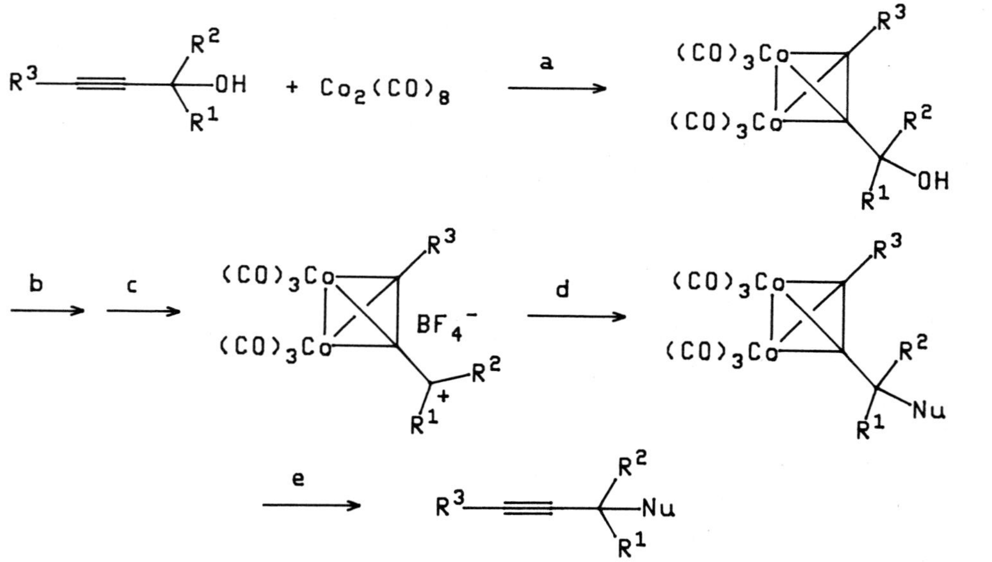

a. rt

b. $[CH_3CH_2C(O)]_2O$

c. $HBF_4 \cdot (CH_3)_2O$

d. NuH

e. [O]

Figure 8.3

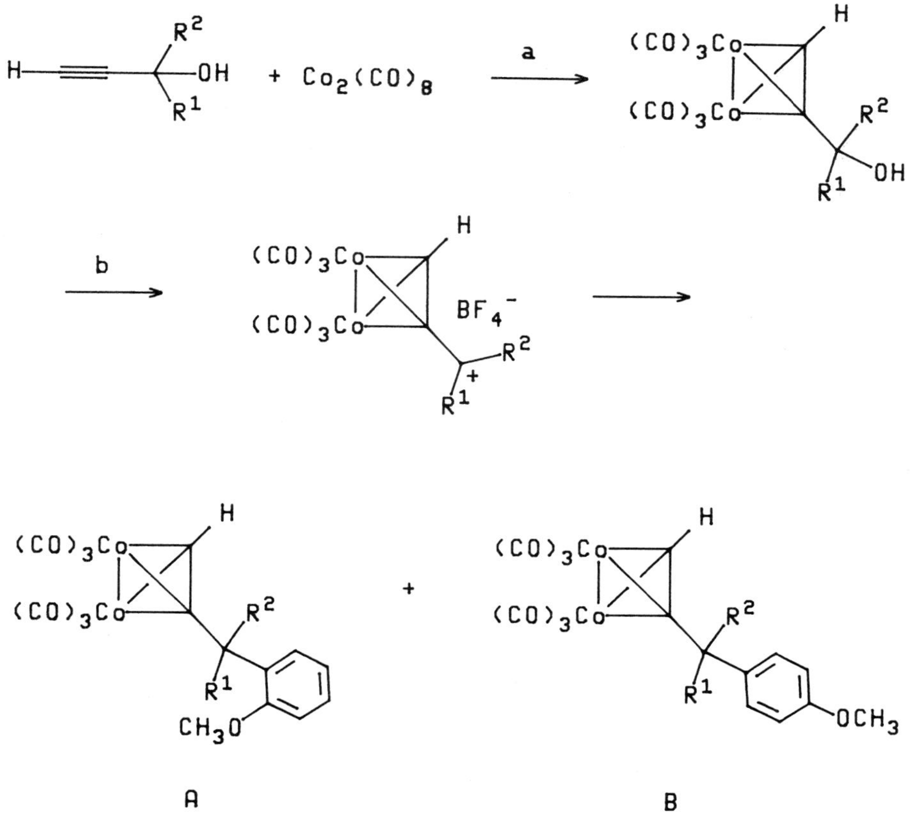

		Yield (%)	
R^1	R^2	A	B
H	H	35	49
CH_3	H	25	53
CH_3	CH_3	0	78

a. rt, petroleum ether

b. 0°C, $HBF_4 \cdot (CH_3)_2O$, anisole

Figure 8.4

utilizes all four of its π electrons in complexation, the complex has drastically reduced reactivity toward electrophiles (see Figure 8.2).

This stability and an efficient decomplexation by mild oxidation [usually $Fe(NO_3)_3 \cdot 9H_2O$ in ethanol or $(NH_4)_2Ce(NO_3)_6$ in acetone at or below room temperature] prompted investigation of the use of dicobalt hexacarbonyl as an alkyne protecting group.

The generation and synthetic utilization of propargyl cations from propargyl alcohols or derived esters is plagued by acid-catalyzed rearrangements. K.M. Nicholas has demonstrated that dicobalt hexacarbonyl complexes of propargyl alcohols can be converted to *remarkably stable cations* that serve as propargyl cation equivalents.[4–6] The sequence of reactions is (1) complex formation with a propargyl alcohol; (2) treatment at low temperature with an acylating agent, then a strong acid with a nonnucleophilic counterion [usually $HBF_4 \cdot (CH_3)_2O$] to generate a salt that is sufficiently stable to be isolated; (3) coupling of the salt with a nucleophile to generate a new carbon–heteroatom or carbon–carbon bond; and (4) decomplexation by mild oxidation (see Figure 8.3). For example, using anisole as a model electron-rich aromatic, a mixture of *ortho*- and *para*-substituted isomers is produced. Increasing the steric bulk of the cation favors the *para* isomer[7] (see Figure 8.4). Excellent yields of coupling products are obtained using 1,3-diketones[8] (see Figure 8.5).

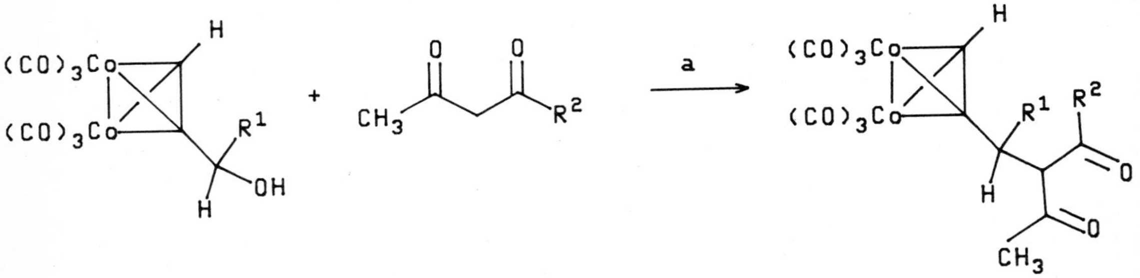

R^1	R^2	Yield (%)
H	CH_3	95
CH_3	CH_3	65
Ph	CH_3	91
H	Ph	90
CH_3	Ph	65
Ph	Ph	95

a. −78 to 0°C, $HBF_4 \cdot (CH_3)_2O$, CH_2Cl_2

Figure 8.5

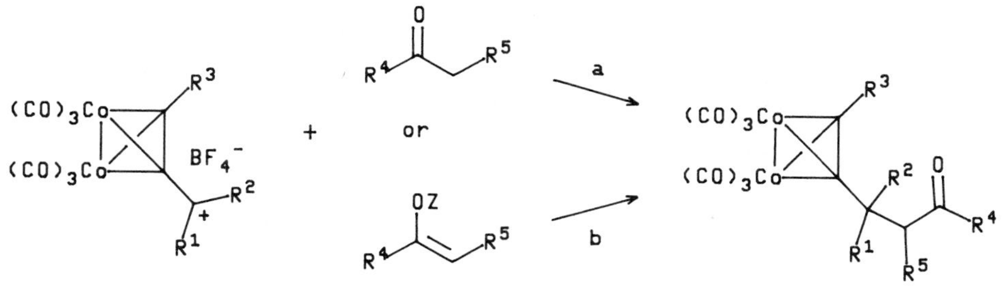

$Z = Ac, Si(CH_3)_3$

62–84% with ketones

51–96% with enol ethers, acetates

a. 0°C (neat ketone)

b. 0°C, CH_2Cl_2

Figure 8.6

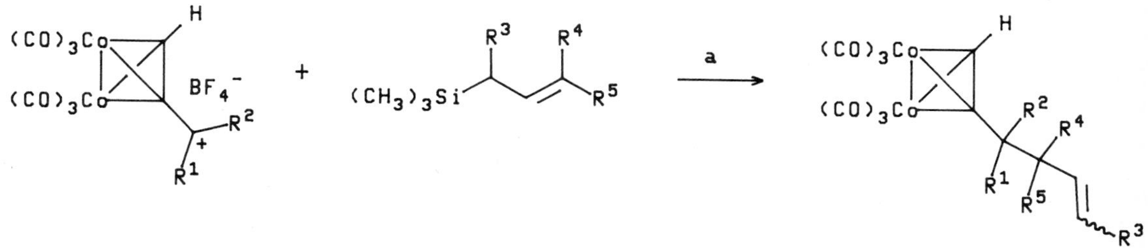

R^1	R^2	R^3	R^4	R^5	Yield (%)
H	H	H	H	H	83
Ph	H	H	H	H	92
CH_3	CH_3	H	H	H	70
H	H	H	CH_3	CH_3	95
Ph	H	H	CH_3	CH_3	91
CH_3	CH_3	H	CH_3	CH_3	0
H	H	$-(CH_2)_3-$		H	79
Ph	H	$-(CH_2)_3-$		H	97
CH_3	CH_3	$-(CH_2)_3-$		H	92

a. 0°C, CH_2Cl_2

Figure 8.7

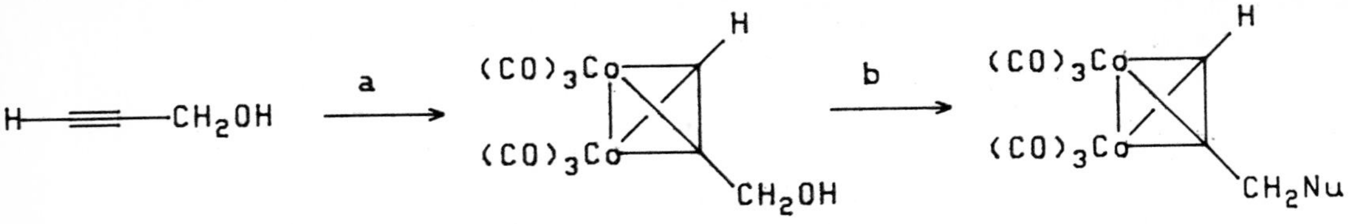

NuH	Yield (%)
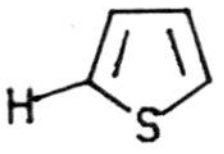 2-thienyl-H (thiophene)	61
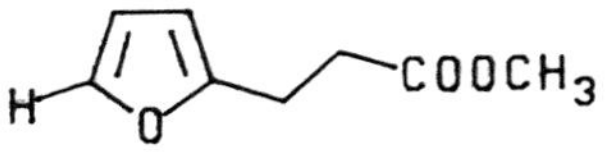 H-furan-$CH_2CH_2COOCH_3$	52
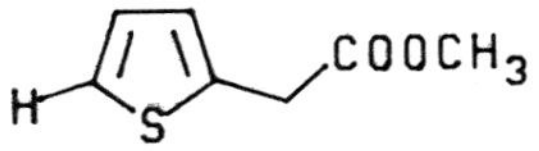 H-thiophene-CH_2COOCH_3	56

a. $Co_2(CO)_8$, petroleum ether; 78%

b. $-78°C$, $HBF_4 \cdot Et_2O$, NuH, CH_2Cl_2

Figure 8.8

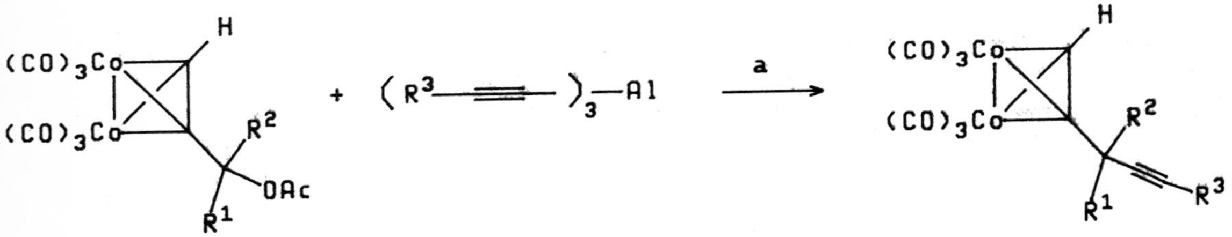

R^1	R^2	R^3	Yield (%)
H	H	n-Bu	66
CH_3	H	n-Bu	61
CH_3	H	n-Bu	94 at −78°C
H	Ph	n-Bu	77
CH_3	CH_3	n-Bu	48
H	Ph	n-hex	46

a. −78 or 0°C, CH_2Cl_2

Figure 8.9

Addition of the complex to a ketone results in coupling via the more stable enol in reasonable yield. To circumvent the requirement for excess ketone, a stoichiometric amount of the corresponding trimethylsilyl enol ether will give the same product in comparable yield[9] (see Figure 8.6). Allylsilanes couple regiospecifically at the position distant from silicon to give 1,5-eneynes. The method appears to be both efficient and general; eneynes with adjacent quaternary carbons cannot be prepared[10] (see Figure 8.7).

Some electron-rich heterocycles (furans and thiophenes) can be functionalized at the 2(5)-position[11] (see Figure 8.8). Using a second method for complex cation formation, 1,4-diynes have been prepared by reaction with trialkynylaluminums[12] (see Figure 8.9).

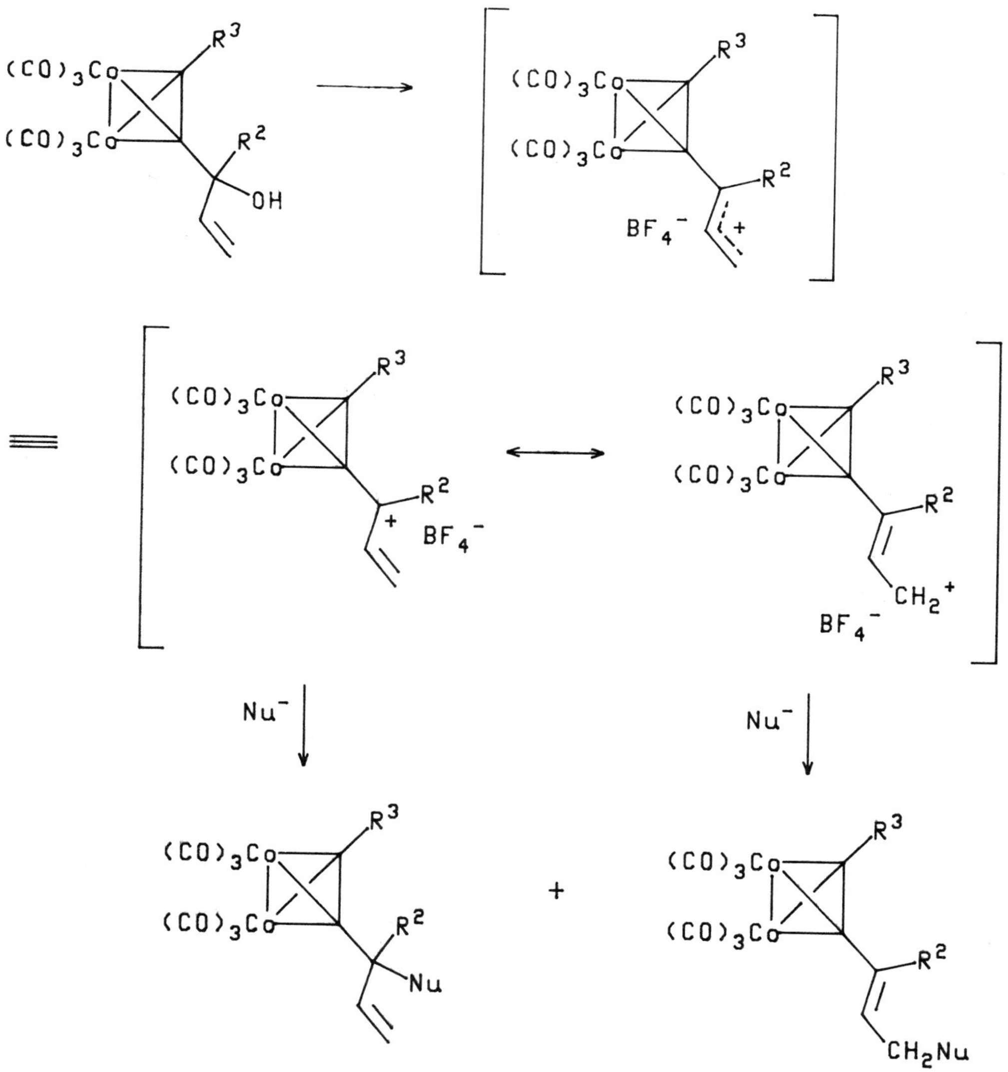

Figure 8.10

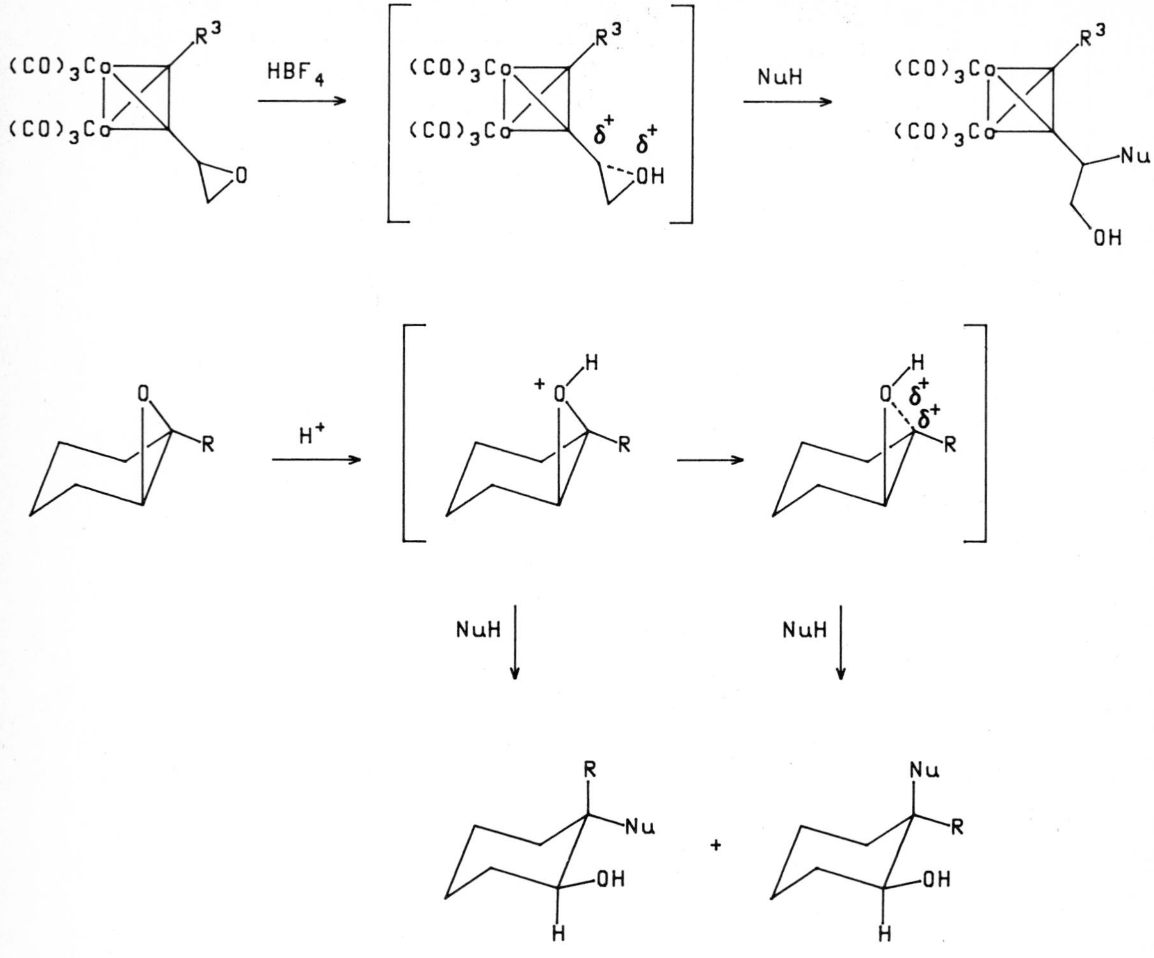

NuH	R	A:B
CH_3OH	$-C\equiv CH$	99:1
H_2O	$-C\equiv CH$	98:2
Cl_3CCOOH	$-C\equiv CH$	48:52
CH_3OH	$-C\equiv CH$ complex	50:50
H_2O	$-C\equiv CH$ complex	41:59
Cl_3CCOOH	$-C\equiv CH$ complex	<5:>95

Figure 8.11

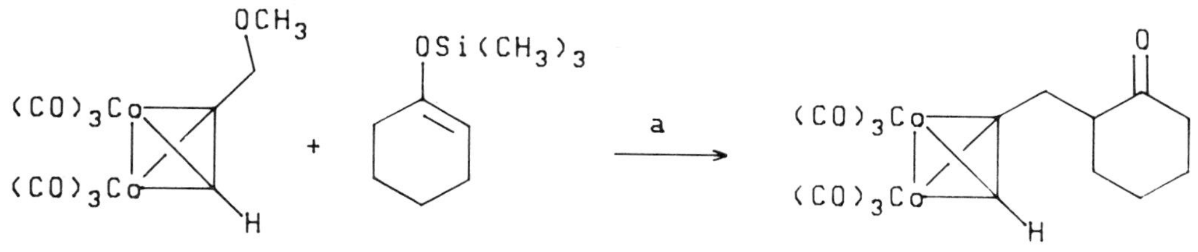

a. −78°C, $BF_3 \cdot Et_2O$, CH_2Cl_2; 92%

Figure 8.12

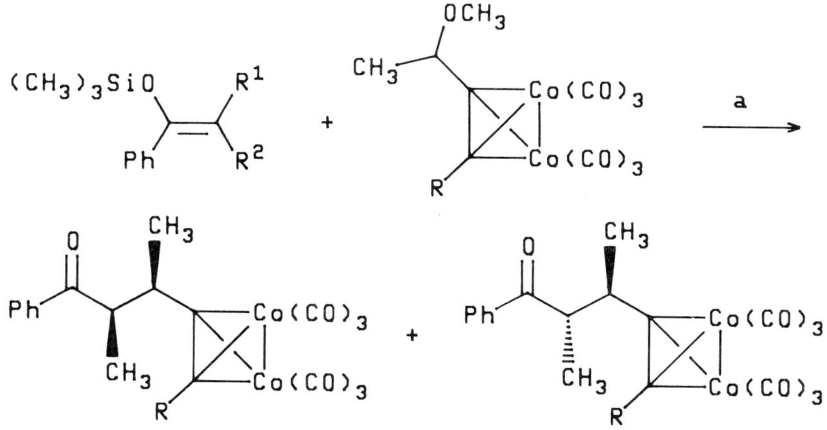

Syn

Anti

R	R^1	R^2	Lewis Acid	Syn:Anti
$(CH_3)_3Si$	CH_3	H	$BF_3 \cdot Et_2O$	15:1
Ph	CH_3	H	$EtAlCl_2$	18:1
Ph	H	CH_3	$EtAlCl_2$	9:1
CH_3	CH_3	H	$BF_3 \cdot Et_2O$	6.8:1
CH_3	H	CH_3	$BF_3 \cdot Et_2O$	3.5:1
H	CH_3	H	$EtAlCl_2$	1.6:1

a. −78°C, Lewis acid, CH_2Cl_2

Figure 8.13 (Reprinted with permission from S. L. Schreiber et al. *J. Am. Chem. Soc.* **1986**, *108* 3128. Copyright 1986, American Chemical Society.)

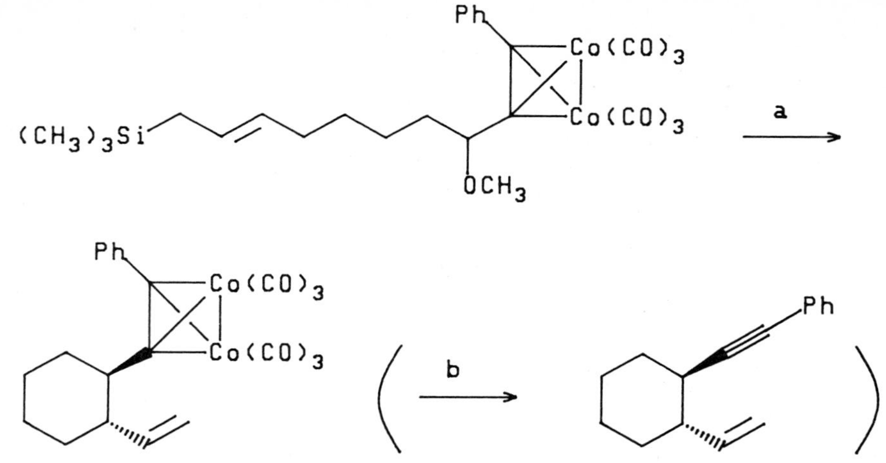

a. $-78°C$, $BF_3 \cdot Et_2O$, CH_2Cl_2; 91%

b. $(CH_3)_3NO$; 89%

Figure 8.14

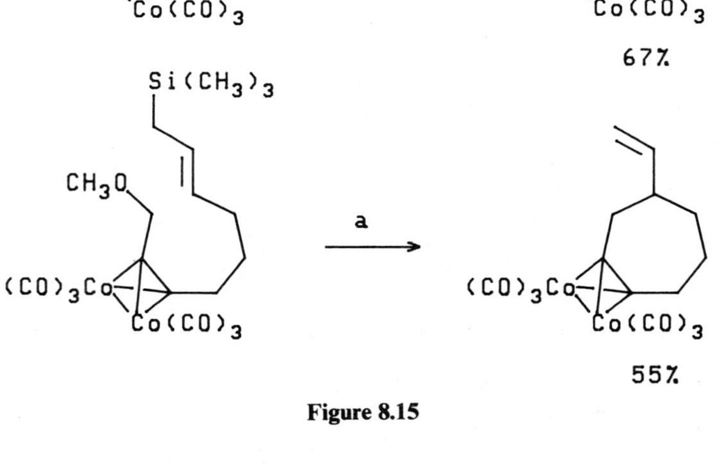

67%

55%

Figure 8.15

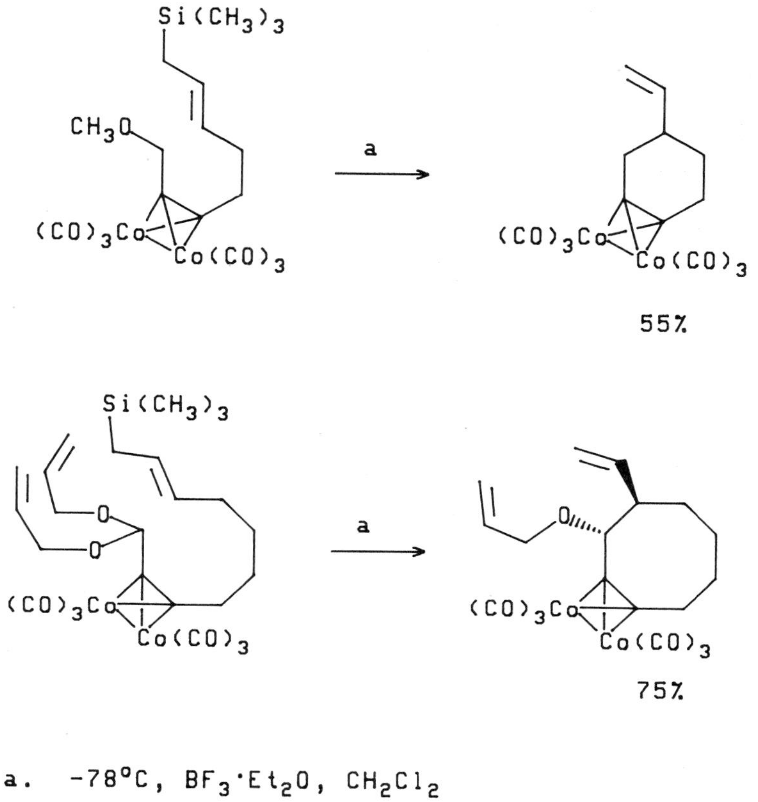

a. $-78^{\circ}C$, $BF_3 \cdot Et_2O$, CH_2Cl_2

Figure 8.15 *(Continued)*

Nucleophilic attack can occur at either terminus of the allylic system when a double bond is in conjugation with the cationic center. Coupling is reversible with oxygen nucleophiles (alcohols, enol acetates); a mixture of regioisomeric coupling products is obtained. With nucleophiles not containing oxygen, irreversible coupling occurs at the least substituted terminus[13] (see Figure 8.10).

Another route to complexed cations is via the acid-catalyzed regiospecific ring opening of epoxides. A comparison of the regiospecificity and stereospecificity of epoxide ring opening using a free and a complexed 1-alkynyl-l-cyclohexene oxide demonstrates the remarkable electron releasing capacity of the complexed alkyne[14] (see Figure 8.11). Propargyl cation utility is further illustrated in Figures 8.12 to 8.15.[15]

8.2 APPLICATION TO CYCLOPENTENONE SYNTHESIS

Coupling of a complexed propargyl cation with 2-acetoxy-1-propene and subsequent decomplexation by mild oxidation are the key steps in a short and efficient synthesis of dihydrojasmone[16] (see Figure 8.16). The alkyne hydration is regiospecific. Carbonyl oxygen participation lowers the energy of one transition state[17] (see Figure 8.17).

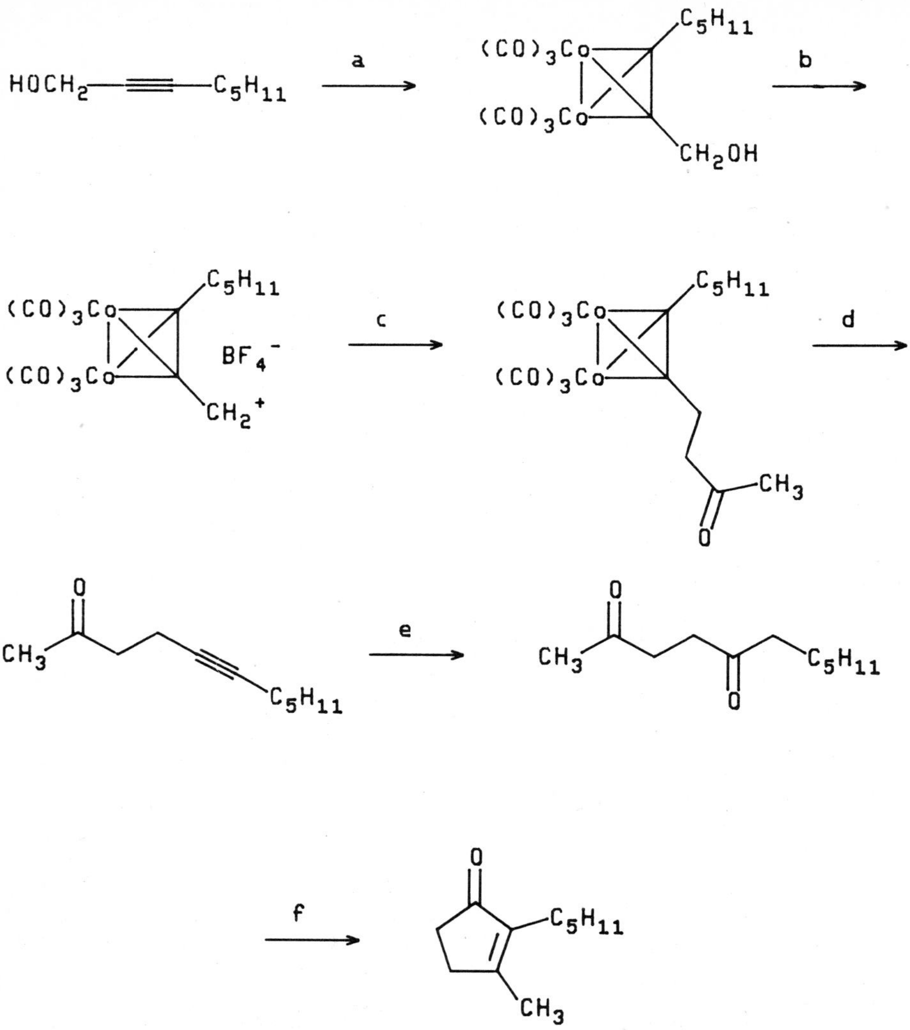

a. rt, $Co_2(CO)_8$, benzene; 86%

b. $-45^{\circ}C$, $[CH_3CH_2C(O)]_2O$, $HBF_4\cdot(CH_3)_2O$, CH_2Cl_2

c. $CH_2=C(OAc)CH_3$, CH_2Cl_2, benzene; 98% for two steps

d. $0^{\circ}C$, $Fe(NO_3)_3\cdot 9H_2O$, 95% EtOH; 94%

e. heat, $HgSO_4$, H_2SO_4, H_2O, CH_3OH; 92%

f. rfx, NaOH, H_2O, EtOH; 93%

Figure 8.16

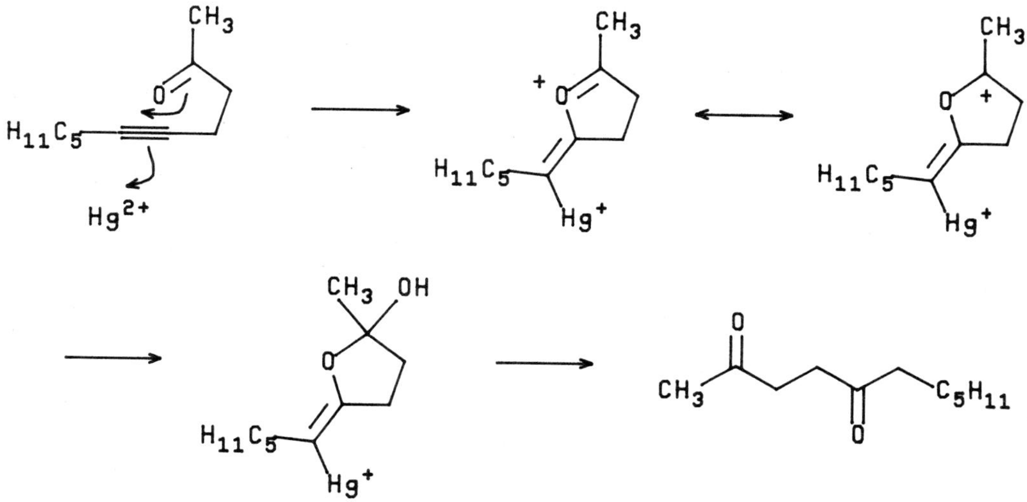

Figure 8.17

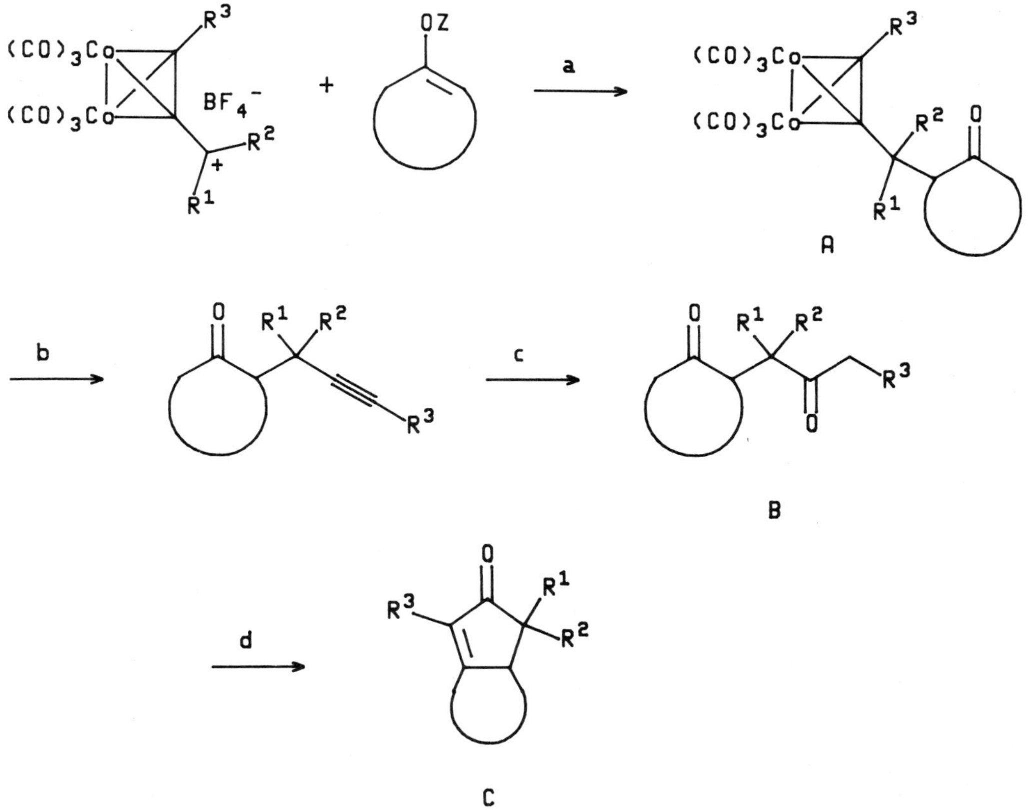

Figure 8.18

				Yield (%)		
Ring Size	R^1	R^2	R^3	A	B	C
5	H	H	H	75	46	73
6	H	H	CH_3	73	67	73
5	Ph	H	H	91	81	63

a. $-78^\circ C$, CH_2Cl_2

b. $-78^\circ C$, $(NH_4)_2Ce(NO_3)_6$, acetone

c. rfx, $HgSO_4$, H_2SO_4, H_2O, CH_3OH

d. rfx, KOH, H_2O

Figure 8.18 *(Continued)*

A similar sequence using specific enolate equivalents of cyclic ketones constitutes a novel method for cyclopentenone annelation[18] (see Figure 8.18).

8.3 TOTAL SYNTHESIS OF (±)-CYCLOCOLORENONE[19,20]

The new, transition-metal-based approach to (±)-cyclocolorenone can be broken down into two operations: (1) stereospecific preparation of a suitably functionalized seven-membered ring and (2) annulation of the cyclopentenone. To prepare the seven-membered ring precursor, two of the three double bonds of tropone[21] are first protected as a diene iron tricarbonyl complex[22] (see Chapter 4). A [3 + 2]-cycloaddition and subsequent thermolysis converts the remaining alkene to a dimethylcyclopropane.[23] The diene iron tricarbonyl complex is then photolyzed in acetic acid, resulting in decomplexation and reduction of the diene.[24] Base converts the β,γ-unsaturated ketone to the α,β-unsaturated isomer. Michael addition of cuprate is stereospecific for the side opposite the dimethylcyclopropane. The enolate can be trapped as a trimethylsilyl enol ether (see Figure 8.19).

The propargyl cation complex is prepared from 2-butyn-1-ol. Coupling with the trimethylsilyl enol ether produces a 3 : 1 mixture of diastereoisomers in moderate yield. Decomplexation by metal oxidation with ceric ammonium nitrate is efficient. Participation by the ketone oxygen provides a mechanistic rationale for subsequent regiospecific hydration of the alkyne. The diastereoisomeric diketones (IIIA, IIIB) can be separated by preparative thin-layer chromatogrphy. The major diastereoisomeric precursors apparently have a *cis*-2,3-geometry since the major ketone (IIIA) is converted to (±)-cyclocolorenone on reaction with ethanolic potassium hydroxide at room temperature.

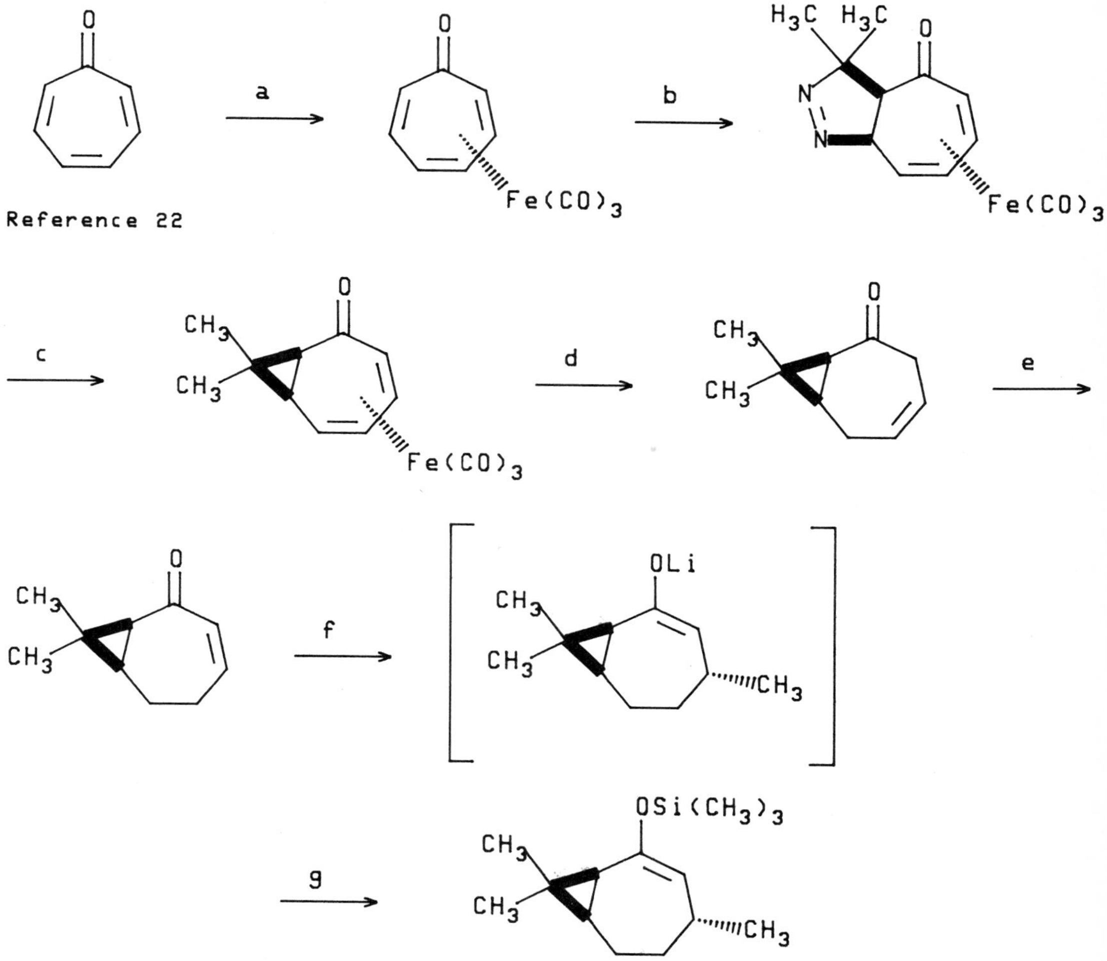

Reference 22

a. rfx, $Fe_2(CO)_9$, Et_2O; 62%

b. $(CH_3)_2C{=}N_2$

c. 80°C

d. hν, AcOH; 40-45% for fours steps from tropone

e. rt, NaOH, H_2O; 91%

f. -78°C, $(CH_3)_2CuLi$, THF

g. -78 to 0°C, $(CH_3)_3SiCl$, THF

Figure 8.19

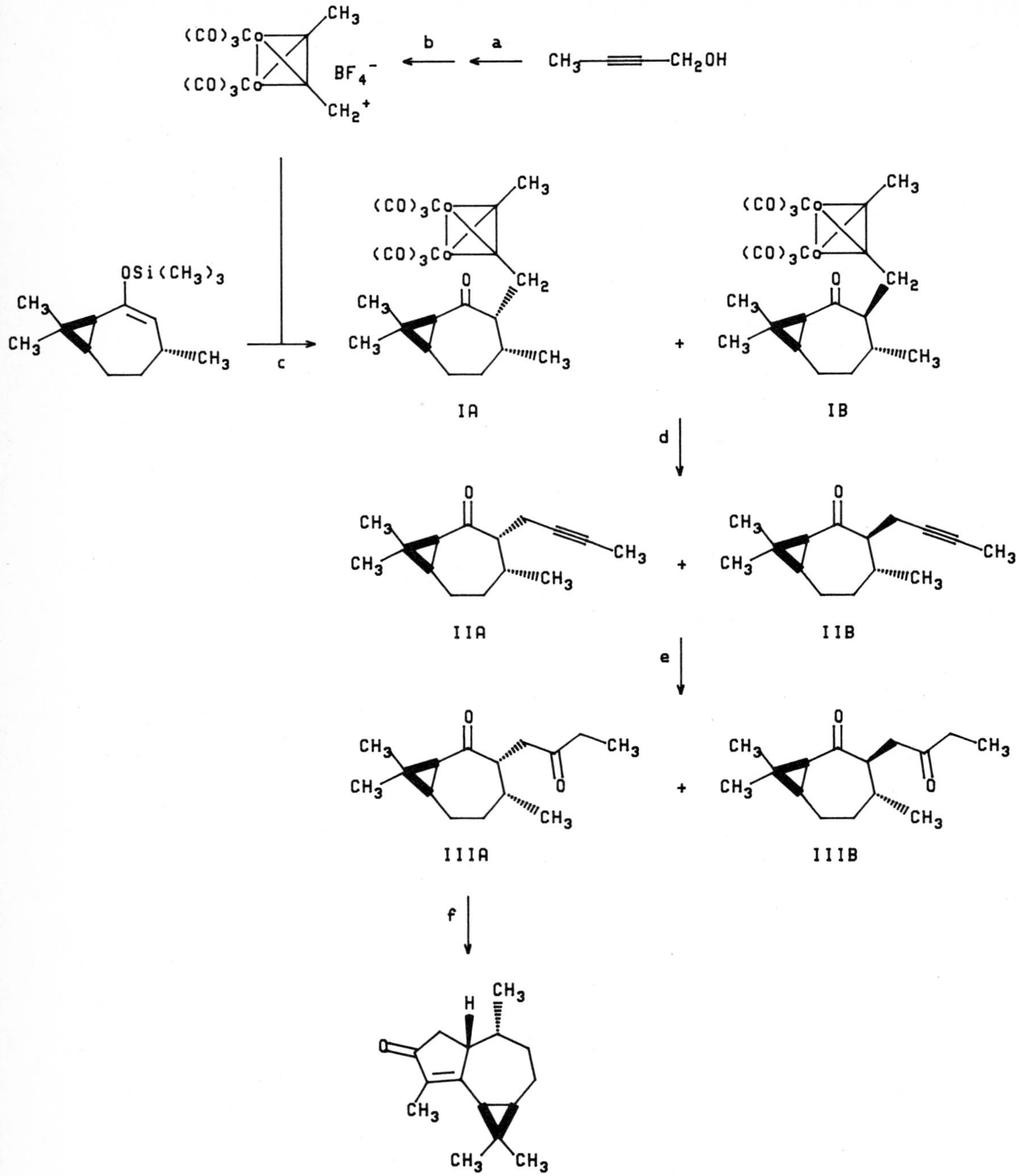

(CO)$_3$Co ... CH$_3$ ... BF$_4^-$ ... CH$_2^+$ ← b ← a CH$_3$—≡—CH$_2$OH

OSi(CH$_3$)$_3$ ... CH$_3$ CH$_3$ ... CH$_3$ → c

IA + IB

d

IIA + IIB

e

IIIA + IIIB

f

(±)-Cyclocolorenone

Figure 8.20

a. rt, $Co_2(CO)_8$, CH_2Cl_2

b. 1) $[CH_3CH_2C(O)]_2O$, 2) $HBF_4 \cdot (CH_3)_2O$

c. -78^oC, CH_2Cl_2; IA:IB, 3:1, 45% yield

d. -78^oC, $(NH_4)_2Ce(NO_3)_6$, acetone; IIA:IIB, 1.5:1, 90% yield

e. rfx, $HgSO_4$, H_2SO_4, H_2O, CH_3OH; IIIA:IIIB, 8:1, 80% yield (separable by preparative TLC)

f. rt, KOH, EtOH; 72%

Figure 8.20 (*Continued*)

The total synthesis of (±)-cyclocolorenone from tropone was completed in 10 steps in an overall yield of 8.4% (assuming a quantitative yield in the cuprate conjugate addition–enolate trap) (see Figure 8.20).

REFERENCES

1. Corbett, R. E.; Speden, R. N. *J. Chem. Soc.* **1958**, 3710.
2. Lee, K-H.; Furukawa, H.; Huang, E-S. *J. Med. Chem.* **1972**, *15*, 609.

3a. Abbaspour, A. *Diss. Abstr. Int. B* **1983**, *43*, 3238; *Chem. Abstr.* **1983**, *99*, 52590f.

3b. Caine, D.; Ingwalson, P. F. *J. Org. Chem.* **1972**, *37*, 3751.

3c. Ingwalson, P. F. *Diss. Abstr. Int. B* **1973**, *34*, 1425; *Chem, Abstr.* **1974**, *80*, B3653c.

3d. Büchi, G.; Kauffman, J. M.; Loewenthal, H. J. E. *J. Am. Chem. Soc.* **1966**, *88*, 3403.

3e. Pesnelle, P.; Ourisson, G. *J. Org. Chem.* **1965**, *30*, 1744.

3f. Büchi, G.; Loewenthal, H. J. E. *Proc. Chem. Soc.* **1962**, 280.

4. See the following references for reviews on the enhanced stability of carbocations adjacent to transition metals:

4a. Haynes, L.; Pettit, R. In *Carbonium Ions*, Olah, G; Schleyer, P. v. R. Ed.; Wiley: New York, 1975; Vol. V.

4b. Vogel, P. *Carbocation Chemistry*; Elsevier: New York, 1985.

5. For a ^{13}C NMR study of (propargyl) dicobalt hexacarbonyl cations, see Padmanabhan, S.; Nicholas, K. M. *J. Organometal. Chem.* **1983**, *268*, C23.
6. For pK_{R^+} measurements on (propargyl) dicobalt hexacarbonyl cations, see Connor, R. E.; Nicholas, K. M. *J. Organometal. Chem.* **1977**, *125*, C45.
7. Lockwood, R. F.; Nicholas, K. M. *Tetrahedron Lett.* **1977**, 4163.
8. Hodes, H. D.; Nicholas, K. M. *Tetrahedron Lett.* **1978**, 4349.
9. Nicholas, K. M.; Mulvaney, M.; Bayer, M. *J. Am. Chem. Soc.* **1980**, *102*, 2508.
10. O'Boyle, J. E.; Nicholas, K. M. *Tetrahedron Lett.* **1980**, 1595.
11. Jaffer, H. J.; Pauson, P. L. *J. Chem. Res.(S)* **1983**, 244.

12. Padmanabhan, S.; Nicholas, K. M. *Tetrahedron Lett.* **1983**, *24*, 2239.
13. Padmanabhan, S.; Nicholas, K. M. *Tetrahedron Lett.* **1982**, *23*, 2555.
14. Saha, M.; Nicholas, K. M. *J. Org. Chem.* **1984**, *49*, 417.
15. Schreiber, S. L.; Sammakia, T; Crowe, T.; Crowe, W. E. *J. Am. Chem. Soc.* **1986**, *108*, 3128.
16. Padmanabhan, S.; Nicholas, K. M. *Syn. Commun.* **1980**, *10*, 503.
17. Stork, G.; Borch, R. *J. Am. Chem. Soc.* **1964**, *86*, 935.
18. Saha, M.; Nicholas, K. M. *Israel J. Chem.* **1984**, *24*, 105.
19. Saha, M.; Muchmore, S.; van der Helm, D.; Nicholas, K. M. *J. Org. Chem.* **1986**, *51*, 1960.
20. Saha, M.; Bagby, B.; Nicholas, K. M. *Tetrahedron. Lett.* **1986**, *27*, 915.
21. Tropone is commercially available (Lancaster Synthesis 89–90 catalog, 5 g = $45.10).
22. Hunt, D. F.; Farrant, G. C.; Rodeheaver, G. T. *J. Organometal. Chem.* **1972**, *38*, 349.
23. Franck-Neumann, M.; Martina, D. *Tetrahedron Lett.* **1975**, 1759.
24. Franck-Neumann, M.; Martina, D.; Brion, F. *Angew. Chem. Int. Ed. Engl.* **1978**, *17*, 690.

CHAPTER 9

(±)-Hirsutic Acid C, (±)-Quadrone, and (±)-Coriolin

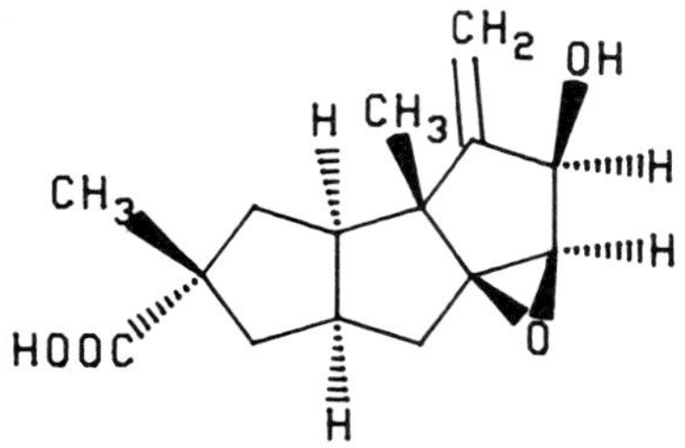

(±)-Hirsutic Acid C

Cyclopenta[4,5]pentaleno[1,6a-b]oxirene-5-carboxylic acid, decahydro-2-hydroxy-3a,5-dimethyl-3-methylene

(1aα, 2β, 3aβ, 3bα, 5α, 6aα, 7aS*)(±)

[55123-33-6]

[1aR-(1aα, 2β, 3aβ, 3bα, 5α, 6aα, 7aS*)]

[3650-17-7]

Figure 9.1

(+)-Hirsutic acid C and a related, uncharacterized compound called hirsutic acid N were first isolated from cultures of Basidiomycetes, *Stereum hirsutum*, in 1947.[1,2] Twenty-six years later hirsutic acid C and a related compound called complicatic acid were isolated from another culture of Basidiomycetes, *Stereum complicatum*[3] (see Figure 9.2).

Although hirsutic acid C has no biological activity[1] (complicatic acid has activity against *Staphylococcus aureus* and *Streptomyces* and *pyogenes*[1,3]), synthetic approaches have been studied by several groups[4] (see Figure 9.3).

(−)-Quadrone was first isolated by a group at W. R. Grace and Company from cultures of *Aspergillus terreus* in 1978.[5] It is the most active of the three targets in this chapter, having antibacterial activity[6], inhibitory activity in vitro against human epidermal carcinoma of the nasopharynx, and inhibitory activity in vivo in mice

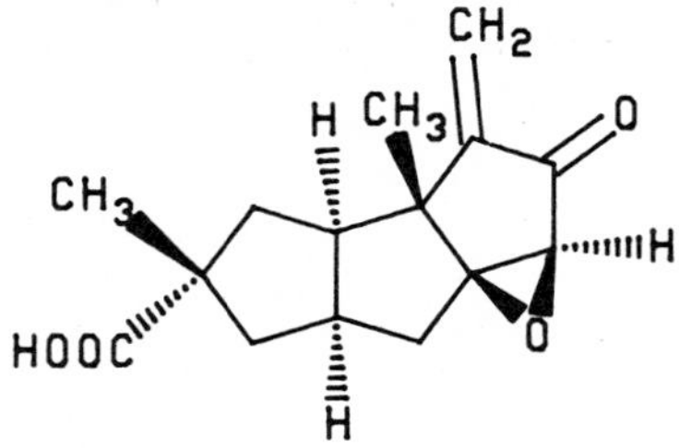

Complicatic Acid

Hirsutic Acid N ?

Figure 9.2

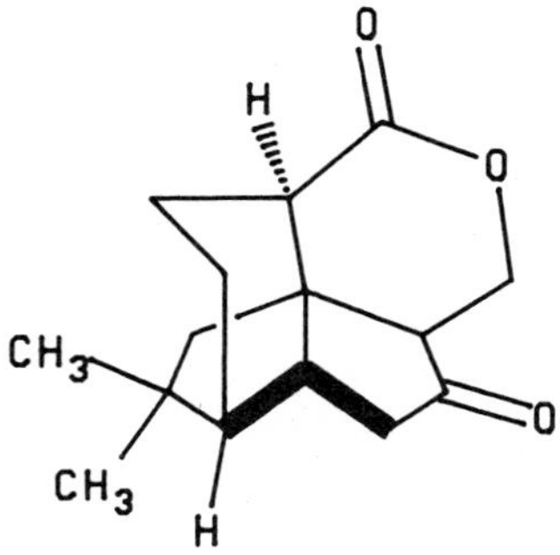

(±)-Quadrone

6,8b-Ethano-8bH-cyclopenta[de]-2-benzopyran-1,4-dione, octahydro-10,10-dimethyl

(3aα, 5aβ, 6α, 8aα, 8bα) (±)

[74807-65-1]

[3aS-(3aα, 5aβ, 6α, 8aα, 8bα)]

[66550-08-1]

Figure 9.3

against P_{388} lymphocytic leukemia and Lewis lung carcinoma.[7] Biological activity may be due to the α-methylene ketone of metabolite terrecyclic acid A. This modest activity coupled with a great interest in five-membered ring synthesis in the 1980s have resulted in several total synthetic efforts[8] (see Figures 9.4 and 9.5).

(−)-Coriolin was first isolated from a cultured both of a Basidiomycetes, *Coriolus consors*, in 1969.[9] Although little biological data have been reported (antibiotic activity,

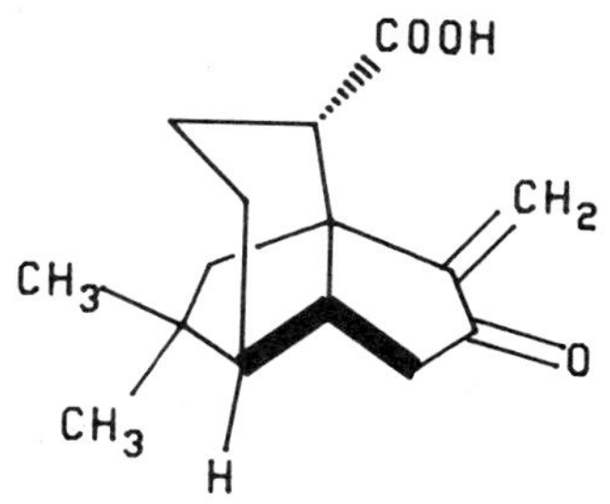

Terrecyclic Acid

Figure 9.4

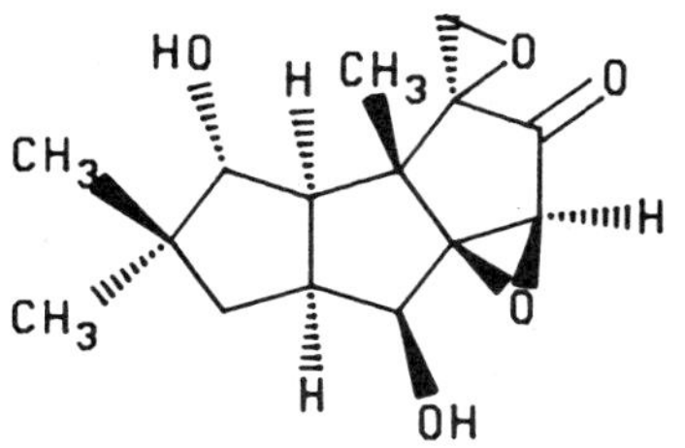

(±)-Coriolin

Spiro[cyclopenta[4,5]pentaleno[1,6a-b]oxirene-3-(3aH), 2'-oxiran]-2-[1aH]one hexahydro-4,7-dihydroxy-3a,5,5-trimethyl

(1aα, 3β, 3aβ, 3bα, 4α, 6aα, 7β, 7aS*)(±)

[74183-95-2]

[1aS-(1aα, 3β, 3aβ, 3bα, 4α, 6aα, 7β, 7aS*)]

[33404-85-2]

Figure 9.5

antitumor activity against L-1210 leukemia in mice[10]), again, a number of total synthetic approaches have been developed in recent years.[11]

9.1 THE KHAND–PAUSON CYCLOPENTENONE SYNTHESIS

9.1.1 Introduction

The first *ortho*-substituted *t*-butylbenzene was prepared by oxidative decomposition of the organometallic complex $Co_2(CO)_4(t\text{-}BuC_2H)_3$.[12] Mills and Robinson later

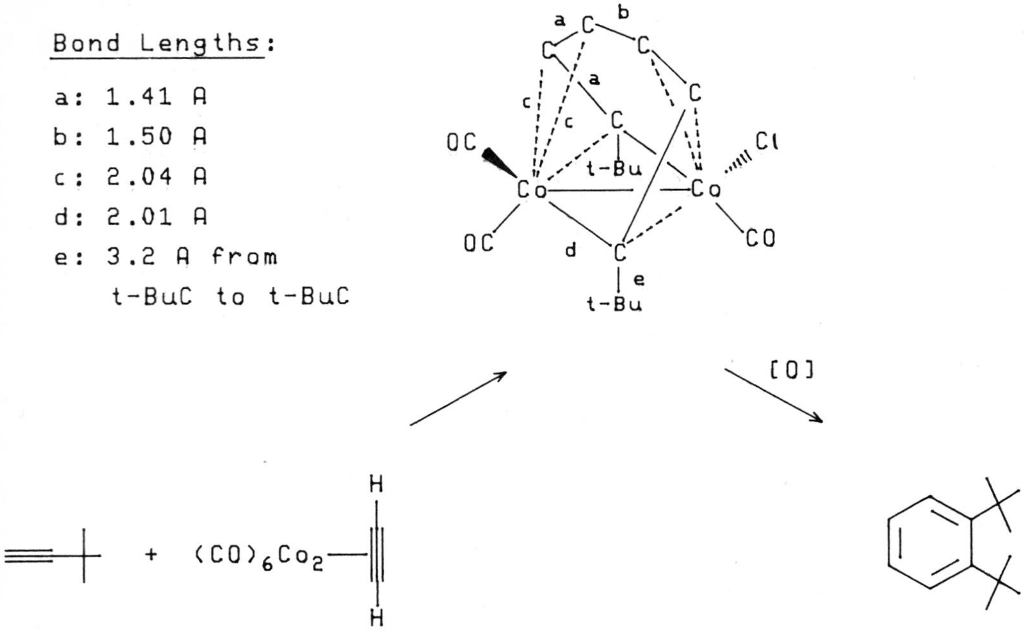

Bond Lengths:

a: 1.41 Å

b: 1.50 Å

c: 2.04 Å

d: 2.01 Å

e: 3.2 Å from t-BuC to t-BuC

Figure 9.6

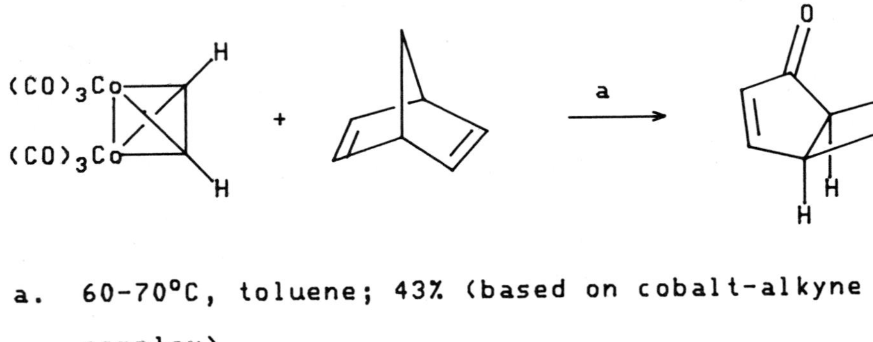

a. 60–70°C, toluene; 43% (based on cobalt-alkyne complex)

Figure 9.7

reported a crystal structure for the related "flyover" complex $Co_2(CO)_4(t\text{-}BuC_2H)_2$-$(C_2H_2)$ and oxidative decomposition to 1,2-di-*t*-butylbenzene[13] (see Figure 9.6).

While investigating a possible analogous insertion of norbornadiene, Khand and Pauson isolated a cyclopentenone with the two alkene carbons derived from the alkyne, two aliphatic carbons from the alkene, and a carbonyl from carbon monoxide[14] (see Figure 9.7). The same product also resulted when acetylene and carbon monoxide were simultaneously passed through a mixture of dicobalt octacarbonyl and norbornadiene in isooctane at 60 to 70°C (14% yield, based on dicobalt octacarbonyl). Research primarily by Pauson and to some extent by Schore, Billington and Schreiber have expanded upon this initial observation, providing information about the range of applicable alkenes, stereoselectivity with respect to alkene substituents, and regioselectivity with respect to alkene and alkyne. This information will be utilized in the development of a plausible mechanism.

9.1.2 Range of Applicable Alkenes

This is the most serious limitation of the Khand–Pauson cyclopentenone synthesis. While strained alkenes such as norbornadiene, norbornene, and cyclobutenes are reactive under relatively mild conditions (60 to 70°C), cyclopentenes, cyclohexenes, and simple acylic alkenes require considerably higher temperature (160°C), limiting the range of applicable alkyne and alkene substituent functionality.

As an alkene becomes more electron deficient, formation of diene by alkene–alkyne coupling without carbon monoxide insertion becomes competitive. Styrene–alkyne couplings afford mixtures of cyclopentenones and dienes in some instances[15] (see Figure 9.8).

Finally, as the number of alkene substituents or steric bulk of the substituents at the carbon atom of attachment increases, carbon–cobalt bond insertion becomes more difficult. Mono- and disubstituted unstrained alkenes and di- and trisubstituted strained alkenes have been successfully employed. When the alkene is not suitable, aromatics produced by alkyne trimerization are commonly observed.

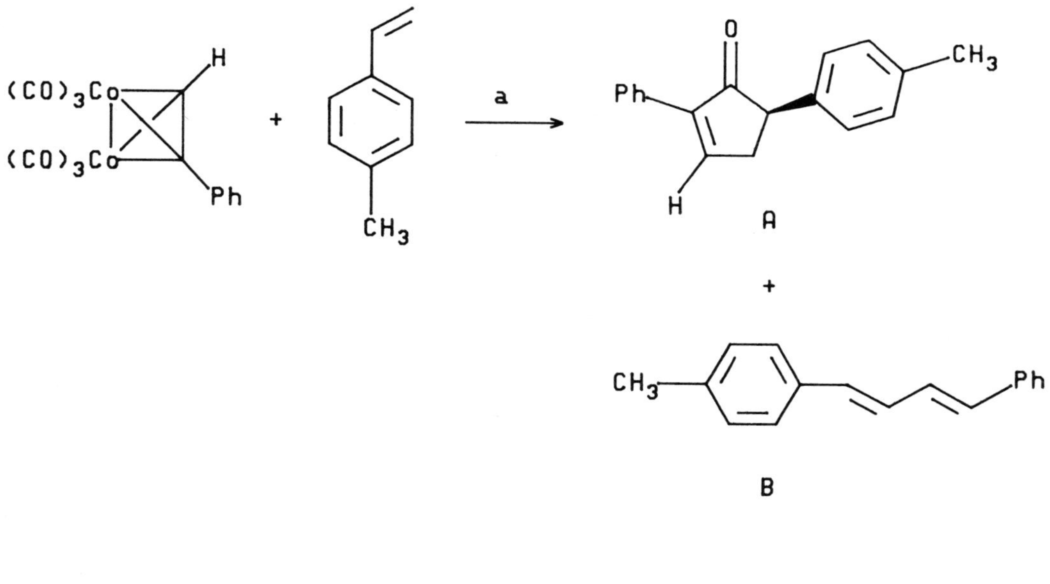

$(CO)_3Co$ H $(CO)_3Co$ Ph + CH_3 —a→ O Ph CH_3 H A + CH_3 Ph B

a. 110°C, toluene; A:13%, B:26% (based on cobalt-alkyne complex)

Figure 9.8

9.1.3 Stereoselectivity with Respect to Alkene Substituents

Annelation of norbornene, norbornadiene, and other similar bicyclic alkenes occurs exclusively on the more accessible *exo* face. A *trans*-disubstituted alkene affords cyclopentenones in which the stereochemical relationship of substituents is maintained[15] (see Figure 9.9).

9.1.4 Regioselectivity with Respect to Alkyne Substituents

In every instance, alkynes afford the product having the larger group adjacent to the carbonyl. Thus, terminal alkynes afford 2-substituted cyclopentenones exclusively. If the steric requirements of the two alkyne substituents are similar (e.g., methyl and ethyl) at the carbon of attachment, a regioisomeric mixture is likely (see Figure 9.10).

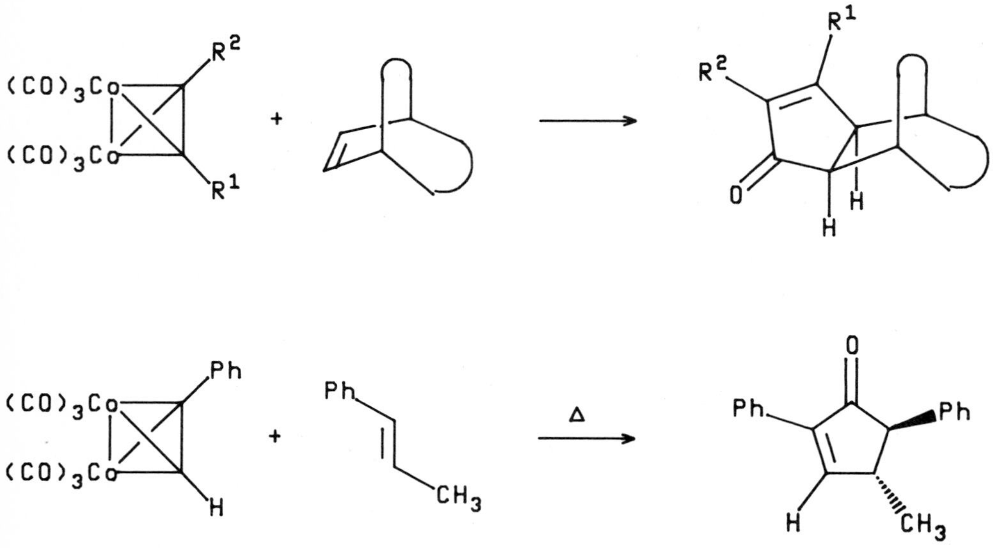

15% (based on cobalt-alkyne complex)

Figure 9.9

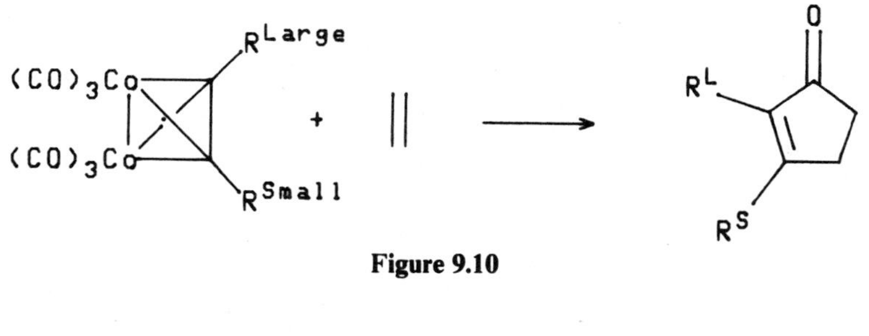

Figure 9.10

9.1.5 Regioselectivity with Respect to Alkene Substituents

Three directing effects, categorized by the distance of the substituents from the alkene carbons, are observed: (1) directly attached, (2) adjacent (allylic), and (3) long range. Using a simple monosubstituted alkene, the substituted carbon becomes C-5 of the cyclopentenone.[15] Using a trisubstituted alkene, the disubstituted carbon atom becomes C-5 of the cyclopentenone[16] (see Figure 9.11).

Using a 1,2-disubstituted alkene in which one substituent is alkyl and the other aryl or vinyl (i.e., a styrene or a 1,3-diene), the cyclopentenone produced has the aryl or vinyl group as a C-5 substituent[17] (see Figure 9.12).

Allylic substitution has a less pronounced effect. The alkene carbon adjacent to a sterically encumbered allylic position becomes C-4 on the major cyclopentenone product and C-5 in the minor cyclopentenone product. One report that apparently contradicts this generalization made only a tentative identification, based on ^{1}H and ^{13}C NMR data, of the single regioisomer obtained[18] (see Figure 9.13).

More distant substituents will sometimes have a significant directing effect, producing a mixture rich in one regioisomer. For example, the reaction of norbornadiene and two equivalents of cobalt–alkyne complex affords a mixture of bis-annelated products in which the *anti* product predominates[14b] (see Figure 9.14).

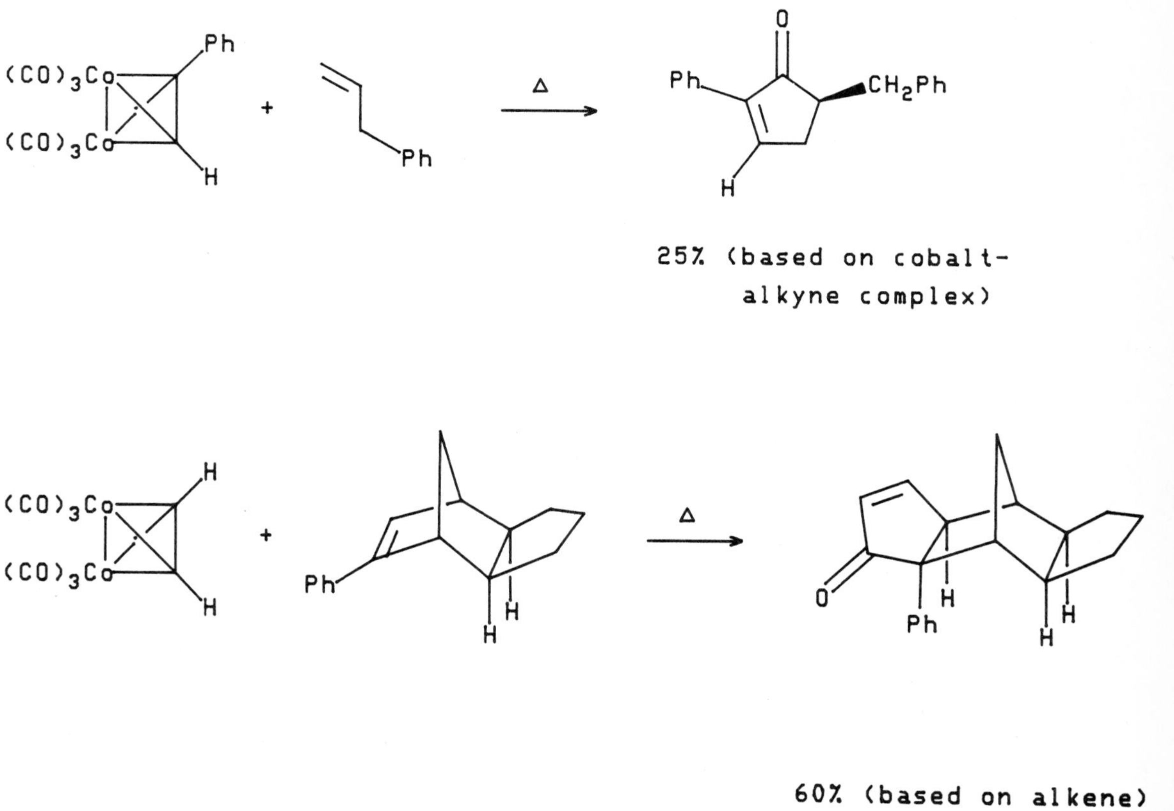

25% (based on cobalt-alkyne complex)

60% (based on alkene)

Figure 9.11

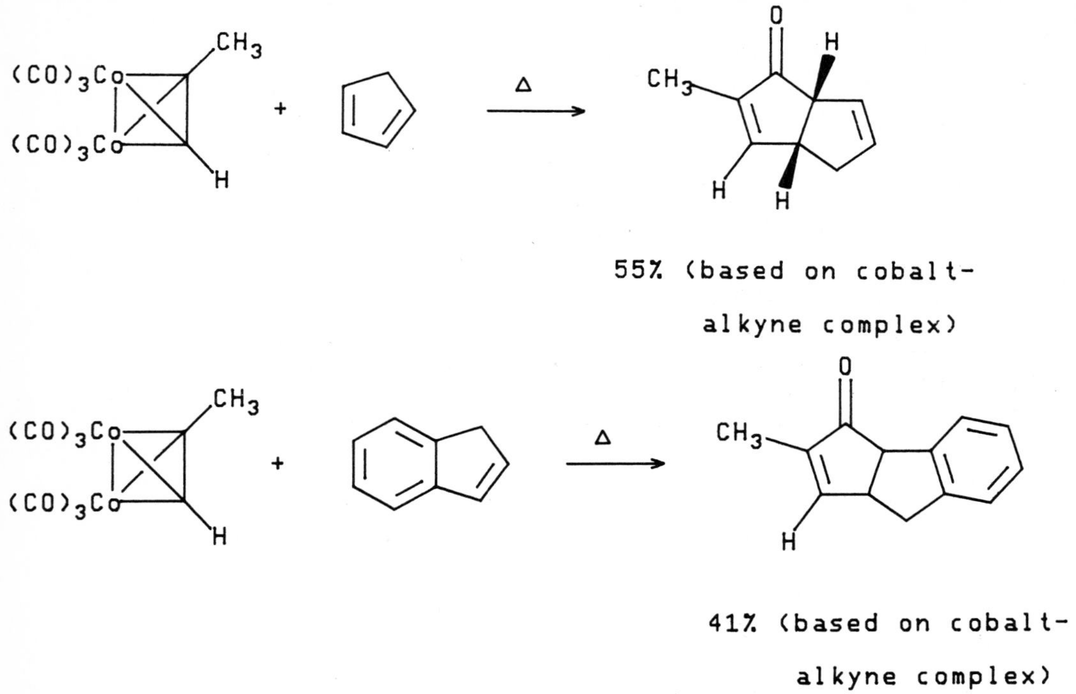

$(CO)_3Co$ CH$_3$ H + $\xrightarrow{\Delta}$ CH$_3$ O H H H

55% (based on cobalt-alkyne complex)

$(CO)_3Co$ CH$_3$ H + $\xrightarrow{\Delta}$ CH$_3$ O H

41% (based on cobalt-alkyne complex)

Figure 9.12

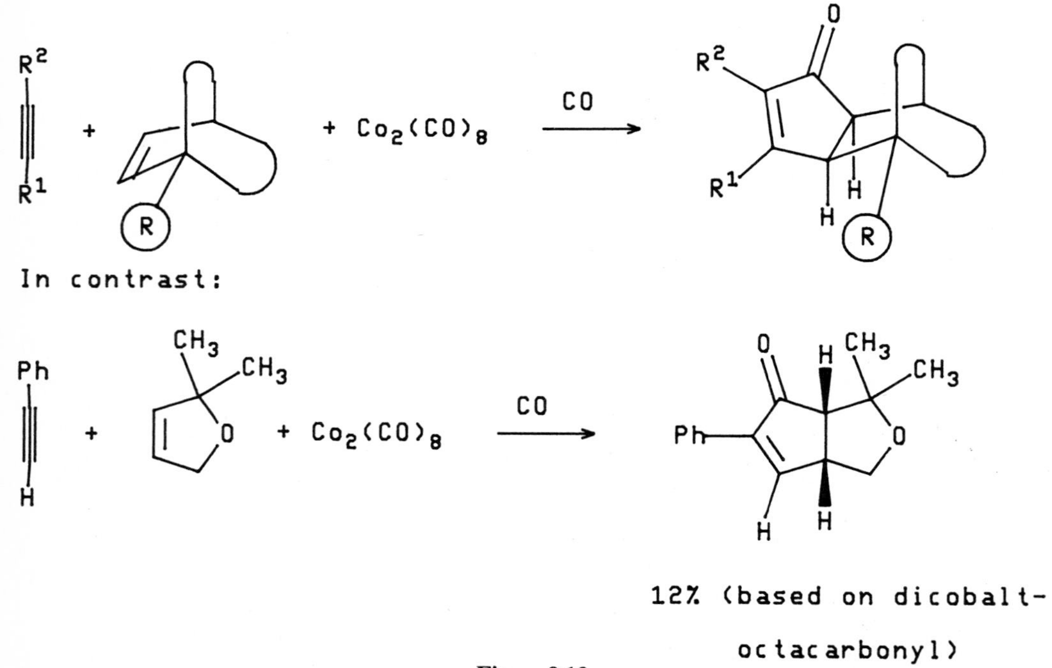

R^2 R^1 + (R) + $Co_2(CO)_8$ $\xrightarrow{CO}$ O R^2 R^1 H H (R)

In contrast:

Ph H + CH$_3$ CH$_3$ O + $Co_2(CO)_8$ $\xrightarrow{CO}$ O H CH$_3$ CH$_3$ Ph O H H

12% (based on dicobalt-octacarbonyl)

Figure 9.13

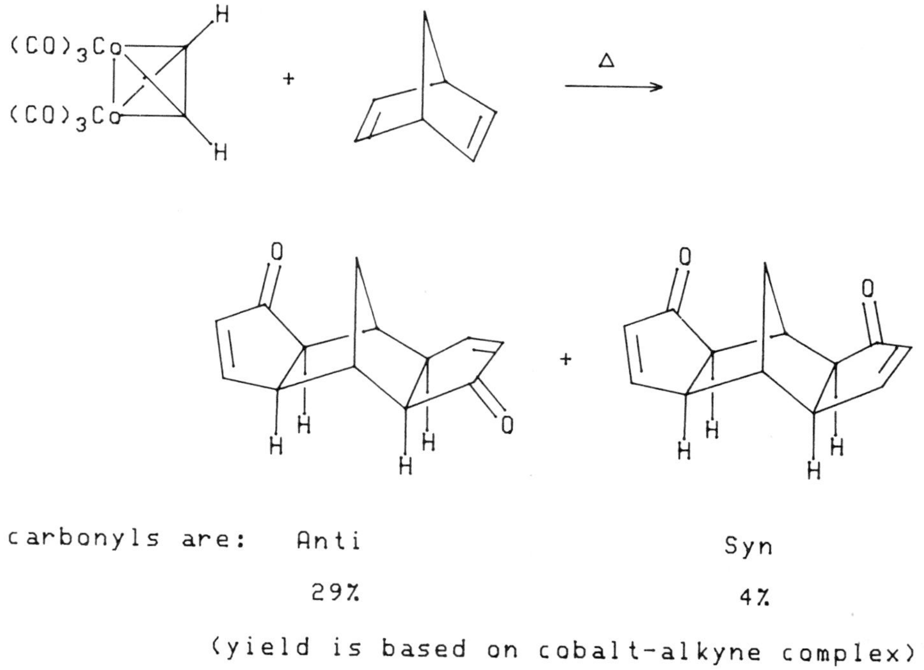

$(CO)_3Co$ $(CO)_3Co$ H H + Δ →

carbonyls are: Anti 29% Syn 4%

(yield is based on cobalt-alkyne complex)

Figure 9.14

9.1.6 Mechanism

A mechanism that adequately accounts for much of the stereo- and regioselectivity observed is illustrated for the case of an unsymmetrical alkyne (R^2 larger than R^1) and monosubstituted alkene. The initial step, insertion of the alkene into a carbon–cobalt bond, can produce eight different flyover complexes (IA-D, IIA-D). Initial carbon–carbon bond formation must involve the most sterically accessible alkyne and alkene carbons (IA, B). Carbon monoxide insertion into the newly formed carbon–cobalt bond, then carbon–carbon bond formation by reductive elimination, closes the five-membered ring. Cleavage of the cobaltacyclopropane affords alkene and dicobalt hexacarbonyl capable of forming a new cobalt–alkyne complex. For the case of a cyclic alkene with one allylic carbon more sterically encumbered than the other, a steric interaction between the allylic substituent and metal carbonyl ligands results in preferential formation of complexes analogous to IC, D.[22] A similar explanation may account for the *preferred* formation of 5-aryl or 5-vinyl cyclopentenones (see Figure 9.15). If R^3 is a bulky substituent on the allylic carbon, a strong steric interaction with the cobalt carbonyls will redirect to ID when $R^1 = H$ (see Figure 9.16).

Finally, when the alkyne substituents, R^1 and R^2, are large, intermediate IA with the alkene substituent on the top face may be preferred to intermediate IB, where there is a possible steric interaction between R^3 and R^2. This directing effect will be discussed in the context of the intramolecular Khand–Pauson cyclopentenone synthesis.

Table 9.1 provides experimental data on the intermolecular cyclopentenone synthesis. Direct comparison of different reports is often difficult since the reaction can

R^2 larger than R^1:

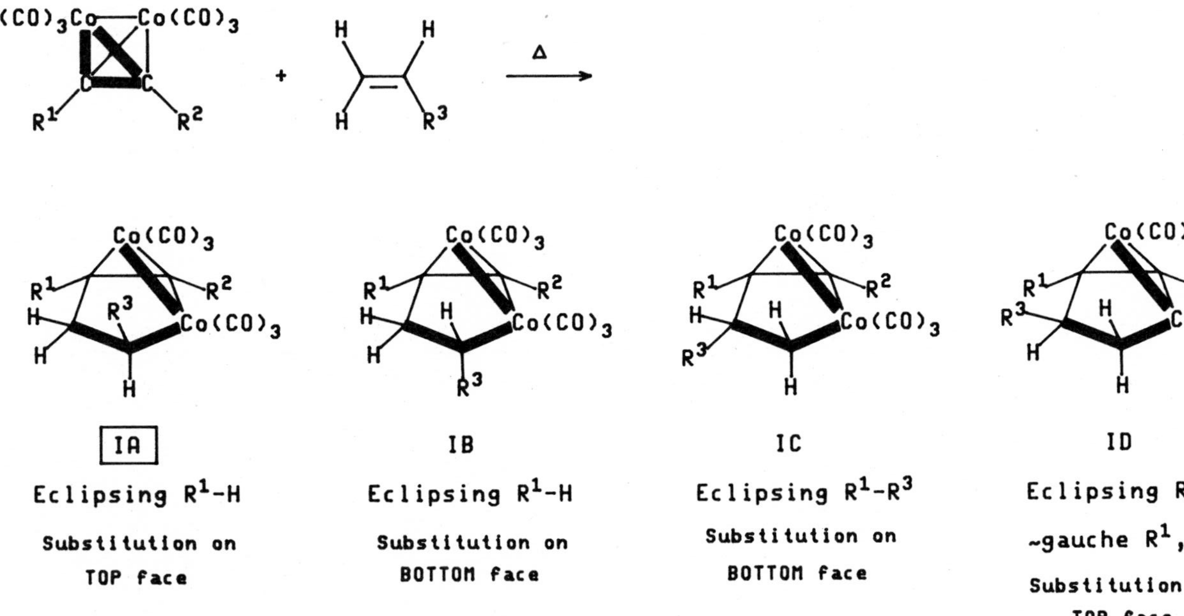

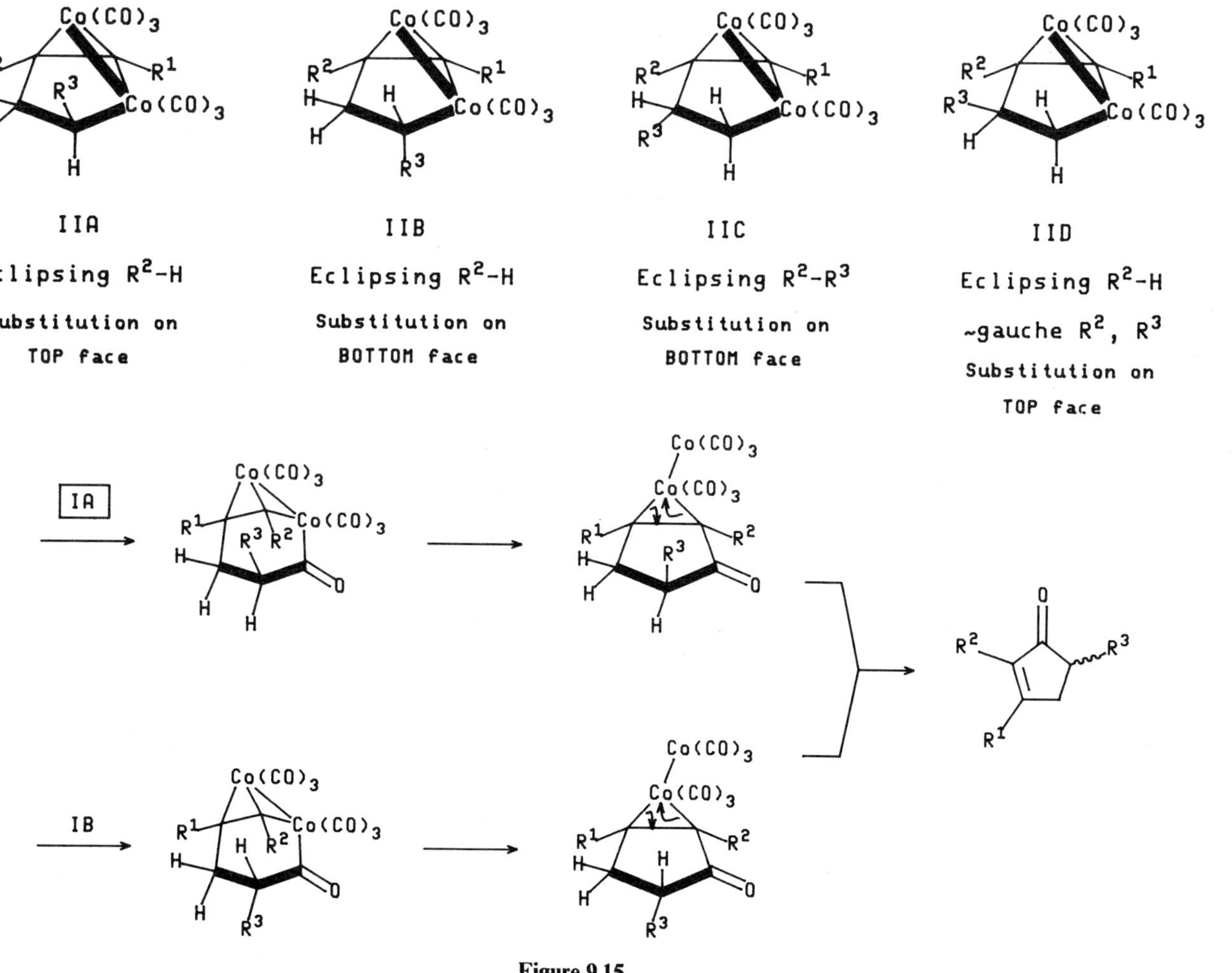
IIA
Eclipsing R2-H
Substitution on
TOP face
IIB
Eclipsing R2-H
Substitution on
BOTTOM face
IIC
Eclipsing R2-R3
Substitution on
BOTTOM face
IID
Eclipsing R2-H
~gauche R2, R3
Substitution on
TOP face
IA
IB

Figure 9.15

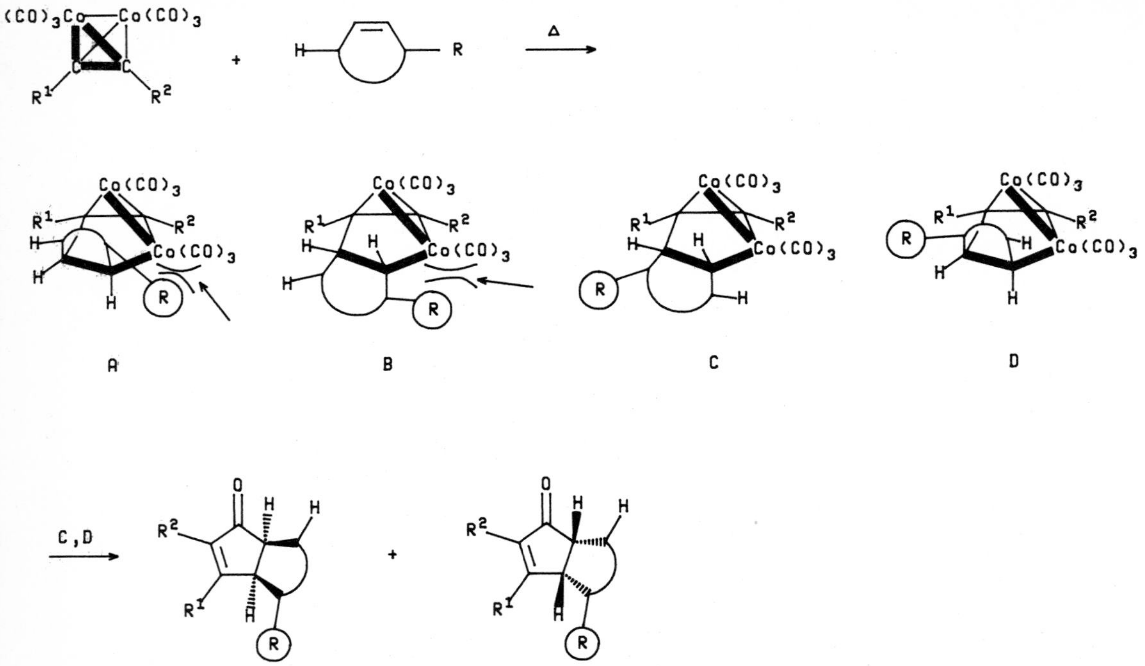

Figure 9.16

TABLE 9.1 The Intermolecular Khand–Pauson Cyclopentenone Synthesis[14b,16–30]

Strained Alkenes

$(CO)_3Co_2(CO)_3(R^1C{\equiv}CR^2)$ + Alkene $\xrightarrow{\text{solvent}}$ Product(s)

R^1	R^2	Alkene	Product(s)	Yield (%)	via:	Reference
H	H		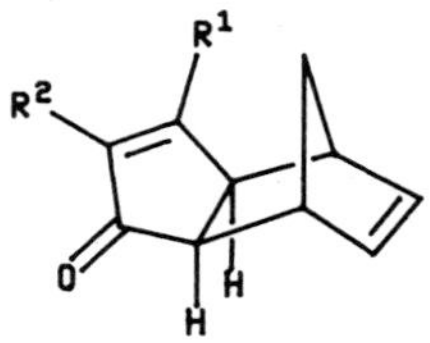	43	A	14b
				47	D	14b
Et	Et			23	A	14b
Ph	Ph			28	A	14b
H	CH_3			33	A	14b
H	Ph			45	A	14b
CH_3	$Si(CH_3)_3$			42	A	16
H	SPh			57	A	25
H	Bu			53	A	31

TABLE 9.1 *(Continued)*

R^1	R^2	Alkene	Product(s)	Yield (%)	via:	Reference
H	H		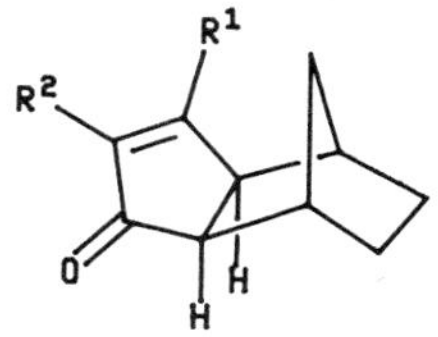	55	A	14b
				74	D	14b
Et	Et			23	A	14b
Ph	Ph			35	A	14b
H	CH_3			33	A	14b
H	Ph			59	A	14b
CH_3	Ph			65	A	16
CH_3	$Si(CH_3)_3$			38	A	16
H	SPh			59	A	25
H	H	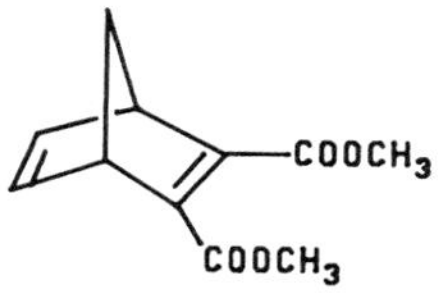	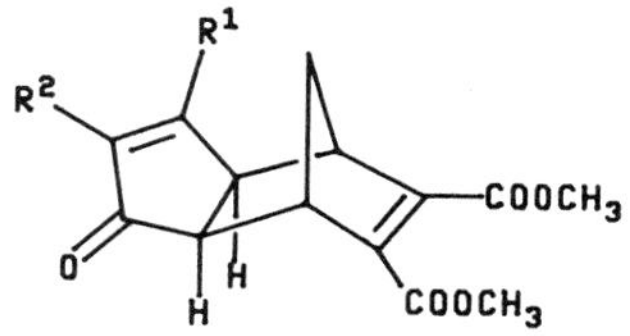	23	A	14b
Ph	Ph			28	A	14b
H	CH_3			38	A	14b
H	Ph			31	A	14b
H	CH_3	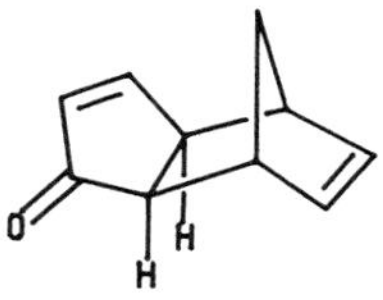	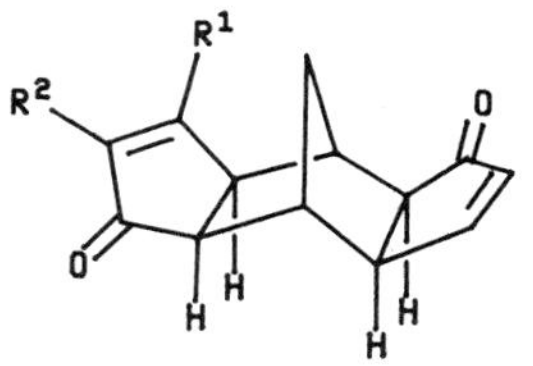	21	C	14b
H	Ph			23	C	14b
H	H	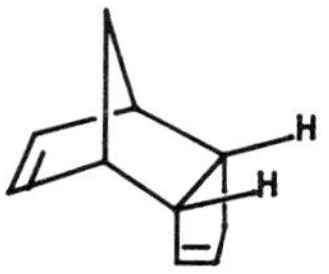	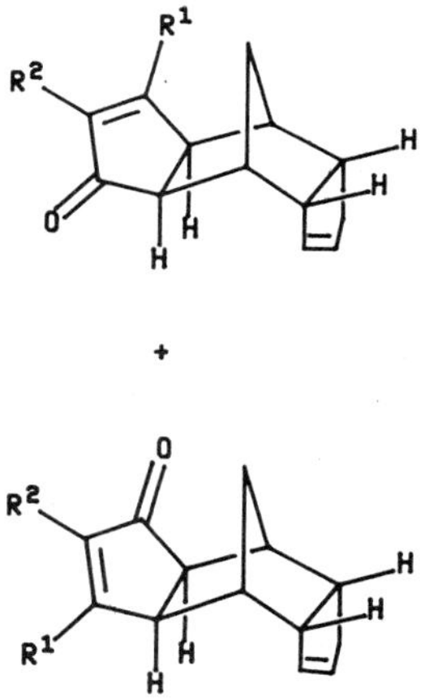	74	C	16
Et	Et			18	C	16
Ph	Ph			23	A	16
H	CH_3			51	C	16
H	Ph			57	A	16
CH_3	Ph			32	A	16

(continued)

TABLE 9.1 (*Continued*)

R^1	R^2	Alkene	Product(s)	Yield (%)	via:	Reference
H	H	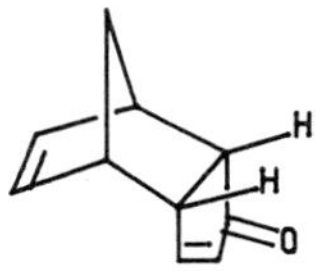	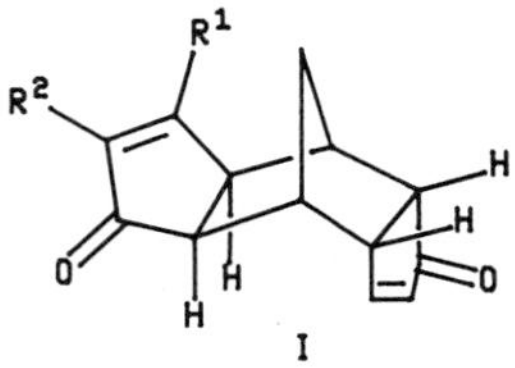 I + 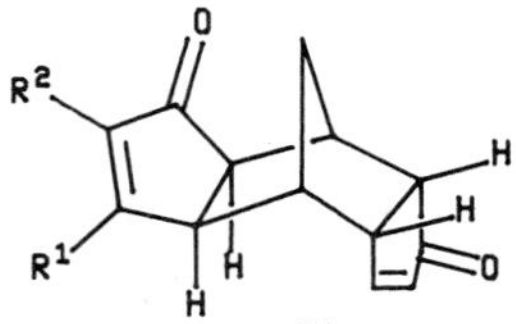 II	I,II = 33,18	C	16
H	CH_3			24,10	C	16
H	Ph			32,9	C	16
H	H	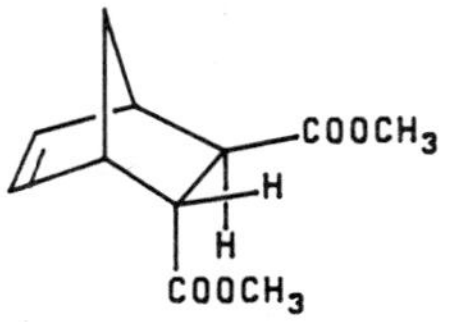 $COOCH_3$, H, H, $COOCH_3$	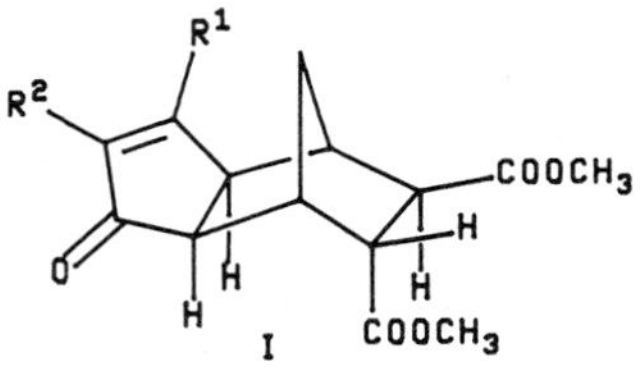 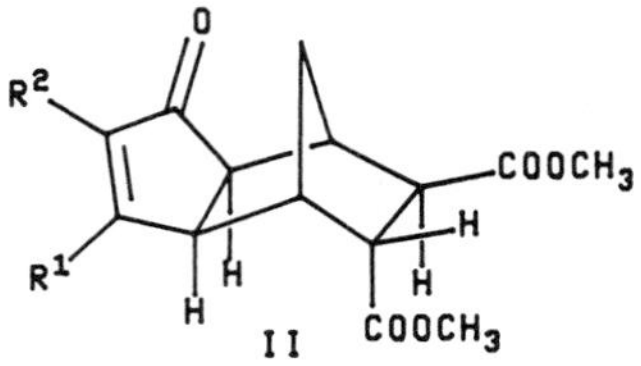 I + II ($COOCH_3$, H, H, $COOCH_3$)	I,II = 25,8	C	16
H	CH_3			53,9	C	16
H	Ph			44,14	C	16
CH_3	Ph			28,14	C	16
H	H	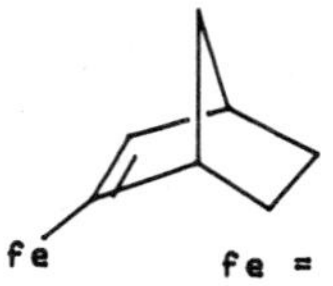 fe; fe = ferrocenyl	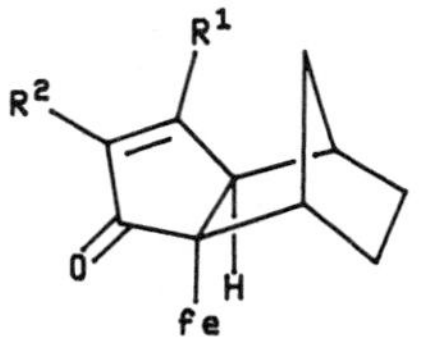 fe	18	C	16

TABLE 9.1 *(Continued)*

R^1	R^2	Alkene	Product(s)	Yield (%)	via:	Reference
H	H	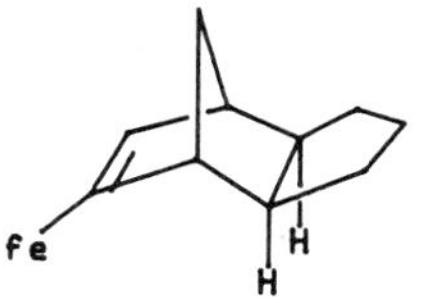 fe = ferrocenyl	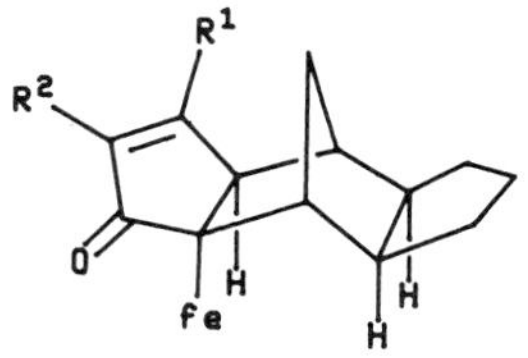	65	C	16
H	H	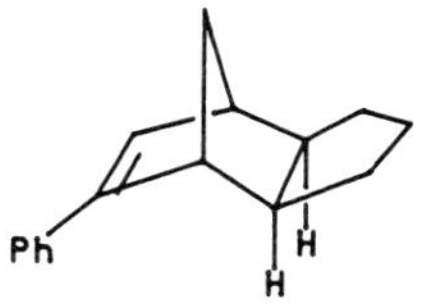	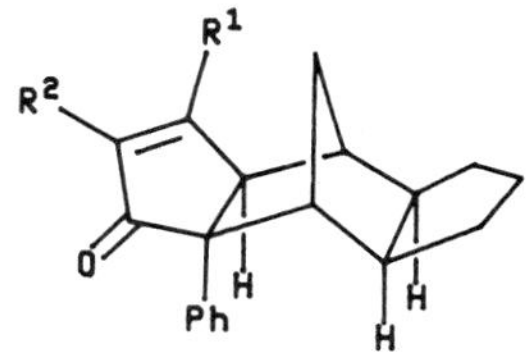	60	C	16
H	CH_3	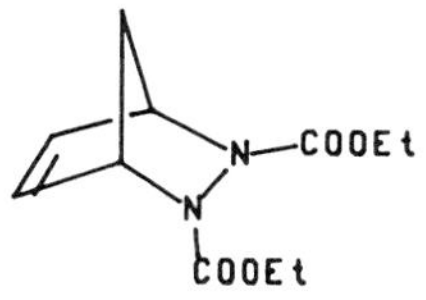	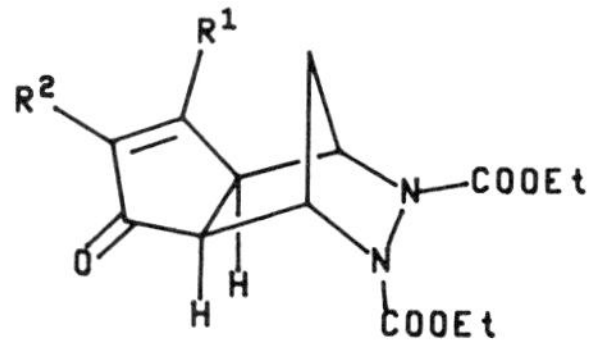	20	C	19
H	Ph			36	C	19
H	H	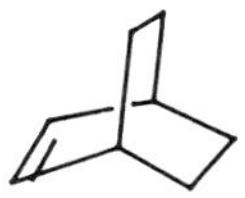	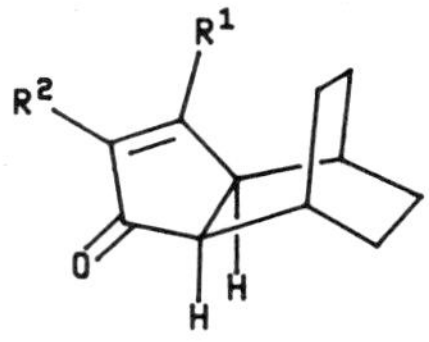	50	C	20
H	CH_3			19	C	20
H	Ph			34	C	20
H	H	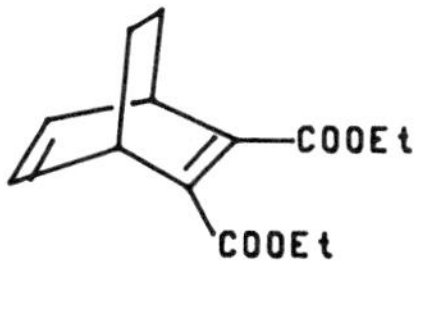	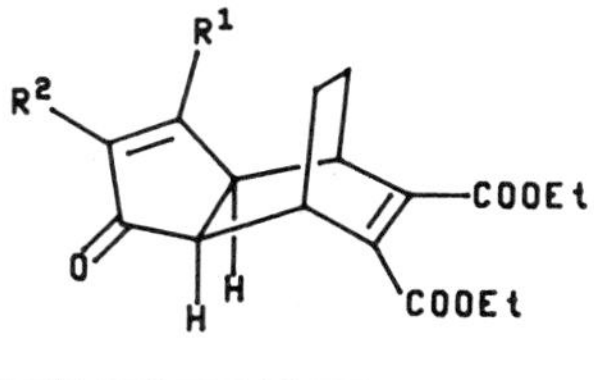	80	C	20
H	CH_3			36	C	20
H	Ph			85	C	20

(continued)

TABLE 9.1 (*Continued*)

R^1	R^2	Alkene	Product(s)	Yield (%)	via:	Reference
H	H		I + II	I,II = 29,4	A	14b
H	CH_3			17,4	A	14b
H	Ph			13,0	A	14b
H	H			No data available		23
Et	Et					23
Ph	Ph					23
H	CH_3					23
H	Ph					23
CH_3	Ph					23
H	H	OCH_3	OCH_3	No data available		23
Et	Et					23
Ph	Ph					23
H	CH_3					23
H	Ph					23
CH_3	Ph					23

Cyclopentenes, Dihydrofurans, Cyclopentadienes

R^1	R^2	Alkene	Product(s)	Yield (%)	via:	Reference
H	$(CH_2)_6COOCH_3$			33	A	24
H	SPh			53	A	25

TABLE 9.1 (*Continued*)

R^1	R^2	Alkene	Product(s)	Yield (%)	via:	Reference
H	Ph	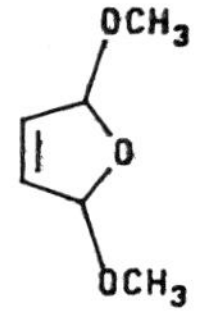 2,5-dimethoxy-2,5-dihydrofuran (OCH_3, O, OCH_3)	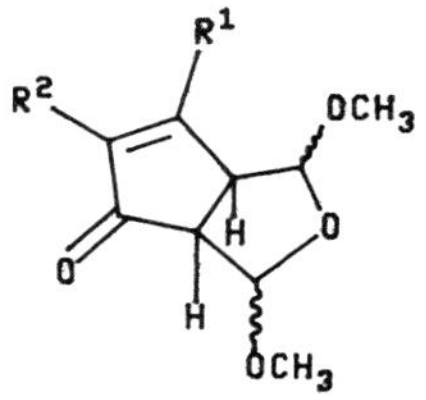 bicyclic enone fused to tetrahydrofuran ring with R^1, R^2, O, H, H, OCH_3, OCH_3	30	A	19
H	Z $CH_2CH{=}CH(CH_2)_3COOCH_3$	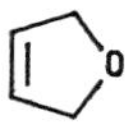 2,5-dihydrofuran (O)	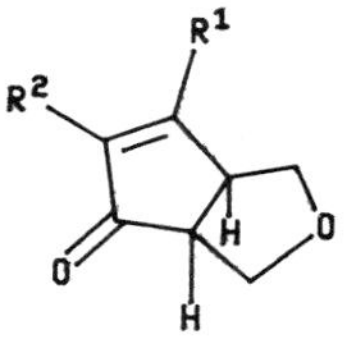 bicyclic enone fused to tetrahydrofuran ring with R^1, R^2, O, H, H, O	65	A	26
H	H			150	D	18
CH_3	CH_3			70	D	18
H	CH_3	 cyclopentene	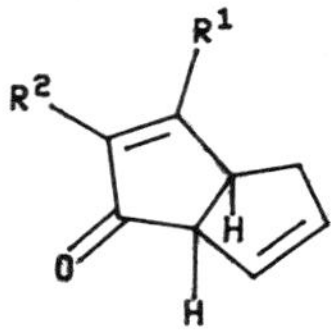 tricyclic enone with R^1, R^2, O, H, H	55	A	17
H	Ph			60	A	17
H	Ph	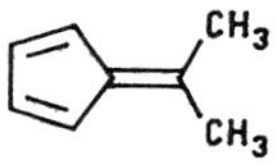 6,6-dimethylfulvene (CH_3, CH_3)	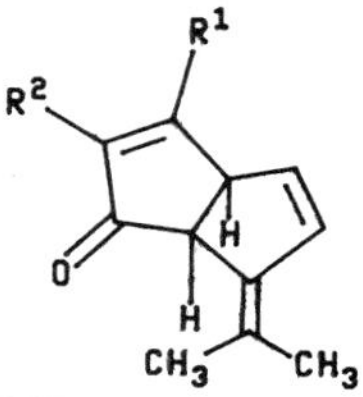 tricyclic enone with R^1, R^2, O, H, H, $=C(CH_3)_2$	23	A	17
H	H	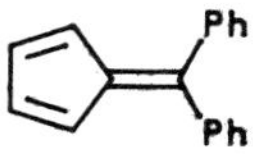 6,6-diphenylfulvene (Ph, Ph)	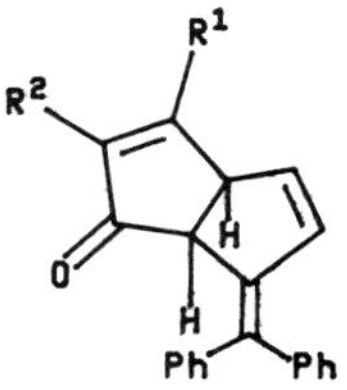 tricyclic enone with R^1, R^2, O, H, H, $=CPh_2$	15	C	17
H	CH_3			15	C	17

(*continued*)

TABLE 9.1 *(Continued)*

R^1	R^2	Alkene	Product(s)	Yield (%)	via:	Reference
H	H	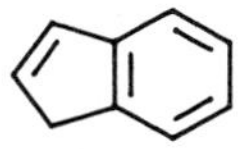	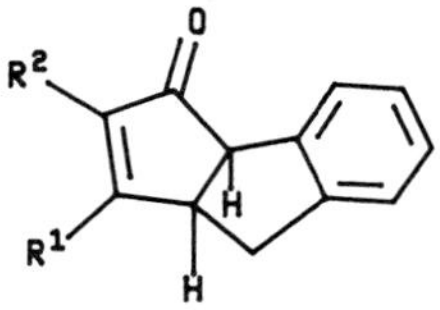	31	A	17
H	CH_3			41	A	17
H	H	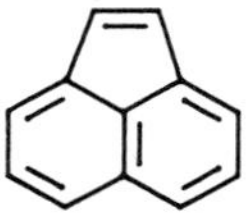	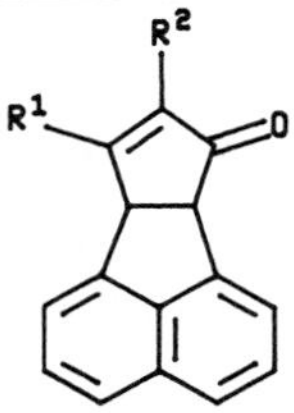	24	A	17
H	CH_3			26	A	17
H	Ph			38	A	17
H	Ph	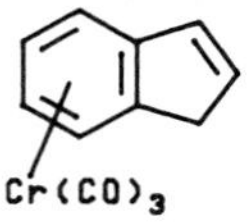 $Cr(CO)_3$	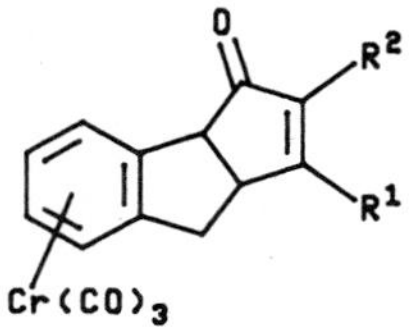 $Cr(CO)_3$	17	C	27
H	H	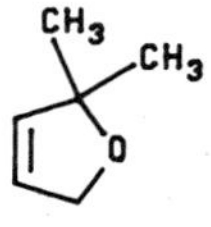	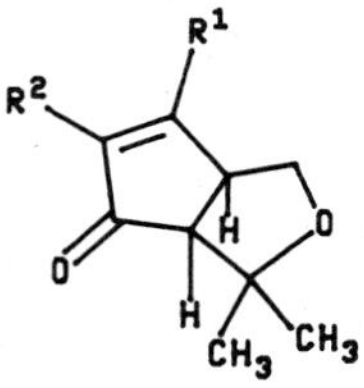	12	D	18

Styrenes, Allylarenes

R^1	R^2	Alkene	Product(s)	Yield (%)	via:	Reference
H	Ph	 C_6H_4X, X = 4-CH_3	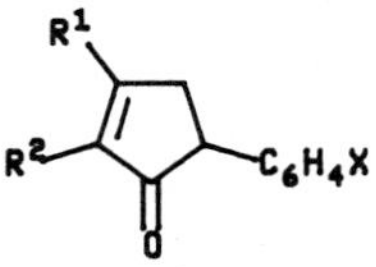 C_6H_4X	13	A	15
		4-F		35	A	15
		4-Cl		8	A	15
		2-Cl		4	A	15
		4-OCH_3		27	A	15

TABLE 9.1 (*Continued*)

R^1	R^2	Alkene	Product(s)	Yield (%)	via:	Reference
H	Ph	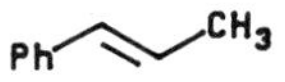 Ph–CH=CH–CH_3	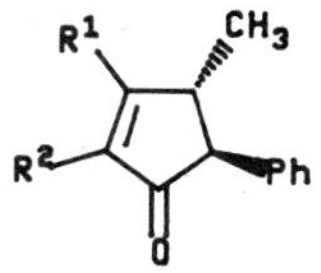 [structure: R^1, R^2, CH_3, Ph, O]	15	A	15
H	Ph	 Ph–CH_2–CH=CH_2	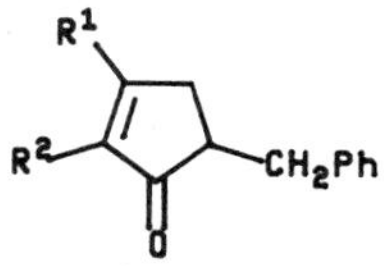 [structure: R^1, R^2, CH_2Ph, O]	25	A	15
H	Ph	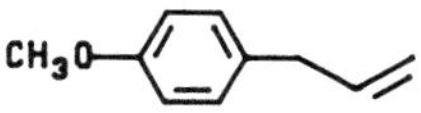 CH_3O–C_6H_4–CH_2–CH=CH_2	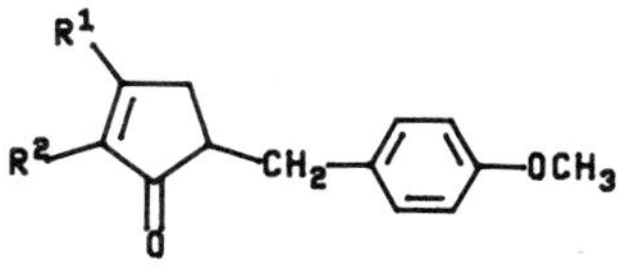 [structure: R^1, R^2, CH_2, OCH_3, O]	29	A	15
H	Ph	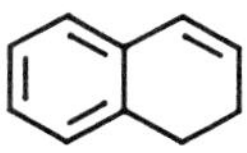 [1,2-dihydronaphthalene]	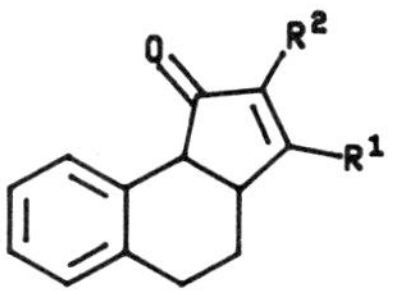 [structure: O, R^2, R^1]	38	A	15
H	CH_3			35	A	15
H	Ph	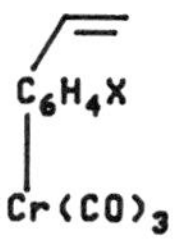 CH_2=CH–C_6H_4X–$Cr(CO)_3$, X = H	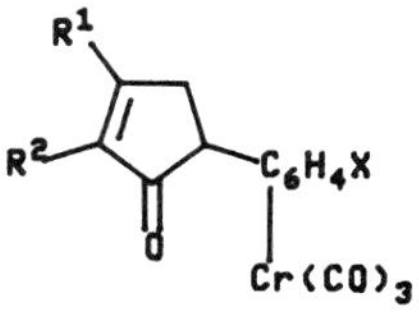 [structure: R^1, R^2, C_6H_4X, $Cr(CO)_3$, O]	37	C	27
		4-CH_3		39	C	27
		4-F		29	C	27
		4-OCH_3		38*	C	27
H	Ph	[allylbenzene–$Cr(CO)_3$]	[structure: R^1, R^2, O, $Cr(CO)_3$]	32	C	27

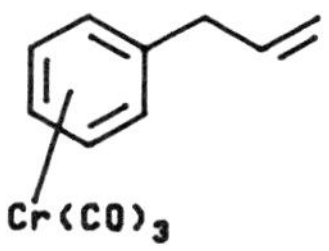

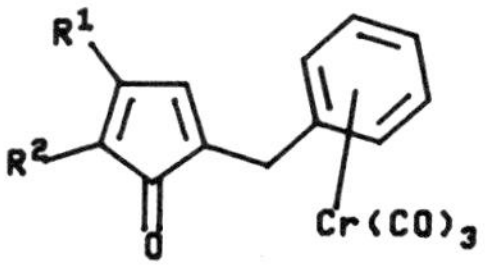

*plus some Cr-free ketone

(*continued*)

TABLE 9.1 (*Continued*)

Simple Alkenes

R^1	R^2	Alkene	Product(s)	Yield (%)	via:	Reference
H	$(CH_2)_6COOCH_3$	$CH_2{=}CH_2$	R^1, R^2, O	46	B	24
H	$CH_2CH\overset{Z}{=}CH(CH_2)_3COOCH_3$		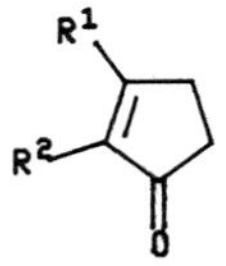	57	B	24
H	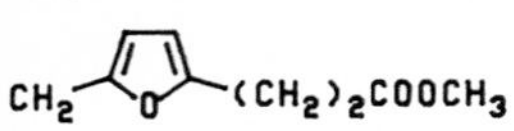 CH_2–(2,5-furandiyl)–$(CH_2)_2COOCH_3$			32	B	28
H	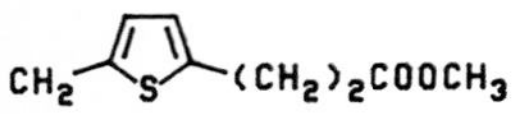 CH_2–(2,5-thiophenediyl)–$(CH_2)_2COOCH_3$			33	B	28
H	SPh			11	A	25
CH_3	CH_3	$CH_2{=}CH{-}CH_2OTHP$	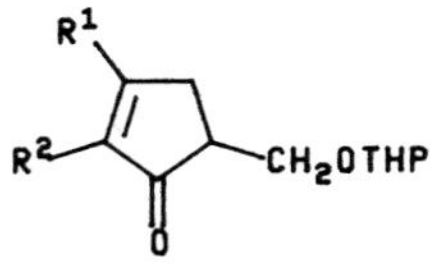 R^1, R^2, CH_2OTHP, O	32	A	29
H	H R = Ac	$CH_2{=}CHOR$	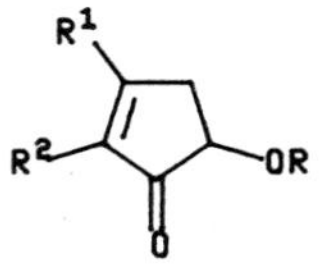 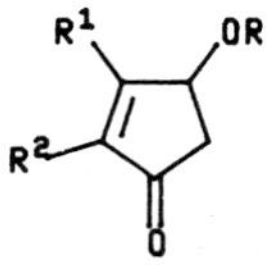 R^1, R^2, OR, O, I + R^1, OR, R^2, O, II	I,II=20,20	D	30
H	H R = t-Bu			20,20	D	30

TABLE 9.1 (*Continued*)

R^1	R^2	Alkene	Product(s)	Yield (%)	via:	Reference
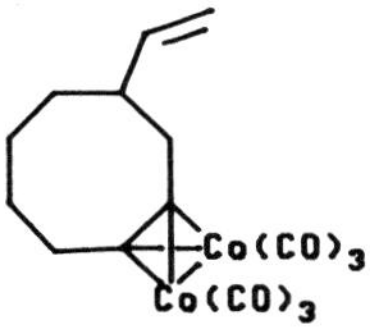 (vinyl-cyclooctyne $Co_2(CO)_6$ complex) +		 norbornene	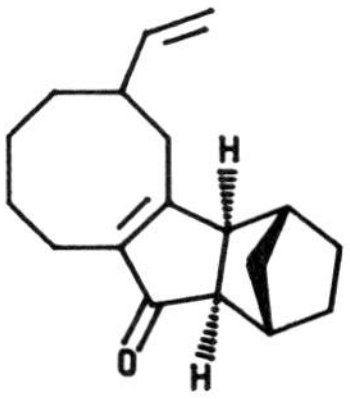 I 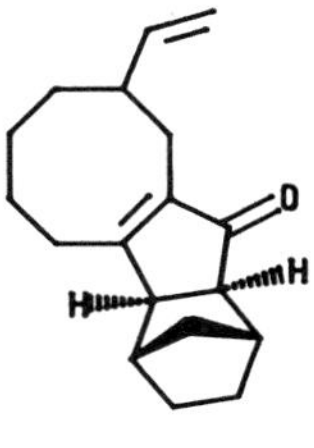 + II	I,II = 33,33	A	21
H	H	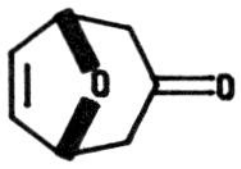	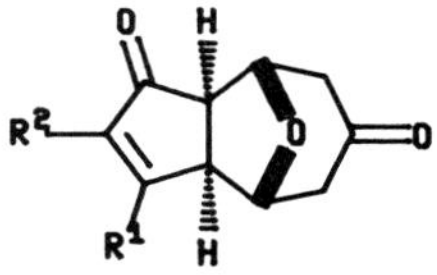	45	E	22
H	Ph			42	E	22
H	Bu			40	E	22
H	H		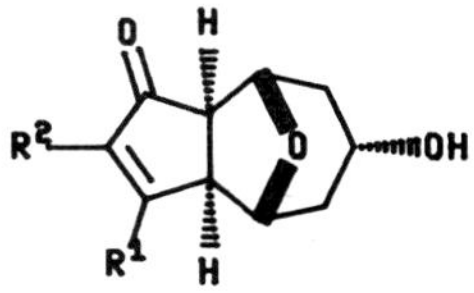	52	E	22
H	Ph			57	E	22
H	Bu			43	E	22

(continued)

TABLE 9.1 (*Continued*)

R^1	R^2	Alkene	Product(s)	Yield (%)	via:	Reference
H	H	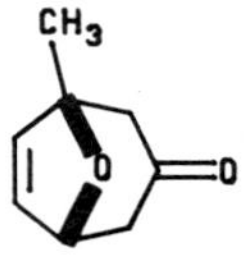	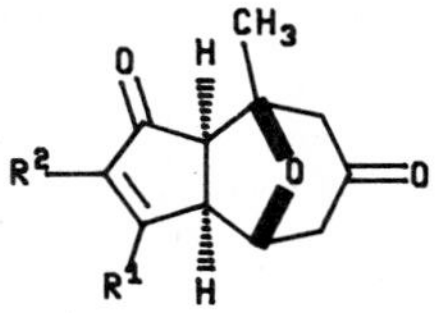 I 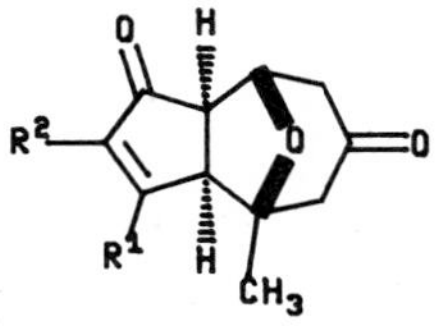 + II	I,II = 20,30	E	22
H	H	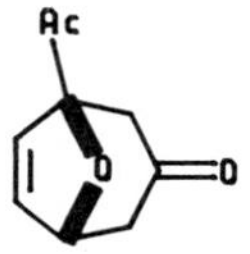	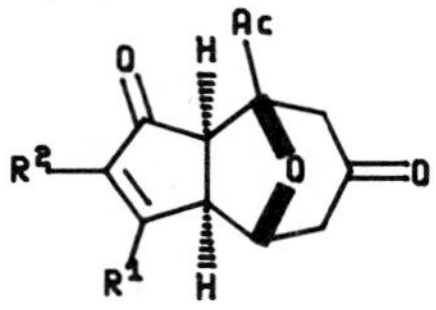 I 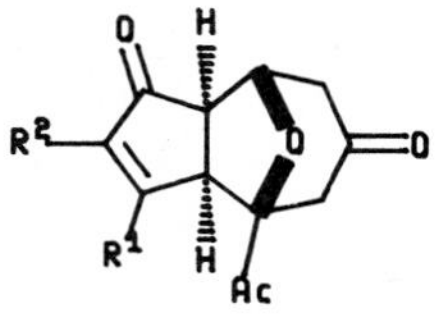 + II	I,II = 6,18	E	22
H	H	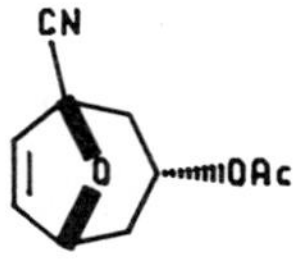	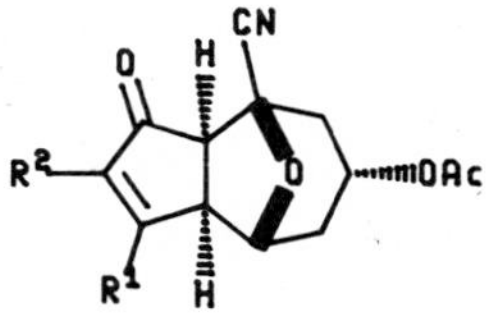 I 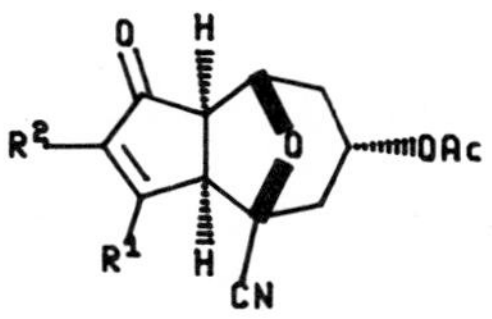 + II	I,II = 23,45	E	22

be run using metal in stoichiometric or catalytic amounts, and the least accessible reagent, either alkene or alkyne–cobalt complex, has been used as the limiting reagent. The following five different methods are used:

A. Under N_2, cobalt–alkyne complex, (excess) alkene; yield based on cobalt–alkyne complex.
B. Under alkene pressure, cobalt–alkyne complex; yield based on cobalt–alkyne complex.
C. Under N_2, (excess) cobalt–alkyne complex, alkene; yield based on alkene.
D. Under CO, $Co_2(CO)_8$, (excess) alkyne, (excess) alkene; yield based on $Co_2(CO)_8$.
E. Under CO, $Co_2(CO)_8$, approximately one equivalent alkyne, approximately one equivalent alkene; yield based on alkene.

9.2 THE INTRAMOLECULAR KHAND–PAUSON CYCLOPENTENONE SYNTHESIS

Two intramolecular versions of the cyclopentenone synthesis have been most extensively studied: (1) synthesis of bicyclo[3.3.0]oct-7-en-6-ones from substrates having alkene and alkyne bridged by three carbons (Table 9.2) and (2) synthesis of 3-oxa-bicyclo[3.3.0]oct-7-en-6-ones from substrates having a C—O—C alkene–alkyne bridge (Table 9.3) (see also Figure 9.17).

As in the intermolecular series, mono-, di-, and trisubstituted alkenes have been successfully employed, and the geometric relationship of alkene substituents is maintained in the product. Cyclization using silicon-substituted alkynes has on occasion been more efficient than cyclization using terminal alkynes. This may be due to a decrease in the propensity for competitive alkyne trimerization (see Chapter 7).

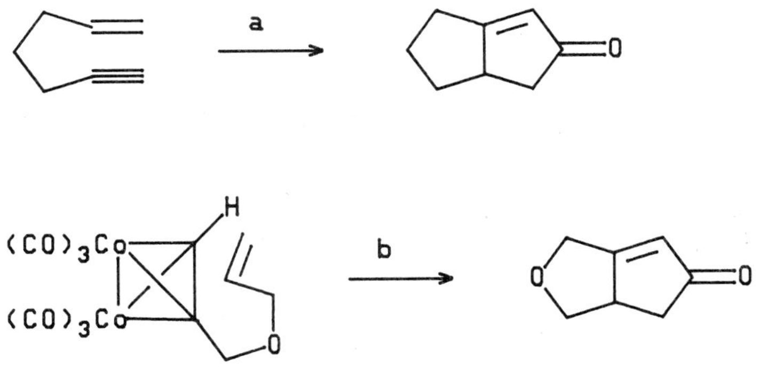

a

O

H

$(CO)_3Co$

$(CO)_3Co$

O

b

O

O

a. 95°C, $Co_2(CO)_8$; 31%; Reference 32

b. 60°C, isooctane; 14%; Reference 40

Figure 9.17

TABLE 9.2 Intramolecular Khand–Pauson Cyclopentenone Synthesis: Three-Carbon Bridge[32–39]

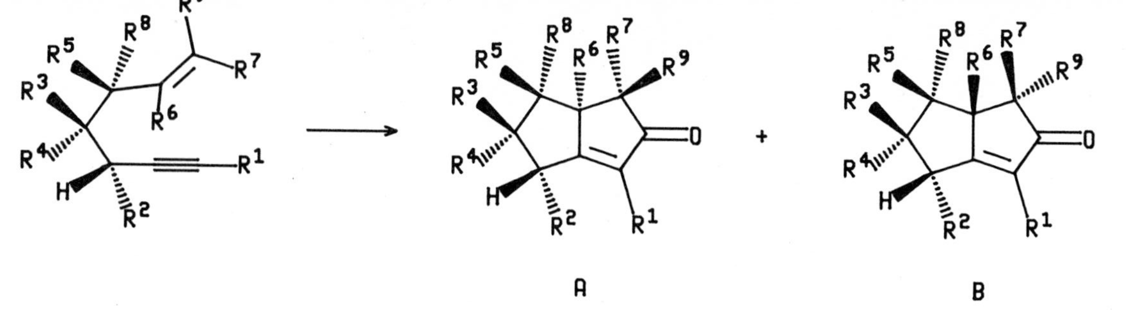

R^1	R^2	R^3	R^4	R^5	R^6	R^7	R^8	R^9	Yield (%)	A : B Ratio	Reference
H	H	H	H	H	H	H	H	H	31	—	32
H	H	H	H	H	$-(CH_2)_3-$		H	H	35	—	33
H	H	CH_3	CH_3	H	$-CHCH_2CH_2-$ | CH_3		OMOM	H	30	—	34
$Si(CH_3)_3$	$OSi(CH_3)_2(t\text{-Bu})$	CH_3	CH_3	H	H	H	H	H	82	26 : 1	36
CH_3	$OSi(CH_3)_2(t\text{-Bu})$	CH_3	CH_3	H	H	H	H	H	65	3 : 1	36
$Si(CH_3)_3$	H	CH_3	COOEt	H	H	H	H	H	86	1.2 : 1	36
$CH_2OSi(CH_3)_2(t\text{-Bu})$	H	CH_3	CH_3	H	H	H	$(CH_2)_2OMOM$	H	45	A only	37
$Si(CH_3)_3$	H	CH_3	CH_3	H	H	H	$(CH_2)_2OMOM$	H	78	A only	35
$Si(CH_3)_3$	H	CH_3	CH_3	H	H	H	OMOM	H	68	A only	35
$Si(CH_3)_3$	H	CH_3	CH_3	H	H		OH	H	25	2.6 : 1	35
$(CH_2)_2CHCH_3$ | OTHP	H	CH_3	CH_3	H	H	H	OMOM	H	20	A only	35
$Si(CH_3)_3$	H	H	CH_3 | $-(CH_2)_2OCO-$ | CH_3		H	H	H	H	51	B only	39
$Si(CH_3)_3$	$OSi(CH_3)_2(t\text{-Bu})$	H	H	H	H	H	$-CHCH_2-$ | $OSi(CH_3)_2(t\text{-Bu})$		76	B only	38

TABLE 9.3 Intramolecular Khand–Pauson Cyclopentenone Synthesis: Carbon–Oxygen–Carbon Bridge[40,41]

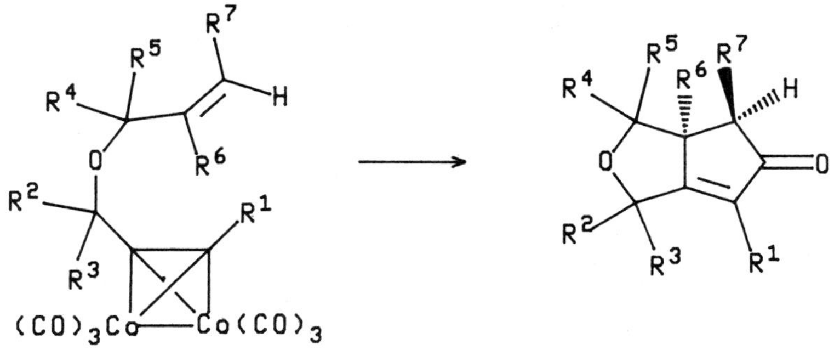

R^1	R^2	R^3	R^4	R^5	R^6	R^7	Yield (%)
H	H	H	H	H	H	H	58
H	CH_3	H	H	H	H	H	29
H	CH_3	CH_3	H	H	H	H	60
CH_3	H	H	H	H	H	H	41
CH_2CH_2OTHP	H	H	H	H	H	H	41
H	H	H	CH_3	CH_3	H	CH_3	92
H	H	H	$—(CH_2)_4—$		H	H	64
H	H	H	$—(CH_2)_5—$		H	H	80
$CH{=}CH_2$	H	H	H	H	H	H	58
H	H	H	H	H	CH_3	H	50
H	H	H	H	H	H	CH_3	46
H	H	H	H	CH_3	H	H	48
H	H	H	CH_3	CH_3	H	H	60
CH_3	H	H	H	H	H	H	60

Recent work on cyclizations to produce 3-oxabicyclo[3.3.0]octenes reveals that reaction rates and thus cyclization efficiency can be significantly increased by bringing the reacting alkene and alkyne ends closer together. This can be accomplished by complexation of the bridge oxygen to silica gel, resulting in a repulsion of the low-polarity alkene and alkyne ends by the polar silica gel surface[41] (see Figure 9.18).

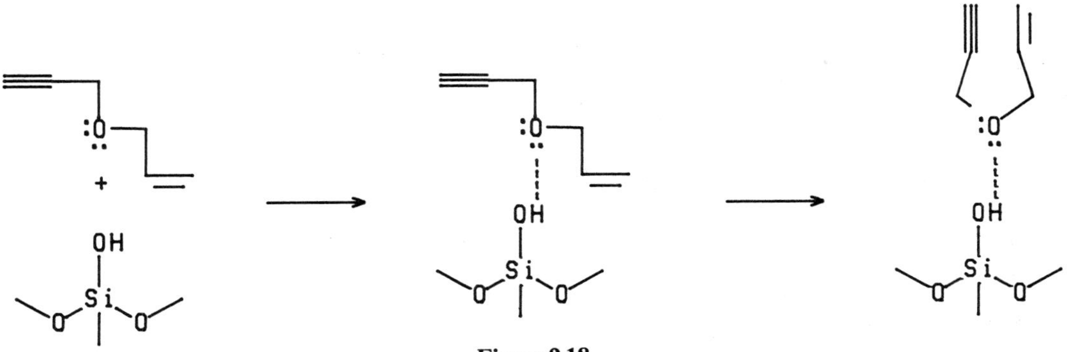

Figure 9.18

Possible Intermediates (3 atom connector)

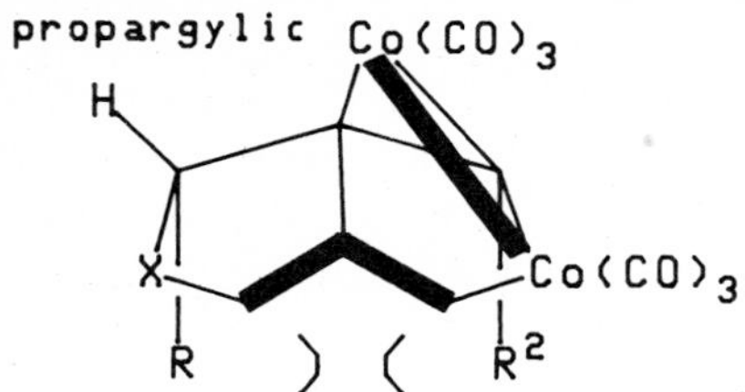

Formation difficult

$R-R^2$ steric interaction

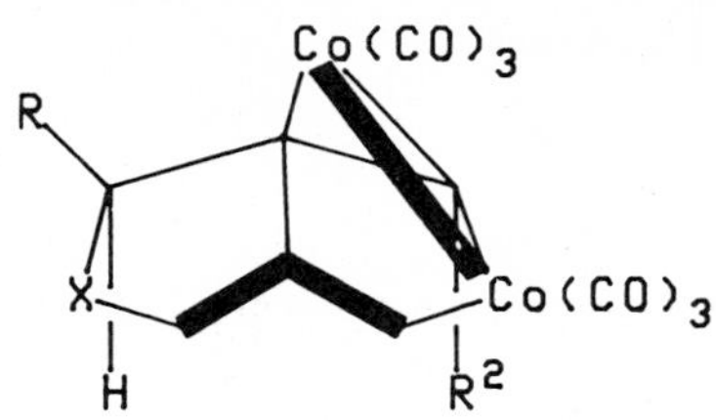

Formation facile

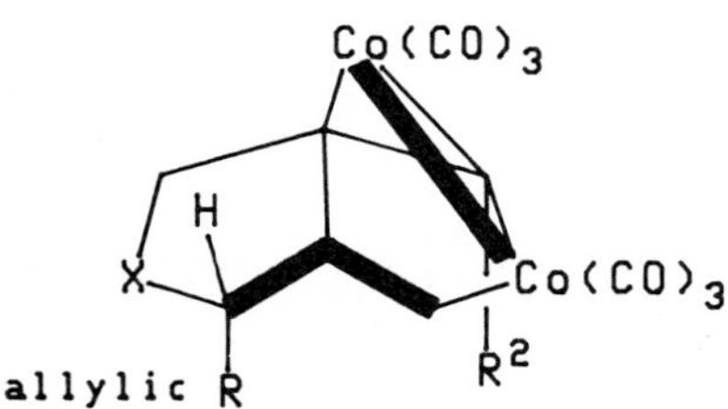

Formation difficult

$R-R^2$ steric interaction

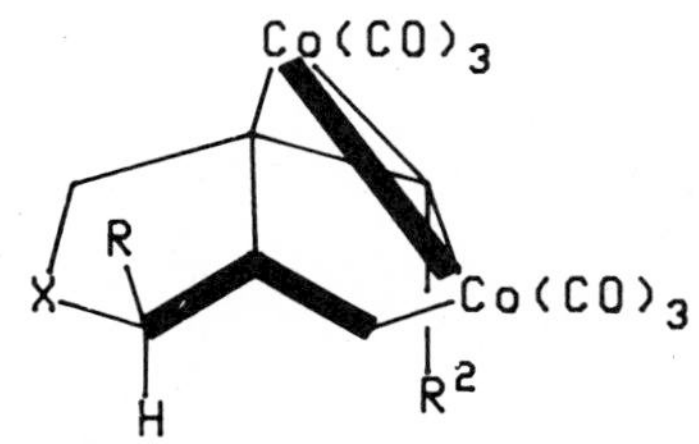

Formation facile

Starting with a cis-alkene:

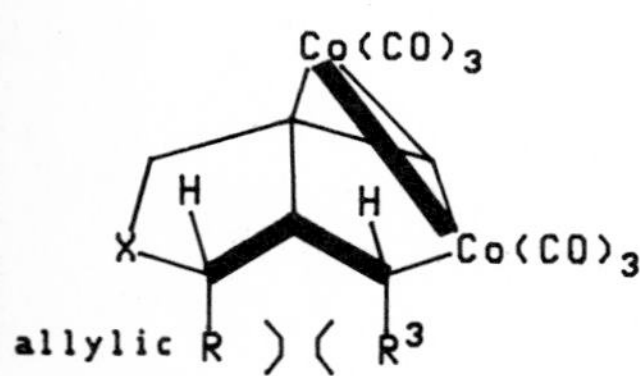

Formation difficult

$R-R^3$ steric interaction

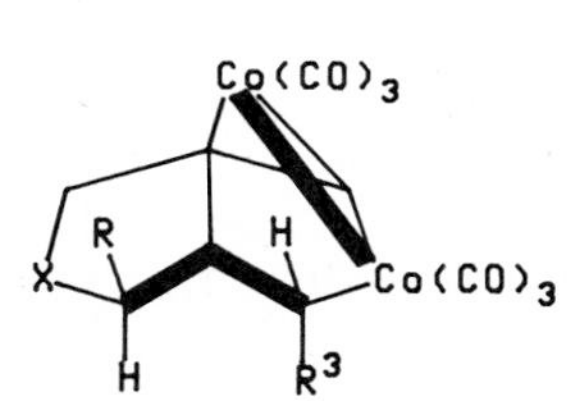

Formation facile

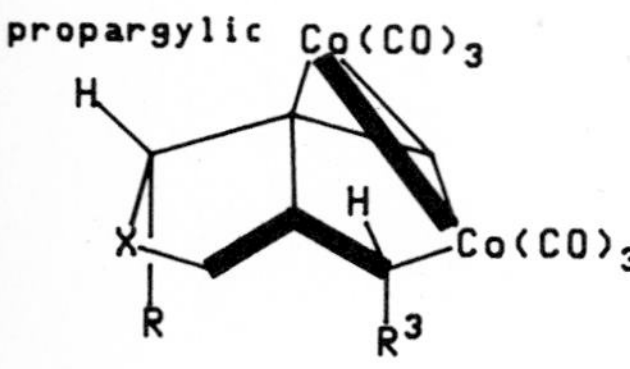

Formation difficult

$R-R^3$ steric interaction

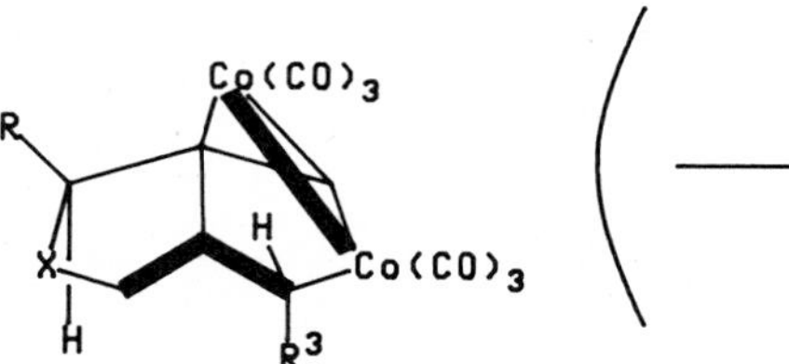

Formation facile

Figure 9.19

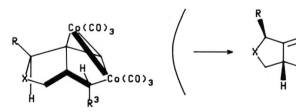

Magnus recognized the rigidity of the cobaltabicyclo[3.3.0]octene flyover intermediates in the intramolecular Khand–Pauson cyclopentenone synthesis as well as the potential for utilization of this rigidity for stereocontrol. A large alkyne substituent (R^2) on the bottom ("concave") face directs substituents on the allylic and propargylic positions of the connecting bridge to the top ("convex") face. A large terminal substituent on a *cis*-substituted alkene may exert a similar directing influence on the allylic and propargylic substituents of the connecting bridge (see Figure 9.19).

In addition, each of the eneynes required for synthesis of the targets contain a quaternary carbon in the bridge. Cyclization should benefit from the Thorpe–Ingold effect.[42]

9.3 TOTAL SYNTHESIS OF (±)-HIRSUTIC ACID C[36]

Synthesis of the requisite eneyne starting from diethyl methylmalonate[43] is completed in just three steps. The alkyne component is derived from propargyl alcohol.[44] The intramolecular Khand–Pauson cyclization affords a nearly 1:1 mixture of diastereoisomeric bicyclo[3.3.0]octenones, which is indicative of the nearly equivalent stabilities of the corresponding flyover intermediates (see Figure 9.20).

Desilylation–hydrolysis of the ester mixture affords an acid mixture enriched in the desired diastereoisomer (bridged proton and carboxyl groups are *cis*). Three additional acid-catalyzed epimerizations convert most of the mixture to the desired diastereoisomer (86% yield). Efficient catalytic hydrogenation is followed by reesterification with diazomethane. The product can be converted to (±)-hirsutic acid C using the procedures of Schuda[4a] and of Greene[4b,d].

The mixture of methyl esters is treated sequentially with an acid catalyst and triethylorthoformate, then with an allylic alcohol. A subsequent Ketal–Claisen reaction is facile at 160°C, affording a γ,δ-unsaturated ketone in excellent yield. Mercury(II)-assisted hydrolysis–elimination of the 4-alkylthio-2-chloroalkene unmasks an α,β-unsaturated methyl ketone. Aldol condensation and dehydration closes the C ring. A highly selective epoxidation of the endocyclic alkene under basic

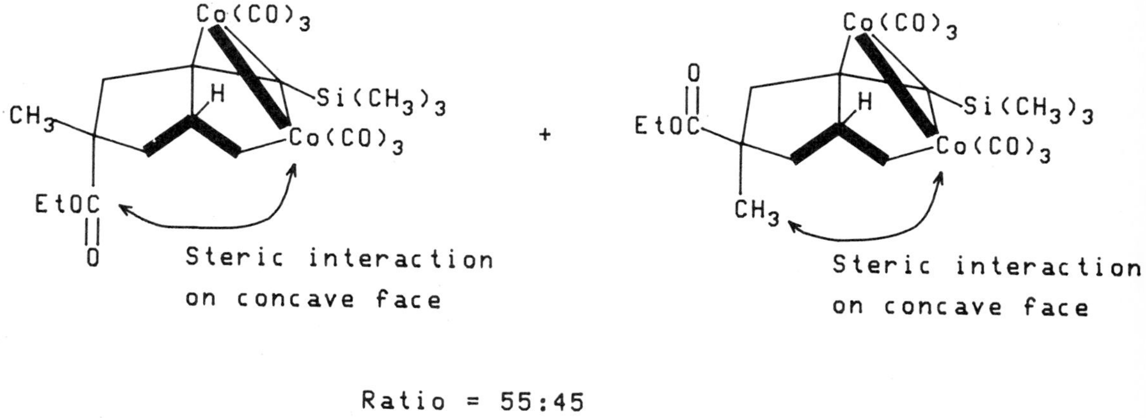

Figure 9.20

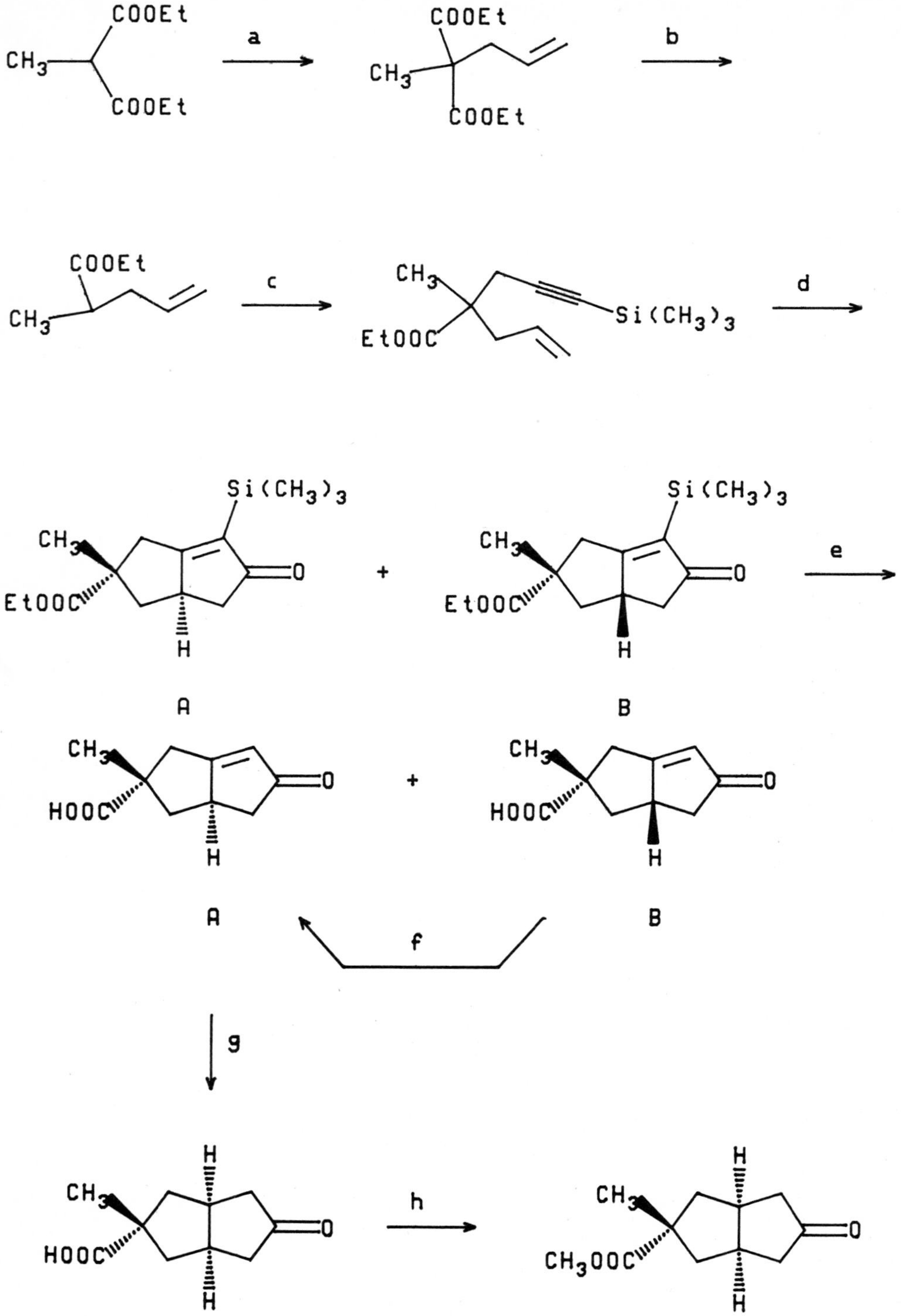
COOEt
CH3
COOEt
a
COOEt
CH3
COOEt
b
COOEt
CH3
c
CH3
EtOOC
Si(CH3)3
d
Si(CH3)3
CH3
EtOOC
O
H
A
+
Si(CH3)3
CH3
EtOOC
O
H
B
e
CH3
HOOC
O
H
A
+
CH3
HOOC
O
H
B
f
g
H
CH3
HOOC
O
H
h
H
CH3
CH3OOC
O
H

Figure 9.21

a. 1) −70 to 0°C, LiN(i-Pr)$_2$, 2) 0 to 20°C, CH_2=CHCH$_2$Br, THF; 99%

b. 180°C, LiI, DMSO, H_2O; 72%

c. 1) 0 to 20°C, LiN(i-Pr)$_2$, 2) 20°C, $(CH_3)_3Si$—≡—CH$_2$OMs*, THF; 71%

*Prepared as follows:

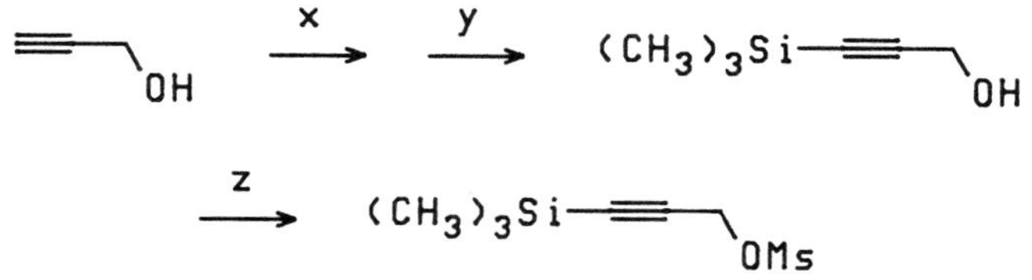

≡—CH$_2$OH —x→ —y→ $(CH_3)_3Si$—≡—CH$_2$OH —z→ $(CH_3)_3Si$—≡—CH$_2$OMs

x. −70°C, n-BuLi, Et_2O

y. −70 to 20°C, $(CH_3)_3SiCl$

z. 0°C, MsCl, CH_2Cl_2; 73%

d. 120°C, $Co_2(CO)_8$, heptane; A:B, 55:45, 86% yield

e. 75°C, CH_3SO_3H, H_2O; A:B, 1.34:1, 96% x 90% stereoselectivity = 86% yield

f. p-TsOH·H_2O

g. H_2, Pd-C, EtOAc; 100%

h. CH_2N_2; ≥ 87% yield

Figure 9.21 (*Continued*)

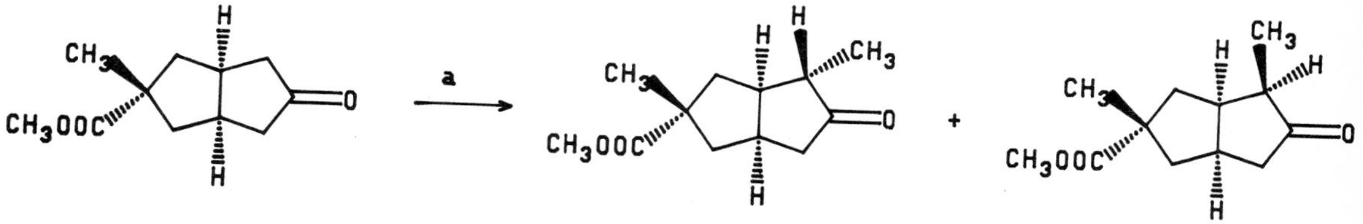

CH$_3$ / CH$_3$OOC / H / H / O —a→ A + B

Figure 9.22

(±)-Hirsutic Acid C

a. 1) −78 to 0°C, $LiN(i\text{-}Pr)_2$, 2) −10°C to rt, CH_3I, THF; A:B, 8:1, 64% yield (based on ketone not recovered)

b. 1) rt, p-TsOH, $HC(OCH_3)_3$, benzene, 2) heat; 84%

c. rt, $Hg(OAc)_2$, $NH_4OC(O)H$, HCOOH, H_2O; 91%

d. −20 to −10°C, t-BuOK, THF, t-BuOH

e. rfx, p-TsOH, benzene; 80%

f. rfx, LiI, DMF; 76% (based on ester not recovered)

g. −35°C, NaOH, H_2O_2, EtOH, H_2O

h. −35 to 0°C, $NaBH_4$; 50%

Figure 9.22 (*Continued*)

conditions is followed by a stereoselective reduction of the ketone with sodium borohydride.

The total synthesis of (±)-hirsutic acid C from diethyl methylmalonate was completed in 16 steps in overall 5.5% yield (assuming a quantitative yield in the esterification using diazomethane) (see Figures 9.21 and 9.22).

9.4 TOTAL SYNTHESIS OF (±)-QUADRONE

In contrast to the short synthesis of the requisite eneyne for the hirsutic acid synthesis, functionality in the allylic position of the connecting bridge necessitates 11 steps in the synthesis of the quadrone eneyne. Ethyl isobutyrate[45] is alkylated with propargyl bromide. An aldehyde is prepared by ester reduction, then alcohol oxidation. Wadsworth–Emmons homologation affords an α,β-unsaturated ester. Reaction of the allylic alcohol (from reduction with di-*i*-butylaluminum hydride) with an acid catalyst and triethylorthoacetate affords the γ,δ-unsaturated ester via a Claisen orthoester rearrangement. The ester is reduced to alcohol and the alcohol protected as a methoxymethyl ether. The alkyne is further functionalized by conversion to the acetylide, then condensed with ethyl chloroformate. Lithium aluminum hydride reduction produces the propargyl alcohol, which is protected as a *t*-butyldimethylsilyl ether (see Figure 9.23).

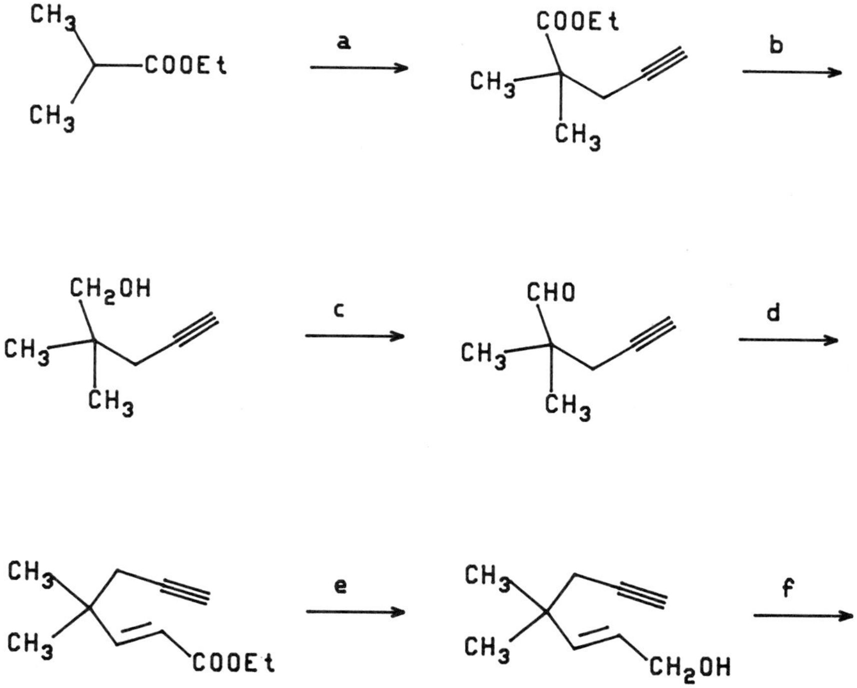

CH_3 CH_3 COOEt —a→ COOEt CH_3 CH_3 —b→

CH_2OH CH_3 CH_3 —c→ CHO CH_3 CH_3 —d→

CH_3 CH_3 COOEt —e→ CH_3 CH_3 CH_2OH —f→

Figure 9.23

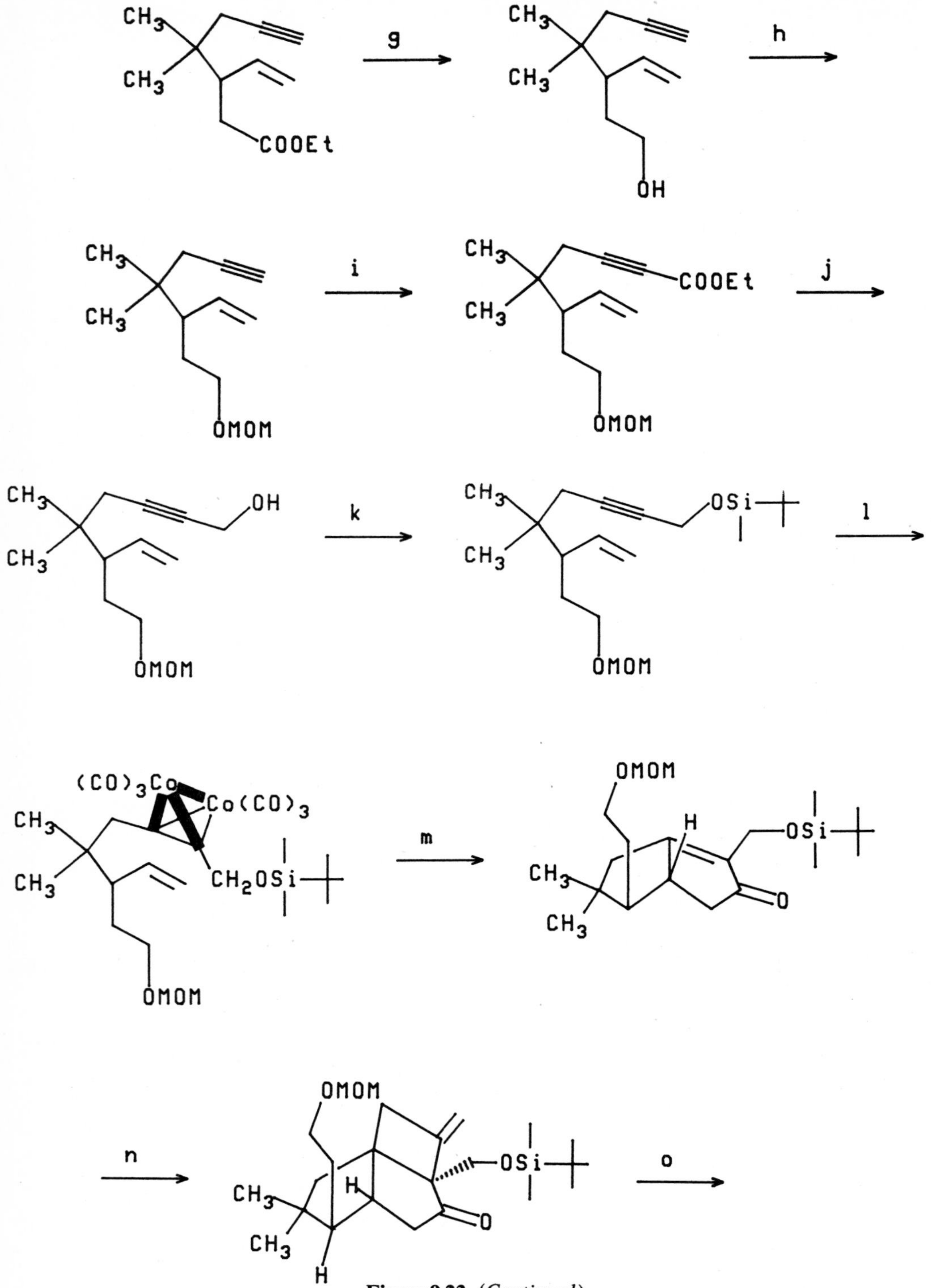

Figure 9.23 (*Continued*)

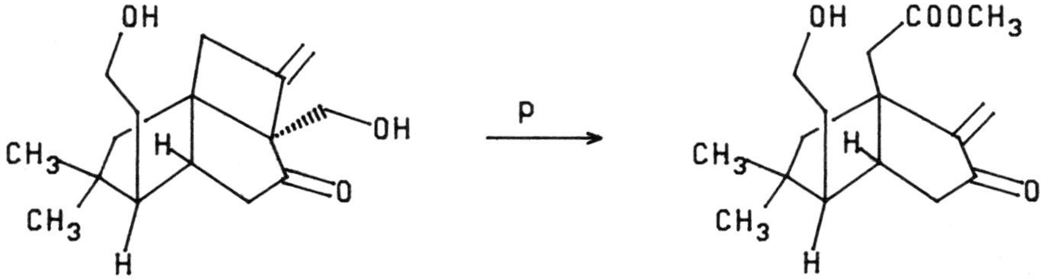

a. 1) −78°C, $LiN(i\text{-}Pr)_2$, 2) $\equiv\!\!-CH_2Br$, THF;

b. $LiAlH_4$; 96%

c. Et_3N, ClC(O)C(O)Cl, DMSO, CH_2Cl_2; 95%

d. $(Et_2O)_2P(O)CH_2COOEt$; 84%

e. $(i\text{-}Bu)_2AlH$, THF; 95%

f. 140°C, cat. EtCOOH, $CH_3C(OEt)_3$; 74%

g. 0°C, $LiAlH_4$, THF; 92%

h. $ClCH_2OCH_3$; 79%

i. 1) −78°C, n-BuLi, hexane, 2) −78 to 0°C, ClCOOEt, THF; 89%

j. −78 to 0°C, $LiAlH_4$, Et_2O; 86%

k. 0°C, Et_3N, $ClSi(CH_3)_2C(CH_3)_3$, CH_2Cl_2; 81%

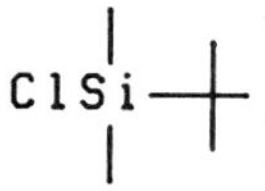

l. 20°C, $Co_2(CO)_8$, heptane; 100%

m. 86°C, CO, 2,6-di-tert-butyl-4-methylpyridine, heptane; 45%

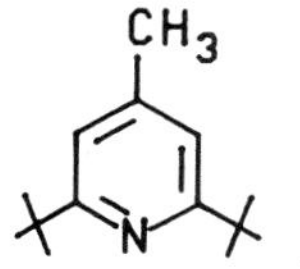

n. 0°C, hν, under N_2, $CH_2{=}C{=}CH_2$, hexane, THF; 84%

o. 65°C, H_2SO_4, HBF_4, THF, H_2O; 96%

p. −78°C, O_3, 4A mol. sieves, CH_2Cl_2, CH_3OH, workup with CH_3SCH_3; 100%

Figure 9.23 (*Continued*)

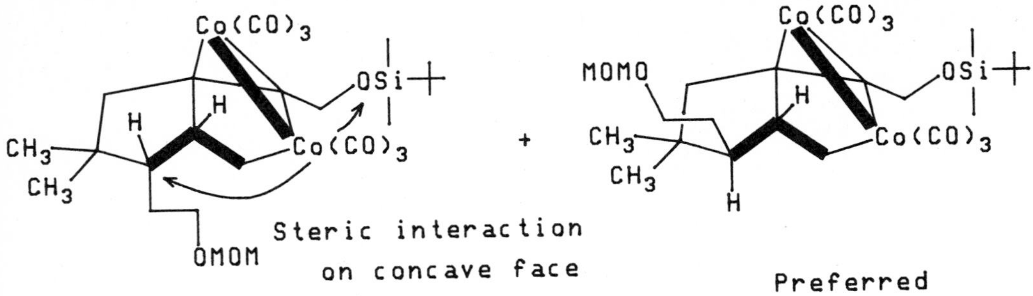

Steric interaction on concave face

Preferred

Figure 9.24

Alkyne–dicobalt hexacarbonyl complex formation occurs on mixing the alkyne and dicobalt octacarbonyl in heptane at room temperature. Heating the mixture at 86°C induces a stereospecific Khand–Pauson cyclization; only the product with the allylic substitution on the convex face is isolated (see Figure 9.24).

Cleavage of the cobalt-coordinated propargyl ether to a cobalt-coordinated propargyl cation may be an efficiency-limiting side reaction (Chapter 8).

A regio- and stereospecific [2 + 2] photochemical cycloaddition with allene is followed by deprotection of the ethers. Ozonolysis in methylene chloride–methanol with reductive workup results in generation and in situ cleavage of a cyclobutanone. This material is converted to (±)-quadrone using procedures developed by Isoe[46] and by Danishefsky[47].

The α-methylene ketone is protected as a ketal by conjugate addition of phenyl selenide, ketalization, and then elimination via the selenoxide. The alcohol is converted to an iodide via the *p*-toluenesulfonate. Intramolecular alkylation of the ester enolate closes the last ring. Sequential deprotection of the acid and ketone affords (±)-terrecyclic acid A. (±)-Terrecyclic acid A cyclizes to (±)-quadrone on heating to 190 to 195°C.

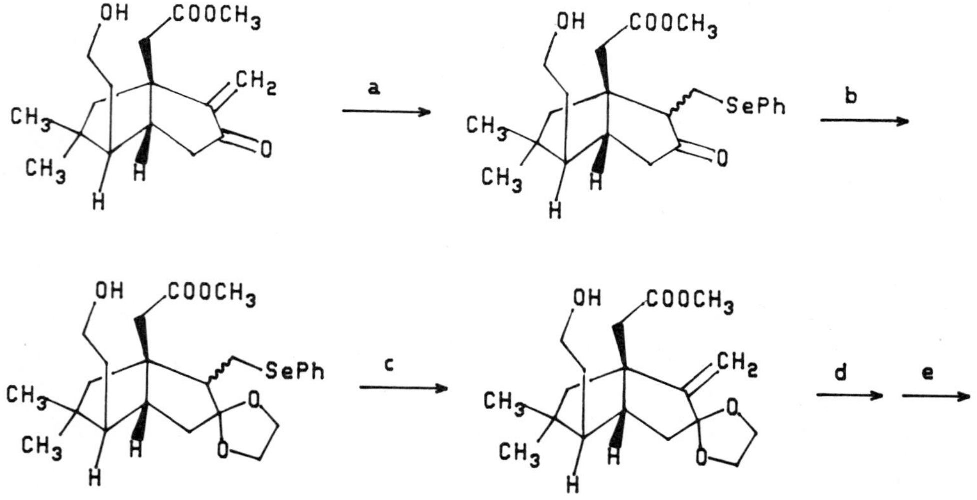

Figure 9.25

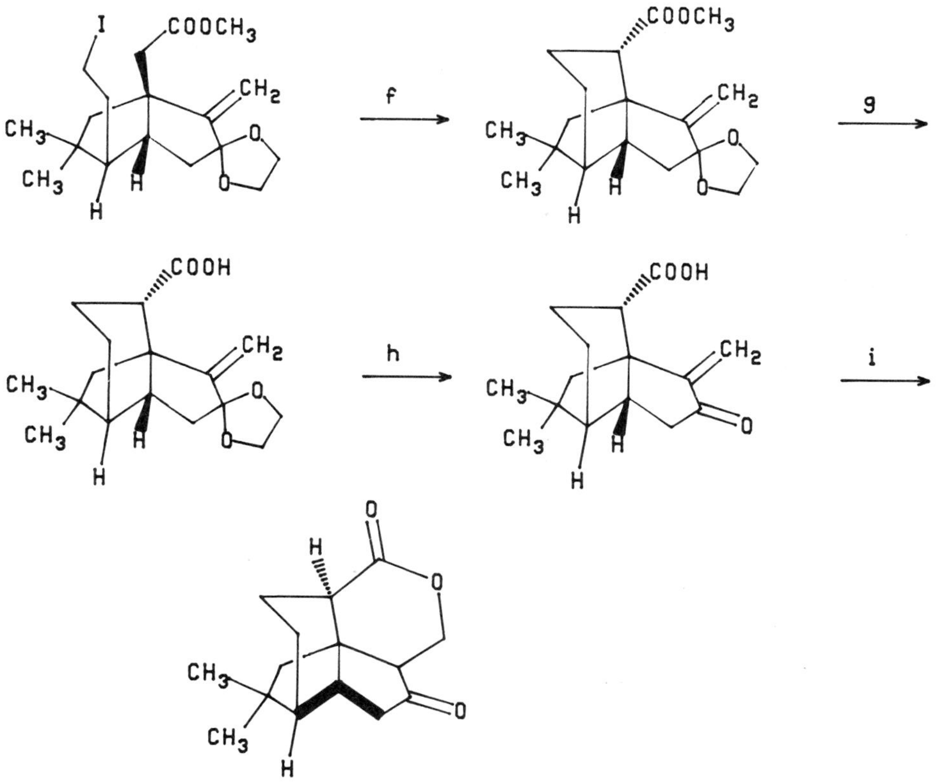

(±)-Quadrone

a. 0°C, $NaBH_4$, PhSeSePh, EtOH

b. p-TsOH, $HOCH_2CH_2OH$, benzene

c. 0 to 70°C, 30% H_2O_2, H_2O, pyridine, CH_2Cl_2; 68% for three steps

d. 0°C, p-TsCl, pyridine

e. 40°C, cat. pyridine, NaI, DMF; "good overall yield"

f. -78 to 60°C, $LiN[Si(CH_3)_3]_2$, THF; 62%

g. 60°C, n-PrSLi, HMPA

h. 0°C, H_3PO_4, H_2O, THF; 87% for two steps

i. 190-195°C; no yield given (51% from precursor β-OH ketone)

Figure 9.25 (*Continued*)

The total synthesis of (±)-quadrone from ethyl isobutyrate was completed in 25 steps in overall 2.9% yield (assuming quantitative conversions of alcohol to iodide by Isoe and (±)-terrecyclic acid A to (±)-quadrone by Danishefsky) (see Figure 9.25).

9.5 TOTAL SYNTHESIS OF (±)-CORIOLIN[36,48]

The synthesis of (±)-coriolin begins with an acid-catalyzed condensation of allyl alcohol with isobutyraldehyde.[49–51] Nucleophilic addition of trimethylsilylacetylide is followed by protection of the propargyl alcohol as a *t*-butyldimethylsilyl ether. The terminal alkyne is unmasked with benzyltriethylammonium chloride–potassium fluoride in refluxing tetrahydrofuran. Alkylation of the acetylide affords the Khand–Pauson eneyne substrate. Cyclization is stereoselective, affording two diastereoisomers in a 3:1 ratio. The major product has the propargylic substituent of the connecting bridge on the convex face (see Figure 9.26).

The ketone produced on catalytic reduction of the major diastereoisomer can be regio- and stereospecifically alkylated on the convex face. Wacker oxidation converts the allyl substituent to an acetonyl group. Aldol condensation then produces an α,β-unsaturated ketone in closing the final ring. Stereospecific γ-hydroxylation is achieved in a three-step sequence: (1) alkene deconjugation, (2) stereospecific epoxidation, and (3) proton abstraction with epoxide ring opening. Silyl ether deprotection using pyridinium polyhydrogen fluoride affords the diol that can be converted to (±)-coriolin using procedures developed by Trost[52], Danishefsky,[53] and Takeuchi.[54]

The ketone is converted to an α-methylene ketone in six steps: (1) protection of the hydroxyl groups, (2) generation of the kinetic enolate and trap with chlorotrimethylsilane, (3) alkylation of the silyl enol ether with dimethylmethylene ammonium iodide, (4) quaternization of the amine, (5) elimination, and (6) diol regeneration with pyridinum polyhydrogen fluoride.

Regio- and stereospecific epoxidation of the endocyclic alkene is achieved with hydrogen peroxide–sodium bicarbonate at 0°C. Introduction of the second epoxide requires three steps: (1) reduction of the ketone to an alcohol, (2) epoxidation of the allylic alcohol under Sharpless conditions, and (3) oxidation of the alcohol back to a ketone.

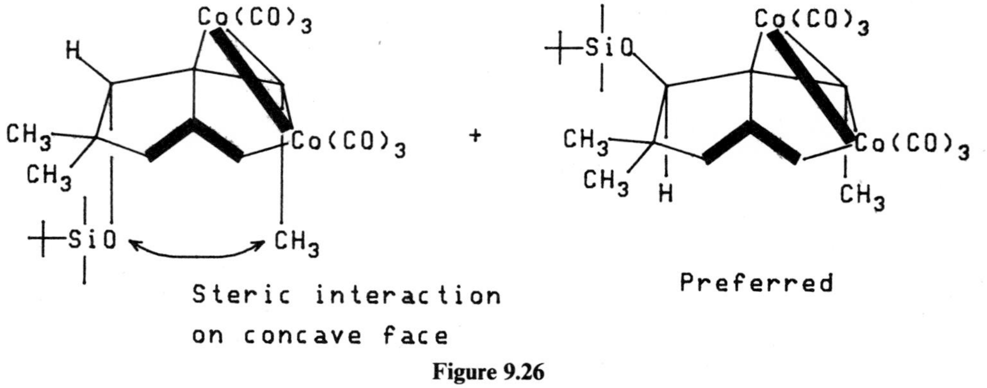

Figure 9.26

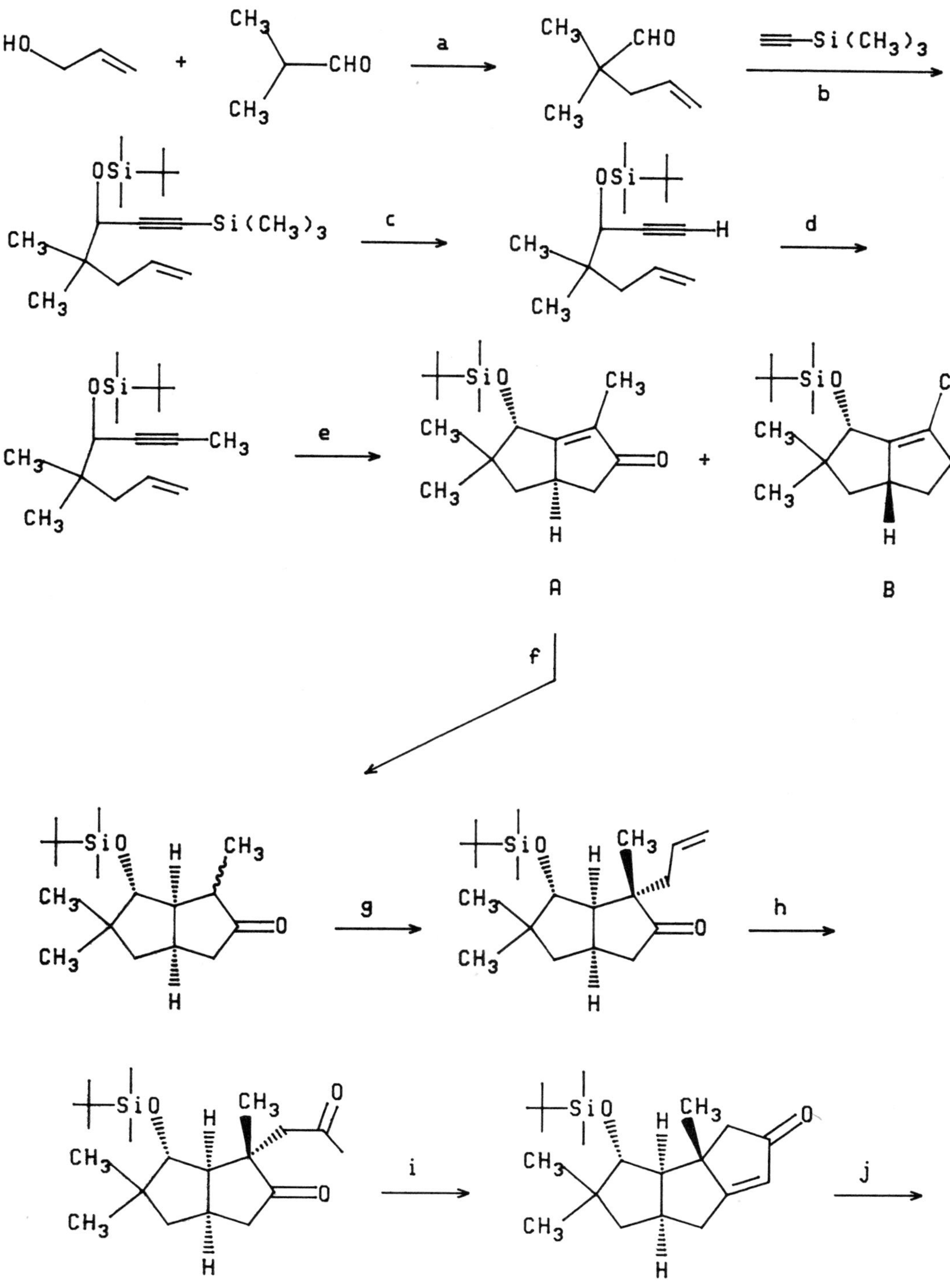

Figure 9.27

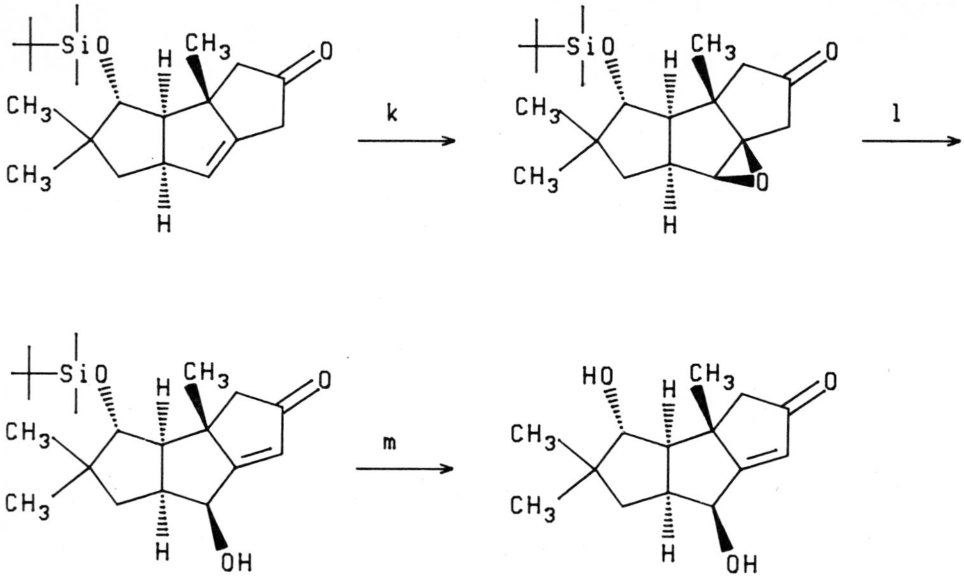

a. rfx, p-TsOH, p-cymene; 93%

b. 1) −70°C, n-BuLi, hexane, 2) −70°C to rfx, ClSi(CH$_3$)$_2$C(CH$_3$)$_3$, THF; 86%

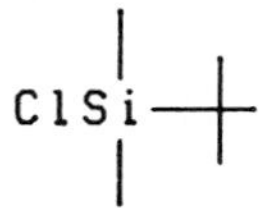

c. rfx, BzNEt$_3$Cl(F), THF; 90%

d. 1) −70°C, n-BuLi, 2) CH_3I, THF; 98%

e. 110°C, $Co_2(CO)_8$, heptane; A:50%, B:15%

f. rt, H_2, Pd-C, 95% EtOH; 92%

g. 1) rfx, NaH, 2) rfx, CH_2=$CHCH_2Br$, DME; 79%

h. rt, O_2, $PdCl_2$, CuCl, DMF, H_2O; 64%

i. rt, t-BuOK, t-BuOH; 74%

j. 1) t-BuOK, 2) AcOH, H_2O, t-BuOH, DME

k. rt, MCPBA, CH_2Cl_2

l. rt, DBU, CH_2Cl_2; 34% for three steps

m. 60°C, pyridinium polyhydrogen fluoride, THF; 74%

Figure 9.27 (*Continued*)

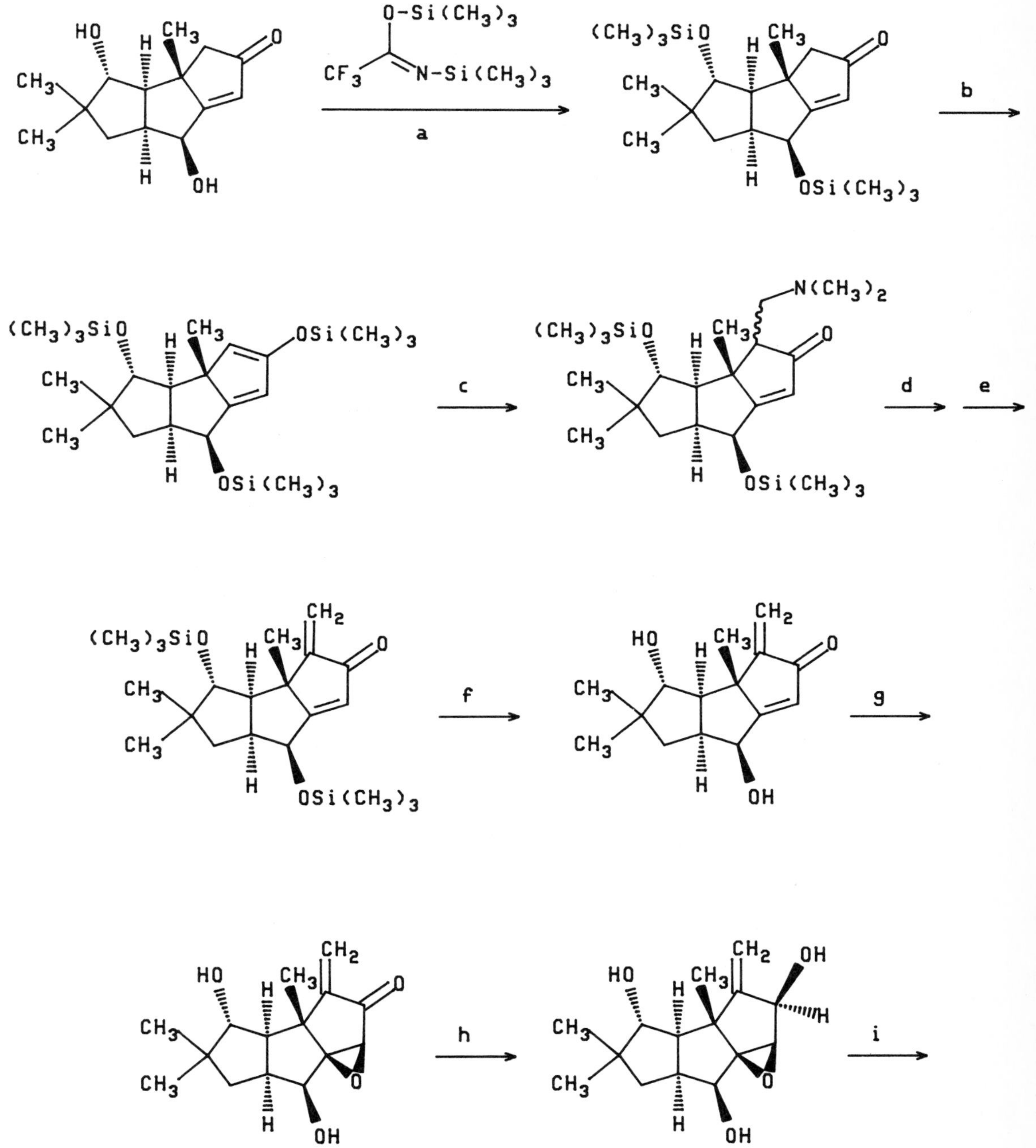

Figure 9.28

(±)-Coriolin

a. 40°C, DMF

b. 1) $LiN(i\text{-}Pr)_2$, 2) $(CH_3)_3SiCl$, THF, HMPA

c. rfx, $(CH_3)_2NCH_2I$, $CHCl_3$

d. rt, CH_3I, Et_2O

e. rt, DBU, CH_2Cl_2; 46% for five steps

f. pyridinium polyhydrogen fluoride, THF; 85%

g. 0°C, $NaHCO_3$, H_2O_2, THF, H_2O

h. 0°C, $NaBH_4$, EtOH; 100%

i. rfx, t-BuOOH, $V(acac)_2$, benzene; 65%

j. rt, CrO_3, pyridine; 40%

Figure 9.28 *(Continued)*

The total synthesis of (±)-coriolin from allyl alcohol and isobutyraldehyde was completed in 23 steps in overall 0.78% yield (assuming a quantitative yield in the last step) (see Figures 9.27 and 9.28).

REFERENCES

1. Heatley, N. G.; Jennings, M. A.; Florey, H. W. *Brit. J. Exptl. Path.* **1947**, *28*, 35.
2. See the following references for structure determination:
2a. Comer, F. W.; McCapra, F.; Qureshi, I. H.; Trotter, J.; Scott, A. I. *Chem. Commun.* **1965**, 310.
2b. Comer, F. W.; Trotter, J. *J. Chem. Soc. (B)* **1966**, 11.
2c. Comer, F. W.; McCapra, F.; Qureshi, I. H.; Scott, A. I. *Tetrahedron* **1967**, *23*, 4761.
3. Mantle, P. G.; Mellows, G. *Trans. Brit. Mycol. Soc.* **1973**, *61* (Pt. 3), 513.

4a. Schuda, P. F.; Phillips, J. L.; Morgan, T. M. *J. Org. Chem.* **1986**, *51*, 2742.
4b. Greene, A. E.; Luche, M-J.; Serra, A. A. *J. Org. Chem.* **1985**, *50*, 3957.
4c. Yamazaki, M.; Shibasaki, M.; Ikegami, S. *J. Org. Chem.* **1983**, *48*, 4402.
4d. Greene, A. E.; Luche, M-J.; Deprés, J-P. *J. Am. Chem. Soc.* **1983**, *105*, 2435.
4e. Shibasaki, M.; Yamazaki, M.; Iseki, K.; Ikegami, S. *Tetrahedron Lett.* **1982**, *23*, 5311.
4f. Yamazaki, M.; Shibasaki, M.; Ikegami, S. *Chem. Lett.* **1981**, 1245.
4g. Venegas, M. G.; Little, R. D. *Tetrahedron Lett.* **1979**, 309.
4h. Trost, B. M.; Shuey, C. D.; DiNinno, F., Jr.; McElvain, S. S. *J. Am. Chem. Soc.* **1979**, *101*, 1284.
4i. Hashimoto, H.; Tsuzuki, K.; Sakan, F.; Shirahama, H.; Matsumoto, T. *Tetrahedron Lett.* **1974**, 3745.
4j. Lansbury, P. T.; Wang, N. Y.; Rhodes, J. E. *Tetrahedron Lett.* **1971**, 1829.
5a. Ranieri, R. L.; Calton, G. J. *Tetrahedron Lett.* **1978**, 499.
5b. Calton, G. J.; Ranieri, R. L.; Espenshade, M. A. *J. Antibiot.* **1973**, *31*, 38.
6. Pickman, A. K.; Towers, G. H. N. *Biochem. Syst. Ecol.* **1983**, *11*, 321.
7a. Calton, G. J.; Espenshade, M. A.; Ranieri, R. L. Belg. 869, 439, 01 Dec 1978; *Chem. Abstr.* **1979**, *90*, P184909r.
7b. Calton, G. J.; Espenshade, M. A.; Ranieri, R. L. U.S. Patent 4 147 798, 03 Apr 1979; *Chem. Abstr.* **1979**, *91*, P18341a.
8a. Funk, R. L.; Abelman, M. M. *J. Org. Chem.* **1986**, *51*, 3247.
8b. Piers, E.; Moss, N. *Tetrahedron Lett.* **1985**, *26*, 2735.
8c. Iwata, C.; Yamashita, M.; Aoki, S.; Suzuki, K.; Takahashi, I.; Arakawa, H.; Imanishi, T.; Tanaka, T. *Chem. Pharm. Bull.* **1985**, *33*, 436.
8d. Wender, P. A.; Wolanin, D. J. *J. Org. Chem.* **1985**, *50*, 4418.
8e. Smith, A. B., III; Konopelski, J. P. *J. Org. Chem.* **1984**, *49*, 4094.
8f. Burke, S. D.; Murtiashaw, C. W.; Saunders, J. O.; Oplinger, J. A.; Dike, M. S. *J. Am. Chem. Soc.* **1984**, *106*, 4558.
8g. Cooper, K.; Pattenden, G.; *J. Chem. Soc., Perkin Trans. 1* **1984**, 799.
8h. Dewanckele, J. M.; Zutterman, F.; Vandewalle, M. *Tetrahedron* **1983**, *39*, 3235.
8i. Paquette, L. A.; Annis, G. D.; Schostarez, H. *J. Am. Chem. Soc.* **1982**, *104*, 6646.
8j. Takeda, K.; Shimono, Y.; Yoshii, E. *J. Am. Chem. Soc.* **1983**, *105*, 563.
9a. Takahashi, S.; Iinuma, H.; Takita, T.; Maeda, K.; Umezawa, H. *Tetrahedron Lett.* **1969**, 4663.
9b. Takeuchi, T.; Iinuma, H.; Iwanaga, J.; Takahashi, S.; Takita, T.; Umezawa, H. *J. Antibiot.* **1969**, *22*, 215.
9c. Takahashi, S.; Iinuma, H.; Takita, T.; Maeda, K.; Umezawa, H. *Tetrahedron Lett.* **1970**, 1637.
9d. Takahashi, S.; Naganawa, H.; Iinuma, H.; Takita, T.; Maeda, K.; Umezawa, H. *Tetrahedron Lett.* **1971**, 1955.
9e. Takeuchi, T.; Iinuma, H.; Takahashi, S.; Umezawa, H. *J. Antibiot.* **1971**, *24*, 631.
10. Umezawa, H.; Umezawa, S.; Tsuchiya, O.; Takeuchi, T.; Nishimura, Y. Jpn. Kokai Tokkyo Koho 79,160,359, 19 Dec 1979.
11a. Mehta, G.; Murthy, A. N.; Reddy, D. S.; Reddy, A. V. *J. Am. Chem. Soc.* **1986**, *108*, 3443.
11b. Demuth, M.; Ritterskamp, P.; Weigt, E.; Schaffner, K. *J. Am. Chem. Soc.* **1986**, *108*, 4149.
11c. Funk, R. L.; Bolton, G. L.; Daggett, J. U.; Hansen, M. M.; Horcher, L. H. M. *Tetrahedron* **1985**, *41*, 3479.

11d. Van Hijfte, L.; Little, R. D. *J. Org. Chem.* **1985**, *50*, 3940.

11e. Ito, T.; Tomiyoshi, N.; Nakamura, K.; Azuma, S.; Izawa, M.; Maruyama, F.; Yanagiya, M.; Shirahama, H.; Matsumoto, T. *Tetrahedron* **1984**, *40*, 241.

11f. Schuda, P. F.; Heimann, M. R. *Tetrahedron* **1984**, *40*, 2365.

11g. Koreeda, M.; Mislankar, S. G. *J. Am. Chem. Soc.* **1983**, *105*, 7203.

11h. Knapp, S.; Trope, A. F.; Theodore, M. S.; Hirata, N.; Barchi, J. J. *J. Org. Chem.* **1984**, *49*, 608.

11i. Wender, P. A.; Howbert, J. J. *Tetrahedron Lett.* **1983**, *24*, 5325.

11j. Iseki, K.; Yamazaki, M.; Shibasaki, M.; Ikegami, S. *Tetrahedron* **1981**, *37*, 4411.

11k. Tatsuta, K.; Akimoto, K.; Kinoshita, M. *Tetrahedron* **1981**, *37*, 4365.

12. Krüerke, U.; Hoogzand, C.; Hübel, W. *Chem. Ber.* **1961**, *94*, 2817.

13. Mills, O. S.; Robinson, G. *Proc. Chem. Soc.* **1964**, 187.

14a. Khand, I. U.; Knox, G. R.; Pauson, P. L.; Watts. W. E. *J. Chem. Soc., Perkin Trans. I* **1973**, 975.

14b. Khand, I. U.; Knox, G. R.; Pauson, P. L.; Watts, W. E.; Foreman, M. I. *J. Chem. Soc., Perkin Trans. I* **1973**, 977.

15. Khand, I. U.; Murphy, E.; Pauson, P. L. *J. Chem. Res. (S)* **1978**, 350.

16. Khand, I. U.; Pauson, P. L. *J. Chem. Soc., Perkin Trans. I* **1976**, 30.

17. Khand, I. U.; Pauson, P. L.; Habib, M. J. A. *J. Chem. Res. (S)* **1978**, 348.

18. Billington, D. C. *Tetrahedron Lett.* **1983**, *24*, 2905.

19. Pauson, P. L.; Khand, I. U. *Ann. N.Y. Acad. Sci.* **1977**, *295*, 2.

20. Khand, I. U.; Pauson, P. L.; Habib, M. J. A. *J. Chem. Res. (S)* **1978**, 346.

21. Schreiber, S. L.; Sammakia, T.; Crowe, W. E. *J. Am. Chem. Soc.* **1986**, *108*, 3128.

22. La Belle, B. E.; Knudsen, M. J.; Olmstead, M. M.; Hope, H.; Yanuck, M. D.; Schore, N. E. *J. Org. Chem.* **1985**, *50*, 5215.

23. Bladon, P.; Khand, I. U.; Pauson, P. L. *J. Chem. Res. (S)* **1977**, 8.

24. Newton, R. F.; Pauson, P. L.; Taylor, R. G. *J. Chem. Res. (S)* **1980**, 277.

25. Daalman, L.; Newton, R. F.; Pauson, P. L.; Wadsworth, A. *J. Chem. Res. (S)* **1984**, 346.

26. Daalman, L.; Newton, R. F.; Pauson, P. L.; Taylor, R. G.; Wadsworth, A. *J. Chem. Res. (S)* **1984**, 344.

27. Khand, I. U.; Mahaffy, C. A. L.; Pauson, P. L. *J. Chem. Res. (S)* **1978**, 352.

28. Jaffer, H. J.; Pauson, P. L. *J. Chem. Res. (S)* **1983**, 244.

29. Billington, D. C.; Pauson, P. L. *Organometallics* **1982**, *1*, 1560.

30. Croudace, M. C.; Schore, N. E. *J. Org. Chem.* **1981**, *46*, 5357.

31. Schore, N. E. *Syn. Commun.* **1979**, *9*, 41.

32. Schore, N. E.; Croudace, M. C. *J. Org. Chem.* **1981**, *46*, 5436.

33. Knudsen, M. J.; Schore, N. E. *J. Org. Chem.* **1984**, *49*, 5025.

34. Schore, N. E.; Knudsen, M. J. *J. Org. Chem.* **1987**, *52*, 569.

35. Magnus, P.; Principe, L. M. *Tetrahedron Lett.* **1985**, *26*, 4851.

36. Magnus, P.; Exon, C.; Albaugh-Robertson, P. *Tetrahedron* **1985**, *41*, 5861.

37. Magnus, P.; Principe, L. M.; Slater, M. J. *J. Org. Chem.* **1987**, *52*, 1483.

38. Carceller, E.; Centellas, V.; Moyano. A.; Pericás, M. A.; Serratosa, F. *Tetrahedron Lett.* **1985**, *26*, 2475.

39. Magnus, P.; Becker, D. P. *J. Am. Chem. Soc.* **1987**, *109*, 7495.

40. Billington, D. C.; Willison, D. *Tetrahedron Lett.* **1984**, *25*, 4041.

41. Simonian, S. O.; Smit, W. A.; Gybin, A. S.; Shashkov, A. S.; Mikaelian, G. S.; Tarasov, V. A.; Ibragimov, I. I.; Caple, R.; Froen, D. E. *Tetrahedron Lett.* **1986**, *27*, 1245.

42a. Eliel, E. L. Stereochemistry of Carbon Compounds, McGraw-Hill: New York, 1962, pp. 197–202.

42b. DeTar, D. F.; Luthra, N. P. *J. Am. Chem. Soc.* **1980**, *102*, 4505.

43. Diethylmethyl malonate is commercially available and inexpensive (Aldrich 88–89 catalog, 500 g for $53.95).

44. Propargyl alcohol is commercially available and inexpensive (Aldrich 88–89 catalog, 500 g for $13.20).

45. Ethyl isobutyrate is commerically available and inexpensive (Lancaster Synthesis 89–90 catalog, 500 g for $25.10).

46. Kon, K.; Ito, K.; Isoe, S. *Tetrahedron Lett.* **1984**, *25*, 3739.

47a. Danishefsky, S.; Vaughan, K.; Gadwood, R. C.; Tsuzuki, K. *J. Am. Chem. Soc.* **1980**, *102*, 4262.

47b. Danishefsky, S.; Vaughan, K.; Gadwood, R.: Tsuzuki, K. *J. Am. Chem. Soc.* **1981**, *103*, 4136.

48. Exon, C.; Magnus, P. *J. Am. Chem. Soc.* **1983**, *105*, 2477.

49. Magnus, P. D.; Nobbs, M. S. *Syn. Commun.* **1980**, *10*, 273.

50. Allyl alcohol is commercially available and inexpensive (Aldrich 88–89 catalog, 1 L, $d = 0.852$ g/mL, for $7.90).

51. Isobutyraldehyde is commercially available and inexpensive (Aldrich 88–89 catalog, 1 L, $d = 0.79$ g/mL, for $7.50).

52. Trost, B. M.; Curran, D. P. *J. Am. Chem. Soc.* **1981**, *103*, 7380.

53. Danishefsky, S.; Zamboni, R.; Kahn, M.; Etheredge, S. J. *J. Am. Chem. Soc.* **1981**, *103*, 3460.

54. Takeuchi, T.; Ishizuka, M.; Umezawa, H.; Nishimura, Y.; Koyama, Y.; Umezawa, S. *J. Antibiot.* **1980**, *33*, 404.

CHAPTER 10

Benzoquinones and Naphthaquinones

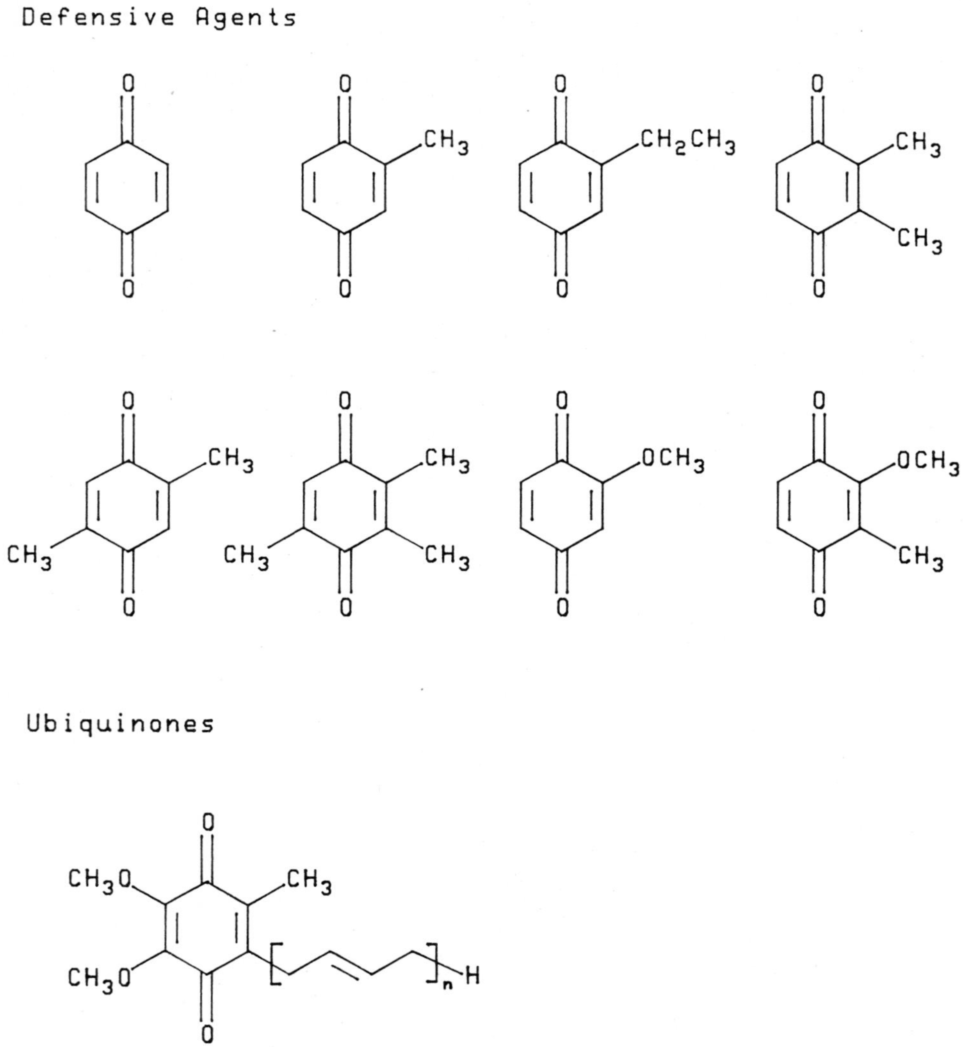

Defensive Agents

O O O O
CH_3 CH_2CH_3 CH_3
CH_3
O O O O

O O O O
CH_3 CH_3 OCH_3 OCH_3
CH_3 CH_3 CH_3 CH_3
O O O O

Ubiquinones

O
CH_3O CH_3
CH_3O n H
O

Figure 10.1

Quinones display a broad spectrum of biological and pharmacological activity. The simplest examples are quinones produced by anthropods as defensive agents against predators.[1–3] Other simple quinones, which occur almost *ubiquitously* in nature, are the ubiquinones (also called coenzymes Q). Ubiquinones play a major role in the respiratory electron transport system[4,5] (see Figure 10.1).

Naturally occurring biologically and pharmacologically active naphthaquinones are far more common. The K vitamins are represented by phylloquinone (K), the menaquinones (MK_n) (involved in bacterial electron transport[5]), demethylmenaquinones (DMK_n), and dihydromenaquinones [$MK_n(II\text{-}H_2)$]. The menaquinone MK_4, the principal vitamin in animals, is believed to be involved in maintenance of adequate

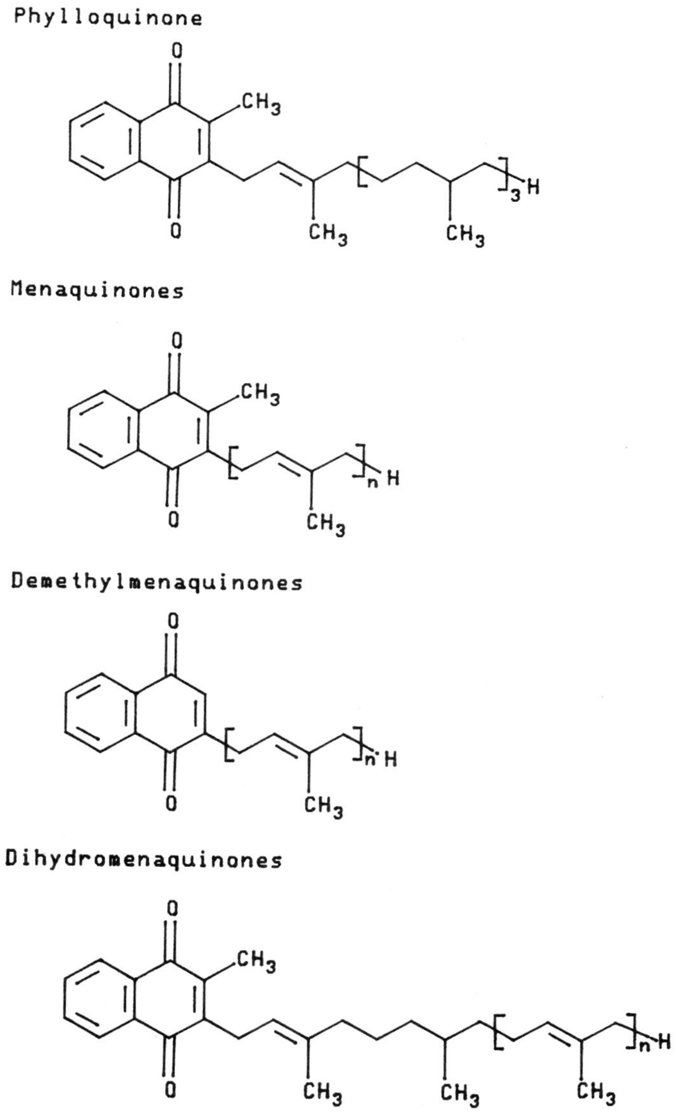

Phylloquinone

Menaquinones

Demethylmenaquinones

Dihydromenaquinones

Figure 10.2

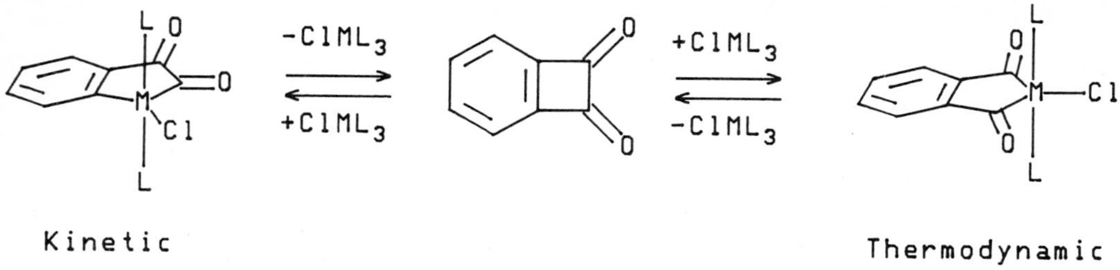

Kinetic

Thermodynamic

L = ligand (commonly PPh_3)

Figure 10.3

levels of prothrombin and other clotting factors (see Figure 10.2). An extensive report on in vivo activity of some 1500 quinones and naphthaquinones against one or more of five animal tumors[6] is indicative of the significant interest in compounds of these classes. Synthetic efforts have mostly been limited to either oxidation of a phenolic precursor or direct functionalization of a quinone.[7] In this chapter we will focus on valence shell electron counting, metal complex geometry, ligand substitution, and chelating ligand geometry in developing an inexpensive quinone/naphthaquinone synthesis based on cobalt–alkyne complexes.

10.1 PHTHALOYL- AND MALEOYLCOBALT COMPLEXES IN QUINONE FORMATION

A low-valent metal can cleave a carbon–carbon bond in a strained ring by oxidative addition. For example, benzocyclobutanediones form phthaloylmetal complexes with rhodium(I), iron(I), and cobalt(I)[8] (see Figure 10.3).

Type 1 complexes, produced from chloro-*tris*-(triphenylphosphine)cobalt, are 16e (coordinatively unsaturated) and are trigonal bipyramidal with the ring and chlorine occupying equatorial positions and the phosphines occupying axial positions. Type 2 complexes, produced from cyclopentadienylcobalt dicarbonyl, are 18e (coordinatively saturated) and are tetrahedral. The preparation of specific phthaloyl- and maleoylcobalt complexes is described in Table 10.1.[8-10]

10.1.1 Method A. From Type 1 Complexes with Silver Tetrafluoroborate

Alkyne coordination to Type 1 complexes is blocked since phosphine dissociation is energetically unfavorable (16e to 14e), and the bulky phosphines block coordination via the open equatorial position. Coordination and subsequent quinone formation do occur when silver tetrafluoroborate is added (method A). The silver salt removes chloride, affording a cationic complex with an easily displaced axial ligand[11] (see Figure 10.4).

Alkyne coordination, alkyne insertion, and reductive elimination result in quinone formation. Since triphenylphosphine is released at the end, two equivalents of silver salt are used. Phosphine–silver complexation prevents phosphine–cobalt complexation

TABLE 10.1 Preparation of Phthaloyl- and Maleoylcobalt Complexes[8–10]

Substrate	Cobalt(I) Complex	Cobalt(III) Complex	Yield (%)
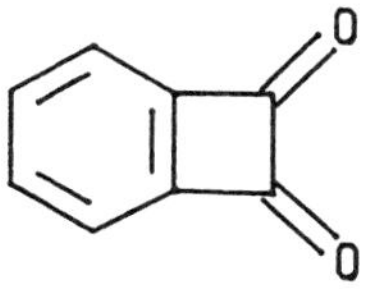	$ClCo(PPh_3)_3$	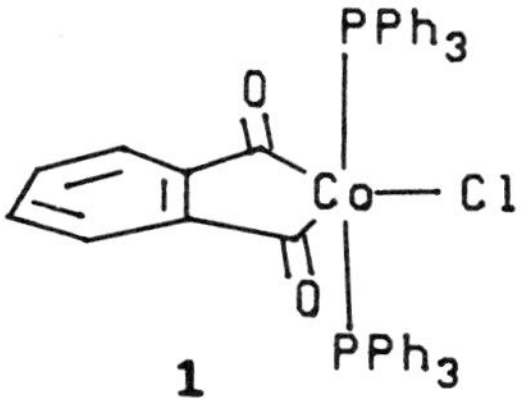 **1**	90
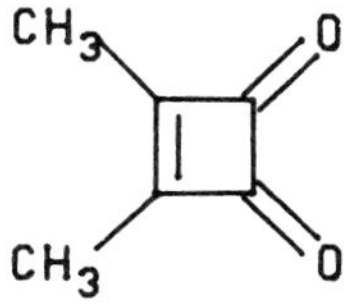	$ClCo(PPh_3)_3$	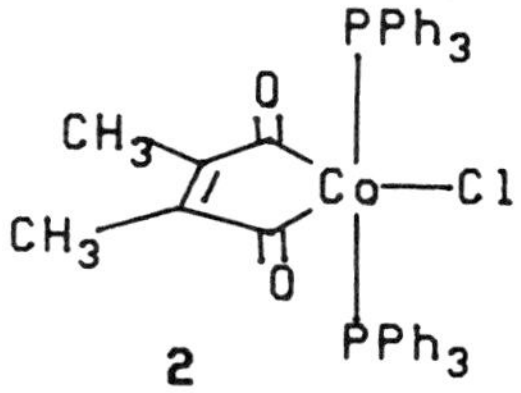 **2**	90
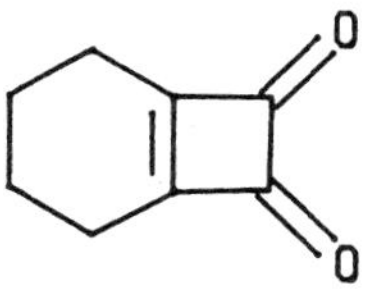	$ClCo(PPh_3)_3$	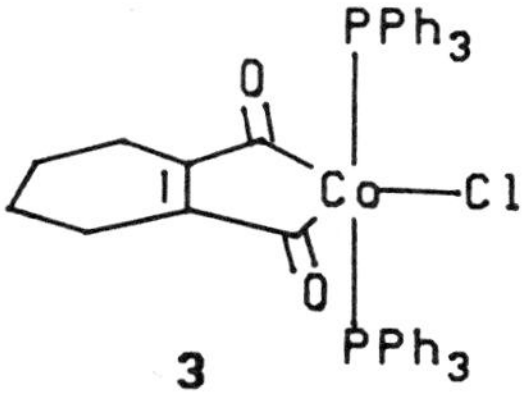 **3**	99
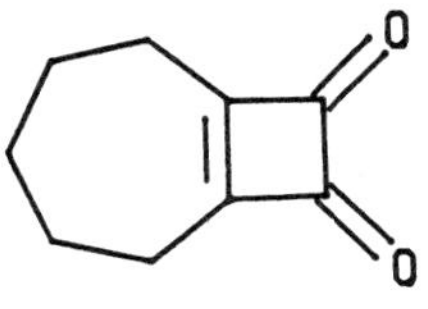	$ClCo(PPh_3)_3$	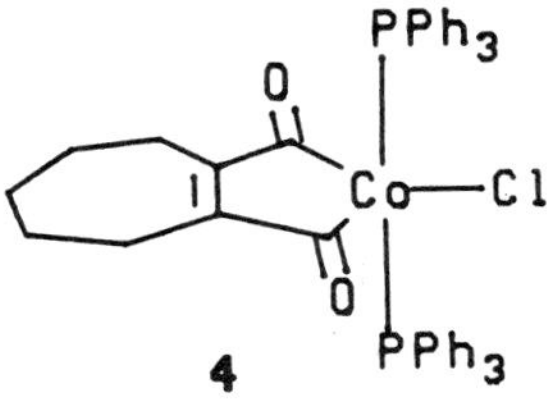 **4**	86
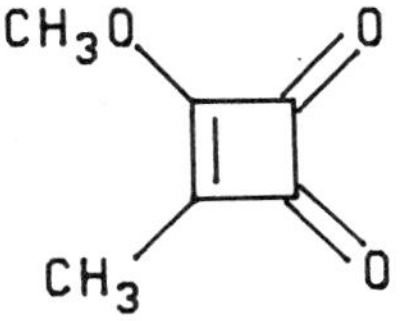	$ClCo(PPh_3)_3$	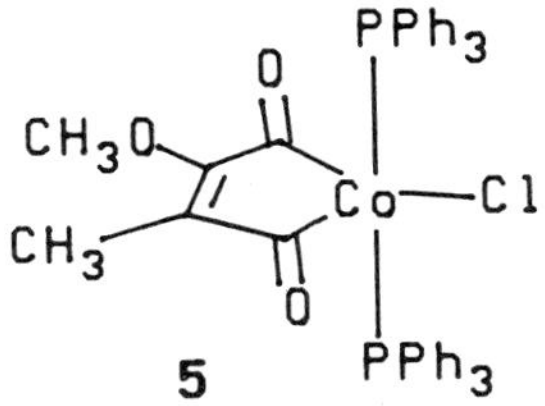 **5**	63

(continued)

TABLE 10.1 (*Continued*)

Substrate	Cobalt(I) Complex	Cobalt(III) Complex	Yield (%)
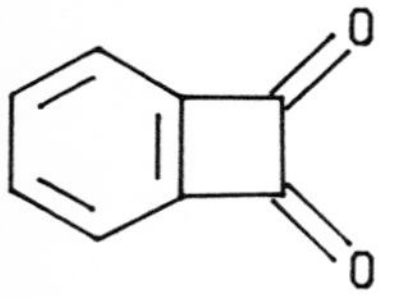	$CpCo(CO)_2$	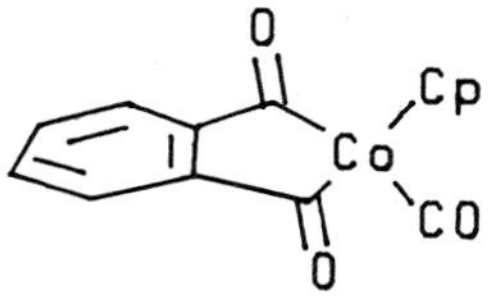 **6**	98
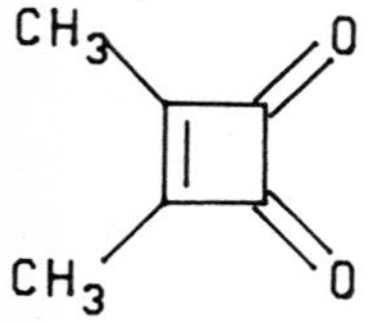	$CpCo(CO)_2$	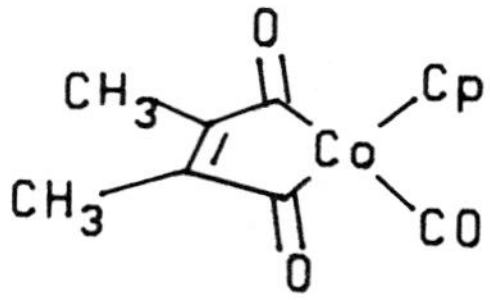 **7**	80

Ⓢ = CH_3CN

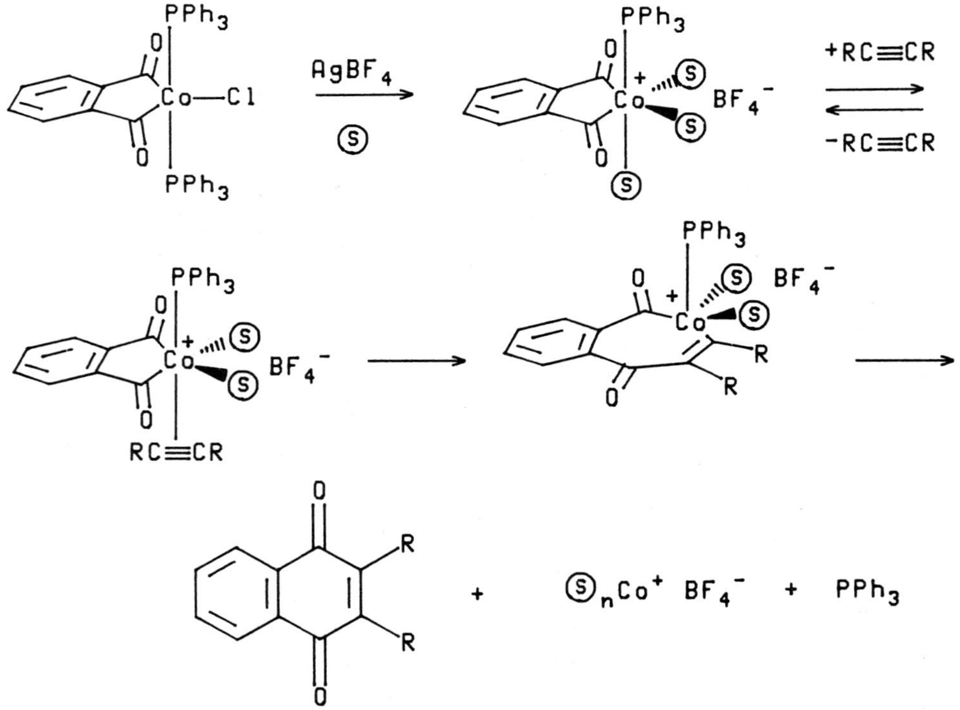

Figure 10.4

and, thus, deactivation of the reactive cobalt(III) intermediate. Quinones can be generated with just one equivalent of silver salt by method B and with no silver salt by methods C and D.

10.1.2 Method B. From Type 1 Complexes with Silver Tetrafluoroborate and Dppe

The chelating ligand bis-(diphenylphosphino)ethane (dppe) can only chelate cobalt in the trigonal bipyramidal complex in an axial-equatorial fashion. Displacement of the triphenylphosphine from the complex produced with silver tetrafluoroborate in acetonitrile results in two equilibrating cationic complexes, each with an easily displaced axial solvent ligand. Reaction with alkyne affords a quinone. A second equivalent of silver salt is not required since the tightly chelating dppe is not prone to elimination from cobalt after reductive elimination[12] (Figure 10.5).

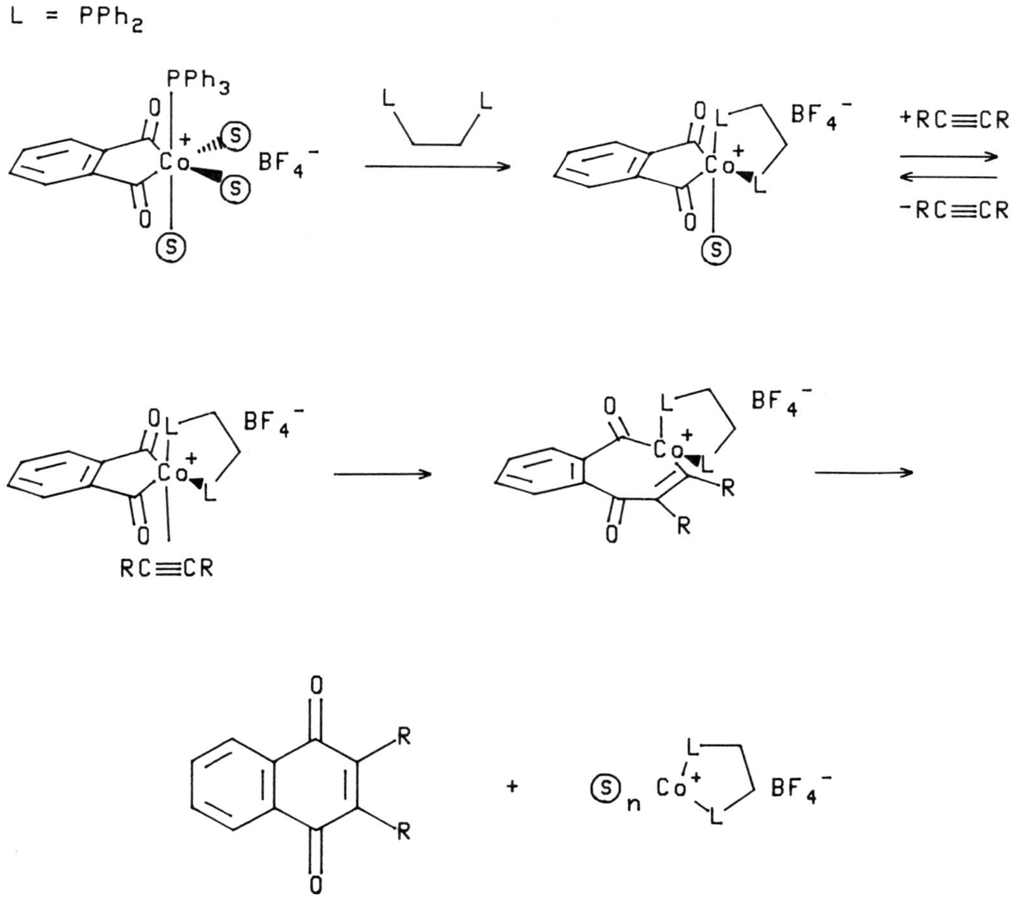

Figure 10.5

10.1.3 Methods C and D. From Type 1 Complexes with Dimethylglyoxime and a Lewis Acid

The chelating ligand dimethylglyoxime (DMG) can only chelate cobalt equatorial-equatorial in an octahedral complex. Thus, displacement of axial phosphine with DMG in pyridine affords an octahedral complex with only an axial ligand free to dissociate. Heating in the presence of cobalt(II) chloride (to complex DMG released) and an alkyne results in loss of pyridine and subsequent quinone formation (method C). Reaction with a Lewis acid and an alkyne results in loss of chloride and again in quinone formation (method D)[13] (see Figure 10.6).

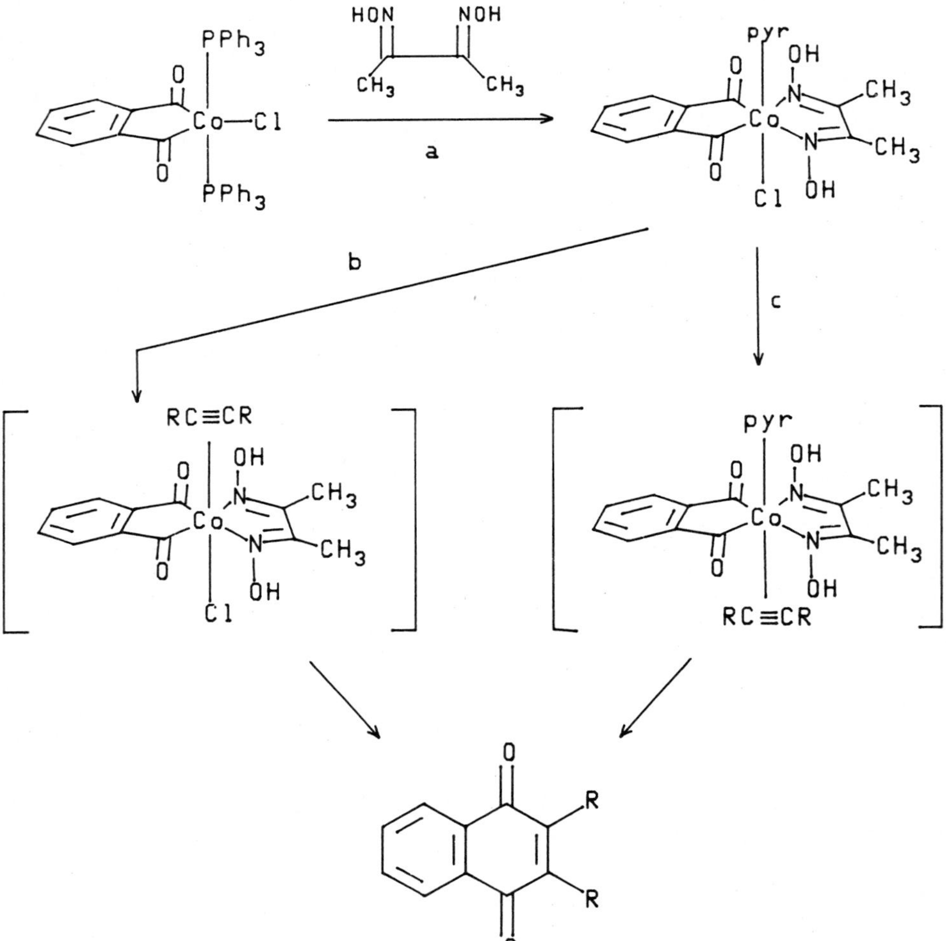

a. pyridine

b. heat, RC≡CR

c. rt, Lewis acid, RC≡CR

Figure 10.6

Figure 10.7

a. $h\nu$

b. Et_2S

c. rt, RC≡CR

d. heat, RC≡CR

e. $h\nu$, Et_2S

f. $(CH_3)_3NO$, (S) = CH_3CN

Figure 10.7 *(Continued)*

10.1.4 Methods E, F, and G. From Type 2 Complexes via Photolysis

Type 2 complexes are coordinatively saturated. Alkyne coordination can only occur after ligand expulsion. Carbon monoxide ligands are usually photochemically labile. Photolysis produces a bis-ketene complex. Addition of alkyne then affords the quinone as its cyclopentadienyl cobalt complex, probably via conversion back to a maleoyl-cobalt complex with alkyne as the additional ligand.[14] Photolysis in the presence of alkyne affords the quinone cyclopentadienyl cobalt complex directly (method E). Alternatively, trapping the 16e intermediate with diethylsulfide affords a new complex with a thermally labile ligand. Heating in presence of an alkyne again produces a benzoquinone as its cyclopentadienyl cobalt complex (method F). Decomplexation is accomplished by cobalt oxidation using ceric ammonium nitrate. A diethylsulfide on a phthaloylcobalt complex is much more difficult to displace. Oxidation with trimethylamine-*N*-oxide in acetonitrile results in replacement of sulfide by solvent. Heating in the presence of an alkyne affords both a naphthaquinone cyclopentadienyl cobalt complex and free naphthaquinone (method G). Some decomplexation under the reaction conditions produces a cobalt complex capable of catalyzing a competing cyclotrimerization of alkynes. Thus, further modification of the method is required before it can be synthetical viable[10] (see Figure 10.7).

10.2 AVAILABILITY OF PRECURSORS

The benzocyclobutanedione is prepared from the corresponding phthalic anhydride. After reaction with hydrazine, oxidation converts a phthalic anhydride to a 1,4-phthalazinedione. The azodicarbonyl group is a potent Diels–Alder dienophile. Thermolysis of the Diels–Alder adduct with anthracene affords the benzocyclobutanedione in excellent yield[15] (see Figure 10.8).

A general synthesis of cyclobutenediones begins with a thioenol ether (from the corresponding ketone). Cycloaddition [2 + 2] with dichloroketene generates the four-membered ring. While treatment with triethylamine is expected to eliminate

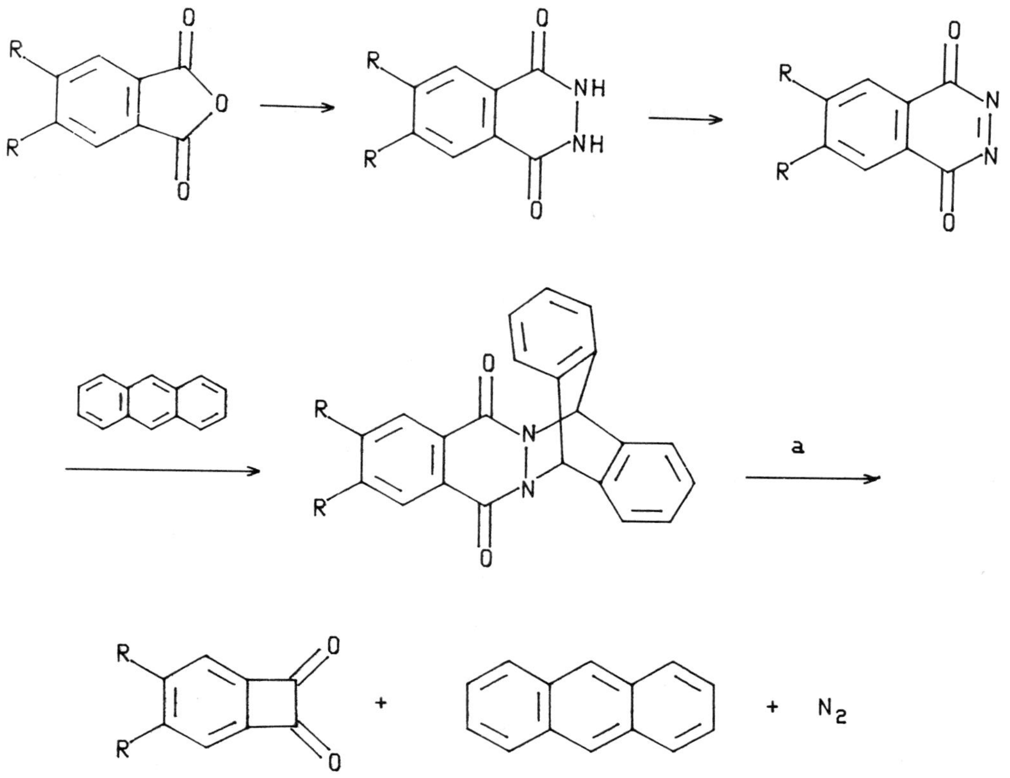

a. high temperature

Figure 10.8

TABLE 10.2 Synthesis of Some Cyclobutenediones[16]

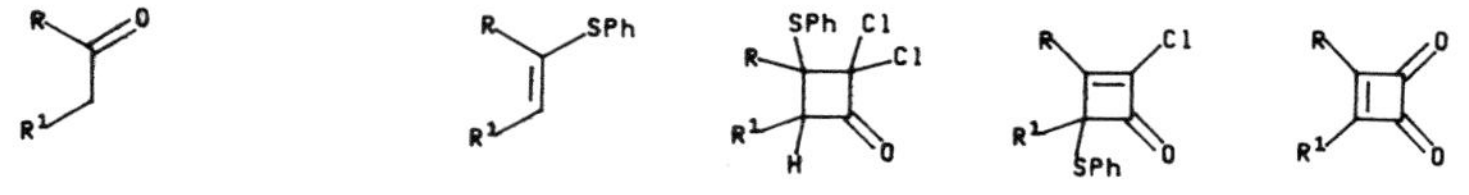

Ketone	Yield (%)			
	Thioenol Ether	[2 + 2] Product	Rearranged Product	Final Product
$R, R^1 = CH_3$	75	—	56[a]	3
$R, R^1 = (CH_2)_4$	75	63	94	94
$R, R^1 = (CH_2)_5$	93	63	87	99
$R = Ph, R^1 = H$	56	62	95	97

[a] For two steps.

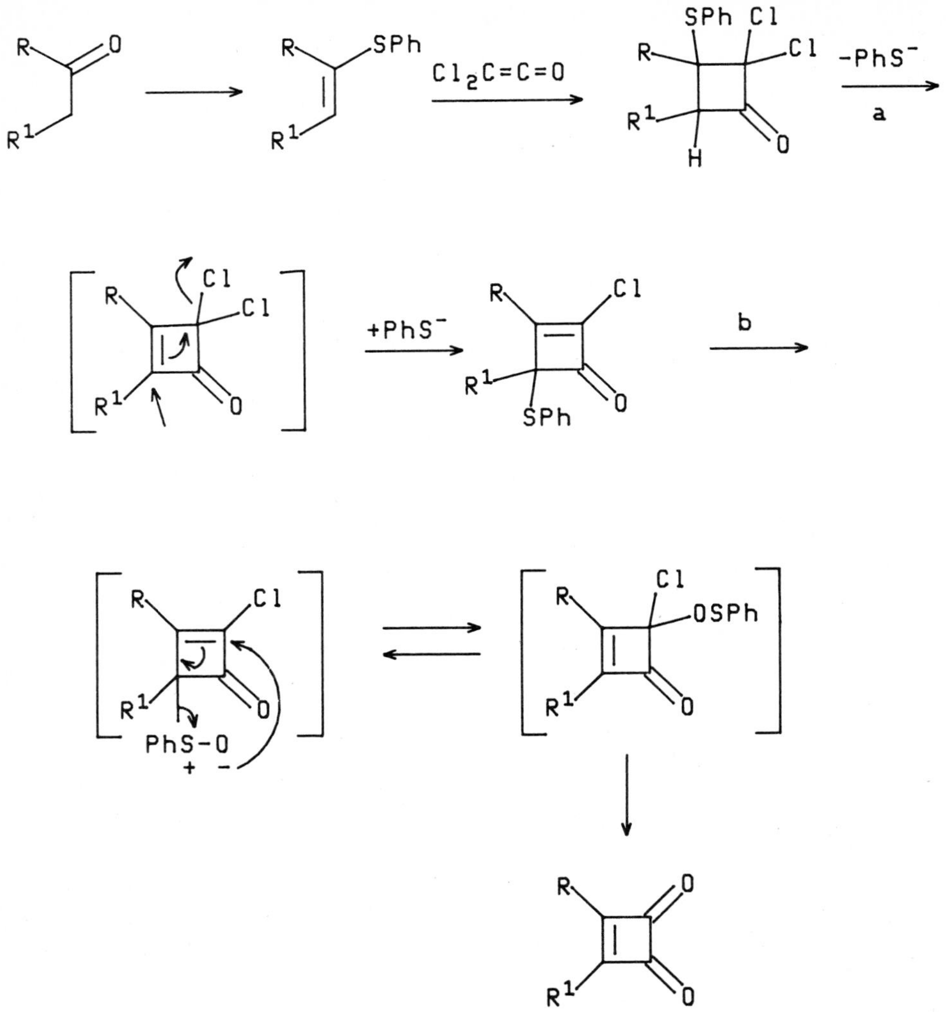

a. 0°C to rt, Et_3N, CH_3CN

b. −78°C to rt, MCPBA, CH_2Cl_2

Figure 10.9

thiophenol, a facile and efficient elimination of HCl gives a rearranged product. Sulfide oxidation with *meta*-chloroperbenzoic acid affords cyclobutenedione via formation of the sulfoxide, [2, 3]-sigmatropic rearrangement, and loss of phenylsulfenyl chloride[16] (see Figure 10.9 and Table 10.2).

Table 10.3 brings together all six methods for quinone formation.

TABLE 10.3 Methods for Quinone Formation[9–13]

Complex	Alkyne $RC{\equiv}CR^1$	Method	Quinone	Yields[a] (%)
1	$R, R^1 = CH_3$	A	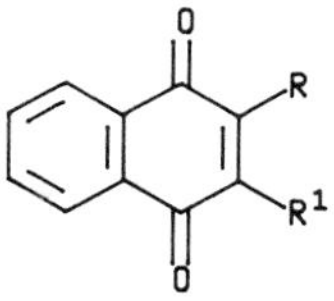 (2-R-3-R¹-1,4-naphthoquinone structure)	73
1	$R, R^1 = Et$	A		90
1	$R, R^1 = Ph$	A		68
1	$R = Ph$, $R^1 = CH_3$	A		78
1	$R = n\text{-Bu}$, $R^1 = H$	A		65
1	$R = Ph$, $R^1 = H$	A		57
1	$R = t\text{-Bu}$, $R^1 = CH_3$	A		72
1	$R = Et$, $R^1 = CH_2CH{=}CH_2$	A		80
1	$R = OEt$, $R^1 = Et$	A(4 eq. $AgBF_4$)		89
1	$R = n\text{-Bu}$, $R^1 = Si(CH_3)_3$	A		68
1	$R = Ph$, $R^1 = CH_2CH_2OH$	A		27
1	$R = COOEt$, $R^1 = CH_3$	A		0
1	$R, R^1 = Et$	B		(93, 95)85
1	$R = n\text{-Bu}$, $R^1 = H$	B		(93, 95)90
1	$R = n\text{-Bu}$, $R^1 = Si(CH_3)_3$	B		(93, 95)86
1	$R = CH_3$, $R^1 = CH_2OEt$	B		(93, 95)63
1	$R = CH_3$, $R^1 = CH(OEt)_2$	B		(93, 95)51
1	$R = Et$, $R^1 = CH_2CH_2OTHP$	B		(93, 95)73
1	$R = n\text{-Bu}$, $R^1 = H$	C		(95)87
1	$R, R^1 = Et$	C		(95)86
1	$R = COOEt$, $R^1 = CH_3$	C		(95)99
1	$R = Et$, $R^1 = OEt$	C		(95)85
1	$R = t\text{-Bu}$, $R^1 = CH_3$	C		(95)72

TABLE 10.3 (*Continued*)

Complex	Alkyne $RC \equiv CR^1$	Method	Quinone	Yields[a] (%)
1	$R = CH_3$ $R^1 = CH_2CH(CH_3)_2$	C		(95)94
1	$R = CH_3$ $R^1 = (E,E) - [CH_2CH{=}C(CH_3)CH_2CH_3]$	C		(95)86
1	$R = n$-Bu $R^1 = H$	D		(95)80
1	$R, R^1 = Et$	D		(95)83
1	$R = COOEt$ $R^1 = CH_3$	D		(95)80
1	$R = Et$ $R^1 = OEt$	D		(95)0
2	$R, R^1 = Et$	C	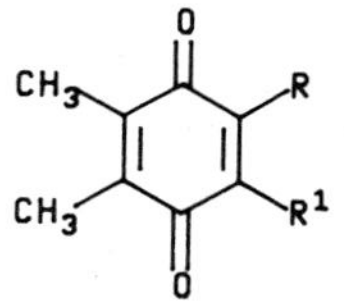	(83)81
2	$R = n$-Bu $R^1 = H$	C		(83)85
3	$R, R^1 = Et$	C	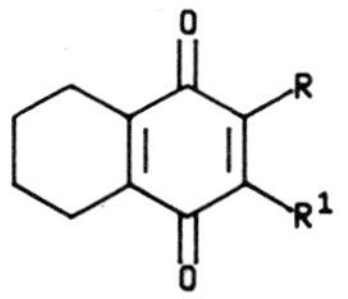	(90)85
3	$R = n$-Bu $R^1 = H$	C		(90)79
4	$R, R^1 = Et$	C	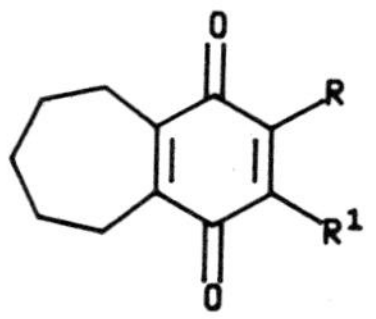	(86)88
4	$R = n$-Bu $R^1 = H$	C		(86)84
5	$R, R^1 = Et$	C[b]	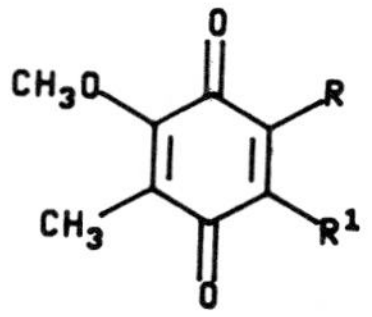	(87)80

TABLE 10.3 *(Continued)*

Complex	Alkyne $RC{\equiv}CR^1$	Method	Quinone	Yields[a] (%)
5	R = H $R^1 = CH_3$	C[b]	+ regioisomer (5:1)	(87)81
5	R = H R^1 = *n*-Bu	C[b]	+ regioisomer (3.7:1)	(87)89
5	R = H R^1 = *t*-Bu	C[b]	+ regioisomer (2.8:1)	(87)73
5	R = COOEt $R^1 = CH_3$	C[b]	+ regioisomer (3.7:1)	(87)64
5	$R = CH_3$ R^1 = OEt	C[b]	+ regioisomer (13.5:1)	(87)81
7	R, R^1 = Et	E	O, CH_3, R, CH_3, R^1, O	94
7	R, R^1 = Et	F		(88, 87)90
7	R = H R^1 = *n*-Bu	F		(88, 85)96
7	R = OEt R^1 = Et	F		(88, 91)97
7	R = Et $R^1 = C(O)CH_3$	F		(88, 84)90
7	$R = CH_3$ $R^1 = C(CH_3){=}CH_2$	F		(88, 84)91

[a] Yield refers to the actual quinone formation step. Yields in parentheses are for additional steps leading from the initial complex to a quinone precursor and, for Type 2 complexes, for subsequent demetallation of the quinone CpCo complexes.

[b] $CoCl_2{\cdot}H_2O$ omitted.

REFERENCES

1. Thomson, R. H. "*Naturally Occurring Quinones;* 2nd ed., Academic: London, 1971.
2. *Biochemistry of Quinones*; Morton, R. A., Ed.; Academic: New York, 1965.
3a. Roth, L. M.; Eisner, T. *Ann. Revs. Entomol.* **1962**, *7*, 107.
3b. Weatherston, J. *Quart., Revs. (London)* **1967**, *21*, 287.
3c. Schlidknecht, V. H. *Angew. Chem.* **1970**, *82*, 17.
4. Redfearn, E. R. *Vitamins Hormones* **1966**, *24*, 465.
5a. Asano, A.; Brodie, A. F. *J. Biol. Chem.* **1964**, *239*, 4280.
5b. White, D. C. *J. Biol. Chem.* **1965**, *240*, 1387.
5c. Crane, F. L.; Low, H. *Physiol. Rev.* **1966**, *46*, 662.
5d. Crane, F. L. In *Biological Oxidations*; Singer, T. P., Ed.; Interscience: New York, 1968.

6. Driscoll, J. S.; Hazard, G. F., Jr.; Wood, H. B., Jr.; Goldin, A. *Cancer Chemother. Rep., Part 2* **1974**, *4*, 1.
7. Thomas, R. H. In *The Chemistry of Quinonoid Compounds, Part I*: Patai, S., Ed.; Wiley: New York, 1974.
8. Liebeskind, L. S.; Baysdon, S. L.; South, M. S.; Blount, J. F. *J. Organometal. Chem.* **1980**, *202*, C73.
9. Liebeskind, L. S.; Leeds, J. P.; Baysdon, S. L.; Iyer, S. *J. Am. Chem. Soc.* **1984**, *106*. 6451.
10. Liebeskind, L. S.; Jewell. C. F., Jr. *J. Organometal. Chem.* **1985**, *285*, 305.
11. Liebeskind, L. S.; Baysdon, S. L.; South, M. S. *J. Am. Chem. Soc.* **1980**, *102*, 7397.
12. Baysdon, S. L.; Liebeskind, L. S. *Organometallics* **1982**, *1*, 771.
13. Liebeskind, L. S.; Baysdon, S. L.; Goedken, V.; Chidambaram, R. *Organometallics* **1986**, *5*, 1086.
14. Jewell, C. F., Jr.; Liebeskind, L. S.; Williamson, M. *J. Am. Chem. Soc.* **1985**, *107*, 6715.
15. McOmie, J. F. W.; Perry, D. H. *J. Chem. Soc., Chem. Commun.* **1973**, 248.
16. Liebeskind, L. S.; Baysdon, S. L. *Tetrahedron Lett.* **1984**, *25*, 1747.

CHAPTER 11

(±)-Acorenone and (±)-Acorenone B

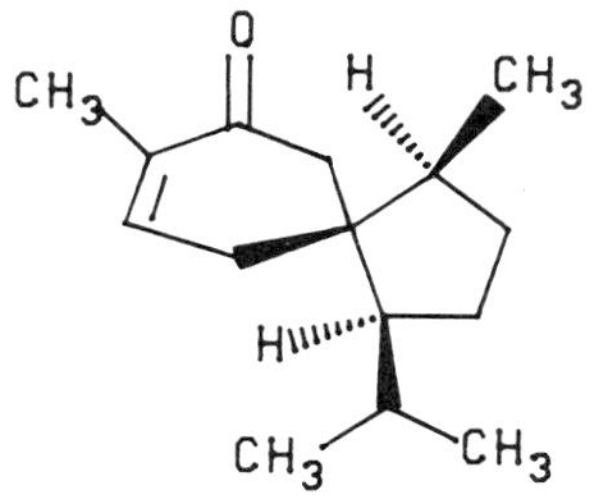

(±)-Acorenone B 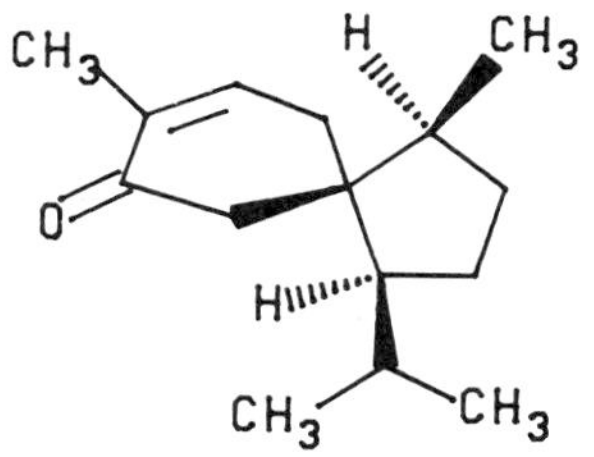   (±)-Acorenone

Spiro[4.5]dec-8-ene-7-one,

4,8-dimethyl-1-(1-methylethyl)

(±)-Acorenone	(1α, 4α, 5α)(±)[63865-67-8]
(±)-Acorenone B	(1α, 4α, 5β)(±)[56362-99-3]
(−)-Acorenone	[1S-(1α, 4α, 5α)][5965-05-8]
(−)-Acorenone B	[1S-(1α, 4α, 5β)][21653-33-8]

Figure 11.1

Both (−)-acorenone, isolated from sweet flag oil (*Acorus calamus L.*)[1] and (−)-acorenone B, isolated from *Bothriochloa intermedia*[2], are spirocyclic sesquiterpenes. The spiro[4.5]decanes of the acorane–alaskane family are important as intermediates in terpene biosynthesis, as constituents of essential oils, and as antifungal agents and stress metabolites.[3] The considerable number of elegant syntheses of both racemic and optically active materials provide an excellent background to Semmelhack's chromium-based approach presented in this chapter.[4]

11.1 η^6-ARENECHROMIUM TRICARBONYL COMPLEXES: PREPARATION

11.1.1 Method A. From Arene and Chromium Hexacarbonyl via Thermolysis

A large number of air-stable, crystalline arenechromium tricarbonyl complexes can be prepared from the arene and chromium hexacarbonyl (Caution: volatile and toxic) by three different methods. Complex formation requires a decarbonylation to open a coordination site on the metal since chromium hexacarbonyl is coordinatively saturated. In method A decarbonylation is achieved by heating the complex at relatively high temperatures (130 to 140°C) in the presence of excess arene in a high boiling, moderately polar solvent such as di-*n*-butyl ether.[5,6]

$$\text{excess arene} + Cr(CO)_6 \xrightarrow{a} \text{areneCr}(CO)_3 + 3\ CO\uparrow \quad (11.1)$$

a. heat, $(n\text{-Bu})_2O$

11.1.2 Method B. From Arene and *tris*-(Ligand)chromium Tricarbonyl

In method B three carbon monoxide ligands of the hexacarbonyl complex are first replaced by more thermally labile ligands such as pyridine, picoline, acetonitrile, or ammonia. A coordination site is then accessed by heating at considerably lower temperatures (60 to 70°C).[7,8]

$$3\ L + Cr(CO)_6 \longrightarrow Cr(CO)_3L_3 + 3\ CO\uparrow \quad (11.2)$$

$$\text{excess arene} + Cr(CO)_3L_3 \xrightarrow{\Delta} \text{areneCr}(CO)_3 + 3\ L \quad (11.3)$$

11.1.3 Method C. From Arene and Chromium Hexacarbonyl via Photolysis

Carbon monoxide can be ejected by photolysis as well as thermolysis. The photochemical method C produces arene complexes from chromium hexacarbonyl at room temperature.[7–9]

$$\text{excess arene} + Cr(CO)_6 \xrightarrow{a} \text{areneCr}(CO)_3 + 3\ CO\uparrow \quad (11.4)$$

a. rt, $h\nu$

11.1.4 Method D. From Arene and Arenechromium Tricarbonyl

Method D is based on an equilibrium established on heating an arenechromium tricarbonyl complex to approximately 200°C in the presence of another arene. An order of arenechromium tricarbonyl complex stability has been established:

$$Ar(CH_3)_6 > Ar(CH_3)_4 > Ar(CH_3)_3 > ArN(CH_3)_2 \geqslant Ar(CH_3)_2 >$$
(most stable)

$$ArCH_3 \approx ArH > ArC(O)CH_3 \approx ArOCH_3 > ArCOOCH_3 > ArCl \approx ArF$$

$$> \text{naphthalene}$$
(least stable)

Thus, the naphthalene complex should be most suitable for preparation of new arene complexes by arene exchange.[10] Complex formation by methods A, B, C, or D is most facile with electron-rich aromatics. Complexes of electron-deficient aromatics (benzaldehyde, benzoic acid, benzonitrile, nitroaromatics) are best prepared by indirect methods, which will be discussed in the section on complex reactivity to follow.

$$\text{arene}_A + \text{arene}_B\text{Cr(CO)}_3 \underset{}{\overset{\Delta}{\rightleftharpoons}} \text{arene}_B + \text{arene}_A\text{Cr(CO)}_3 \quad (11.5)$$

11.1.5 Method E. From Chromium Pentacarbonyl Carbenes and Alkynes

There is one other established method for complex preparation *that does not start with the arene.* Method E produces complexes from the condensation of alkynes with chromium pentacarbonyl carbenes and is discussed in detail in Chapter 12 (see Figure 11.2).

11.2 η^6-ARENECHROMIUM TRICARBONYL COMPLEXES: REACTIVITY[11–13]

Figure 11.3 summarizes the reactivity of the arene complexes relative to the reactivity of the free arene.[11]

The valence orbitals for arenechromium tricarbonyl complexes are derived from combination of π-orbitals of the arene with those of chromium tricarbonyl. Extended Hückel Theory (EHT) calculations predict a net transfer of electron density from arene to metal, resulting in enhanced acidity of ring protons, enhanced acidity of α-protons, enhanced acidity of β-protons, decreased reactivity of the ring with electrophiles, and increased reactivity of the ring with nucleophiles.[14]

11.2.1 Enhanced Acidity of Ring Protons

Strongly basic anions such as methyl- or butyllithium can abstract a ring proton. Abstraction is regioselective for the *ortho* position when the substituent is methoxy, chloro, or fluoro. Trap of the lithiated arene complex by electrophiles provides an indirect route to more highly functionalized arene complexes[15] (see Figure 11.4).

Regioselectivity of lithiation of the arene complex is often different from that of the free arene. For example, 3-methoxybenzyl alcohol lithiates at the 2-position; the arene complex lithiates at the 4-position (see Figure 11.5).

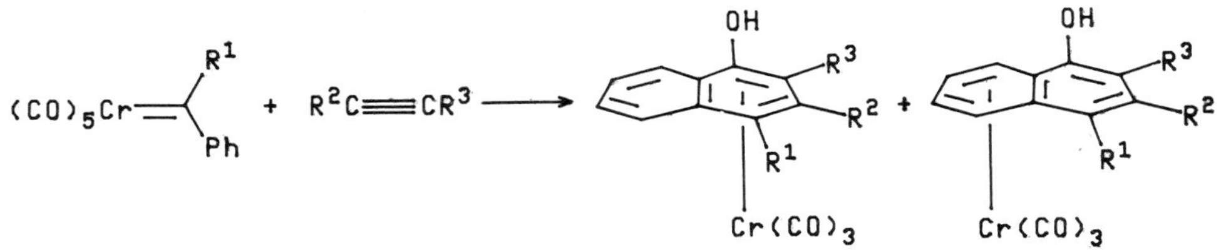

$$(\text{CO})_5\text{Cr}{=}\text{C}(\text{R}^1)(\text{Ph}) + \text{R}^2\text{C}{\equiv}\text{CR}^3 \longrightarrow \text{[OH, R}^1\text{, R}^2\text{, R}^3\text{-naphthalene]Cr(CO)}_3 \text{ (substituted ring)} + \text{[OH, R}^1\text{, R}^2\text{, R}^3\text{-naphthalene]Cr(CO)}_3 \text{ (unsubstituted ring)}$$

Figure 11.2

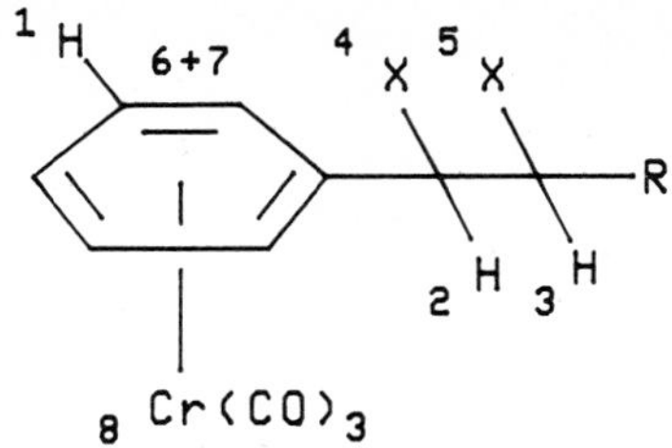

1) enhanced acidity of ring protons

2) enhanced acidity of α-protons

3) enhanced acidity of β-protons

4) enhanced rate of solvolysis of α-leaving groups

5) enhanced rate of solvolysis of β-leaving groups

6) decreased reactivity of the ring with electrophiles

7) increased reactivity of the ring with nucleophiles

8) stereoselectivity due to the chromium tricarbonyl group

Figure 11.3

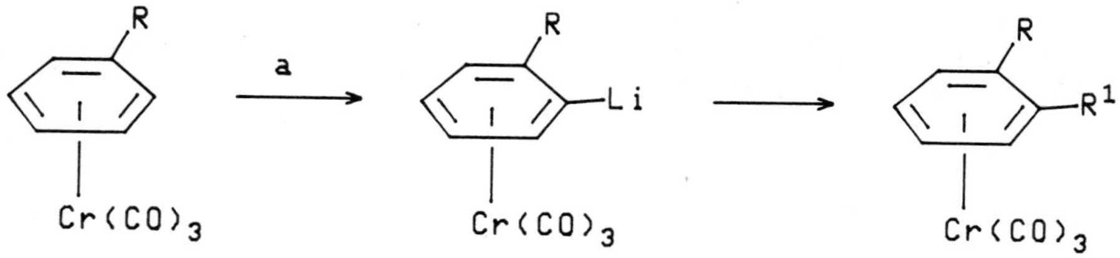

R = H, OCH_3, F, Cl

R^1 = $COOCH_3$, CH_3, Ph, CH_2OH, TMS, $(CH_3)_2COH$

a. −78°C, n-BuLi, THF, TMEDA

Figure 11.4

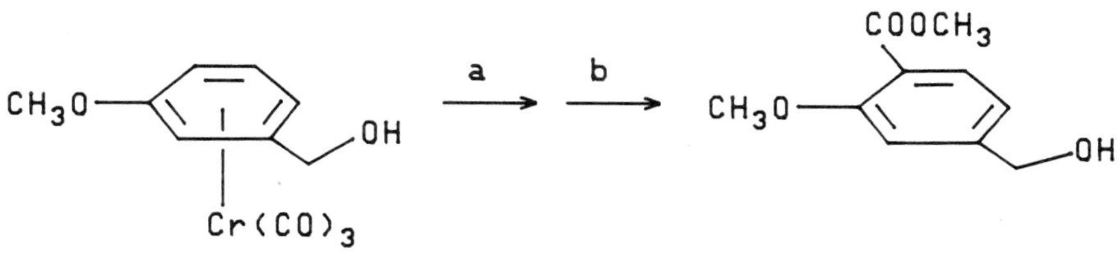

a. −78°C, n-BuLi, THF, TMEDA

b. 1) CO_2, 2) hν, O_2, 3) CH_2N_2; 77% for two steps

Figure 11.5

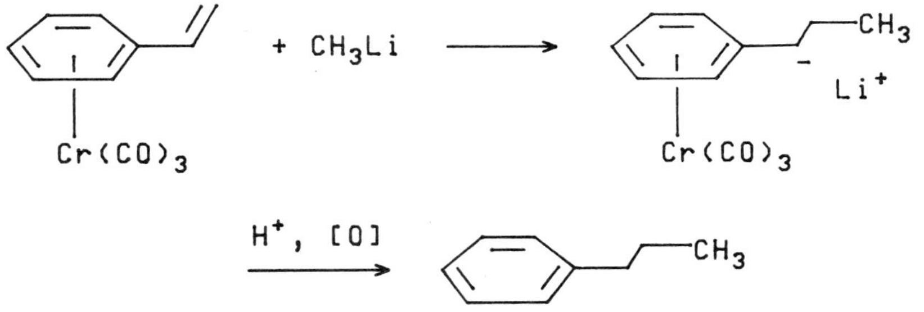

H^+, [O]

Figure 11.6

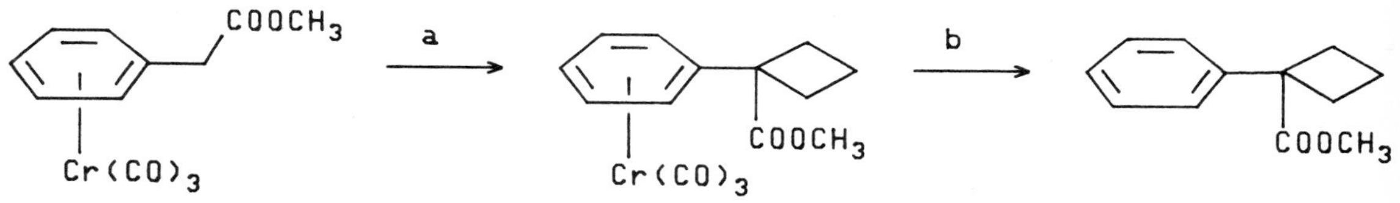

a. 1) rt, NaH, 2) $Br(CH_2)_3Br$, DMF

b. hν, O_2; 87% for two steps

Figure 11.7

11.2.2 Enhanced Acidity of α-Protons

α-Anion stabilization should make possible the use of complexed styrenes as Michael acceptors. While the concept has been demonstrated, efficiency is limited by competing subsequent reactions of the anionic intermediate[16] (see Figure 11.6).

Jaouen later provided clear and simple evidence of α-anion stabilization. While no reaction occurs on mixing methyl phenylacetate with sodium hydride and an alkylating agent in *N,N*-dimethylformamide, facile dialkylation occurs with the corresponding arene complex[17] (see Figure 11.7).

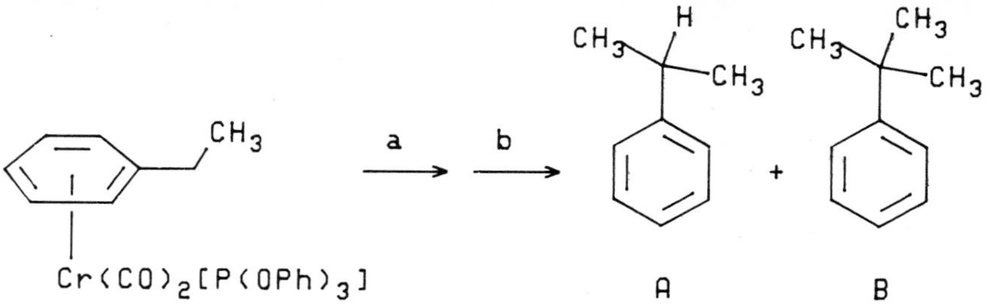

a. 1) rt, t-BuOK, DMSO, 2) CH_3I

b. hν, O_2; A:71%, B:8%

Figure 11.8

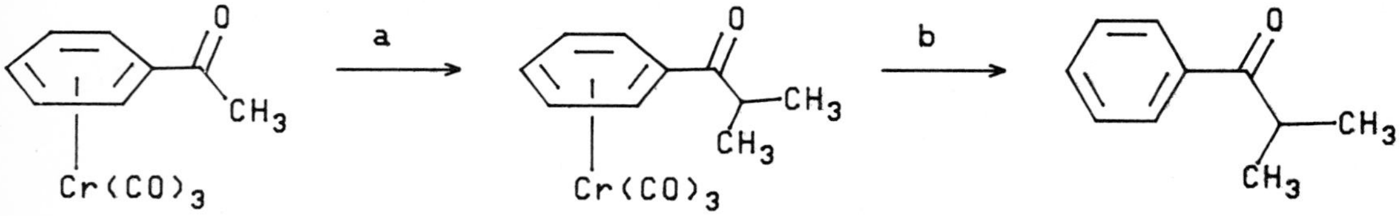

a. NaH, CH_3I, DMF; 90%

b. hν, O_2

Figure 11.9

The extent of α-anion stabilization and thus the ratio of α-monoalkylated to α,α-dialkylated product can be attenuated by proper choice of metal ligands. Replacement of one carbon monoxide ligand on η^6-(ethylbenzene)chromium tricarbonyl by a superior donor, a phosphite, results in increased electron density on the metal, increased electron density on the ring, and thus decreased ability of the complex to stabilize an α-carbanion. Reaction of this modified complex with a strong base followed by an alkylating agent results in moderately selective monoalkylation (see Figure 11.8).

11.2.3 Enhanced Acidity of β-Protons

While no reaction occurs on mixing acetophenone with sodium hydride and an alkylating agent in dimethylformamide, the arene complex undergoes facile dialkylation (see Figure 11.9).

11.2.4 Enhanced Rate of Solvolysis of α-Leaving Groups

Two observations lead to the conclusion that α-cations on arene complexes are more stable than α-cations on free arenes.

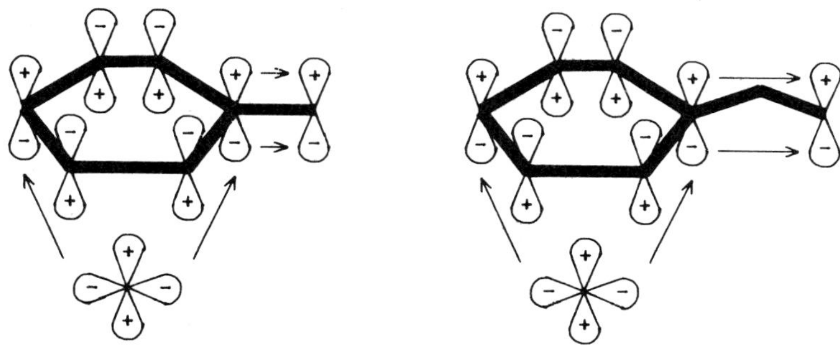

Figure 11.10

Observation 1. The S_N1 solvolyses of benzyl, benzhydryl, and cumyl chloride complexes are 10^5, 10^3, and 28 times faster, respectively, than solvolyses of the free arenes.[18,19]

Observation 2. The chromium complex of benzylthiocyanate rapidly isomerizes to the chromium complex of benzylisothiocyanate.[20]

While net electron density transfer from the arene to the metal explains much of the chemistry of arenechromium tricarbonyl complexes, transfer in the opposite direction would explain enhanced stability of α- and β-cations.[21] Both direct electron donation from the metal to the α-carbon and $\delta \rightarrow \pi^*$ indirect donation through a ring carbon are possible (see Figure 11.10).

11.2.5 Enhanced Rate of Solvolysis of β-Leaving Groups

Two observations lead to the conclusion that β-cations on arene complexes are more stable than β-cations on the free arenes (see Table 11.1).

TABLE 11.1 Comparison of the Rate of S_N1 Solvolyses of Complexed and Uncomplexed β-Arylalkylmethane Sulfonates[22]

R	R^1	Solvent	Temperature (°C)	Rate Enhancement on Complexation
H	H	AcOH	90	10,000
H	H	HCOOH	86.6	7,900
CH_3	H	AcOH	90	3,200
CH_3	H	HCOOH	86.6	1,900
CH_3	CH_3	AcOH	75	1,600

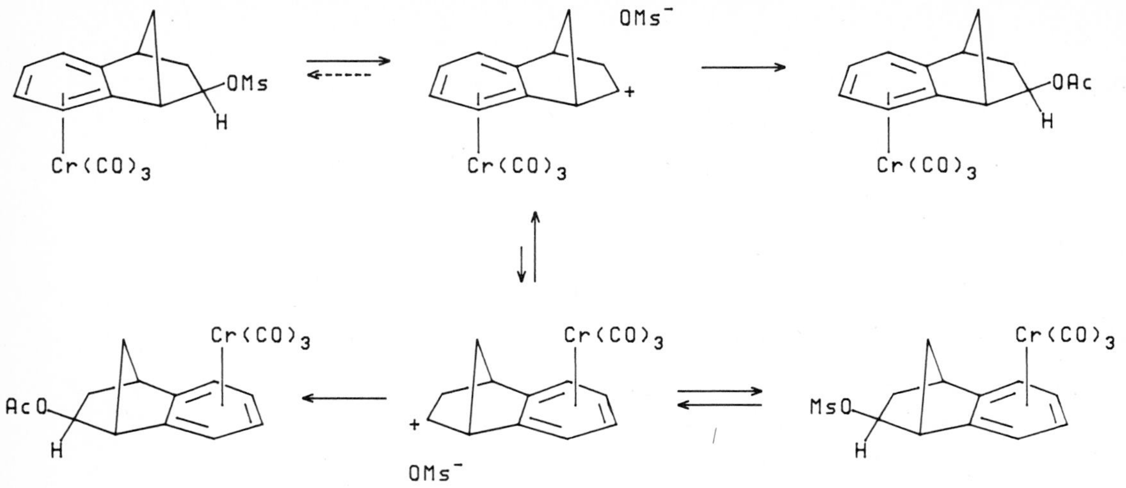

Figure 11.11

Observation 1. The S_N1 solvolyses of complexed β-arylalkylmethane sulfonates are considerably faster than solvolyses of the free arenes.[22]

Observation 2. The acetolysis of an *exo*-mesylate on an *endo*-complexed π-benzonorbornene is 300 times faster than on an *exo*-complexed π-benzonorbornene. Both result in the thermodynamically less stable *endo*-complexed π-benzonorbornene *exo*-acetate[23,24] (see Figure 11.11).

11.2.6 Decreased Reactivity of the Ring with Electrophiles

The functionalization of arenes in arene complexes by electrophilic substitution is limited by the oxidative instability of η^6-arenechromium tricarbonyl complexes. Of the large number of synthetically significant methods for electrophilic aromatic substitution of arenes, only Friedel–Crafts acetylation[25,26] and mercuration[27–29] have been successful with the arene complexes. The decrease in ring reactivity toward electrophiles is accompanied by an altered regioselectivity of attack. Alkylbenzenes undergo nearly exclusive *para*-acetylation; η^6-(alkylbenzene)chromium tricarbonyl complexes are attacked at *ortho* and *meta* positions, with the proportion of *meta* product increasing as the steric bulk of the alkyl substituent increases (see Figure 11.12).

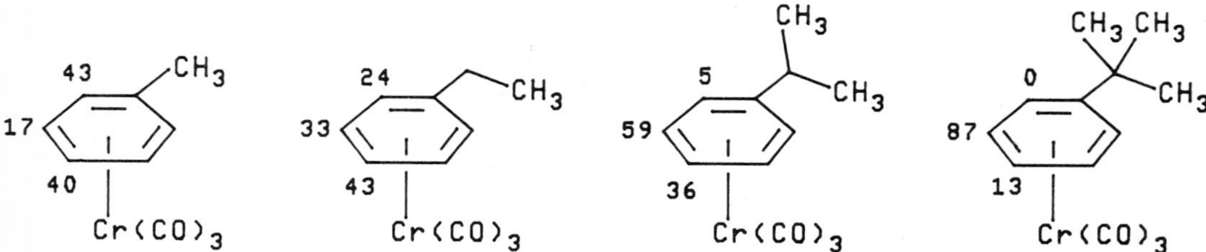

Figure 11.12 Regioselectivity of attack by electrophiles on η^6-arenechromium tricarbonyl complexes.

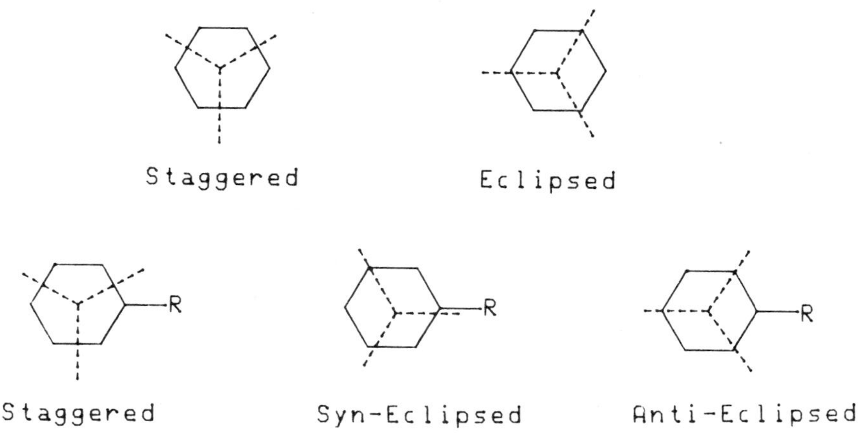

Staggered

Eclipsed

Staggered

Syn-Eclipsed

Anti-Eclipsed

Figure 11.13

Carpenter has rationalized these results by considering the stabilities of rotational conformers or rotamers of the arene complexes.[14] η^6-Benzenechromium tricarbonyl can exist in two different rotational conformers. These are named for the spatial relationship of metal carbonyl ligands to aromatic ring carbons as either staggered or eclipsed. η^6-(Monosubstituted benzene)chromium tricarbonyl can exist in three different rotational conformers: staggered, *syn*-eclipsed (which has the substituent and metal carbonyl ligand in close proximity), and *anti*-eclipsed (which has the substituent and metal carbonyl ligand distant) (see Figure 11.13).

The staggered benzenechromium tricarbonyl rotamer is more stable than the eclipsed, but the difference is small (0.3 kcal/mol), and interconversion of rotamers is facile. In the staggered rotamer all six carbons are equivalent. In the eclipsed rotamer three carbons are staggered and three are eclipsed. EHT calculations provide charges for the ring carbons in the eclipsed rotamer and lead to the conclusion that electrophiles will attack staggered ring carbons.

For monosubstituted arene complexes, the *syn*-eclipsed rotamer is more stable when the substituent is electron-releasing, and the *anti*-eclipsed is more stable when the substituent is electron-withdrawing. In addition, as the steric bulk of an electron-releasing substituent increases, the *syn*-eclipsed rotamer becomes less stable, less favored, and the *anti*-eclipsed more favored. Again, the energy differences are relatively small and interconversion of rotamers is facile. EHT calculations again lead to the conclusion that, in the *absence of electronic effects from ring substituents*, electrophiles will attack staggered arene carbons. This translates into *ortho–para* attack when the substituent is electron releasing, *meta* attack when the substituent is electron-withdrawing, and *meta* attack when the substituent is both electron-releasing and bulky. These predictions are supported by studies of the regioselectivity of Friedel–Crafts acetylation.[25,26]

11.2.7 Increased Reactivity of the Ring with Nucleophiles

A method for nucleophilic substitution of a leaving group or a hydrogen on an aromatic ring would nicely complement the well-established methodology for

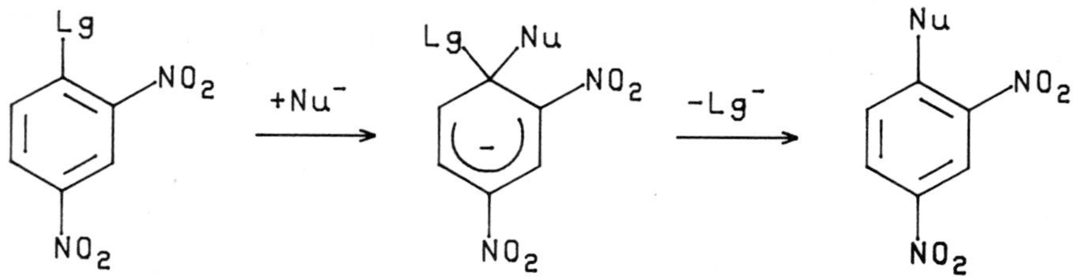

Figure 11.14

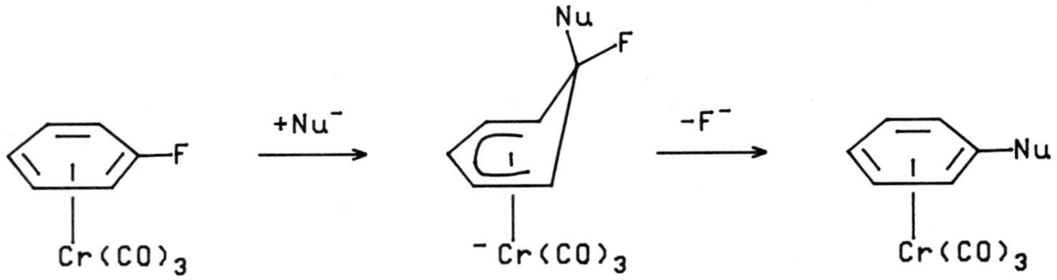

Figure 11.15

electrophilic substitution. Nucleophilic substitution of a leaving group is possible when the ring possesses *ortho*- and/or *para*-situated nitro substituents that stabilize the intermediate Meisenheimer complex (see Figure 11.14).

The reaction of η^6-chloro(or fluoro)benzenechromium tricarbonyl with sodium methoxide in methanol similarily produces η^6-anisolechromium tricarbonyl in excellent yield. Amines, cyanide, and some carbanions also displace fluoride efficiently.[30–34] From kinetic rate data an apparently general two-step mechanism can be proposed (see Figure 11.15).

Step 1. An η^5-cyclohexadienyl complex is formed by rapid nucleophilic attack from the side opposite the metal, the *exo* side.

Step 2. The η^5-cyclohexadienyl complex slowly fragments by loss of halide from the same side as the metal, the *endo* side.

Although several of the successful carbanions are strongly basic, an alternative elimination–addition (aryne) mechanism is unlikely. While *p*-chlorotoluenechromium tricarbonyl would be expected to produce some *meta*- as well as *para*-substituted products via the aryne, only *para*-subsitution is observed with isobutyronitrile anion[34] (see Figure 11.16).

The range of carbon nucleophiles that cleanly displace chloride is disappointingly limited. Grignards, cuprates, and mercuric chlorides are unreactive below room temperature and give only products of decomposition on heating. In addition, several lithium anions react rapidly at −78°C but fail to efficiently substitute for chlorine even at higher temperatures. Both applicable carbanions and these inapplicable lithium carbanions apparently attack the ring with little regioselectivity. Only the adduct from

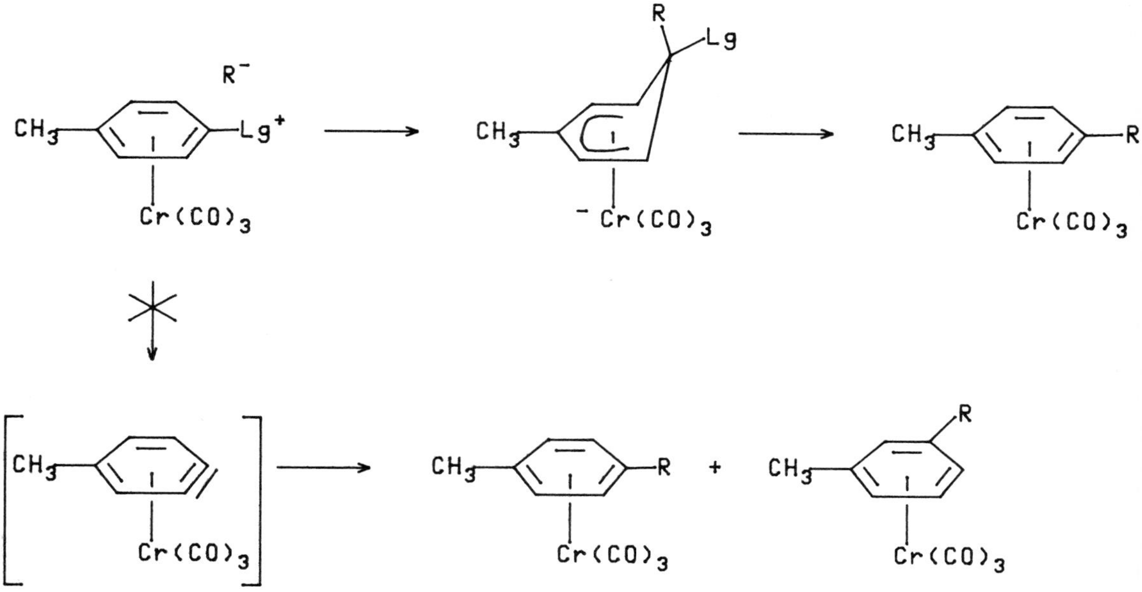

Figure 11.16

attack on the carbon bearing chlorine can lead to substitution products. Attack by applicable carbanions is reversible; return to the η^6-arene complex eventually results in efficient conversion. Attack by inapplicable carbanions is irreversible. Only the small portion of adduct from attack on carbon bearing chlorine is converted to product. When a mixture of η^5-cyclohexadienyl complexes is subjected to oxidation–oxidative demetallation with iodine, products resulting from overall nucleophilic substitution for ring hydrogens are obtained. Interestingly, little or no product from substitution for the hydrogen *para* to the leaving group is observed (see Figure 11.17).

The halogen substituent is clearly not required for nucleophilic attack. A similar nucleophilic substitution for hydrogen can be accomplished using η^6-benzenechromium tricarbonyl (see Figure 11.18).

The η^5-cyclohexadienyl anion that results from the reaction of 2-lithio-1,3-dithiane was isolated and a crystal structure obtained. Nucleophilic attack is stereospecifically *exo* and the carbon bearing the nucleophile is displaced 38.6° from the plane of the η^5-system. Again, the range of applicable carbanions is somewhat limited. Grignards, cuprates, mercuric chlorides, and anions more stable than ester enolates ($pKa < 25$) are all unreactive, while several strongly basic anions such as methyl- or butyllithium abstract a ring proton, as discussed earlier.[35,36] The regioselectivity of attack on η^6-(monosubstituted benzene)chromium tricarbonyl complexes has received considerable attention (see Table 11.2).

Several patterns emerge from these results: (1) rings with electron-releasing substituents are attacked *met*; (2) as the bulk of the electron-releasing substituents increases, there is increasing redirection of attack from *meta* to *para*; and (3) rings with electron-withdrawing substituents are attacked *para*. These three patterns neatly fit the rotamer stability rationalization for regioselectivity of attack by electrophiles discussed

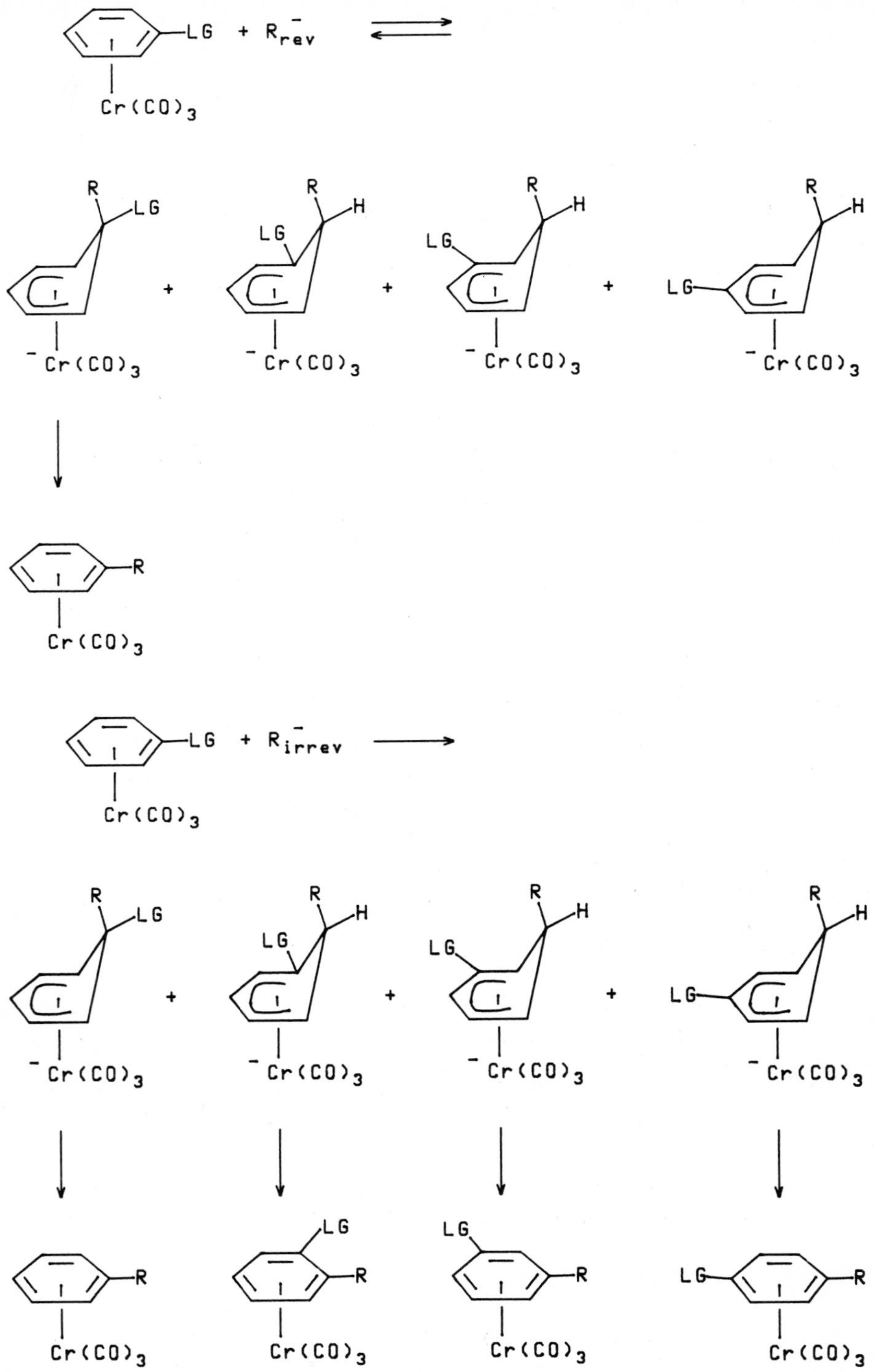

Figure 11.17

LG = Cl

R_{rev} = $CH(COOCH_3)_2$, $C(CH_3)_2CN$, CN, $C(CH_3)_2COOEt$, PhC(CN)(OR), $C(CH_3)_2COO^-$

R_{irrev} = CH_2COOCH_3, $C(CH_3)_2COOCH_3$, $CH_2COOt\text{-}Bu$, CH_2CN,

C≡CH, HC(S(CH₂)₃S) [1,3-dithian-2-yl]

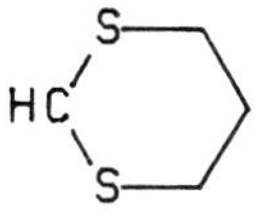

Li^+ is counterion in all cases.

	Yield (%)			
	ortho	meta	para	ipso
R_{irrev} = $CH_2COOt\text{-}Bu$	71	21	0	0
R_{rev} = $C(CH_3)_2CN$	2	56	0	19
$C(CH_3)_2CN$*	0	0	0	100

*At longer reaction time

Figure 11.17 (*Continued*)

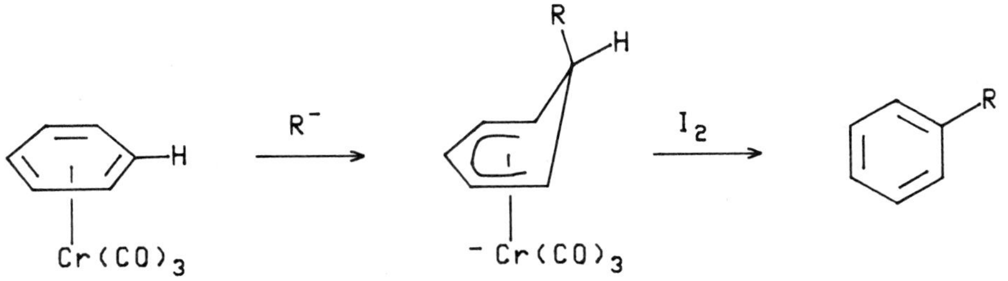

R_{appl} = $C(CH_3)_2CN$, CH_2CN, $CH[S(CH_2)_3S]$, C(CN)C(OR)R, $CH(SPh)_2$, $C(CH_3)_3$, $p\text{-}CH_3\text{-}C_6H_4$, CH_2COOR, $CH(CH_3)COOR$, $C(CH_3)_2COOR$

Li^+ is counterion in all cases.

Figure 11.18

TABLE 11.2 Regioselectivity of Nucleophilic Attack on η^6-(Monosubstituted benzene)-chromium Tricarbonyl Complexes[37–39]

Substituent X	RLi Carbination R	Product Ratios *Ortho*	*meta*	*para*	Combined Yield XArR (%)
CH_3	CH_2COO*t*-Bu	28	72	0	89
OCH_3	CH_2COO*t*-Bu	4	96	0	86
Cl	CH_2COOR^1	54	45	1	98
Cl	$CH(CH_3)COOR^1$	53	46	1	88
Cl	$C(CH_3)_2COOR^1$	5	95	1	84
Cl	CH_2COO*t*-Bu	70	24	0	87
CH_3	CH_2CN	35	63	2	88
t-Bu	CH_2CN	28	48	24	51
CH_3	$C(CH_3)_2CN$	1	97	2	95
t-Bu	$C(CH_3)_2CN$	0	55	45	78
CH(*t*-Bu)$_2$	$C(CH_3)_2CN$	0	0	100	63
$Si(CH_3)_3$	$C(CH_3)_2CN$	0	2	98	65
$N(CH_3)_2$	$C(CH_3)_2CN$	1	99	0	92
Cl	$C(CH_3)_2CN$	10	89	1	84
CH_3	$C(OR^2)(CH_3)CN$	0	96	4	75
CH_3CH_2	$C(OR^2)(CH_3)CN$	0	94	6	89
i-Pr	$C(OR^2)(CH_3)CN$	0	80	20	88
t-Bu	$C(OR^2)(CH_3)CN$	0	35	65	86
CH(*t*-Bu)$_2$	$C(OR^2)(CH_3)CN$	0	0	100	17
CF_3	$C(OR^2)(CH_3)CN$	0	30	70	33
CH_3	1,3-dithianyl	52	46	2	94
Cl	1,3-dithianyl	46	53	1	56
CH_3	CH_2SPh	52	46	2	96
i-Pr	CH_2SPh	47	46	7	86
t-Bu	CH_2SPh	45	32	23	88
	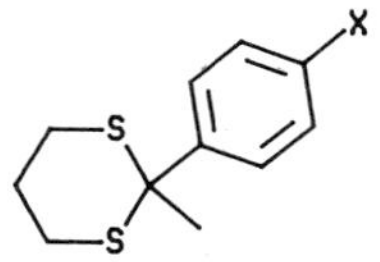 [2-methyl-2-(4-X-phenyl)-1,3-dithiane]				
t-Bu	$X = O^-$	0	73	27	84
t-Bu	$X = N(CH_3)_2$	0	63	37	80
t-Bu	$X = OCH_3$	0	54	46	71
t-Bu	$X = CH_3$	0	43	57	75
t-Bu	$X = H$	0	28	72	75
t-Bu	$X = Cl$	0	21	79	80

R^1 is presumably methyl or ethyl.
$R^2 = CH(CH_3)OCH_2CH_3$.

previously. EHT calculations lead to the conclusion that, in the absence of electronic effects from arene substituents, electrophiles will attack staggered arene carbons and nucleophiles will attack eclipsed arene carbons. This translates into nucleophilic attack *meta* when R is electron-releasing, *ortho–para* when R is electron-withdrawing, and *ortho–para* when R is bulky and electron-releasing.

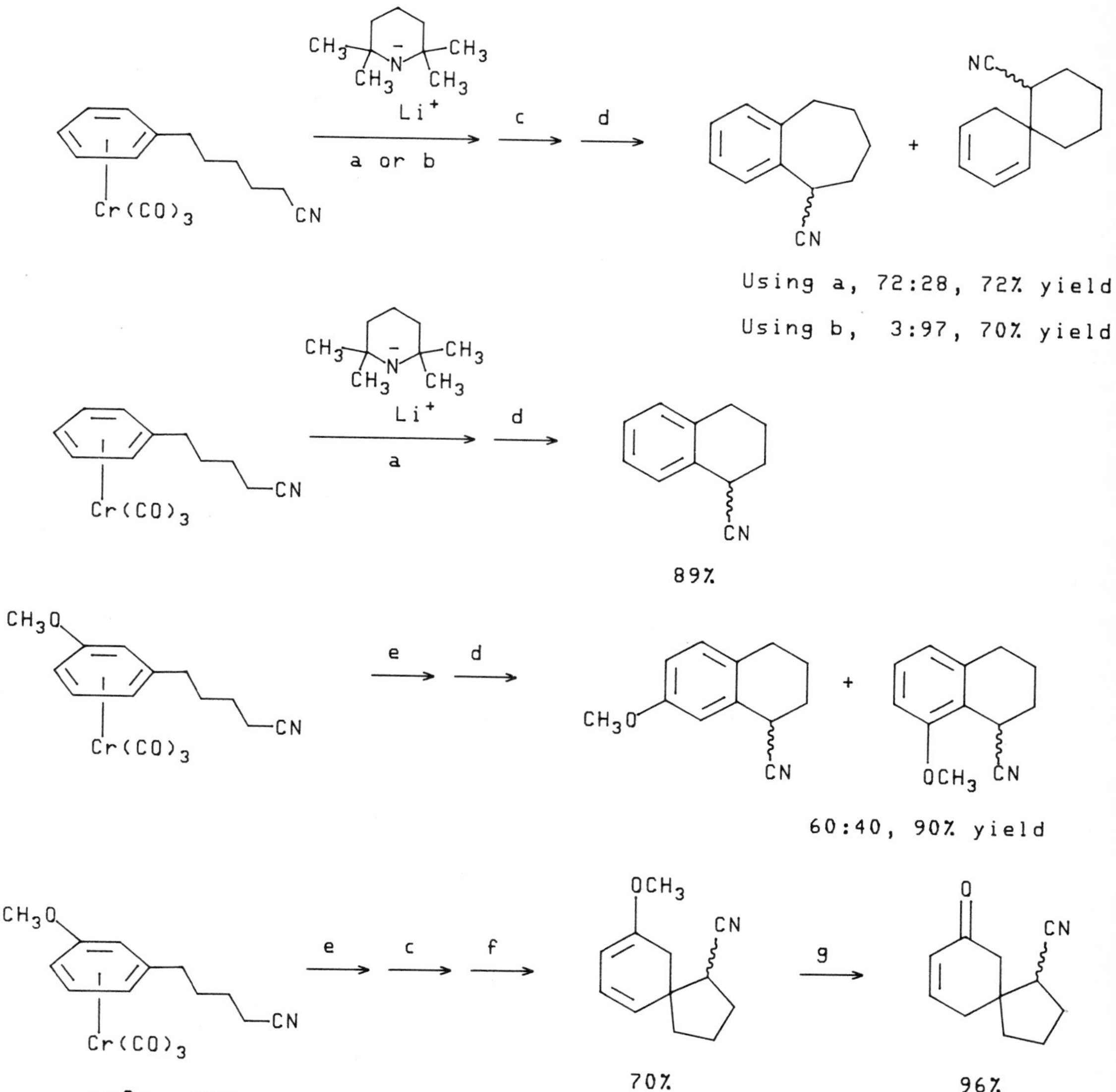

Using a, 72:28, 72% yield

Using b, 3:97, 70% yield

89%

60:40, 90% yield

70%

96%

a. $-78^{\circ}C$, THF

b. $0^{\circ}C$, THF

c. $-78^{\circ}C$, CF_3COOH

d. I_2

e. $0^{\circ}C$, $LiN(i\text{-}Pr)_2$

f. NH_4OH

g. $100^{\circ}C$, 5 M HCl, H_2O

Figure 11.19

This one-parameter analysis does not account for several additional patterns:

1. The important orbital interaction is between the lowest energy arene-centered unoccupied molecular orbital (UMO) and the highest occupied molecular orbital (HOMO) for the carbanion.
2. The HOMO for the anion is at a lower energy level than the UMO for the complex. A more electron-releasing aryl substituent on the nucleophile translates into a less stable anion and a higher energy nucleophile HOMO. Thus, with the less stable anion, the nucleophile HOMO and complex UMO are closer in energy, and frontier orbital overlap dominates addition. *Ortho/meta* addition is favored over *para*. Conversely, a more electron-withdrawing aryl substituent on the nucleophile translates into a more stable anion and a lower energy nucleophile HOMO. With the more stable anion the frontier orbital interaction is weak; charge control, neatly rationalized in terms of rotamer stability, dominates.
3. Intramolecular attack by carbanions followed by trap with acid and oxidative demetallation can produce annelated aromatics and/or spiro-fused cyclohexadienes. The length of the chain connecting arene with nucleophile, the temperature of anion generation and nucleophilic attack, and effects due to other ring substituents are all important factors in determining product ratios[40] (see Figure 11.19).
4. There is a difference in regioselectivity for *ortho* and *para* positions most clearly seen in reactions involving toluene and chlorobenzene complexes.
5. Regioselectivity changes considerably with changes in carbanion. A more recent rationalization is based on both orbital control and charge control. EHT calculations are consistent with orbital-controlled substitutions at the *ortho* and *meta* positions. This expanded analysis is best illustrated with the results obtained with the series of 2-(*para*-substituted)phenyl-2-lithio-1,3-dithianes.

As an alternative to oxidative demetallation, which converts the intermediate η^5-cyclohexadienyl anion to a substituted aromatic, reaction with electrophiles and

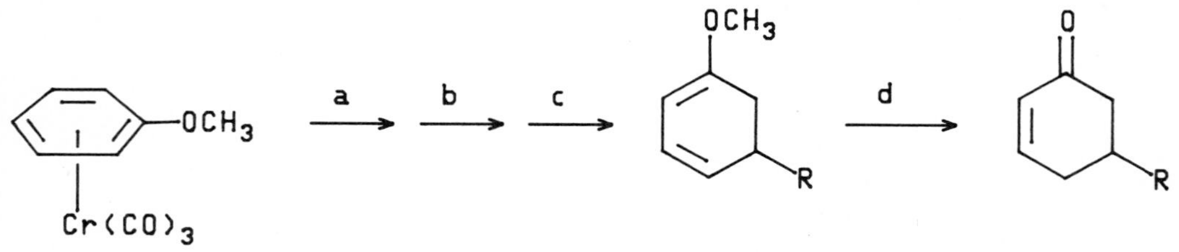

a. −78°C, RLi, THF

b. −78°C, CF_3COOH

c. NH_4OH

d. 100°C, HCl, H_2O

Figure 11.20

subsequent demetallation was recognized as a potential route to 5-nucleophile-6-electrophile-1,3-cyclohexadienes. Cleavage of the η^5-cyclohexadienyl anion complexes with acid and subsequent demetallation provides cyclohexadienes. η^6-Anisolechromium tricarbonyl is similarly converted to 5-nucleophile-cyclohex-2-en-1-one.[41] Some success with other electrophiles is achieved only after recognition of the importance of reversibility of nucleophilic attack[13] (see Figure 11.20).

When nucleophilic addition is reversible, reaction of the η^5-cyclohexadienyl anion with an alkylating agent affords only starting η^6-arenechromium tricarbonyl and alkylated nucleophile. When nucleophilic addition is irreversible (e.g., with methyldithianelithium, phenyllithium, or *t*-butyllithium), alkylation of the η^5-cyclohexadienyl anion and subsequent oxidative demetallation with iodine or simple ligand substitution with triphenylphosphine affords the targeted 4,5-disubstituted cyclohexadienes (see Figure 11.21).

Carbon monoxide insertion is most efficient when the reaction is carried out in a carbon monoxide atmosphere or in the presence of triphenylphosphine. Suitable alkylating agents are primary iodides and sulfates and allylic and benzylic bromides. Ethyl bromide reacts only on heating. The reactivity profile is consistent with an S_N2

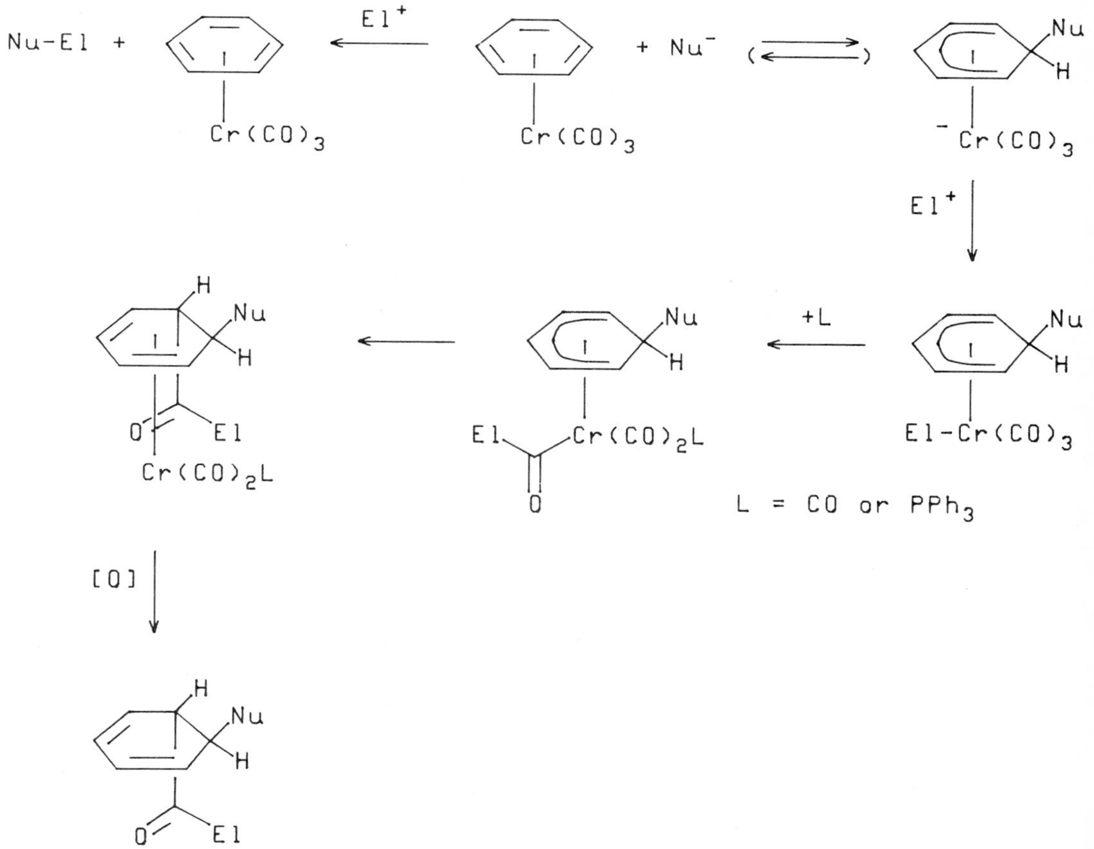

Nu-El + Cr(CO)3 ← El+ Cr(CO)3 + Nu− (⇌) Nu H −Cr(CO)3

El+ ↓

Nu H El-Cr(CO)3 → +L → Nu H El Cr(CO)2L O → Nu H H O El Cr(CO)2L

L = CO or PPh3

[O] ↓

H Nu H O El

Figure 11.21

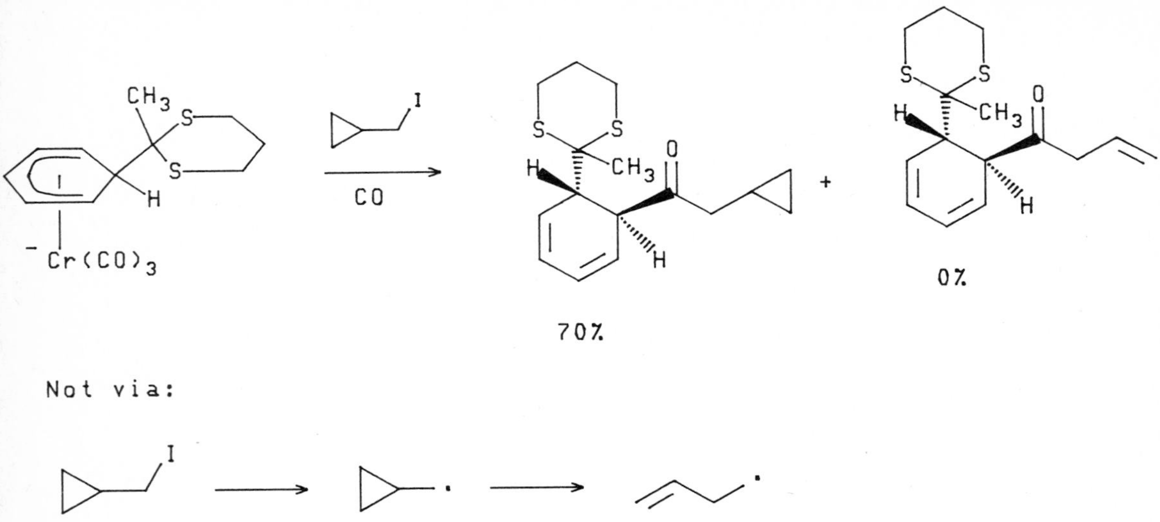

70%

0%

Not via:

Figure 11.22

displacement of leaving group by the metal anion:

$$\text{benzyl, allyl} > \text{primary} \gg \text{secondary}$$

$$\text{OTf} > \text{OMs} \approx \text{I} > \text{Br} \gg \text{Cl}$$

The large rate difference for iodides and bromides prompted an investigation of a possible change from a redical to a polar mechanism. No homoallyl product was isolated from the reaction of an η^5-cyclohexadienyl anion with cyclopropylmethyl iodide (see Figure 11.22).

11.2.8 Stereoselectivity Due to the Chromium Tricarbonyl Group

The stereoselective *exo*-mode of attack by nucleophiles on η^6-arenechromium tricarbonyl complexes has already been discussed. The steric bulk of chromium with its carbon monoxide ligands can also be utilized to effect an asymmetric synthesis on a ring side chain.[42]

Stereoisomers can be produced by complexation of certain arenes to chromium. Only single complexes are produced by complexation of benzene, monosubstituted benzenes and *para*-disubstituted benzenes. The eclipsed rotamers of these complexes all possess a plane of symmetry containing two ring carbons, the substituent atoms attached directly to the ring, chromium, and one carbon monoxide. Mirror images are rapidly interconverted by rotation of the chromium tricarbonyl moiety. Only single complexes are produced by complexation of *ortho*- and *meta*-disubstituted benzenes when the substituents are identical. While the eclipsed rotamers of these complexes have no plane of symmetry, the mirror images are rapidly interconverted by rotation of chromium tricarbonyl. Finally, complexation of *ortho*- and *meta*-disubstituted benzenes when the two substituents are different can produce pairs of enantiomers. The

Figure 11.23

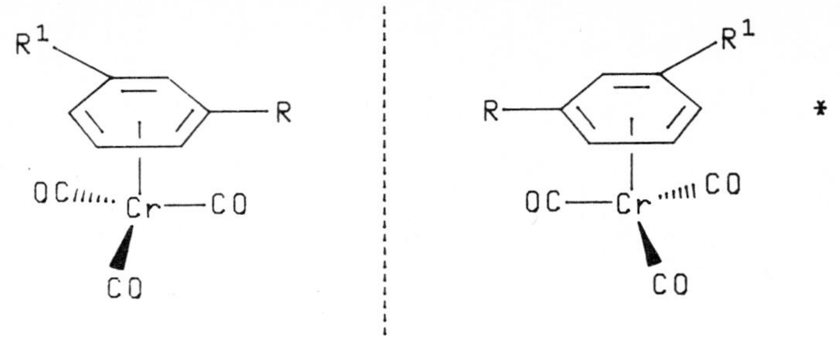

*Non-superimposable mirror images

Figure 11.23 (*Continued*)

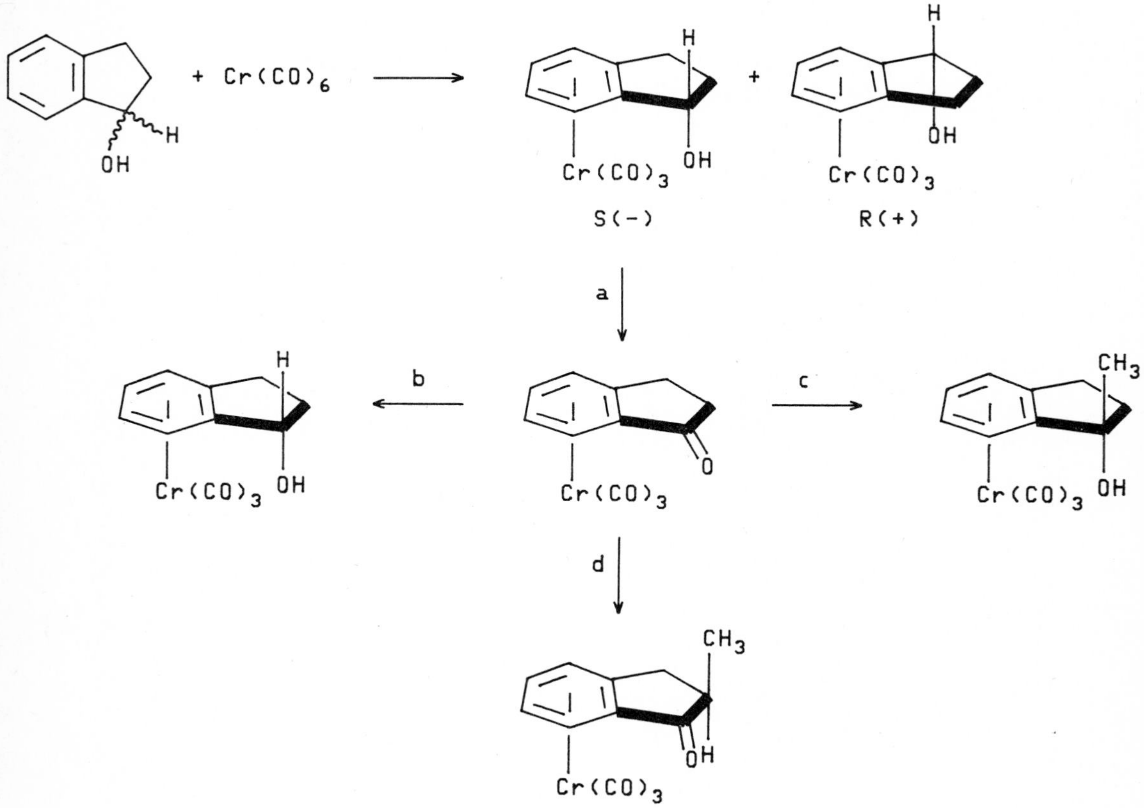

S(−)

R(+)

a. [O]

b. rfx, $LiAlH_4$, Et_2O

c. rt or rfx, CH_3MgBr, Et_2O

d. 1) rt, NaH, 2) CH_3I, DMF

Figure 11.24

eclipsed rotamers of these complexes have no plane of symmetry, and the mirror images cannot be interconverted by rotation of chromium tricarbonyl (see Figure 11.23).

To illustrate, complexation of racemic 1-indanol can produce as many as four diastereoisomers. In fact, complexation occurs exclusively on the same face as the hydroxyl group. The pair of diastereoisomeric complexes are separable. Oxidation of the (*S*)-isomer affords a single complex of indanone. Attack on the ketone by hydride reducing agents and Grignard reagents and alkylation of the corresponding enolate with methyl iodide all occur exclusively on the *exo* side opposite the metal[43–47] (see Figure 11.24).

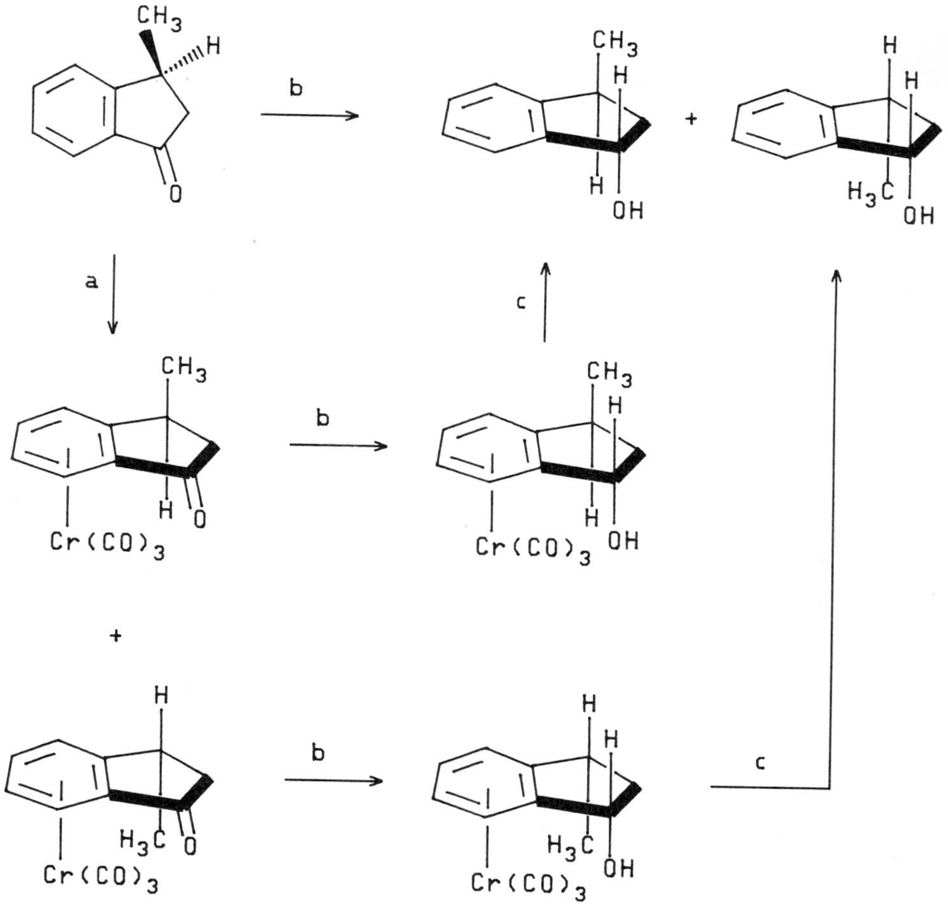

a. rfx, $Cr(CO)_6$, $(n\text{-}Bu)_2O$; separable mixture

b. rt, $NaBH_4$, i-PrOH

c. rt, hν, Et_2O

Figure 11.25

While little asymmetric induction is observed in hydride reduction of 3-methylindanone, complexation affords separable diastereoisomers that undergo stereospecific *exo* hydride reduction and demetallation en route to pure diastereoisomeric alcohols[48–51] (see Figure 11.25).

Stereoselectivity of nucleophilic attack on a carbonyl and alkylation of an enolate can be quite high in complexes of open-chain systems.[52–54] In addition, the reactivity of a complexed acetophenone can be elegantly attenuated by changing to chromium ligands with superior electron-donating capability[55] (see Figure 11.26).

11.3 TOTAL SYNTHESES OF (±)-ACORENONE AND (±)-ACORENONE B[56]

Several of the themes presented in the organometallic background section are illustrated in the total synthesis of (±)-acorenone and (±)-acorenone B:

1. The ease of complexation of electron-rich aromatics via method A
2. The regioselectivity of attack by a nucleophile on an η^6-anisolechromium tricarbonyl complex

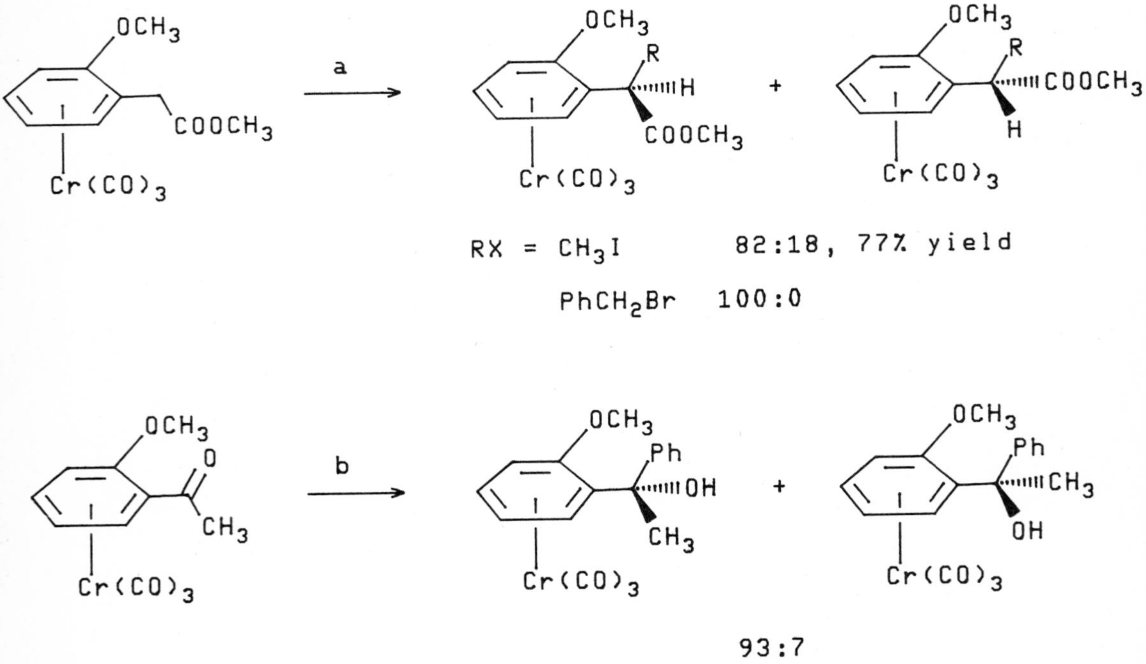

a. 1) rt, NaH, 2) RX, DMF

b. rt or rfx, PhMgBr, Et_2O

Figure 11.26

OCH_3 CH_3 —a→ CH_3O CH_3 $Cr(CO)_3$ —b→ (CN, CH_3, Li, CH_3, O, O, CH_3, CH_3)

R H CH_3O CH_3 $^-Cr(CO)_3$ —c→ OCH_3 CH_3 R —d→ —e→

OCH_3 CH_3 O CH_3 CH_3 —f→ —g→ OCH_3 OH CH_3 CH_3 CH_3

—h→ OCH_3 CH_3 * CH_3 CH_3 —i→ Br OCH_3 CH_3 * CH_3 CH_3

—j→ CN OCH_3 CH_3 * CH_3 CH_3 —k→

Figure 11.27

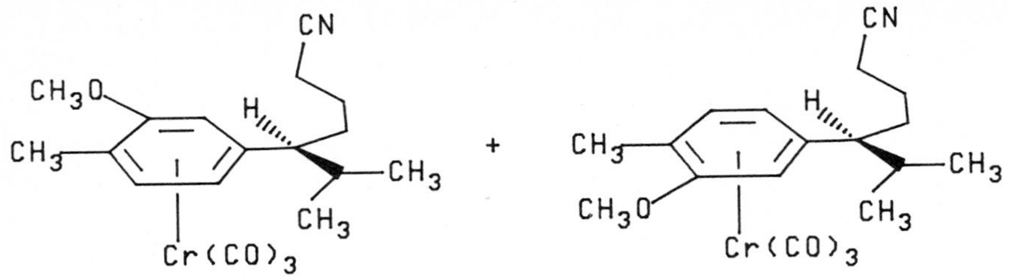

a. rfx, $Cr(CO)_6$, dioxane; (95%)

b. THF

c. I_2

d. rt, 5% H_2SO_4, H_2O, CH_3OH

e. rt, NaOH, H_2O; 77% for five steps

f. CH_2=$CHCH_2MgBr$

g. H_3O^+

h. Et_3SiH, CF_3COOH

i. hν, HBr

j. KCN; 77% for four steps

k. 1) rfx, $Cr(CO)_6$, dioxane, 2) rt, CO, THF;

A:50%, B:34%

Figure 11.27 *(Continued)*

3. The *exo*-stereospecific nucleophilic attack on an η^6-arenechromium tricarbonyl complex
4. The recovery of nucleophile-substituted arene from an η^5-cyclohexadienyl anionic complex by oxidation–oxidative demetallation using iodine
5. The generation of separable diastereoisomeric arenechromium tricarbonyl complexes from an arene with a chiral center in the side chain
6. The addition of hexamethylphosphoric triamide to facilitate nucleophilic attack by a nitrile-stabilized carbanion on an η^6-arene complex
7. The formation of a 5-nucleophile-cyclohex-2-en-1-one by nucleophilic attack on an η^6-anisole complex followed by quench with acid, demetallation with ammonium hydroxide, and aqueous acid hydrolysis

Heating *o*-methylanisole[57] with chromium hexacarbonyl in dioxane affords the η^6-arene complex in excellent yield (95%). Regioselective *meta* attack by methoxycyano-hydrin acetal anion[58] and subsequent oxidation–oxidative demetallation with iodine affords the nucleophile-substituted aromatic, contaminated with only a trace amount of a regioisomer. Pure material is available from a careful fractional distillation. The cyanohydrin acetal is converted to the corresponding ketone by acetal deprotection with aqueous acid, then cyanohydrin deprotection with aqueous base. Reaction with allyl Grignard is followed by reduction of the tertiary-benzylic alcohol by "ionic hydrogenation".[59] Hydrogen bromide is added in anti-Markovnikov fashion to the alkene. The primary bromide is replaced (S_N2) by cyanide. The aromatic, now with a chiral center in the side chain, is again complexed with chromium, affording a separable 60:40 mixture of diastereoisomers in 84% yield after treatment with high pressure of carbon monoxide to cleave a nitrile–$Cr(CO)_5$ complex (see Figure 11.27).

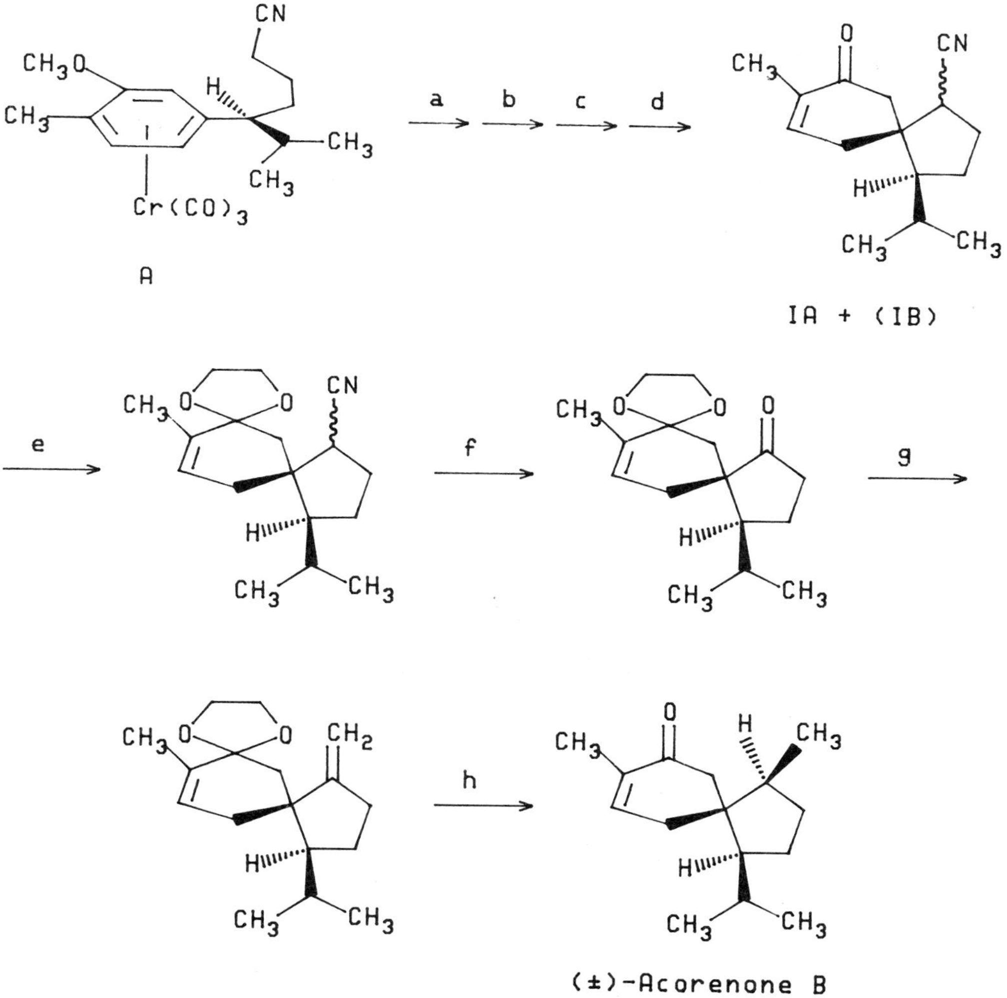

A

IA + (IB)

(±)-Acorenone B

Figure 11.28

a. −78°C, $LiN(i\text{-}Pr)_2$, HMPA, THF

b. −78°C, CF_3SO_3H, THF

c. −30°C, NH_4OH, Et_2O, THF

d. 80°C, HCl, H_2O, CH_3OH; IA:40%, IB:5% for four steps

e. H^+, $HOCH_2CH_2OH$

f. 1) −78°C, $LiNEt_2$, 2) O_2, THF, 3) $(CH_3)_2S$, 4) aqueous base

g. $Ph_3P{=}CH_2$

h. aqueous acid

i. H_2, $(Ph_3P)_3RhCl$; 45% for four steps from IA

Figure 11.28 *(Continued)*

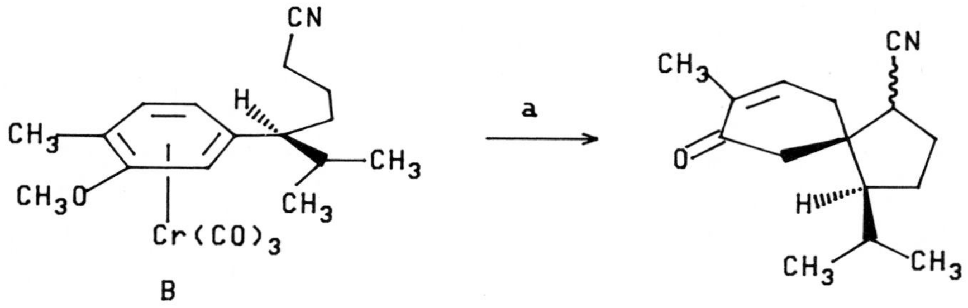

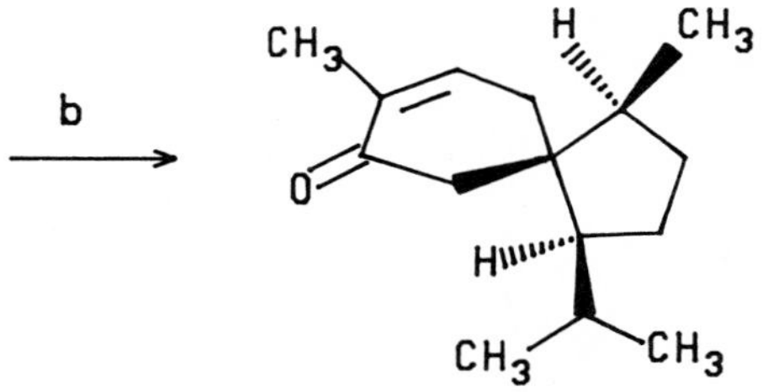

(±)-Acorenone

a. Same as for A, Steps a–d; 15% for four steps

b. Same as for A, Steps e–h; 40% for four steps

Figure 11.29

The major diastereoisomer (A) is converted to the nitrile-stabilized carbanion with lithium diisopropylamide in tetrahydrofuran at −78 °C. Addition of hexamethylphosphoric triamide solvates the lithium cation, increasing the reactivity of the carbanion, and thus facilitates a regioselective nucleophilic attack on the complexed arene. After 4 h at −78 °C, trifluoromethanesulfonic acid is added to trap the desired η^5-cyclohexadienyl anion as the spiro-fused cyclohexadiene. Demetallation by workup with concentrated aqueous ammonium hydroxide affords the enol ether, which is then hydrolyzed with acid. Careful chromatographic separation affords noncyclized arene (30%), fused ring isomers (20%), and two spiro-cyclohexanones (40 and 5%). The spiro-cyclohexanones differ only in the orientation of the nitrile substituent relative to the isopropyl; both are converted to the same ketone using an oxidative decyanation procedure.[60] Wittig olefination is followed by ketal deprotection. Finally, stereospecific catalytic hydrogenation of the *exo*-methylene using Wilkinson's catalyst affords (±)-acorenone B (see Figure 11.28).

The minor diastereoisomer (B) is converted to (±)-acorenone by an analogous sequence. Spirocyclization is more stereoselective but considerably less efficient.

(±)-Acorenone B was prepared from *o*-methylanisole in 15 steps in overall 5.3% yield. The preparation and purification of diastereoisomeric complex A is the crucial efficiency-limiting operation. (±)-Acorenone was prepared from *o*-methylanisole in 15 steps in overall 1.2% yield. In this case both the preparation of diastereoisomeric complex B and spirocyclization are efficiency limiting. Stereospecific syntheses of A and B would improve both syntheses significantly (see Figure 11.29).

REFERENCES

1. Vrkoc, J.; Herout, V.; Sorm, F. *Coll. Czech. Chem. Commun.* **1961**, *26*, 3183.
2. McClure, R. J.; Schorno, K. S.; Bertrand, J. A.; Zalkow, L. H. *Chem. Commun.* **1968**, 1135.
3. Marshall, J. A.; Brady, F.; Andersen, N. H. *Fortschr. Chem. Org. Naturst.* **1974**, *31*, 283.

4a. Iwata, C.; Nakamura, S.; Shinoo, Y.; Fusaka, T.; Kishimoto, M.; Uetsuji, H.; Maezaki, N.; Tanaka, T. *J. Chem. Soc., Chem. Commun.* **1984**, 781.

4b. Baldwin, S. W.; Fredericks, J. E. *Tetrahedron Lett.* **1982**, *23*, 1235.

4c. Lange, G. L.; Neidert, E. E.; Orrom, W. J.; Wallace, D. J. *Can. J. Chem.* **1978**, *56*, 1628.

4d. Lange, G. L.; Orrom, W. J.; Wallace, D. J. *Tetrahedron Lett.* **1977**, 4479.

4e. Pesaro, M.; Bachman, J-P. *J. Chem. Soc., Chem. Commun.* **1978**, 203.

4f. Naegeli, P. *Tetrahedron Lett.* **1978**, 2127.

4g. Oppolzer, W.; Mahalanabis, K. K.; Bättig, K. *Helv. Chim. Acta* **1977**, *60*, 2388.

4h. Oppolzer, W.; Mahalanabis, K. K. *Tetrahedron Lett.* **1975**, 3411.

4i. Rascher, W.; Wolf, H. *Tetrahedron* **1977**, *33*, 575.

4j. Wolf, H.; Kolleck, M.; Rascher, W. *Chem. Ber.* **1976**, *109*, 2805.

4k. Wolf, H.; Kolleck, M. *Tetrahedron Lett.* **1975**, 451.

4l. Trost, B. M.; Hiroi, K.; Holy, N. *J. Am. Chem. Soc.* **1975**, *97*, 5873.

5. Mahaffy, C. A. L.; Pauson, P. L. *Inorg. Syn.* **1979**, *19*, 154.
6. Nicholls, B.; Whiting, M. C. *J. Chem. Soc.* **1959**, 551.

7. *Comprehensive Organometallic Chemistry*; Wilkinson, G., Ed.; Vol. 3, Pergamon: New York, 1982; pp 1001–1021.
8. Rausch, M. D. *J. Org. Chem.* **1974**, *39*, 1787.
9. Deckelmann, K.; Werner, H. *Helv. Chim. Acta* **1970**, *53*, 139.
10. Mahaffy, C. A. L.; Pauson, P. L. *J. Chem. Res. (S)* **1979**, 126.
11. Semmelhack, M. F. *Ann. N.Y. Acad. Sci.* **1977**, *295*, 36.
12. Semmelhack, M. F. *Pure Appl. Chem.* **1981**, *53*, 2379.
13. Kündig, E. P. *Pure Appl. Chem.* **1985**, *57*, 1855.
14. Albright, T. A.; Carpenter, B. K. *Inorg. Chem.* **1980**, *19*, 3092.
15. Semmelhack, M. F.; Bisaha, J.; Czarny, M. *J. Am. Chem. Soc.* **1979**, *101*, 768.
16. Knox, G. R.; Leppard, D. G.; Pauson, P. L.; Watts, W. E. *J. Organometal. Chem.* **1972**, *34*, 347.
17. Jaouen, G.; Meyer, A.; Simonneaux, G. *J. Chem. Soc., Chem. Commun.* **1975**, 813.
18. Holmes, J. D.; Jones, D. A. K.; Pettit, R. *J. Organometal. Chem.* **1965**, *4*, 324.
19. Gubin, S. P.; Khandkarova, V. S.; Kreindlin, A. Z. *J. Organometal. Chem.* **1974**, *64*, 229.
20. Ceccon, A. *J. Organometal. Chem.* **1971**, *29*, C19.
21. Davis, R. E.; Simpson, H. D.; Grice, N.; Pettit, R. *J. Am. Chem. Soc.* **1971**, *93*, 6688.
22. Bly, R. S.; Mateer, R. A.; Tse, K-K.; Veazey, R. L. *J. Org. Chem.* **1973**, *38*, 1518.
23. Bly, R. S.; Maier, T. L. *J. Org. Chem.* **1978**, *43*, 614.
24. Bly, R. S.; Strickland, R. C. *J. Am. Chem. Soc.* **1970**, *92*, 7459.
25. Herberich, G. E.; Fischer, E. O. *Chem. Ber.* **1962**, *95*, 2803.
26. Jackson, W. R.; Jennings, W. B. *J. Chem. Soc. (B)* **1969**, 1221.
27. Magomedov, G. K.-I.; Syrkin, V. G.; Frenkel', A. S. *J. Gen. Chem. USSR* **1972**, *42*, 2443.
28a. Magomedov, G. K.-I.; Syrkin, V. G.; Frenkel', A. S.; Zakharchenko, O. A. *J. Gen. Chem. USSR* **1975**, *45*, 2484.
28b. Magomedov, G. K.-I.; Syrkin, V. G.; Frenkel', A. S.; Nekrasov, Yu. S.; Zavina, T. A. *J. Gen. Chem. USSR* **1975**, *45*, 2487.
29. Razuvaev, G. A.; Petuchov, G. G.; Artemov, A. N.; Sirotkin, N. I. *J. Organometal. Chem.* **1972**, *37*, 313.
30. Brown, D. A.; Raju, J. R. *J. Chem. Soc. (A)* **1966**, *40*, 1617.
31. Bunnett, J .F.; Hermann, H. *J. Org. Chem.* **1971**, *36*, 4081.
32. Rosca, S. I.; Rosca, S. *Rev. Chim.* **1974**, *25*, 461.
33. Mahaffy, C. A. L.; Pauson, P. L. *J. Chem. Res. (S)* **1979**, 128.
34. Semmelhack, M. F.; Hall, H. T. *J. Am. Chem. Soc.* **1974**, *96*, 7091 and 7092.
35. Semmelhack, M. F.; Hall, H. T.; Yoshifuji, M.; Clark, G. *J. Amer. Chem. Soc.* **1975**, *97*, 1247.
36. Semmelhack, M. F.; Hall, H. T., Jr.; Farina, R.; Yoshifuji, M.; Clark, G.; Bargar, T.; Hirotsu, K.; Clardy, J. *J. Am. Chem. Soc.* **1979**, *101*, 3535.
37. Semmelhack, M. F.; Clark, G. *J. Am. Chem. Soc.* **1977**, *99*, 1675.
38. Semmelhack, M. F.; Clark, G. R.; Farina, R.; Saeman, M. *J. Am. Chem. Soc.* **1979**, *101*, 217.
39. Semmelhack, M. F.; Garcia, J. L.; Cortes, D.; Farina, R.; Hong, R.; Carpenter, B. K. *Organometallics* **1983**, *2*, 467.
40. Sommelhack, M. F.; Thebtaranonth, Y.; Keller, L. *J. Am. Chem. Soc.* **1977**, *99*, 959.
41. Semmelhack, M. F.; Harrison, J. J.; Thebtaranonth, Y. *J. Org. Chem.* **1979**, *44*, 3275.
42. For an analysis of the general question of asymmetric induction in reactions of η^6-arene chromium tricarbonyl complexes, see Besancon, J.; Tainturier, G.; Tirouflet, J. *Bull. Soc. Chim. Fr.* **1971**, 1804.

43. Falk, H.; Schlögl, K.; Steyrer, W. *Monatsch. Chem.* **1966**, *97*, 1029.
44. Jackson, W. R.; Mitchell, T. R. B. *J. Chem. Soc. (B)* **1969**, 1228.
45. Jaouen, G.; Dabard, R. *Tetrahedron Lett.* **1971**, 1015.
46. Jaouen, G.; Caro, B.; Le Bihan, J. Y. *C.R. Hebd. Seances Acad. Sci.* **1972**, *274*, 902.
47. Meyer, A.; Jaouen, G. *J. Chem. Soc., Chem. Commun.* **1974**, 787.
48. Jaouen, G.; Meyer, A. *J. Am. Chem. Soc.* **1975**, *97*, 4667.
49. Caro, B.; Jaouen, G. *J. Chem. Soc., Chem. Commun.* **1976**, 655.
50. Caro, B.; Gentric, E.; Grandjean, D.; Jaouen, G. *Tetrahedron Lett.* **1978**, 3009.
51. Gentric, E.; Le Borgne, G.; Grandjean, D. *J. Organometal. Chem.* **1978**, *155*, 207.
52. des Abbayes, H.; Boudeville, M-A. *J. Org. Chem.* **1977**, *42*, 4104.
53. Meyer, A.; Dabard, R. *J. Organometal. Chem.* **1972**, *36*, C38.
54. Besancon, J.; Tirouflet, J.; Card, A.; Dusausoy, Y. *J. Organometal. Chem.* **1973**, *59*, 267.
55. Boyer, B.; Lamaty, G.; Caro, B.; Jaouen, G. *J. Organometal. Chem.* **1981**, *216*, C51.
56. Semmelhack, M. K.; Yamashita, A. *J. Am. Chem. Soc.* **1980**, *102*, 5924.
57. 2-Methylanisole is commercially available (Aldrich 88–89 catalog, 100 g for $48.50).
58. Stork, G.; Maldonado, L. *J. Am. Chem. Soc.* **1971**, *93*, 5286.
59. Kursanov, D. N.; Parnes, Z. N.; Loim, N. M. *Synthesis* **1974**, 633.
60. Selikson, S. J.; Watt, D. S. *J. Org. Chem.* **1975**, *40*, 267.

CHAPTER 12

(±)-Daunomycinone and (±)-Adriamycinone

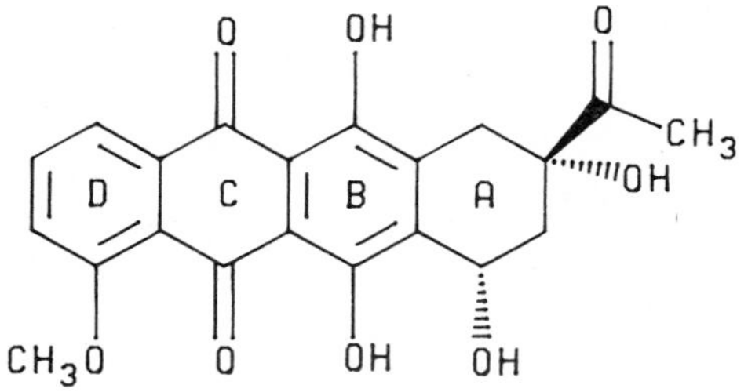

Daunomycinone, Daunorubicinone,

Leukaemomycinone C

5,12-Naphthacenedione,8-acetyl-7,8,9,10-tetrahydro-6,8,10,11-tetrahydroxy-1-methoxy

(8S-cis)[21794-55-8]

cis(±)[59367-20-3]

Figure 12.1

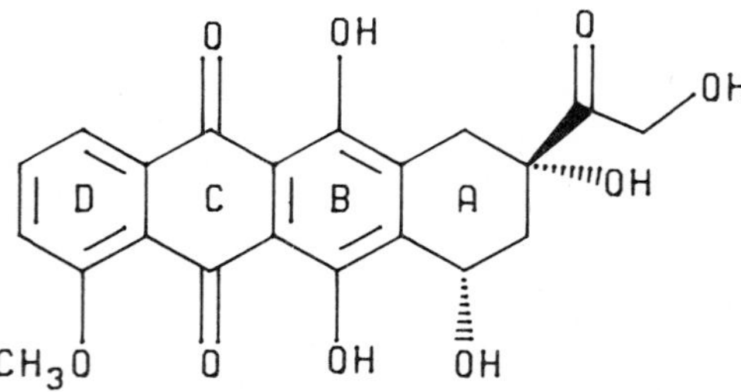

Adriamycinone, Doxorubicinone,

5,12-Naphthacenedione,7,8,9,10-tetrahydro-6,8,10,11-tetrahydroxy-8-(hydroxyacetyl)-1-methoxy

(8S-cis)[24385-10-2]

cis(±)[89164-17-0]

Figure 12.2

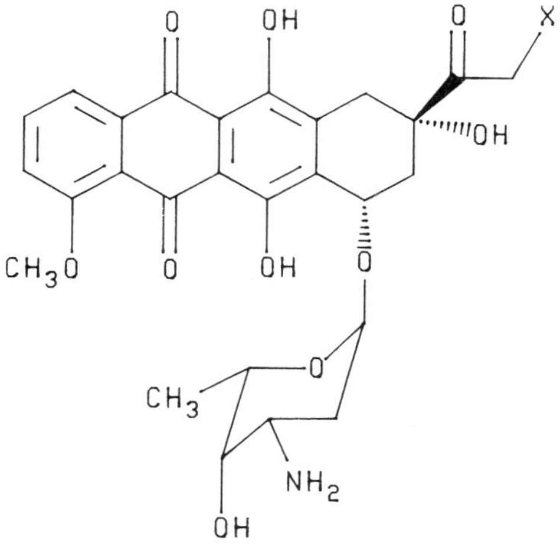

X = H, Daunomycin

OH, Adriamycin

Figure 12.3

Daunomycinone and adriamycinone are the aglycones of the potent anthracycline antitumor antibiotics daunomycin and adriamycin. Daunomycinone was first isolated from *Streptomyces peucetius* and shown to have antileukemic activity in 1964.[1] Adriamycin was subsequently isolated from mutant strains of *S. peucetius*[2,3] (see Figure 12.3).

Daunomycin is the most active agent for the treatment of acute nonlymphocytic leukemia, has significant activity against childhood and adult acute lymphocytic leukemia, and some measurable activity against refractory or recurrent neuroblastoma. Adriamycin has an uncommonly broad spectrum of activity including activity against "hematologic malignancies, metastables or inoperable sarcomas, breast and lung cancer, some genitourinary tumors, hepatoma, and a variety of childhood solid tumors."[4] In common with other antineoplastic agents, both are "profoundly toxic to bone marrow, mucous membranes, and hair follicles."[4] A more unusual and serious problem is irreversible cardiac toxicity resulting in congestive heart failure. Fortunately, there is now some evidence that this risk may be reduced by modification of treatment schedule and dosage.[5]

While the mechanisms of antitumor activity of daunomycin and adriamycin remain uncertain, three possibilities have been extensively explored: DNA intercalation, free radical formation via a metabolic reduction, and some direct mechanism of action involving the cell membrane.[6] Of these, DNA intercalation and metabolic reduction provide a framework for understanding activity changes with structural modification of the anthracycline aglycone.

DNA Intercalation. Most of the adriamycin within a treated cell is bound to DNA. Binding may involve insertion of the planar tetracyclic framework into the DNA helix parallel to the base pairs, resulting in some local unwinding of the sugar phosphate backbone.[7] Insertion impairs the ability of DNA to act as a template for DNA and RNA polymerases and induces some single-strand scission, perhaps mediated by oxygen radicals.

Free Radical Formation. Oxygen radicals are key intermediates in a pathway leading from one electron reduction of anthracyclines to cell destruction. The anthracycline radical converts oxygen to superoxide, superoxide produces hydroxyl radicals and hydrogen peroxide, and hydroxyl radicals destroy cells by DNA strand scission.[8] The strand scission chain can be blocked by blocking hydroxyl radical/hydrogen peroxide availability using superoxide dismutase, catalase, or sodium benzoate.[9]

The significant antitumor activity, a developing "handle" on an aglycone structure–anthracycline activity relationship, and the clinical utilization of anthracyclines have prompted several total synthesis of both daunomycinone[10] and adriamycinone.[11]

12.1 CARBENE COMPLEXES

Since the first transition metal carbene complex was prepared by E. O. Fischer in 1964,[12a] the interest in synthetic utilization of carbene complexes as replacements for free carbenes has led to the synthesis and study of carbene complexes of nearly all the transition metals. Ironically, the chemistry of carbene complexes studied to date bears little resemblance to that of free carbenes; the metal plays a crucial mechanistic role in almost every case. Carbene complexes can be divided into two general categories: the Fischer-type carbenes with an electrophilic carbene carbon (metals from Groups VI to VIII in low oxidation states) and the Schrock-type carbenes with a nucleophilic carbene carbon (early transition metals) (see Figure 12.4).

Three different processes employing carbene complexes have been exploited synthetically: cocyclization of Fischer-type carbenes with alkynes to give hydroquinones (this chapter), cocyclization of Fischer-type carbenes with imines to give β-lactams, and alkene metathesis using Schrock-type carbenes (Chapter 13).

12.2 PENTACARBONYLCHROMIUM CARBENES: PREPARATION AND ELABORATION

The first and still most common method for preparation of heteroatom-stabilized chromium carbenes is by nucleophilic attack on coordinated carbon monoxide of

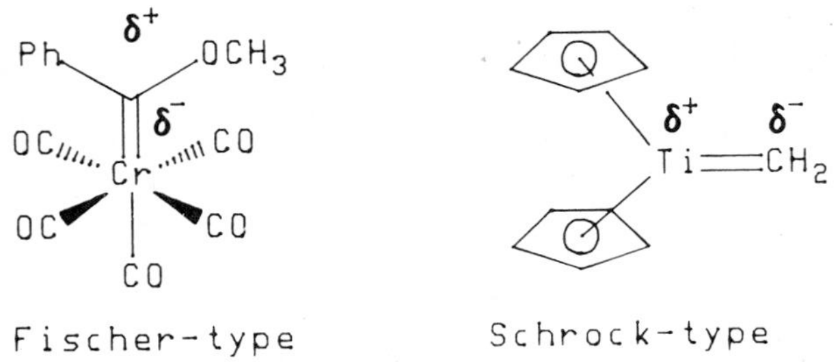

Figure 12.4

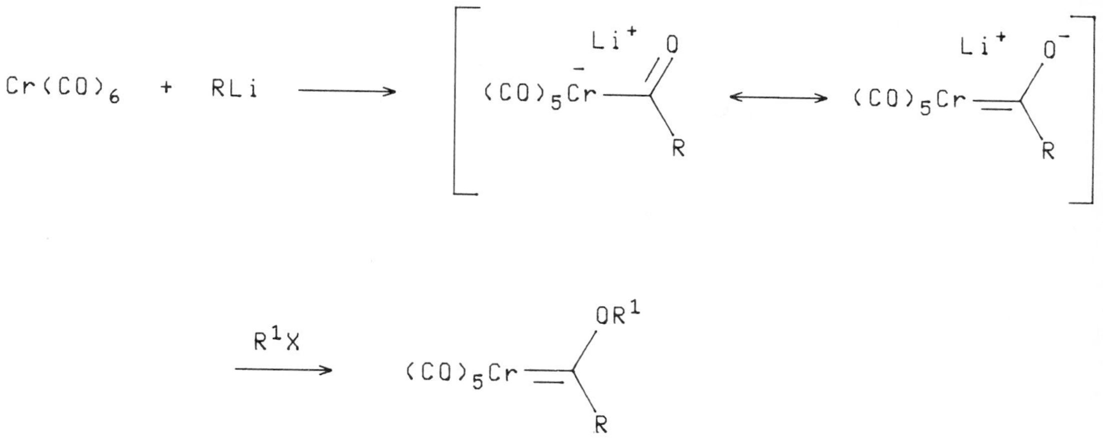

$$Cr(CO)_6 + RLi \longrightarrow \left[(CO)_5\bar{Cr}\text{—}C(=O)R \;\; Li^+ \longleftrightarrow (CO)_5Cr{=}C(O^-)R \;\; Li^+ \right] \xrightarrow{R^1X} (CO)_5Cr{=}C(OR^1)R$$

Figure 12.5

chromium hexacarbonyl followed by trap of the acylmetallate with an alkylating agent. Organolithium reagents are suitable nucleophiles; oxonium salts or fluoroalkanesulfonates alkylate the acylmetallate most efficiently[12] (see Figure 12.5).

More highly functionalized carbene complexes can be prepared from these simple carbenes by several methods: nucleophilic attack on the carbene carbon with replacement of alkoxy, nucleophilic attack on the carbene carbon with insertion of the nucleophile, generation of a carbene α-anion and capture by electrophiles, Michael addition to vinyl carbenes, and Diels–Alder cycloaddition using vinyl or alkynyl carbenes as dienophiles.

12.2.1 Nucleophilic Attack on the Carbene Carbon with Replacement of Alkoxy

Nucleophilic attack on pentacarbonylchromium carbenes could occur either on the carbon of a coordinated carbon monoxide or on the carbene carbon. In fact, attack by amines,[13–17] mercaptans,[13,18] and organolithium reagents[19] occurs on the carbene carbon. This is not due to charge control but rather to frontier orbital control: Casey's "molecular orbital calculations on $(CO)_5CrC(OCH_3)CH_3$ reveal that the LUMO is energetically isolated and spatially localized on the carbene carbon atom."[20] Replacement of alkoxy efficiently produces new carbene complexes under quite mild conditions (see Figure 12.6).

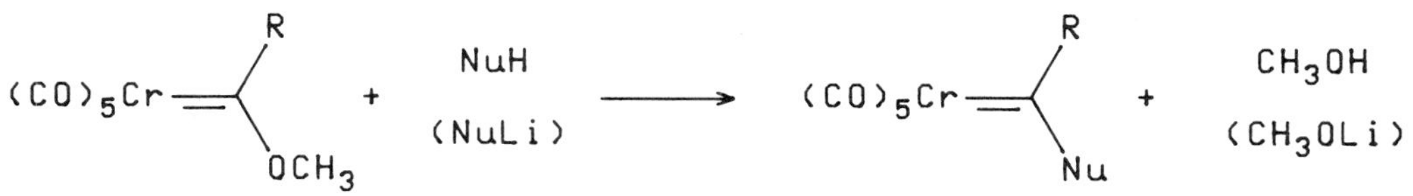

$$(CO)_5Cr{=}C(R)(OCH_3) + \underset{(NuLi)}{NuH} \longrightarrow (CO)_5Cr{=}C(R)(Nu) + \underset{(CH_3OLi)}{CH_3OH}$$

Figure 12.6

12.2.2 Nucleophilic Attack on the Carbene Carbon with Insertion of the Nucleophile

Attack by multiply bonded nucleophilic systems such as ynamines,[21,22] cyanamides,[23] and enamines[24] also initially occurs on the carbene carbon. Instead of alkoxy elimination, the adducts ring close to metallocycles. Ring opening then produces the observed insertion products. With ynamines, the alkenyl carbene is formed exclusively in the *E*-configuration in a highly stereospecific process. Insertion of enamines is considerably less efficient due, in part, to competing metallocycle reductive elimination to cyclopropanes (see Figure 12.7).

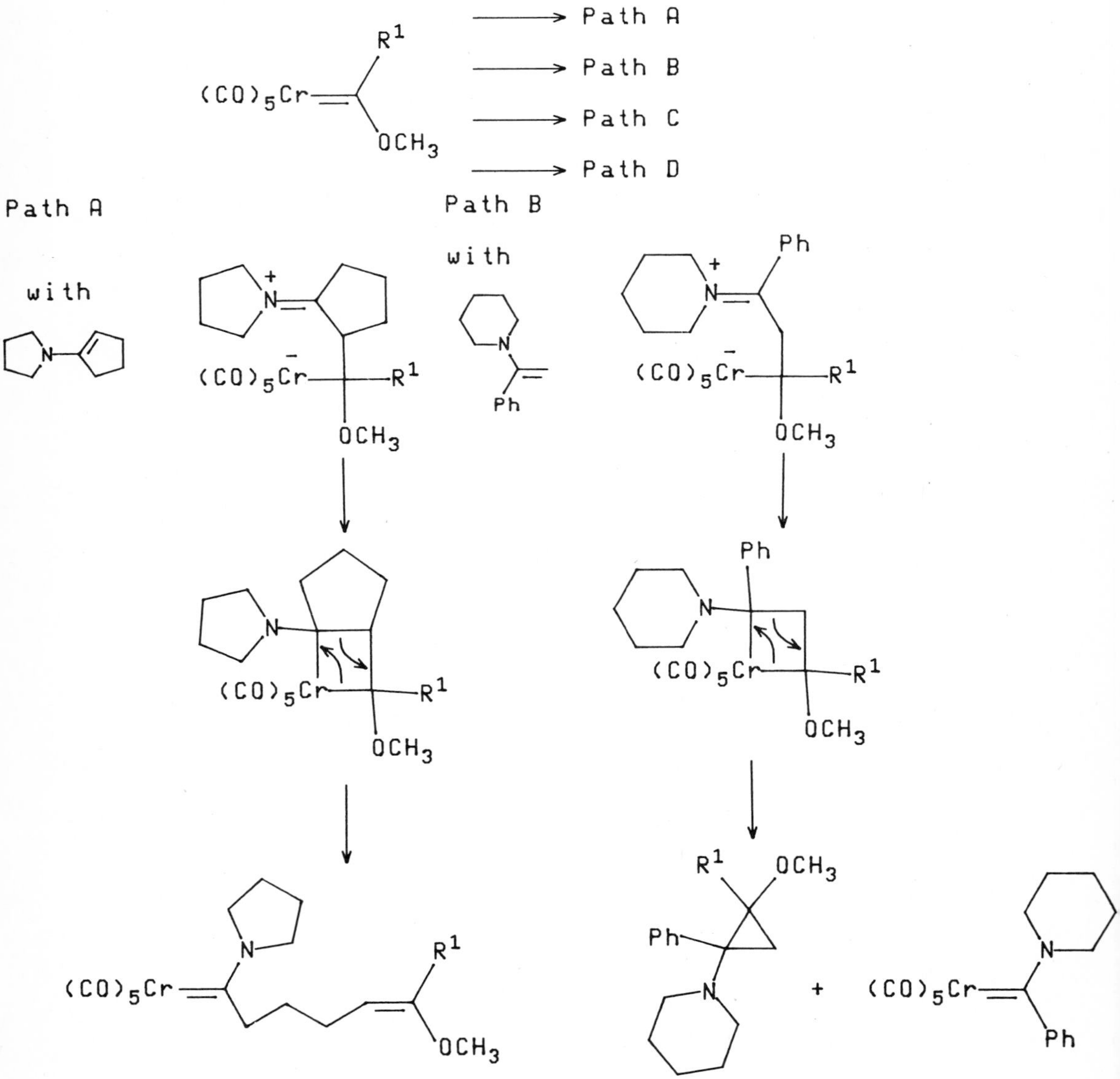

Figure 12.7

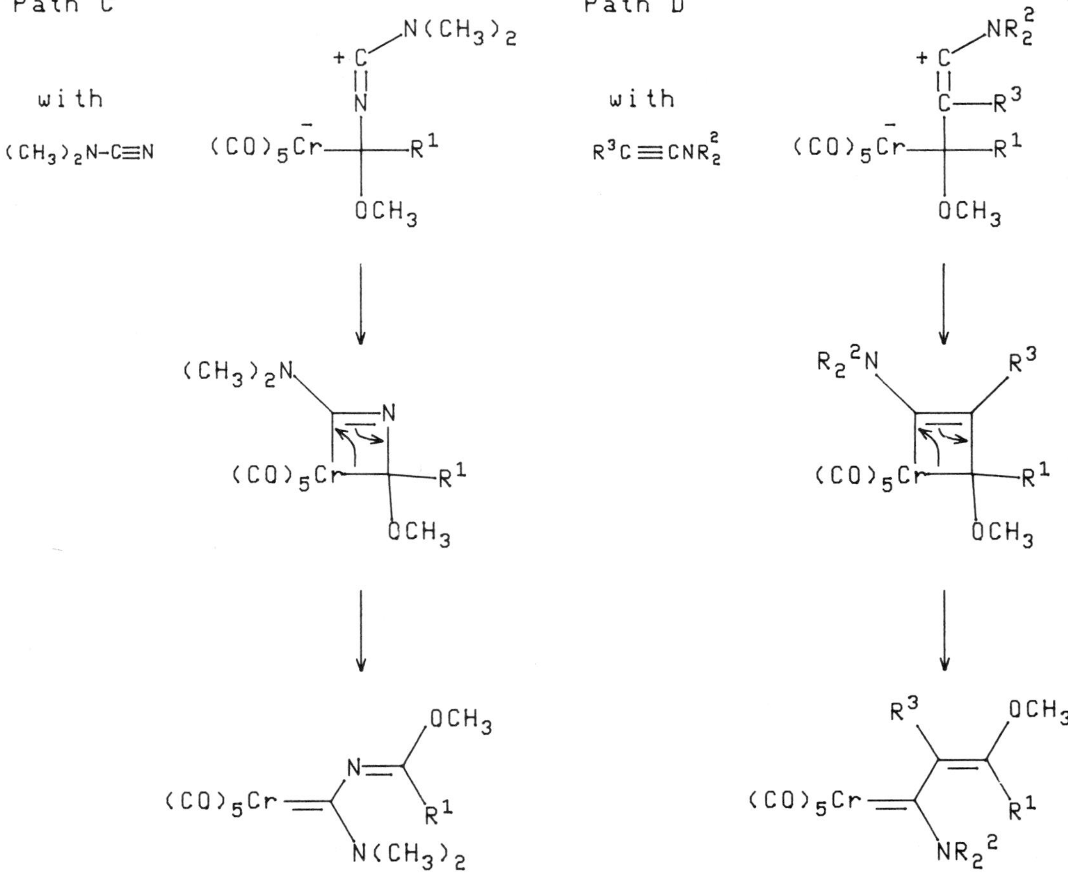

Figure 12.7 (*Continued*)

12.2.3 Generation of a Carbene α-Anion and Capture by Electrophiles

Protons α to the carbene carbon are acidic; the resulting α-anion is stabilized by charge delocalization onto the metal. For example, protons on the methyl group of (methylmethoxycarbene)pentacarbonylchromium(0) have a pKa of approximately 8; this is one of the most acidic methyl groups known.[25] The highly stabilized anion is a weak nucleophile only capable of synthetically useful alkylation by a few potent electrophiles, in particular, epoxides,[26] α-bromoesters,[26] α-chloroethers,[27] aldehydes,[27] and Michael acceptors.[28] For example, reaction with ethylene oxide produces an alkoxide capable of intramolecular displacement of alkoxy from the carbene carbon. The (2-oxacyclopentylidene)pentacarbonylchromium(0) produced can be converted to an α-anion with *n*-butyllithium, then (1) alkylated with chloromethyl methyl ether, (2) condensed with aldehydes, or (3) alkylated with methyl vinyl ketone. The ratio of monoalkylated to dialkylated products can be contrôlled to some extent by reaction conditions (see Figure 12.8).

W. D. Wulff has looked at two approaches to increasing the versatility of the carbene functionalization methodology: (1) increase the reactivity of the electrophiles and (2) increase the reactivity (i.e., decrease the stability) of the carbene α-anion. Aldehydes, ketones, and acetals undergo aldol condensation with carbene α-anions when activated by Lewis acid complexation. Ketone condensation is most efficient with boron trifluoride etherate; aldehyde condensation is most efficient with titanium tetrachloride. A diastereoselective condensation is observed using (methoxyethylcarbene)pentacarbonylchromium(0) and benzaldehyde-titanium tetrachloride. The 6.0:1.0 mixture of *anti* and *syn* isomers obtained does not necessarily represent diastereoselectivity limits; stereoselectivity of the α-anion formation was not determined[29] (see Figure 12.9).

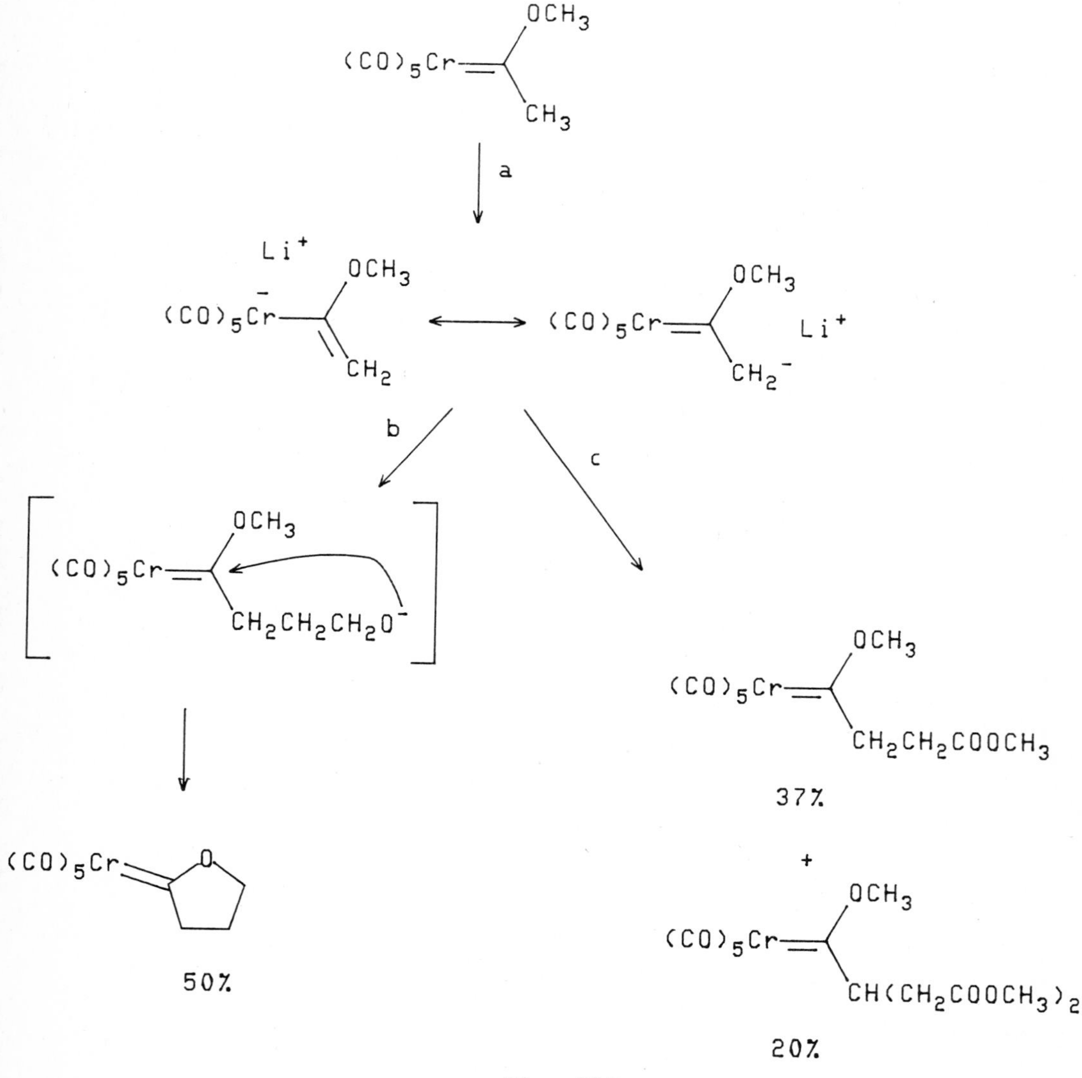

Figure 12.8

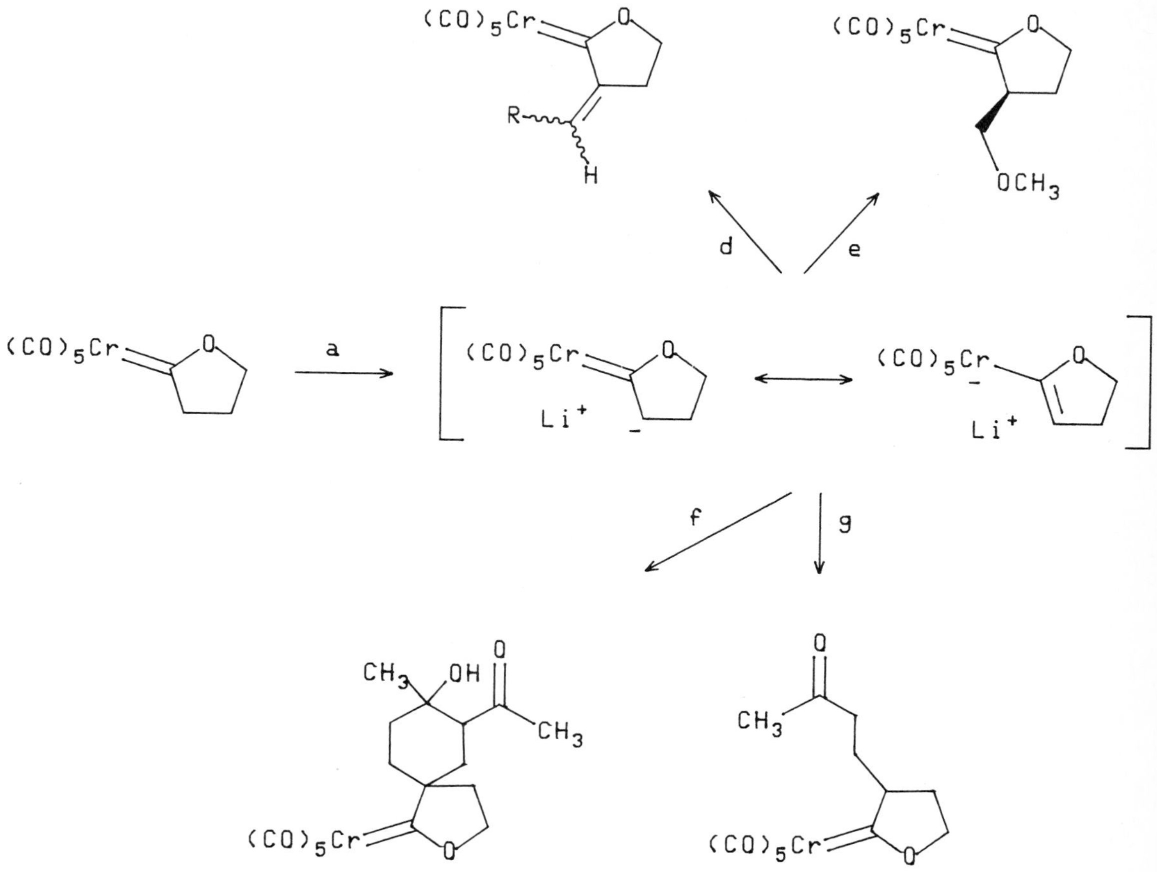

a. $-78^{\circ}C$, n-BuLi, THF

b.  (ethylene oxide)

c. $BrCH_2COOCH_3$

d. 1) $-78^{\circ}C$, RCHO, 2) HCl, Et_2O

e. $0^{\circ}C$, CH_3OCH_2Cl

f. 1) $-78^{\circ}C$, cat. n-BuLi, THF, 2) -78 to $0^{\circ}C$, $CH_2{=}CHC(O)CH_3$, 3) HCl, Et_2O; 36%

g. 1) $-78^{\circ}C$, cat. n-BuLi, THF, 2) $0^{\circ}C$, $CH_2{=}CHC(O)CH_3$, 3) HCl, Et_2O; 41%

Figure 12.8 *(Continued)*

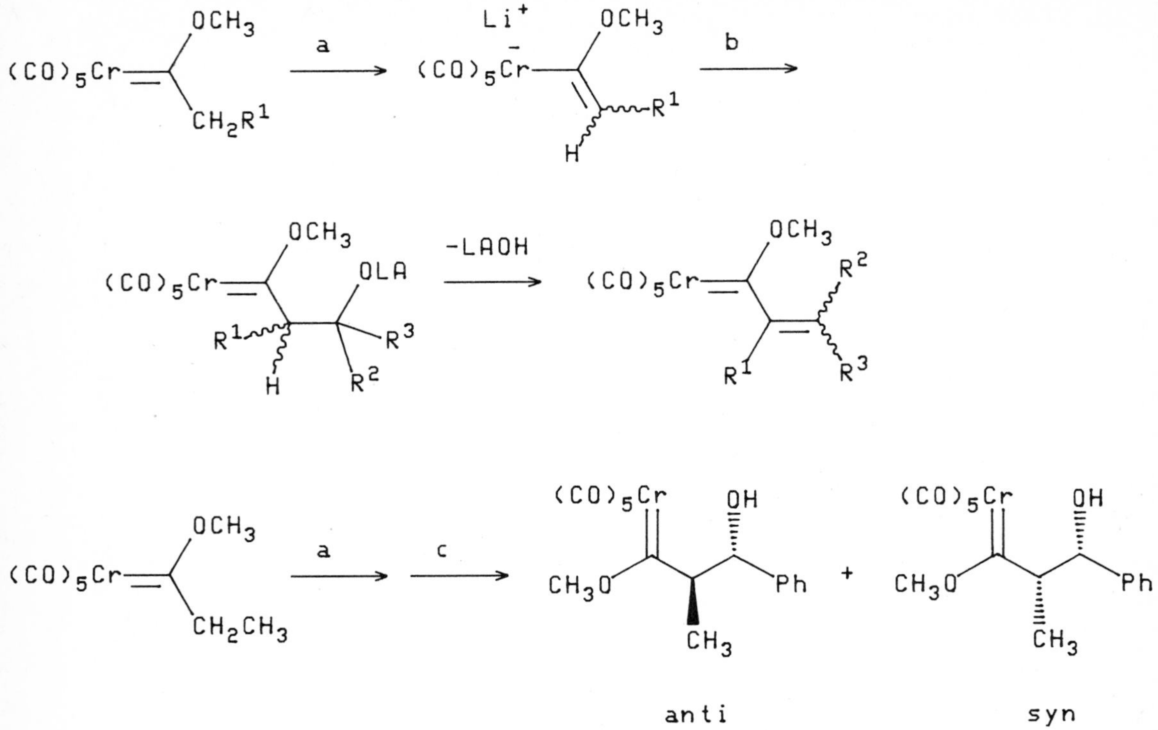

a. $-78°C$, n-BuLi, Et_2O

b. $-78°C$, $R^2C(O)R^3$, Lewis acid (LA), CH_2Cl_2

c. $-78°C$, PhCHO, $TiCl_4$, CH_2Cl_2; anti:syn, 6:1, 55% for two steps

Figure 12.9

Modification or replacement of ancillary ligands is a common way to modulate the stability/reactivity of a transition metal complex. Replacement of one carbon monoxide by tri-*n*-butylphosphine or triphenylphosphine results in significantly decreased acidity of the carbene α-hydrogen and decreased stability/increased reactivity of the resulting α-anion. Methyl iodide, primary halides, primary alkyl-, allyl-, and propargylfluorosulfonates, and benzyl halides all give clean alkylation products in high yield. Conversion back to pentacarbonyl complexes after alkylation would be desirable since the monophosphine complexes are not as useful as the pentacarbonyl complexes in the key synthetic steps of the (±)-daunomycinone and (±)-adriamycinone syntheses. Replacement of triphenylphosphine is achieved by stirring in tetrahydrofuran at room temperature under a high pressure of carbon monoxide[30] (see Figure 12.10).

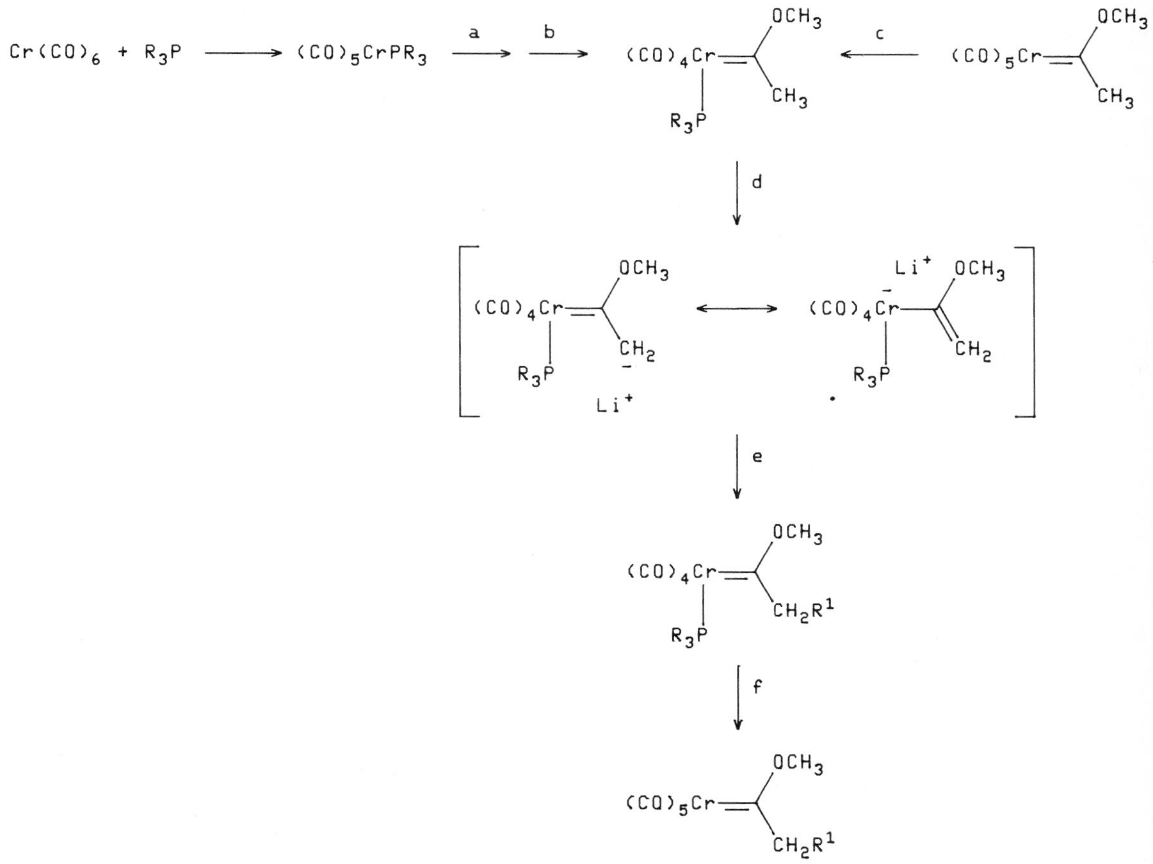

$Cr(CO)_6 + R_3P \longrightarrow (CO)_5CrPR_3$ ... a, b ... $(CO)_4Cr{=}C(OCH_3)(CH_3)(R_3P)$ ← c ← $(CO)_5Cr{=}C(OCH_3)CH_3$

a. 0°C to rt, CH_3Li, Et_2O

b. 0°C to rt, CH_3OSO_2F

c. heat or hν, R_3P

d. -78°C, n-BuLi, THF

e. R^1X

f. R = Ph, rt, high pressure CO, THF

Figure 12.10

12.2.4 Michael Addition to Vinyl Carbenes

The same stability that makes pentacarbonyl carbene α-anions poor nucleophiles can be used to drive carbon–carbon bond formation by Michael addition in vinyl carbene complexes. Casey prepared the requisite vinyl carbenes by two methods: (1) nucleophilic attack of a vinylmagnesium bromide on chromium hexacarbonyl, then alkylation of the acylmetallate or (2) aldol condensation of a carbene α-anion with an aldehyde.

High yields of Michael adducts are obtained with isobutyrophenone and cyclopentanone enolates. Two possible mechanisms are (1) direct nucleophilic attack by the enolate anion on the terminal vinyl carbon or (2) initial electron transfer from enolate to Michael acceptor, then coupling of the radical ions. The products of condensation with acetone enolate are demetallated vinyl ethers, which apparently result from initial attack at the carbene carbon [31] (see Figure 12.11).

12.2.5 Diels–Alder Cycloaddition Using Vinyl or Alkynyl Carbenes as Dienophiles

The strongly electron-withdrawing carbene carbon activates vinyl carbenes to Michael addition. Vinyl carbenes are also excellent dienophiles. Diels–Alder cycloadditions occur more than 10^4 times faster with (vinylmethoxycarbene)pentacarbonylchromium(0) than with methylacrylate and have regioselectivities and stereoselectivities

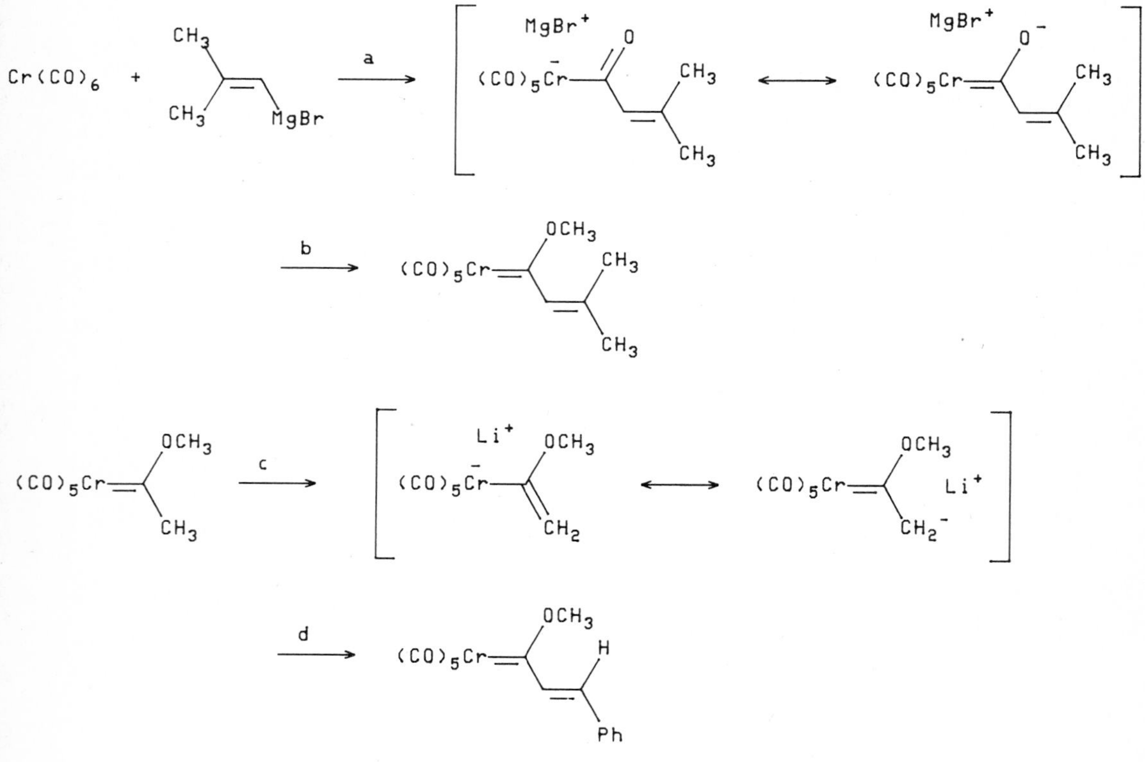

a. rt, THF

b. rt, CH_3OSO_2F; 11% for two steps

c. −78°C, n-BuLi, Et_2O

d. −78 to 0°C, PhCHO; 52% for two steps

Figure 12.11

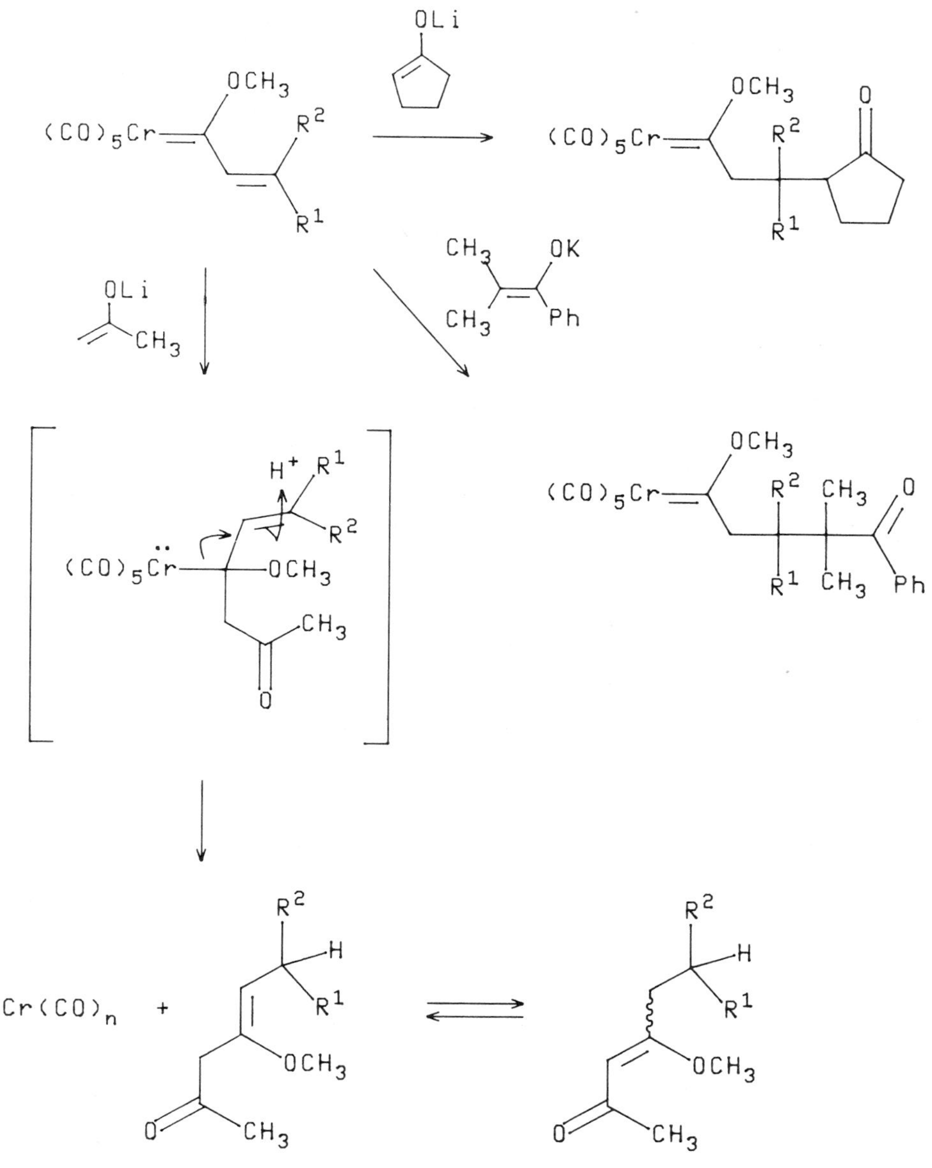

Figure 12.11 (*Continued*)

usually associated with Lewis-acid-catalyzed processes[32,33] (see Figures 12.12 and 12.13).

Diels–Alder cycloaddition using an alkynyl carbene provides a new and efficient approach to vinyl carbenes. The alkynyl carbenes are prepared by the standard Fischer method[34] (see Figure 12.14).

Pentacarbonylchromium carbenes are generally yellow, crystalline solids (red if α,β-unsaturated). They are stable (1) during short periods of exposure during handling, (2)

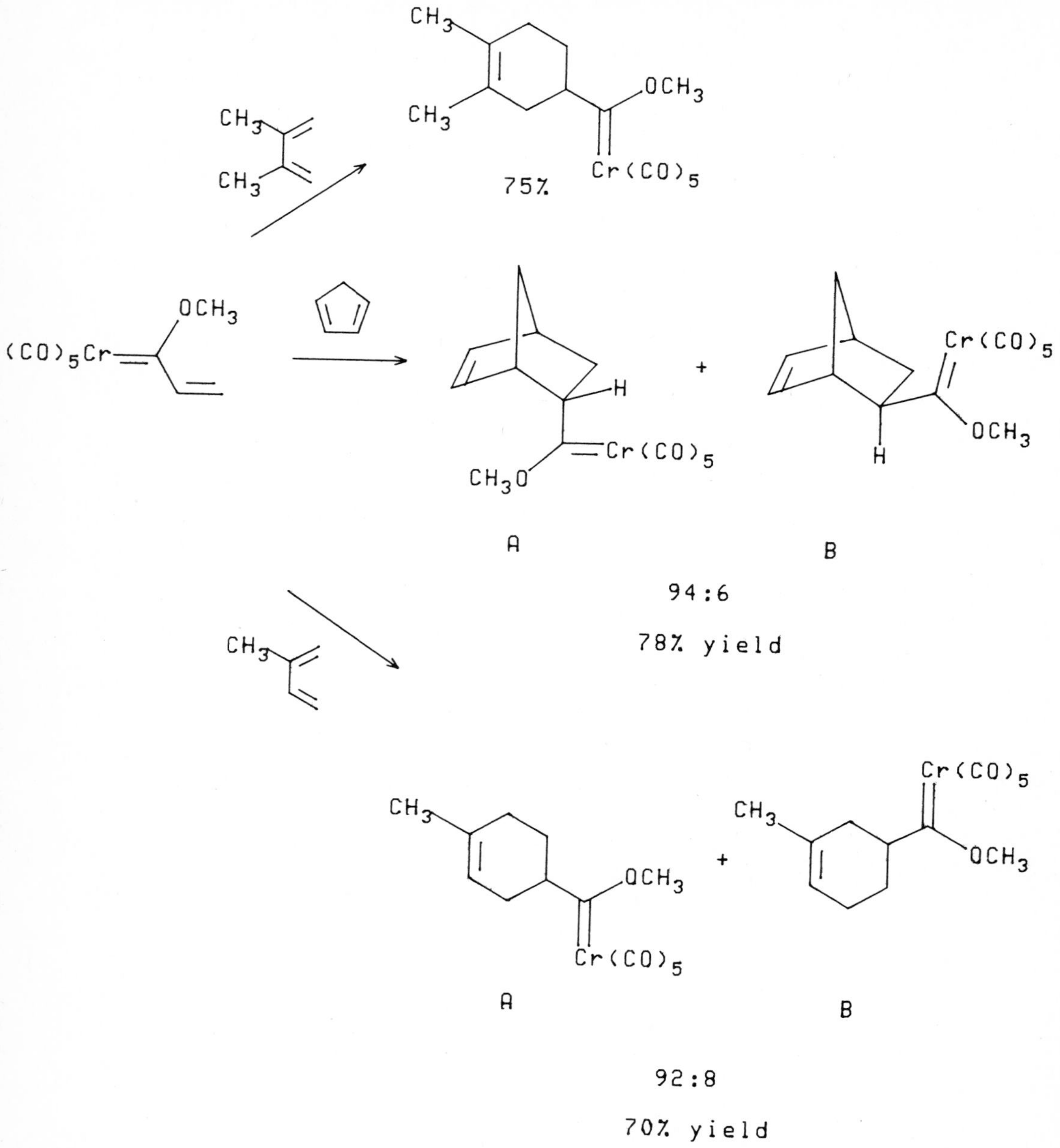

CH_3 CH_3 OCH_3 $Cr(CO)_5$ 75%

$(CO)_5Cr$ OCH_3

CH_3O $Cr(CO)_5$ H + $Cr(CO)_5$ OCH_3 H

A B

94 : 6

78% yield

CH_3

CH_3 OCH_3 $Cr(CO)_5$ + CH_3 $Cr(CO)_5$ OCH_3

A B

92 : 8

70% yield

Figure 12.12

to acids and bases under appropriate low-temperature conditions, and (3) at temperatures in excess of 100 °C. Despite their pronounced reactivity and strong metal–carbene carbon bond polarization, they can be chromatographed on silica gel. Carbenes without polar functional groups will elute with nonpolar solvents such as hexanes.

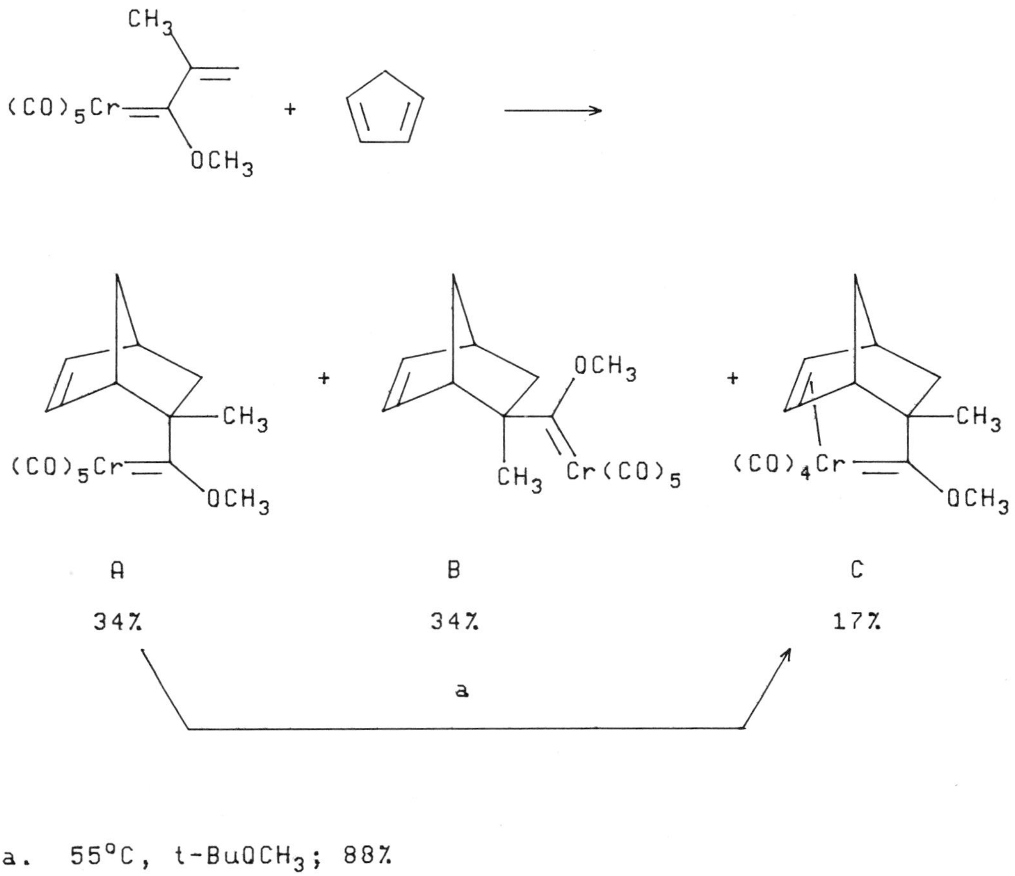

(CO)5Cr … + cyclopentadiene →

A 34% + B 34% + C 17%

a. 55°C, t-BuOCH3; 88%

Figure 12.13

12.3 CARBENE–ALKYNE COCYCLIZATION

Pentacarbonylchromium carbenes undergo a great variety of reactions including replacement of the metal by oxygen,[27,32,35] alkene metathesis[36] (see Chapter 13), and [2 + 1]-cycloaddition.[37] Replacement by oxygen is of analytical value as a means of characterizing heteroatom-stabilized carbenes by conversion to carboxylic acid derivatives. Since alkene metathesis is more efficient with titanium carbenes, and [2 + 1]-cycloadditions can be accomplished with more traditional carbene sources, chromium carbenes did not find a synthetic niche until Dötz discovered that vinyl and aryl carbenes undergo a unique cyloaddition with alkynes[38,39] (see Figure 12.15).

12.3.1 Mechanism of Carbene–Alkyne Cocyclization

Two reasonable mechanisms for arene formation have been proposed. Both have as the initial step (step 1) a rate-limiting dissociation of a *cis*-carbon monoxide from the

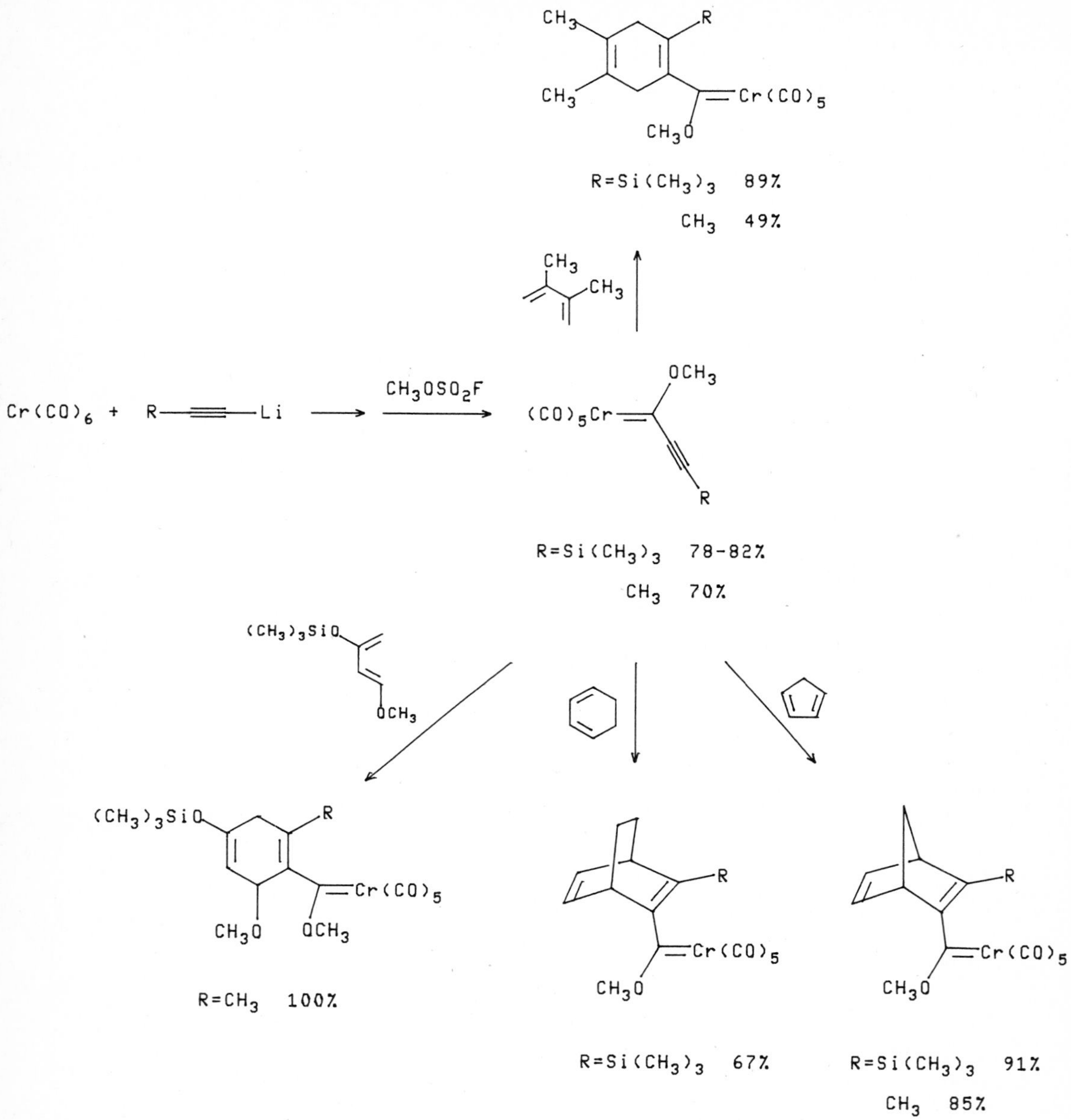

Figure 12.14

octahedral carbene complex. Arene formation can be suppressed by the presence of carbon monoxide.[40] In addition, a detailed kinetic study provides a rate law for the reaction of (phenylmethoxycarbene)pentacarbonylchromium(0) with alkynes in di-*n*-butyl ether. The relatively high activation enthalpy ($\Delta H^{\dagger} = 25.8 \pm 0.5$ kcal mol^{-1}) and positive activation entropy ($\Delta S^{\dagger} = 6.2 \pm 1.4$ cal mol^{-1} K^{-1}) are characteristic of carbonyl carbene complex substitutions that involve dissociation of a carbon monoxide.[41] The

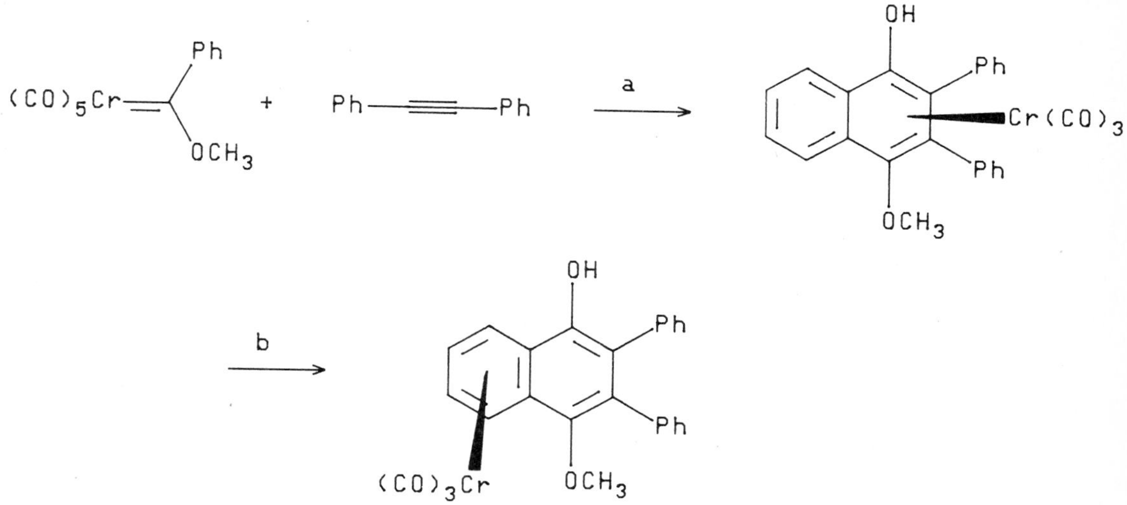

a. 45°C, $(n\text{-}Bu)_2O$; 62%

b. 70°C, $(n\text{-}Bu)_2O$

Figure 12.15

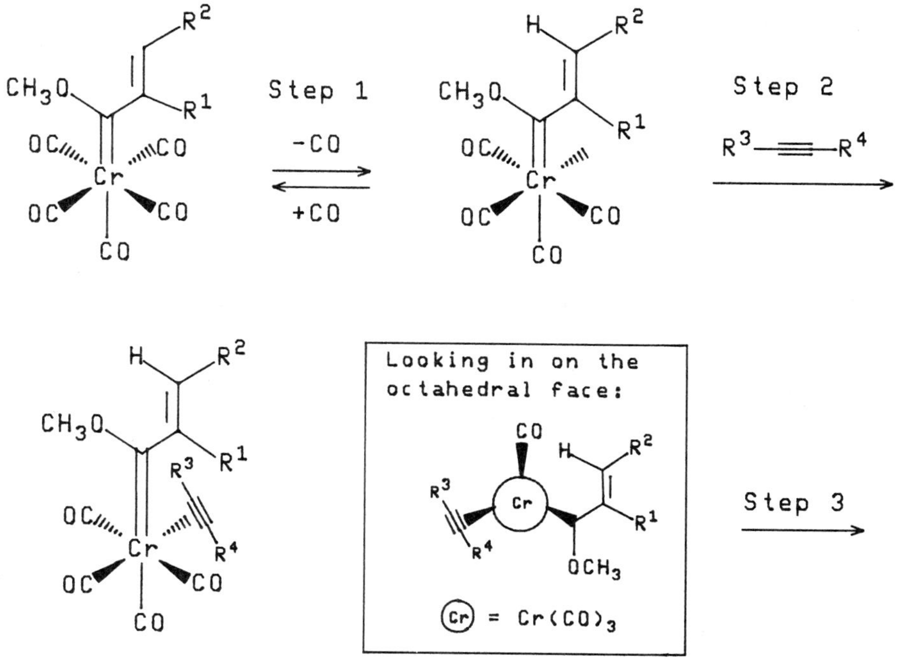

Step 1

-CO

+CO

Step 2

Looking in on the octahedral face:

Ⓒr = $Cr(CO)_3$

Step 3

Figure 12.16

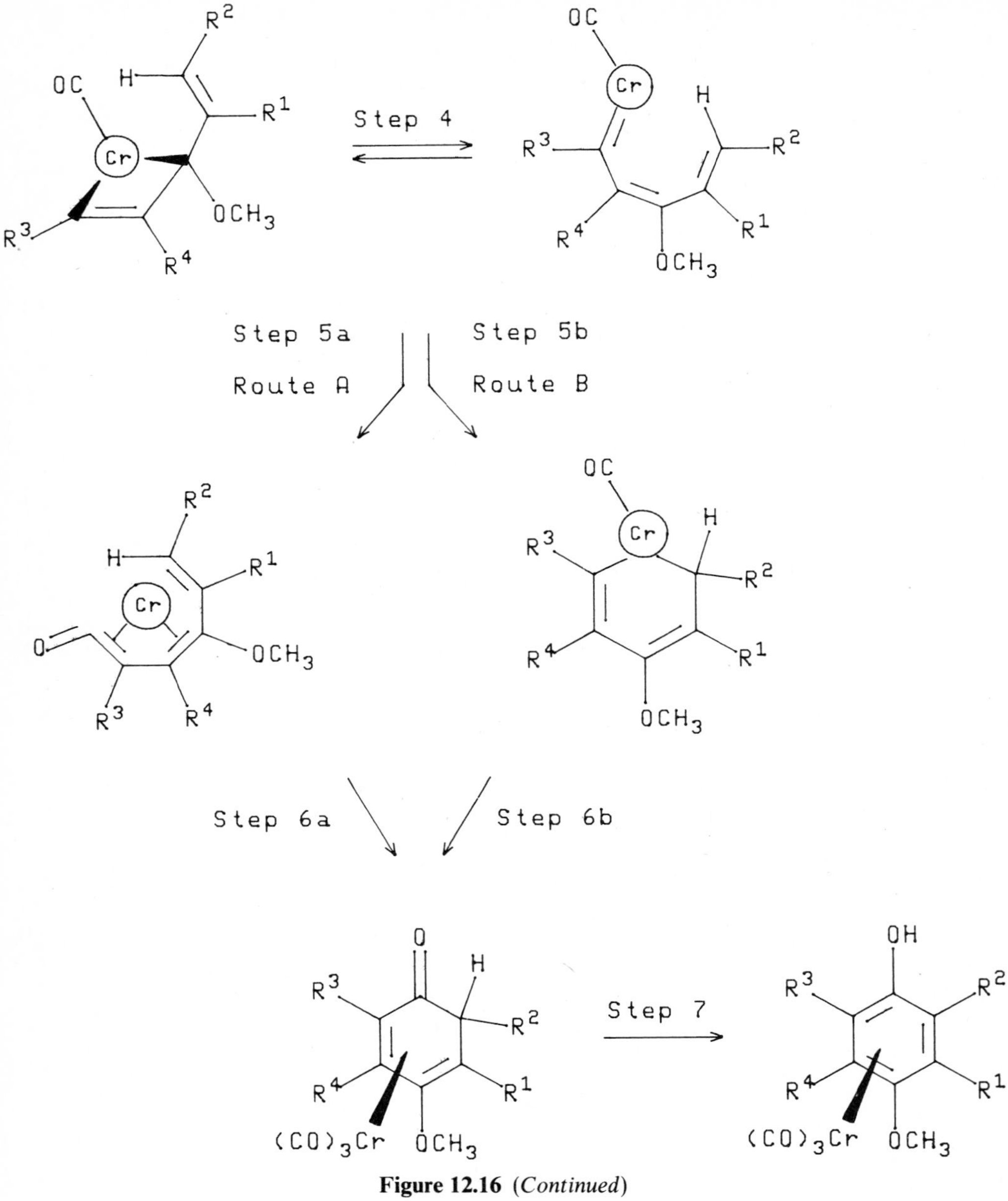

Step 4

Step 5a Route A

Step 5b Route B

Step 6a

Step 6b

Step 7

Figure 12.16 (*Continued*)

anticipated loss of a *cis*-ligand was demonstrated by Casey: selective ^{13}CO exchange with a *cis*-carbon monoxide on (phenylmethoxycarbene)pentacarbonylchromium(0) was observed by ^{13}C NMR spectroscopy.[42]

Also common to both mechanisms is subsequent alkyne coordination (step 2), positioning all three components used in arene formation on one face of the octahedron. The rate of alkyne coordination and rate of arene formation increases with increasing electron density on the triple bond.[41] The carbon–chromium double bond and alkyne form a chromacyclobutene (step 3) that ring opens to a chromabutadiene (step 4). In route A the carbene inserts carbon monoxide to give a chromium–complexed vinyl ketene (step 5a). This chromium-complexed 1,6-hexatriene then ring closes (step 6a). Keto-enol tautomerization to aromatize (step 7) then affords the η^6-chromium tricarbonyl complex of the arene. In route B the chromabutadiene (which is also a chromahexatriene) ring closes (step 6b). Carbon monoxide insertion (step 6b) and keto-enol tautomerization to aromatize (step 7) then affords the same η^6-chromium tricarbonyl complex of the arene (see Figure 12.16).

Evidence for Route A: Route A is supported by the following observations: (1) isolation of a product possibly resulting from trap of a vinyl ketene with a nucleophile[43,44] (see Figure 12.17), (2) isolation of a vinyl ketene that could not form arene,[45] and (3) redirection of the carbene–alkyne cocyclization to produce cyclobutenone.[46] Reaction of (methoxyphenylcarbene)pentacarbonylchromium(0) with tolan in the presence of alcohols or amines affords products expected from nucleophilic trap of the vinyl ketene.[43]

Cocyclizations of (arylmethoxycarbene)pentacarbonylchromium(0) or (alkenylmethoxycarbene)pentacarbonylchromium(0) complexes with 3-butyn-1-ol produce an intermediate capable of either arene formation or intramolecular nucleophilic trap of the vinyl ketene. The preferred route is dictated by the choice of the aryl or alkenyl group. The rates are comparable with *p*-tolylcarbene. The *E*-2-butenylcarbene is converted to arene while the 2-naphthylcarbene is converted to lactone[44] (see Figure 12.18).

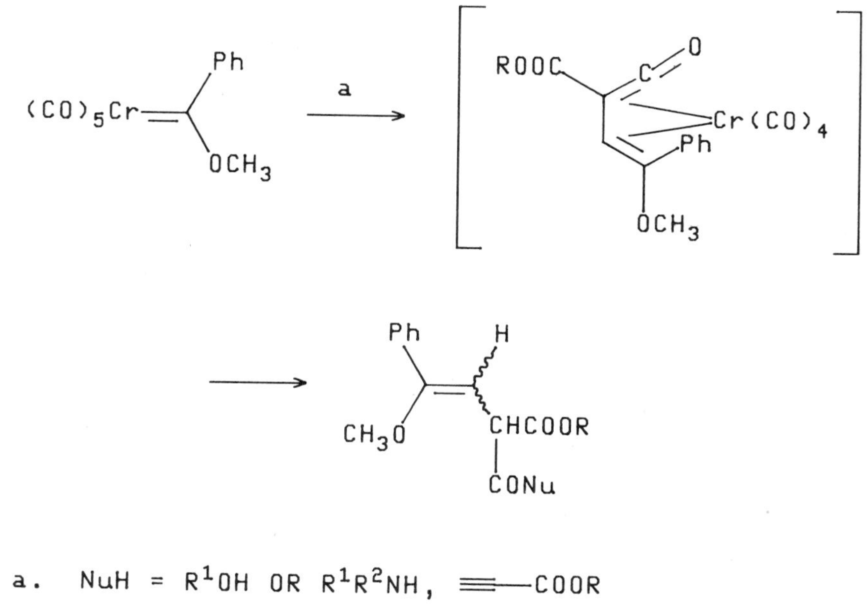

$(CO)_5Cr{=}C(Ph)(OCH_3)$ —a→ [ROOC–CH=C(–C=O)… $Cr(CO)_4$, Ph, OCH_3] → $CH_3O(Ph)C{=}C(H)CH(COOR)CONu$

a. NuH = R^1OH OR R^1R^2NH, ≡—COOR

Figure 12.17

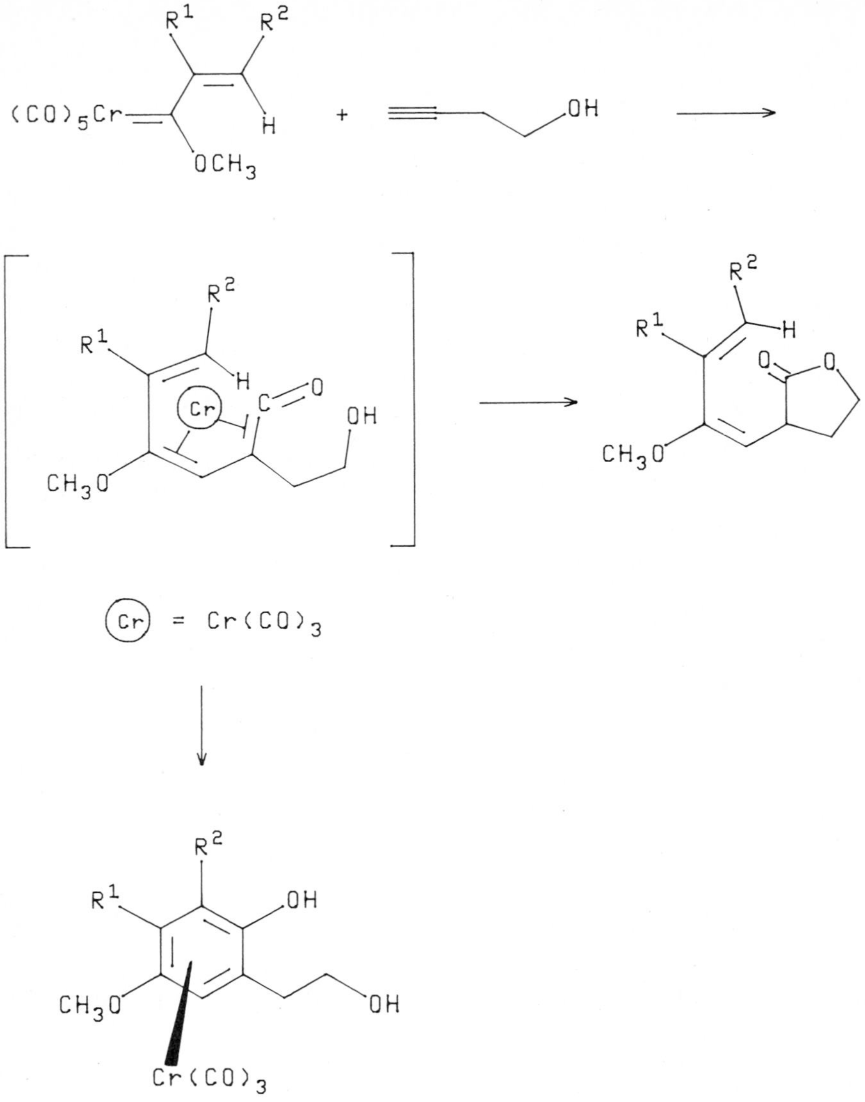

Figure 12.18

Cocyclization of (phenylmethoxycarbene)pentacarbonylchromium(0) with bis-trimethylsilylacetylene proceeds via the chromacyclobutene. Ring opening produces a chromabutadiene that is forced out of the diene configuration necessary for arene formation because of an unfavorable steric interaction between the two trimethylsilyl substituents. Carbon monoxide insertion and metal migration affords a stable vinyl ketene as the η^6-arenechromium tricarbonyl complex.[45a,b] With trimethylsilylacetylene only a small amount of vinyl ketene is isolable[45c] (see Figure 12.19).

Cocyclization of (methylmethoxycarbene)pentacarbonylchromium(0) with tolan proceeds via the chromacyclobutene to a cyclobutenone. Cyclobutenone formation

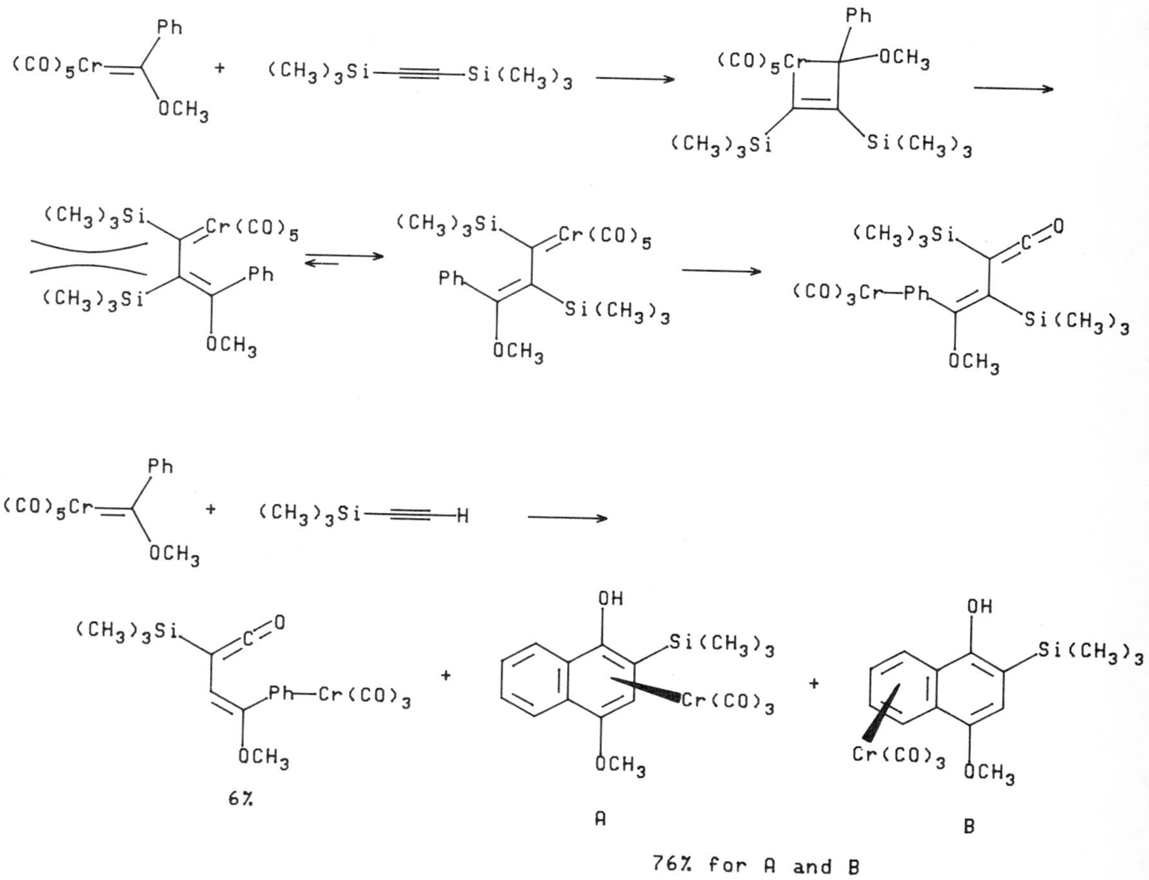

Figure 12.19

may involve either carbon monoxide insertion at the chromacyclobutene stage or electrophilic ring closure of the vinyl ketene[46] (see Figure 12.20).

Evidence for Route B: Redirection of the carbene–alkyne cocyclization to produce indenes provides some evidence for route B.[47] Cocyclization of [(2,6-dimethylphenyl)-methoxycarbene]pentacarbonyl chromium(0) with tolan proceeds via an intermediate that reductively eliminates to form indene rather than insert carbon monoxide. A subsequent 1,5-sigmatropic methyl shift rearomatizes the ring (see Figure 12.21).

12.3.2 Scope and Limitations of Carbene–Alkyne Cocyclization

Despite the relatively recent discovery of the process, the concentrated efforts of Dötz and of Wulff have largely determined the scope, limitations, and reliability of carbene–alkyne cocyclization. Cocyclization is most often run in donor solvents such as diethyl ether, tetrahydrofuran, di-*n*-butyl ether, and *t*-butyl methyl ether at temperatures ranging from 35 to 70 °C. Both nonstabilized and heteroatom-stabilized carbenes can

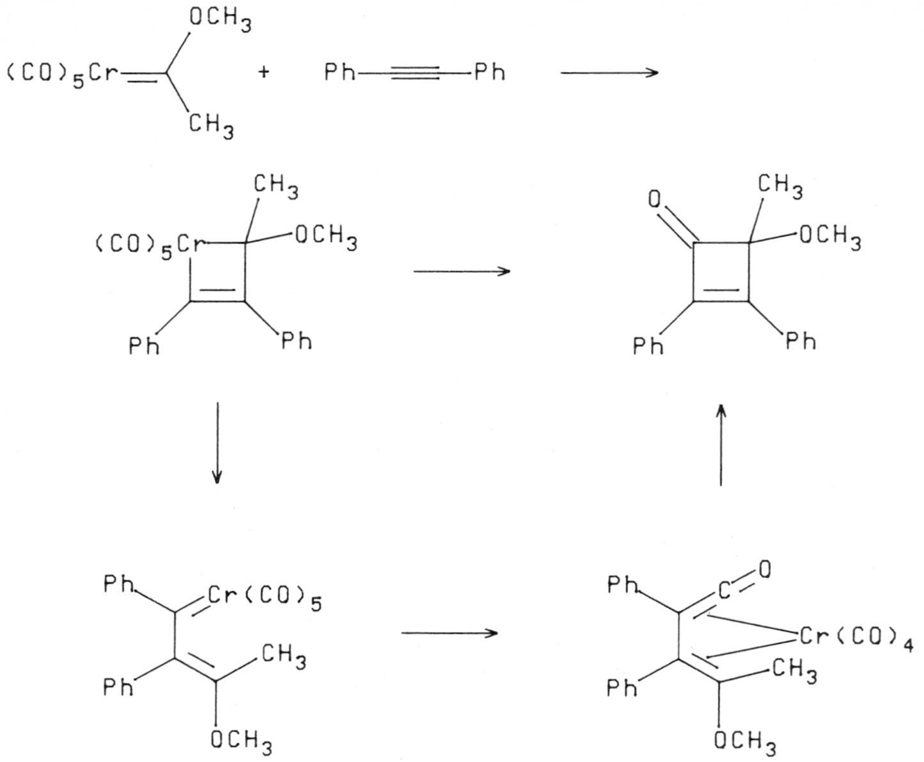

Figure 12.20

be employed. The vinyl portion can be a simple alkene, cyclopentene, cyclohexene, cycloheptene, or vinyl ether or can be incorporated in aryl, naphthyl, furanyl, or thienyl rings. When the terminal carbon of the vinyl carbene is dialkyl substituted, cocyclization affords the cyclohexadienone. When the terminal carbon of the vinyl carbene is alkyl and trimethylsilyl substituted, the higher migratory aptitude of silicon results in conversion of the intermediate cyclohexadienone to a hydroquinone methyl ether trimethylsilyl ether complex[34] (see Figure 12.22).

Finally, reaction of (methylmethoxycarbene)pentacarbonylchromium(0) with an alkyne affords a chromacyclobutene that ring opens to a chromabutadiene. The chromabutadiene, a vinyl carbene, is capable of cocyclization with a second equivalent of alkyne. In situ reduction of the intermediate cyclohexadienone by chromium(0) affords a phenol. This two-alkyne annelation is considerably more efficient when cocyclization is intramolecular[48] (see Figure 12.23).

Attempted cocyclization using electron-rich, nucleophilic alkynes such as ynamines results in attack on the carbene carbon and alkyne insertion.[21,22] Cocyclization using electron-poor alkynes, ynones and propiolates, is not efficient. Alkyl-, aryl-, and trimethylsilyl-substituted internal alkynes work well. Alkene, alcohol, ester, and amide functionality on alkyl side chains and ether, halogen, and trifluoromethyl substituents on aryl rings are compatible.

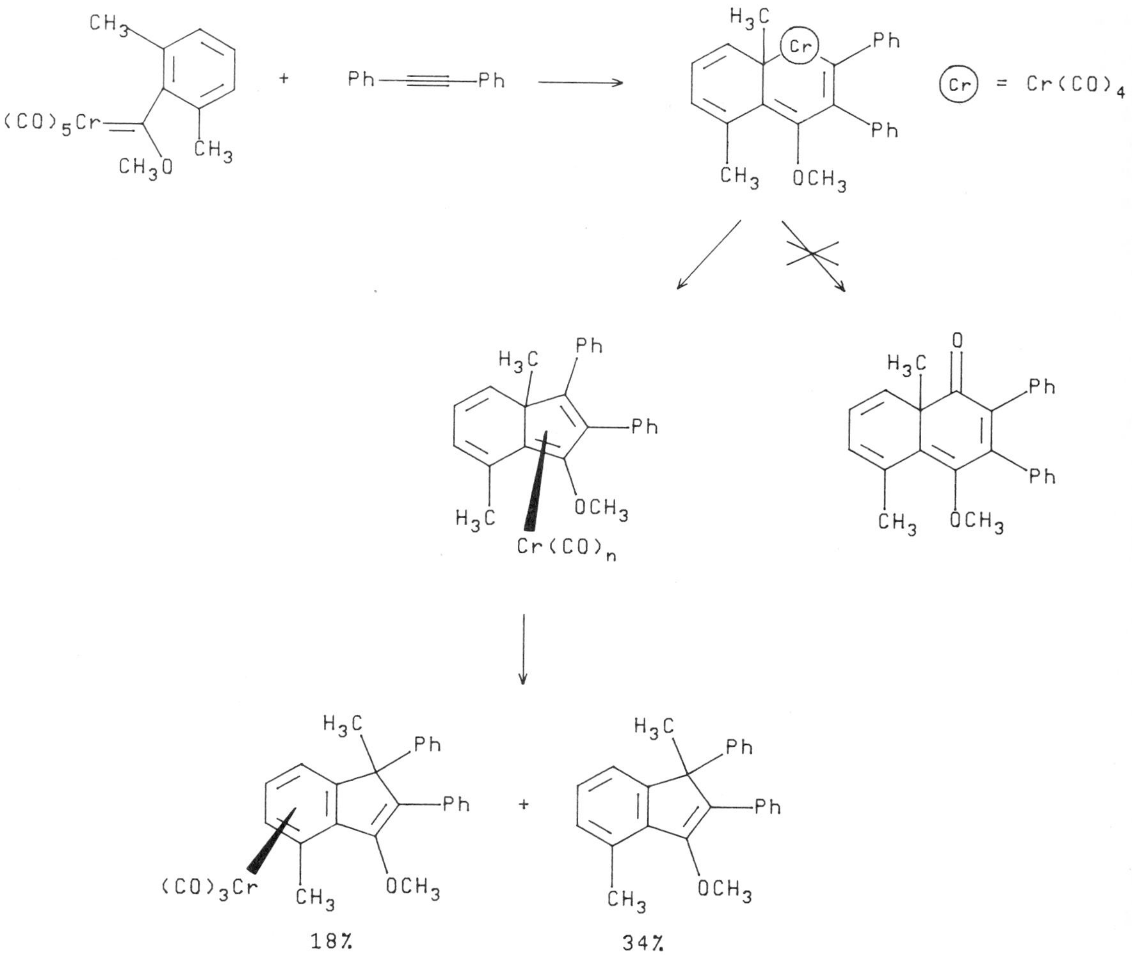

Figure 12.21

Cocyclization with unsymmetrical alkynes can produce two regioisomers. Terminal alkynes regioselectively afford the product with alkyne substituent adjacent to hydroxy. Internal alkynes afford mixtures; the product with the more sterically demanding alkyne substituent adjacent to hydroxy predominates. Regioselectivity increases with increase in the difference in substituent steric bulk adjacent to the alkyne carbons[49] (see Figure 12.24). Regioselectivity can be traced back to chromacyclobutene formation when the initial bond formation is between the carbene carbon and the less sterically encumbered alkyne carbon.

Other regioselectivity questions arise from use of aryl, naphthyl, and furyl carbenes. With *meta*-substituted arylmethoxycarbenes, both the positions *ortho* and *para* to the ring substituent can be involved in cocyclization; *para* is favored. With 2-naphthylmethoxycarbene, cocyclization can involve the 1-position, producing a phenanthrene, or 3-position, producing an anthracene. Cocyclization affords only the phenanthrene. With 3-furylmethoxycarbene, cocyclization can involve the 2-position, producing a

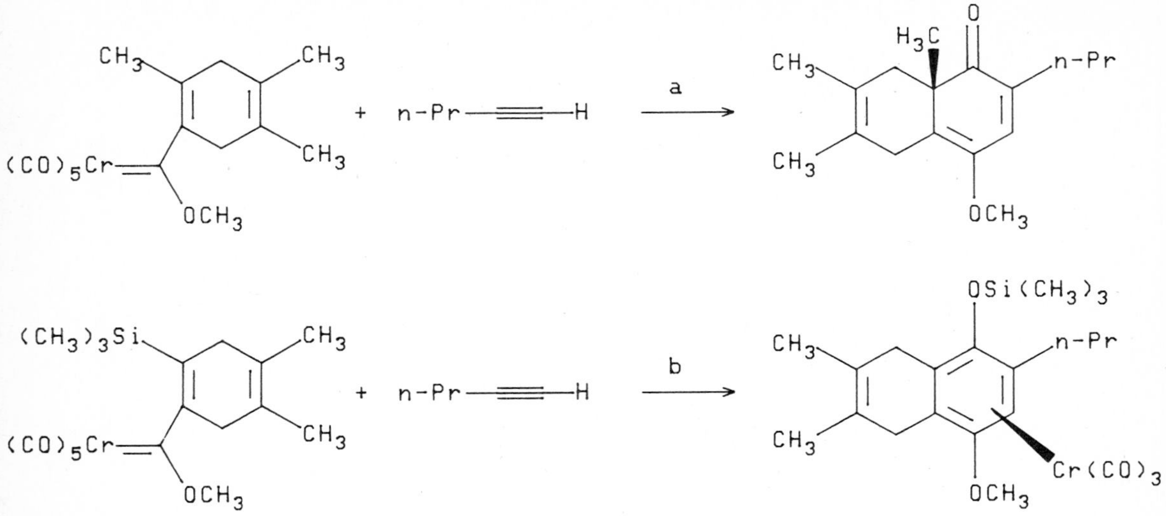

a. 50°C, THF; 67%

b. 45°C, THF; 97%

Figure 12.22

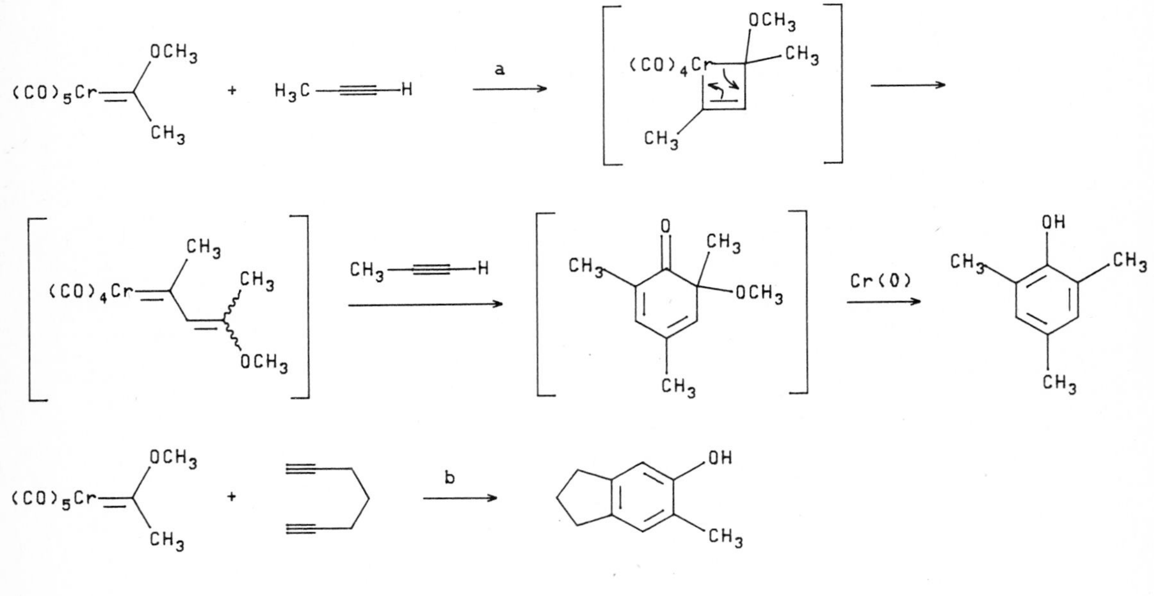

a. 45°C, THF; 33%

b. 70°C, THF; 57%

Figure 12.23

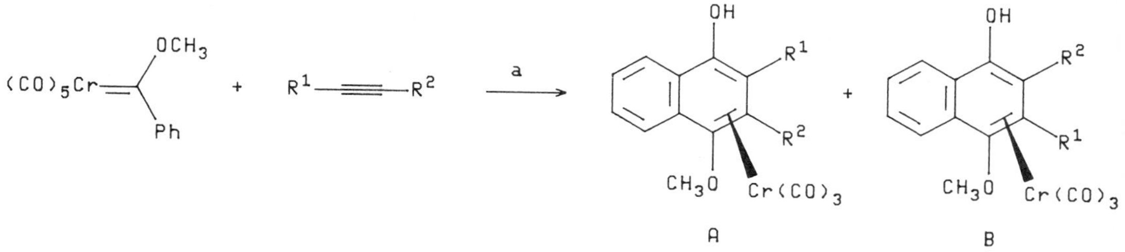

R^1	R^2	A:B Ratio	Yield (%)
Et	CH_3	2.2:1	56
t-Bu	CH_3	11.5:1	59
Ph	CH_3	A only	73

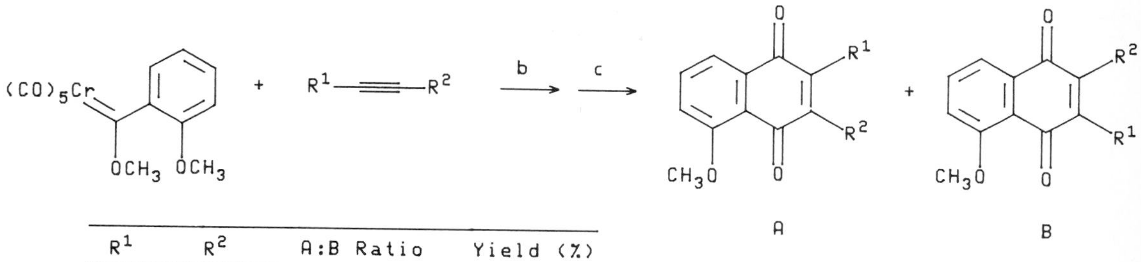

R^1	R^2	A:B Ratio	Yield (%)
Et	CH_3	1.5:1	81
n-Pr	CH_3	2.9:1	64
i-Pr	CH_3	4.8:1	61
Ph	CH_3	A only	78

a. 45°C, $(n\text{-}Bu)_2O$

b. 45°C, THF

c. $(NH_4)_2Ce(NO_3)_6$, H_2O

Figure 12.24

benzofuran, or 4-position, producing an isobenzofuran. Cocyclization affords only the benzofuran. With the nonstabilized phenylfuryl- and phenylnaphthylcarbenes, cocyclization utilizes the phenyl ring only. Finally, with a nonstabilized diarylcarbene, one ring *p*-trifluoromethylphenyl, the other *p*-tolyl, cocyclization utilizes the *p*-trifluoromethylphenyl ring.

One of the major concerns in the synthetic application of carbene–alkyne cocyclization is stability of the product, a chromium tricarbonyl complex of a hydroquinone monomethyl ether. With arylmethoxycarbenes, heating the cocyclization mixture for extended periods or at higher temperatures can result in metal migration to give an alternative arene complex. Attempted chromatographic purification of the hydroquinone monomethyl ether complex on silica gel can result in demetallation and ring oxidation, affording the quinone. Often a workup is chosen to remove the metal, affording the hydroquinone monomethyl ether [high-pressure CO, I_2, Fe(III), or air], or to remove the metal and oxidize to quinone [moist silica and air, Ce(IV) and H_2O, DDQ and H_2O, Ag_2O, or HNO_3 in AcOH], or to remove the metal and oxidize to quinone monoketal [Ce(IV) and CH_3OH]. Thus, efficiency of the cocyclization may be masked in some cases by an inefficient workup procedure. Wulff's four different workup procedures for one cocyclization product is illustrative[50] (see Figure 12.25).

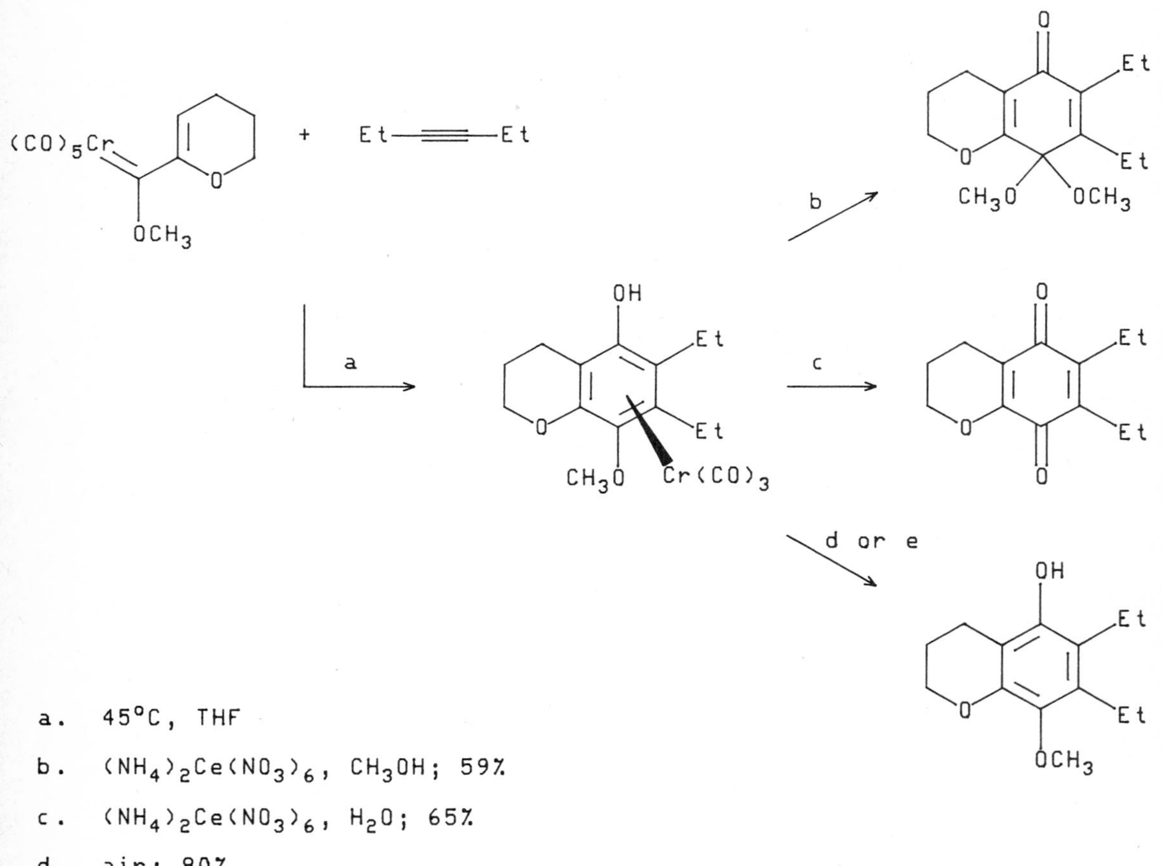

Figure 12.25

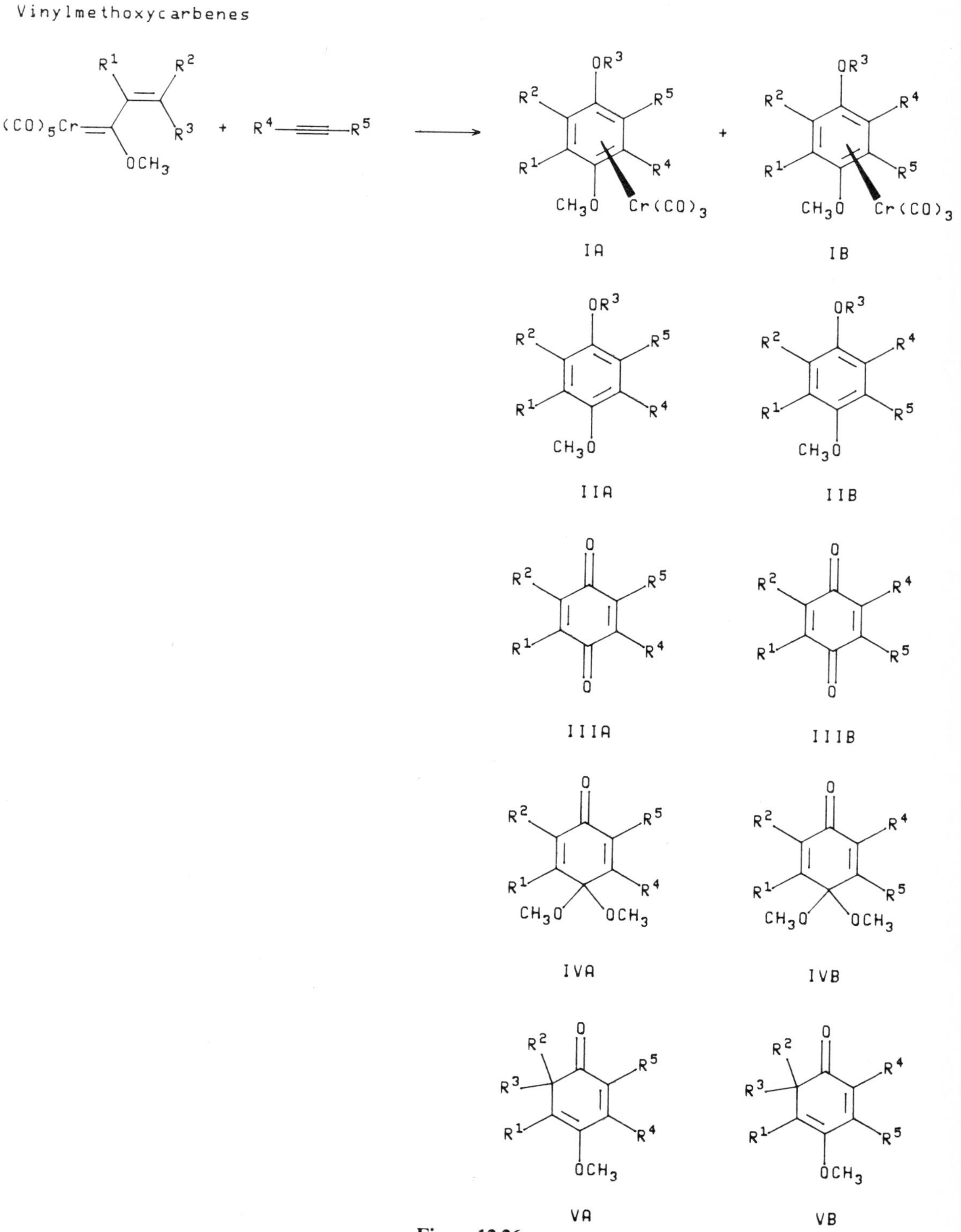
Vinylmethoxycarbenes
IA
IB
IIA
IIB
IIIA
IIIB
IVA
IVB
VA
VB

Figure 12.26

TABLE 12.1 Vinylmethoxycarbene–Alkyne Cocyclization

R^1	R^2	R^3	R^4	R^5	Product	Yield (%)	Reference
CH_3	CH_3	H	CH_3	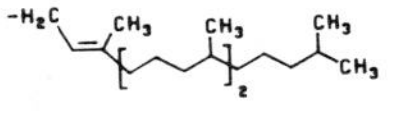 A	IA, B[f]	59	51
$(CH_2)_3$		H	Ph	Ph	I	26	52
H	CH_3	CH_3	H	Ph	VA	55(THF) 81(CH_3CN)	45c
H	CH_3	CH_3	H	$Si(CH_3)_3$	VA	50(THF) 70(CH_3CN)	45c
H	CH_3	CH_3	H	*n*-Bu	VA	60(THF) 80(CH_3CN)	45c
H	CH_3	CH_3	Et	Et	V	39(THF) 42(CH_3CN)	45c
H	CH_3	CH_3	H	CH_3	VA	59(THF) 62(CH_3CN)	45c
H	CH_3	CH_3	H	CH_2OAc	VA	55(CH_3CN)	45c
$—CH(CH_3)(CH_2)_3—$		CH_3	H	Ph	VA	58[a]	45c
$—CH(CH_3)(CH_2)_3—$		CH_3	H	$Si(CH_3)_3$	VA	75[b]	45c
$—CH(CH_3)(CH_2)_3—$		CH_3	H	*n*-Bu	VA	78[c]	45c
$—CH(CH_3)(CH_2)_3—$		CH_3	H	CH_3	VA	44[d]	45c
$—(CH_2)_4—$		H	H	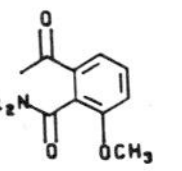 B	IIA	17(THF) 30(CH_3CN)	53
$—(CH_2)_4—$		H	H	C(O)Ph	IIA	17	53
$—(CH_2)_4—$		H	H	$COOCH_3$	IIA	22	53
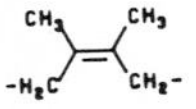 C		$Si(CH_3)_3$	H	*n*-Pr	IA	97	34
C		$Si(CH_3)_3$	Et	Et	II	52	34
C		CH_3	H	*n*-Pr	VA	67	34
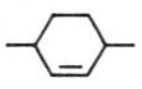 D		$Si(CH_3)_3$	H	*n*-Pr	IIA	95	34
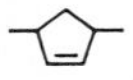 E		$Si(CH_3)_3$	H	*n*-Pr	IIA	56	34
E		CH_3	H	*n*-Pr	VA	47	34
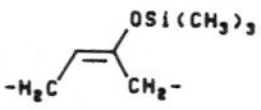 F		$Si(CH_3)_3$	H	*n*-Pr	II	58	34

TABLE 12.1 *(Continued)*

R^1	R^2	R^3	R^4	R^5	Product	Yield (%)	References
—$O(CH_2)_3$—		H	Et	Et	II	80	50
CH_3	H	H	CH_3	CH_3	I	68	54
CH_3	CH_3	H	H	*n*-Pr	IA	75	54
CH_3	CH_3	H	CH_3	*n*-Pr	IA, B[g]	65	54
H	Ph	H	Ph	Ph	I	90	54
—$(CH_2)_3$—		H	Ph	Ph	II	36	50
—$(CH_2)_3$—		H	H	Ph	IIA	76	50
—$(CH_2)_3$—		H	Et	Et	III	37	50
—$(CH_2)_3$—		H	H	*n*-Pr	IIA	54	50
—$(CH_2)_4$—		H	Et	Et	III	65	50
—$(CH_2)_4$—		H	Et	Et	II	64	50
—$(CH_2)_4$—		H	H	*n*-Pr	IIIA	61	50
—$(CH_2)_4$—		H	H	$Si(CH_3)_3$	IIA	71	50
—$(CH_2)_4$—		H	H	(Z)$CH{=}CHOCH_3$	IIA	68	50
—$(CH_2)_4$—		H	H	$COOCH_3$	IIA	22	50
—$(CH_2)_4$—		H	H	C(O)Ph	IIA	17	50
—$(CH_2)_4$—		H	$COOCH_3$	$COOCH_3$	II	8	50
—$(CH_2)_5$—		H	H	*n*-Pr	IIA	66	50
—$(CH_2)_5$—		H	H	$CH(CH_3)CH{=}CH_2$	IIA	57	50
—$O(CH_2)_3$—		H	Et	Et	II	80	50
—$O(CH_2)_3$—		H	Et	Et	III	73	50
—$O(CH_2)_3$—		H	H	COOEt | CH_2CNHAc | COOEt	IIIA	71	50
—$O(CH_2)_3$—		H	H	*n*-Pr	IIIA	67	50
—$O(CH_2)_2O$—		H	Et	Et	III	58	50
—$O(CH_2)_2O$—		H	H	*n*-Pr	IIIA	38	50
EtO	CH_3[e]	H	Ph	Ph	II	67	50
EtO	CH_3[e]	H	H	*n*-Pr	IIIA	40	50
EtO	H	H	H	$CH_2CH{=}C(CH_3)_2$	IIA	23	50
CH_3	H	H	H	*n*-Pr	IIIA, B[h]	51	50
H	CH_3	H	H	*n*-Pr	IIIA, B[i]	75	50
H	CH_3	H	H	*n*-Pr	IIA	60	50
CH_3	CH_3	H	H	CH_2CH_2OH	IA	76	44

[a] Methyls *trans–cis*, 95:5.
[b] Methyls *trans–cis*, 92:8.
[c] Methyls *trans–cis*, 90:10.
[d] Methyls *trans*–cis, 91:9.
[e] 1.6:1 mixture of *E* and *Z* alkenes.
[f] A:B = 70:30.
[g] A:B = 80:20.
[h] A:B = 93:7.
[i] A:B = 99:1.

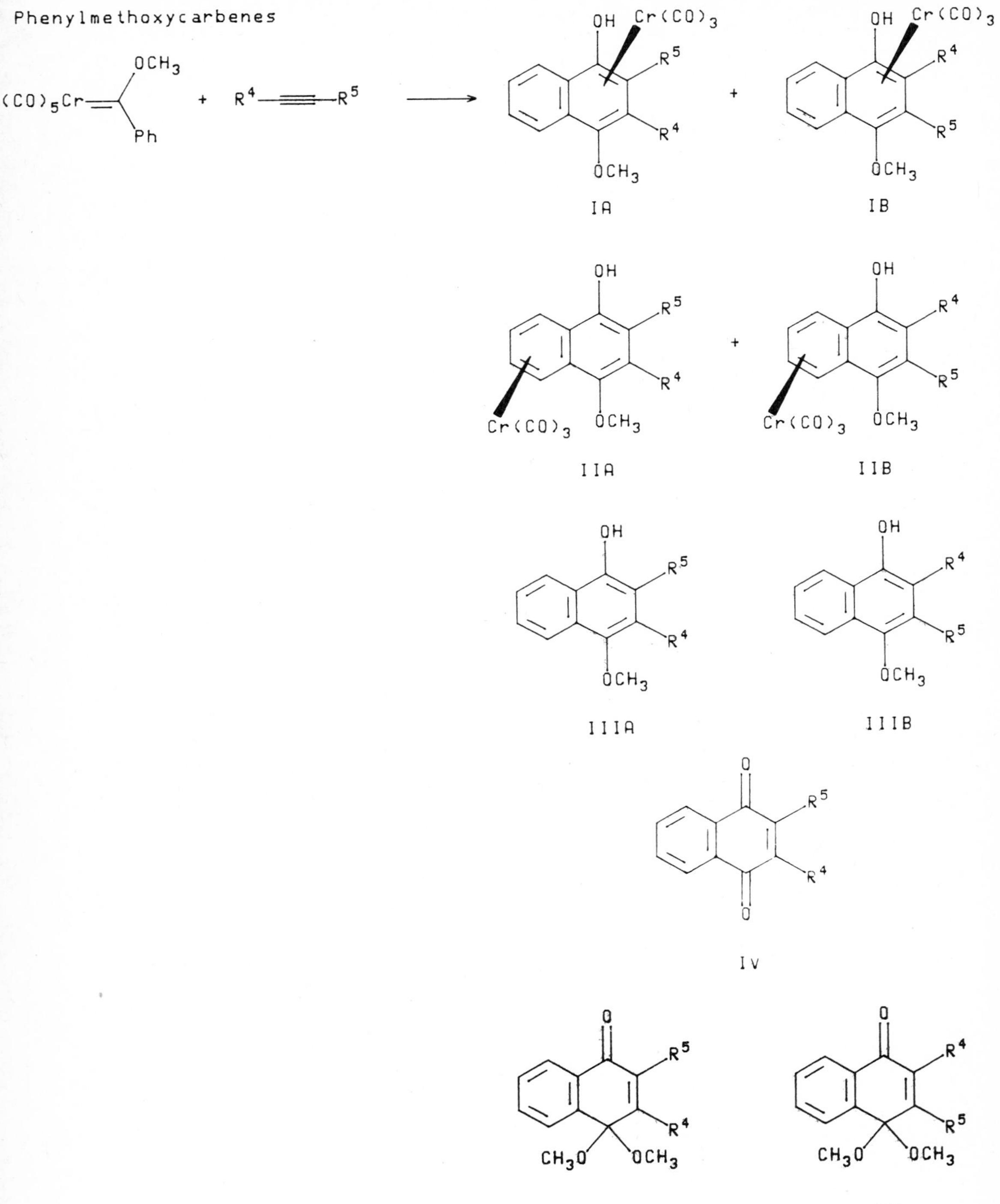

Phenylmethoxycarbenes

$(CO)_5Cr{=}C(OCH_3)Ph$ + $R^4{-}C{\equiv}C{-}R^5$ ⟶

IA + IB

IIA + IIB

IIIA IIIB

IV

VA VB

Figure 12.27

TABLE 12.2 Phenylmethoxycarbene–Alkyne Cocyclization

R^4	R^5	Product	A : B ratio	Yield (%)	Reference
CH_3	CH_3	I	—	68	40
Et	Et	I	—	65	40
Ph	Ph	I, II	—	20, 18	47c
H	CH_3	IV	—	66	56
H	Et	I	A only	35	40
H	*n*-Pr	I	A only	45	40
H	*n*-Bu	I	A only	45	40
H	*t*-Bu	I	A only	79	57
H	*n*-C_8H_{17}	I	A only	59	57
H	$Si(CH_3)_3$	I, II	A only	76	45b
H	$CH_2CH{=}C(CH_3)_2$	I	ND	60	55
CH_3	Et	I	69 : 31	56	57
CH_3	*n*-Pr	I	ND	80	57
CH_3	*t*-Bu	I	92 : 8	59	57
CH_3	$CH_2CH{=}C(CH_3)_2$	I	ND	85	55
CH_3	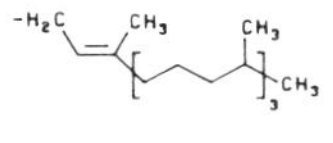 A	I	63 : 37	92	55
CH_3	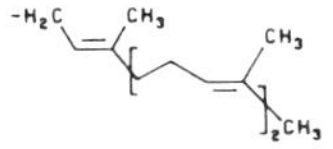 B	I	ND	90	55
CH_3	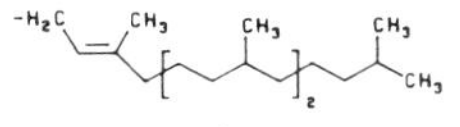 C	I	63 : 37	92	57
CH_3		IV	—	48	56
CH_3	Ph	I	A only	73	57
CH_3	*p*-$CH_3C_6H_4$	I	A only	73	41
Et	*n*-Pr	I	ND	62	57
Et	*t*-Bu	I	ND	66	57
n-Pr	*n*-Bu	I	ND	62	57
Ph	*p*-$CH_3C_6H_4$	I	53 : 47	52	57
Ph	*p*-$CF_3C_6H_4$	I	56 : 44	66	57
p-$CH_3C_6H_4$	*p*-$CH_3C_6H_4$	I	54 : 46	62	57

ND = not determined.

(Substituted phenyl)methoxycarbenes

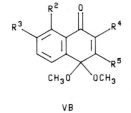

$(CO)_5Cr=C(OCH_3)Ar + R^4-C \equiv C-R^5 \longrightarrow$ IA + IB

IA

IB

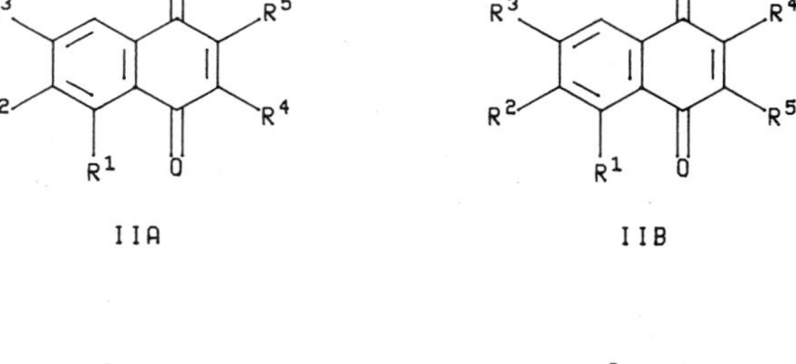

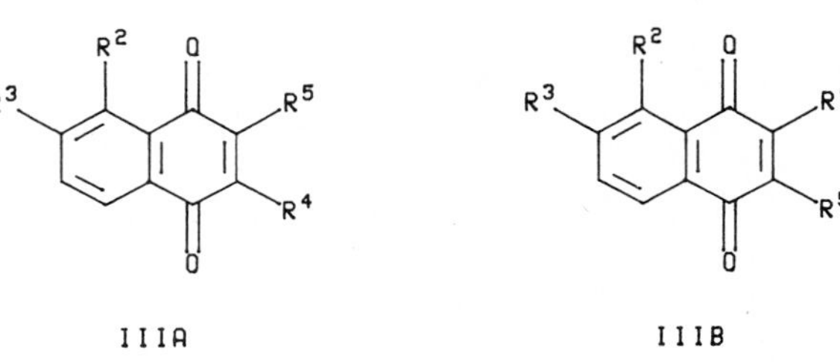

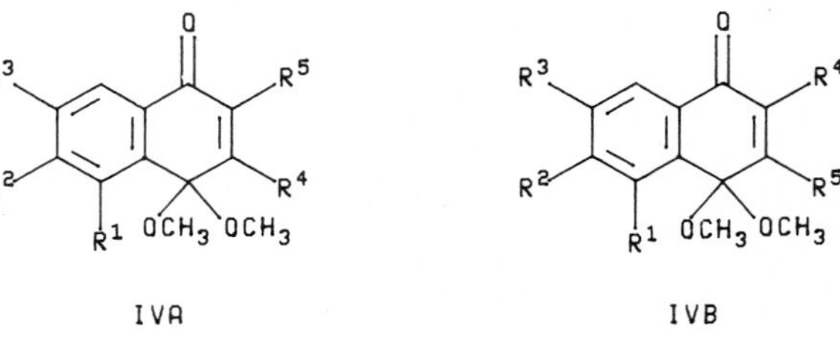

VA

VB

Figure 12.28

TABLE 12.3 (Substituted phenyl)methoxycarbene–Alkyne Cocyclization[44,49,52]

R^1	R^2	R^3	R^4	R^5	Product	A:B Ratio	Yield (%)
H	H	CH_3	Ph	Ph	I	—	40
H	H	CF_3	Ph	Ph	I	—	25
H	H	CH_3	H	CH_2OH	I	A only	60
H	H	CH_3	H	$(CH_2)_2OH$	I	A only	17
H	H	CH_3	H	$(CH_2)_3OH$	I	A only	44
H	H	CH_3	H	$(CH_2)_4OH$	I	A only	75
H	CH_3	H	Et	Et	IV, V (IV : V, 54 : 46)	—	79
H	OCH_3	H	Et	Et	IV, V (IV : V, 68 : 32)	—	81
H	OCH_3	H	H	*n*-Pr	II, III (II : III, 91 : 9)	A only	76
H	F	H	Et	Et	IV, V (IV:V, 88:12)	—	70
H	CF_3	H	Et	Et	IV, V (IV : V, 96 : 4)	—	55
OCH_3	H	H	CH_3	Et	II	60:40	81
OCH_3	H	H	CH_3	*n*-Pr	II	74:26	64
OCH_3	H	H	CH_3	*i*-Pr	II	83:17	61
OCH_3	H	H	CH_3	Ph	II	A only	78
OCH_3	H	H	H	*n*-Pr	II	A only	74
OCH_3	H	H	H	$(CH_2)_3COOt$-Bu	II	A only	66
OCH_3	H	H	H	$(CH_2)_3CONHt$-Bu	II	A only	70

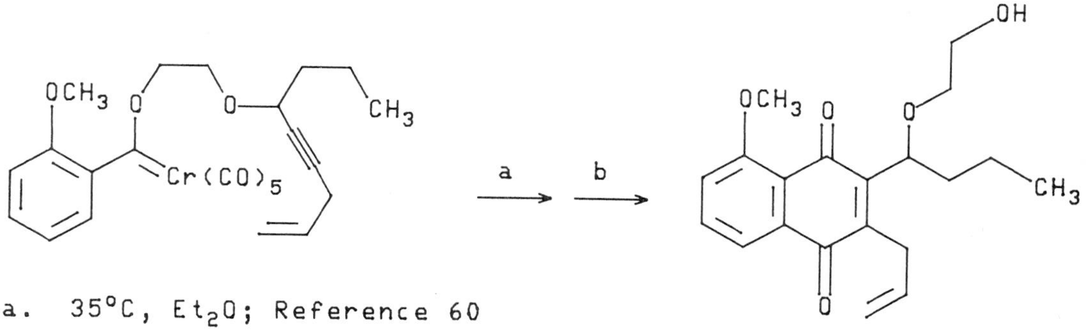

a. 35°C, Et_2O; Reference 60

b. DDQ, CH_3CN, H_2O; 51%

Figure 12.29

TABLE 12.4 (Naphthyl- and Heteroaryl) methoxycarbene–Alkyne Cocyclization[52]

$(CO)_5Cr{=}C(OCH_3)Ar$ + $R^4{-}C{\equiv}C{-}R^5$ ⟶ Product

Ar	R^4	R^5	Product	Yield (%)
1-naphthyl	Ph	Ph	CH_3O, Ph, Ph, OH, $Cr(CO)_3$	22
2-naphthyl	Ph	Ph	HO, Ph, Ph, OCH_3, $Cr(CO)_3$	19
2-furyl	Ph	Ph	OH, Ph, Ph, OCH_3, $Cr(CO)_3$	19

(continued)

TABLE 12.4 *(Continued)*

Ar	R^4	R^5	Product	Yield (%)
2-furyl	H	n-Pr	OH, n-Pr, $Cr(CO)_3$, H, OCH_3	23
2-thienyl	Ph	Ph	OH, Ph, $Cr(CO)_3$, Ph, OCH_3	28
2-thienyl	H	n-Pr	OH, n-Pr, $Cr(CO)_3$, H, OCH_3	40
3-furyl	Ph	Ph	OCH_3, Ph, $Cr(CO)_3$, Ph, OH	62

TABLE 12.5 Diarylcarbene–Alkyne Cocyclization[40,52]

$$(CO)_5Cr{=}C(Ar^1)(Ar^2) + R^4{-}C{\equiv}C{-}R^5 \longrightarrow \text{Product}$$

Ar^1	Ar^2	R^4	R^5	Product	Yield (%)
Ph	2-naphth	Ph	Ph	1-OH-2,3-Ph-4-(2-naphth)naphthalene $Cr(CO)_3$	11
4-$CH_3C_6H_4$	4-$CF_3C_6H_4$	Ph	Ph	1-OH-2,3-Ph-4-(4-$CH_3C_6H_4$)-7-CF_3-naphthalene $Cr(CO)_3$	18
Ph	2-furyl	Ph	Ph	1-OH-2,3-Ph-4-(2-furyl)naphthalene $Cr(CO)_3$	22
Ph	Ph	CH_3	CH_3	1-OH-2-R^5-3-R^4-4-Ph-naphthalene $Cr(CO)_3$	12
Ph	Ph	H	n-Pr		11
Ph	Ph	H	Ph		27
Ph	Ph	Ph	Ph		20
Ph	Ph	COOEt	Ph		7

TABLE 12.6 Intramolecular Phenylalkoxycarben–Alkyne Cocyclization[59]

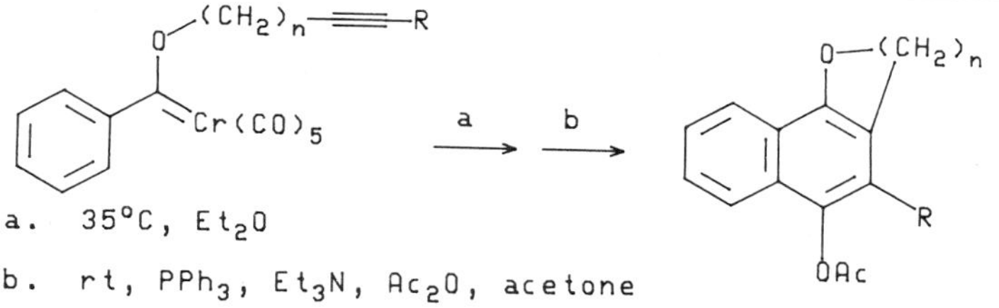

a. 35°C, Et_2O

b. rt, PPh_3, Et_3N, Ac_2O, acetone

n	R	Yield (%)
2	H	16
2	CH_3	81
3	H	18
3	CH_3	62
4	H	38
4	CH_3	62

TABLE 12.7 Two-Alkyne Annulations via in situ Generated Vinylcarbene Complexes[48]

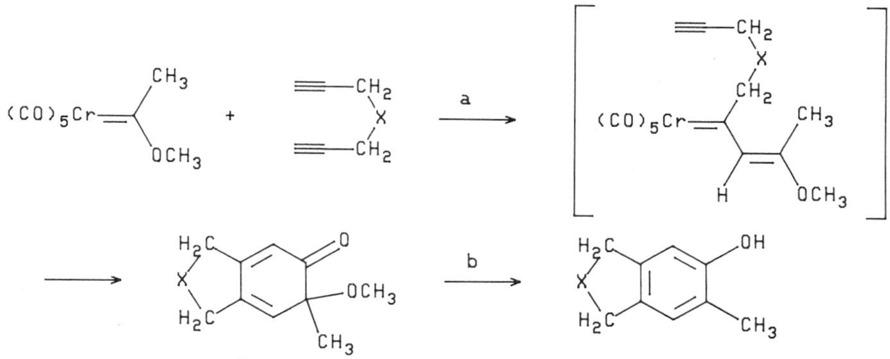

I

II

a. 70°C, THF

b. [Cr(0)]

X	Product(s)	Yield(s) (%)
CH_2	II	57
CH_2	I,II*	26,39
O	II	31
$C(COOEt)_2$	II	57
CH_2CH_2	II	32

*in CH_3CN

Of the many methods for workup, demetallation with high-pressure carbon monoxide (in diethyl ether at room temperature) is particularly attractive since chromium is recovered as the recyclable hexacarbonyl complex.

The scope and limitations of carbene–alkyne cocyclization are detailed in Figures 12.26–12.29 and Tables 12.1–12.7.

12.4 TOTAL SYNTHESIS OF (±)-DAUNOMYCINONE

Two different cocyclization-based approaches to (±)-daunomycinone have been successful. *Both* utilize a carbene–alkyne cocyclization to construct the B ring in the key step. Wulff's approach (routes A and B) starts with intact A and D rings, constructs the B ring by cocyclization, then closes the C ring by intramolecular ring acylation. Dötz's approach (route C) starts with intact C and D rings, constructs the B ring by cocyclization, then closes the A ring by intramolecular ring acylation (see Figure 12.30).

12.4.1 Route A. From a Cyclohexenylmethoxycarbene Prepared by Fischer's Method[53]

The cyclohexenylmethoxycarbene is contructed using Fischer's method.[12] A vinyl anion is prepared from 1,4-cyclohexadione.[61] One ketone is protected as a ketal,[62] the other converted to a 2,4,6-triisopropylbenzenesulfonylhydrazone. The hydrazone affords a vinyl anion on treatment with *n*-butyllithium at $-78°C$ (Shapiro reaction[63,64]). The acylmetallate obtained with chromium hexacarbonyl is first isolated as the tetramethylammonium salt, then alkylated with methyltrifluoromethanesulfonate (see Figure 12.31).

A redesign of the route A alkyne is necessary since (1) the key cocyclization of carbene with an electron-deficient alkyne would be inefficient and (2) subsequent intramolecular ring acylation to close the C ring would require harsh conditions incompatible with sensitive A ring functionality. The alkyne chosen is available from *o*-anisic acid[65] via the *N*,*N*-diethylamide.[66] *Ortho*lithiation of the amide and trap with propargyl aldehyde affords a lactone[67] (see Figure 12.32).

The carbene–alkyne cocyclization using this terminal alkyne is regiospecific, affording only the product with alkyne substituent adjacent to the hydroxy group. Demetallation with iron(III) affords the hydroquinone monomethyl ether in 76% yield. Hydroxyl protection, C—O bond reduction of the benzylic ester, a facile intramolecular ring acylation, and oxidation up to the quinone affords the tetracyclic daunomycinone framework (see Figure 12.33).

Deprotection of the hydroquinone ethers and ketal is best achieved in four operations: removal of the MOM protection (TFA in CH_2Cl_2), oxidation of the hydroquinone monomethyl ether to quinone, B-ring quinone reduction to hydroquinone, and then ketal deprotection (see Figure 12.34).

The A ring is functionalized using Kende's procedure.[68] Alkynyl Grignard condensation with the ketone affords the tertiary alcohol. Oxymercuration–demercuration converts the alkyne to a methyl ketone. Free radical benzylic bromination and subsequent bromide solvolysis introduces the 10-hydroxyl group, affording a 5:2 mixture of (±)-epidaunomycinone and (±)-daunomycinone. The epi form is converted to (±)-daunomycinone in trifluoroacetic acid at room temperature.

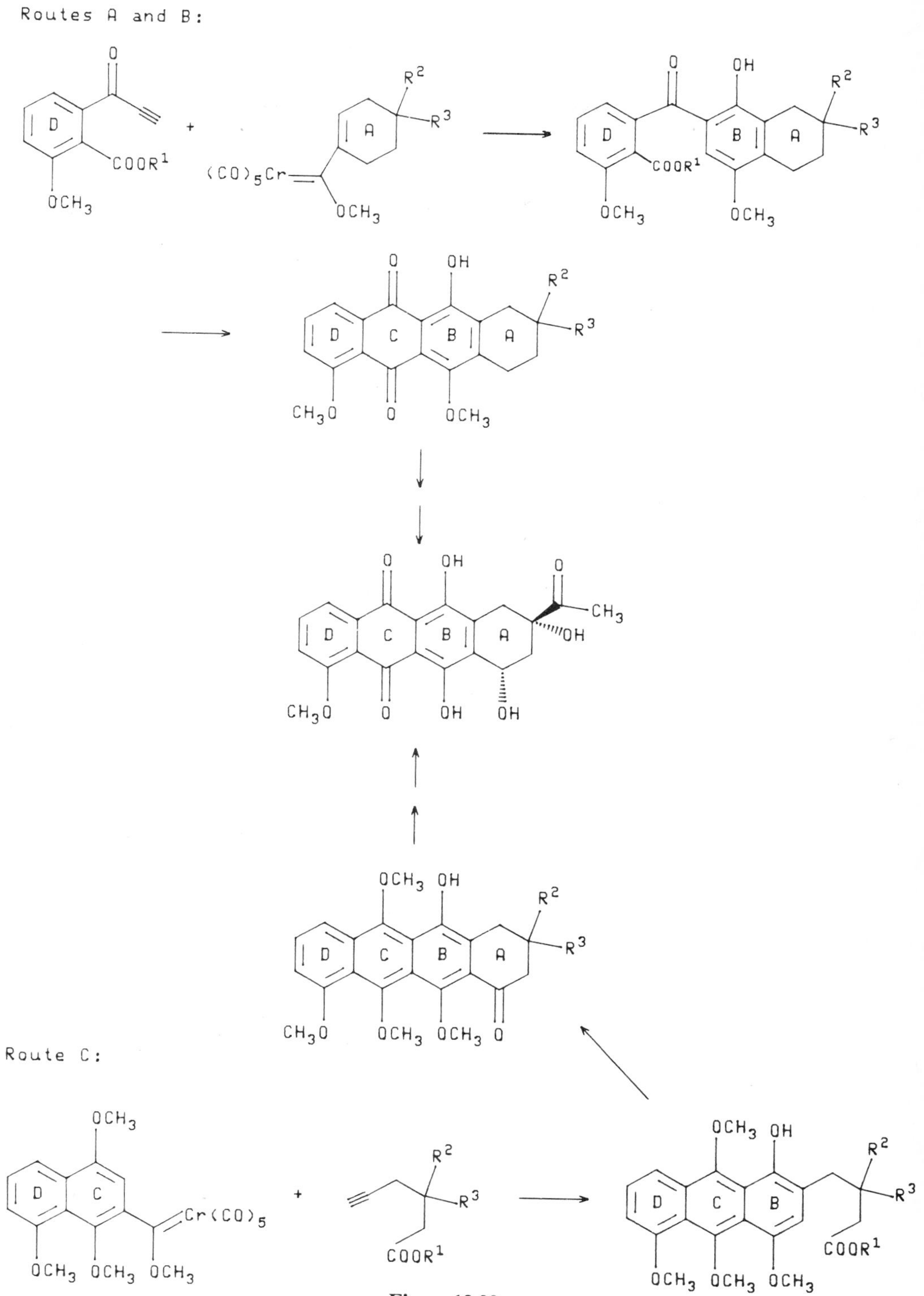

Routes A and B:

Route C:

Figure 12.30

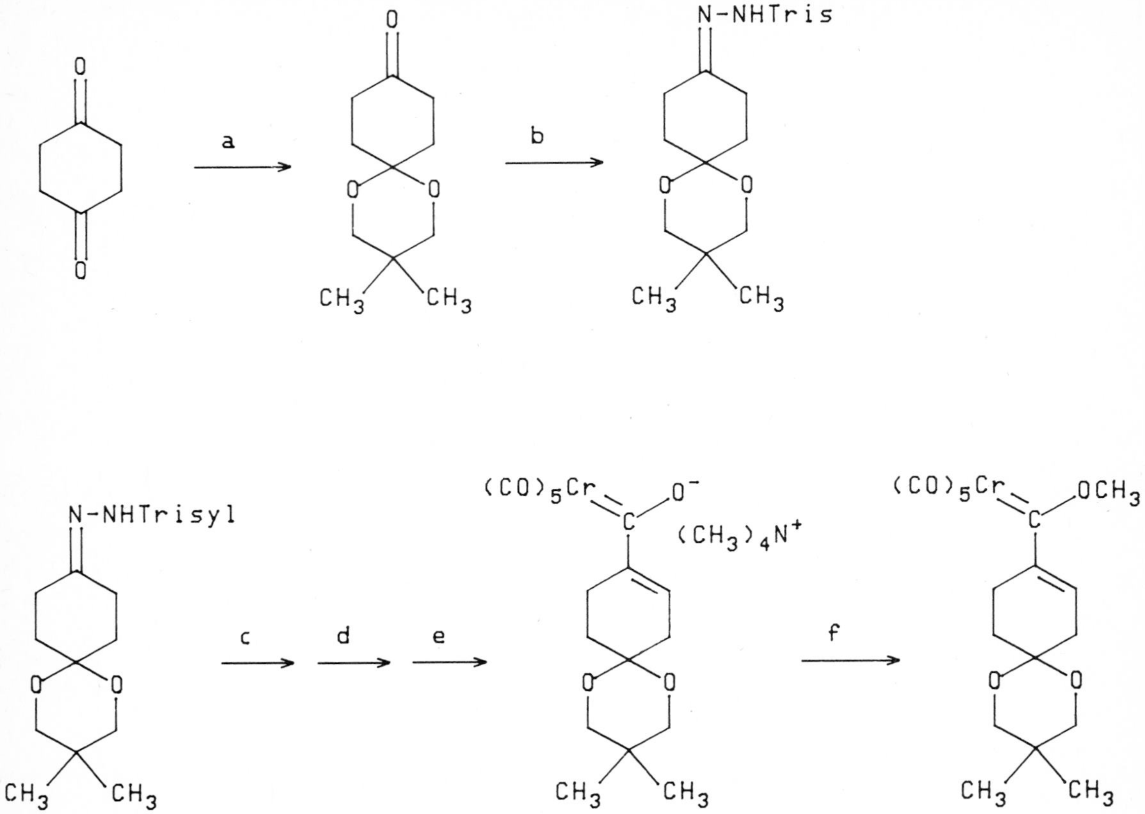

a. $HOCH_2C(CH_3)_2CH_2OH$; 85%, >94% pure; Reference 61c

b. rt, $TrisNHNH_2$, CH_3OH; 93%

c. -78°C, n-BuLi, THF, hexane

d. rt, $Cr(CO)_6$, Et_2O

e. $(CH_3)_4NBr$; 41% for three steps

f. rt, CH_3OSO_2F, CH_2Cl_2; 88%

Figure 12.31

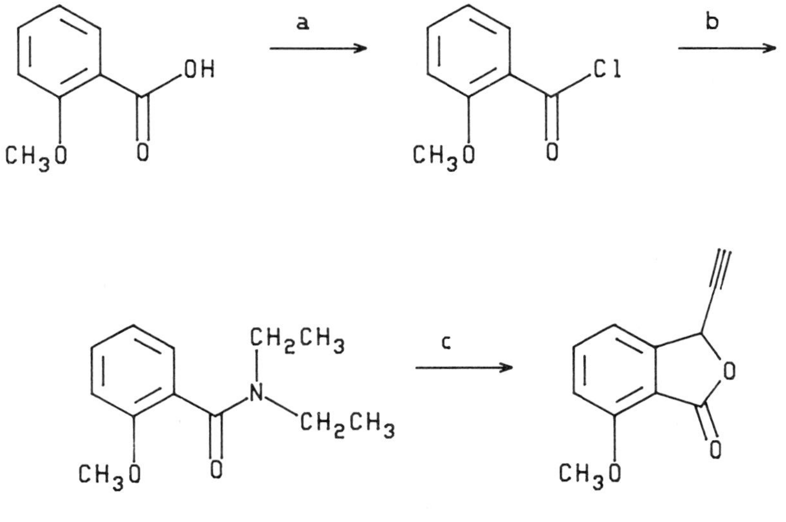

a. heat, $SOCl_2$

b. Et_2NH, Et_2O; 87%

c. 1) −78°C, s-BuLi, TMEDA, 2) ≡—C(O)H; 42%

Figure 12.32

The total synthesis of (±)-daunomycinone from 1,4-cyclohexanedione required 19 steps and afforded an overall 0.93% yield (from *o*-anisic acid, 17 steps, overall 1.3% yield). The synthesis could be improved significantly by development of an alternative to the Kende procedure. The tetracyclic ketone synthesis from 1,4-cyclohexanedione required 14 steps and afforded an overall 9.5% yield (from *o*-anisic acid, 13 steps, overall 12.1% yield) (see Figure 12.35).

12.4.2 Route B. From a Cyclohexenylmethoxycarbene Prepared by Diels–Alder Cycloaddition[34]

A route A-type cyclohexenylmethoxycarbene can also be prepared by Diels–Alder cycloaddition using an alkynylmethoxycarbene as the dienophile. The alkynylmethoxycarbene is prepared by Fischer's method.[12,69] The Diels–Alder cycloaddition[70] and carbene–alkyne cocyclization are carried out in one pot. Ketalization of the sensitive trimethylsilyl enol ether, hydrolysis of the hydroquinone trimethylsilyl ether, and demetallation affords the route A intermediate hydroquinone monomethyl ether in 23% unoptimized yield.

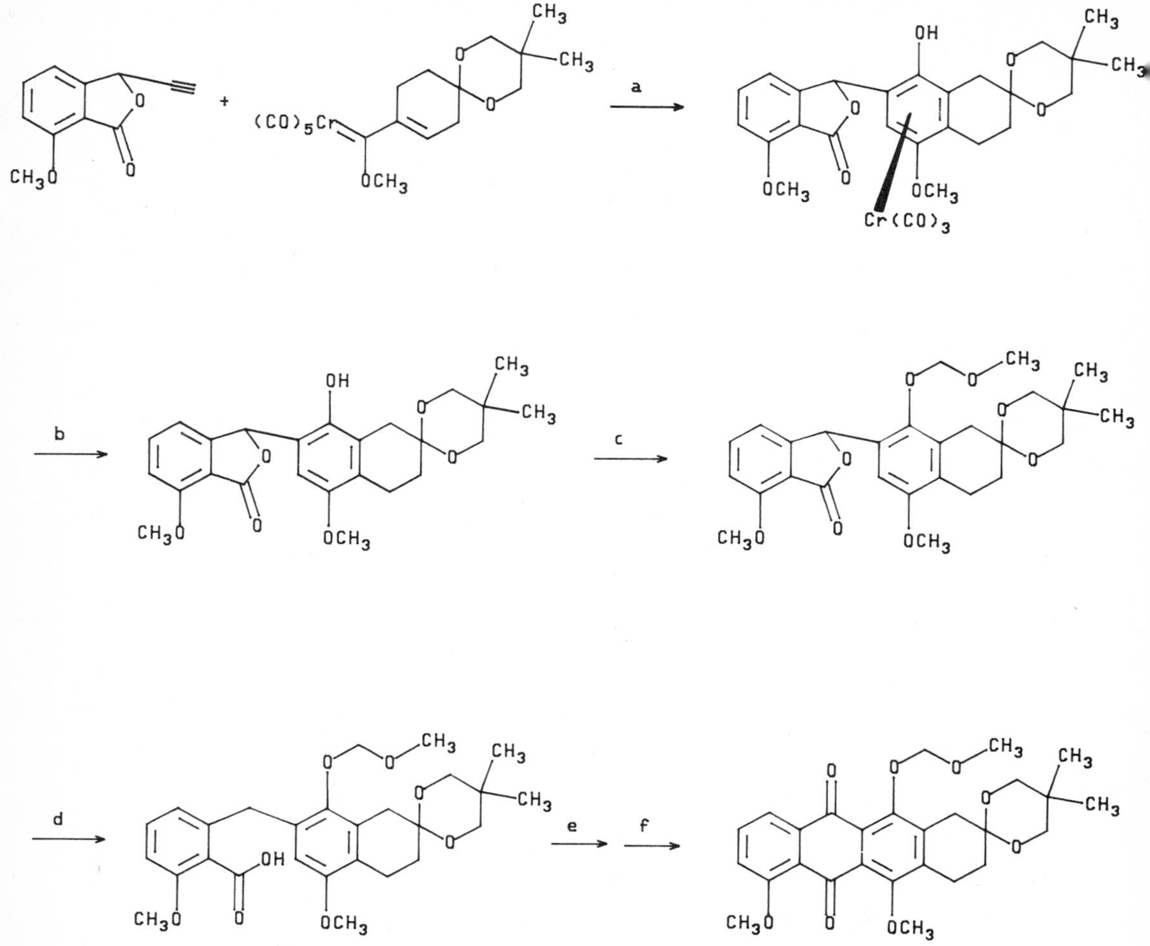

a. 45°C, THF

b. $[Fe(DMF)_3Cl_2][FeCl_4]$, THF; 76% for two steps

c. rt, $ClCH_2OCH_3$, i-$PrNEt_2$, CH_2Cl_2; 91%

d. Zn, $CuSO_4$, KOH, pyridine; 95%

e. 0°C to rt, TFAA, t-Bu–(pyridine)–t-Bu, CH_2Cl_2

f. rt, O_2, Triton B, CH_3OH; 69% for two steps

Figure 12.33

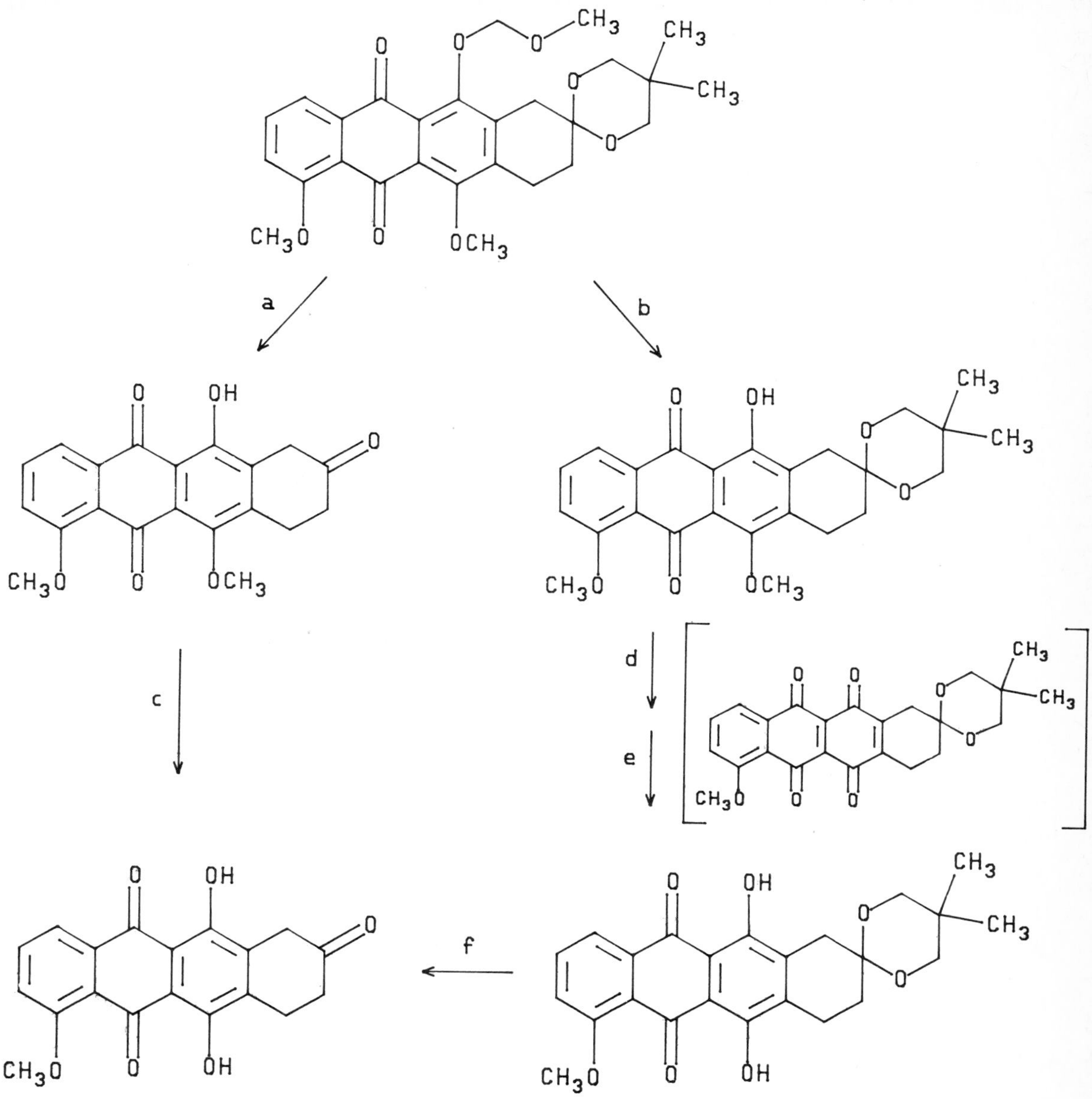

a. 56°C, H_2SO_4, acetone; 100%

b. rt, CF_3COOH, CH_2Cl_2; 99%

c. 55°C, 48% HBr, AcOH; 65%

d. rt, AgO, HNO_3, acetone, H_2O

e. rt, $NaHSO_3$, H_2O; 96% (crude) for two steps

f. 45°C, H_2SO_4, acetone, EtOAc, H_2O; 77%

Figure 12.34

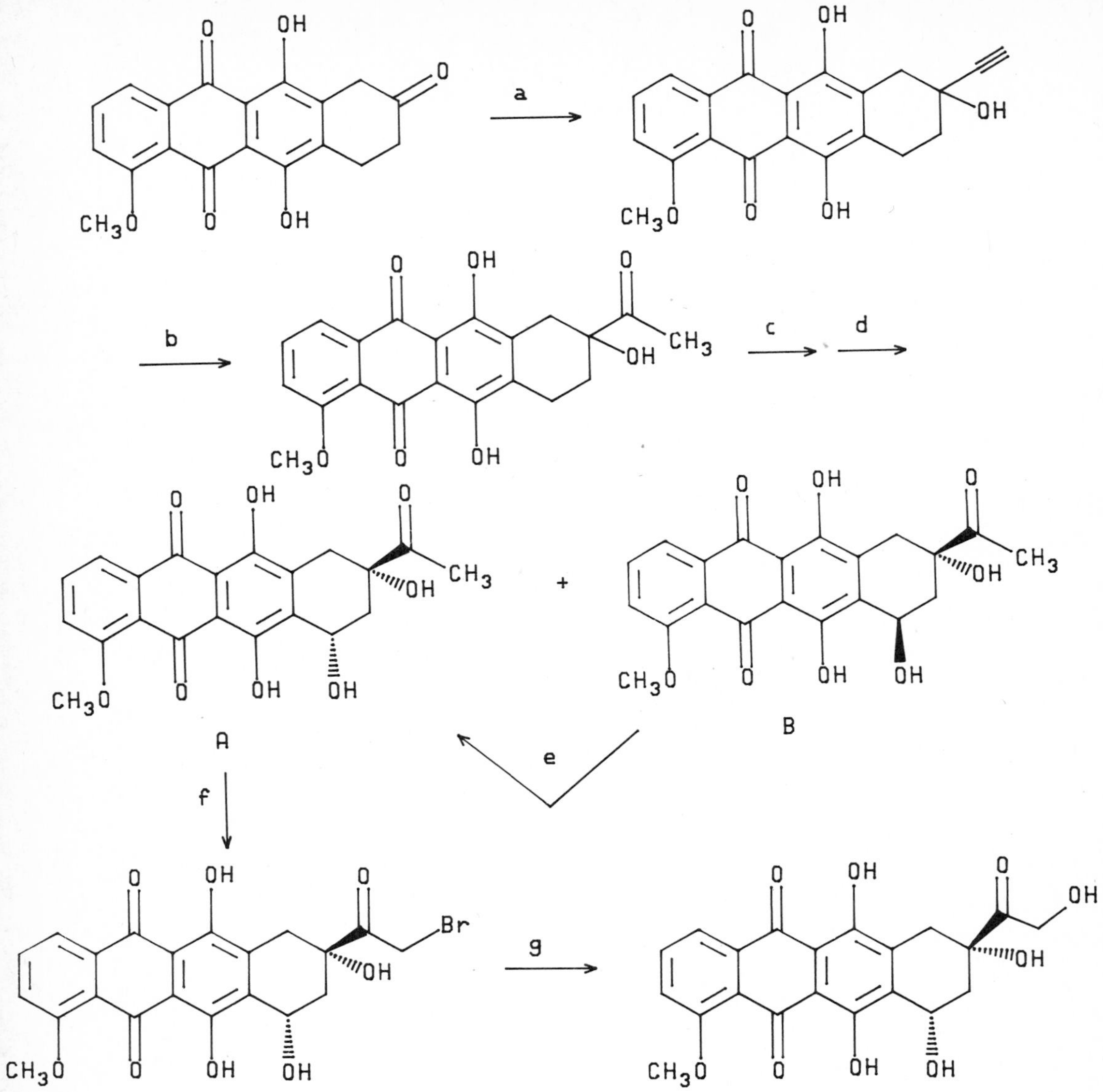

(±)-Adriamycinone

a. ≡—MgBr, THF; 52%

b. 70°C, H_2SO_4, HgO, H_2O; 40%

c. hν, Br_2, CCl_4

d. H_2O; A:B, 5:2, 50% yield of A for three steps

e. 25°C, TFA; 76%

f. Br_2, $CHCl_3$

g. NaOH, acetone, H_2O; 87% for two steps

Figure 12.35

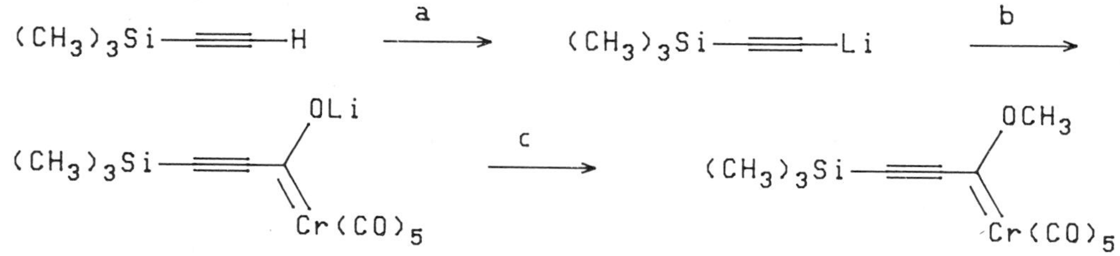

$(CH_3)_3Si—C≡C—H$ →(a) $(CH_3)_3Si—C≡C—Li$ →(b)

$(CH_3)_3Si—C≡C—C(OLi)=Cr(CO)_5$ →(c) $(CH_3)_3Si—C≡C—C(OCH_3)=Cr(CO)_5$

a. $-78^{o}C$, n-BuLi, Et_2O

b. $0^{o}C$, $Cr(CO)_6$, Et_2O

c. CH_3OSO_2F, Et_2O, THF; 77%

Figure 12.36

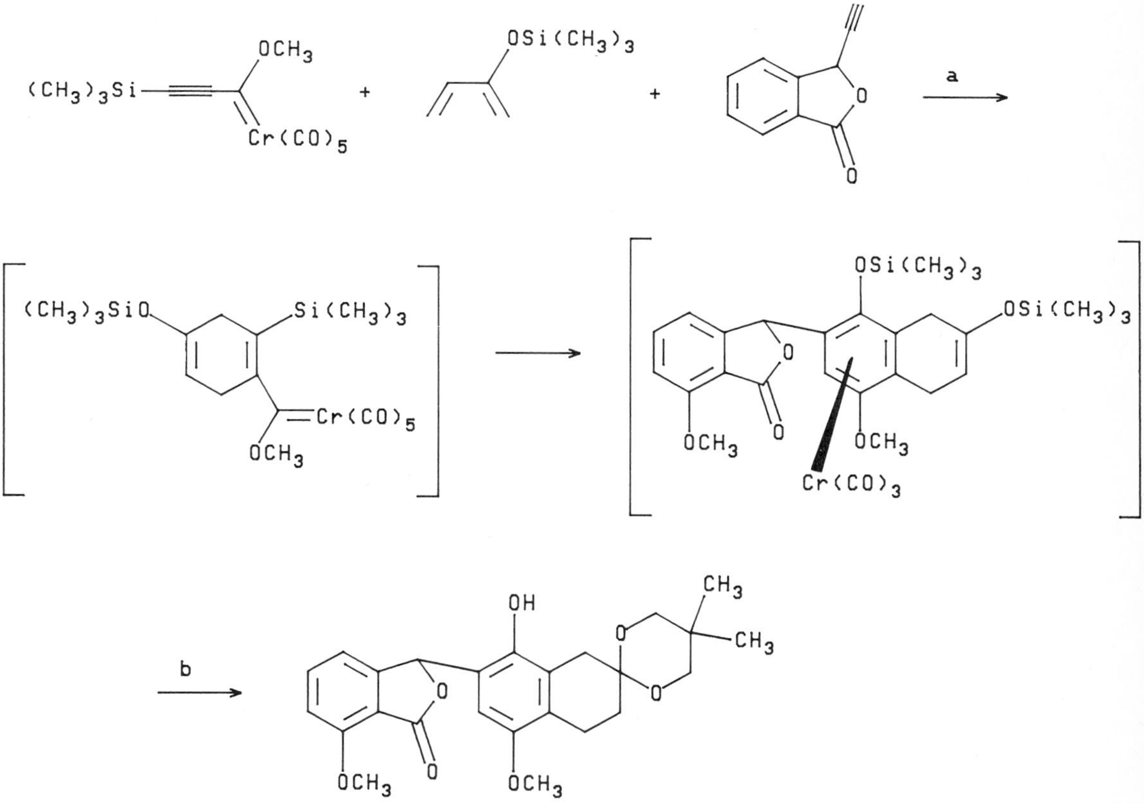

$(CH_3)_3Si—C≡C—C(OCH_3)=Cr(CO)_5$ + $OSi(CH_3)_3$ + O, O →(a)

[$(CH_3)_3SiO$, $Si(CH_3)_3$, $=Cr(CO)_5$, OCH_3] → [$OSi(CH_3)_3$, $OSi(CH_3)_3$, O, O, OCH_3, OCH_3, $Cr(CO)_3$]

→(b) OH, CH_3, O, CH_3, O, O, O, OCH_3, OCH_3

a. $50^{o}C$, THF

b. $HOCH_2C(CH_3)_2CH_2OH$, cat. $(CH_3)_3SiCl$; 23%

Figure 12.37

The total synthesis of (±)-daunomycinone from trimethylsilylacetylene required 16 steps and afforded an overall 0.18% yield (from 2-trimethylsiloxy-1,3-butadiene, 15 steps, overall 1.0% yield) (see Figures 12.36 and 12.37).

12.4.3 Route C. From a Naphthylmethoxycarbene[70]

The naphthylmethoxycarbene is constructed using Fischer's method.[12] Reaction of 1,5-dihydroxynaphthalene[72] with dimethylsulfate affords 1,5-dimethoxynaphthalene.[73] Regioselective Vilsmeier formylation followed by Baeyer–Villiger oxidation of the aldehyde introduces a hydroxyl group. Regioselective bromination adjacent to the hydroxyl, then methylation, affords 2-bromo-1,4,8-trimethoxynaphthalene.[61b,74] Lithium–halogen exchange produces the naphthyllithium required for carbene preparation. This carbene, which possesses a methoxy substituent on the ring position adjacent to the carbene carbon, readily eliminates a carbon monoxide. Thus carbene–alkyne cocyclization in route C will not involve an initial rate-determining loss of a carbon monoxide ligand (see Figure 12.38).

The alkyne can be constructed in a number of ways. The shortest and most efficient approach begins with methyl acetoacetate.[75] Deprotonation with methoxide affords the stabilized anion, which is alkylated with propargyl bromide. Subsequent deprotonation with sodium hydride in tetrahydrofuran affords a stabilized anion, which is alkylated with methyl bromoacetate. When some methanol is present in the alkylation medium, simultaneous deacylation of the β-ketoester affords 3-carbomethoxy-5-hexynoic acid methyl ester directly (see Figure 12.39).

Carbene–alkyne cocyclization with this terminal alkyne is regiospecific, affording the hydroquinone monomethyl ether complex with alkyne substituent adjacent to the hydroxyl in 72% isolated yield. Demetallation with high-pressure carbon monoxide quantitatively produces the hydroquinone monomethyl ether and recyclable chromium hexacarbonyl. O-Methylation, then facile intramolecular ring acylation produces tetracyclic material (see Figure 12.40).

The tetracyclic material is converted to (±)-daunomycinone using the procedures of Hauser,[76] Sih,[77] and Kende.[68] Hauser reduced the ketone to a methylene in two steps (borohydride reduction to alcohol, triethylsilane reduction to methylene). The ester is best converted to the daunomycinone methyl ketone by hydrolysis to the acid, then nucleophilic acyl substitution with methyllithium. The ketone is protected; then the C ring is oxidized to a quinone with a nitrogen oxide and cerium(IV) in aqueous acetonitrile. Selective demethylation of the B-ring hydroquinone dimethyl ether and subsequent deketalization affords (±)-8,10-dideoxydaunomycinone.

The 8-hydroxyl group is introduced using Sih's procedure.[77] Heating with acetic anhydride and *p*-toluenesulfonic acid converts the ketone to an enol acetate and protects the hydroquinone. Oxidation with *m*-chloroperbenzoic acid affords the epoxide. Ring opening to an α-hydroxy ketone and liberaton of the hydroquinone are effected by aqueous base, then aqueous acid hydrolyses. Conversion of (±)-10-deoxydaunomycinone to (±)-daunomycinone by free radical benzylic bromination and hydrolysis of the bromide[68] has been described in route A.

Total synthesis of (±)-daunomycinone from 1,5-dihydroxynaphthalene required 27 steps and afforded an overall 3.2% yield (from methyl acetoacetate, 21 steps, overall 5.3% yield). Route C is the most efficient, although longer primarily due to the

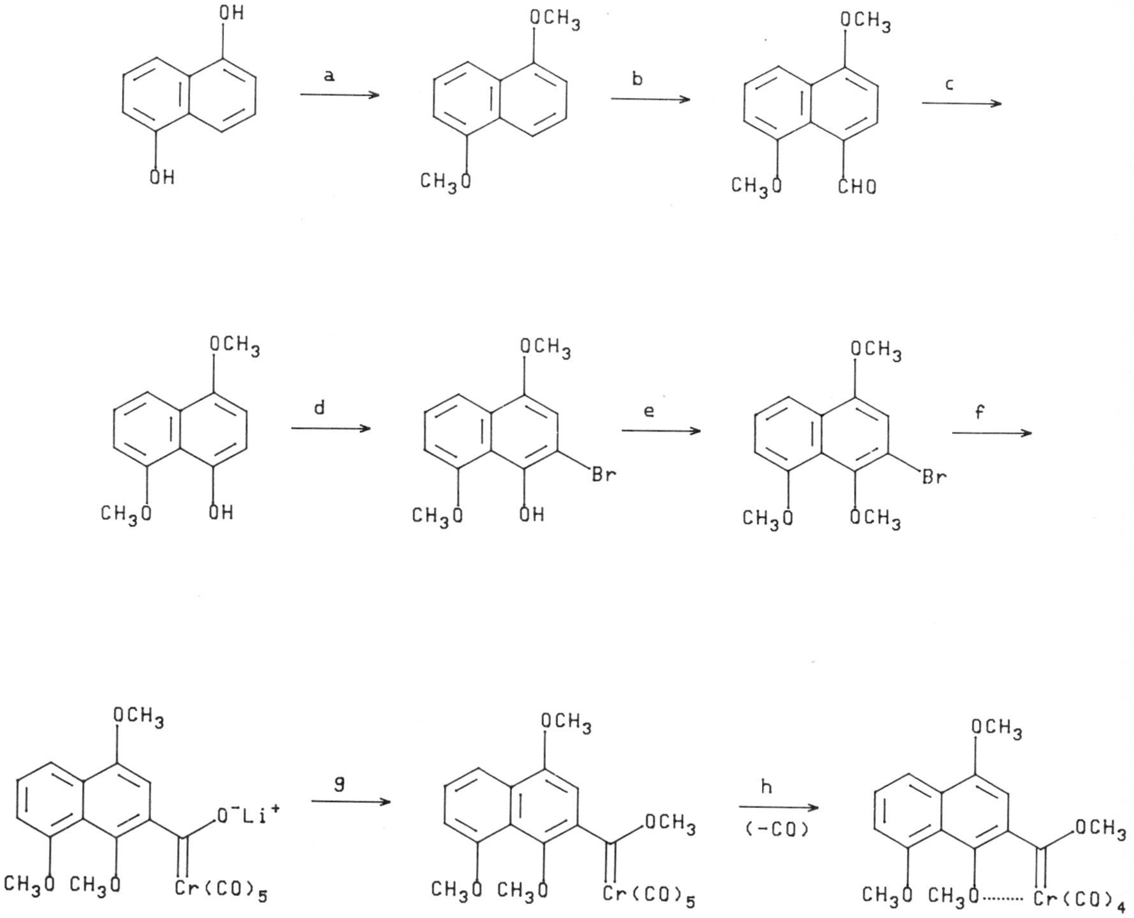

a. ≤70°C, NaOH, H_2O, $CH_3OSO_2OCH_3$; 100%

b. 0°C to rfx, $POCl_3$-DMF, toluene; 93%

c. rt, MCPBA, CH_2Cl_2; 77%

d. rt, Br_2, CCl_4; 98%

e. 1) NaOH, 2) $CH_3OSO_2OCH_3$, EtOH; 79%

f. 1) n-BuLi, 2) $Cr(CO)_6$

g. $(CH_3)_3OBF_4$; 79% for two steps

h. 55°C, t-$BuOCH_3$; 97%

Figure 12.38

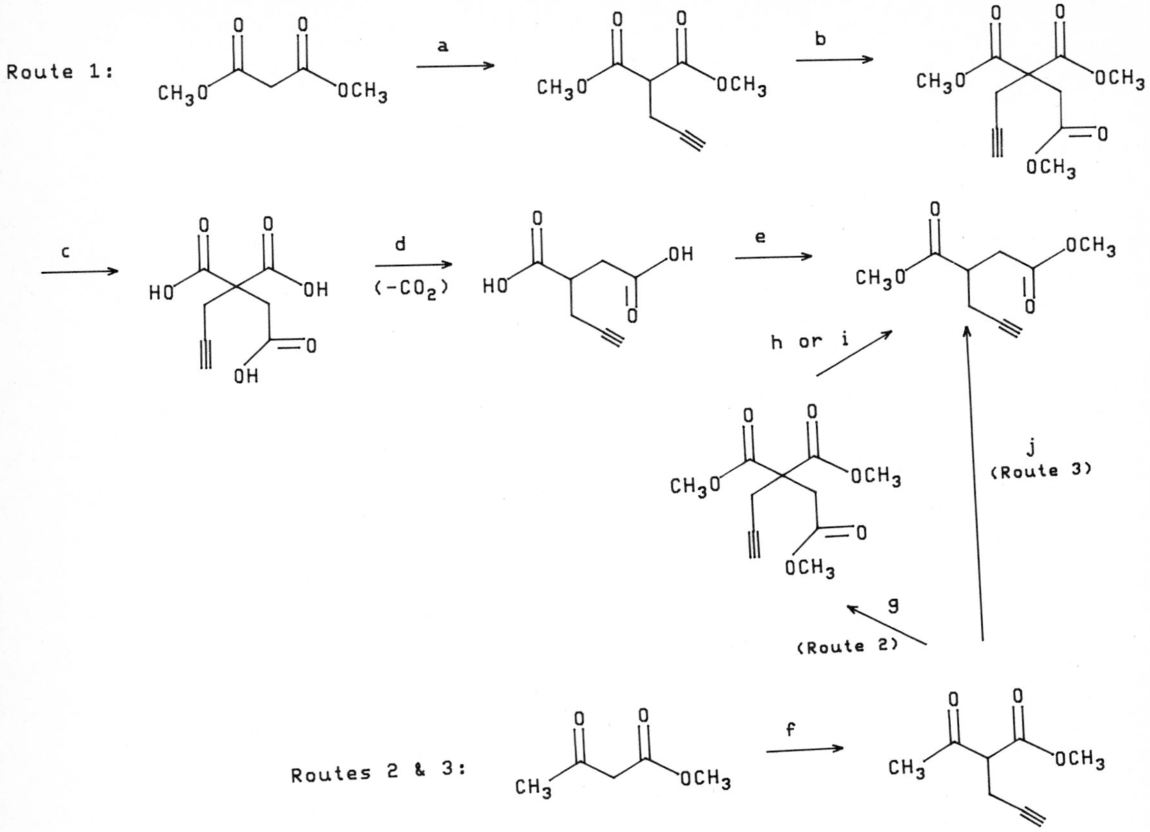

Route 1:

a. 1) 0°C, $NaOCH_3$, CH_3OH, 2) 0°C, ≡—CH_2Br; 52%

b. 1) rt, NaH, THF, 2) rt to rfx, $BrCH_2COOCH_3$; 98%

c. 1) rfx, KOH, H_2O, EtOH, 2) HCl; 98%

d. 160°C; 51%

e. rt, CH_2N_2, Et_2O; 75%

f. 1) 0°C, $NaOCH_3$, CH_3OH, 2) 0°C to rfx, ≡—CH_2Br; 76%

g. 1) rt, NaH, THF, 2) rt to rfx, $BrCH_2COOCH_3$; 98%

h. 1) 90°C, KOH, H_2O, EtOH, 2) HCl, 3) CH_2N_2; 68%

i. 1) HCl, H_2O, EtOH, 2) CH_2N_2, Et_2O; 78%

j. 1) rt, NaH, THF, 2) rt to rfx, $BrCH_2COOCH_3$,
3) CH_3OH; 92%

Figure 12.39

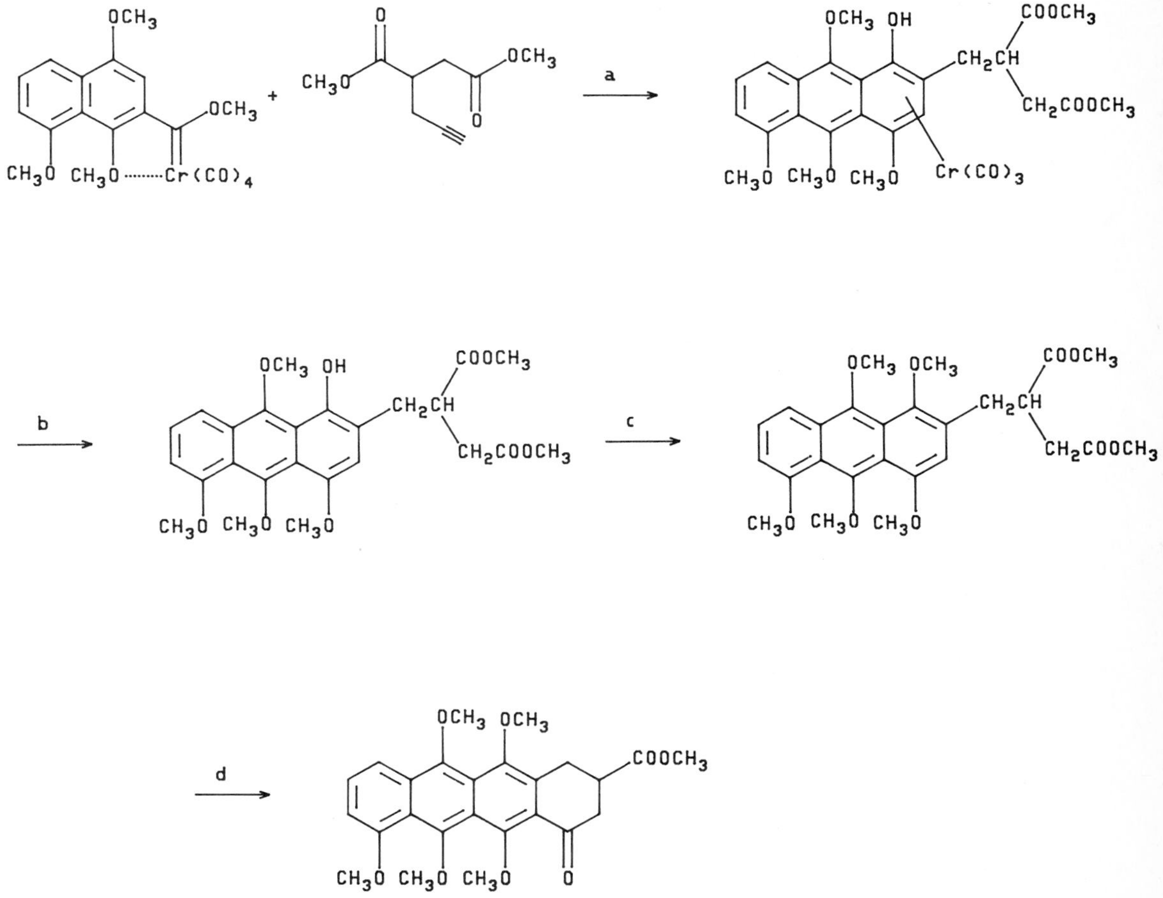

a. 50°C, t-$BuOCH_3$; 72%

b. 50°C, CO, Et_2O; 100%

c. $CH_3OSO_2OCH_3$

d. 0°C, 2:1 TFAA:TFA, CH_2Cl_2; 64% for two steps

Figure 12.40

Figure 12.41

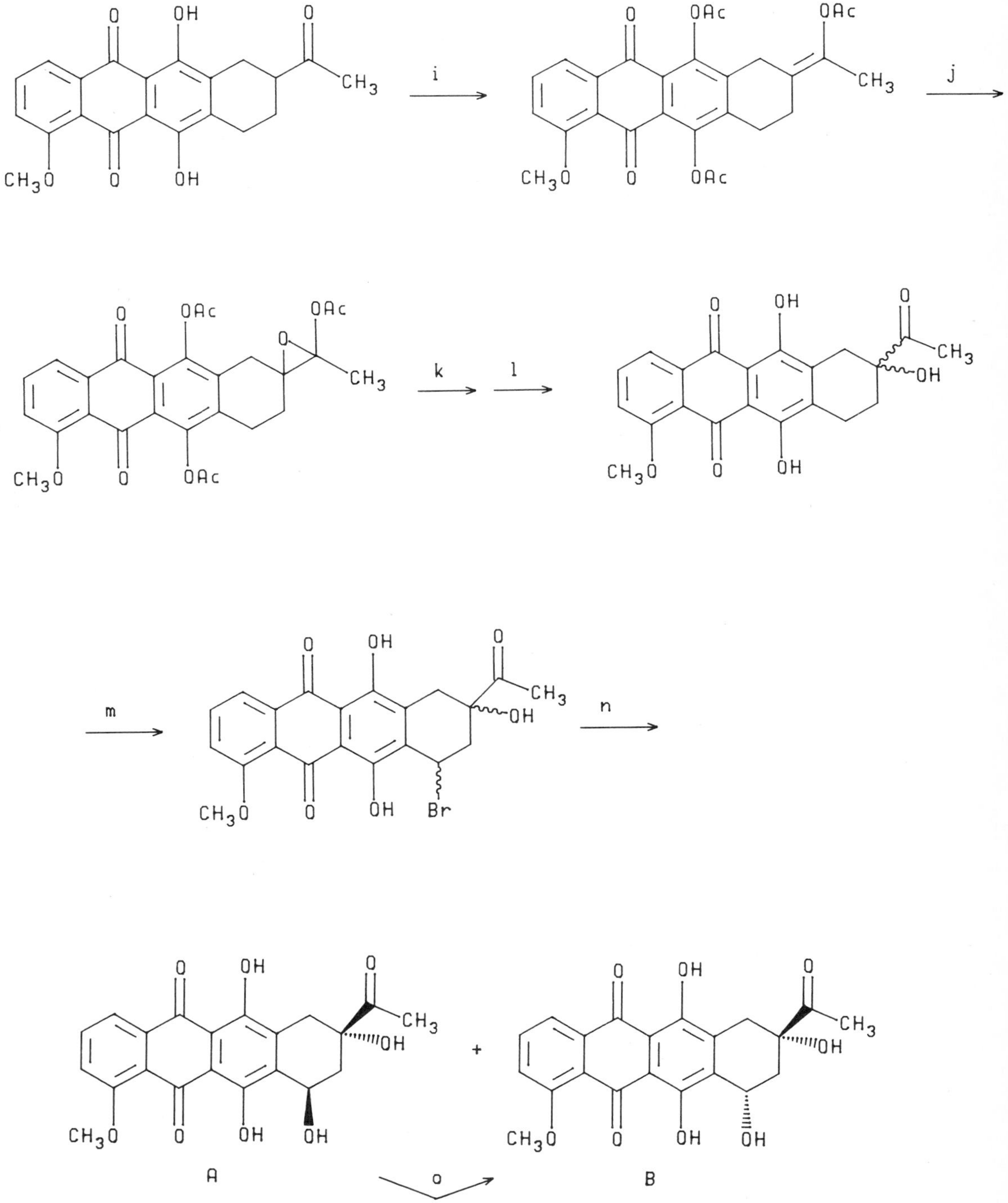

Figure 12.41 (*Continued*)

a. −10°C, $NaBH_4$, THF, H_2O; 96%

b. −20°C, Et_3SiH, TFA, CH_2Cl_2; 89%

c. 100°C, KOH, H_2O; 98%

d. 0°C, LiH, then CH_3Li, DME; 94%

e. rfx, p-TsOH·H_2O, $HOCH_2CH_2OH$, benzene; 97%

f. 0°C, N-oxide, $(NH_4)_2Ce(NO_3)_6$, H_2O, CH_3CN; 96%

g. rt, AgO, acetone; 95%

h. rt, TFA, HCl, DMF, H_2O; 94%

i. rfx, p-TsOH·H_2O, Ac_2O

j. 25°C, MCPBA

k. rt, NaOH, H_2O, EtOH

l. rt, AcOH, H_2SO_4, H_2O; 50% for four steps

m. hν, Br_2, CCl_4

n. H_2O

o. rt, TFA; A:B, 5:2, 50% yield for three steps

Figure 12.41 *(Continued)*

unappealing necessity for removal and reintroduction of oxygen functionality at the 10-position (see Figure 12.41).

12.5 CONVERSION OF (±)-DAUNOMYCINONE TO (±)-ADRIAMYCINONE

The ketone of (±)-daunomycinone is converted to the α-hydroxy ketone of (±)-adriamycinone by careful α-bromination, then solvolysis of the bromide.[78]

REFERENCES

1. Di Marco, A.; Gaetani, M.; Orezzi, P.; Scarpinato, B. M.; Silvestrini, R.; Soldati, M.; Dasdia, T.; Valentini, L. *Nature (London)* **1964**, *201*, 706.
2. Arcamone, F.; Franceschi, G.; Penco, S.; Selva, A. *Tetrahedron Lett.* **1969**, 1007.
3. Bonadonna, G.; Monfardini, S.; De Lena, M.; Fossati-Bellani, F.; Beretta, G. *Cancer Res.* **1970**, *30*, 2572.

4. *Anthracyclines: Current Status and New Developments*; Crooke, S. T.; Recih, S. D., Eds; Academic: New York, 1980.
5. Weiss, A. J.; Manthel, R. W. *Cancer* **1977**, *40*, 2046.
6. *Cancer Chemotherapy, 1979, Annual 1*; Pinedo, H. M., Ed. Elsevier: New York, 1979.

7a. Calendi, E.; Di Marco, A.; Reggiani, M.; Scarpinato, B.; Valentini, L. *Biochim. Biophys. Acta* **1965**, *103*, 25.

7b. Di Marco, A.; Zunino, F.; Silvestrini, R.; Gambarucci, C.; Gambetta, R. A. *Biochem. Pharmacol.* **1971**, *20*, 1323.

7c. Pigram, W. J.; Fuller, W.; Hamilton, L. D. *Nature New Biol.* **1972**, *235*, 17.

8. Fridovich, I. *Science* **1978**, *201*, 875.
9. Lown, J. W.; Sim, S-K.; Majumdar, K. C.; Chang, R-Y. *Biochem. Biophys. Res. Commun.* **1977**, *76*, 705.
10. See the following references for the most recent synthetic work on daunomycinone:

10a. Rao, A. V. R.; Mehendale, A. R.; Reddy, K. B. *Indian J. Chem.* **1984**, *23B*, 1154.

10b. Keay, B. A.; Rodrigo, R. *Tetrahedron* **1984**, *40*, 4597.

10c., Kelly, T. R.; Ananthasubramanian, L.; Borah, K.; Gillard, J. W.; Goerner, R. N., Jr.; King, P. F.; Lyding, J. M.; Tsang, W.-G.; Vaya, J. *Tetrahedron* **1984**, *40*, 4569.

10d. Swenton, J. S.; Freskos, J. N.; Morrow, G. W.; Sercel, A. D. *Tetrahedron* **1984**, *40*, 4625.

10e. Sibi, M. P.; Altintas, N.; Snieckus, V. *Tetrahedron* **1984**, *40*, 4593.

10f. Tamariz, J.; Schwager, L.; Stibbard, J. H. A.; Vogel, P. *Tetrahedron Lett.* **1983**, *24*, 1497.

10g. Roa, A. V. R.; Reddy, K. B.; Mehendale, A. R. *J. Chem. Soc., Chem. Commun.* **1983**, 564.

10h. Keay, B. A.; Rodrigo, R. *Can. J. Chem.* **1983**, *61*, 637.

10i. Broadhurst, M. J.; Hassall, C. H. *J. Chem. Soc. Perkin Trans 1* **1982**, 2227.

10j. Kimball, S. D.; Kim, K. S.; Mohanty, D. K.; Vanotti, E.; Johnson, F. *Tetrahedron Lett.* **1982**, *23*, 3871.

11a. Horton, D.; Priebe, W.; Turner, W. R. *Carbohydrate Res.* **1981**, *94*, 11.

11b. Smith, T. H.; Fujiwara, A. N.; Lee, W. W.; Wu, H. Y.; Henry, D. W. *J. Org. Chem.* **1977**, *42*, 3653.

11c. Raynolds, P. W.; Manning, M. J.; Swenton, J. S. *Tetrahedron Lett.* **1977**, 2383.

11d. Smith, T. H.; Fujiwara, A. N.; Henry, D. W.; Lee, W. W. *J. Am. Chem. Soc.* **1976**, *98*, 1969.

12a. Fischer, E. O.; Maasböl, A. *Angew. Chem. Int. Ed. Engl.* **1964**, *3*, 580.

12b. Fischer, E. O.; Schubert, U.; Kleine, W,; Fischer, H. *Ilorg. Syn.* **1949**, *19*, 164.

13. Klabunde, U.; Fischer, E. O. *J. Am. Chem. Soc.* **1967**, *89*, 7141.
14. Connor, J. A.; Fischer, E. O. *Chem. Commun.* 1967, 1024.
15. Fischer, E. O.; Leupold, M. *Chem. Ber.* **1972**, *105*, 599.
16. Brunner, H.; Doppelberger, J.; Fischer, E. O; Lappus, M. *J. Organometal. Chem.* **1976**, *112*, 65.
17. Werner, H.; Fischer, E. O.; Heckl, B.; Kreiter, C. G. *J. Organometal. Chem.* **1971**, *28*, 367.
18. Fischer, E. O.; Leupold, M.; Kreiter, C. G.; Müller, J. *Chem. Ber.* **1972**, *105*, 150.
19. Fischer, E. O.; Held, W.; Kreissel, F. R.; Frank, A.; Huttner, G. *Chem. Ber.* **1977**, *110*, 656.
20. Block, T. F.; Fenske, R. F.; Casey, C. P. *J. Am. Chem. Soc.* **1976**, *98*, 443.
21. Dötz, K. H. *Chem. Ber.* **1977**, *110*, 78.
22. Fischer, H.; Dötz, K. H. *Chem. Ber.* **1980**, *113*, 193.
23. Fischer, E. O.; Schubert, U.; Märkl, R. *Chem. Ber.* **1981**, *114*, 3412.
24. Dötz, K. H.; Pruskil, I. *Chem. Ber.* **1981**, *114*, 1980.
25. Casey, C. P.; Anderson, R. L. *J. Am. Chem. Soc.* **1974**, *96*, 1230.

26. Casey, C. P.; Anderson, R. L. *J. Organometal. Chem.* **1974**, *73*, C28.
27. Casey, C. P.; Brunsvold, W. R. *J. Organometal. Chem.* **1975**, *102*, 175.
28. Casey, C. P.; Brunsvold, W. R.; Scheck, D. M. *Inorg. Chem.* **1977**, *16*, 3059.
29. Wulff, W. D.; Gilbertson, S. R. *J. Am. Chem. Soc.* **1985**, *107*, 503.
30. Xu, Y-C.; Wulff, W. D. *J. Org. Chem.* **1987**, *52*, 3263.
31. Casey, C. P.; Brunsvold, W. R. *Inorg. Chem.* **1977**, *16*, 391.
32. Wulff, W. D.; Yang, D. C. *J. Am. Chem. Soc.* **1983**, *105*, 6726.
33. Dötz, K. H.; Kuhn, W.; Müller, G.; Huber, B.; Alt. H. G. *Angew. Chem. Int. Ed. Engl.* **1986**, *25*, 812.
34. Wulff, W. D.; Yang, D. C. *J. Am. Chem. Soc.* **1984**, *106*, 7565.
35a. Lukehart, C. M.; Zeile, J. V. *J. Organometal. Chem.* **1975**, *97*, 421.
35b. Lukehart, C. M.; Zeile, J. V. *Inorg. Chim. Acta* **1976**, *17*, L7.
35c. Dötz, K. H.; Fügen-Köster, B.; Neugebauer, D. *J. Organometal. Chem.* **1979**, *182*, 489.
36a. Fischer, E. O.; Dötz, K. H. *Chem. Ber.* **1972**, *105*, 3966.
36b. Fischer, E. O.; Dorrer, B. *Chem. Ber.* **1974**, *107*, 1156.
37a. Fischer, E. O.; Dötz, K. H. *Chem. Ber.* **1970**, *103*, 1273.
37b. Dötz, K. H.; Fischer, E. O. *Chem. Ber.* **1972**, *105*, 1356.
37c. Dorrer, B.; Fischer, E. O.; Kalbfus, W. *J. Organometal. Chem.* **1974**, *81*, C20.
38. Dötz, K. H. *Angew. Chem. Int. Ed. Engl.* **1975**, *14*, 644.
39a. Dötz, K. H. *Pure Appl. Chem.* **1983**, *55*, 1689.
39b. Dötz, K. H. *Angew. Chem. Int. Ed. Engl.* **1984**, *23*, 587.
40. Dötz, K. H.; Dietz, R. *Chem. Ber.* **1977**, *110*, 1555.
41. Fischer, H.; Mühlemeier, J.; Märkl, R.; Dötz, K. H. *Chem. Ber.* **1982**, *115*, 1355.
42. Casey, C. P.; Cesa, M. C. *Organometallics* **1982**, *1*, 87.
43. Yamashita, A.; Scahill, T. A. *Tetrahedron Lett.* **1982**, *23*, 3765.
44. Dötz, K. H.; Sturm, W. *J. Organometal. Chem.* **1985**, *285*, 205.
45a. Dötz, K. H. *Angew. Chem. Int. Ed. Engl.* **1979**, *18*, 954.
45b. Dötz, K. H.; Fügen-Köster, B. *Chem. Ber* **1980**, *113*, 1449.
45c. See also Tang, P-C.; Wulff, W. D. *J. Am. Chem. Soc.* **1984**, *106*, 1132.
46. Dötz, K. H.; Dietz, R. *J. Organometal Chem.* **1978**, *157*, C55.
47a. Dötz, K. H.; Dietz, R.; Kappenstein, C.; Neugebauer, D.; Schubert, U. *Chem. Ber.* **1979**, *112*, 3682.
47b. For other reports of indene formation, see Dötz, K. H.; Neugebauer, D. *Angew. Chem. Int. Ed. Engl.* **1978**, *17*, 851.
47c. Dötz, K. H. *J. Organometal. Chem.* **1977**, *140*, 177.
48. Wulff, W. D.; Kaesler, R. W.; Peterson, G. A.; Tang, P-C. *J. Am. Chem. Soc.* **1985**, *107*, 1060.
49. Wulff, W. D.; Tang, P. C.; McCallum, J. S. *J. Am. Chem. Soc.* **1981**, *103*, 7677.
50. Wulff, W. D.; Chan, K-S.; Tang, P-C. *J. Org. Chem.* **1984**, *49*, 2293.
51. Dötz, K. H.; Kuhn, W. *Angew. Chem. Int. Ed. Engl.* **1983**, *22*, 732.
52. Dötz, K. H.; Dietz, R. *Chem. Ber.* **1978**, *111*, 2517.
53. Wulff, W. D.; Tang, P-C. *J. Am. Chem. Soc.* **1984**, *106*, 434.
54. Dötz, K. H.; Kuhn, W. J. *Organometal. Chem.* **1983**, *252*, C78.
55. Dötz, K. H.; Pruskil, I.; Mühlemeier, J. *Chem. Ber.* **1982**, *115*, 1278.
56. Dötz, K. H.; Pruskil, I. J. *Organometal Chem.* **1981**, *209*, C4.
57. Dötz, K. H.; Mühlemeier, J.; Schubert, U.; Orama, O. *J. Organometal. Chem.* **1983**, *247*, 187.

58. Dötz, K. H.; Dietz, R.; von Imhof, A.; Lorenz, H.; Huttner, G. *Chem. Ber.* **1976**, *109*, 2033.
59. Semmelhack, M. F.; Bozell, J. J. *Tetrahedron. Lett.* **1982**, *23*, 2931.
60. Semmelhack. M. F.; Bozell, J. J.; Sato, T.; Wulff, W.; Spiess, E.; Zask, A. *J. Am. Chem. Soc.* **1982**, *104*, 5850.
61. 1,4-Cyclohexanedione is commercially available (Lancaster Synthesis 89–90 catalog, 100 g for $71.30)
62a. Haslanger, M.; Lawton, R. G. *Syn. Comm.* **1974**, *4*, 155.
62b. Mussini, P.; Orsini, F.; Pelizzoni, F. *Syn. Comm.* **1975**, *5*, 283.
62c. Marshall, J. A.; Flynn, G. A. *Syn. Comm.* **1979**, *9*, 123.
63. Adlington, R. M.; Barrett, A. G. M. *Acc. Chem. Res.* **1983**, *16*, 55.
64. Chamberlin, A. R.; Stemke, J. E.; Bond, F. T. *J. Org. Chem.* **1978**, *43*, 147.
65. *o*-Anisic acid is commercially available and inexpensive (Fluka 88–89 catalog, 100 g for $18.00).
66. McCabe, E. T.; Barthel, W. F.; Gertler, S. I.; Hall, S. A. *J. Org. Chem.* **1954**, *19*, 493.
67a. Beak, P.; Brown, R. A. *J. Org. Chem.* **1982**, *47*, 34.
67b. Potts, K. T.; Bhattacharjee, D.; Walsh, E. B. *J. Org. Chem.* **1986**, *51*, 2011.
68. Kende, A. S.; Tsay, Y-G.; Mills, J. E. *J. Am. Chem. Soc.* **1976**, *98*, 1967.
69. Trimethylsilylacetylene is commercially available (Lancaster Synthesis 89–90 catalog, 25 g for $91.60).
70. 2-Trimethylsiloxy-1,3-butadiene is commercially available (Aldrich 88–89 catalog, 5 g for $50.90).
71a. Dötz, K. H.; Popall, M. *Tetrahedron* **1985**, *41*, 5797.
71b. Dötz, K. H. Ger. Offen. DE 3, 407, 632 12 Sept 1985 (Appl. 01 Mar 1984); *Chem. Abstr.* **1986**, *104*, 168270c.
72. 1,5-Dihydroxynaphthalene is commercially available and inexpensive (Aldrich 88–89 catalog, 500 g for $28.80).
73. Naylor, C. A., Jr.; Gardner, J. H. *J. Am. Chem. Soc.* **1931**, *53*, 4109.
74. Hannan, R. L.; Barber, R. B.; Rapoport, H. *J. Org. Chem.* **1979**, *44*, 2153.
75. Methyl acetoacetate is commercially available and inexpensive (Kodak 87–88 catalog, 3 kg for $26.30).
76. Hauser, F. M.; Prasanna, S. *J. Am. Chem. Soc.* **1981**, *103*, 6378.
77. Gleim, R. D.; Trenbeath, S.; Mittal, R. S. D.; Sih, C. J. *Tetrahedron. Lett.* **1976**, 3385.
78a. Smith, T. H.; Fujiwara, A. N.; Henry, D. W.; Lee, W. W. *J. Am. Chem. Soc.* **1976**, *98*, 1969.
78b. Smith, T. H.; Fujiwara, A. N.; Lee, W. W.; Wu, H. Y.; Henry, D. W. *J. Org. Chem.* **1977**, *42*, 3653.

CHAPTER 13

$(\pm)$-$\Delta^{9(12)}$-Capnellane

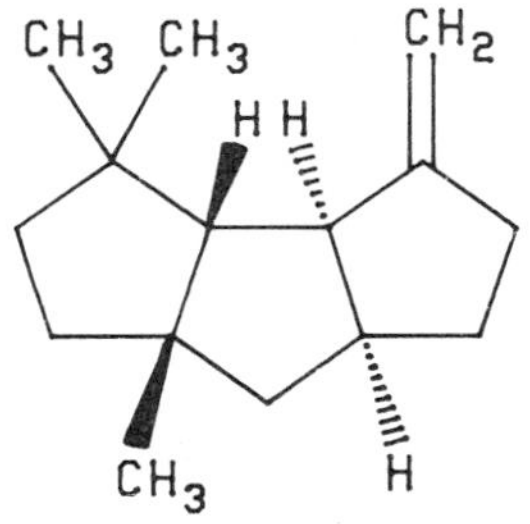

(±)-$\Delta^{9(12)}$-Capnellane

1H-cyclopenta[a]pentalene,

decahydro-3,3,7a-trimehyl-4-methylene

(3aα, 3bβ, 6aβ, 7aα)(±) [81370-78-7]

Figure 13.1

The sesquiterpene (±)-$\Delta^{9(12)}$-capnellane, isolated from the soft coral *Capnella imbricata* (Quoy and Gaimard, 1833), is the presumed biosynthetic precursor of the capnellenols. The capnellane group may serve as a component of the coral reef defense mechanism that inhibits the growth of microorganisms and settlement of larvae.[1] This property and a structural similarity to the pharmacologically significant hirsutanes (antibiotic, antitumor)[2] have prompted total synthetic efforts by several groups.[3] In this chapter, R. H. Grubbs' synthesis elegantly illustrates recent advances in the synthetic application of titanium carbene chemistry.

13.1 TITANIUM CARBENE CHEMISTRY

Certain transition metals can catalyze the pairwise exchange of alkenes in a process known as alkene metathesis (see Figure 13.2).

The currently accepted mechanism for alkene metathesis is a carbene chain reaction involving initiation and propagation steps. In the initiation step a reactive metal

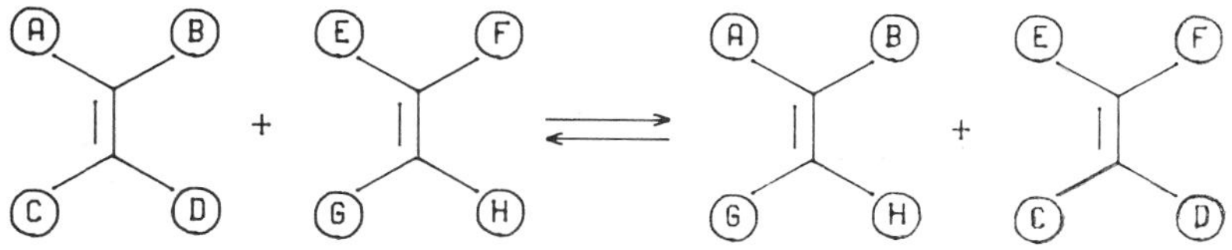

Figure 13.2

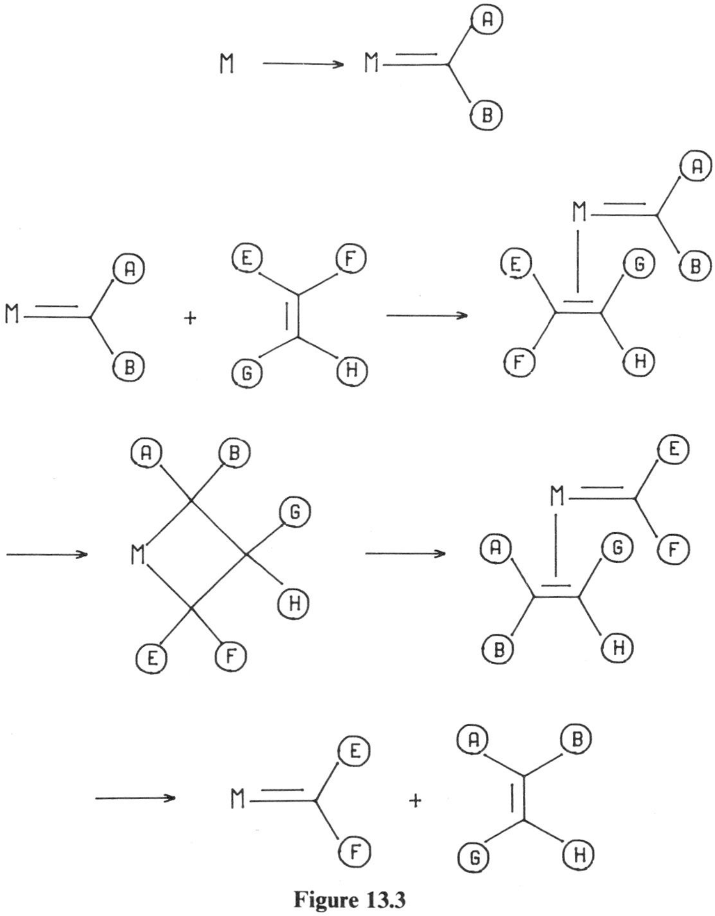

Figure 13.3

carbene is generated. In the propagation steps a metal carbene is utilized and a new metal carbene is produced. In principle, total scrambling of the methylene units will result in a statistical mixture of alkene products (see Figure 13.3).

We will be focusing on a metathesis catalyst where M = titanium.[4,5] The Tebbe reagent catalyst is generated at low temperature by reaction of titanocene dichloride with two equivalents of trimethylaluminum. It decomposes at −40°C to provide a titanium carbene (see Figure 13.4).

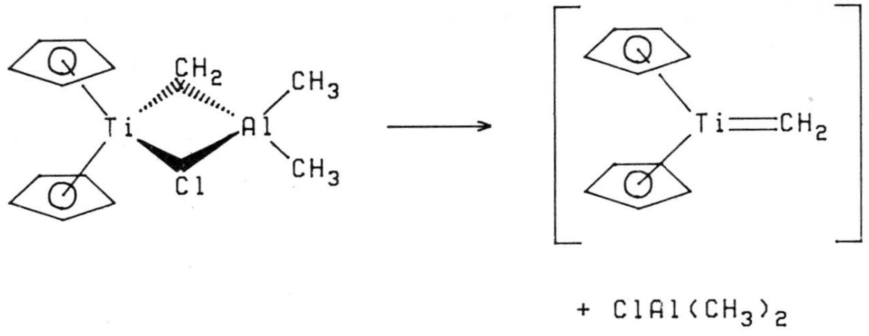
CH2
CH3
Ti
Al
CH3
Cl
Ti=CH2
+ ClAl(CH3)2

Figure 13.4

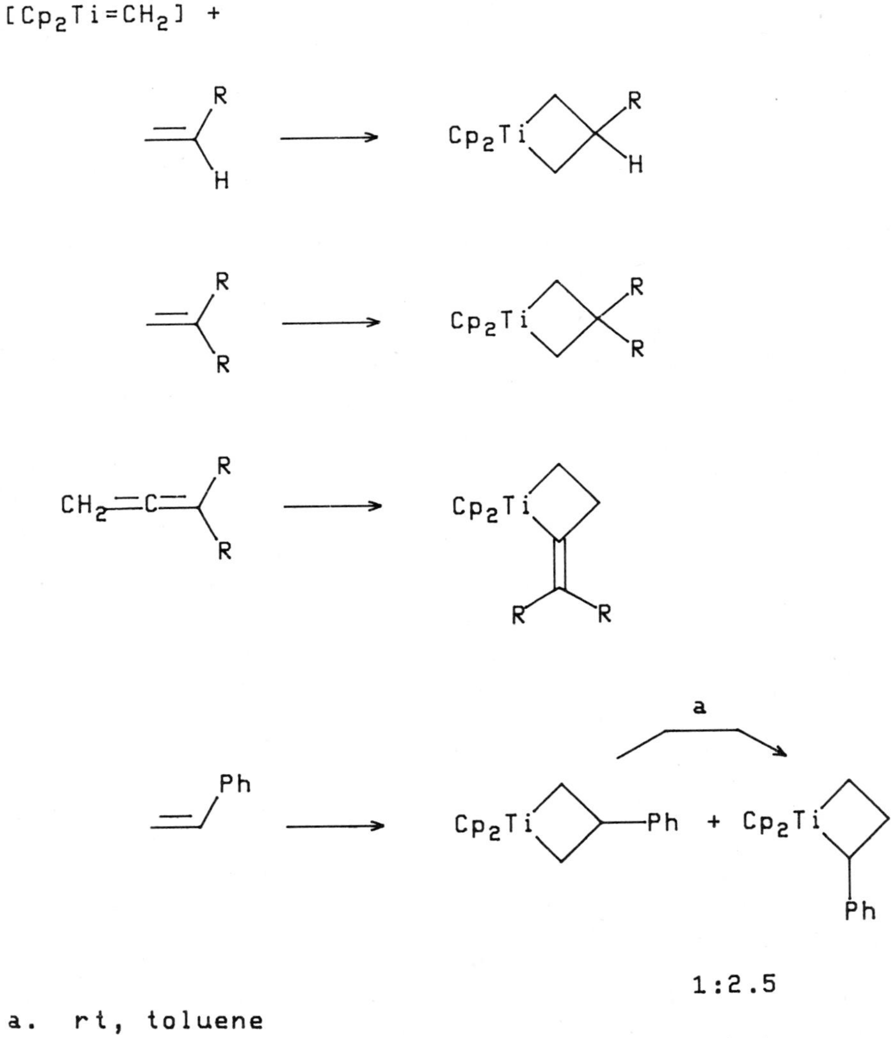
[Cp2Ti=CH2] +
R
H
Cp2Ti
R
H
R
R
Cp2Ti
R
R
CH2=C=
R
R
Cp2Ti
R
R
a
Ph
Cp2Ti
Ph
+ Cp2Ti
Ph
1:2.5
a. rt, toluene

Figure 13.5

Attempted utilization of the Tebbe reagent in alkene metathesis results in the isolation of moderately stable titanacyclobutanes.[4,5] A base is required since titanacyclobutane formation is accompanied by release of a strong Lewis acid, dimethylaluminum chloride. Pyridine, 4-dimethylaminopyridine, and polymer-bound analogs are commonly used. Mono- or 1,1-disubstituted alkenes are conveniently converted to products with ring substituents distant from the metal. Allenes are converted to products with the alkene adjacent to the metal. Titanacyclobutane formation is not regiospecific. With styrene, a 1 : 2.5 ratio of β to α isomers is produced, and the situation is further complicated by isomerization of β to α in toluene at room temperature (see Figure 13.5).

These titanacyclobutanes are stable at temperatures required for carbene generation from the Tebbe reagent (−40°C) but revert to carbene and alkene at higher temperatures. Thus, isolation of an intermediate metallocyclobutane and demonstration of conversion to metal carbene and alkene provide support for the proposed metal carbene chain mechanism.

13.2 REACTIONS OF TITANACYCLOBUTANES

Titanacyclobutanes can be cleaved by oxidation (see Figure 13.6). The Type 2 mechanism of oxidative cleavage was elucidated using the titanacyclobutane from *cis*-2-butene (see Figure 13.7). The major cleavage product results from initial attack on the less sterically encumbered carbon–titanium bond. Carbon–titanium bond cleavage

Type 1. Carbon-Titanium Bond Cleavage by Electrophiles

$$Cp_2Ti(CH_2)_2CR_2 \xrightarrow{2\ Br_2} Cp_2TiBr_2 + (BrCH_2)_2CR_2$$

$$Cp_2Ti(CH_2)_2CR_2 \xrightarrow{2\ HCl} Cp_2TiCl_2 + (HCH_2)_2CR_2$$

Type 2. Carbon-Titanium Bond Cleavage by Iodine

$$Cp_2Ti(CH_2)_2CR_2 \xrightarrow{I_2} Cp_2TiI_2 + \text{1,1-}R_2\text{-cyclopropane}$$

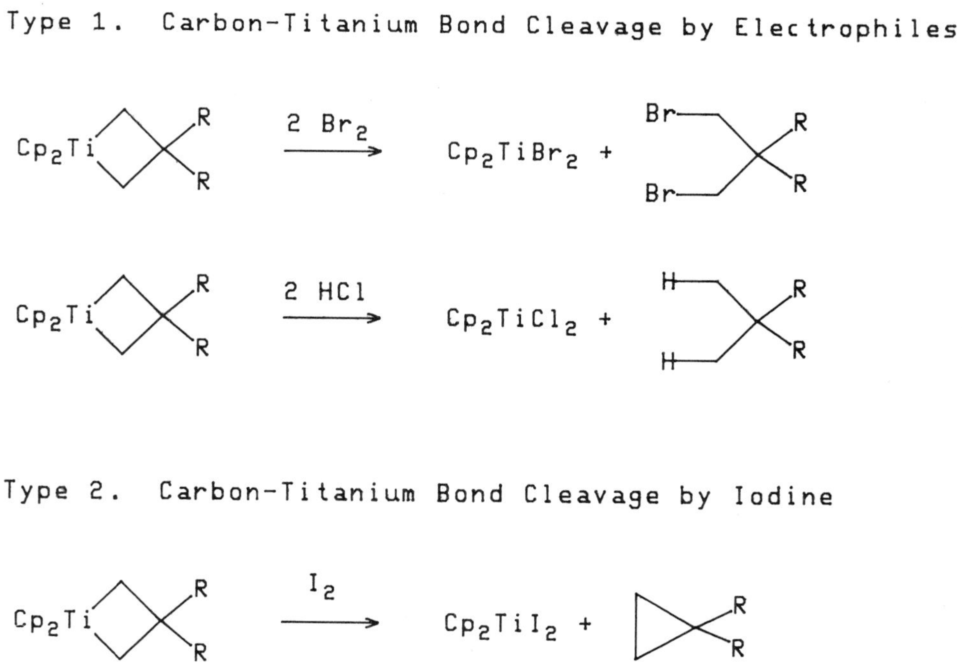

Figure 13.6

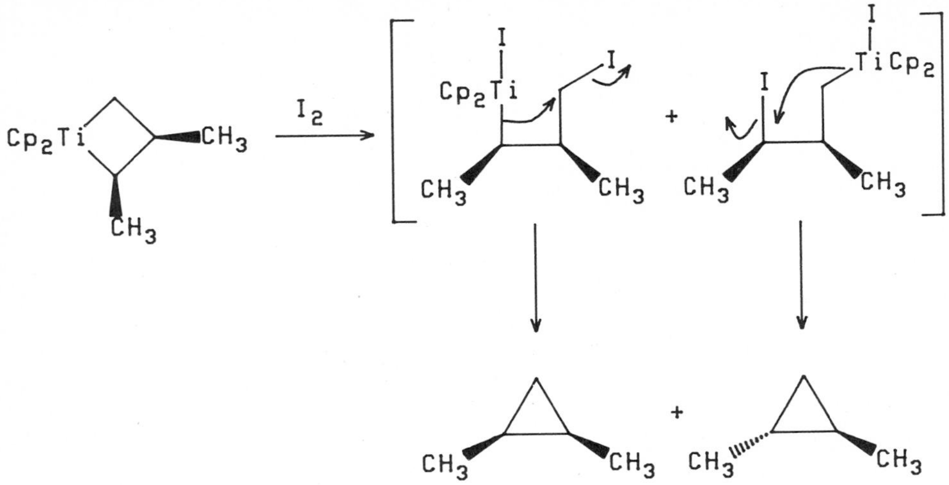

Figure 13.7

Type 3. Thermal Decomposition to Metal Carbene and Alkene and Subsequent Capture of Carbene by C=C and C=O

$[Cp_2Ti{=}CH_2]$ +

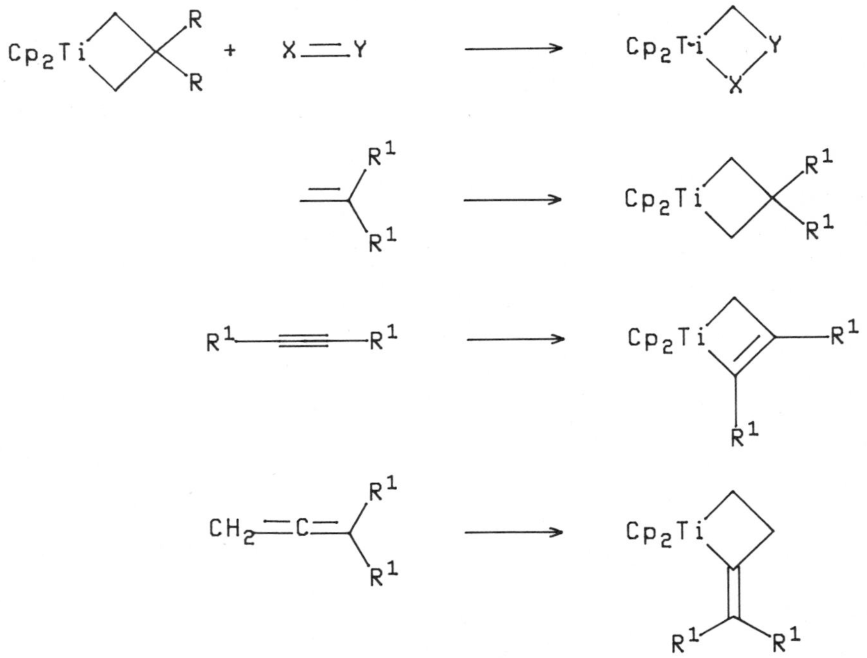

Figure 13.8

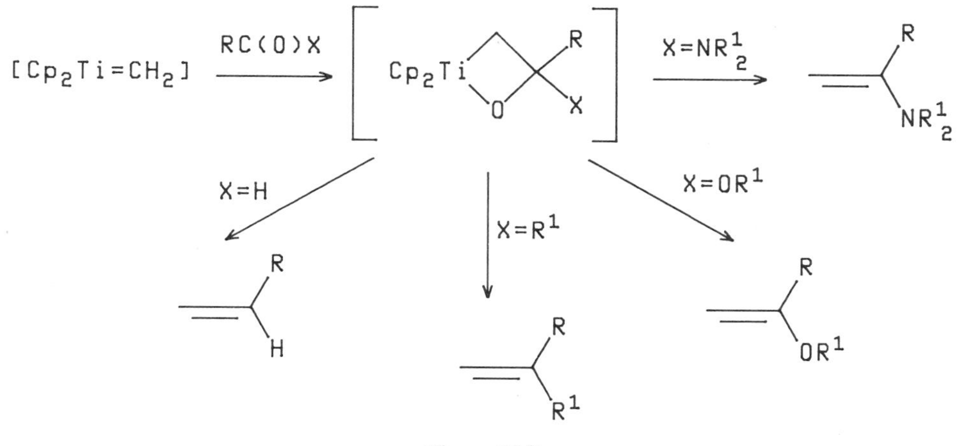

Figure 13.9

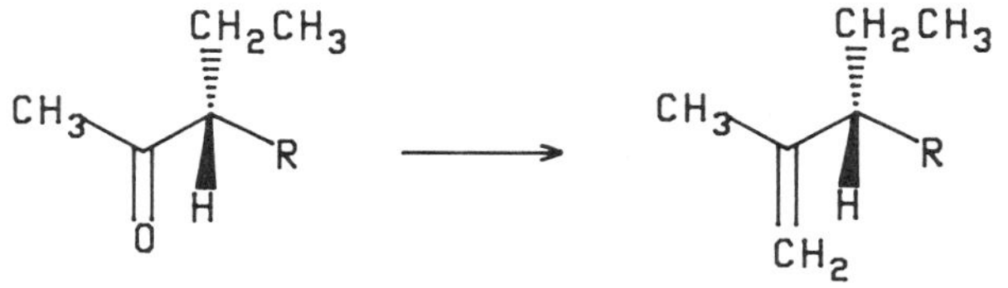

with $Ph_3P{=}CH_2$ 20–30% yield of racemized product

with $Cp_2Ti{=}CH_2$ 60–70% yield of optically active product

Figure 13.10

with retention of stereochemistry is followed by S_N2 displacement of iodide *with inversion of stereochemistry.*[6]

Thermal decomposition of a titanacyclobutane affords a metal carbene and an alkene. The carbenes from the Tebbe reagent or from a titanacyclobutane can be used as Wittig reagent equivalents (see Figure 13.8).

	$Cp_2Ti{=}CH_2$	$Ph_3P{=}CH_2$
	$\overset{\delta+}{Cp_2Ti}{-}\overset{\delta-}{CH_2}$	$\overset{\delta+}{Ph_3P}{-}\overset{\delta-}{CH_2}$
forms	$Cp_2{=}TiO$	$Ph_3P{=}O$

In both the Wittig and titanium carbene cases, there is significant negative charge density on carbon and a high affinity for oxygen by the atom at the cationic center (P, Ti). While the Wittig reaction is limited to aldehydes and ketones, the equivalent condensation with titanium carbene is applicable to aldehydes, ketones, esters, and amides producing mono- and 1,1-disubstituted alkenes, enol ethers, and enamines, respectively (see Figure 13.9). The low basicity of the titanium carbene permits its use

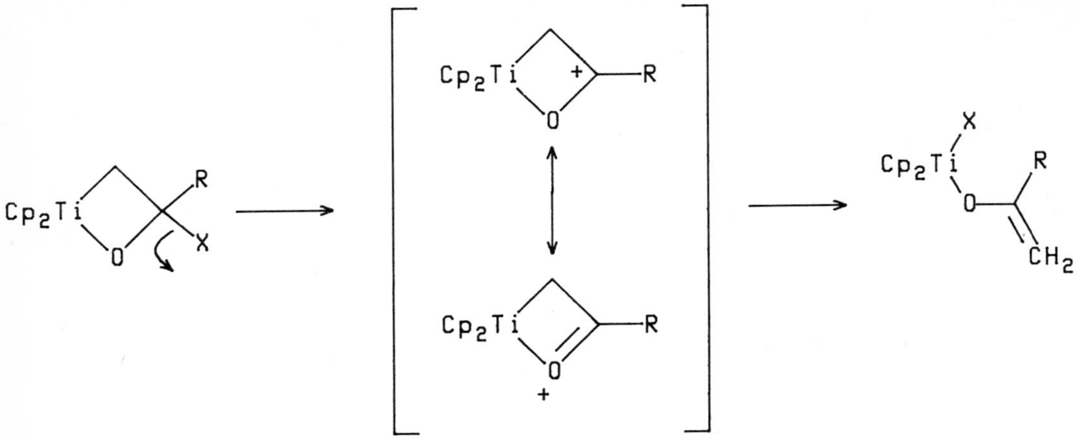

Figure 13.11

TABLE 13.1 Temperatures Required for Carbene Generation from Titanacyclobutanes[a]

Titanacyclobutane	Temperature
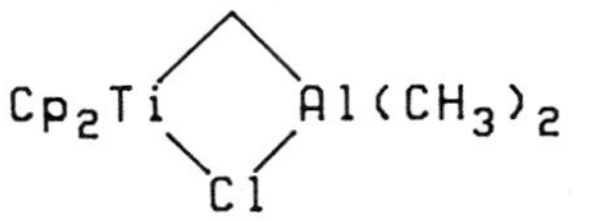 Cp_2Ti … $Al(CH_3)_2$, Cl	−40°C
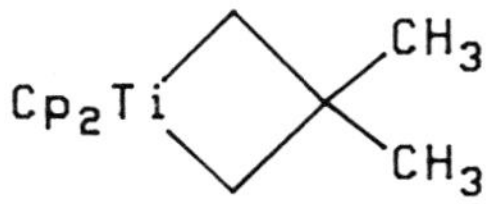 Cp_2Ti … CH_3, CH_3	5°C
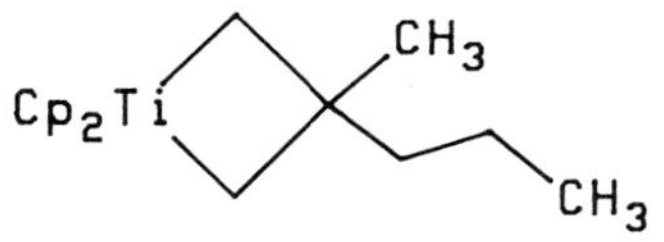 Cp_2Ti … CH_3, CH_3	0°C
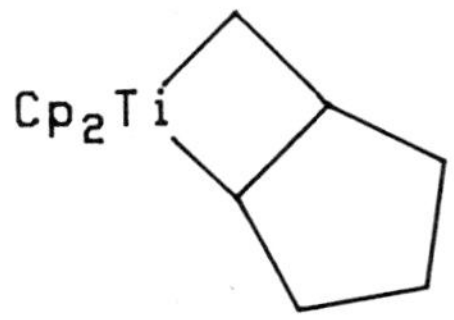 Cp_2Ti (fused cyclopentane)	45°C
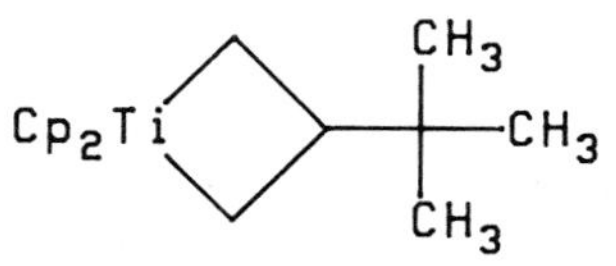 Cp_2Ti … CH_3, CH_3, CH_3	60°C

[a]Used with permission from IUPAC.

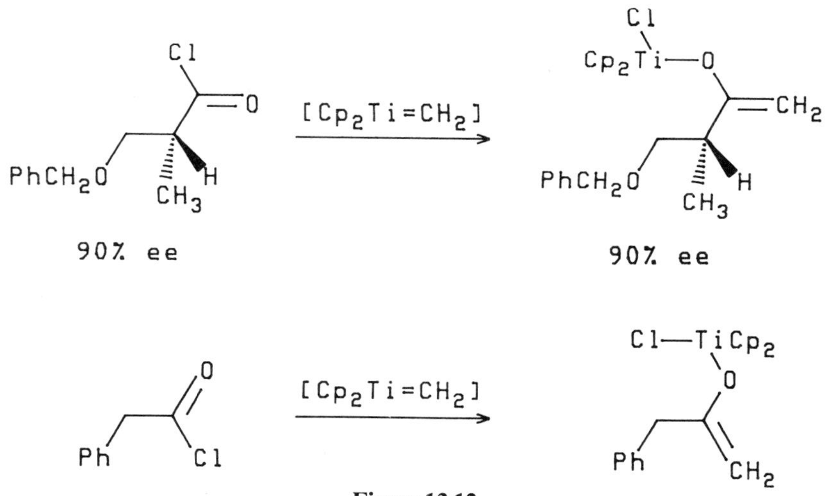

$$\text{PhCH}_2\text{O-CH}_2\text{-CH(CH}_3\text{)-C(=O)Cl} \ (90\%\ ee) \xrightarrow{[\text{Cp}_2\text{Ti=CH}_2]} \text{PhCH}_2\text{O-CH}_2\text{-CH(CH}_3\text{)-C(=CH}_2\text{)-O-Ti(Cl)Cp}_2 \ (90\%\ ee)$$

$$\text{PhCH}_2\text{C(=O)Cl} \xrightarrow{[\text{Cp}_2\text{Ti=CH}_2]} \text{PhCH}_2\text{C(=CH}_2\text{)-O-TiCp}_2\text{Cl}$$

Figure 13.12

with carbonyl compounds possessing chiral centers at easily epimerizable positions α to carbonyl (see Figure 13.10).

The acid sensitivity of enol ether and enamine products necessitates base complexation of the Lewis acid released on condensation when using the Tebbe reagent. Alternative utilization of a titanacyclobutane as the carbene precursor has distinct advantages:

1. While the Tebbe reagent is air and moisture sensitive, the titanacyclobutanes are crystalline and air stable.
2. No base is necessary since acid is not a by-product.
3. The temperature required for carbene generation can be modulated by variation of the titanacyclobutane ring substituents (see Table 13.1). In an optimal condensation, carbene is generated at a temperature where subsequent reaction of the carbene is rapid.

In condensations with more reactive carboxylic acid derivatives such as acid chlorides or anhydrides, an alternative cleavage provides a stable titanium enolate of the corresponding methyl ketone[7] (see Figure 13.11). The potential utility of this method of specific enolate generation is readily apparent (see Figure 13.12).

13.3 EXTENSION TO MORE ELABORATE CARBENES

With the titanacyclobutanes substituted α to the metal, will cleavage to carbene and alkene be *productive* or regenerate the starting materials? One series of titanacyclobutanes substituted α to the metal are obtained from cyclic alkenes (see Figure 13.13).

Starting with relatively unstrained cyclic alkenes, cleavage results in a return to starting materials. With 3,3-dimethylcyclopropene, cleavage is *redirected by relief of ring strain* and constitutes a key step in a novel 1,4-diene synthesis[8] (see Figure 13.14).

$[Cp_2Ti{=}CH_2]$ +

Figure 13.13

$[Cp_2Ti{=}CH_2]$ +

+ $[Cp_2Ti{=}O]_n$

Figure 13.14

$[Cp_2Ti{=}CH_2]$ + $CH_2{=}C{=}CR_2$

Figure 13.15

A second series of α-substituted titanacyclobutanes is obtained from allenes. Here, again, cleavage is productive and constitutes the key step in a novel allene synthesis[9] (see Figure 13.15).

13.4 TOTAL SYNTHESIS OF (±)-$\Delta^{9(12)}$-CAPNELLANE[10]

The first substrate for titanium carbene chemistry is prepared in six steps from γ-butyrolactone[11] (see Figure 13.16). The relative stereochemistry at all four chiral centers is established in a Diels–Alder cycloaddition.

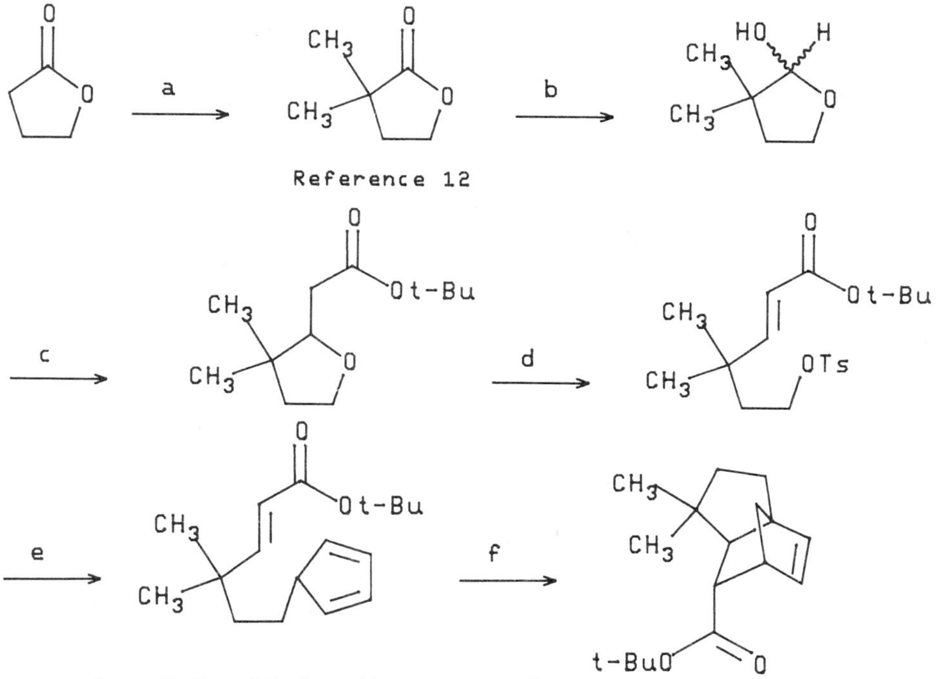

a. rfx, NaH, CH_3I, dioxane; 65%

b. −78°C, $(i\text{-}Bu)_2AlH$

c. 25°C, $(EtO)_2P(O)CH_2COOt\text{-}Bu$, NaH, benzene; 89% for two steps

d. 1) −78°C, $LiN(i\text{-}Pr)_2$; 2) −78 to 0°C, p-TsCl, THF; 83% for two steps

e. 25°C, CpMgCl, THF

f. 75°C, benzene; 81% for two steps

Figure 13.16

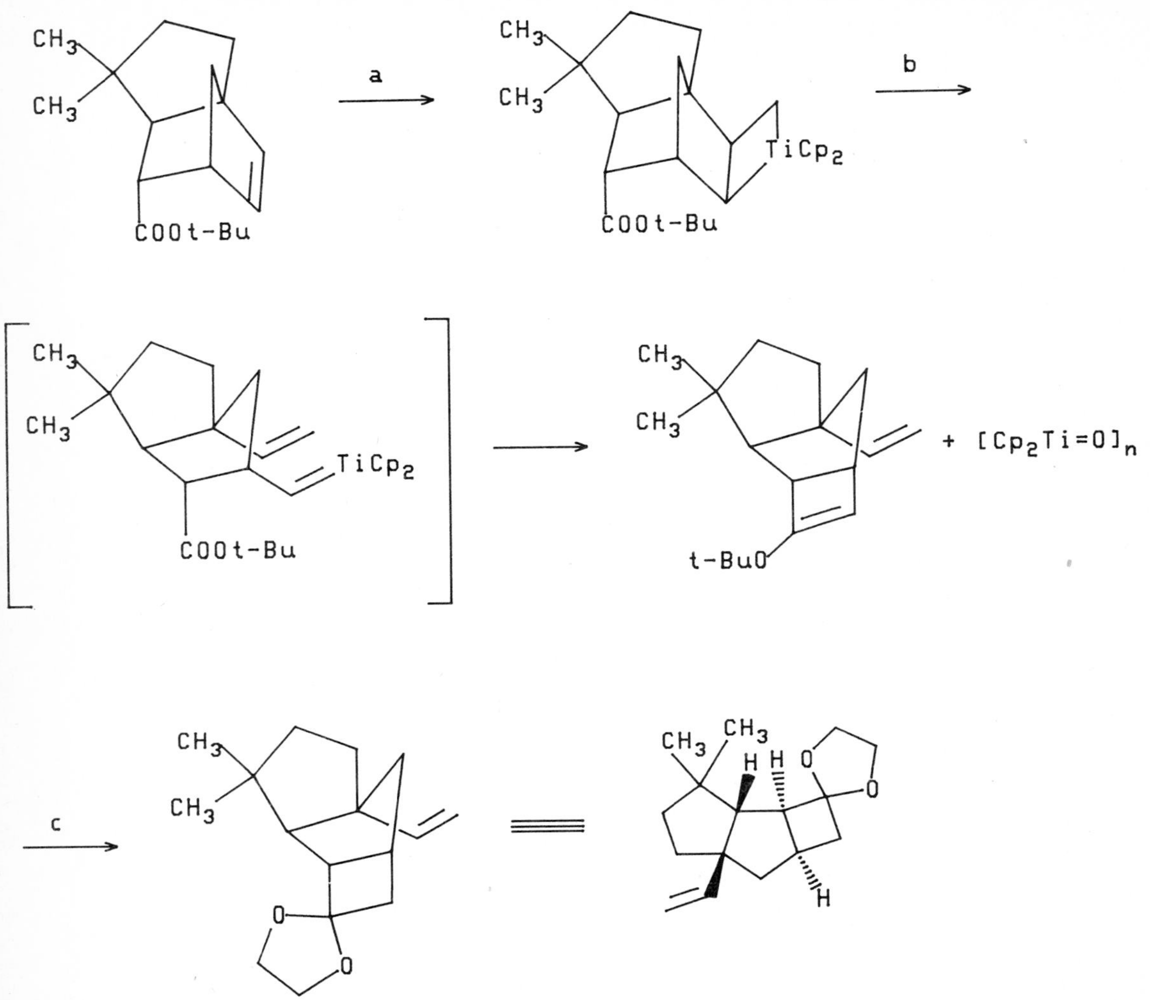

a. 25°C, [$Cp_2Ti=CH_2$], benzene

b. 90°C, benzene

c. rfx, p-TsOH·H_2O, $HOCH_2CH_2OH$, benzene; 81% for three steps

Figure 13.17

A carbene is generated by addition of the Tebbe reagent to a solution of 4-dimethylaminopyridine in benzene at room temperature. Condensation with alkene produces a titanacyclobutane both regio- and stereospecifically. Competing Wittig-like attack on the *t*-butyl ester is not a problem under the mild reaction conditions employed. The carbene enters on the *exo*, more sterically accessible, face with the metal

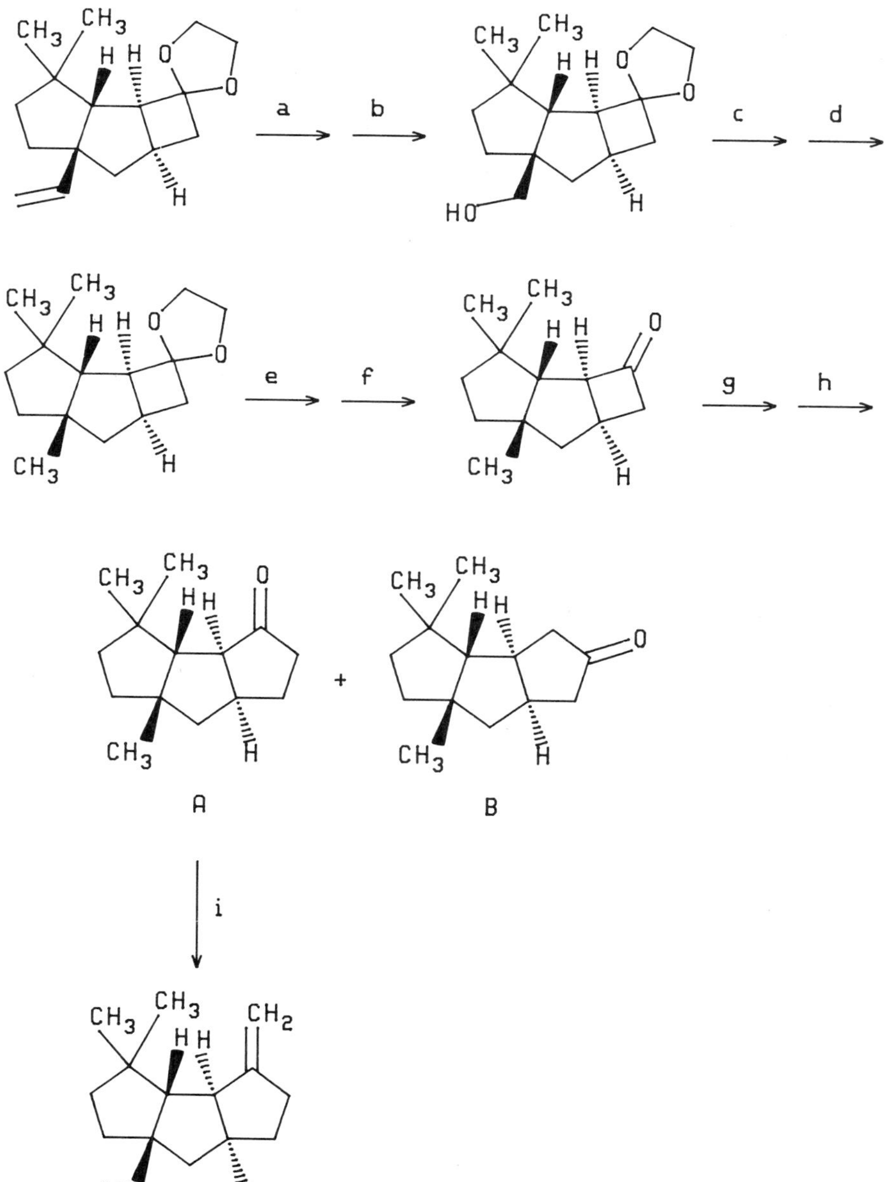

Figure 13.18

a. −78°C, O_3, CH_3OH, CH_2Cl_2

b. −78 to 25°C, $NaBH_4$; 91% for two steps

c. 1) n-BuLi, 2) $(Me_2N)_2P(O)Cl$, Et_3N, DME

d. −50 to −40°C, Li, t-BuOH, $EtNH_2$, THF

e. rfx, p-TsOH·H_2O, acetone, benzene

f. 25°C, 0.15 eq PDC, CH_2Cl_2; 68% for four steps

g. −78°C, BF_3·Et_2O, $N_2CHCOOEt$, Et_2O

h. 150°C, NaCl, DMSO, H_2O; A:B, 83:17, 73% yield for two steps (mixture separable by flash chromatography)

i. −40 to 25°C, $[Cp_2Ti{=}CH_2]$; 93%

Figure 13.18 (*Continued*)

and cyclopentadienyl ligands distant from the more congested bridge position. Ring opening of the titanacyclobutane at 90°C, directed by relief of bicyclic ring strain, is productive. The resulting carbene reacts in an intramolecular Wittig-like condensation with the *t*-butyl ester to afford an enol ether. The hydrolytic instability of the enol ether necessitates protection as a ketal. This three-step sequence is accomplished in a remarkable 81% yield! (See Figure 13.17.)

Completion of the synthesis requires (1) conversion of vinyl to methyl, (2) ring expansion from four- to five-membered, and (3) Wittig condensation with the ketone. The vinyl substituent is transformed to methyl by ozonolysis and reduction to give a hydroxymethyl. The alcohol is converted to a tetramethylphosphorodiamidate ester, and the ester is reduced with lithium metal.[13] Although literature precedent indicates possible regiospecific ring expansion,[14] an 83:17 mixture of cyclopentanones is obtained. These are readily separated by flash chromatography. Finally, Wittig-like condensation is best achieved using the titanium carbene generated by addition of the Tebbe reagent to a solution of 4-dimethylaminopyridine in diethyl ether at −40°C.

The total synthesis of (±)-$\Delta^{9(12)}$-capnellane was completed in 18 steps in 17% overall yield from α,α-dimethyl-γ-butyrolactone (19 steps in overall 11% yield from γ-butyrolactone (19 steps in overall 11% yield from γ-butyrolactone) (see Figure 13.18).

REFERENCES

1. Sheikh, Y. M.; Singy, G.; Kaisin, M.; Eggert, H.; Djerassi, C.; Tursch, B.; Daloze, D.; Braekman, J. C. *Tetrahedron*, **1976**, *32*, 1171 and references cited therein.

2a. Takeuchi, T.; Iinuma, H.; Iwanaga, J.; Takahashi, S.; Takita, T.; Umezawa, H. *J. Antibiot.* **1969**, *22*, 215.

2b. Takeuchi, T.; Takahashi, S.; Iinuma, H.; Umezawa, H. *J. Antibiot.* **1971**, *24*, 631.
3a. Mehta, G.; Murthy, A. N.; Reddy, D. S.; Reddy, A. V. *J. Am. Chem. Soc.* **1986**, *108*, 3443.
3b. Liu, H. J.; Kulkarni, M. G. *Tetrahedron Lett.* **1985**, *26*, 4847.
3c. Curran, D. P.; Chen, M-H. *Tetrahedron Lett.* **1985**, 26, 4991.
3d. Paquette, L. A.; Stevens, K. E. *Can. J. Chem.* **1984**, 62, 2415.
3e. Crisp, G. T.; Scott, W. J.; Stille, J. K. *J. Am. Chem. Soc.* **1984**, *106*, 7500.
3f. Mehta, G.; Reddy, D. S.; Murty, A. N. *J. Chem. Soc., Chem. Commun.* **1983**, 824.
3g. Piers, E.; Karunaratne, V. *Can. J. Chem.* **1984**, *62*, 629.
3h. Paquette, L. A. *Top. Curr. Chem.* **1984**, *119*, 1.
4. For an excellent review on titanium carbene complexes, see Brown-Wensley, K. A.; Buchwald, S. L.; Cannizzo, L.; Clawson, L.; Ho, S.; Meinhardt, D.; Stille, J. R.; Straus, D.; Grubbs, R. H. *Pure & Appl. Chem.* **1983**, *55*, 1733.
5. For an excellent review on related tungsten carbene complexes, see Schrock, R. R. *J. Organometal. Chem.* **1986**, *300*, 249.
6. Ho, S. C. H.; Straus, D. A.; Grubbs, R. H. *J. Am. Chem. Soc.* **1984**, *106*, 1533.
7a. Stille, J. R.; Grubbs, R. H. *J. Am. Chem. Soc.* **1983**, *105*, 1664.
7b. Chou, T-S.; Huang, S-B. *Tetrahedron Lett.* **1983**, *24*, 2169.
8. Gilliom, L. R.; Grubbs, R. H. *Organometallics* **1986**, *5*, 721.
9. Buchwald, S. L.; Grubbs, R. H. *J. Am. Chem. Soc.* **1983**, *105*, 5490.
10. Stille, J. R.; Grubbs, R. H. *J. Am. Chem. Soc.* **1986**, *108*, 855.
11. γ-Butyrolactone is commercially available and inexpensive (Lancaster Synthesis 88–89 catalog, 1 kg for $14.40).
12. Baas, J. L.; Davies-Fidder, A.; Huisman, H. O. *Tetrahedron* **1966**, *22*, 285.
13. Ireland, R. E.; Muchmore, D. C.; Hengartner, U. *J. Am. Chem. Soc.* **1972**, *94*, 5098.
14. Liu, H. J.; Ogino, T. *Tetrahedron Lett.* **1973**, 4937.

CHAPTER 14

Prostaglandins, 5-(*E*) Analogs: (±)-5-(*E*)- $PGF_{2\alpha}$

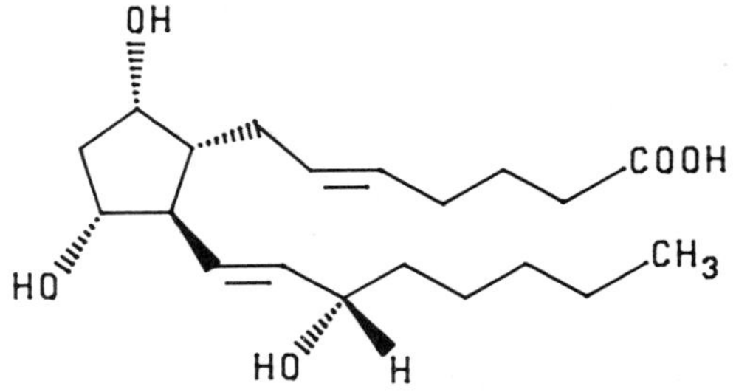

Prosta-5,13-dien-1-oic acid, 9,11,15-trihydroxy

(5E,9α,11α,13E,15S)[36150-01-3]

(5E,9α,11α,13E,15S)(±)[59727-54-7]

Figure 14.1

Prostaglandin research began with several seemingly unrelated observations of diverse biological effects:

1. Kurzok and Lieb observed that human semen contained a factor capable of causing either strong contraction or relaxation of a human uterus.[1]
2. Von Euler and Goldblatt demonstrated independently that human semen and sheep vesicular glands contain a vasodepressor and muscle stimulation factor.[2,3]
3. Burr and Burr reported that a deficiency state in rats on fat-free diets could be prevented by addition of polyunsaturated fatty acids.[4]

Some 30 years passed before prostaglandin structures were assigned to the factors (PGE_1, PGE_2, and $PGF_{2\alpha}$ were the first prostaglandins identified) and the polyunsaturated fatty acids, particularly arachidonic acid, demonstrated to be prostaglandin precursors. Since the early 1960s, the prostaglandin field has developed exponentially, highlighted most recently by research on prostacyclins, prostaglandin endoperoxides, leukotrienes, and thromboxanes.[5]

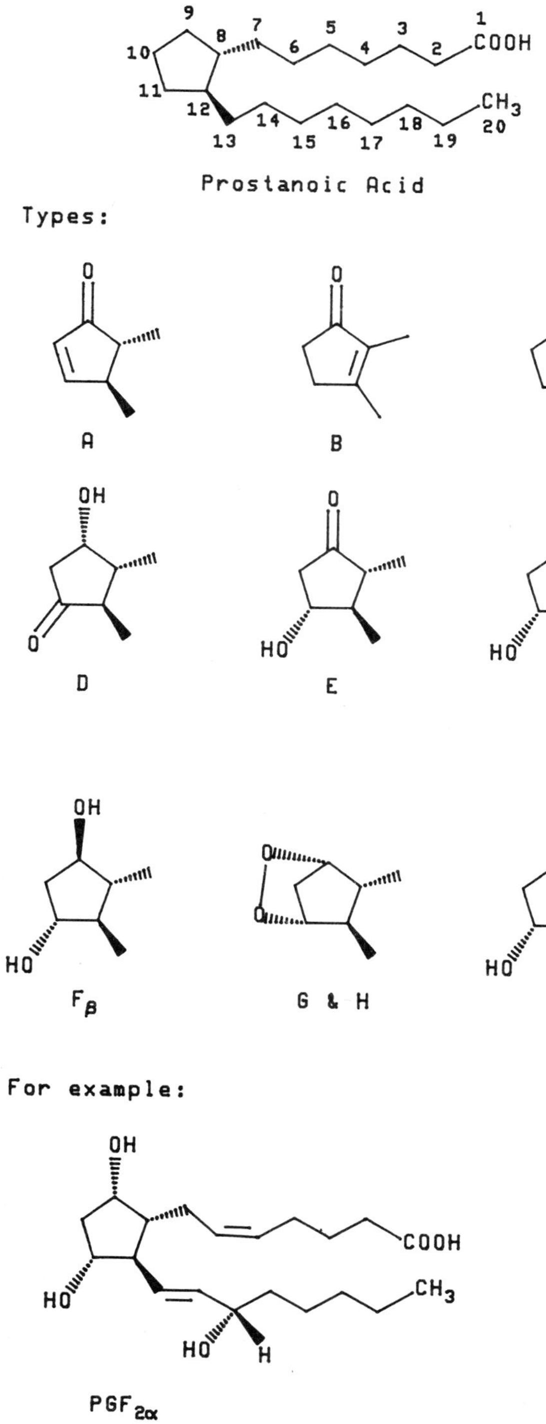

Prostanoic Acid

Types:

A B C

D E F_α

F_β G & H I

For example:

$PGF_{2\alpha}$

Figure 14.2

OR

base

CH_3

acyl transferase

phospholipase A_2

CoA

CH_3

hydrolase

synthetase

OH

CH_3

Arachidonic Acid

cyclooxygenase

OH

CH_3

HO—O

OOH

CH_3

PGG_2

peroxidase

OH

CH_3

HO H

PGH_2

PGI_2 PGE_2 ⟶ $PGF_{2\alpha}$ ⟵ PGD_2

Figure 14.3

A systematic nomenclature is based on prostanoic acid.[6] The initial descriptor, X ($PGX_{n\alpha}$ or $PGX_{n\beta}$) specifies functionality on the cyclopentane ring. The subscript n species the number of carbon–carbon double bonds in the side chains. The subscript α or β specifies the 9-hydroxyl group as down or up, respectively. While considerable variation is possible in the cyclopentane portion, there are only three common series of prostaglandins ($n = 1, 2, 3$) where

$n = 1$ has (E)C=C at 13, 14
$n = 2$ has (E) C=C at 13, 14
(Z) C=C at 5, 6
$n = 3$ has (E) C=C at 13, 14
(Z) C=C at 5, 6
(Z) C=C at 17, 18

Note that the 15-(S) hydroxyl group appears to be important for biological activity (see Figure 14.2).

The biosynthesis of $PGF_{2\alpha}$ from arachidonic acid is mediated by cyclooxygenase (arachidonic acid $\rightarrow PGG_2$), peroxidase ($PGG_2 \rightarrow PGH_2$), and an enzyme yet to be isolated ($PGH_2 \rightarrow PGF_{2\alpha}$).[7] The rate-determining step is apparently hydrolysis of the glycerophospholipid to free arachidonic acid since the concentration of free arachidonic acid in unstimulated cells is very low. Alternative routes to $PGF_{2\alpha}$ involve reduction of the ketone carbonyl of PGE_2 and PGD_2 (see Figure 14.3).

The wide range of biological effects attributed to prostaglandins has been a serious limitation to clinical utilization. Thus, analogs with high specific activity yet minimal side effects are unusual. For example, $PGF_{2\alpha}$ (1) contracts human uteral smooth muscle in vitro, (2) inhibits the "readsorption of water and electrolytes in the small intestine, leading to interpooling,"[8] and (3) contracts bronchial smooth muscles in vitro in many species and in man. It was believed that $PGF_{2\alpha}$ had an important role in allergic response since release of $PGF_{2\alpha}$ occurs during allergen-provoked asthma attacks. However, more recent work has shown the endoperoxides and thromboxane TXA_2 to be considerably more potent bronchoconstrictors.

The analog 5-(E)-$PGF_{2\alpha}$ has been shown to prevent nosebleeds and to have useful activity for the treatment of asthma.[8]

Since prostaglandins are only available in minute quantities from natural sources, quantities necessary for pharmacological evaluation must be synthesized. Many synthetic approaches are now well established.[9] This chapter highlights Schwartz's modification of Corey's $PGF_{2\alpha}$ synthesis for the preparation of ($\pm$)-5-(E)-$PGF_{2\alpha}$.[10]

14.1 ORGANOZIRCONIUM(IV) COMPLEXES

A general scheme for equilibrium between metal hydride, metal hydride–π-alkene complex, and σ-alkyl metal complex is illustrated in Figure 14.4. Hoffmann proposed that the position for the second equilibrium (K_2) should depend on the electron density of the metal.[11] High electron density, such as in d^8-palladium, stabilizes the π-alkene complex by back donation into the empty π^* orbital and thus drives the equilibrium to

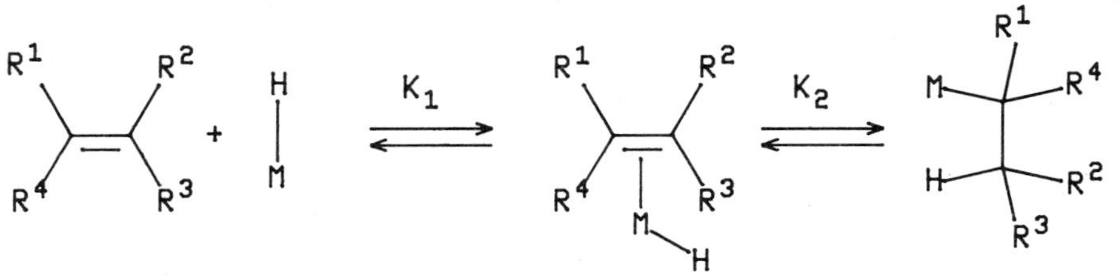

Figure 14.4

the left. Low electron density, such as in d^0-zirconium, favors the σ-alkyl complex. Indeed, much of the beauty of organopalladium chemistry lies in the design to ultimately produce a σ-alkylpalladium complex that readily β-hydride eliminates to afford alkene products. The synthesis and utilization of organozirconium(IV) complexes that do not β-hydride eliminate is the subject of this chapter.

Wailes and Weigold were the first to prepare stable alkyl- and vinylzirconium complexes by *hydrozirconation.*[12] Utilization of organozirconium(IV) complexes for organic synthesis, pioneered by Schwartz,[13] required development of methods for selective replacement of the metal. Poor reactivity of the organozirconium(IV) complexes with carbon electrophiles prompted the development of transmetallation methods, in this instance, the replacement of zirconium by another metal with a well-precedented reaction manifold, including metal replacement by carbon. The pioneering research efforts on transmetallation by both Schwartz and Negishi[14] prompted transformation of a process that was largely unknown just 10 years ago into a versatile and reliable synthetic method.

14.2 PREPARATION OF ORGANOZIRCONIUM(IV) COMPLEXES FROM ZIRCONOCENE DICHLORIDE

Organozirconium(IV) complexes are both air and moisture sensitive. Preparation, characterization, and utilization in synthesis must be performed under a dry, inert atmosphere.

14.2.1 Method A. Tri-*t*-butoxyaluminum Hydride

Vinyl and alkylzirconium(IV) complexes are most often prepared by *syn*-hydrozirconation of alkynes and alkenes, respectively, using zirconocene hydride chloride. The air- and moisture-sensitive hydride is prepared by reduction of zirconocene dichloride with lithium tri-*t*-butoxyaluminum hydride in tetrahydrofuran[12b] (see Figure 14.5).

Bis-hydrozirconation of the alkyne is not a problem since alkynes are 70 to 100 times more reactive than similarly substituted alkenes. The bulky metal center is initially introduced regioselectively at the less encumbered carbon of the alkyne or alkene. With alkynes, the regioselectivity is controlled by both the difference in steric bulk of alkyne substituents and the ratio of alkyne to zirconocene hydride chloride. Using an excess of

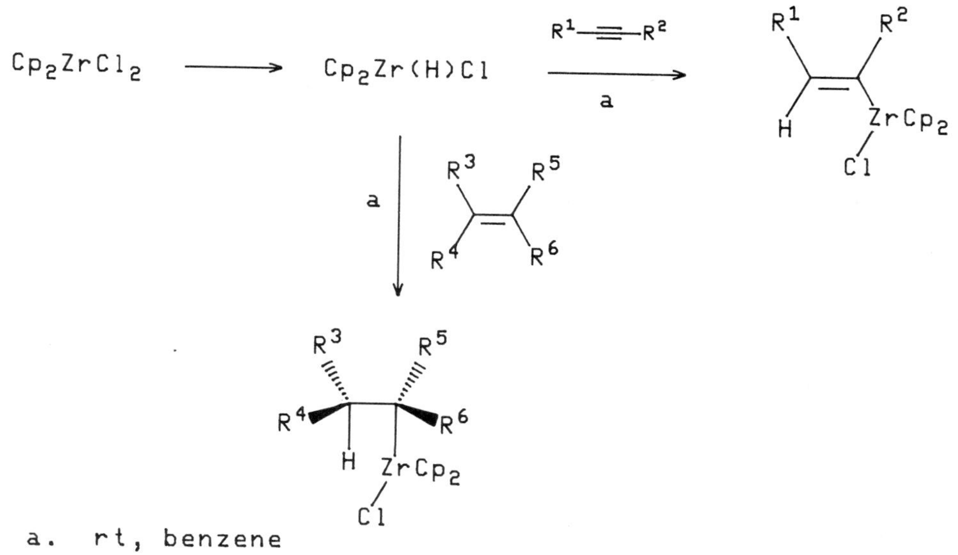

Figure 14.5

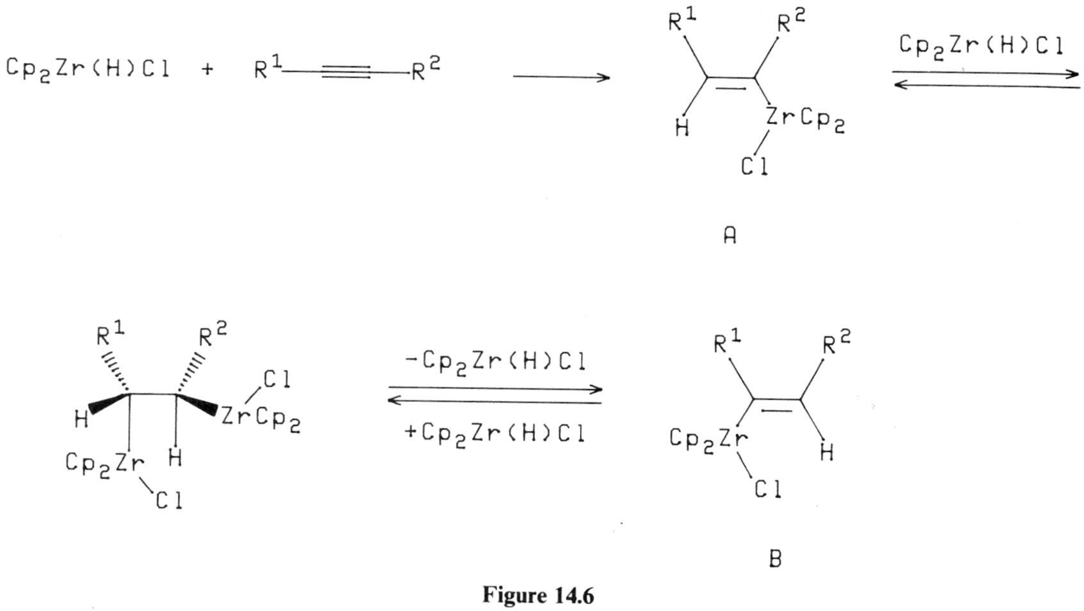

Figure 14.6

the hydride, the initial kinetic ratio is improved to the thermodynamic ratio by a hydrozirconation–dehydrozirconation sequence[15] (see Table 14.1 and Figure 14.6).

Alkylzirconium complexes formed from internal alkenes undergo metal migration by a β-hydride elimination-metal hydride readdition sequence. The metal migrates to a position that has minimal steric interaction between the organic ligand and the bulky metal center[16] (see Figure 14.7).

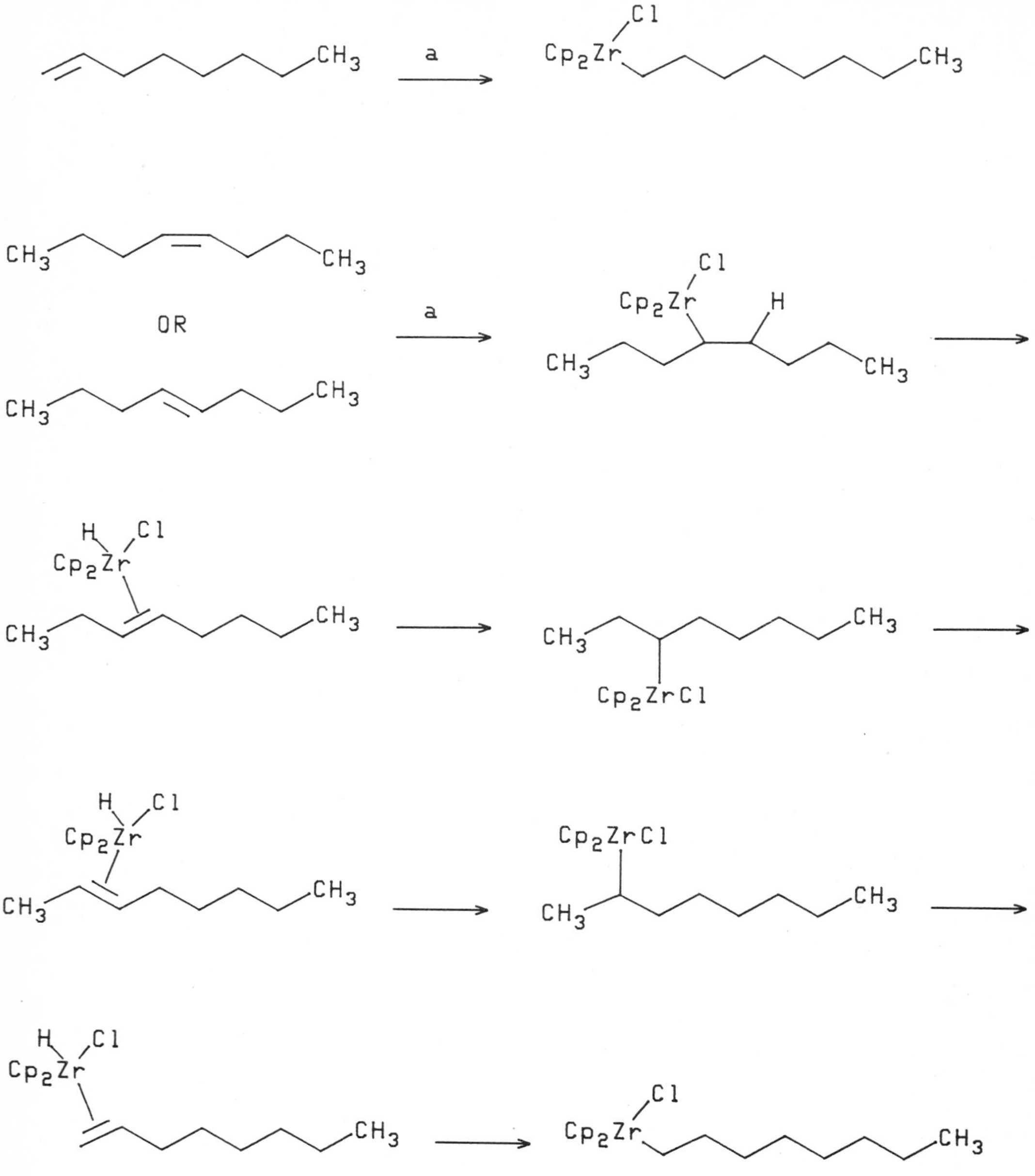

a

a

a. rt, $Cp_2Zr(H)Cl$, benzene

Figure 14.7

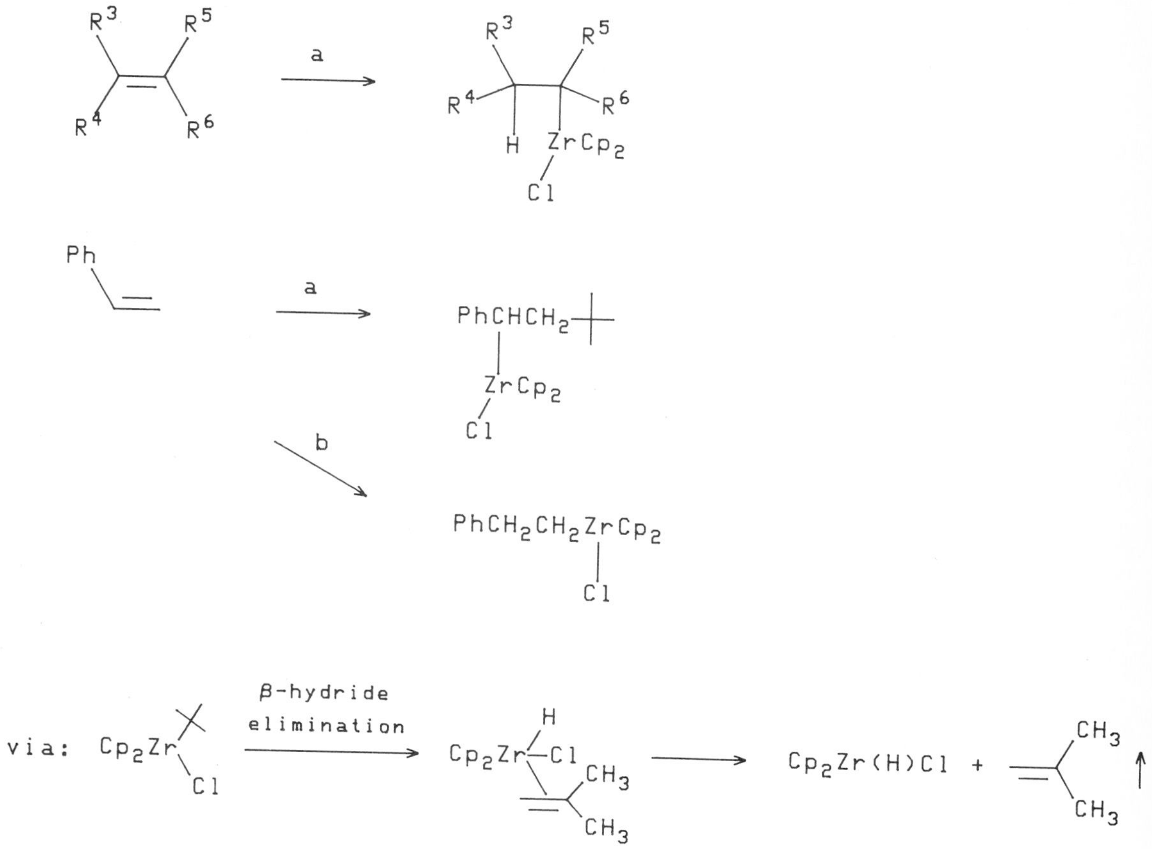

a. rt, Cp_2ZrCl_2, t-BuMgCl, benzene, Et_2O

b. rt, Cp_2ZrCl_2, t-BuMgCl, benzene

Figure 14.8

Hydrozirconation is successful using mono-, di-, and trisubstituted alkenes, mono- and disubstituted alkynes, and 1,3-dienes. An unusual mode of addition is observed with 1,3-dienes: hydrozirconation occurs via 1,2-addition to the more accessible alkene to give γ,δ-unsaturated complexes.[17] Compatible functionality is limited by the hydridic nature of the zirconocene hydride chloride; ethers (alkyl, silyl, and tetrahydropyranyl) and *t*-butyl esters have been successfully incorporated.

14.2.2 Method B. *t*-Butylmagnesium Chloride

The zirconocene hydride chloride can be prepared in situ with *t*-butylmagnesium chloride and zirconocene dichloride.[18] With conjugated alkenes, a novel *t*-butylzirconation is observed in benzene-diethyl ether while hydrozirconation predominates in benzene alone (see Figure 14.8).

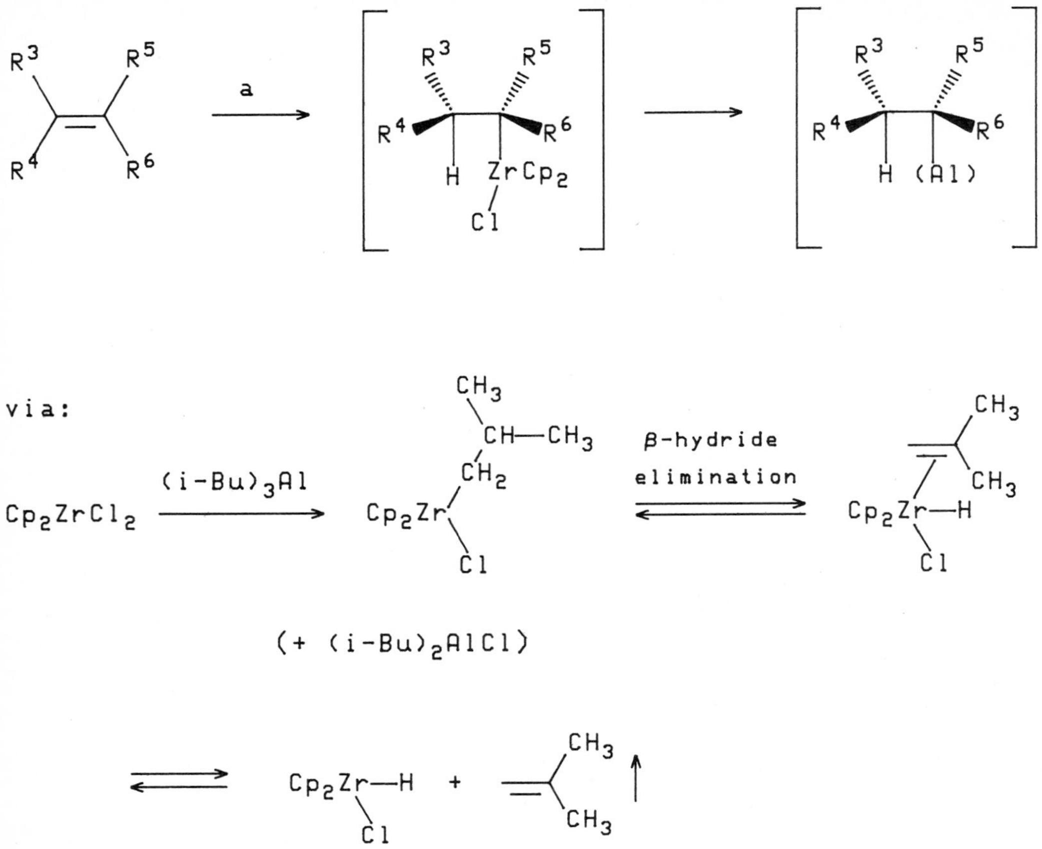

via:

(+ (i-Bu)$_2$AlCl)

a. rt, (i-Bu)$_3$Al, Cp$_2$ZrCl$_2$, ClCH$_2$CH$_2$Cl

Figure 14.9

14.2.3 Method C. Tri-*i*-butylaluminum

In situ generation of zirconocene hydride chloride is also possible usng tri-*i*-butylaluminum and zirconocene dichloride.[19] Hydrozirconation produces an alkylaluminum complex after transmetallation (see Figure 14.9).

14.2.4 Method D. Tri-*i*-propylaluminum and Trimethylaluminum

π-Complexation of zirconocene hydride chloride with propene should be stronger than with isobutylene. Thus, reaction of zirconocene dichloride with tri-*i*-propylaluminum produces both $Cp_2Zr(n\text{-}Pr)Cl$ and zirconocene hydride chloride. In the presence of an alkyne, both *n*-propylzirconation and hydrozirconation are observed.[19] Reaction of

TABLE 14.1 Hydrozirconation of Unsymmetrical Alkynes[a]

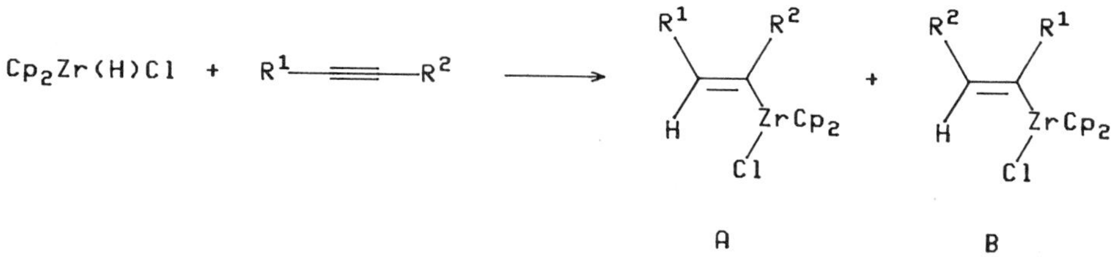

$Cp_2Zr(H)Cl$ + $R^1—≡—R^2$ ⟶ A + B

R^1	R^2	A:B Ratio	A:B Ratio after Treatment with Additional Hydride
$CH_3CH_2CH_2CH_2$	H	>98:2	—
CH_3CH_2	CH_3	55:45	89:11
$CH_3CH_2CH_2$	CH_3	69:31	91:9
$(CH_3)_2CHCH_2$	CH_3	55:45	>95:<5
$(CH_3)_2CH$	CH_3	84:16	>98:<2
$(CH_3)_3C$	CH_3	>98:2	—

[a]Reprinted with permission from D. W. Hart, et al., *J. Am. Chem. Soc.*, **1975**, 97, 679. Copyright 1975, American Chemical Society.

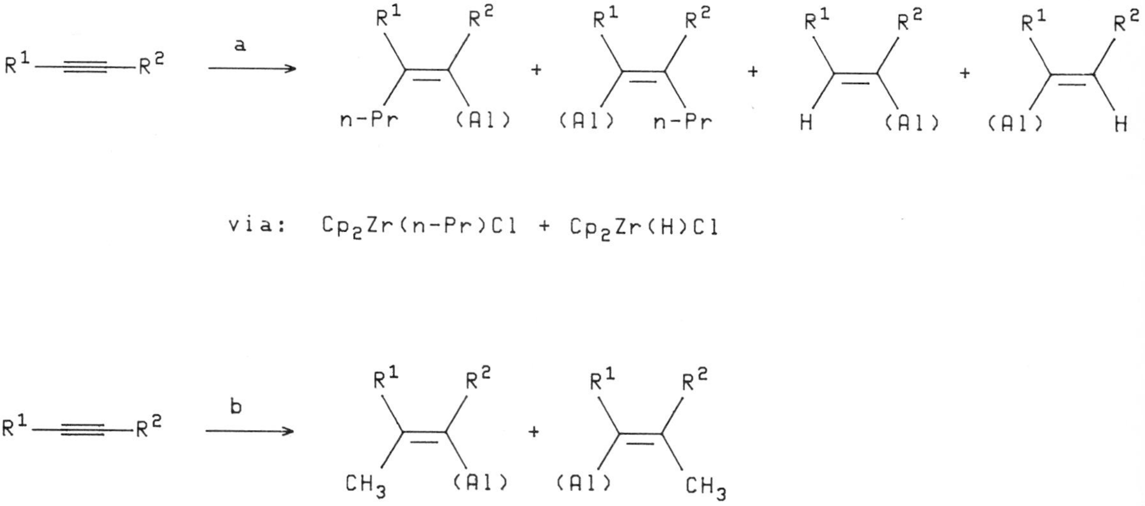

via: $Cp_2Zr(n\text{-}Pr)Cl$ + $Cp_2Zr(H)Cl$

a. rt, $(n\text{-}Pr)_3Al$, Cp_2ZrCl_2, $ClCH_2CH_2Cl$

b. rt, $(CH_3)_3Al$, Cp_2ZrCl_2, $ClCH_2CH_2Cl$

Figure 14.10

zirconocene dichloride with trimethylaluminum can afford only $Cp_2Zr(CH_3)Cl$, leading to a clean methylzirconation. Again, the vinylzirconium complex transmetallates to a vinylaluminum complex[20] (see Figure 14.10).

14.3 ZIRCONIUM REPLACEMENT BY HYDROGEN, HALOGEN, OXYGEN, AND CARBON

Organozirconium(IV) complexes can be directly demetallated with acids, dihalogen, *N*-halosuccinimide, oxidizing agents, and acyl chlorides (see Table 14.2). Coordination and insertion of carbon monoxide converts organozirconium(IV) complexes to acyl complexes that can be analogously demetallated with acids, dihalogen, and oxidizing agents[21] (see Figure 14.11). Replacement of the metal in a vinylzirconium(IV) complex occurs *with retention of configuration.*[15]

Erythro and *threo* $(CH_3)_3$ CCHDCHD zirconium(IV) complexes provide a "handle"

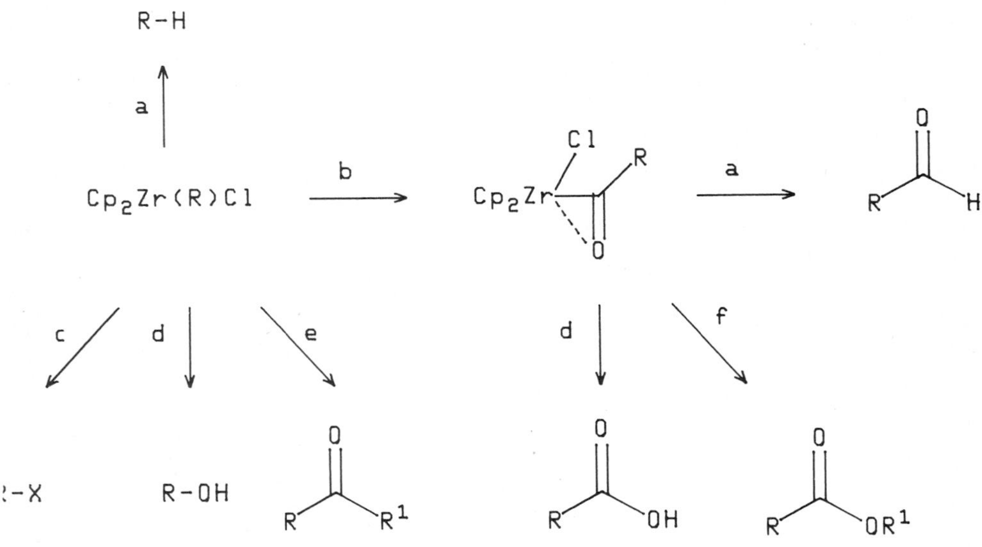

a. HX

b. CO

c. NXS or X_2

d. [O]

e. $R^1C(O)Cl$

f. X_2, R^1OH

Figure 14.11

TABLE 14.2 Oxidative Demetallation of Organozirconium (IV) Complexes with *N*-Halosuccinimides[15]

R^1	R^2	Complex IA:IB	X	Vinyl Halide IIA:IIB	Yield (VPC,%)
H	$CH_3CH_2CH_2CH_2$	>98:<2	Br	>98:<2	75
			Cl	>98:<2	100
CH_3	CH_3CH_2	72:28	Br	72:28	97
			Cl	71:29	53
CH_3	$(CH_3)_2CHCH_2$	>95:<5	Br	>95:<5	75
			Cl	>95:<5	71
CH_3	$(CH)_2CH$	85:16	Br	83:17	95

for determination of the stereoselctivity of C—Zr bond cleavage in alkylzirconium(IV) complexes.[22] Due to the size of both the *t*-butyl and Cp_2ZrCl substituents, these *erythro* and *threo* forms will be essentially "locked" in conformations having these groups in an *anti* relationship. The *erythro* form has a dihedral angle for the vicinal hydrogens of 180°, and the H—H coupling constant is large ($J_{H,H} = 12.9$ Hz). The *threo* form has a dihedral angle for the vicinal hydrogens of 60°, and the H—H coupling constant is small ($J_{H,H} = 4.2$ Hz). Metal replacement using bromine or *N*-bromosuccinimide occurs *with retention of configuration* (see Figure 14.12).

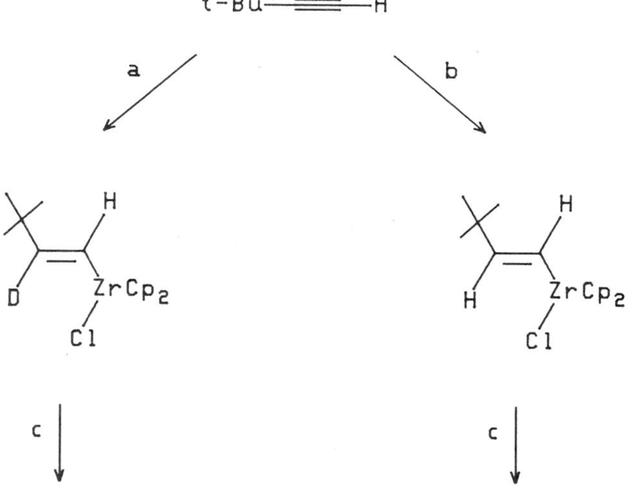

Figure 14.12

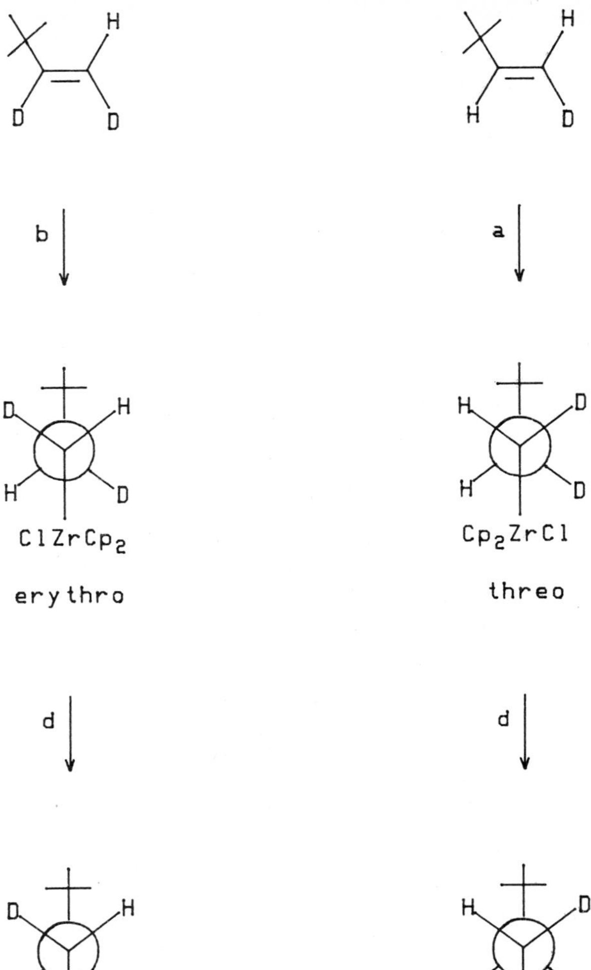

a. rt, $Cp_2Zr(D)Cl$, benzene

b. rt, $Cp_2Zr(H)Cl$, benzene

c. dilute D_2SO_4, D_2O

d. 10°C, Br_2, benzene

Figure 14.12 (*Continued*)

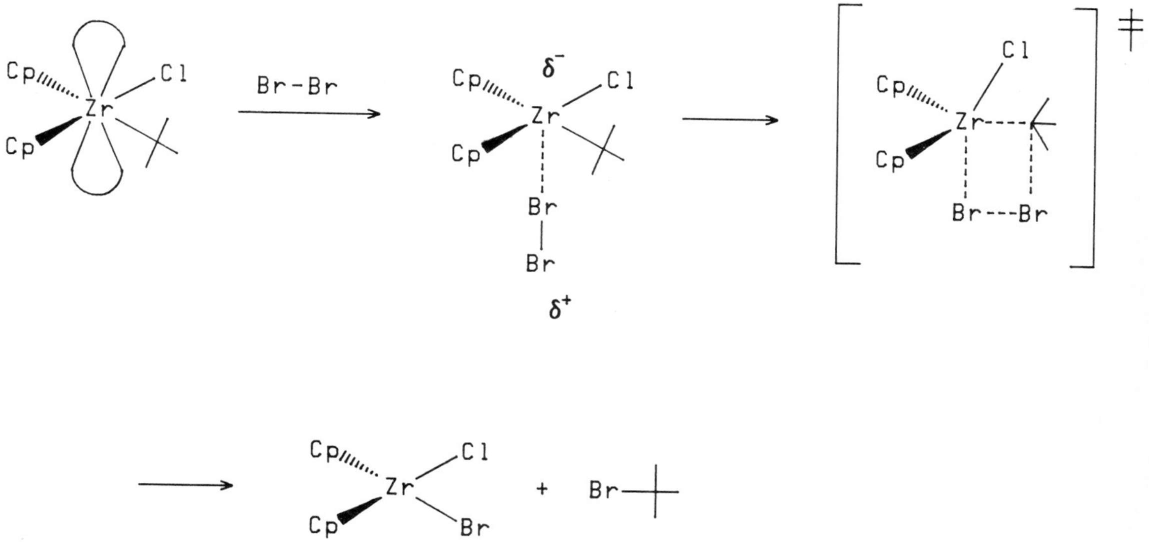

Figure 14.13

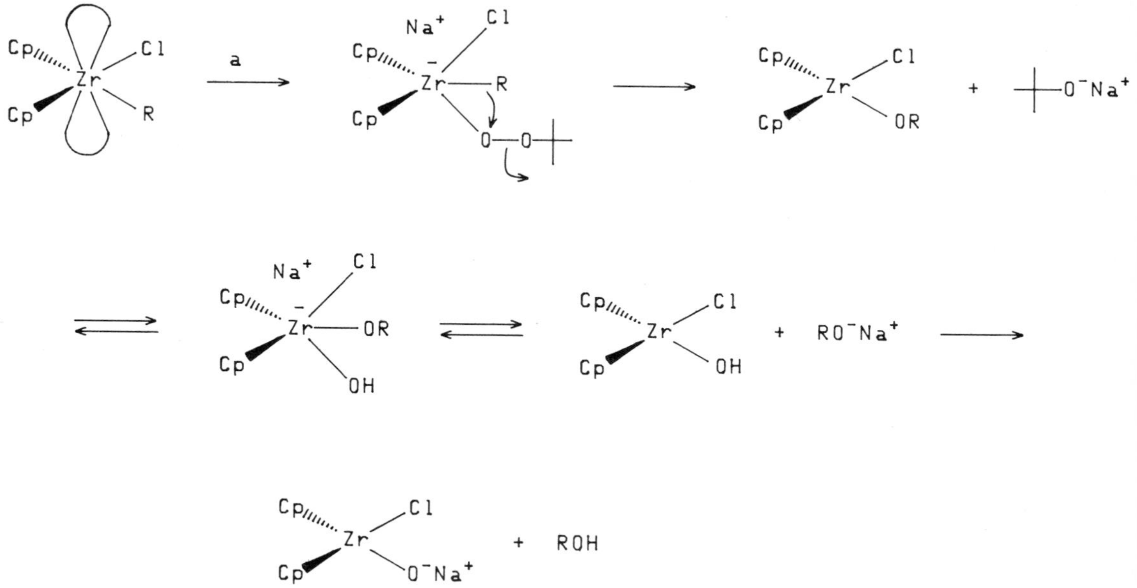

R = 1-octyl: 72%

a. rt, t-BuOOH, benzene

Figure 14.14

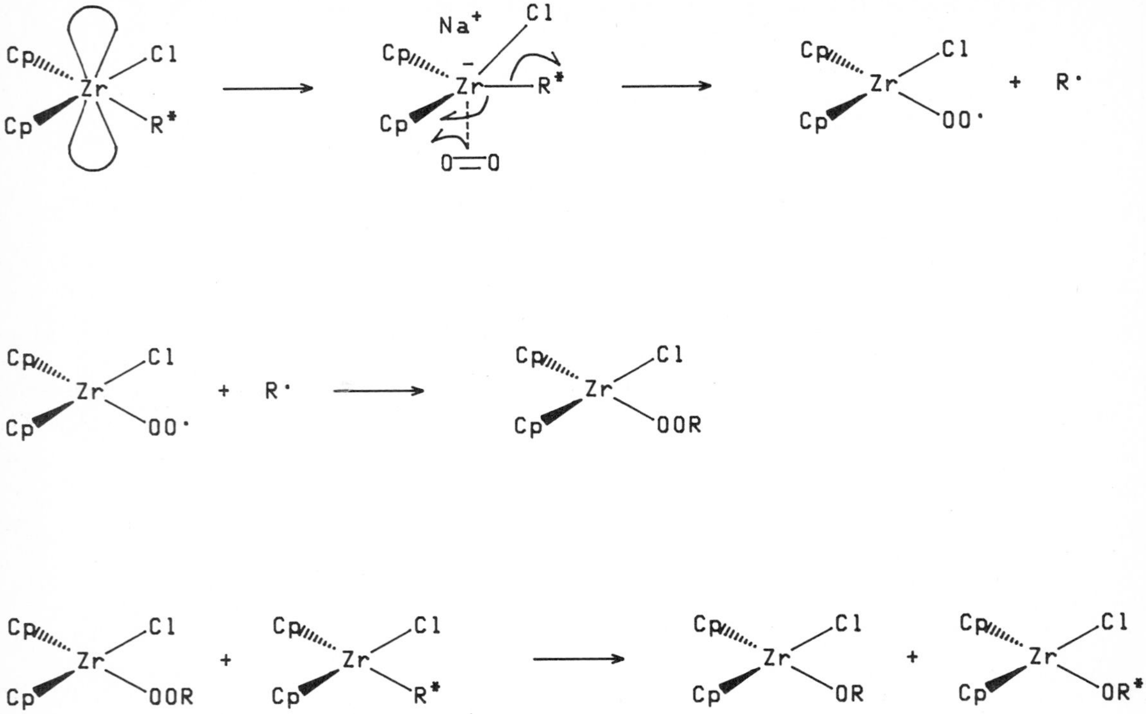

R* - optically active

R - racemic

Figure 14.15

The vast majority of alkyl metal complexes are demetallated with inversion[23] by electrophiles. An appropriate mechanism involves initial oxidation of the metal, then nucleophilic displacement of the metal (S_N2) from the backside. Since zirconium(IV) is d^0, it is not prone to the initial oxidation necessary for backside displacement. The organozirconium(IV) complex has a LUMO, dy^2 in character, located in the *x*-*y* plane containing the two non-Cp ligands. A frontside attack is facilitated by an initial complexation of the organozirconium(IV) complex, via this LUMO, and the cleavage reagent. Polarization is followed by a cyclic transition state having, to illustrate with bromine, simultaneous Zr—C and Br—Br bond cleavage and Zr—Br and C—Br bond formation (see Figure 14.13).

Replacement of zirconium by oxygen can be achieved using a variety of electrophilic reagents.[24] Protic oxidizing agents (peracids, peroxides) afford alcohols directly. Reaction with molecular oxygen produces a metal alkoxide that affords alcohol on hydrolysis. Peroxides and peracids convert the *erythro* and *threo* $(CH_3)_3CHDCHDCZrCp_2(Cl)$ complexes to alcohol with *retention of configuration.* The mechanism may involve formation of a metal peroxide followed by alkyl migration with simultaneous cleavage of the weak O—O bond (see Figure 14.14).

Reaction with molecular oxygen, then hydrolysis, converts these *erythro* and *threo* complexes to alcohols with *50% retention and 50% racemization*, which is indicative of an sp^2 carbon in an intermediate. Monohydrozirconation of 1,5-hexadiene followed by cleavage with oxygen and hydrolysis affords 5-hexen-1-ol (80%) and cyclopentylmethanol (3%). The small quantity of cyclized material is indicative of a short-lived radical intermediate. A suggested mechanism accounts for 50% retention–50% racemization and racemization via a radical intermediate. Initial metal–dioxygen complexation is followed by carbon–metal bond homolysis to produce a carbon radical and metal peroxide radical. Recombination affords a metal alkyl peroxide. This metal alkyl peroxide inserts oxygen *with retention* into a second alkyl metal complex. Thus, for each carbon racemized, one carbon retains configuration (see Figure 14.15).

Few carbon electrophiles are suitable for this frontside attack–metal replacement scheme. Electrophiles with the necessary polarization, alkyltrifluoromethylsulfonates, for example, are not suited for the initial complexation to the metal, a prerequisite to the facile frontside attack. Electrophiles that are able to complex, alkyl chlorides, for example, are not sufficiently polarized. Only acyl chlorides have both the ability to form the initial complex *and* sufficient reactivity to cleave the carbon–metal bond. Even with acyl chlorides, the cleavage is slow and applicable only to alkylzirconium(IV) complexes.[25]

The low reactivity of organozirconium(IV) complexes with carbon electrophiles prompted efforts to *transmetallate* to produce organometallic complexes capable of carbon–carbon bond formation.

14.4 TRANSMETALLATION

14.4.1 To Aluminum

The established mechanism for electrophilic cleavage of the C—Zr bond can be extended to cleavage by a Lewis acidic metal halide such as aluminum chloride. In fact, transmetallation of the organic fragment occurs rapidly and quantitatively at 0°C in dichloromethane. While the organozirconium(IV) complexes react slowly and often inefficiently with acid chlorides, carbon–metal cleavage of an organoaluminum

TABLE 14.3 Comparison of Direct Cleavage and Transmetallation–Cleavage in the Oxidative Demetallation of Organozirconium(IV) Complexes with Acid Chlorides[25]

R	R^1	VPC Yield (%)	
		Zr only	Zr to Al
t-$BuCH_2CH_2$	CH_3	60 (29 h)	98 (30 min)
$CH_3(CH_2)_7$	CH_3	80 (> 2 days)	98 (60 min)
cyclohexyl	CH_3	8 (48 h)	—
cyclohexyl	Ph	—	64[a] (45 min)
(*E*)-(*t*-Bu)CH=CH	CH_3	—	97 (45 min)
(*E*)-(*i*-Pr)CH=CH	CH_3	—	98 (48 min)

[a]Isolated yield.

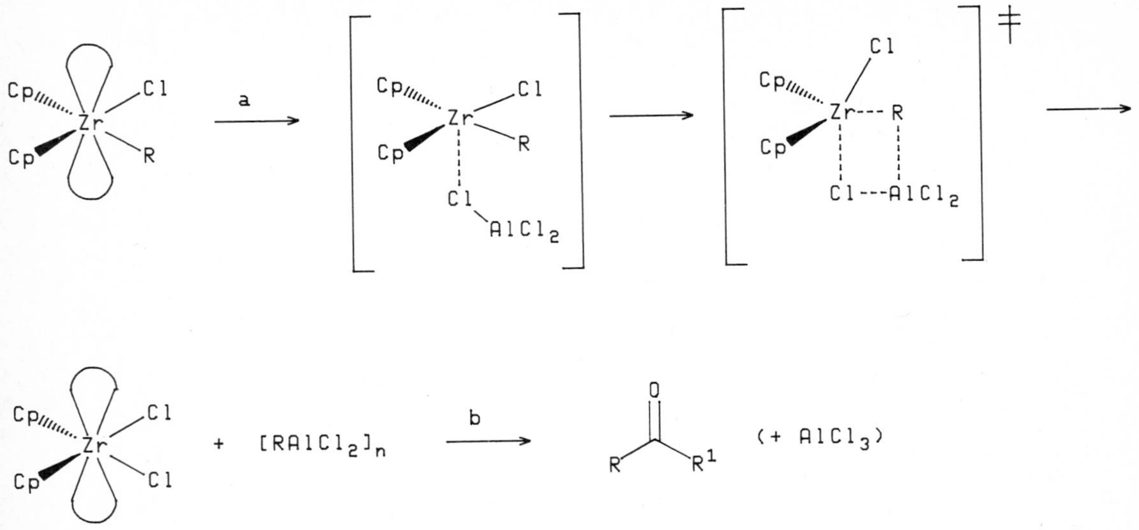

a. 0°C, $AlCl_3$, CH_2Cl_2

b. −30°C, $R^1C(O)Cl$, CH_2Cl_2

Figure 14.16

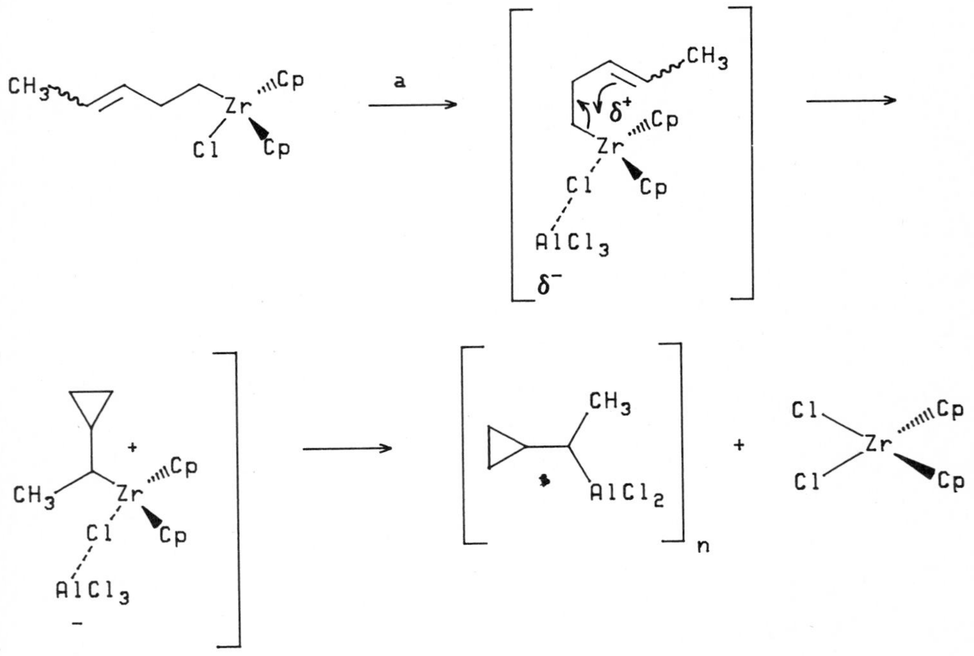

Figure 14.17

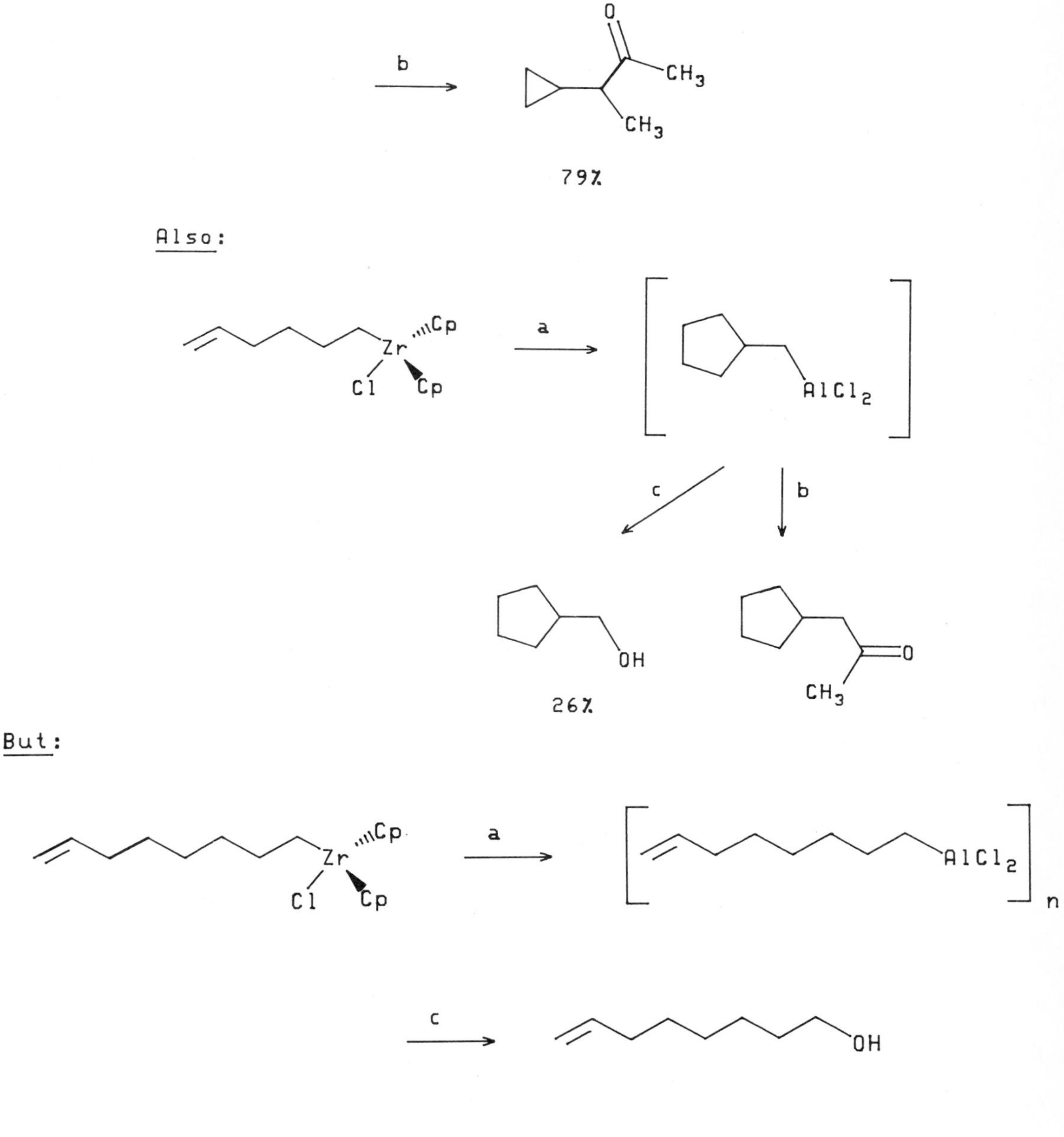

b

79%

Also:

a

c

b

26%

But:

a

n

c

a. −30°C, $AlCl_3$, CH_2Cl_2

b. −30°C, $CH_3C(O)Cl$, CH_2Cl_2

c. 1) O_2, 2) H_2O

Figure 14.17 (*Continued*)

dichloride is complete in less than 1 h at −30°C.[25] The scheme for transmetallation and cleavage is, in principle, catalytic in the Lewis acid, but, in fact, acid–base complexation with the ketone oxygen necessitates use of a stoichiometric quantity of aluminum chloride. Table 14.3 provides a comparison of direct cleavage and transmetallation cleavage (see Figure 14.16).

When the organozirconium(IV) complex contains a remote double bond, transmetallation can be accompanied by cyclization prior to introduction of the acyl chloride. Although an organoaluminum dichloride containing a remote double bond can cyclize, the facility of cyclization here is perhaps indicative of zirconium participation[26] (see Figure 14.17).

Transmetallation of an acylzirconium(IV) complex affords an acylaluminum

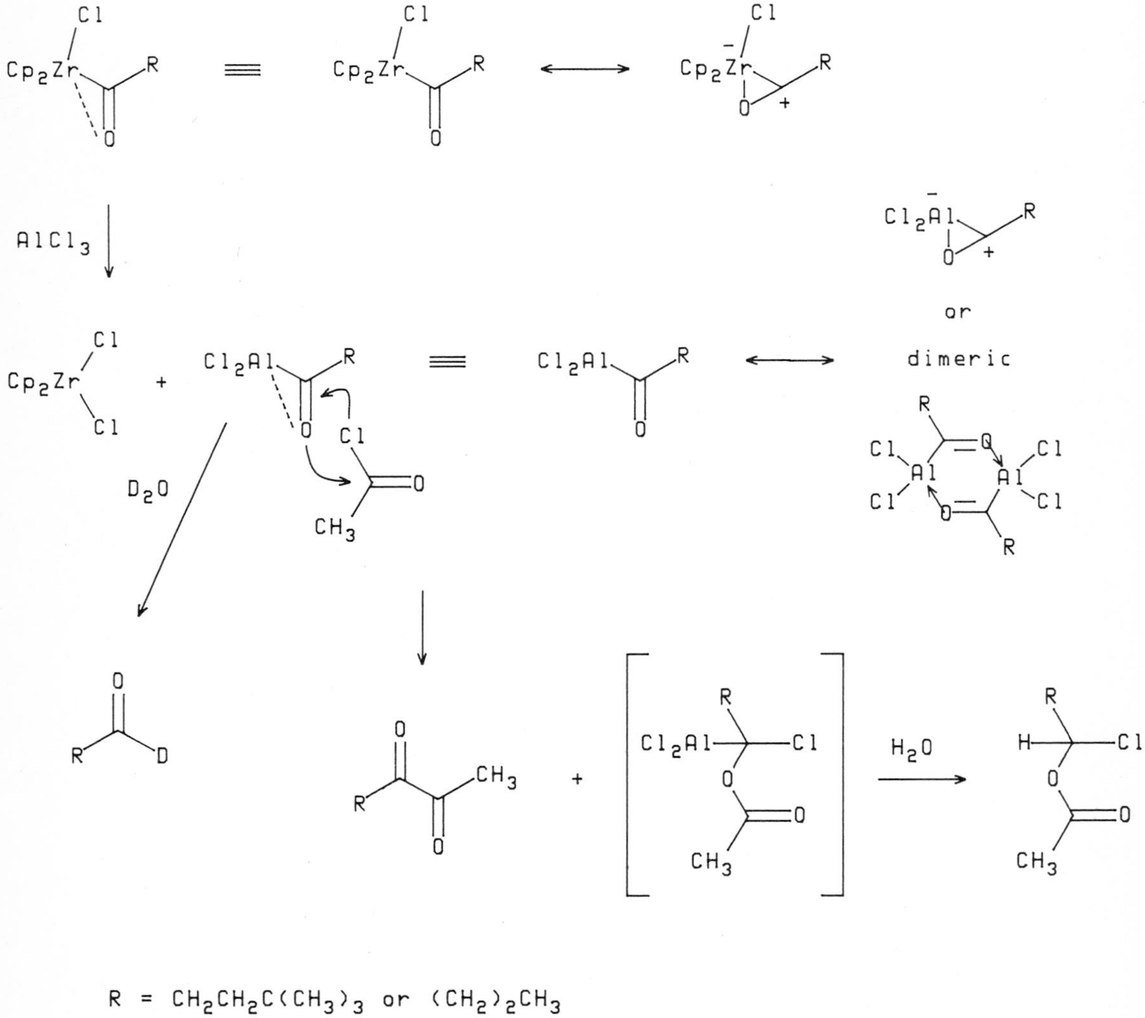

$R = CH_2CH_2C(CH_3)_3$ or $(CH_2)_2CH_3$

Figure 14.18

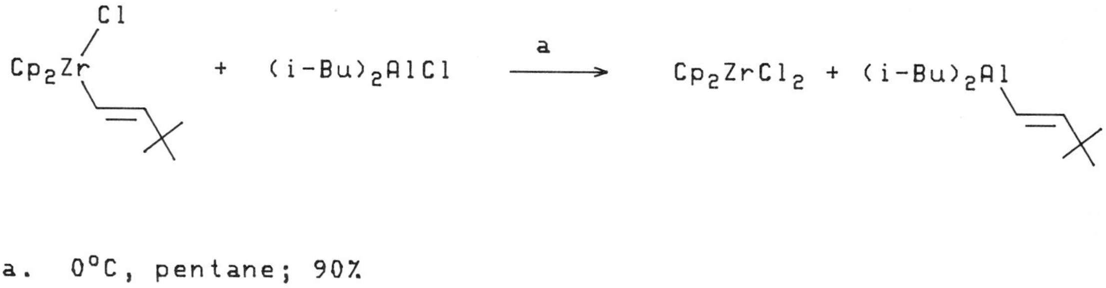

$Cp_2Zr(Cl)(CH{=}CHC(CH_3)_3) + (i\text{-}Bu)_2AlCl \xrightarrow{a} Cp_2ZrCl_2 + (i\text{-}Bu)_2AlCH{=}CHC(CH_3)_3$

a. 0°C, pentane; 90%

Figure 14.19

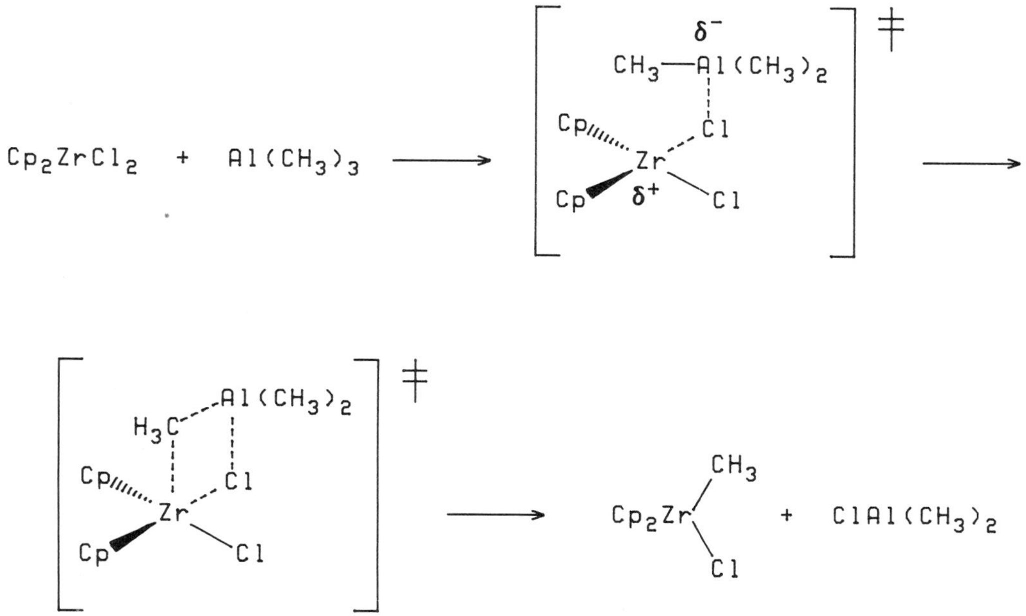

$Cp_2ZrCl_2 + Al(CH_3)_3 \longrightarrow$ [δ^- CH_3—$Al(CH_3)_2$ ⋯ Cl—Zr(δ^+)(Cp)(Cp)—Cl]‡ $\longrightarrow$

[H_3C⋯$Al(CH_3)_2$ ⋯ Cl; H_3C⋯Zr(Cp)(Cp)(Cl)]‡ $\longrightarrow$ $Cp_2Zr(CH_3)Cl + ClAl(CH_3)_2$

Figure 14.20

dichloride that has been characterized by IR and NMR. Particularly informative is the IR carbonyl bond stretch at 1530 cm^{-1}, indicative of aluminum oxygen bonding; the IR carbonyl band of the starting zirconium complex appears at 1550 cm^{-1}. Further evidence for the acylaluminum dichloride was obtained by (1) deuterolysis and (2) cleavage with acetyl chloride. Deuterolysis affords an aldehyde that is at least 80% C(O)D. Deuterolysis of the alternative aluminum enolate would lead to significant deuterium incorporation into the α-position. Cleavage with acetyl chloride affords both an α-diketone and an α-chloroalkyl acetate[27] (see Figure 14.18.)

Several general conclusions were derived from this early transmetallation work. For effective transmetallation, the entering metal with its full complement of ligands should be more electronegative than the departing zirconium. In addition, the facility of

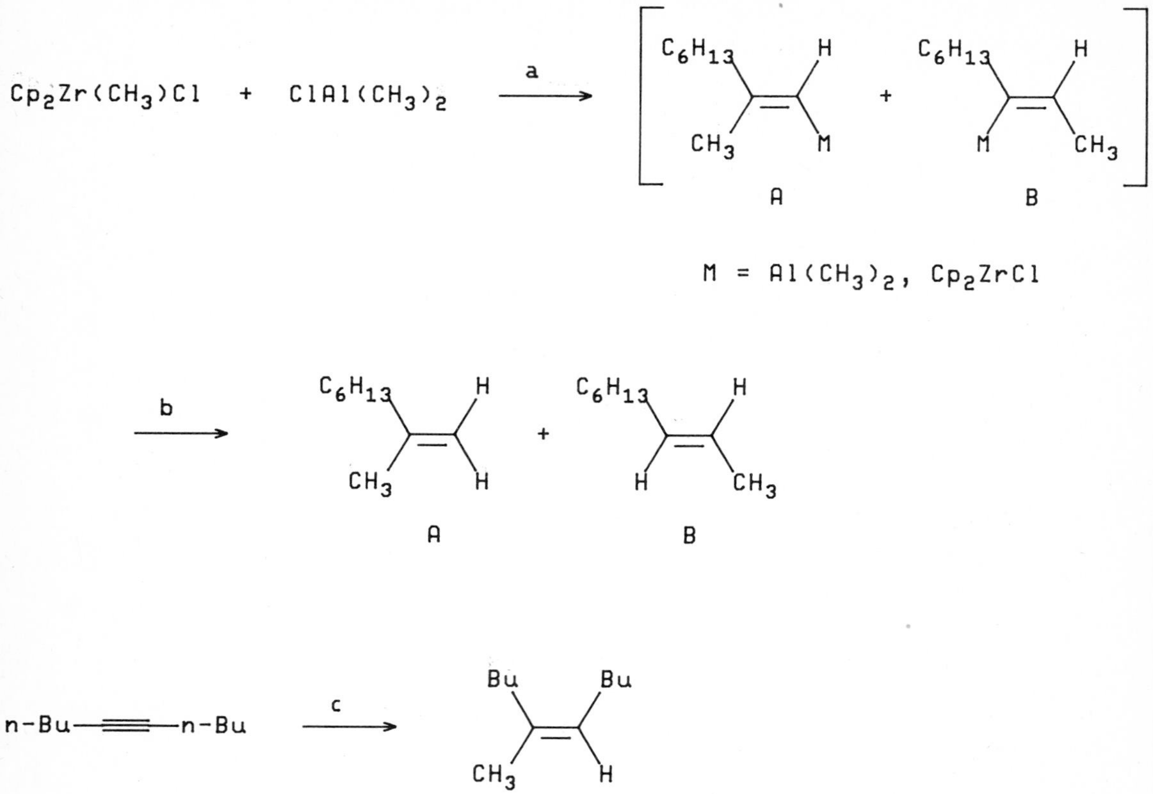

a. rt, ≡—C_6H_{13}, $ClCH_2CH_2Cl$; A:B, 20:1

b. 3 N HCl; A:B, 95:5, 100% yield (vpc)

c. 1) 50°C, $Cp_2Zr(CH_3)Cl$ + $ClAl(CH_3)_2$, 2) H_2O; 89% (vpc)

Figure 14.21

transmetallation is apparently dependent on the hybridization of the mobile carbon. It is known that vinylaluminum compounds "exhibit a greater tendency toward association through alkenyl 'bridges' than do alkyl derivatives (through alkyl 'bridges')."[26,28] Extending this observation, one might expect that the transition state for transmetallation with a bridging vinyl will be lower in energy than one with a bridging alkyl. In fact, only vinyl ligands are transferrable in Zr to Cu, Zr to Pd, Zr to Zn, and Zr to Ni transmetallations.

Transfer of a vinyl group from zirconium(IV) to the less electronegative aluminum of di-*i*-butylaluminium chloride is also rapid at 0°C[29] (see Figure 14.19).

In contrast, Negishi reported that zirconocene dichloride reacts with trimethylaluminum in a reverse-sense transmetallation.[30] Note, again, the bridging chloride in the initial complex and transition state (see Figure 14.20).

Perhaps the vinylaluminum complex is favored in the first case due to the greater thermodynamic stability of a bridged vinylaluminum complex in comparison with that of an alkylaluminum. Negishi's $Cp_2Zr(CH_3)Cl\text{-}ClAl(CH_3)_2$ system rapidly exchanges methyl groups at low temperature and will "methylzirconate" 1-octyne stereospecifically and regioselectively (see Figure 14.21). Transfer of the vinyl group from zirconium to aluminum is reversible but strongly favors aluminum [$M = Al(CH_3)_2 : Cp_2ZrCl = 20:1$]. Protic workup affords a 95:5 mixture of 2-methyl-1-octene and *trans*-2-nonene in quantitative yield. Methylzirconation of internal alkynes is similarly efficient. More vigorous conditions are required.

The method is efficient only for introduction of alkyl groups lacking β-hydrogen. *n*-Propylzirconation of 1-octyne followed by protic workup affords, in addition to the expected alkenes, 20% 1-octene formed by a β-hydride elimination–hydrozirconation sequence. Use of a catalytic amount of zirconocene dichloride favors hydrozirconation[19] (see Figure 14.22).

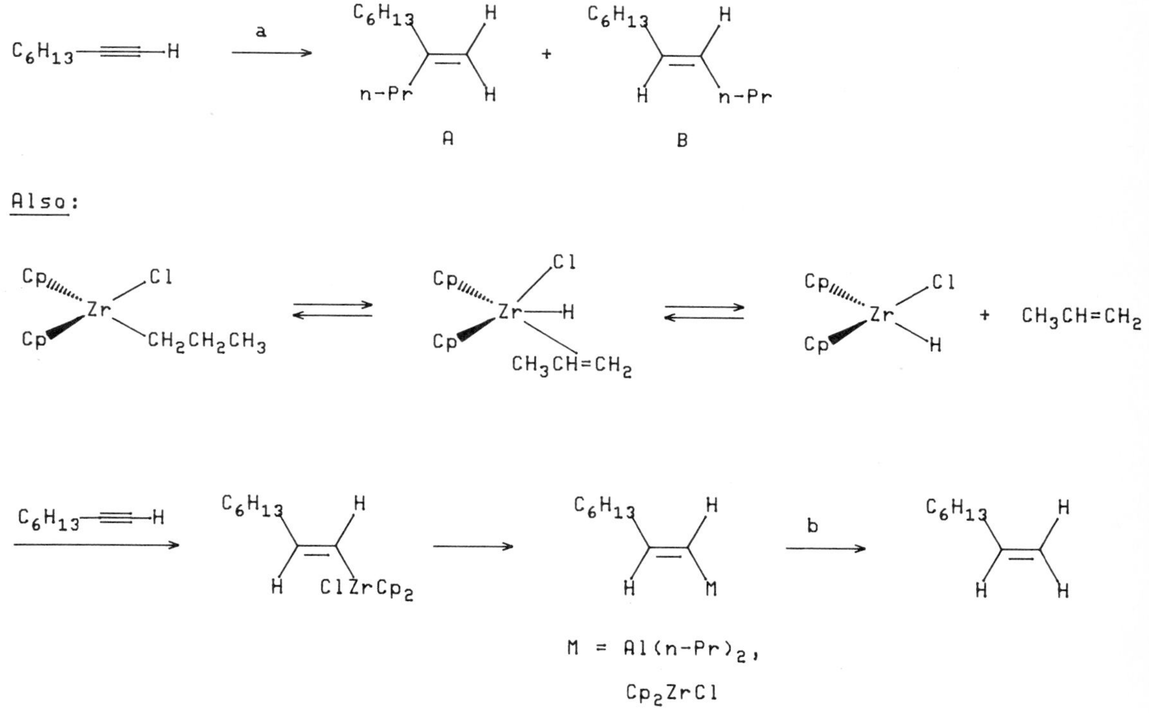

Also:

a. 1) 0°C, $Cp_2Zr(n\text{-}Pr)Cl$, $ClAl(n\text{-}Pr)_2$, 2) H_2O; A:B, 80:20, 75% yield; with catalytic Zr, A:29%, B:8%

b. H_2O; 20%; with catalytic Zr, 47%

Figure 14.22

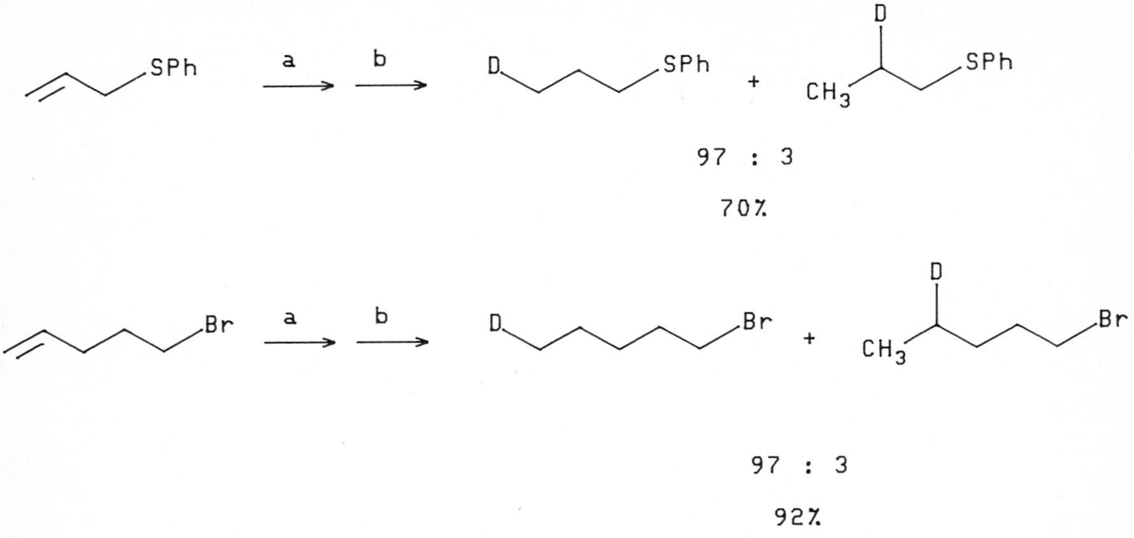

SPh $\xrightarrow{a}$ $\xrightarrow{b}$ D–(CH2)3–SPh + CH3–CH(D)–CH2–SPh

97 : 3

70%

Br $\xrightarrow{a}$ $\xrightarrow{b}$ D–(CH2)5–Br + CH3–CH(D)–(CH2)3–Br

97 : 3

92%

a. 0°C, 10% Cp_2ZrCl_2, $(i\text{-}Bu)_3Al$, $ClCH_2CH_2Cl$

b. D_2O

Figure 14.23

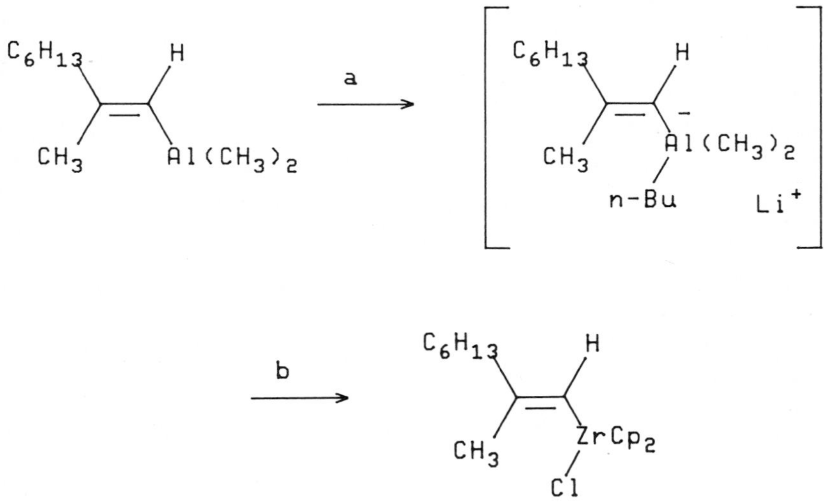

C_6H_{13}, H, CH_3, $Al(CH_3)_2$ $\xrightarrow{a}$ [C_6H_{13}, H, CH_3, $\bar{A}l(CH_3)_2$, n-Bu, Li^+]

$\xrightarrow{b}$ C_6H_{13}, H, CH_3, $ZrCp_2$, Cl

a. n-BuLi, hexane

b. –78°C to rt, Cp_2ZrCl_2, THF; 95% (by NMR)

Figure 14.24

The equilibrium between alkylzirconium and zirconium hydride complexes can be driven to hydride by cleavage of the zirconium–alkene π-complex. An increase in the number of alkene substituents should weaken the π-complex and facilitate cleavage. Thus, a mixture of tri-*i*-butylaluminum and a catalytic amount of zirconocene dichloride efficiently produces a vinylaluminum complex by hydrozirconation and transmetallation[19] (see Figure 14.23). This catalytic mixture efficiently cyclizes 6-trimethylsilylhex-1-en-5-yne to trimethylsilylmethylenecyclopentene. The method appears to be limited to the synthesis of five- and six-membered exocyclic alkenes.[29]

Finally, while Zr to Al transmetallation occurs with alkenylzirconium complexes and dialkylaluminum chlorides, the reverse Al to Zr transmetallation can be driven by prior formation of an ate complex[30] (see Figure 14.24).

14.4.2 To Copper

Transmetallation of vinylzirconium(IV) complexes to cuprous chloride would provide an attractive alternative route to vinylcopper complexes (usually prepared from relatively inaccessible vinyllithiums). Transmetallation is slow at $-78\,°C$ in tetrahydrofuran. The vinylcopper complex can be trapped as an ate complex, then used in Michael addition to α,β-unsaturated ketones[31] (see Figure 14.25).

The low thermal stability of vinylcopper complexes results in homocoupling of the vinyl ligand during transmetallation at $0\,°C$. Diene formation, inefficient utilization of

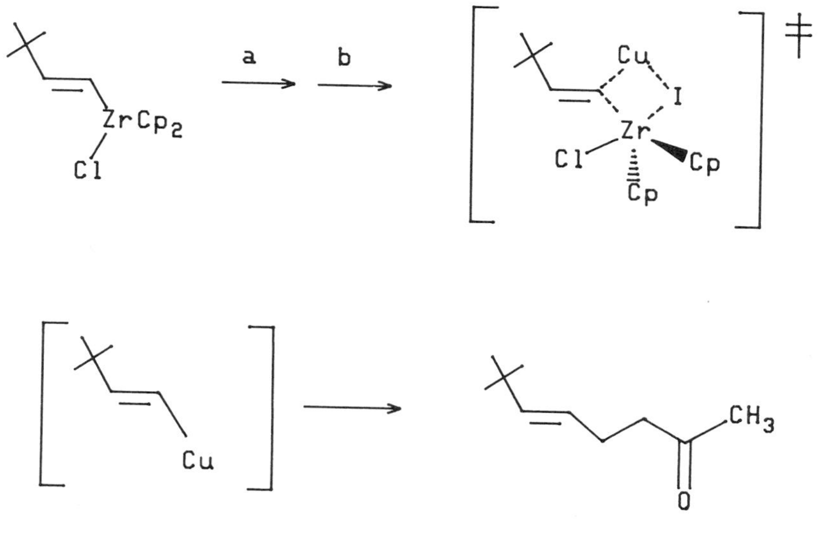

a. $-78°C$, LiI, $CH_2{=}CHC(O)CH_3$

b. $CuOSO_2CF_3$, THF; 73% (vpc)

Figure 14.25

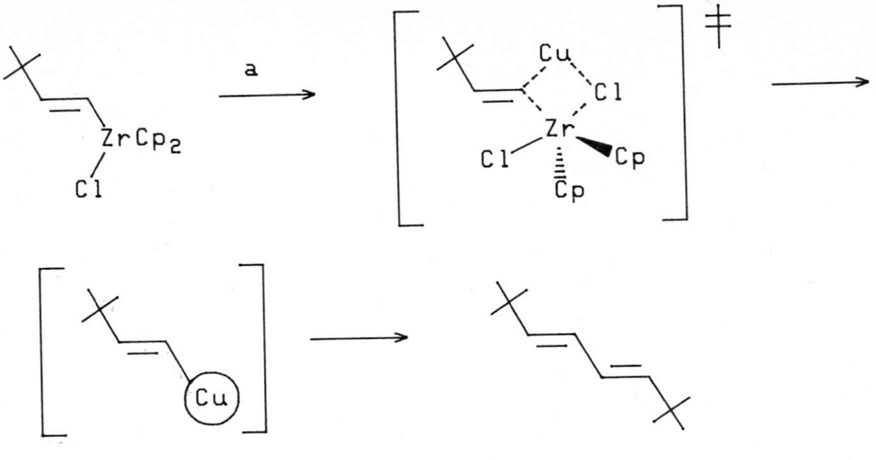

a. 0°C, CuCl, THF; 90% (vpc)

Figure 14.26

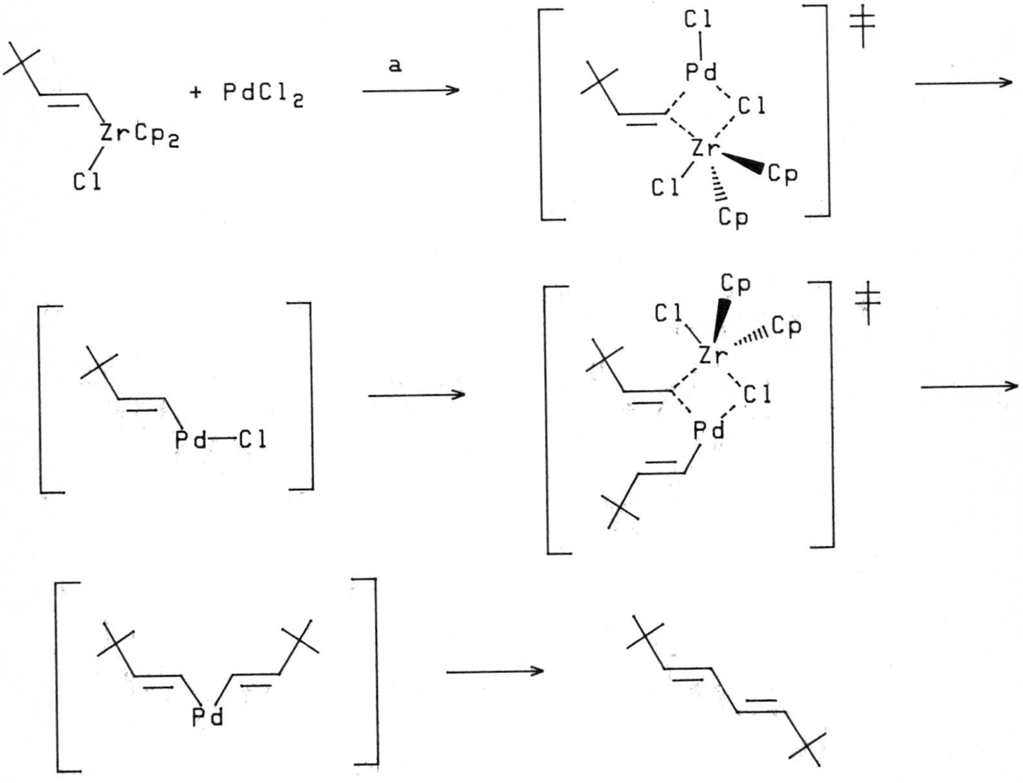

a. rt, sulfolane; 62%

Figure 14.27

a

b

c

81%

trace

d

e

A

B

f, g

Figure 14.28

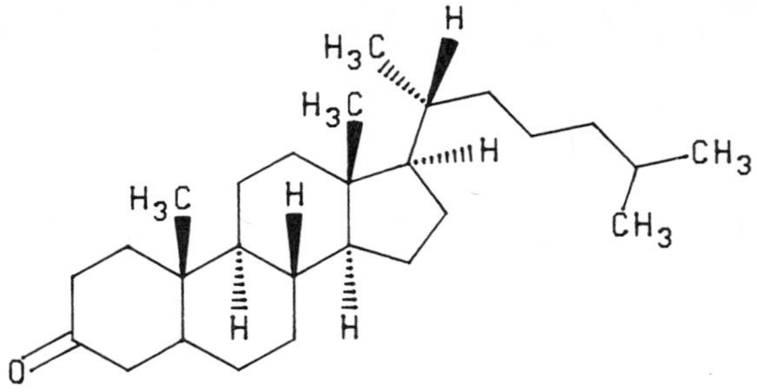

20(R)-Cholestane-3-one

---

a. rfx, $HOCH_2CH_2OH$, p-TsOH·H_2O, benzene; 95%

b. rt, NaOAc, PCC, CH_2Cl_2; 97%

c. rfx, $CH_3CH{=}PPh_3$, THF

d. rfx, NaCl, Na_2CO_3, 15-crown-5, CH_2Cl_2, $PdCl_2(CH_3CN)_2$; 96%

e. $-78^{\circ}C$, maleic anhydride (ancillary ligand), THF; 96% combined yield for A and B, 78% isolated for A

f. H_2, PtO, EtOAc

g. H_3O^+; 94% for two steps

Figure 14.28 (*Continued*)

the organic ligand in homocuprates, and the relatively low reactivity of heterocuprates limit the synthetic utility of Zr to Cu transmetallation (see Figure 14.26).

14.4.3 To Palladium

Transmetallation from Zr to Pd links hydrozirconation to the manifold of carbon–carbon bond formations mediated by palladium. For example, reaction of a vinylzirconium(IV) complex with palladium(II) chloride affords diene via Zr to Pd transmetallation, a second Zr to Pd transmetallation, and reductive elimination (see Figure 14.27).

14.4.4 To π-Allylpalladium Complexes

Reductive elimination will follow a single transmetallation when the palladium(II) complex already bears an organic substituent. For example, coupling with a preformed

π-allylpalladium chloride is the key step in a stereo- and regioselective introduction of a steroid side chain of 20(*R*)-cholestan-3-one[32,33] and 25-hydroxycholesterol.[34] Palladium is replaced with retention of configuration (see Figures 14.28 and 14.29).

Regioselectivity is controlled by choice of the ancillary ligand on palladium. In the presence of maleic anhydride, high selectivity for coupling at the less hindered (exocyclic) terminus is observed. An explanation is based on the donor–acceptor capabilities of ligands and the square planar geometry of palladium. The terminal carbons of an unsymmetrical allyl in a π-allylpalladium complex will have different Pd—C distances: shorter for the more sterically accessible (better donor) carbon and longer for the less sterically accessible (weaker donor) carbon. The thermodynamically favored arrangement of ligands in the square planar metal sphere will have the stronger

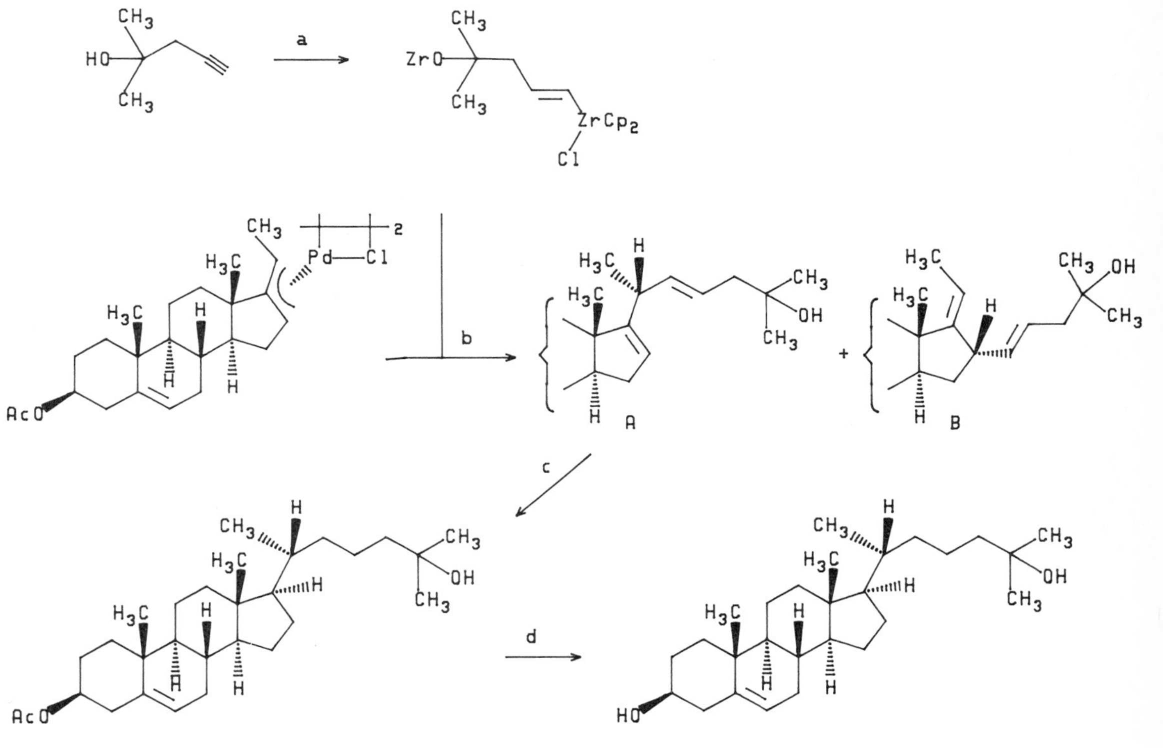

25-Hydroxycholesterol

a. rt, 2 $Cp_2Zr(H)Cl$, benzene; 45%

b. -78°C, maleic anhydride (ancillary ligand), THF;
A:70%, B:18%

c. H_2, PtO, EtOAc; 97%

d. OH^-, H_2O; 97%

Figure 14.29

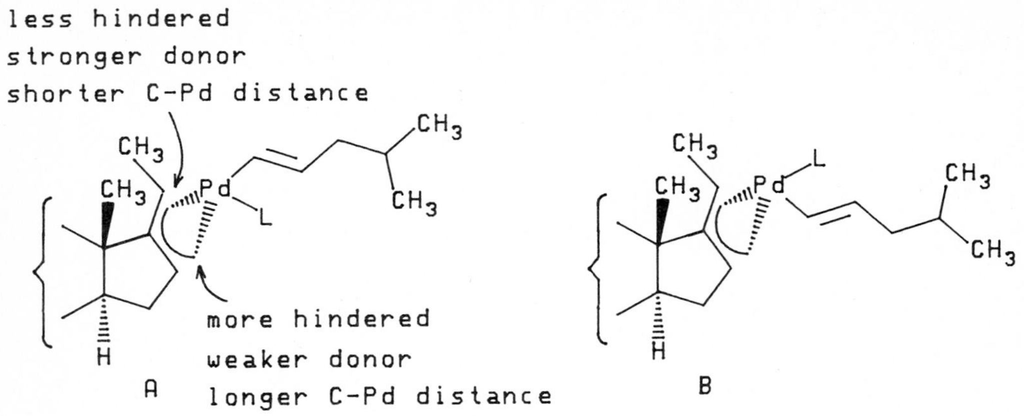
less hindered
stronger donor
shorter C-Pd distance
CH_3
CH_3
Pd
L
CH_3
CH_3
H
more hindered
weaker donor
longer C-Pd distance
A
CH_3
CH_3
Pd
L
CH_3
CH_3
H
B

A favored when L = maleic anhydride

B favored when L = PPh_3

Figure 14.30

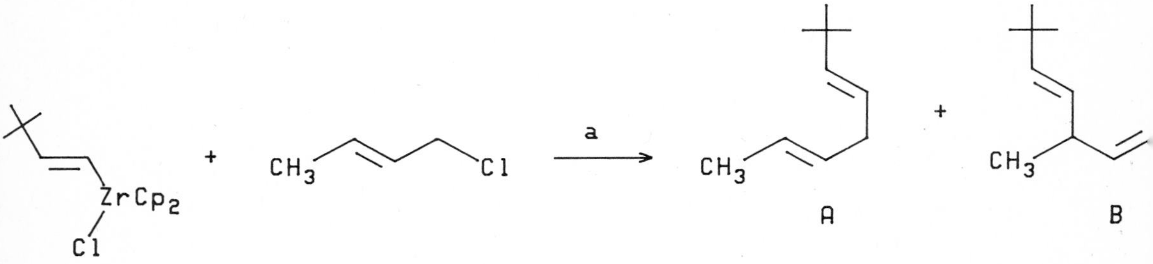
$ZrCp_2$
Cl
+
CH_3
Cl
a
CH_3
A
+
CH_3
B

9 eq PPh_3	A:B =	10:90 (vpc)
3 eq maleic anhydride		92:8 (vpc)

a. rt, 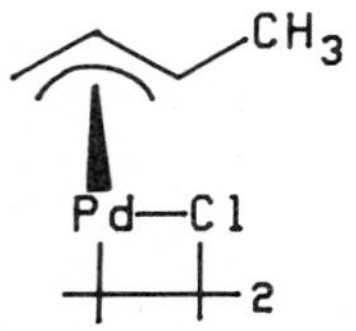 catalyst, THF
CH_3
Pd—Cl
2

Figure 14.31

acceptor ligand *trans* to the better donor carbon. Since carbons being coupled in reductive elimination must be *syn*, coordination of maleic anhydride, favoring complex A over complex B, leads to coupling at the less hindered terminus while coordination of phosphine, favoring complex B over complex A, leads to coupling at the more hindered terminus (see Figure 14.30).

The requirement for a stoichiometric quantity of palladium might be eliminated by generation of the π-allylpalladium complex in situ by oxidative addition of an allylic chloride or acetate to palladium(0). However, *tetrakis*-(triphenylphosphine)palladium(0) is required (see Chapter 3) for facile oxidative addition; the phosphine requirement would eliminate the ancillary ligand "handle" on regioselectivity of coupling. Schwartz found that a highly reactive "naked" palladium(0) catalyst is released when a π-allylpalladium chloride dimer couples with an alkenylzirconium(IV) complex. Again, the presence of maleic anhydride strongly favors the coupling at the least hindered terminus[36] (see Figure 14.31).

Ancillary ligand control apparently cannot overcome the steric problems associated with formation of a quaternary carbon. Coupling of an alkenylzirconium(IV) complex with isoprenyl chloride using *tetrakis*-(triphenylphosphine)palladium(0) occurs exclusively at the most accessible carbon[36] (see Figure 14.32).

14.4.5 To σ-Vinyl- and σ-Arylpalladium(II) Halides

Organopalladium(II) bromides and iodides are also prepared under mild conditions by mild oxidative addition of vinyl and aryl halides to palladium(0). Electron-rich or neutral vinyl halides oxidatively add to *tetrakis*-(triphenylphosphine)palladium(0). Electron-deficient vinyl halides require a bis-phosphine complex, prepared by reduction of bis-(triphenylphosphine)palladium dichloride with di-*i*-butylaluminum hydride. Coupling of vinylpalladium(II) halides with monosubstituted vinylzirconium(IV) complexes is rapid; only a single conjugated diene, retaining the original configuration of the vinylzirconium(IV) complex and vinyl halide, is isolated[37] (see Table 14.4).

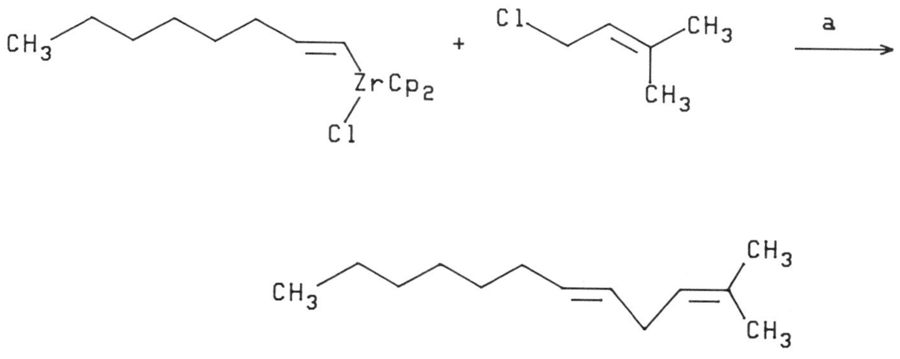

CH_3 ... $ZrCp_2$, Cl + Cl ... CH_3, CH_3 $\xrightarrow{a}$ CH_3 ... CH_3, CH_3

a. rt, L_4Pd, $ClCH_2CH_2Cl$; 97%

Figure 14.32

TABLE 14.4 Coupling of Vinylpalladium(II) Halides with Monosubstituted Vinylzirconium(IV) Complexes[37]

a. rt, catalyst, THF

X	R^1	R^2	R^3	R^4	Catalyst	Yield (%)
I	*n*-C_5H_{11}	H	*n*-C_4H_9	H	$Pd(PPh_3)_4$	91
Br	*n*-C_4H_9	H	$COOCH_3$	CH_3	$Pd(PPh_3)_2$	75
Br	CH_2OTHP	H	$COOCH_3$	CH_3	$Pd(PPh_3)_2$	70
Br	CH_2OTHP	H	H	H	$Pd(PPh_3)_4$	77
Br	*n*-C_4H_9	—$(CH_2)_3C(O)$—		H	$Pd(PPh_3)_2$	60

TABLE 14.5 Coupling of Organopalladium(II) Halides with Di- and Trisubstituted Vinylzirconium(IV) Complexes via a Zinc Shuttle[30,38]

a. rt, $ZnCl_2$, $Pd(PPh_3)_4$, THF

R^1	R^2	R^3	R	X	Yield (%) VPC	Yield (%) Isolated
C_4H_9	H	C_4H_9	Ph	I	—	65
C_2H_5	H	C_2H_5	Ph	I	85	72
C_2H_5	H	C_2H_5	(*E*)-$C_4H_9CH{=}CH$	I	86	71
C_2H_5	H	C_2H_5	$CH_2{=}CH$	Br	79	62
C_6H_{13}	CH_3	H	Ph	I	89	—

14.4.6 To Zinc, Then to Palladium

Coupling of organopalladium(II) halides with *di*substituted and *tri*substituted vinylzirconium(IV) complexes is considerably less efficient due to slow Zr to Pd transmetallation. A complex with metal electronegativity similar to palladium chloride but with a lower steric requirement ($ZnCl_2$, $CdCl_2$) can serve as a "shuttle" between

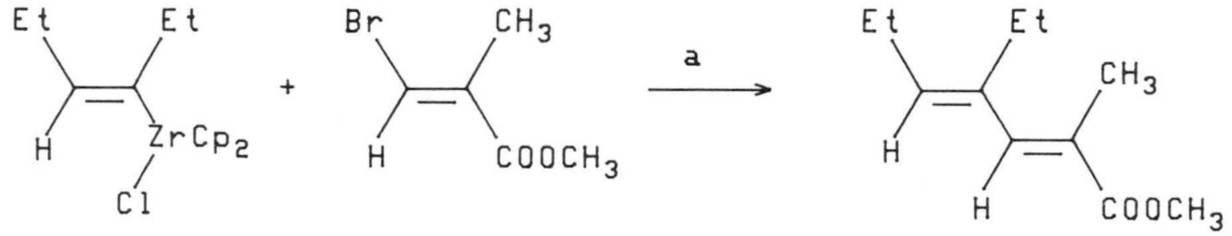

a. rt, $Pd(PPh_3)_4$, THF; <2% after 6 h,

82% after 1 h with $ZnCl_2$

With Pd, via:

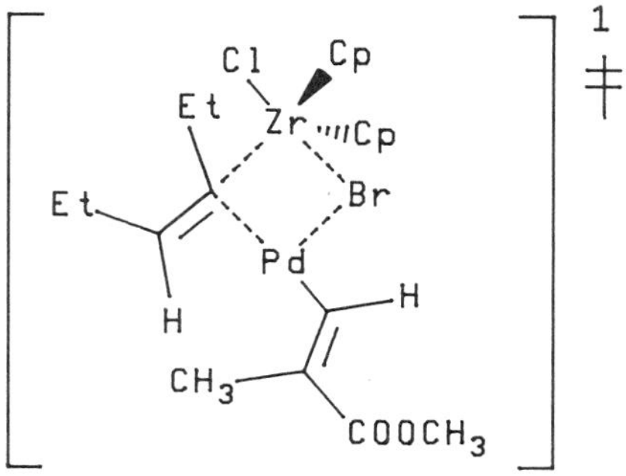

With Pd and $ZnCl_2$, via:

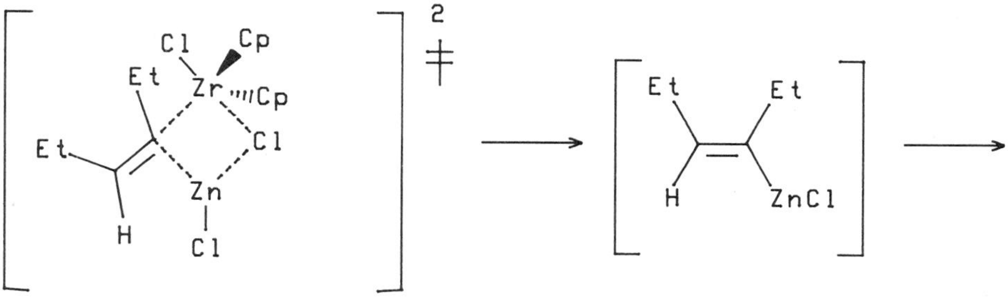

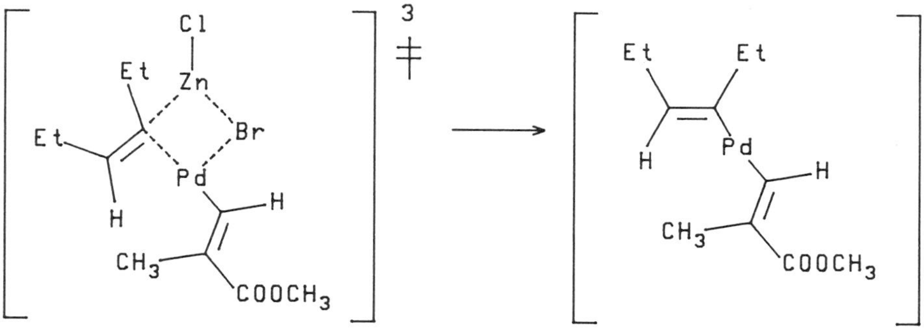

Transition State 1 is higher in energy than States 2 or 3

Figure 14.33

zirconium and palladium. To illustrate, essentially no coupling (< 2% product) occurs at room temperature in 6 h in the palladium-catalyzed reaction of a disubstituted vinylzirconium(IV) complex with a vinyl bromide. The same mixture with one equivalent of zinc chloride added affords 82% coupling product at room temperature in just 1 h. Other examples of the Zr to Zn to Pd shuttling process are shown in Table 14.5[30,38] (see Figure 14.33).

14.4.7 To σ-Alkylnickel(III)

The low thermal stability of vinylcopper complexes from Zr to Cu transmetallation, inefficient utilization of organic ligands in homocuprates, and low reactivity of

TABLE 14.6 Nickel-Catalyzed Michael Addition of Vinylzirconium(IV) Complexes to α,β-Unsatuated ketones[39,42]

a. −78 to 0°C, 10% $Ni(acac)_2$, THF

b. rt, NH_4Cl, H_2O

R^1	R^2	R^3		R^4	Yield (%) VPC	Yield (%) Isolated
t-Bu	H	CH_3		H	>95	—
t-Bu	H		$—(CH_2)_3—$		77	73
t-Bu	H		$—(CH_2)_2—$		78	—
n-hex	H		$—(CH_2)_2—$		—	49
n-C_5H_{11}–CH(H)–O–CH(CH_3)–O–CH_2CH_3	H		$—(CH_2)_2—$		—	59
t-Bu	CH_3		$—(CH_2)_3—$		—	10
Et	Et		$—(CH_2)_3—$		—	10
t-Bu	H		—CH_2CH(O*t*-Bu)—		77	—
t-Bu	H		—CH_2CH(OPh)—		—	80[a]
Et	Et		$—(CH_2)_3—$		—	52[a]
n-C_5H_{11}–CH(H)–OSi(CH$_3$)$_2$(*t*-Bu)	H		—$CH_2CH[OC(CH_3)_2Ph]$—		—	84[a]

[a]Catalyst = $Ni(acac)_2$ + $(i\text{-}Bu)_2AlH$ (1:1).

heterocuprates in Michael addition prompted a search for alternative metals to mediate the Michael addition of organozirconium(IV) complexes. Loots and Schwartz reported efficient nickel-catalyzed Michael addition of monosubstituted vinylzirconium(IV) complexes.[39] Michael addition of di- and trisubstituted vinylzirconium(IV) complexes is inefficient, presumably due to slow Zr to Ni transmetallation. Some symmetrical diene from coupling of two vinylzirconium(IV) complexes is observed *prior* to the start of conjugate addition. Apparently the nickel catalyst is a reduced form released in a reductive elimination that generates diene. An alternative catalytic system is generated from nickel acetylacetonate reduction with di-*i*-butylaluminum hydride (1 molar equivalent based on nickel). The nickel species involved in transmetallation is a σ-alkylnickel(III) formed by reduction of Ni(III) to Ni(I), electron transfer from Ni(I) to enone to give caged Ni(II) and enone radical anion,

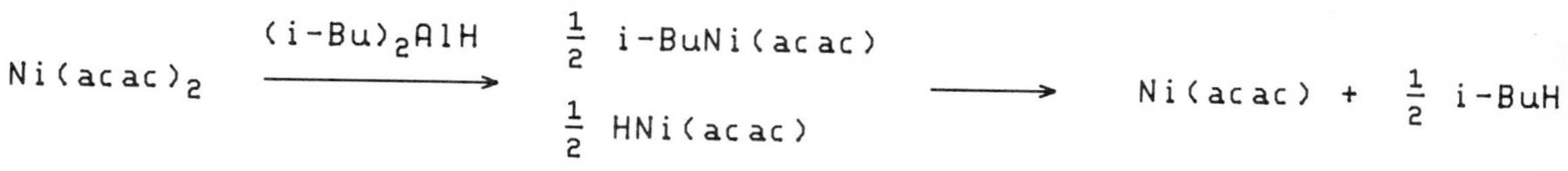

$$\mathrm{Ni(acac)_2} \xrightarrow{(i\text{-Bu})_2\mathrm{AlH}} \tfrac{1}{2}\, i\text{-BuNi(acac)} + \tfrac{1}{2}\,\mathrm{HNi(acac)} \longrightarrow \mathrm{Ni(acac)} + \tfrac{1}{2}\, i\text{-BuH}$$

Figure 14.34

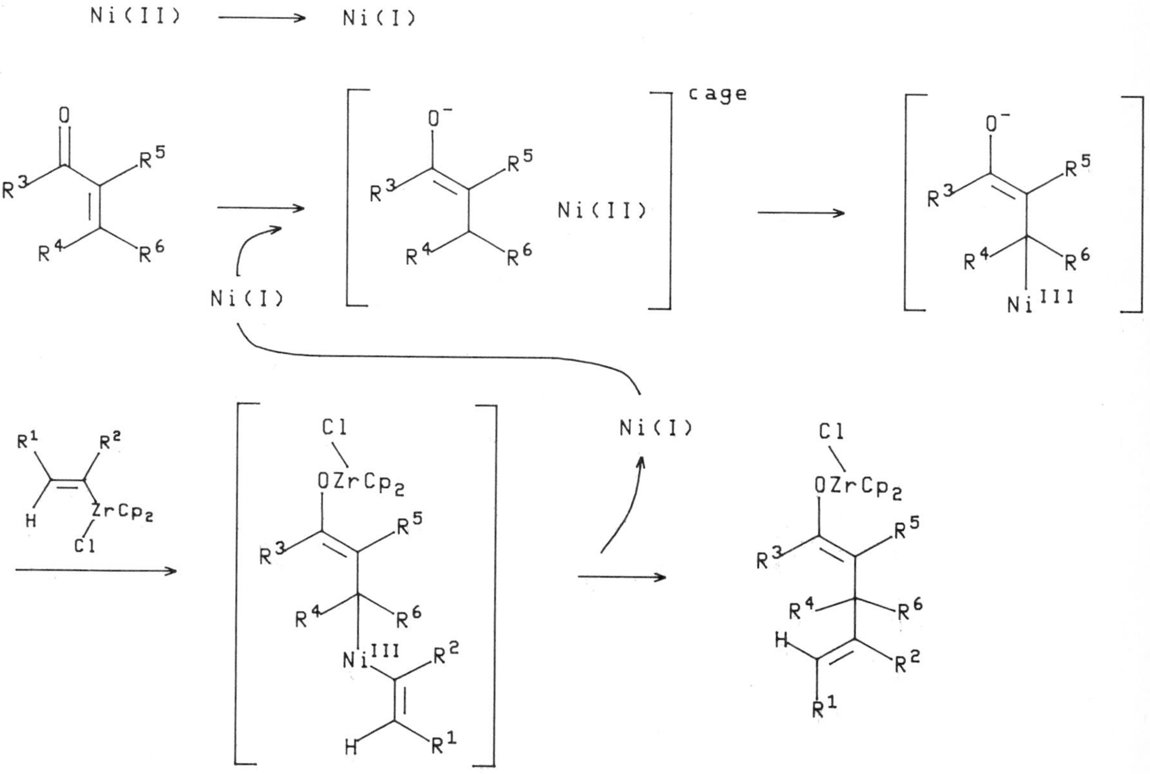

Figure 14.35

and then collapse to generate a C—Ni σ bond (see Figure 14.35).[40–42] Both gas evolution and cyclic voltammetry studies support this mechanism:

1. Evolution of 0.5 equivalent of isobutene and appearance of catalytic activity are observed on mixing nickel acetylacetonate with di-*i*-butylaluminum hydride at 0°C. Activity is negligible after 24 h [complete reduction to Ni(0)] (see Figure 14.34).
2. Cyclic voltammetry of the catalyst system reveals at least three distinct anodic waves (Ep_a = − 1.66, − 1.16, − 0.49 V), each coupled to a large cathodic wave. The three anodic waves correspond to three independent oxidizable species, each able to catalyze the process with a unique efficiency, each with a unique lifetime at 23 °C (for − 1.66 V, 20 min; for − 0.49 V, 3–4 h; for − 1.16 V, a low level of activity still present after 24 h) (see Table 14.6 and Figure 14.36).

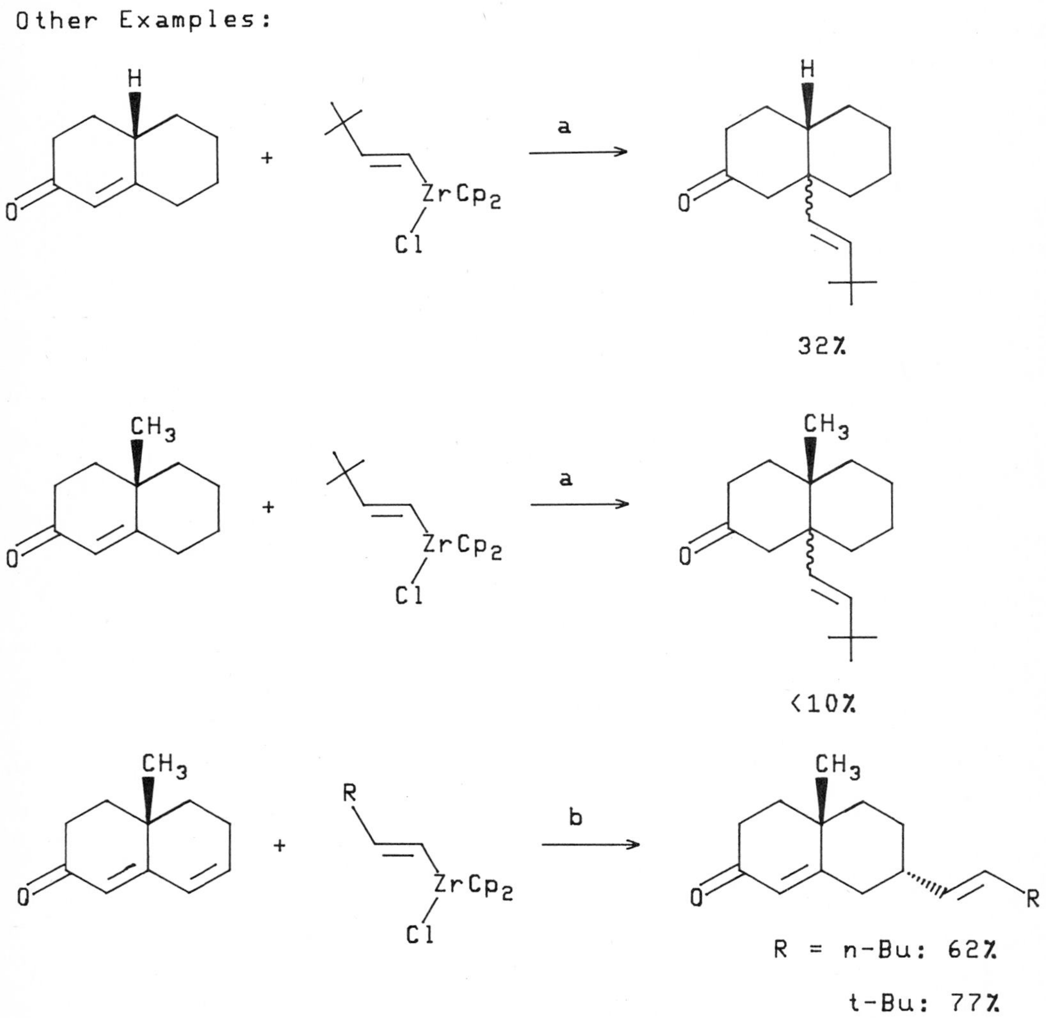

Other Examples:

Figure 14.36

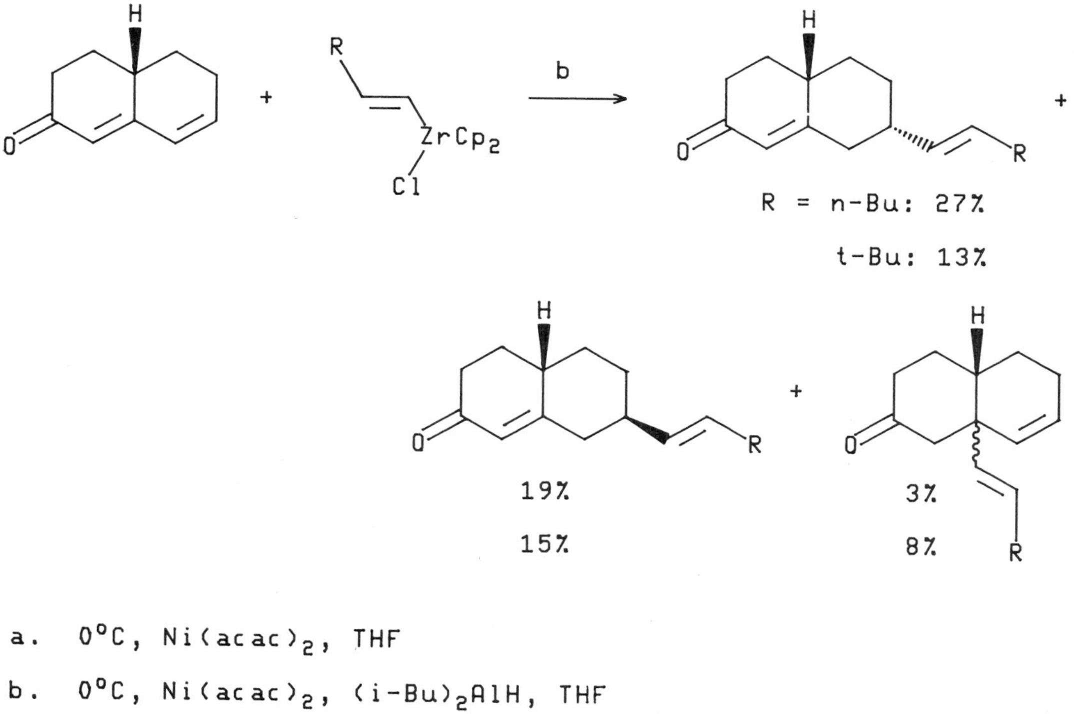

R = n-Bu: 27%

t-Bu: 13%

19%

15%

3%

8%

a. 0°C, Ni(acac)$_2$, THF

b. 0°C, Ni(acac)$_2$, (i-Bu)$_2$AlH, THF

Figure 14.36 (*Continued*)

14.4.8 To σ-Arylnickel(II) Halides

Vinylzirconium(IV) complexes will also efficiently transmetallate with σ-arylnickel(II) halides produced by oxidative addition of aryl halides (halogen = I, Br) to nickel(0). The coupling with monosubstituted vinylzirconium(IV) complexes is rapid at room temperature. Chloride, ether, nitrile, and ester functionalities do not interfere[41,43] (see

TABLE 14.7 Coupling of σ-Arylnickel(II) Halides with Vinylzirconium(IV) Complexes[43]

R	Ar	X	Yield (VPC, %)
n-C_5H_{11}	Ph	I	96
n-C_5H_{11}	1-naphthyl	Br	70
$EtOCH_2CH_2$	Ph	I	99
$THPOCH_2CH_2$	Ph	I	84
n-C_4H_9	p-ClC_6H_4	I	95
n-C_4H_9	p-$CH_3OC_6H_4$	I	80
n-C_4H_9	p-NCC_6H_4	Br	92
n-C_4H_9	p-$CH_3OOCC_6H_4$	I	92

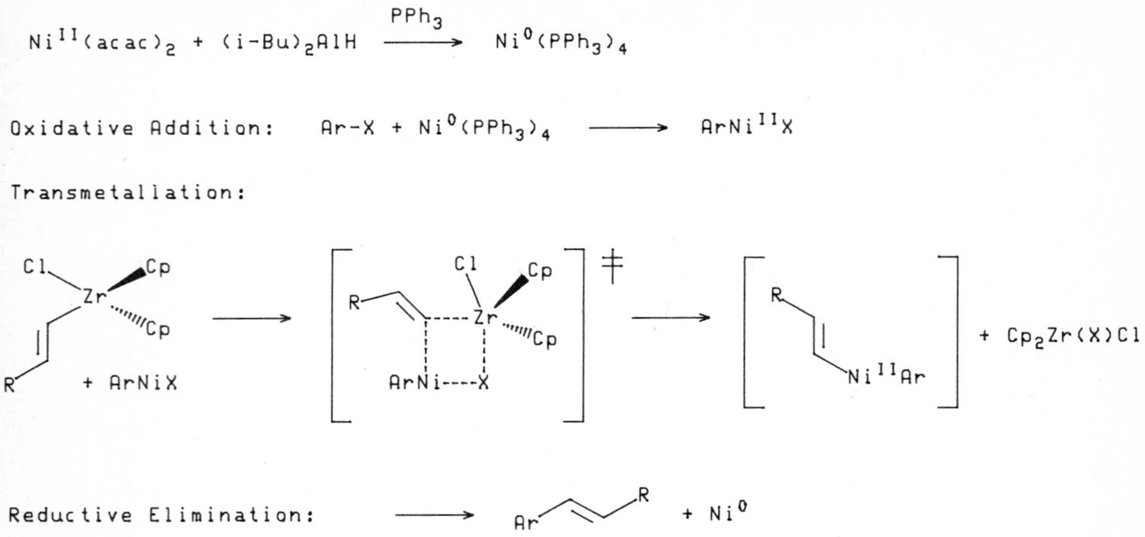

$$Ni^{II}(acac)_2 + (i\text{-}Bu)_2AlH \xrightarrow{PPh_3} Ni^0(PPh_3)_4$$

Oxidative Addition: $Ar\text{-}X + Ni^0(PPh_3)_4 \longrightarrow ArNi^{II}X$

Transmetallation:

Cl, Cp, Zr, Cp, R, + ArNiX → [R, Cl, Cp, Zr, Cp, ArNi----X]‡ → [R, $Ni^{II}Ar$] + $Cp_2Zr(X)Cl$

Reductive Elimination: → Ar, R + Ni^0

Figure 14.37

Figure 14.37 and Table 14.7). Selective replacement of an iodide in the presence of a bromide is the key step in total synthesis of (±)-aurantioclavine[44] (see Chapter 2).

14.4.9 To Zinc, Then to Nickel

Biaryl formation is competitive in coupling of σ-arylnickel(II) halides with di- and trisubstituted vinylzirconium(IV) complexes due to slow transmetallation.[43] A zinc "shuttle" significantly increases rate and efficiency (see Figure 14.38).

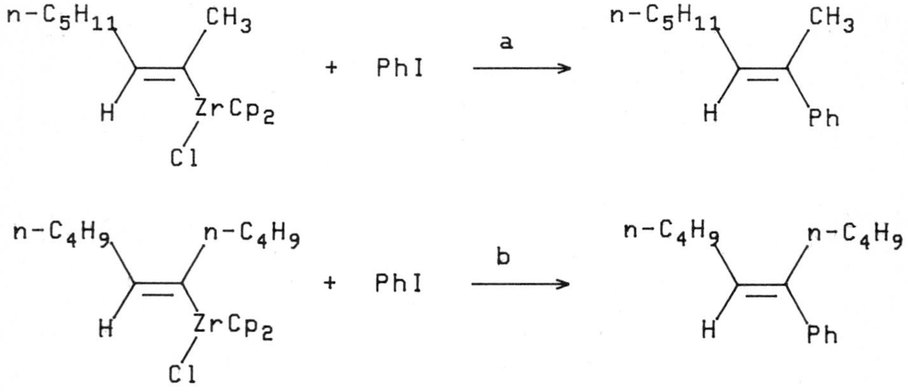

n-C_5H_{11}, CH_3, H, $ZrCp_2$, Cl + PhI —a→ n-C_5H_{11}, CH_3, H, Ph

n-C_4H_9, n-C_4H_9, H, $ZrCp_2$, Cl + PhI —b→ n-C_4H_9, n-C_4H_9, H, Ph

a. rt, $Ni(PPh_3)_4$, THF; 35% (vpc)

b. rt, $Ni(PPh_3)_4$, $ZnCl_2$, THF; 80% (vpc)

Figure 14.38

14.5 TOTAL SYNTHESIS OF 5-(*E*)-PGF$_{2\alpha}$

First, a functionalized cyclopentenone is prepared from cyclopentadiene in three steps.[45] 1,4-Addition of two protected hydroxyls (one as cumyloxy, the other as acetate) is followed by acetate deprotection by basic hydrolysis. Jones oxidation affords 4-cumyloxy-2-cyclopenten-1-one in 48% overall yield.[46] 1-Octyn-3-ol[47] is protected as a benzyl methyl ether[48] and hydrozirconated regio- and stereospecifically to give an (*E*)-vinylzirconium(IV) reagent (see Figures 14.39 and 14.40).

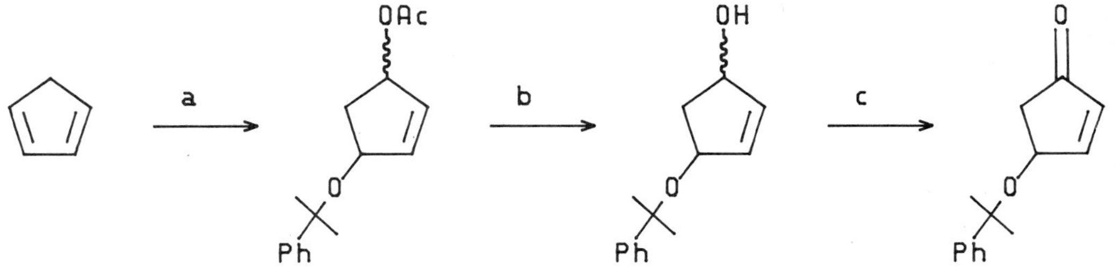

a. 0°C, $FeSO_4$, $Cu(OAc)_2$, Ph–C(CH₃)₂–OO–C(CH₃)₂–Ph, AcOH, H_2O

b. rt, 10% KOH, CH_3OH, H_2O

c. CrO_3, H_2SO_4, acetone; 48% for three steps

Figure 14.39

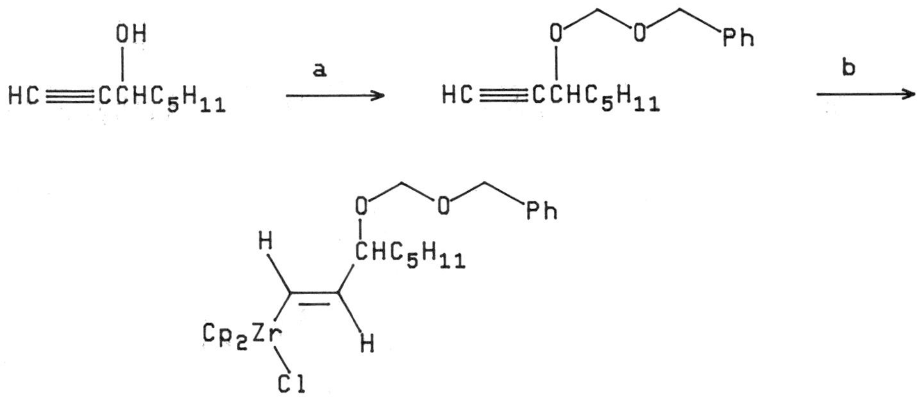

a. rt, $PhCH_2OCH_2Cl$, $(i\text{-}Pr)_2NEt$; no yield reported

b. rt, $Cp_2Zr(H)Cl$, benzene; 80–95%

Figure 14.40

Route A

Route B

Route C

(2-chloromethyl)tetrahydrofuran ring with Cl

a ↓

$HC \equiv C(CH_2)_3OH$

$Br(CH_2)_3Cl$

b ↓

f ↓

$HC \equiv C(CH_2)_3OTs$

$HC \equiv C(CH_2)_3Cl$

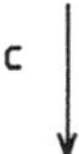

c ↓

g ↙

$HC \equiv C(CH_2)_4OH$

$HC \equiv C(CH_2)_3CN$

e ↘

d ↓

$HC \equiv C(CH_2)_3COOH$

h ↓

$HC \equiv C(CH_2)_3C(=O)Cl$

i ↓

$HC \equiv C(CH_2)_3C(=O)O{-}C(CH_3)_3$ —j→ $Cp_2Zr(Cl)CH{=}CH(CH_2)_3CO{-}C(CH_3)_3$

a. 1) $NaNH_2$, NH_3, 2) NH_4Cl; 85%

b. rt, p-TsCl, benzene; 83%

Figure 14.41

c. rfx, KCN, EtOH, H_2O; 75%

d. rfx, KOH, H_2O; 90%

e. CrO_3, H_2SO_4, acetone; 82%

f. NaC≡CH, NH_3; 55%

g. rfx, NaCN, H_2O, EtOH; 70%

h. rt, ClC(O)C(O)Cl, DMF, benzene; 82%

i. rfx, t-BuOH, N,N-dimethylaniline, Et_2O

j. rt, $Cp_2Zr(H)Cl$, benzene; 80-95%

Figure 14.41 (*Continued*)

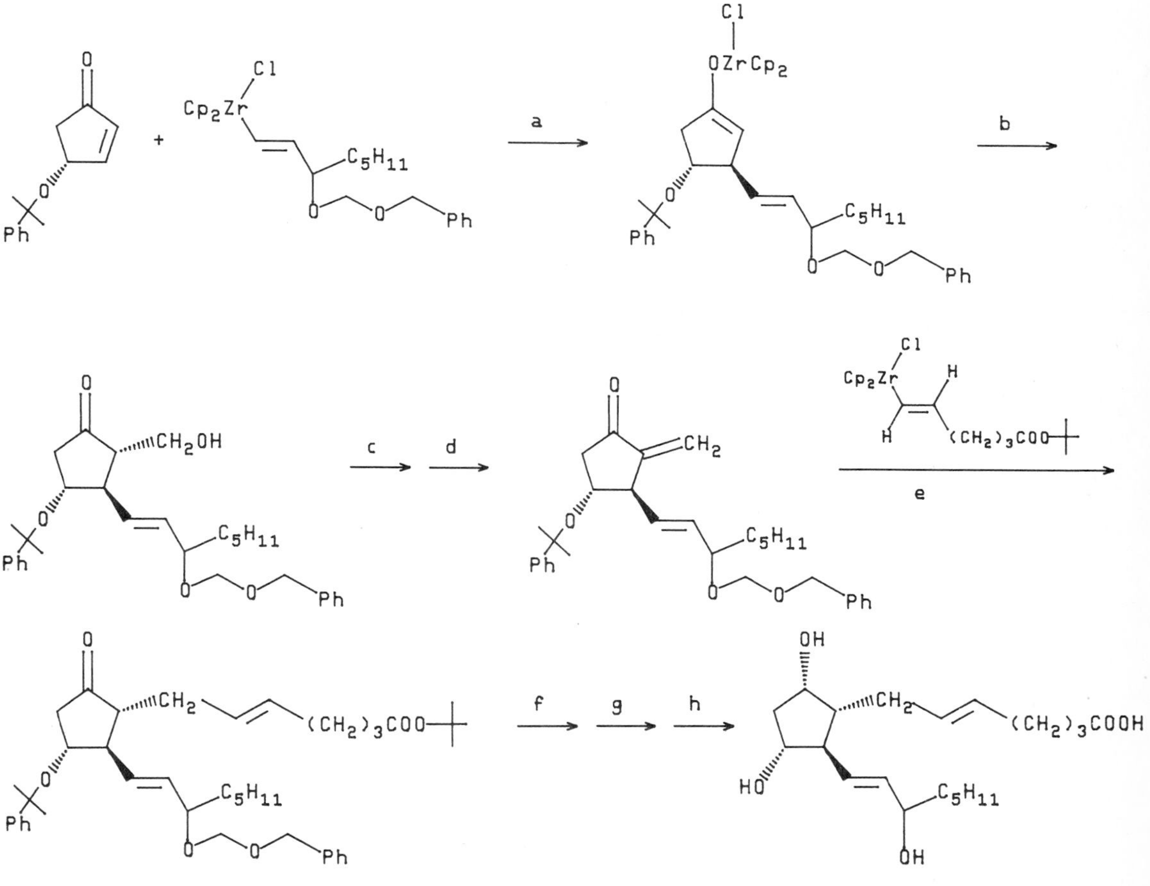

a b c d e f g h

(±)-5-(E)-$PGF_{2\alpha}$

Figure 14.42

a. 0°C, $Ni(acac)_2$, $(i\text{-}Bu)_2AlH$, THF

b. CH_2O, Et_2O; 69% for two steps, 1.5:1 mixture of C_{15} diastereoisomers

c. 0°C, CH_3OSO_2Cl, pyridine

d. $(i\text{-}Pr)_2NEt$, Et_2O; 80%

e. rt, $Ni(acac)_2$, $(i\text{-}Bu)_2AlH$, THF; 66%

f. H_3O^+

g. −78°C, $(s\text{-}Bu)_3BHLi$, THF

h. Na, NH_3, EtOH

Figure 14.42 (*Continued*)

Regio- and stereospecific hydrozirconation of *t*-butyl-5-hexynoate affords another (*E*)-vinylzirconium(IV) reagent. Three unique approaches to precursor 5-hexynoic acid can be considered. Jones oxidation (route A) of 5-hexyn-1-ol[49] affords the acid directly.[50] Alternatively, tetrahydrofurfuryl chloride[51] can be ring opened to 4-pentyn-1-ol with sodium amide.[52] Conversion of the alcohol to a tosylate, displacement with cyanide, then basic hydrolysis afford the acid in overall 48% yield[53] (four steps, route B). Finally, a moderately selective displacement of bromide in 1-bromo-3-chloropropane[54] by sodium acetylide[55] affords 5-chloro-1-pentyne. Nucleophilic displacement by cyanide followed by basic hydrolysis affords the acid in overall 29% yield[56] (three steps, route C). The *t*-butyl ester is prepared via the acid chloride[49,57] (see Figure 14.41).

Nickel-catalyzed conjugate addition affords a zirconium enolate that is trapped with dry formaldehyde in diethyl ether. Dehydration is achieved by conversion to the mesylate, then *E*-2 elimination with base. A second nickel-catalyzed conjugate addition is followed by standard aqueous workup to afford a fully functionalized cyclopentanone. The ketone can be reduced stereoselectivity with lithium tri-*sec*-butylborohydride. Deprotection of the *t*-butyl ester followed by the removal of the benzyloxymethyl and cumyl ethers with sodium in ammonia–ethanol would afford (±)-5-(*E*)-$PGF_{2\alpha}$.[46,48]

The fully functionalized cyclopentenone precursor was synthesized from cyclopentadiene in six steps in overall 17.5% yield (from 5-hexyn-1-ol in seven steps in overall 23.3% yield, assuming quantitative yield in the formation of the *t*-butyl ester).

After deprotection of the *t*-butyl ester, stereospecific ketone reduction and ether deprotection, both demonstrated on the 5-*Z* isomer, would complete the synthesis (60% yield of the C-15 epimeric methyl esters after treatment with diazomethane) (see Figure 14.42).

REFERENCES

1. Kurzok, R.; Lieb, C. C. *Proc. Soc. Exp. Biol. Med.* **1930**, *28*, 268.
2. Goldblatt, M. W. *J. Physiol. (London)* **1935**, *84*, 208.

3a. von Euler, U. S. *Arch. Exptl. Path. Pharmakol.* **1934**, *175*, 78.

3b. von Euler, U. S. *Arch. Exptl. Path. Pharmakol.* **1936**, *181*, 181.

4a. Burr. G. O.; Burr, M. M. *J. Biol. Chem.* **1930**, *86*, 587.

4b. Burr, G. O.; Burr, M. M. *J. Biol Chem.* **1929**, *82*, 345.

5. See the following references for recent reviews:

5a. *Prostaglandin Research*; Crabbe, P., Ed.; Academic: New York, 1977.

5b. Pace-Asciak, C.; Granström, E. Eds.; *Prostaglandins and Related Substances*; Elsevier: New York, 1983.

5c. Moore, P. K. *Prostanoids: Pharmacological, Physiological and Clinical Relevance*; Cambridge University Press: New York, 1985.

5d. *Prostaglandins in Clinical Research;* Sinzinger, W.; Schrör, K., Eds.; Alan R. Liss: New York, 1986.

6. Nelson, N. A. *J. Med. Chem.* **1974**, *17*, 911.
7. "The activity to synthesize $PGF_{2\alpha}$ from PGH_2 is found in the microsomes of cow and guinea pig uterus." Wlodawer, P.; Kindahl, H.; Hamberg, M. *Biochim. Biophys. Acta* **1976**, *431*, 603.
8. Lincoln, F. H., Jr.; Pike, J. E. Ger. Offen. 2, 144, 048, 16 Mar 1972; *Chem. Abstr.* **1972**, *77*, P47938s.
9. Mitra, A. *The Synthesis of Prostaglandins*, Wiley: New York, 1977.
10. See following references for previous synthetic work:

10a. Nicolaou, K. C.: Sipio, W. J.; Magolda, R. L.; Claremon, D. A. *J. Chem. Soc., Chem. Commun.* **1979**, 83.

10b. Nicolaou, K. C.; Barnette, W. E.; Magolda, R. L. *J. Am. Chem. Soc.*, **1978**, *100*, 2567.

10c. Corey, E. J.; Szekely, I.; Shiner, C. S. *Tetrahedron Lett.* **1977**, 3529.

10d. Schneider, W. P.; Bundy, G. L.; Lincoln, F. H.; Daniels, E. G.; Pike, J. E. *J. Am. Chem. Soc.* **1977**, *99*, 1222.

10e. Bundy, G. L.; Daniels, E. G.; Lincoln, F. H.; Pike, J. E. *J. Am. Chem. Soc.* **1972**, *94*, 2124.

11. Lauher, J. W.; Hoffmann, R. *J. Am. Chem. Soc.* **1976**, *98*, 1729.

12a. Kautzner, B.; Wailes, P. C.; Weigold, H. *Chem. Commun.* **1969**, 1105.

12b. Wailes, P. C.; Weigold, H. *J. Organometal. Chem.* **1970**, *24*, 405.

12c. Wailes, P. C.; Weigold, H.; Bell, A. P. *J. Organometal. Chem.* **1971**, *27*, 373.

12d. Wailes, P. C.; Weigold, H.; Bell, A. P. *J. Organometal. Chem.* **1972**, *43*, C32.

13a. Schwartz, J.; Labinger, J. A. *Angew. Chem. Int. Ed. Engl.* **1976**, *15*, 333.

13b. Carr, D. B.; Yoshifuji, M.; Shoer, L. I.; Gell, K. I.; Schwartz, J. *Ann. N.Y. Acad. Sci.* **1977**, *295*, 127.

13c. Schwartz, J. *Pure Appl. Chem.* **1980**, *52*, 733.

14a. Negishi, E-i. *Pure Appl. Chem.* **1981**, *53*, 2333.

14b. Negishi, E-i. *Acc. Chem. Res.* **1982**, *15*, 340.

14c. Negishi, E-i.; Takahashi, T. *Aldrichimica Acta* **1985**, *18*, 31.

15. Hart, D. W.; Blackburn, T. F.; Schwartz, J. *J. Am. Chem. Soc.* **1975**, *97*, 679.
16. Hart, D. W.; Schwartz, J. *J. Am. Chem. Soc.* **1974**, *96*, 8115.
17. Bertelo, C. A.; Schwartz, J. *J. Am. Chem. Soc.* **1976**, *98*, 262.

18. Negishi, E-i.; Miller, J. A.; Yoshida, T. *Tetrahedron Lett.* **1984**, *25*, 3407.
19. Negishi, E-i.; Yoshida, T. *Tetrahedron Lett.* **1980**, *21*, 1501.
20. Van Horn, D. E.; Negishi, E-i. *J. Am. Chem. Soc.* **1978**, *100*, 2252.
21. Bertelo, C. A.; Schwartz, J. *J. Am. Chem. Soc.* **1975**, *97*, 228.
22. Labinger, J. A.; Hart, D. W.; Seibert, W. E., III; Schwartz, J. *J. Am. Chem. Soc.* **1975**, *97*, 3851.
23. Bock, P. L.; Boschetto, D. J.; Rasmussen, J. R.; Demers, J. P.; Whitesides, G. M. *J. Am. Chem. Soc.* **1974**, *96*, 2814.
24. Blackburn, T. F.; Labinger, J. A.; Schwartz, J. *Tetrahedron Lett.* **1975**, 3041.
25. Carr, D. B.; Schwartz, J. *J. Am. Chem. Soc.* **1977**, *99*, 638.
26. Carr, D. B.; Schwartz, J. *J. Am. Chem. Soc.* **1979**, *101*, 3521.
27. Carr, D. B.; Schwartz, J. *J. Organometal. Chem.* **1977**, *139*, C21.
28a. Visser, H. D.; Oliver, J. P. *J. Organometal. Chem.* **1972**, *40*, 7.
28b. Mason, R.; Mingos, D. M. P. *J. Organometal. Chem.* **1973**, *50*, 53.
29. Miller, J. A.; Negishi, E-i. *Israel J. Chem.* **1984**, *24*, 76.
30. Negishi, E-i.; Boardman, L. D. *Tetrahedron Lett.* **1982**, *23*, 3327.
31. Yoshifuji, M.; Loots, M. J.; Schwartz, J. *Tetrahedron Lett.* **1977**, 1303.
32. Temple, J. S.; Schwartz, J. *J. Am. Chem. Soc.* **1980**, *102*, 7381.
33. Temple, J. S.; Riediker, M.; Schwartz, J. *J. Am. Chem. Soc.* **1982**, *104*, 1310.
34. Reidiker, M.; Schwartz, J. *Tetrahedron Lett.* **1981**, *22*, 4655.
35. Hayasi, Y.; Riediker, M.; Temple, J. S.; Schwartz, J. *Tetrahedron Lett.* **1981**, *22*, 2629.
36. Matsushita, H.; Negishi, E-i. *J. Am. Chem. Soc.* **1981**, *103*, 2882.
37. Okukado, N.; Van Horn, D. E.; Klima, W. L.; Negishi, E-i. *Tetrahedron Lett.* **1978**, 1027.
38. Negishi, E-i.; Okudado, N.; King, A. O.; Van Horn, D. E.; Spiegel, B. I. *J. Am. Chem. Soc.* **1978**, *100*, 2254.
39. Loots, M. J.; Schwartz, J. *J. Am. Chem. Soc.* **1977**, *99*, 8045.
40. Dayrit, F. M.; Gladkowski, D. E.; Schwartz, J. *J. Am. Chem. Soc.* **1980**, *102*, 3976.
41. Dayrit, F. M.; Schwartz, J. *J. Am. Chem. Soc.* **1981**, *103*, 4466.
42. Schwartz, J.; Loots, M. J.; Kosugi, H. *J. Am. Chem. Soc.* **1980**, *102*, 1333.
43. Negishi, E-i.; Van Horn, D. E. *J. Am. Chem. Soc.* **1977**, *99*, 3168.
44. Hegedus, L. S.; Toro, J. L.; Miles, W. H.; Harrington, P. J. *J. Org. Chem.* **1987**, *52*, 3319.
45. Dicyclopentadiene is commercially available and inexpensive (Aldrich 88–89 catalog, 3 kg for $13.25).
46. Stork, G.; Isobe, M. *J. Am. Chem. Soc.* **1975**, *97*, 6260.
47. 1-Octyn-3-ol is commercially available (Fluka 88–89 catalog, 100 mL, $d = 0.866$ g/mL, for $45.00).
48. Stork, G.; Isobe, M. *J. Am. Chem. Soc.* **1975**, *97*, 4745.
49. 5-Hexyn-1-ol is commercially available (Farchan Laboratories, 50 g for $96.00).
50. Earl, R. A.; Vollhardt, K. P. C. *J. Org. Chem.* **1984**, *49*, 4786.
51. Tetrahydrofurfuryl chloride is commercially available (Fluka 88–89 catalog, 1 L, $d = 1.11$ g/mL, for $19.80).
52. Jones, E. R. H.; Eglinton, G.; Whiting, M. C. Org. Syn., Coll. **1963**, *4*, 755.
53a. Eglinton, G.; Whiting, M. C. *J. Chem. Soc.* **1950**, 3650.
53b. Eglinton, G.; Whiting, M. C. *J. Chem. Soc.* **1953**, 3052.

54. 1-Bromo-3-chloropropane is commercially available and inexpensive (Lancaster Synthesis 89–90 catalog, 500 g for $16.10).

55. Sodium acetylide is commercially available and inexpensive (Fluka 88–89 catalog, 500 mL of an 11–15% suspension in xylene, $d = 1.199$ g/mL, for $63.00).

56. Ferrier, R. J.; Tedder, J. M. *J. Chem. Soc.* **1957**, 1435.

57. Sandler, S. R.; Karo, W. *Organic Functional Group Preparations*; Academic: New York, 1968; Vol. 1, p 250.

CHAPTER 15

$(\pm)$-$\Delta^{9(12)}$-Capnellane

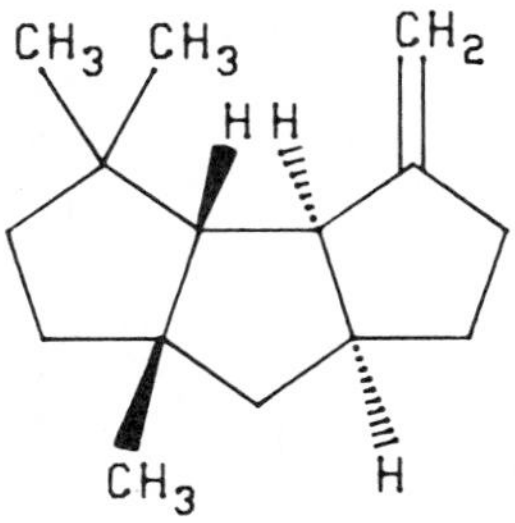

CH3 CH3 CH2
H H
CH3 H

(±)-$\Delta^{9(12)}$-Capnellane

1H-cyclopenta[a]pentalene,

decahydro-3,3,7a-trimethyl-4-methylene

(3aα, 3bβ, 6aβ, 7aα)(±) [81370-78-7]

Figure 15.1

The sesquiterpene (±)-$\Delta^{9(12)}$-capnellane, isolated from the soft coral *capnella imbricata* (Quoy and Gaimard, 1833), is the presumed biosynthetic precursor of the capnellenols. The capnellane group may serve as a component of a coral reef defense mechanism which inhibits the growth of microorganisms and settlement of larvae.[1] This property and structural similarity to the pharmacologically significant hirsutanes (antibiotic, anticancer)[2] have prompted total synthetic efforts by several groups.[3] One successful approach using the unique reactivity of titanium carbene complexes was described in Chapter 13.

15.1 TRANSMETALLATION: FROM TIN(IV) TO PALLADIUM(II)[4]

Carbon–carbon bond formation between organic electrophiles and organotin(IV) complexes is catalyzed by palladium(0). In the initial step, *oxidative addition* of the electrophile to palladium(0) produces a palladium(II) species. This undergoes facile *transmetallation* of an organic ligand from tin(IV) to palladium(II). The resulting diorganopalladium(II) species then rearranges and *reductively eliminates* to generate

Oxidative Addition

$$R{-}X + Pd(0) \longrightarrow R{-}Pd^{II}{-}X$$

Transmetallation

$$R{-}Pd^{II}{-}X + R^1{}_4Sn^{IV} \longrightarrow R{-}Pd^{II}{-}R^1 + R^1{}_3Sn^{IV}X$$

Rearrangement in the Coordination Sphere of the Metal

$$R{-}Pd^{II}{-}R^1 \longrightarrow \begin{array}{l} R{-}Pd^{II} \\ \quad\;\; | \\ \quad\;\; R^1 \end{array}$$

Reductive Elimination

$$\begin{array}{l} R{-}Pd^{II} \\ \quad\;\; | \\ \quad\;\; R^1 \end{array} \longrightarrow R{-}R^1 + Pd(0)$$

OVERALL

$$R{-}X + R^1{}_4Sn \xrightarrow{Pd(0)} R{-}R^1 + R^1{}_3SnX$$

Figure 15.2

the new carbon–carbon bond and regenerate the palladium(0) catalyst. Coupling is typically done in tetrahydrofuran at reflux under an inert atmosphere, using either *tetrakis*-(triphenylphosphine)palladium(0) or a palladium(II) complex as the palladium(0) source (see Figure 15.2).

If this sequence is carried out under a carbon monoxide atmosphere, the initial organopalladium(II) intermediate can coordinate and insert carbon monoxide. Transmetallation, rearrangement, and reductive elimination produce a ketone. Carbonylative coupling is typically done in tetrahydrofuran at 40 to 50 °C (higher temperatures result in competitive direct coupling) under carbon monoxide using *tetrakis*-(triphenylphosphine)palladium(0) (see Figure 15.3).

Tetraorganotin(IV) transfers an organic ligand about one hundred times more rapidly than the by-product triorganotin(IV) halide. Using a tin reagent with four identical organic ligands, three are discarded. Such ligand waste severely limits the utility of the method. The problem and solution are similar to the problem and solution in the use of cuprates for Michael addition reactions. Various organic ligands transfer from tin(IV) to palladium(II) at significantly different rates.

$$RC{\equiv}C > RCH{=}CH > Ar > RCH{=}CH_2{-}CH_2,\ ArCH_2 \gg C_nH_{2n+1}$$

$$(sp > sp^2 > sp^3)$$

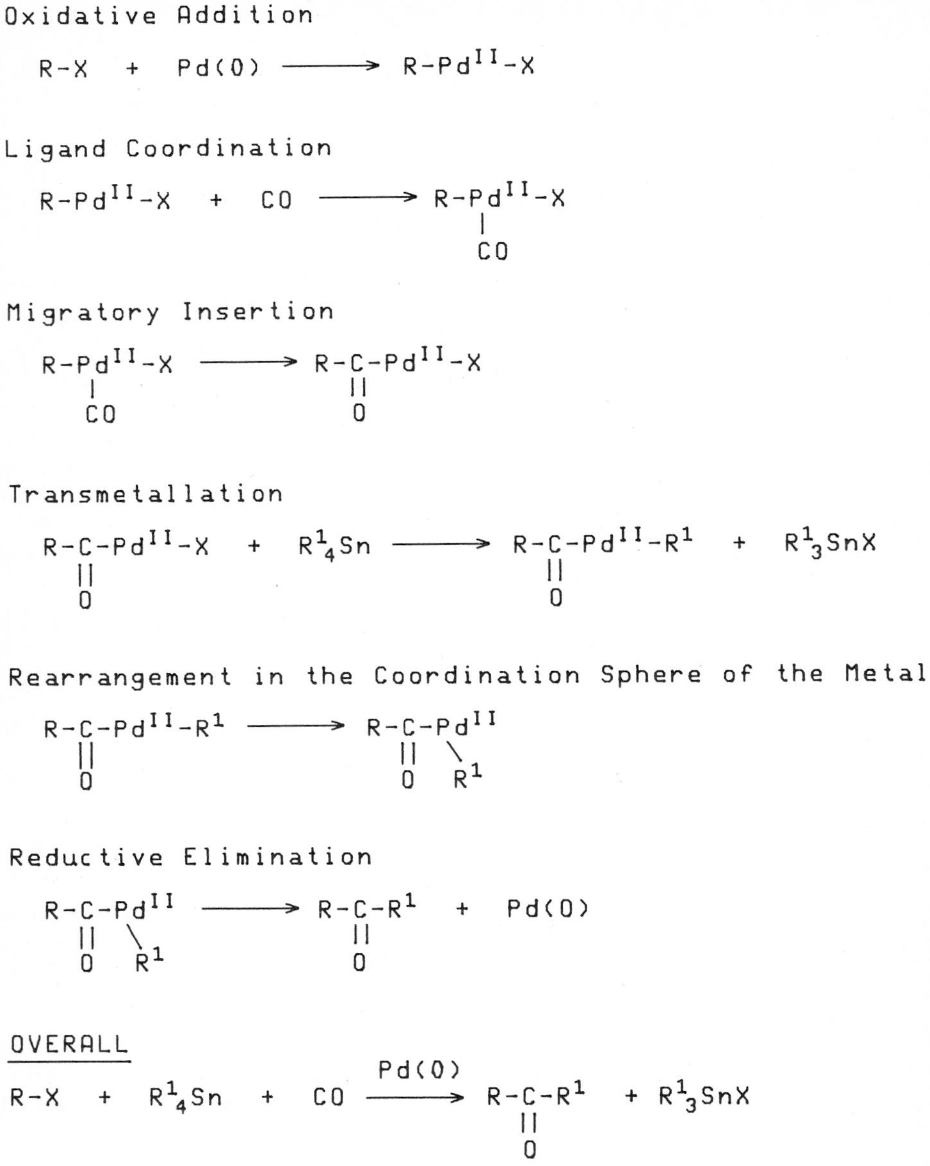

Oxidative Addition

$$R{-}X + Pd(0) \longrightarrow R{-}Pd^{II}{-}X$$

Ligand Coordination

$$R{-}Pd^{II}{-}X + CO \longrightarrow R{-}\underset{\displaystyle CO}{\underset{|}{Pd^{II}}}{-}X$$

Migratory Insertion

$$R{-}\underset{\displaystyle CO}{\underset{|}{Pd^{II}}}{-}X \longrightarrow R{-}\underset{\displaystyle O}{\underset{||}{C}}{-}Pd^{II}{-}X$$

Transmetallation

$$R{-}\underset{\displaystyle O}{\underset{||}{C}}{-}Pd^{II}{-}X + R^1{}_4Sn \longrightarrow R{-}\underset{\displaystyle O}{\underset{||}{C}}{-}Pd^{II}{-}R^1 + R^1{}_3SnX$$

Rearrangement in the Coordination Sphere of the Metal

$$R{-}\underset{\displaystyle O}{\underset{||}{C}}{-}Pd^{II}{-}R^1 \longrightarrow R{-}\underset{\displaystyle O}{\underset{||}{C}}{-}Pd^{II} \;(\backslash R^1)$$

Reductive Elimination

$$R{-}\underset{\displaystyle O}{\underset{||}{C}}{-}Pd^{II} \;(\backslash R^1) \longrightarrow R{-}\underset{\displaystyle O}{\underset{||}{C}}{-}R^1 + Pd(0)$$

OVERALL

$$R{-}X + R^1{}_4Sn + CO \xrightarrow{Pd(0)} R{-}\underset{\displaystyle O}{\underset{||}{C}}{-}R^1 + R^1{}_3SnX$$

Figure 15.3

Since alkyl groups transfer at the slowest rate, they are suitable "dummy" ligands for tin(IV). Thus, a trialkyl(alkynyl)tin(IV) complex will transfer alkynyl, a trialkyl(vinyl)tin(IV) complex will transfer vinyl, a trialkyl(aryl)tin(IV) complex will transfer aryl, and so on.

If the organic electrophile used in oxidative addition has β-hydrogens, the relative rates of β-hydride elimination and transmetallation (or carbon monoxide insertion)

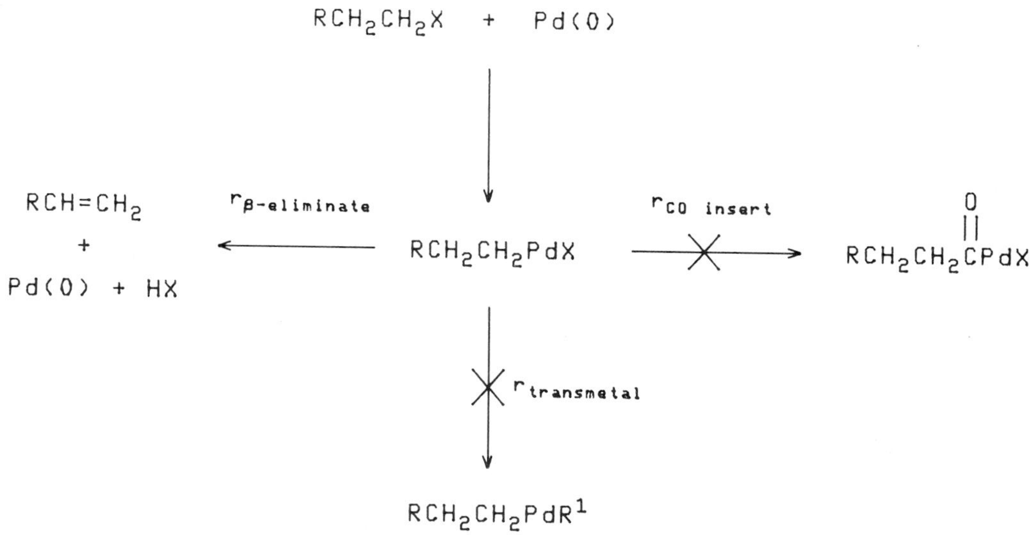

$$RCH_2CH_2X + Pd(0)$$

$$\downarrow$$

$$RCH=CH_2 + Pd(0) + HX \xleftarrow{r_{\beta\text{-eliminate}}} RCH_2CH_2PdX \xrightarrow[\times]{r_{CO\ insert}} RCH_2CH_2C(=O)PdX$$

$$RCH_2CH_2PdX \xrightarrow[\times]{r_{transmetal}} RCH_2CH_2PdR^1$$

Figure 15.4

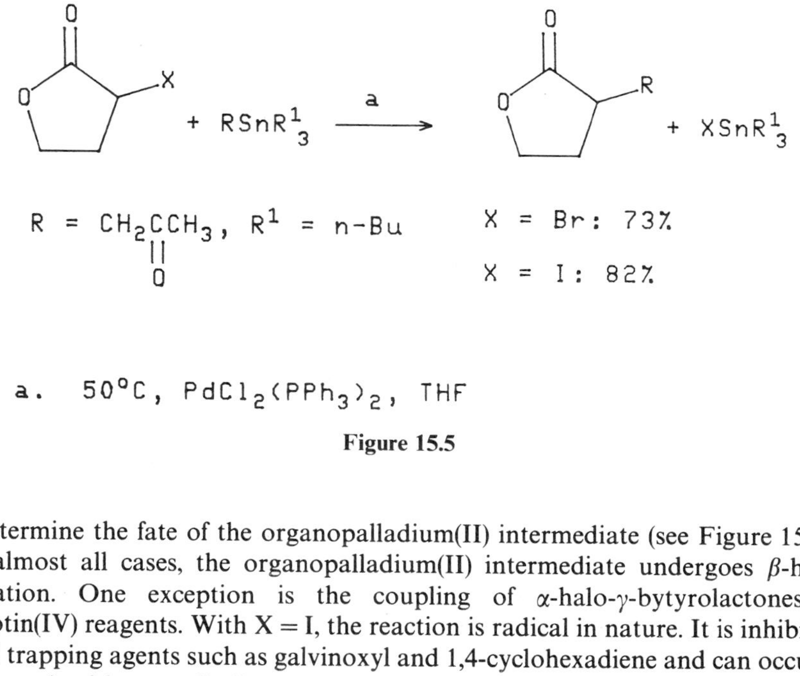

$$\alpha\text{-X-}\gamma\text{-butyrolactone} + RSnR^1_3 \xrightarrow{a} \alpha\text{-R-}\gamma\text{-butyrolactone} + XSnR^1_3$$

$R = CH_2C(=O)CH_3$, $R^1 = n\text{-Bu}$ $X = Br$: 73%

$X = I$: 82%

a. 50°C, $PdCl_2(PPh_3)_2$, THF

Figure 15.5

will determine the fate of the organopalladium(II) intermediate (see Figure 15.4).

In almost all cases, the organopalladium(II) intermediate undergoes β-hydride elimination. One exception is the coupling of α-halo-γ-bytyrolactones with organotin(IV) reagents. With $X = I$, the reaction is radical in nature. It is inhibited by radical trapping agents such as galvinoxyl and 1,4-cyclohexadiene and can occur (at a slower rate) without palladium catalysis. With $X = Br$, the coupling is apparently palladium catalyzed[5] (see Figure 15.5).

To circumvent this limitation, organic groups with β-hydrogens can be coupled if the organic group comes from tin since rearrangement and reductive elimination are rapid (see Figure 15.6).

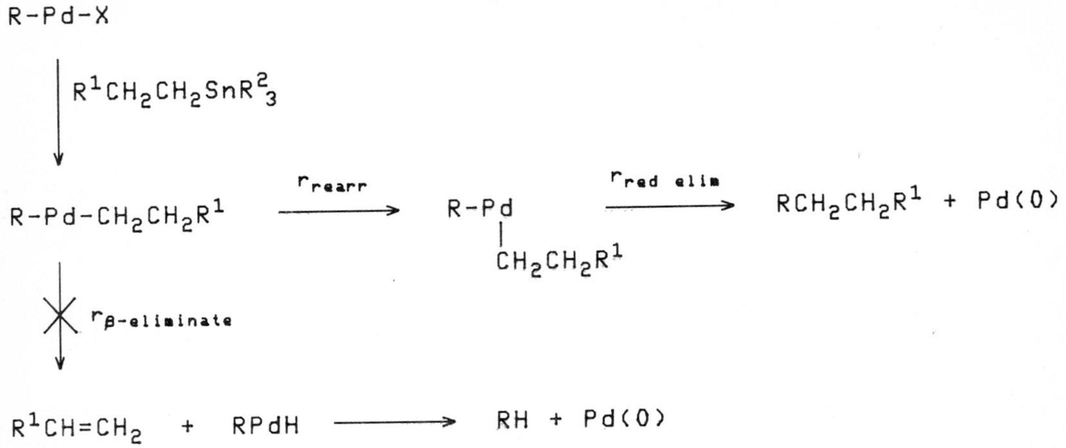

$$R\text{-}Pd\text{-}X \xrightarrow{R^1CH_2CH_2SnR^2_3} R\text{-}Pd\text{-}CH_2CH_2R^1 \xrightarrow{r_{rearr}} R\text{-}Pd(CH_2CH_2R^1) \xrightarrow{r_{red\ elim}} RCH_2CH_2R^1 + Pd(0)$$

$$R\text{-}Pd\text{-}CH_2CH_2R^1 \not\xrightarrow{r_{\beta\text{-}eliminate}} R^1CH{=}CH_2 + RPdH \longrightarrow RH + Pd(0)$$

Figure 15.6

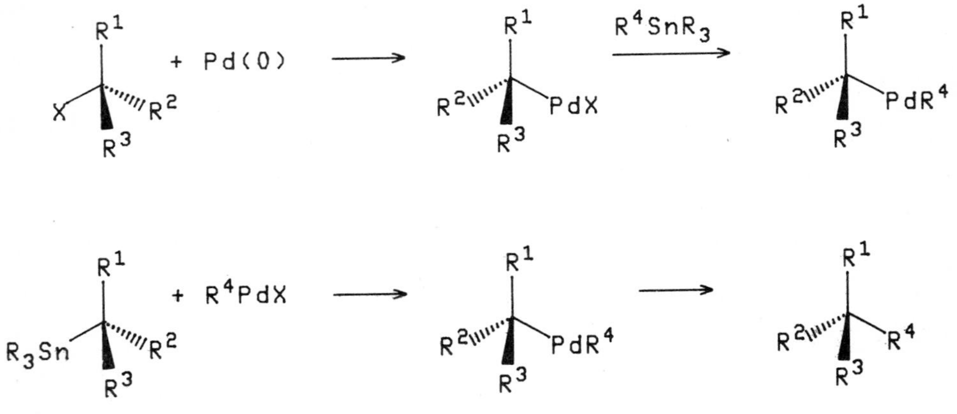

$$R^1R^2R^3C\text{-}X + Pd(0) \longrightarrow R^1R^2R^3C\text{-}PdX \xrightarrow{R^4SnR_3} R^1R^2R^3C\text{-}PdR^4$$

$$R^1R^2R^3C\text{-}SnR_3 + R^4PdX \longrightarrow R^1R^2R^3C\text{-}PdR^4 \longrightarrow R^1R^2R^3C\text{-}R^4$$

Figure 15.7

The palladium-catalyzed coupling of an organic halide with an organotin reagent occurs under mild conditions (temperature is determined by the ease of halide oxidative addition) and tolerates a wide range of functionality. The organic halide can be an acid chloride, allyl halide (Cl, Br), benzyl halide (Cl, Br), aryl halide (Br, I), vinyl halide (I), or α-halocarboxylic acid or acid derivative (Br, I). The organotin reagent can transfer hydride, alkynyl, vinyl, aryl, allyl, or alkyl groups. The coupling is stereospecific. Carbons (sp^3) bound to tin or halogen are inverted in the product since (1) oxidative addition proceeds with inversion, (2) transmetallation proceeds with inversion of the transferred tin(IV) ligand, and (3) rearrangement and reductive elimination proceed with retention (see Figure 15.7).

15.1.1 Coupling with Acid Chlorides

Coupling of a vinyltin reagent with an acid chloride is the key step in a total synthesis of the antibiotic (±)-pyrenophorin.[6] The synthesis is greatly simplified by noting the symmetry and disconnecting into two identical pieces. One coupling partner is prepared by ring opening of γ-valerolactone, protecting the alcohol as the silyl ether, and converting the acid to acid chloride. The other partner is prepared by azo bis-isobutyronitrile-initiated addition of a tributyltin radical to a propargyl ester[7] (see Figures 15.8 and 15.9).

(±)-Pyrenophorin

a. KOH, H_2O

b. $Ph_2Si(t\text{-}Bu)Cl$

c. KOH, $(n\text{-}Bu)_4NOH$, H_2O; 74% for three steps

d. $ClC(O)C(O)Cl$, CH_2Cl_2

Figure 15.8

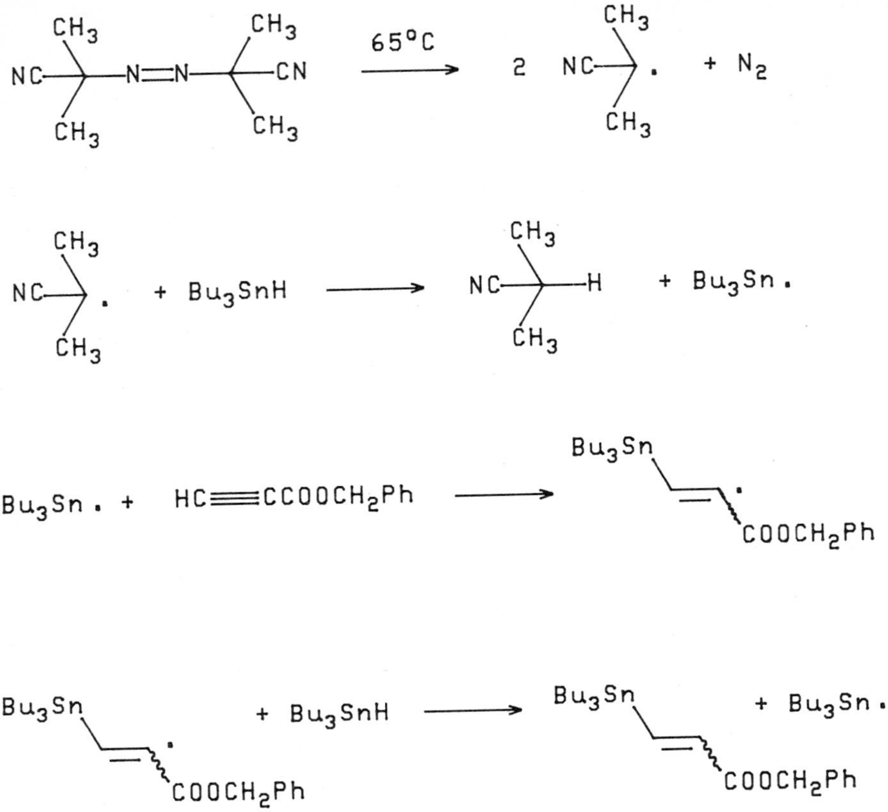

Figure 15.9

The coupling is most efficient at 65 °C using $PhCH_2Pd(PPh_3)_2Cl$ as the palladium(0) source and under a carbon monoxide atmosphere to prevent a potentially competitive decarbonylative coupling. The ketone is protected as the ketal. Deprotection of the ester and silyl ether affords the alcohol acid. Macrocycle formation using triphenylphosphine–diethylazodicarboxylate followed by ketal deprotection affords (±)-pyrenophorin[8,9] (see Figures 15.10 and 15.11).

15.1.2 Coupling with Allyl Halides

Allyl halides oxidatively add to palladium(0) to produce σ-alkylpalladium complexes that rapidly collapse to π-allyl complexes. While coupling with an organotin reagent can occur at either allyl terminus, coupling at the most sterically accessible terminus is apparently preferred[10] (see Figure 15.12).

15.1.3 Carbonylative Coupling with Allyl Halides

Allyltin reagents generally couple with predominant allylic transposition. However, in a carbonylative coupling with an allyl halide, coupling takes place (1) without allylic

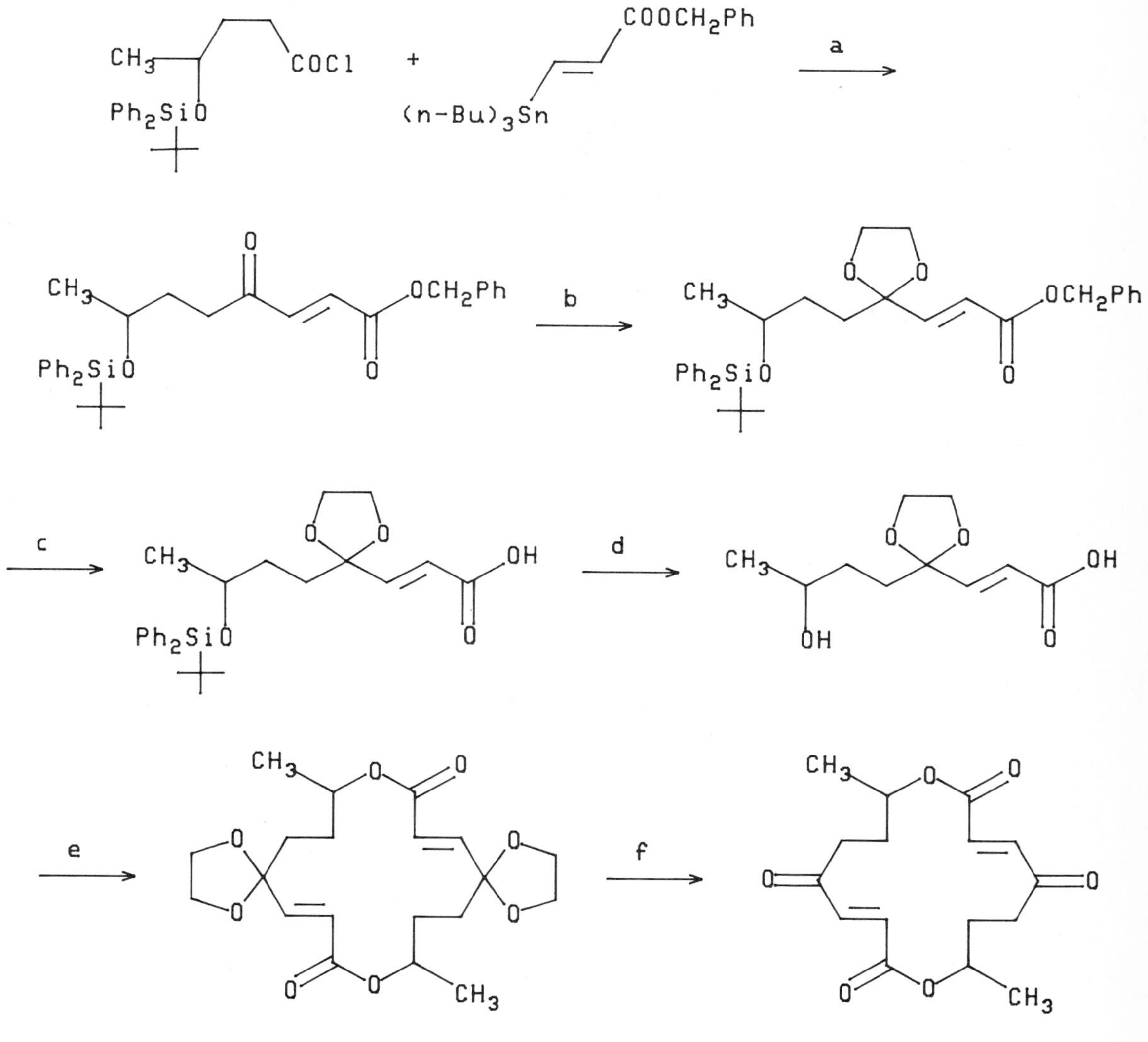

(±)-Pyrenophorin

a. 65°C, CO, $PhCH_2Pd(PPh_3)_2Cl$, $CHCl_3$; 59%

b. heat, $BF_3 \cdot Et_2O$, $HOCH_2CH_2OH$, benzene; 76%

c. heat, KOH, $(n\text{-}Bu)_4NOH$, THF; 92%

d. $(n\text{-}Bu)_4NF$; 76%

e. −35 to 15°C, PPh_3, DEAD, toluene, CCl_4; 24%

f. rt, $p\text{-}TsOH \cdot H_2O$, acetone; 40%

Figure 15.10

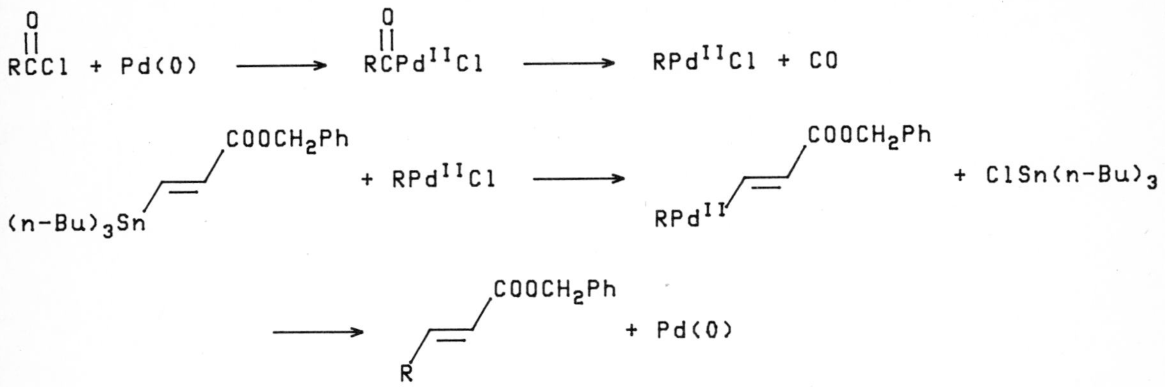

RCCl + Pd(0)
RCPd^IICl
RPd^IICl + CO
COOCH2Ph
(n-Bu)3Sn
+ RPd^IICl
RPd^II
+ ClSn(n-Bu)3
+ Pd(0)
R

Figure 15.11 Mechanism for decarbonylative coupling of acid chlorides with vinyltin reagents

R1
R2
R3
X
+ Pd(0)
Pd^IIX
R4SnR3
R4
and
or

For example:

CH3
CHO
Cl
+ (n-Bu)3Sn
OCH3
a

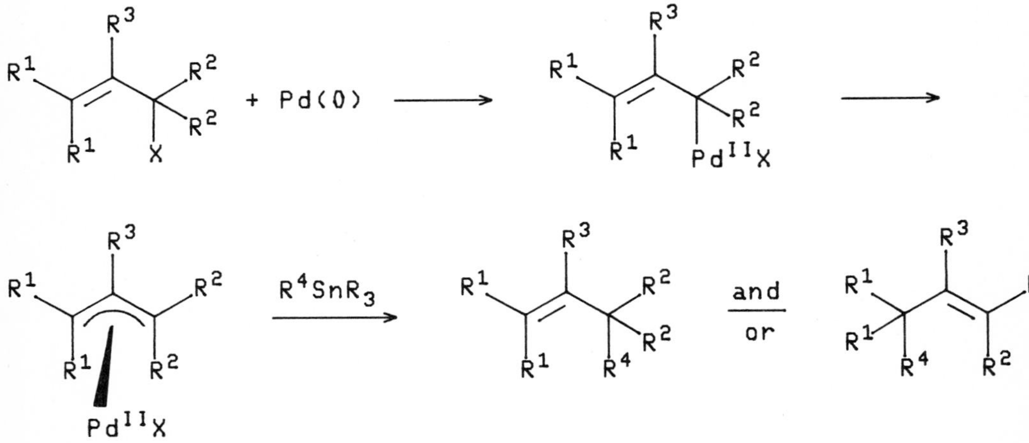

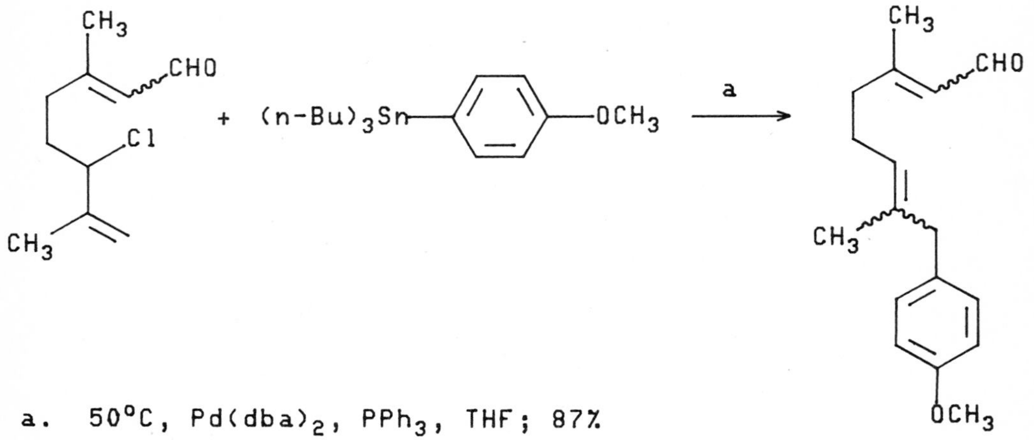

a. 50°C, Pd(dba)$_2$, PPh$_3$, THF; 87%

Figure 15.12

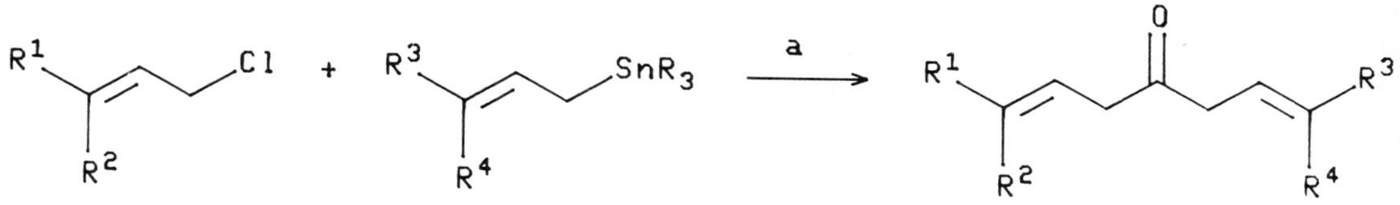

R^1, R^2 (allyl chloride, –Cl) + R^3, R^4 (allyl SnR_3) $\xrightarrow{a}$ R^1, R^2, R^3, R^4 (diallyl ketone, O)

a. 25°C, 3 atm CO, $Pd(dba)_2$, THF; 20-62%

Figure 15.13

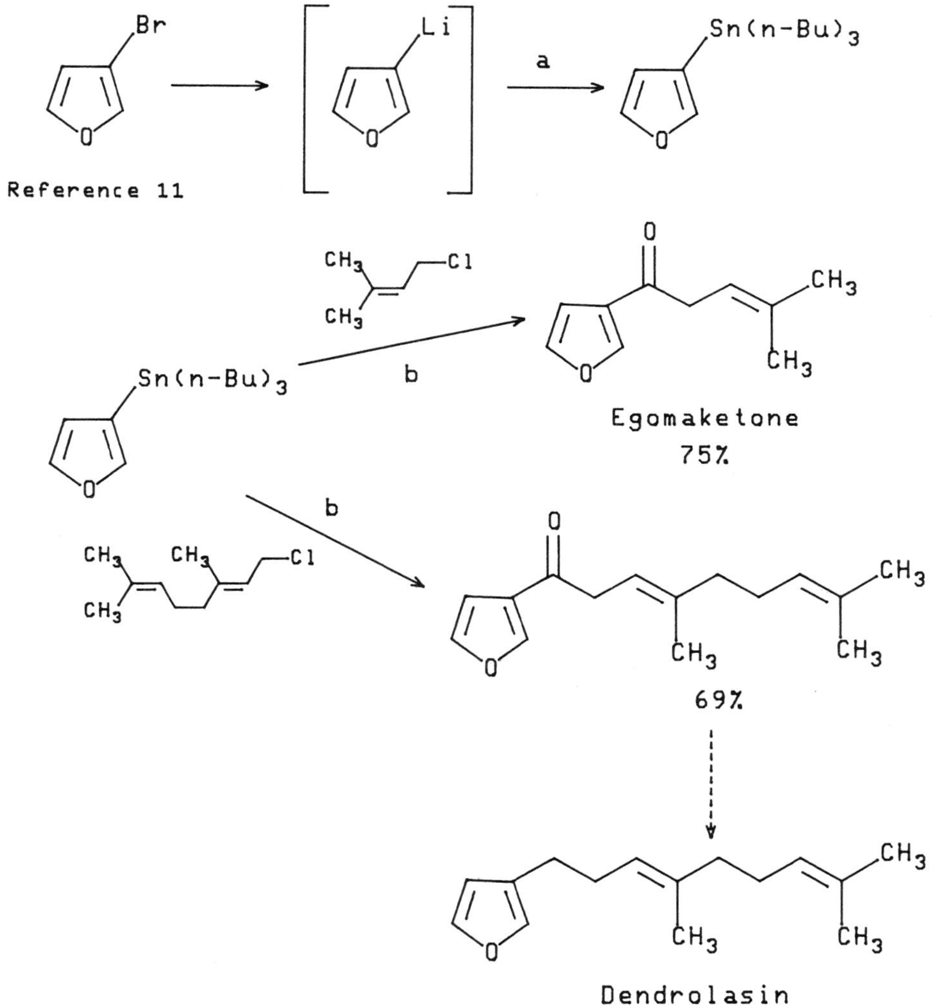

Br (3-bromofuran) ⟶ [Li (3-furyllithium)] $\xrightarrow{a}$ $Sn(n\text{-}Bu)_3$ (3-furyl)

Reference 11

$Sn(n\text{-}Bu)_3$ (3-furyl) + CH_3, CH_3, –Cl $\xrightarrow{b}$ O, CH_3, CH_3

Egomaketone
75%

$Sn(n\text{-}Bu)_3$ (3-furyl) + CH_3, CH_3, CH_3, –Cl $\xrightarrow{b}$ O, CH_3, CH_3, CH_3

69%

⇣

CH_3, CH_3, CH_3

Dendrolasin

a. −78°C, $(n\text{-}Bu)_3SnCl$, Et_2O; 79%

b. 50°C, CO, $PhCH_2Pd(PPh_3)_2Cl$ or $Pd(PPh_3)_4$

Figure 15.14

transposition of the tin fragment and, as expected, (2) at the least hindered terminus of the π-allyl (see Figure 15.13).

A similar coupling between a heteroaryltin reagent and allyl halide is utilized in short syntheses of egomaketone and a dendrolasin precursor. (Dendrolasin is a major constituent of the mandibular gland secretion of the Formicine ant, *Lasius faliginosis*) (see Figure 15.14).

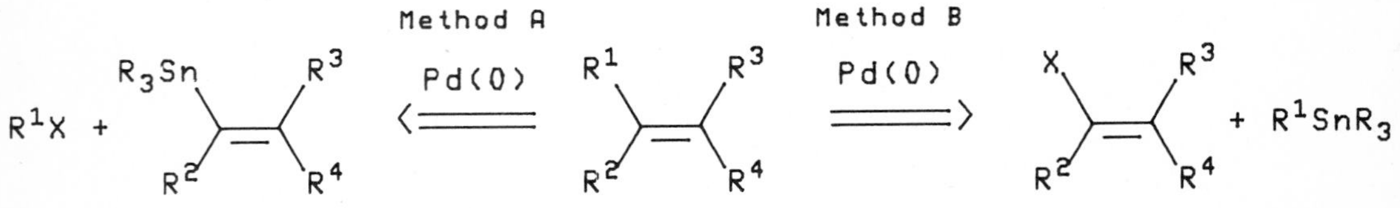

Method A: Pd(0); Method B: Pd(0); + R^1SnR_3

Figure 15.15

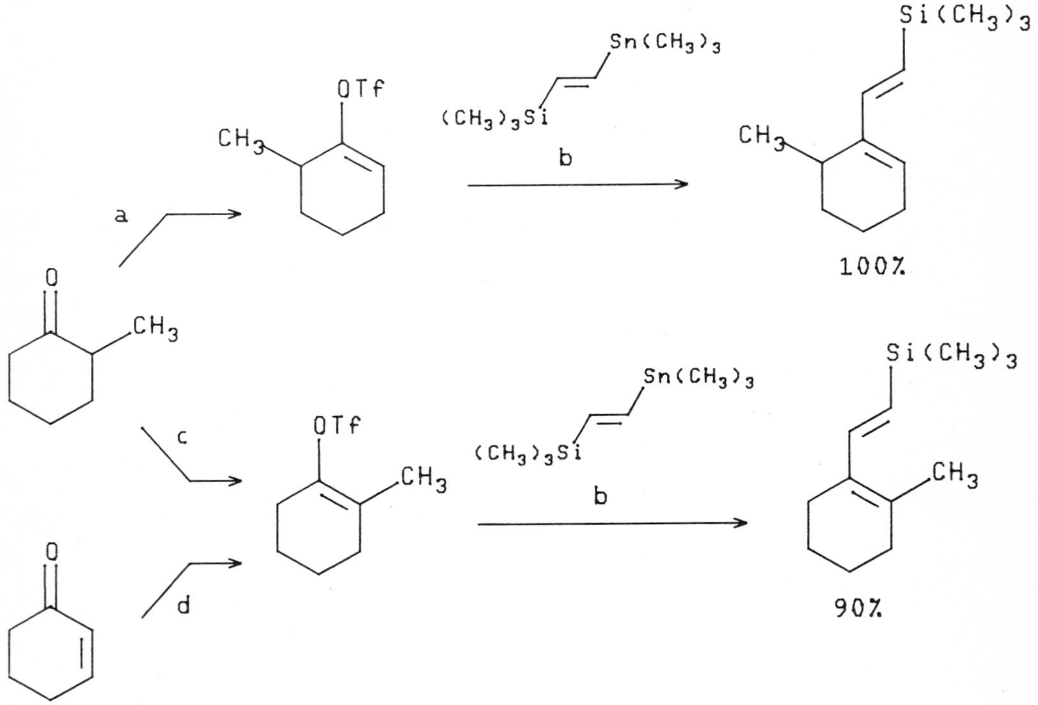

a. 1) LiN(i-Pr)$_2$, 2) Tf$_2$NPh; 91%

b. rfx, LiCl, Pd(PPh$_3$)$_4$, THF

c. 1) (i-Pr)$_2$NMgBr, 2) Tf$_2$NPh; 63%

d. 1) L-Selectride®, 2) Tf$_2$NPh; 90%

Figure 15.16

15.1.4 Coupling with Vinyl Halides and Vinyl Triflates

There are two methods for alkene synthesis using palladium-catalyzed coupling of organic electrophiles with organotin reagents. Method A is somewhat limited by the inaccessibility of regio- and stereochemically pure vinyltin reagents. In addition, while alkene geometry is retained in coupling using a vinyltin *E*-isomer, geometry is scrambled using a vinyltin *Z*-isomer. Method B is severely limited by the inaccessibility of regio- and stereochemically pure vinyl iodides (see Figure 15.15).

Scott and Stille found that vinyl triflates will oxidatively add to palladium(0). Since vinyl triflates are accessible in regio- and stereochemically pure form from the corresponding ketone enolates, vinyl triflates can be substituted for vinyl iodides in method B. Lithium chloride is required for efficient tin(IV) to palladium(II) transmetallation (see Figure 15.16). This methodology is applied to the synthesis of the unusual carbon skeleton of pleraplysillin-1, a new type of sesquiterpene from the sponge *Pleraplysilla spinifera*[12] (see Figure 15.17).

15.1.5 Carbonylative Coupling with Vinyl Triflates

Vinyl triflates could also be coupled with activated alkenes[15] and with organotin(IV) reagents under carbonylative conditions. Some less reactive organotin reagents require

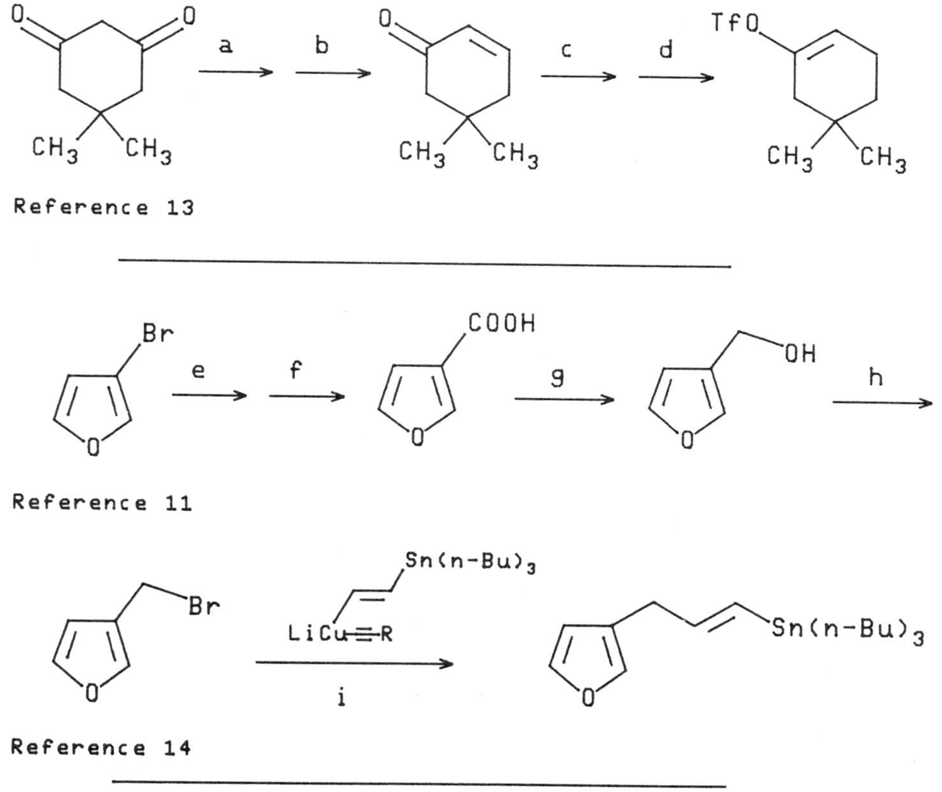

Reference 13

Reference 11

Reference 14

Figure 15.17

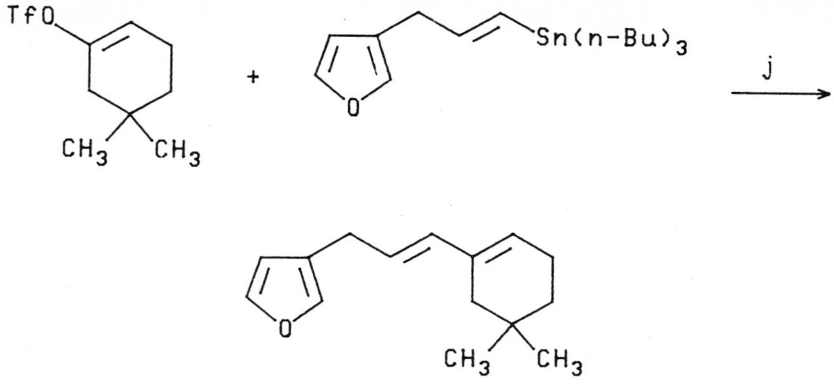

TfO ... + ... $Sn(n\text{-}Bu)_3$ —j→

CH_3 CH_3

CH_3 CH_3

Pleraplysillin-1

a. H^+, $TsNHNH_2$, CH_3OH

b. heat, K_2CO_3, H_2O; 52%

c. L-Selectride®

d. Tf_2NPh; 93%

e. Li

f. CO_2; 97%

g. $LiAlH_4$, Et_2O; 86-91%

h. PBr_3, pyridine, Et_2O

i. 73%

j. rfx, LiCl, $Pd(PPh_3)_4$, THF; 75%

Figure 15.17 (*Continued*)

addition of zinc chloride for coupling at the low temperatures necessary to prevent direct coupling. In these cases catalysis may involve sequential ligand transfer from Sn(IV) to Zn(II) to Pd(II) (see Figures 15.18 and 15.19).

Carbonylative coupling of a vinyl triflate and a silicon-directed Nazarov cyclization are the key steps in Stille's iterative approach to construction of the tricyclic ring system of (±)-$\Delta^{9(12)}$-capnellane.

15.2 THE SILICON-DIRECTED NAZAROV CYCLIZATION

Ferric chloride-catalyzed conversion of a diene-one to a cyclopentenone is known as the Nazarov cyclization. With unsymmetrical diene-ones, two cyclopentenones are

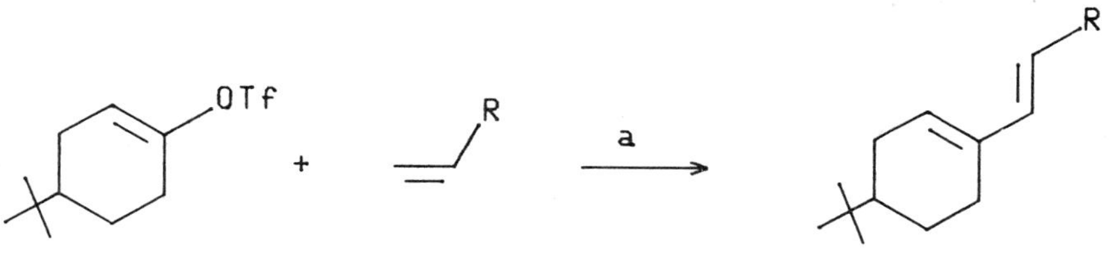

OTf + R → (a) R

R = $COOCH_3$ 91%

$C(O)CH_3$ 89%

CHO 86%

*Ph 3%

CN 99% (mixture of cis, trans)

*with $Pd(PPh_3)_4$ catalyst, 57%

a. 75°C, $PdCl_2(PPh_3)_2$, Et_3N, THF

Figure 15.18

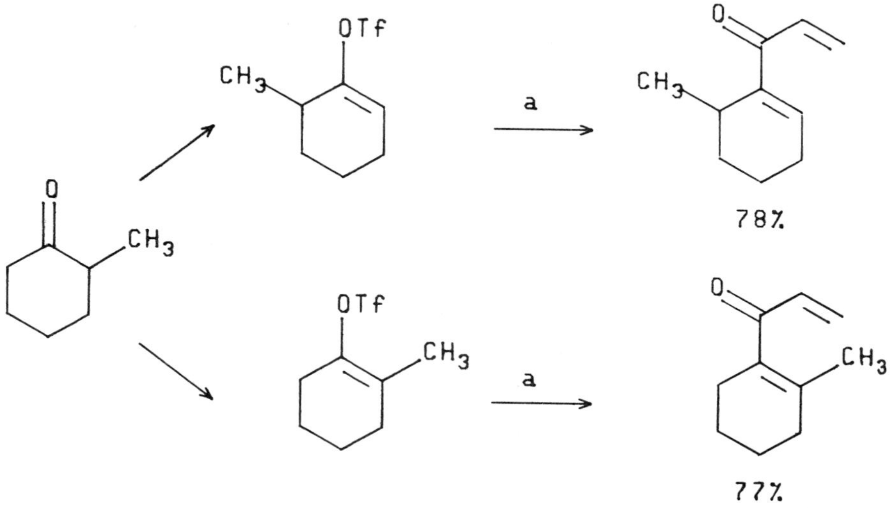

O, CH_3 → OTf, CH_3 → (a) O, CH_3 78%

O, CH_3 → OTf, CH_3 → (a) O, CH_3 77%

a. 55°C, CO, LiCl, $Pd(PPh_3)_4$, $(CH_3)_3SnCH{=}CH_2$, THF

Figure 15.19

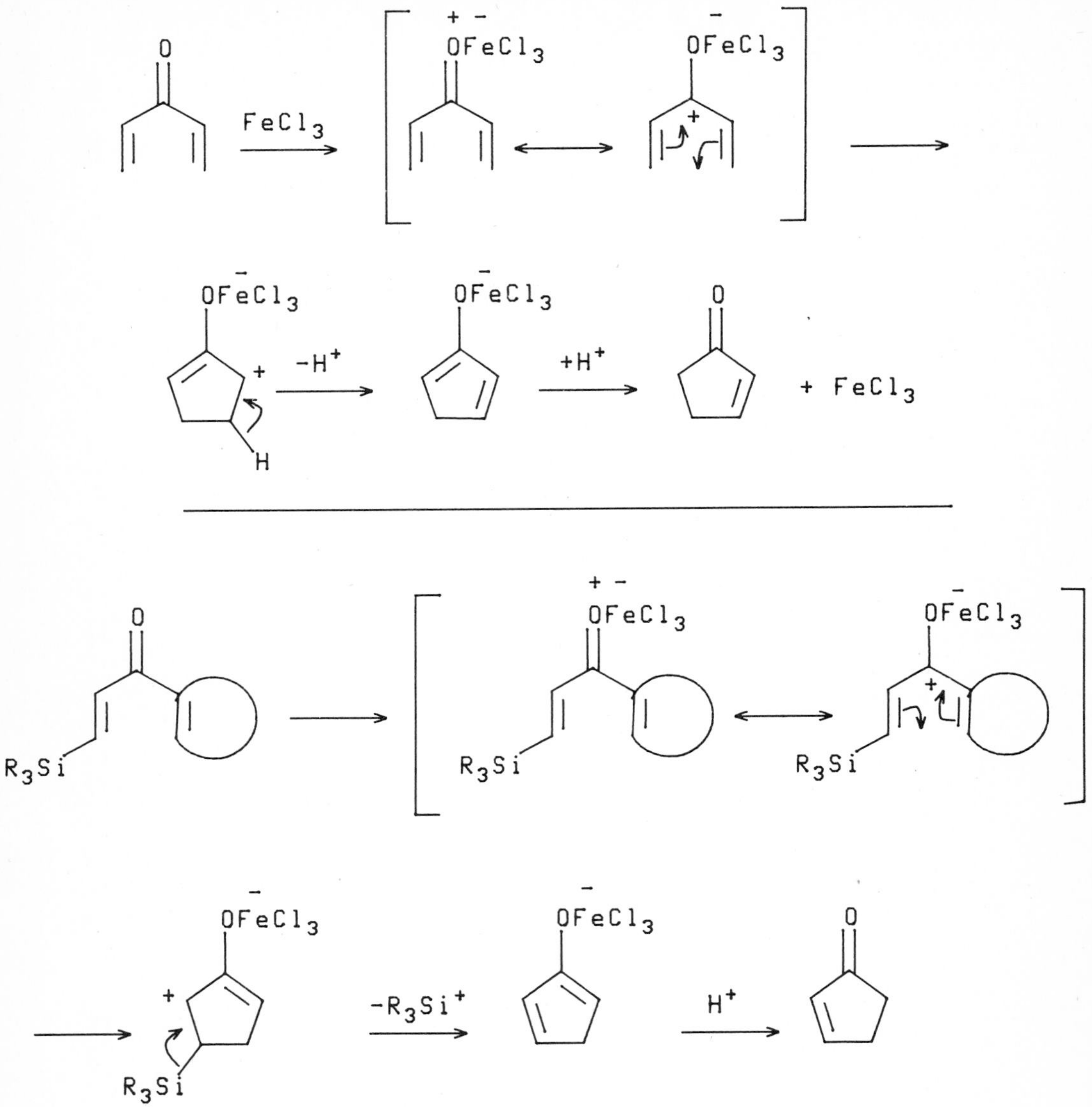

Figure 15.20

produced. If one alkene possesses a silicon β to the carbonyl carbon, the greater stabilization of one intermediate carbocation (β to silicon) results in formation of only one cyclopentenone product (see Figure 15.20).

15.3 TOTAL SYNTHESIS OF (±)-$\Delta^{9(12)}$-CAPNELLANE[16]

Dieckman cyclization of diethyl adipate[17] followed by exhaustive methylation and hydrolysis–decarboxylation affords 2,2,5-trimethylcyclopentanone.[18] The vinyl

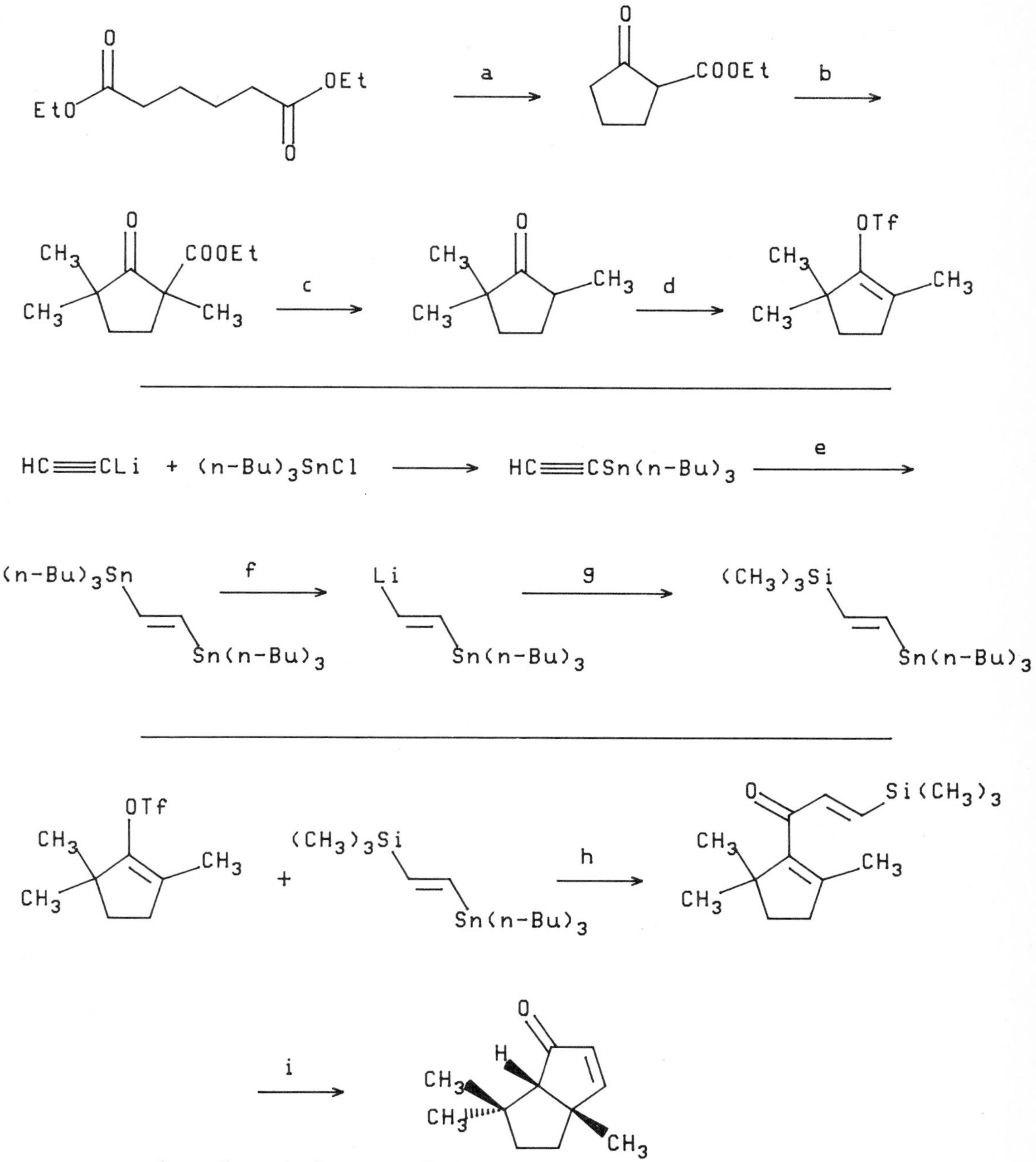

a. rfx, Na, toluene; 72%

b. 1) $NaNH_2$, 2) CH_3I; 45%

c. rfx, HCl, H_2O; 63%

Figure 15.21

d. 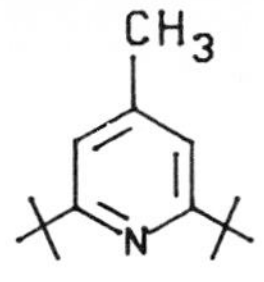 (2,6-di-tert-butyl-4-methylpyridine), Tf_2O, CH_2Cl_2; 85%

e. $(n\text{-}Bu)_3SnH$

f. −78°C, n-BuLi, THF

g. $(CH_3)_3SiCl$; 60% for three steps

h. 55°C, CO, LiCl, $Pd(PPh_3)_4$, THF; 87%

i. 100°C, $BF_3 \cdot Et_2O$, toluene; 70%

Figure 15.21 *(Continued)*

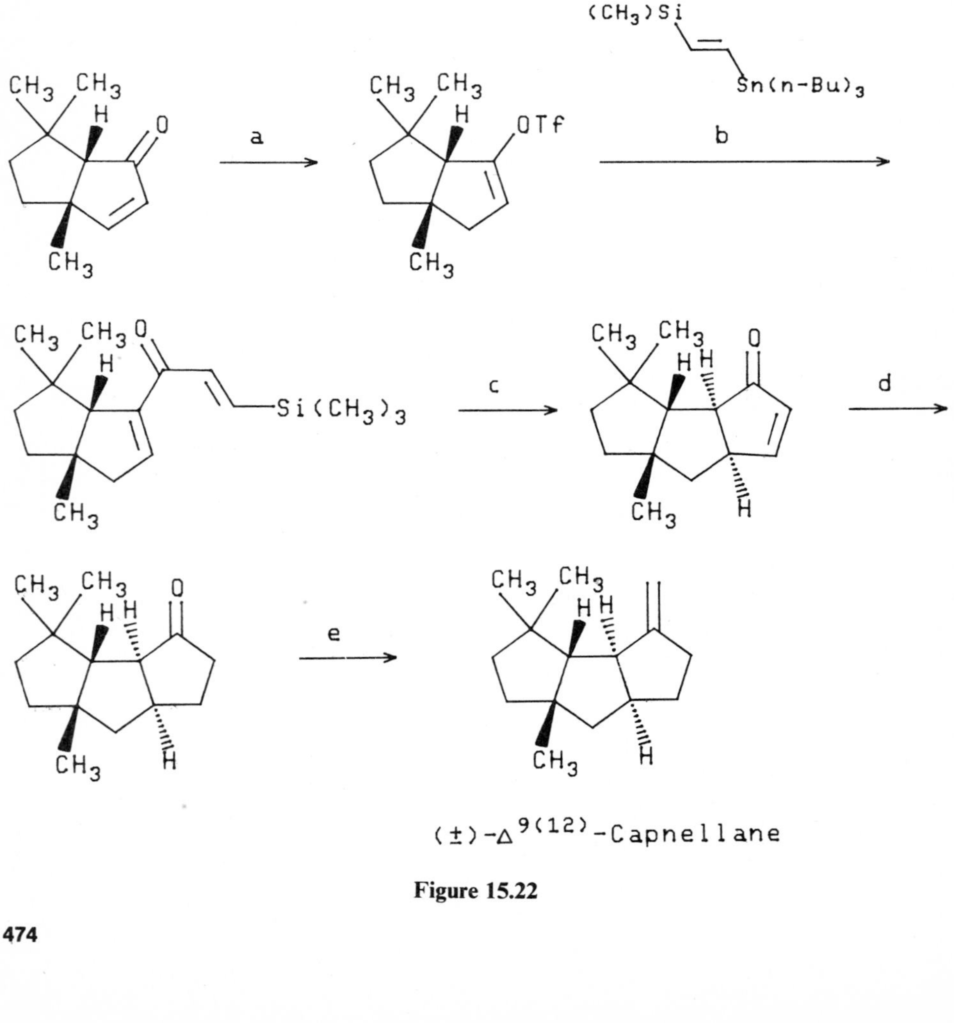

a, b: $(CH_3)Si$–CH=CH–$Sn(n\text{-}Bu)_3$, c, d, e

(±)-$\Delta^{9(12)}$-Capnellane

Figure 15.22

a. 1) L-Selectride®, 2) Tf_2NPh; 76%

b. 55°C, CO, LiCl, $Pd(PPh_3)_4$, THF; 86%

c. 25°C, $BF_3 \cdot Et_2O$, toluene; 88%

d. rt, H_2, Pd-C, EtOAc

e. rfx, $Ph_3P{=}CH_2$, benzene; 60% for two steps

Figure 15.22 (*Continued*)

triflate is best prepared using trifluoromethanesulfonic anhydride. Tri-*n*-butylstannylacetylene is prepared from lithium acetylide and tri-*n*-butyltin chloride. Addition of tri-*n*-butyltin hydride gives 1,2-bis-tri-*n*-butylstannylethylene. Lithium–metal exchange produces a vinyl anion that is coupled with trimethylsilyl chloride.[19,20] Palladium-catalyzed coupling of the vinyl triflate and organotin reagent under carbon monoxide affords a diene-one set for silicon-directed Nazarov cyclization. Boron trifluoride etherate gives yields superior to those obtained with the usual Nazarov catalyst, ferric chloride (see Figure 15.21).

The enolate from conjugate reduction is trapped as the vinyl triflate.[21] A second carbonylative coupling again gives a diene-one set for silicon-directed Nazarov cyclization. Cyclization produces the desired (and most stable) *cis–anti–cis* tricyclic system. Double-bond reduction and Wittig condensation completes the synthesis (see Chapter 13 for a discussion of the relative merits of the Wittig reagent and the titanium carbene for this alkene formation).

The synthesis of ($\pm$)-$\Delta^{9(12)}$-capnellane was completed in 10 steps from diethyl adipate in overall 3.6% yield (7 steps from 2,2,5-trimethylcyclopentanone in overall 18% yield) (see Figure 15.22).

REFERENCES

1. Sheikh, Y. M.; Singy, G.; Kaisin, M.; Eggert, H.; Djerassi, C.; Tursch, B.; Daloze, D.; Braekman, J. C. *Tetrahedron* **1976**, *32*, 1171 and references cited therein.

2a. Takeuchi, T.; Iinuma, H.; Iwanaga, J.; Takahashi, S.; Takita, T.; Umezawa, H. *J. Antibiot.* **1969**, *22*, 215.

2b. Takeuchi, T.; Takahashi, T.; Iinuma, H.; Umezawa, H. *J. Antibiot.* **1971**, *24*, 631.

3a. Mehta, G.; Murthy, A. N.; Reddy, D. S.; Reddy, A. V. *J. Am. Chem. Soc.* **1986**, *108*, 3443.

3b. Liu, H. J.; Kulkarni, M. G. *Tetrahedron Lett.* **1985**, *26*, 4847.

3c. Curran, D. P.; Chen, M-H. *Tetrahedron Lett.* **1985**, *26*, 4991.

3d. Paquette, L. A.; Stevens, K. E. *Can. J. Chem.* **1984**, **62**, 2415.

3e. Stille, J. R.; Grubbs, R. H. *J. Am. Chem. Soc.* **1986**, *108*, 855.

3f. Mehta, G.; Reddy, D. S.; Murty, A. N. *J. Chem. Soc., Chem. Commun. 1983*, 824.

3g. Piers, E.; Karunaratne, V. *Can. J. Chem.* **1984**, *62*, 629.

3h. Paquette, L. A. *Top. Curr. Chem.* **1984**, *119*, 1.

4. For an excellent recent review, see Stille, J. K. *Pure Appl. Chem.* **1985**, *57*, 1771.

5. Simpson, J. H.; Stille, J. K. *J. Org. Chem.* **1985**, *50*, 1759.
6a. Labadie, J. W.; Stille, J. K. *Tetrahedron Lett.* **1983**, *24*, 4283.
6b. Labadie, J. W.; Tueting, D.; Stille, J. K. *J. Org. Chem.* **1983**, *48*, 4634.
7. Piers, E.; Chong, J. M.; Morton, H. E. *Tetrahedron Lett.* **1981**, *22*, 4905.
8. Gerlach, H.; Oertle, K.; Thalmann, A. *Helv. Chim. Acta* **1977**, *60*, 2860.
9. For preparation of chiral pyrenophorin, see Seebach, D.; Seuring, B.; Kalinowski, H-O.; Lubosch, W.; Renger, B. *Angew. Chem. Int. Ed. Engl.* **1977**, *16*, 264.
10a. Sheffy, F. K.; Stille, J. K. *J. Am. Chem. Soc.* **1983**, *105*, 7173.
10b. Sheffy, F. K.; Godschalx, J. P.; Stille, J. K. *J. Am. Chem. Soc.* **1984**, *106*, 4833.
11. Fukuyama, Y.; Kawashima, Y.; Miwa, T.; Tokoroyama, T. *Synthesis* **1974**, 443.
12. Scott, W. J.; Crisp, G. T.; Stille, J. K. *J. Am. Chem. Soc.* **1984**, *106*, 4630.
13. Hiegel, G. A.; Burk, P. *J. Org. Chem.* **1973**, *38*, 3637.
14. Parker, K. A.; Johnson, W. S. *Tetrahedron Lett.* **1969**, 1329.
15. Scott, W. J.; Peña, M. R.; Swärd, K.; Stoessel, S. J.; Stille, J. K. *J. Org. Chem.* **1985**, *50*, 2302.
16. Crisp, G. T.; Scott, W. J.; Stille, J. K. *J. Am. Chem. Soc.* **1984**, *106*, 7500.
17. Diethyl adipate is commercially available and inexpensive (Aldrich 88–89 catalog, 500 g for $13.05).
18. Dubois, J.-E.; Ford, J.-F. *Tetrahedron* **1972**, *28*, 1653.
19. Seyferth, D.; Vick, S. C. *J. Organometal. Chem.* **1978**, *144*, 1.
20. Nesmeyanov. A. N.; Borisov, A. E. *Dokl. Akad. Nauk SSSR* **1967**, *174*, 96.
21. Mc Murry, J. E.; Scott, W. J. *Tetrahedron Lett.* **1983**, *24*, 979.

INDEX